STUDENT SOLUTIONS MANUAL
Chapters 1–23

Chemistry: An Atoms-Focused Approach

Second Edition

Chemistry: An Atoms-Focused Approach

Second Edition

Thomas R. Gilbert, Rein V. Kirss, Natalie Foster,
Stacey Lowery Bretz

Karen S. Brewer
HAMILTON COLLEGE

W. W. NORTON & COMPANY

NEW YORK · LONDON

W. W. Norton & Company has been independent since its founding in 1923, when William Warder Norton and Mary D. Herter Norton first published lectures delivered at the People's Institute, the adult education division of New York City's Cooper Union. The Nortons soon expanded their program beyond the Institute, publishing books by celebrated academics from America and abroad. By mid-century, the two major pillars of Norton's publishing program—trade books and college texts—were firmly established. In the 1950s, the Norton family transferred control of the company to its employees, and today—with a staff of four hundred and a comparable number of trade, college, and professional titles published each year—W. W. Norton & Company stands as the largest and oldest publishing house owned wholly by its employees.

Media Editor: Chris Rapp
Associate Media Editor: Julia T. Sammaritano, Arielle Holstein
Assistant Media Editor: Doris Chiu
Production Manager: Eric Pier-Hocking
Composition by codeMantra
Manufacturing by Sterling Pierce

ISBN 978-0-393-60382-8

W. W. Norton & Company, Inc., 500 Fifth Avenue, New York, NY 10110

www.wwnorton.com

W. W. Norton & Company, Ltd., 15 Carlisle Street, London W1D 3BS

1 2 3 4 5 6 7 8 9 0

Contents

Preface

Students often begin an introductory course in chemistry with a little trepidation and want to know the secret formula for success. The simple answer (admittedly hard to put into practice) is for you to engage fully with the course during lectures, in laboratories, and in working problems. We hope that this solutions manual will help you as you work on the problems assigned outside the class.

In the textbook, you are introduced to the COAST method as an approach to solving problems and answering conceptual questions. This method encourages you to assemble relevant information and to plan an approach for the problems before you start calculating the answer. After solving the problem, COAST guides you to think further about the problem and to extend your knowledge. The COAST method, as summarized below, is used throughout this *Manual*:

COLLECT AND ORGANIZE	Restatement of the problem to delineate exactly what information has been provided and what is being asked.
ANALYZE	Strategy to solve the problem including which formulas are relevant and which unit conversions are necessary.
SOLVE	Application of the strategy to solve a problem numerically or to answer a conceptual question.
THINK ABOUT IT	Reminders to check the answer or consider whether the answer makes sense. In this manual, this step is sometimes used to add factual information extending the context of the question.

This *Student Solutions Manual* contains the worked-out solutions to every odd-numbered problem at the end of each chapter of the textbook. Generally speaking, the conventions for significant figures as outlined in Chapter 1 of the textbook are used, except when an additional significant figure in the answer would clarify the difference that you might see on your calculator compared to what the strict adherence to significant figure rules would give as an answer. For most of the calculations presented in this *Manual*, all of the significant digits were kept in the calculator for intermediate values in a multistep calculation.

As you use this *Solutions Manual*, keep in mind that there are often several valid approaches to solving a particular problem in chemistry. Neither your professor nor you will always solve the problem in exactly the same way as it is presented here. Also, in class, you may be shown additional approaches to solving the many problems that you will encounter in introductory chemistry. Regardless of the strategy you employ, be mindful of writing down your calculation in each step. Paying attention to the information presented in the problem and the units at each stage of calculation will help you arrive at the correct solution.

CHAPTER 1 | Matter and Energy: An Atomic Perspective

1.1. **Collect and Organize**

Figure P1.1(a) shows "molecules," each consisting of one red sphere and one blue sphere, and Figure P1.1(b) shows separate blue spheres and red spheres. For each figure we are to determine whether the substance(s) depicted is a solid, liquid, or gas and whether the figures show an element, a compound, a mixture of elements, or a mixture of compounds.

Analyze

An element is composed of all the same type of atom, and a compound is composed of two or more types of atoms. Solids have a definite volume and a highly ordered arrangement where the particles are close together. Liquids also have a definite volume but have a disordered arrangement of particles that are close together. Gases have disordered particles that fill the volume of the container and are far apart from each other.

Solve

(a) Because each particle in Figure P1.1(a) consists of one red sphere and one blue sphere, all the particles are the same—that is a compound. The particles fill the container and are disordered, so those particles are in the gas phase.

(b) Because it shows a mixture of red and blue spheres, Figure P1.1(b) depicts a mixture of blue elemental atoms and red elemental atoms. The blue spheres fill the container and are disordered, so those particles are in the gas phase. The red spheres have a definite volume and are ordered, so those particles are in the solid phase.

Think About It

Remember that both elements and compounds may be either pure or present in a mixture.

1.3. **Collect and Organize**

In this question we are to consider whether the reactants, as depicted, undergo a chemical reaction and/or a phase change.

Analyze

Chemical reactions involve the breaking and making of bonds in which atoms are combined differently in the products than in the reactants. When we consider a possible phase change, remember the following: Solids have a definite volume and a highly ordered arrangement where the particles are close together. Liquids also have a definite volume but have a disordered arrangement of particles that are close together. Gases have disordered particles that fill the volume of the container and are far apart from one another.

Solve

In Figure P1.3, two pure elements (red–red and blue–blue) in the gas phase recombine to form a compound (red–blue) in the solid phase (ordered array of molecules). Therefore, answer b describes the reaction shown.

Think About It

A phase change does not necessarily accompany a chemical reaction. We will learn later that the polarity of the product will determine whether a substance will be in the solid, liquid, or gaseous state at a given temperature.

1.5. **Collect and Organize**

Given a space-filling model of formic acid pictured in Figure P1.5, we are to write the molecular formula of formic acid.

Analyze

To determine the identity of the atoms in the space-filling model, we use the Atomic Color Palette (inside back cover of your textbook): black is carbon, red is oxygen, and white is hydrogen. To determine the formula, we need to count each type of atom in the structure.

Solve

The structure of formic acid includes two hydrogen atoms, two oxygen atoms, and one carbon atom. Therefore, its molecular formula is CH_2O_2.

Think About It

Writing the molecular formula as "2HC2O" is confusing and wrong. We indicate the number of atoms in a molecular formula with a subscript within the molecular formula, not with a coefficient. Also, you will find that it is a convention for organic molecular formulas to list C first, followed by H.

1.7. **Collect and Organize**

For this question we are asked to differentiate "hypothesis" from "scientific theory."

Analyze

These terms are part of the scientific method and result from different aspects of the process of observing and explaining a natural phenomenon.

Solve

A hypothesis is a tentative explanation of an observation or set of observations, whereas a scientific theory is a concise explanation of a natural phenomenon that has been extensively tested and explains why certain phenomena are always observed.

Think About It

Notice that a hypothesis might become a theory after much experimental testing.

1.9. **Collect and Organize**

In this question we consider how Dalton's atomic theory supported his law of multiple proportions.

Analyze

Dalton's law of multiple proportions states that when two elements combine to make two (or more) compounds, the ratio of the masses of one of the elements, which combine with a given mass of the second element, is always a ratio of small whole numbers. His atomic theory states that matter in the form of elements and compounds is made up of small, indivisible units—atoms.

Solve

Dalton's atomic theory explained the small, whole-number mass ratios in his law of multiple proportions because the compounds contained small, whole-number ratios of atoms of different elements per molecule or formula unit.

Think About It

Dalton's theory is not strictly true–atoms are divisible into electrons, protons, and neutrons (and even further into subatomic quarks) and some compounds do not have whole-number ratios of atoms. For most matter and most compounds we encounter in chemistry, though, his theory is true.

1.11. **Collect and Organize**

In this question we are asked to explain why scientists opposed Proust's law of definite proportions when he proposed it.

Analyze

The law of definite proportions states that the ratio of elements in a compound is always the same.

Solve

Proust's law needed to have corroborating evidence to fully support it. At the time, experiments to prepare a compound of tin with oxygen yielded various compositions. The compounds they prepared, later turned out to be mixtures of two compounds of tin oxide.

Think About It

Tin can form either tin(II) oxide, SnO, or tin(IV) oxide, SnO_2. What do you think the ratio of the elements would be for a 50–50 mixture of these two compounds? Of a 25–75 mixture?

1.13. Collect and Organize

We are to define *theory* as used in everyday conversation and differentiate it from its use in science.

Analyze

Theory in everyday conversation has a quite different meaning from its meaning in science.

Solve

Whereas *theory* in normal conversation means someone's idea or opinion that is open to speculation, a scientific theory is a concise and testable explanation of natural phenomena based on observation and experimentation that can accurately predict the results of experiments.

Think About It

Theory in normal conversation is more akin to a *hypothesis* or a *guess* that may or may not be testable.

1.15. Collect and Organize

For the foods listed, we are to determine which are heterogeneous.

Analyze

A heterogeneous mixture has visible regions of different composition.

Solve

Clear regions of different composition are evident in a Snickers bar (b) and in an uncooked hamburger (d), but not in bottled water (a) or in grape juice (c).

Think About It

Bottled water contains homogeneous mixtures of small amounts of dissolved minerals such as salts of sodium, magnesium, and calcium that give water its flavor.

1.17. Collect and Organize

For the foods listed, we are to determine which are heterogeneous.

Analyze

A heterogeneous mixture has visible regions of different composition.

Solve

Clear regions of different composition are evident in orange juice (with pulp) (d) and tomato juice (e), but not in apple juice, cooking oil, or solid butter, or tomato juice (a–c).

Think About It

When butter melts, you notice milk solids and clear regions that are definitely discernible. Therefore, homogeneous solid butter becomes heterogeneous when heated.

1.19. Collect and Organize

We are asked to consider whether distillation would be effective in removing suspended soil particles from water.

Analyze

In distillation, evaporation of a liquid and subsequent condensation of the vapor is used to separate substances of different volatilities.

Solve

Soil particles are not volatile, but water is; we can boil water, but not the soil. Therefore, yes, distillation can be used to remove soil particles from water. That is not a widely used process to purify water because boiling water is energy- and time-intensive. Filtration would be both cheaper and faster.

Think About It

In this distillation process we would collect pure water through condensation, and the soil particles would be left behind in the distillation flask.

1.21. **Collect and Organize**

For this question we are to list some chemical and physical properties of gold.

Analyze

A chemical property is seen when a substance undergoes a chemical reaction, thereby becoming a different substance. A physical property can be seen without any transformation of one substance into another.

Solve

One chemical property of gold is its resistance to corrosion (oxidation). Gold's physical properties include its density, color, melting temperature, and electrical and thermal conductivity.

Think About It

Another metal that does not corrode (or rust) is platinum. Platinum and gold, along with palladium, are often called "noble metals."

1.23. **Collect and Organize**

We are asked in this question to name three properties to distinguish among table sugar, water, and oxygen.

Analyze

We can distinguish among substances by using either physical properties (such as color, melting point, and density) or chemical properties (such as chemical reactions, corrosion, and flammability).

Solve

We can distinguish among table sugar, water, and oxygen by examining their physical states (sugar is a solid, water is a liquid, and oxygen is a gas) and by their densities, melting points, and boiling points.

Think About It

These three substances are also very different at the atomic level. Oxygen is a pure element made up of diatomic molecules; water is a liquid compound made up of discrete molecules of hydrogen and oxygen (H_2O); and table sugar is a solid compound made up of carbon, hydrogen, and oxygen atoms.

1.25. **Collect and Organize**

From the list of properties of sodium, we are to determine which are physical and which are chemical properties.

Analyze

Physical properties are those that can be observed without transforming the substance into another substance. Chemical properties are observed only when one substance reacts with another and therefore is transformed into another substance.

Solve

Density, melting point, thermal and electrical conductivity, and softness (a–d) are all physical properties, whereas tarnishing and reaction with water (e and f) are both chemical properties.

Think About It

Because the density of sodium is less than that of water, a piece of sodium will float on water as it reacts.

1.27. **Collect and Organize**

We are to explain whether an extensive property can be used to identify a substance.

Analyze

An extensive property is one that, like mass, length, and volume, is determined by size or amount.

Solve

Extensive properties will change with the size of the sample and therefore cannot be used to identify a substance.

Think About It

We could, for example, have the same mass of feathers and lead, but their mass alone will not tell us which mass measurement belongs to which—the feathers or the lead.

1.29. **Collect and Organize**

We are to explain whether the extinguishing of fires by carbon dioxide, CO_2, is a result of its chemical or physical properties (or both).

Analyze

Physical properties are those that can be observed without transforming the substance into another substance. Chemical properties are observed only when one substance reacts with another and therefore is transformed into another substance.

Solve

Carbon dioxide is a nonflammable gas (a chemical property; it does not burn) and it is denser than air (a physical property; it smothers the flames by excluding oxygen from the fuel). Therefore, CO_2's fire-extinguishing properties are due to both its physical and its chemical properties.

Think About It

Some metals, such as magnesium, will burn in carbon dioxide; you cannot extinguish those fires with a CO_2 fire extinguisher.

1.31. **Collect and Organize**

We are asked to compare the arrangement of water molecules in water as a solid (ice) and water as a liquid.

Analyze

Figure A1.31 shows the arrangement of the water molecules in both those phases.

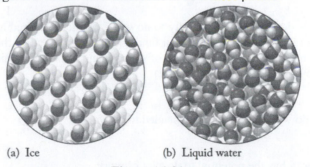

(a) Ice   (b) Liquid water

Figure A1.31

Solve

Water molecules in both the ice and liquid forms contain hydrogen bonds that link individual molecules together closely. In those arrangements, the hydrogen atoms in a water molecule point to an oxygen atom of a neighboring water molecule. The oxygen atom, in turn, points toward a hydrogen atom of a neighboring molecule. The hydrogen bonds in liquid water have no long-range structure and can move around each other, whereas in ice the molecules are arranged in a rigid hexagonal arrangement.

Think About It

The structure of ice is more open than the structure of liquid water. That is why, when water freezes, it expands.

1.33. **Collect and Organize**

We are to determine which phase (solid, liquid, or gas) has the greatest particle motion and which has the least.

Analyze

Gases have particles much separated from each other; these particles, therefore, have a wide range of movement. Particles in solids and liquids are close to one another, and therefore the particle motion in both phases is restricted. Solids hold their particles in rigid arrays.

Solve

Because of their freedom of movement, gases have the greatest particle motion; because of the restriction of their solid lattice, solids have the least particle motion.

Think About It

Heating a solid or liquid can melt or vaporize a substance. During these phase changes with the addition of heat, particle motion increases.

1.35. **Collect and Organize**

We are asked to identify the process that results in snow disappearing, but not melting, on a sunny but cold winter day.

Analyze

The cold air temperature does not allow the snow to melt, but the sunny day does add warmth to the solid snow.

Solve

The snow, instead of melting, sublimes: going directly from the solid ice crystals in snow to water vapor.

Think About It

A more familiar example of sublimation is that of "dry ice," which is solid CO_2 and at ambient temperature and pressure it sublimes, rather than melts, to give a "fog" for stage shows.

1.37. **Collect and Organize**

We are asked whether energy and work are related.

Analyze

In this context, energy is defined as the capacity to do work. Work is defined as moving an object with a force over some distance. Energy also is thought to be a fundamental component of the universe. The Big Bang theory postulates that all matter originated from a burst of energy, and Albert Einstein proposed that $m = E/c^2$ (mass equals energy divided by the speed of light squared).

Solve

Energy is the ability to do work and must be expended to do work.

Think About It

A system with high energy has the potential to do a lot of work.

1.39. **Collect and Organize**

From three statements about heat, we are asked to choose those that are true.

Analyze

Heat is defined as the transfer of energy between objects or regions of different temperature and the energy flows from high thermal energy to low thermal energy. Heat is an extrinsic property; the amount of heat in a substance depends on the quantity of the substance and it is measured by taking the temperature.

Solve
All these statements (a, b, and c) are true.

Think About It
Later, in this course, we will be able to quantify the heat in a substance or the heat released or absorbed by a physical change or a chemical reaction.

1.41. **Collect and Organize**
We are asked to compare the kinetic energy of a subcompact car (1400 kg) with that of a dump truck (18,000 kg) when they are traveling at the same speed.

Analyze
We can compare the kinetic energies by using the equation

$$KE = \frac{1}{2}mu^2$$

Solve
For the subcompact car:

$$KE = \frac{1}{2}(1400 \text{ kg})u_1^{\,2}$$

For the dump truck:

$$KE = \frac{1}{2}(18,000 \text{ kg})u_2^{\,2}$$

Because $u_1 = u_2$, we can replace u_1 in the first expression with u_2. The ratio of the kinetic energies is

$$\frac{KE_2}{KE_1} = \frac{\frac{1}{2}(18,000 \text{ kg})u_2^{\,2}}{\frac{1}{2}(1400 \text{ kg})u_2^{\,2}} = \frac{18,000 \text{ kg}}{1400 \text{ kg}} = 13$$

When traveling at the same speed, the dump truck has 13 times more kinetic energy than the subcompact car.

Think About It
The same dump truck has more kinetic energy when traveling faster because kinetic energy depends on the velocity of an object, as well as its mass.

1.43. **Collect and Organize**
We are to compare SI units with U.S. customary units.

Analyze
SI units are based on a decimal system to describe basic units of mass, length, temperature, energy, and so on. U.S. customary units vary.

Solve
SI units, which were based on the original metric system, can be easily converted into a larger or smaller unit by multiplying or dividing by multiples of 10. U.S. customary units are more complicated to manipulate. For example, to convert miles to feet you have to know that 1 mile is 5280 feet, and to convert gallons to quarts you have to know that 4 quarts is in 1 gallon.

Think About It
Once you can visualize a meter, a gram, and a liter, using the SI system is quite convenient.

1.45. **Collect and Organize**
For this question we are to think about why scientists might prefer the Celsius scale over the Fahrenheit scale.

Analyze
The Celsius scale is based on a 100-degree range between the freezing point and boiling point of water, whereas the Fahrenheit scale is based on the 100-degree range between the freezing point of a concentrated salt solution and the average internal human body temperature.

Solve

Scientists might prefer the Celsius scale because it is based on the phase changes (freezing and boiling) for a pure common solvent (water).

Think About It

Because the difference in the freezing and boiling point of water on the Fahrenheit scale is 180 degrees compared with 100 degrees for the Celsius scale, one degree Fahrenheit is smaller than one degree Celsius. Notice in Figure A1.45 that a 10°C range is much larger than a 10°F range.

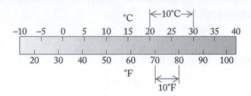

Figure A1.45

1.47. **Collect and Organize**

In this question we are to define the *absolute* temperature scale.

Analyze

The Kelvin scale is the absolute temperature scale, and its lowest temperature is 0 K.

Solve

The absolute temperature scale (Kelvin scale) has no negative temperatures, and its zero value is placed at the lowest possible temperature.

Think About It

Because the Kelvin scale has no negative temperatures, it will often be used in equations when using a negative temperature (in Celsius) that would result in a nonsensical answer.

1.49. **Collect and Organize**

For this problem we need to convert the distance of the Olympic mile (1500 m) in meters to miles and then to feet. Then we need to compare that distance with an actual mile by using a ratio and then converting that to a percentage.

Analyze

To convert the distance, we can use the following conversions:
$$\frac{1\ km}{1000\ m}, \quad \frac{0.6214\ mi}{1\ km}, \quad and \quad \frac{5280\ ft}{1\ mi}$$
To determine the percentage the Olympic mile distance is compared with the actual mile, we will use
$$\%\ distance = \frac{Olympic\ mile\ distance\ in\ feet}{5280\ ft} \times 100$$

Solve

$$1500\ m \times \frac{1\ km}{1000\ m} \times \frac{0.6214\ mi}{1\ km} = 0.9321\ mi$$

$$\%\ distance = \frac{0.9321\ mi}{1\ mi} \times 100 = 93.2\%$$

Think About It

This calculation shows that the Olympic mile is just a little bit shorter than the conventional mile.

1.51. **Collect and Organize**

This problem asks for a simple conversion of length: from meters to miles.

Analyze

The conversions that we need include meters to kilometers and kilometers to miles:

$$\frac{1 \text{ km}}{1000 \text{ m}} \quad \text{and} \quad \frac{0.6214 \text{ mi}}{1 \text{ km}}$$

Solve

$$4.0 \times 10^3 \text{ m} \times \frac{1 \text{ km}}{1000 \text{ m}} \times \frac{0.6214 \text{ mi}}{1 \text{ km}} = 2.5 \text{ mi}$$

Think About It

The answer is reasonable because 4000 m would be a little over 2 mi when estimated. For a natural piece of silk to be that long, though, is surprising.

1.53. **Collect and Organize**

To determine the Calories burned by the wheelchair marathoner in a race, we can first find the number of hours the race will be for the marathoner at the pace of 13.1 miles per hour. We can then calculate the Calories burned from that value and the rate at which the marathoner burns Calories.

Analyze

The time for the marathoner to complete the race will be given by

$$\text{time to complete the marathon} = \frac{\text{distance of the marathon}}{\text{pace of the marathoner}}$$

The Calories burned will be computed by

$$\text{Calories burned} = \frac{\text{Calories burned}}{\text{hr}} \times \text{length of the marathon race}$$

Solve

$$\text{time to complete the marathon} = \frac{26.2 \text{ mi}}{13.1 \text{ mi/hr}} = 2.00 \text{ hr}$$

$$\text{Calories burned} = \frac{665 \text{ Cal}}{\text{hr}} \times 2.00 \text{ hr} = 1330 \text{ Cal}$$

Think About It

We could solve this problem without touching a calculator. Because the marathoner takes 2.00 hr to complete the race, the Calories she burns are simply twice the number of Calories she burns in 1 hr.

1.55. **Collect and Organize**

A light-year is the distance light travels in 1 year. To determine the distance of 4.3 light-years in kilometers, we will first have to convert 4.3 years into seconds and then use the speed of light to determine the distance the light travels over that time.

Analyze

We can find the length of time of 4.3 years in seconds by using the following conversions:

$$\frac{1 \text{ yr}}{365.25 \text{ d}}, \quad \frac{1 \text{ d}}{24 \text{ hr}}, \quad \frac{1 \text{ hr}}{60 \text{ min}}, \quad \text{and} \quad \frac{1 \text{ min}}{60 \text{ s}}$$

We can find the distance of 4.3 light-years in meters from the speed of light:

distance of 4.3 light-years = speed of light (in meters/second) × 4.3 yr (in seconds)

We can convert this value into kilometers by using

$$\frac{1 \text{ km}}{1000 \text{ m}}$$

Solve

$$4.3 \text{ yr} \times \frac{365.25 \text{ d}}{1 \text{ yr}} \times \frac{24 \text{ hr}}{1 \text{ d}} \times \frac{60 \text{ min}}{1 \text{ hr}} \times \frac{60 \text{ s}}{1 \text{ min}} = 1.36 \times 10^8 \text{ s}$$

$$\text{distance to Proxima Centauri} = (1.36 \times 10^8 \text{ s}) \times \frac{2.998 \times 10^8 \text{ m}}{\text{s}} \times \frac{1 \text{ km}}{1000 \text{ m}} = 4.1 \times 10^{13} \text{ km}$$

Think About It

This is a very large distance since light travels so fast. The light-year, being such a large distance, is an ideal unit for expressing astronomical distances.

1.57. Collect and Organize

To solve this problem, we need to know the volume of water in liters that is to be removed from the swimming pool. Using that volume and the rate at which the water can be siphoned, we can find how long removing the water will take.

Analyze

The volume of water to be removed in cubic meters can be found from

length of pool (in meters) × width of pool (in meters) × depth of water to be removed (in meters)

This volume will have to be converted to liters through the conversion

$$\frac{1 \text{ L}}{1 \times 10^{-3} \text{ m}^3}$$

The time to siphon the water is determined by the rate at which the siphon pump operates:

$$\text{time to siphon the water} = \frac{\text{volume of water to be siphoned in liters}}{\text{rate at which the water can be siphoned in liters per second}}$$

Solve

The volume of the water to be siphoned out of the pool is

$$50.0 \text{ m} \times 25.0 \text{ m} \times \left(3.0 \text{ cm} \times \frac{1 \text{ m}}{100 \text{ cm}} \right) = 37.5 \text{ m}^3$$

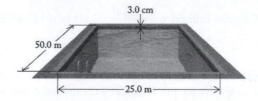

Converting this into liters,

$$37.5 \text{ m}^3 \times \frac{1 \text{ L}}{1 \times 10^{-3} \text{ m}^3} = 3.75 \times 10^4 \text{ L}$$

The amount of time to siphon this water is

$$\text{time to siphon the water} = \frac{3.75 \times 10^4 \text{ L}}{5.2 \text{ L/s}} = 7210 \text{ s}$$

$$7210 \text{ s} \times \frac{1 \text{ min}}{60 \text{ s}} = 120 \text{ min, or } 2.0 \text{ hr}$$

Think About It

This may be a surprisingly long time to siphon only 3.0 cm of water from the pool, but the total volume to be siphoned is quite large because of the pool's size.

1.59. **Collect and Organize**

We can solve this by converting the time for the runner to complete the race into seconds and then converting the distance of the race from kilometers into meters. The runner's speed is the ratio of distance (meters) to time (seconds).

Analyze

Conversions for time that we need are

$$\frac{60 \text{ s}}{1 \text{ min}}$$

To convert kilometers into miles, we need

$$\frac{1000 \text{ m}}{1 \text{ km}}$$

Solve

The time for the runner to complete the race is

$$\left(41 \text{ min} \times \frac{60 \text{ s}}{1 \text{ min}}\right) + 23 \text{ s} = 2483 \text{ s}$$

The distance of the race in kilometers is

$$10.0 \text{ km} \times \frac{1000 \text{ m}}{\text{km}} = 1.00 \times 10^4 \text{ m}$$

The runner's average speed then is

$$\frac{1.00 \times 10^4 \text{ m}}{2483 \text{ s}} = 4.03 \text{ m/s}$$

Think About It

The answer makes sense because a walking speed is around 3 mph, or 1.3 m/s. Running could easily be imagined at 9 mph, or 4.0 m/s.

1.61. **Collect and Organize**

In this problem we need to use the density of magnesium to find the mass of a specific size block of the metal.

Analyze

Density is defined as the mass of a substance per unit volume. Appendix 3 gives the density of magnesium as 1.738 g/cm^3. We have to find the volume of the block of magnesium by multiplying the length by the height by the depth (the result will be in cubic centimeters). We can then find mass through the following formula:

$$\text{mass (g)} = \text{density (g/cm}^3) \times \text{volume (cm}^3)$$

Solve

The volume of the block of magnesium is

$$2.5 \text{ cm} \times 3.5 \text{ cm} \times 1.5 \text{ cm} = 13 \text{ cm}^3$$

Therefore, the mass of the block is

$$13 \text{ cm}^3 \times 1.738 \text{ g/cm}^3 = 23 \text{ g}$$

Think About It

The mass of a sample depends on the amount of substance present. Here we have about 23 g. For a quick estimate, a block of magnesium of about 10 cm^3 would weigh more than 1.7 times that of 1 cm^3, or 17 g. Because we have more than 10 cm^3 of this sample and the density is a little greater than 1.7 g/cm^3, our answer of 23 g is reasonable.

1.63. **Collect and Organize**

In this problem we use the density to find the volume of sulfuric acid the chemist needs. This situation uses the definition density = mass/volume.

Analyze

We can easily solve this problem by rearranging the density equation:

$$\text{density} = \frac{\text{mass}}{\text{volume}} \quad \text{or} \quad \text{volume} = \frac{\text{mass}}{\text{density}}$$

Solve

$$\text{volume needed} = \frac{35.0 \text{ g}}{1.84 \text{ g/mL}} = 19.0 \text{ mL}$$

Think About It

With a density of about 2 g/cm^3 to get a mass of about 40 g, we might estimate that the chemist would need 20 mL. This estimate shows that our answer is reasonable.

1.65. **Collect and Organize**

This problem asks us to convert weights from ounces to grams and then to kilograms.

Analyze

Conversions for weight (mass) that we need are

$$\frac{1 \text{ oz}}{28.35 \text{ g}} \quad \text{and} \quad \frac{1 \text{ kg}}{1000 \text{ g}}$$

Solve

$$0.934 \text{ oz} \times \frac{28.35 \text{ g}}{1 \text{ oz}} = 26.5 \text{ g}$$

$$26.5 \text{ g} \times \frac{1 \text{ kg}}{1000 \text{ g}} = 0.0265 \text{ kg}$$

Think About It

Because the silver dollar weighs just under an ounce, its mass will be slightly less than 28.35 g, so our answer of 26.5 g makes sense.

1.67. **Collect and Organize**

To answer this question we need to use the density of copper to compute the mass of the copper sample that is 125 cm^3 in volume. Next, we use that mass to find out how much volume (in cubic centimeters) that mass of gold would occupy.

Analyze

We need the density both of copper and of gold from Appendix 3 to convert from volume to mass (for copper) and then from mass to volume (for gold). These densities are 8.96 g/mL for copper and 19.3 g/mL for gold. One milliliter is equivalent to 1 cm^3, so the densities are 8.96 g/cm^3 and 19.3 g/cm^3, respectively. The density formulas that we need are

$$\text{mass of copper} = \text{density of copper} \times \text{volume}$$

$$\text{volume of gold} = \frac{\text{mass}}{\text{density of gold}}$$

Solve

$$\text{mass of copper} = 8.96 \text{ g/cm}^3 \times 125 \text{ cm}^3 = 1120 \text{ g}$$

$$\text{volume of gold} = \frac{1120 \text{ g}}{19.3 \text{ g/cm}^3} = 58.0 \text{ cm}^3$$

Think About It

Because gold is more than twice as dense as copper, we would expect the volume of a gold sample to have about half the volume of that of the same mass of copper.

1.69. Collect and Organize

Using the density of mercury, we can find the volume of 1.00 kg of mercury.

Analyze

Appendix 3 gives the density of mercury as 13.546 g/mL. Because this property is expressed in grams per milliliter, not kilograms per milliliter, we have to convert kilograms into grams by using the conversion factor

$$\frac{1000 \text{ g}}{1 \text{ kg}}$$

Once we have the mass in grams, we can use the rearranged formula for density to find volume:

$$\text{volume of mercury (mL)} = \frac{\text{mass of mercury (g)}}{\text{density of mercury (g/mL)}}$$

Solve

$$1.00 \text{ kg} \times \frac{1000 \text{ g}}{1 \text{ kg}} = 1.00 \times 10^3 \text{ g}$$

$$\text{volume of mercury} = \frac{1.00 \times 10^3 \text{ g}}{13.546 \text{ g/mL}} = 73.8 \text{ mL}$$

Think About It

That result is a fairly small amount that weighs 1 kg. That value is the result of mercury's relatively high density.

1.71. Collect and Organize

Because we are not directly given the mass and volume of the two planets, Earth and Venus, we have to use their relative masses and volumes to find the density of Venus compared with that of Earth.

Analyze

The relative masses and volumes of the two planets can be expressed as

$$\text{mass of Venus} = 0.815 \times \text{mass of Earth}$$

$$\text{volume of Venus} = 0.88 \times \text{volume of Earth}$$

To find the density of Venus, we will have to rearrange these into

$$\frac{\text{mass of Earth}}{\text{volume of Earth}} \times \frac{\text{volume of Earth}}{\text{volume of Venus}} \times \frac{\text{mass of Venus}}{\text{mass of Earth}} = \frac{\text{mass of Venus}}{\text{volume of Venus}}$$

or

$$\text{density of Earth} \times \frac{100}{88} \times \frac{81.5}{100} = \text{density of Venus}$$

Solve

$$\frac{5.5 \text{ g}}{\text{cm}^3} \times \frac{100}{88} \times \frac{81.5}{100} = 5.1 \text{ g/cm}^3$$

Think About It

With Earth being larger than Venus, and more massive, immediately predicting whether Venus would be more or less dense than Earth is hard. However, because the difference in the mass (18.5%) between Earth and Venus is greater than the difference in volume (12%), it makes sense that the density of Venus is lower than that of Earth.

1.73. Collect and Organize

To determine whether a cube made of high-density polyethylene (HDPE) will float on water, we need to compare the density of the HDPE with that of water. If HDPE's density is less than water's, the cube will float.

Analyze

To compare the densities of the two substances (water and HDPE), we need to have them in the same units. We can approach this in either of two ways—convert the seawater density to kilograms per cubic meter or convert the HDPE density to grams per cubic centimeter. Let's do the latter, using the following conversions:

$$\frac{100 \text{ cm}}{1 \text{ m}} \quad \text{and} \quad \frac{1000 \text{ g}}{1 \text{ kg}}$$

To calculate the density of the HDPE sample, we must divide the mass of the cube of HDPE in grams by the volume in cubic centimeters.

Solve

$$\text{volume of the HDPE cube} = \left(1.20 \times 10^{-2} \text{ m}\right)^3 \times \left(\frac{100 \text{ cm}}{1 \text{ m}}\right)^3 = 1.728 \text{ cm}^3$$

$$\text{mass of the HDPE cube in grams} = 1.70 \times 10^{-3} \text{ kg} \times \frac{1000 \text{ g}}{1 \text{ kg}} = 1.70 \text{ g}$$

$$\text{density of the HDPE cube} = \frac{1.70 \text{ g}}{1.728 \text{ cm}^3} = 0.984 \text{ g/cm}^3$$

That density is less than the density of the seawater (1.03 g/cm^3), so the cube of HDPE will float on water.

Think About It

Certainly boats are made of other materials (such as iron) that are denser than water. Those boats float because the mass of the water they displace is greater than their mass.

1.75. Collect and Organize

Given that the mass of the Golden Jubilee diamond is 545.67 carats (with 1 carat = 0.200 g), we are to calculate the mass of the diamond in grams and in ounces.

Analyze

First, we can use the fact that 1 carat = 0.200 g to convert the mass of the diamond from carats to grams. Then we can use the fact that 1 ounce = 28.35 g to convert the result in grams to ounces.

Solve

$$\text{mass of the diamond in grams} = 545.67 \text{ carats} \times \frac{0.200 \text{ g}}{1 \text{ carat}} = 109 \text{ g}$$

$$\text{mass of the diamond in ounces} = 109 \text{ g} \times \frac{1 \text{ ounce}}{28.35 \text{ g}} = 3.85 \text{ ounces}$$

Think About It

From these conversion factors we can determine the number of ounces in 1 carat:

$$1 \text{ carat} \times \frac{0.200 \text{ g}}{\text{carat}} \times \frac{1 \text{ ounce}}{28.35 \text{ g}} = 7.05 \times 10^{-3} \text{ ounce}$$

1.77. Collect and Organize

For the values given, we are asked to choose those that contain three significant figures.

Analyze

Writing all the values in scientific notation will help determine the number of significant figures in each.

(a) 7.02
(b) 6.452
(c) $302 = 3.02 \times 10^2$
(d) 6.02×10^{23}
(e) $12.77 = 1.277 \times 10^1$
(f) 3.43

Solve

The values that have three significant figures are (a) 7.02, (c) 302, (d) 6.02×10^{23}, and (f) 3.43.

Think About It

Remember that a zero between two other digits is always significant.

1.79. Collect and Organize

We are to express the result of each calculation to the correct number of significant figures.

Analyze

Section 1.8 in the textbook gives the rules regarding the significant figures that carry over in calculations. Remember to operate on the weak-link principle.

Solve

(a) The least well-known value has two significant figures, so the calculator result of 9.225×10^2 is reported as 9.2×10^2.
(b) The sum results in the least well-known digit in the hundredths place to give the sum known to four significant digits, which determines the significant digits to be four for the multiplication and division steps, so the calculator result of 1.29334×10^{-16} is reported as 1.293×10^{-16}.
(c) The numerator is known only to the tenths place for three significant digits and the denominator is known to four significant digits. The least well-known value, then, has three significant figures, so the calculator result of 1.5336×10^{-23} is reported as 1.53×10^{-23}.
(d) The numerator is known only to the tenths place for three significant digits and the denominator is known to four significant digits. The least well-known value, then, has three significant figures, so the calculator result of 3.726×10^{-6} is reported as 3.73×10^{-6}.

Think About It

Indicating the correct number of significant figures for a calculated value indicates the level of confidence we have in our calculated value. Reporting too many significant figures would indicate a higher level of precision in our number than we actually have.

1.81. Collect and Organize

We are asked in this problem to convert from kelvin to degrees Celsius.

Analyze

The relationship between the Kelvin temperature scale and the Celsius temperature scale is given by

$$K = {}^{\circ}C + 273.15$$

Rearranging this gives the equation to convert Kelvin to Celsius temperatures:

$${}^{\circ}C = K - 273.15$$

Solve

$${}^{\circ}C = 4.2\ K - 273.15 = -269.0\,{}^{\circ}C$$

Think About It
Because 4.2 K is very cold, we would expect that the Celsius temperature would be very negative. It should not, however, be lower than –273.15 K, since that is the lowest temperature possible.

1.83. **Collect and Organize**
Given the boiling point of ethyl chloride in degrees Celsius, we are to compute the boiling point in °F and K.

Analyze
The relationship between the Kelvin temperature scale and the Celsius temperature scale is given by
$$K = °C + 273.15$$
The relationship between the Celsius and Fahrenheit temperature scales is given by
$$°C = \frac{5}{9}\left(°F - 32\right)$$
We will have to rearrange this expression to find °F from °C:
$$°F = \frac{9}{5}\left(°C\right) + 32$$

Solve
The boiling point of ethyl chloride in the Fahrenheit and Kelvin scales is
$$K = 12.3°C + 273.15 = 285.4 \text{ K}$$
$$°F = \frac{9}{5}\left(12.3°C\right) + 32 = 54.1°F$$

Think About It
Notice that the answer is reported to four significant figures for the temperature in Kelvin and to two significant figures for the temperature in Fahrenheit because of the addition and multiplication rule.

1.85. **Collect and Organize**
This question asks us to convert the coldest temperature recorded on Earth from Fahrenheit to Celsius degrees and K.

Analyze
Since the Celsius and Kelvin scales are similar (offset by 273.15°), once we convert from Fahrenheit to Celsius, finding the Kelvin temperature will be straightforward. The equations we need are
$$°C = \frac{5}{9}\left(°F - 32\right)$$
$$K = °C + 273.15$$

Solve
$$°C = \frac{5}{9}\left(-128.6°F - 32\right) = -89.2°C$$
$$K = -89.2°C + 273.15 = 183.9 \text{ K}$$

Think About It
This temperature is cold on any scale!

1.87. **Collect and Organize**
We are asked to compare the critical temperature (T_c) of three superconductors. The critical temperatures, however, are given in three different temperature scales, so for the comparison, we will need to convert them to a single scale.

Analyze

Which temperature scale we use as the common one does not matter, but since the critical temperatures are low, expressing all the temperatures in kelvin might be easiest. The equations we will need are

$$K = {}^{\circ}C + 273.15 \quad \text{and} \quad {}^{\circ}C = \frac{5}{9}\left({}^{\circ}F - 32\right)$$

Solve

The T_c for $YBa_2Cu_3O_7$ is already expressed in kelvin, $T_c = 93.0$ K.
The T_c of Nb_3Ge is expressed in degrees Celsius and can be converted to kelvin by

$$K = -250.0\,{}^{\circ}C + 273.15 = 23.2 \text{ K}$$

The T_c of $HgBa_2CaCu_2O_6$ is expressed in Fahrenheit degrees. To get this temperature in kelvin, first convert to Celsius degrees:

$$ {}^{\circ}C = \frac{5}{9}\left(-231.1\,{}^{\circ}F - 32\right) = -146.2\,{}^{\circ}C $$

$$ K = -146.2\,{}^{\circ}C + 273.15 = 127.0 \text{ K} $$

The superconductor with the highest T_c is $HgBa_2CaCu_2O_6$ with a T_c of 127.0 K.

Think About It

The superconductor with the lowest T_c is Nb_3Ge with a T_c of 23.2 K, more than 100 K lower than the T_c of $HgBa_2CaCu_2O_6$.

1.89. **Collect and Organize**

In this question we are to determine the number of suspect data points that can be identified by using Grubbs' test.

Analyze

Grubbs' test is a test to statistically detect whether a particular data point is an outlier in a data set.

Solve

Because we test through Grubbs' test whether a particular data point is an outlier, we test only one data point at a time.

Think About It

If a data point is determined to be an outlier through Grubbs' test, it can be removed from the data set.

1.91. **Collect and Organize**

We are to decide whether the measure of mean ± standard deviation or a 95% confidence interval has the greater variability.

Analyze

The equation to determine the confidence interval is

$$\mu = \bar{x} \pm \frac{ts}{\sqrt{n}}$$

where μs is the true mean value, $\bar{x}$ is the mean, t is the value for a particular confidence level, s is the standard deviation, and n is the number of values in the data set.

Solve

Mean standard deviation is slightly greater because the span of 95% confidence interval for seven data points is

$$\pm\left(\frac{ts}{\sqrt{n}}\right) = \pm\left(\frac{2.447 \times s}{\sqrt{7}}\right) = 0.9249s$$

Think About It

Remember that built into the calculation of standard deviation and confidence levels is the assumption that the data vary randomly. The conclusion would be reversed when $n < 7$.

1.93. **Collect and Organize**

Given the data from three manufacturers of circuit boards for copper line widths, we are to calculate the mean and standard deviation, determine which of the data sets would include the data point of 0.500 µm in the 95% confidence interval, and decide which manufacturer was "precise and accurate" and which was "precise but not accurate."

Analyze

To calculate the mean, we sum all the values of the data set and divide by the number of data points in that data set. To calculate the standard deviation, we use the following formula

$$s = \sqrt{\frac{\sum_i \left(x_i - \bar{x}\right)^2}{n-1}}$$

To calculate the 95% confidence intervals for the data sets we use

$$\mu = \bar{x} \pm \frac{t\, s}{\sqrt{n}}$$

Solve

(a) For each manufacturer, the mean and standard deviations are:

Manufacturer 1

$$\bar{x} = \frac{0.512 + 0.508 + 0.516 + 0.504 + 0.513}{5} = 0.5106$$

$$s = \sqrt{\frac{(0.512-0.511)^2 + (0.508-0.511)^2 + (0.516-0.511)^2 + (0.504-0.511)^2 + (0.513-0.511)^2 +}{5-1}} = 0.0047$$

Manufacturer 2

$$\bar{x} = \frac{0.514 + 0.513 + 0.514 + 0.514 + 0.512}{5} = 0.5134$$

$$s = \sqrt{\frac{(0.514-0.513)^2 + (0.513-0.513)^2 + (0.514-0.513)^2 + (0.514-0.513)^2 + (0.512-0.513)^2 +}{5-1}} = 0.0009$$

Manufacturer 3

$$\bar{x} = \frac{0.500 + 0.501 + 0.502 + 0.502 + 0.501}{5} = 0.5012$$

$$s = \sqrt{\frac{(0.500-0.501)^2 + (0.501-0.501)^2 + (0.502-0.501)^2 + (0.502-0.501)^2 + (0.501-0.501)^2 +}{5-1}} = 0.0008$$

(b) For each manufacturer, the 95% confidence interval is:

Manufacturer 1

$$\mu = 0.5106 \pm \frac{2.776 \times 0.0047}{\sqrt{5}} = 0.5106 \pm 0.0058$$

This range would be 0.5048–0.5164 µm and would not include the 0.500 µm data point.

Manufacturer 2

$$\mu = 0.5134 \pm \frac{2.776 \times 0.0009}{\sqrt{5}} = 0.5134 \pm 0.0011$$

This range would be 0.5123–0.5145 µm and would not include the 0.500 µm data point.

Manufacturer 3

$$\mu = 0.5012 \pm \frac{2.776 \times 0.0008}{\sqrt{5}} = 0.5012 \pm 0.0010$$

This range would be 0.5002–0.5022 μm and would include the 0.500 μm data point for three significant digits.

(c) The manufacturer that is both precise (smallest spread of values) and accurate (closest average to 0.500 μm) is Manufacturer 3. The manufacturer that is precise (smallest spread of values) but not accurate (far from the average to 0.500 μm) is Manufacturer 2.

Think About It

In manufacturing electronic circuit boards the specifications must be very strictly adhered to. Manufacturer 3, which prints boards with the highest precision and accuracy, will win the contract.

1.95. Collect and Organize

In this question we will use Grubbs' test to determine whether the value 3.41 should be considered an outlier in a data set.

Analyze

To determine whether a data point is an outlier, we use the formula

$$Z = \frac{|x_i - \bar{x}|}{s}$$

where x_i is the data point being tested, $\bar{x}$ is the mean of the data, and s is the standard deviation. If the value of Z is greater than 1.887 ($n = 6$ for this data set) then that data point is an outlier.

Solve

The mean and standard deviations for this data set are

$$\bar{x} = \frac{3.15 + 3.03 + 3.09 + 3.11 + 3.12 + 3.41}{6} = 3.15$$

$$s = \sqrt{\frac{(3.15 - 3.15)^2 + (3.03 - 3.15)^2 + (3.09 - 3.15)^2 + (3.11 - 3.15)^2 + (3.12 - 3.15)^2 + (3.41 - 3.15)^2}{6 - 1}} = 0.1327$$

Applying Grubbs' test:

$$Z = \frac{|x_i - \bar{x}|}{s} = \frac{|3.41 - 3.15|}{0.1327} = 1.959$$

That value is greater than 1.887, so that data point is an outlier.

Think About It

This data point then can be discarded from the data set.

1.97. Collect and Organize

This question considers the runoff of nitrogen every year into a stream caused by a farmer's application of fertilizer. We must consider that not all the fertilizer contains nitrogen and not all the fertilizer runs off into the stream. We must also account for the flow of the stream in taking up the nitrogen runoff.

Analyze

First, we have to determine the amount of nitrogen in the fertilizer (10% of 1.50 metric tons, or 1500 kg since 1 metric ton = 1000 kg). Then we need to find how much of that nitrogen gets washed into the stream (15% of the mass of N in the fertilizer). Our final answer must be in milligrams of N, so we can convert the mass of N that gets washed into the stream from kilograms to milligrams.

$$\text{mass of fertilizer in kg} \times 0.10 = \text{mass of N in fertilizer in kg}$$

$$\text{mass of N in fertilizer in kg} \times 0.15 = \text{mass of N washed into the stream in kg}$$

$$\text{mass of N that washes into the stream in kg} \times \frac{1000\ \text{g}}{1\ \text{kg}} \times \frac{1000\ \text{mg}}{1\ \text{g}} = \text{mass of N that washes into the stream in mg}$$

Next, we need to know how much water flows through the farm each year via the stream. To find this, we must convert the rate of flow in cubic meters per minute to liters per year. We can convert this through one line by using dimensional analysis with the following conversions:

$$\frac{1000 \text{ L}}{1 \text{ m}^3}, \quad \frac{1 \text{ hr}}{60 \text{ min}}, \quad \frac{1 \text{ d}}{24 \text{ hr}}, \quad \text{and} \quad \frac{1 \text{ yr}}{365.25 \text{ d}}$$

Solve

The amount of N washed into the stream each year is

$$1500 \text{ kg} \times 0.10 = 150 \text{ kg of N in the fertilizer}$$

$$150 \text{ kg} \times 0.15 = 22.5 \text{ kg of N washed into the stream in one year}$$

$$22.5 \text{ kg} \times \frac{1000 \text{ g}}{1 \text{ kg}} \times \frac{1000 \text{ mg}}{1 \text{ g}} = 2.25 \times 10^7 \text{ mg of N washed into the stream in one year}$$

The amount of stream water flowing through the field each year is

$$\frac{1.4 \text{ m}^3}{1 \text{ min}} \times \frac{1000 \text{ L}}{1 \text{ m}^3} \times \frac{60 \text{ min}}{1 \text{ hr}} \times \frac{24 \text{ hr}}{1 \text{ d}} \times \frac{365.25 \text{ d}}{1 \text{ yr}} = 7.36 \times 10^8 \text{ L/yr}$$

The additional concentration of N added to the stream by the fertilizer is

$$\frac{2.25 \times 10^7 \text{ mg of N/yr}}{7.36 \times 10^8 \text{ L/yr}} = 0.031 \text{ mg/L}$$

Think About It

The calculated amount of nitrogen added to the stream seems reasonable. The concentration is relatively low because the stream is moving fairly swiftly and the total amount of nitrogen that washes into the stream over the year is not too great. The problem, however, does not tell us whether this amount would harm the plant and animal life in the stream.

1.99. ### Collect and Organize

In this problem we need to express each mixture of chlorine and sodium as a ratio. The mixture closest to the ratio for chlorine to sodium will be the one with the desired product, leaving neither sodium nor chlorine left over.

Analyze

First, we must calculate the ratio of chlorine to sodium in sodium chloride. This is a simple ratio of the masses of those two substances:

$$\frac{\text{mass of chlorine}}{\text{mass of sodium}} = \text{ratio of the two components}$$

We can compare the ratios of the other mixtures by making the same calculations.

Solve

In sodium chloride, the mass ratio of chlorine to sodium is

$$\frac{1.54 \text{ g of chlorine}}{1.00 \text{ g of sodium}} = 1.54$$

Repeating this calculation for the four mixtures, we obtain the ratio of chlorine to sodium:

$$\frac{17.0 \text{ g}}{11.0 \text{ g}} = 1.55 \text{ for mixture a} \qquad \frac{12.0 \text{ g}}{6.5 \text{ g}} = 1.8 \text{ for mixture c}$$

$$\frac{10.0 \text{ g}}{6.5 \text{ g}} = 1.5 \text{ for mixture b} \qquad \frac{8.0 \text{ g}}{6.5 \text{ g}} = 1.2 \text{ for mixture d}$$

Both mixtures a and b react so that neither sodium nor chlorine is left over.

Think About It

Mixture c has leftover chlorine and mixture d has leftover sodium after the reaction is complete.

1.101. **Collect and Organize**

This problem asks us to compute the percentages of the two ingredients in trail mix as manufactured on different days.

Analyze

Because we compare each day's percentage of peanuts in the trail mix bags with the ideal range of 65%–69%, we have to compute each day's percentage of peanuts from the data given. Each day has a total of 82 peanuts plus raisins, so the percentage of the mix in peanuts for each day is calculated from the equation

$$\% \text{ peanuts} = \frac{\text{number of peanuts in mix}}{82} \times 100$$

Solve

For each day, the percentage of peanuts is

$$\frac{50}{82} \times 100 = 61\% \text{ peanuts, Day 1} \qquad \frac{48}{82} \times 100 = 59\% \text{ peanuts, Day 21}$$

$$\frac{56}{82} \times 100 = 68\% \text{ peanuts, Day 11} \qquad \frac{52}{82} \times 100 = 63\% \text{ peanuts, Day 31}$$

The only day that met the specifications for the percentage of peanuts in the trail mix was Day 11.

Think About It

On Days 1, 21, and 31, too few peanuts were in the trail mix.

1.103. **Collect and Organize**

Given the correct dosage of phenobarbital per day and the details of the drug given to a patient over 3 days, we are to determine how many times over the prescribed dose was given to an overdose patient.

Analyze

We can use a common unit of milligrams of the drug to compare the prescribed amount with the overdose amount. To do so we will have to convert 0.5 grains into milligrams and multiply by the 3 days the drug was given. The actual amount given to the patient was four times 130 mg. We can then compare these two dosages in a ratio.

Solve

Amount of phenobarbital prescribed for 3 days in milligrams:

$$\frac{0.5 \text{ grains}}{\text{day}} \times \frac{64.79891 \text{ mg}}{\text{grain}} \times 3 \text{ days} = 97.1984 \text{ mg}$$

Actual amount given to patient in four doses over three days:

$$\frac{130 \text{ mg}}{\text{dose}} \times 4 \text{ doses} \times 3 \text{ d} = 1560 \text{ mg}$$

Ratio of actual dose to prescribed dose for 3 days:

$$\frac{1560 \text{ mg}}{97.1984 \text{ mg}} = 16 \text{ times too much phenobarbital was given}$$

Think About It

An overdose of such a powerful sedative as phenobarbital can be fatal. Symptoms include shallow breathing, extreme sleepiness, and blurry vision.

1.105. **Collect and Organize**

Given the temperatures for the freezing point and boiling point of water measured using three digital hospital thermometers, we are to determine which ones could detect a 0.1°C increase in temperature and which would give an accurate reading of normal body temperature of 36.8°C.

Analyze

(a) To detect a 0.1°C temperature rise, the scale on the thermometer would not have to be expanded over the range so that the 0.1°C could be detected (the thermometers all read only to a tenth of a degree). However, if the temperature scale for the thermometer is contracted, it will detect the 0.1˚C temperature change because its intervals of 0.1˚C are smaller.

(b) To determine whether any of the thermometers can accurately measure a temperature of 36.8˚C, we need to consider the calibration curves (constructed by comparing the measured freezing and boiling points with the actual—that is, by plotting the correct temperatures vs. the measured temperatures). From the equation for the line, we can solve for the reading on the thermometer when the actual temperature is 36.8˚C.

Solve

(a) We can't tell from the data given. All reading for these thermometers end in an even number, which may mean that the minimum detectable change in temperature is 0.2˚C.

(b) The calibration curves for all three thermometers are shown below. In each graph, the slope is derived from the actual range of the freezing point and boiling point of water (100˚C–0˚C, Δy) and the range for the particular thermometer for these points (Δx).

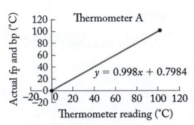 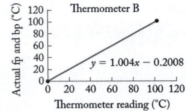 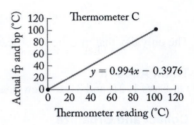

The equation of the line gives the calibration equation for us to use in the calculation of the temperature each thermometer would read (x) for the actual temperature of 36.8˚C (y).

For thermometer A:

$$36.8˚C = 0.998x + 0.7984$$
$$x = 36.1˚C$$

For thermometer B:

$$36.8˚C = 1.004x - 0.2008$$
$$x = 36.9˚C$$

For thermometer C:

$$36.8˚C = 0.994x - 0.3976$$
$$x = 37.4˚C$$

None of these thermometers can accurately read the patient's temperature as 36.8˚C within ±0.1˚C.

Think About It

Thermometer B comes closest in measuring a temperature of 36.8˚C.

1.107. Collect and Organize

We are asked to convert the speed (the heliocentric velocity) of the *New Horizons* spacecraft from kilometers per second to miles per hour.

Analyze

To convert from kilometers to miles, we use the conversion 1 mile = 1.609 km; to convert from seconds to hours we use the conversions 60 s = 1 min and 60 min = 1 hour.

Solve

$$\frac{14.51 \text{ km}}{s} \times \frac{1 \text{ mi}}{1.609 \text{ km}} \times \frac{60 \text{ s}}{1 \text{ min}} \times \frac{60 \text{ min}}{1 \text{ hr}} = 3.246 \times 10^4 \text{ mi/hr}$$

Think About It

With the speed of light being 6.706×10^8 miles/hour this spacecraft is traveling at only 0.005% the speed of light.

CHAPTER 2 | Atoms, Ions, and Molecules: The Building Blocks of Matter

2.1. Collect and Organize

This question asks us to identify the element in Figure P2.1 that has the fewest protons in its nucleus.

Analyze

An element is defined by the number of protons in its nucleus; as we move to higher elements in the periodic table, the number of protons increases.

Solve

The lightest, lowest atomic number element highlighted is hydrogen (purple). Hydrogen therefore has the fewest protons (one) in its nucleus.

Think About It

The other elements highlighted have more protons in their nucleus: helium (blue) has 2, fluorine (green) has 9, sulfur (yellow) has 16, and arsenic (red) has 33.

2.3. Collect and Organize

We are asked to identify which shaded element in Figure P2.1 is stable and yet has no neutrons in its nucleus.

Analyze

The presence of neutrons in nuclei helps to overcome the repulsive forces of more than one proton in the nucleus.

Solve

The only shaded element that does not have more than one proton in its nucleus and therefore would be stable without neutrons is the element shaded purple, hydrogen.

Think About It

Without the neutrons, the nuclei with more than one proton would fly apart.

2.5. Collect and Organize

From among the shaded elements in the periodic table shown in Figure P2.4, we are to identify a transition metal, an alkali metal, and a halogen.

Analyze

Transition metals occupy the middle of the periodic table and reside in the *d*-block, alkali metals occupy the first column of the periodic table in the *s*-block, and halogens occupy the second-to-last column on the right, next to the noble gases.

Solve

(a) The transition metal element is shaded green and is gold (Au).
(b) The alkali metal is shaded blue and is sodium (Na).
(c) The halogen is shaded lilac and is chlorine (Cl).

Think About It

A few other regions of the periodic table also have names. Can you identify the alkaline earth metals, the noble gases, and the chalcogens?

2.7. **Collect and Organize**

Using Figure P2.7 and our knowledge of the charges and the masses of alpha and beta particles, we are to determine which arrow (red, blue, or green) represents each particle's behavior as it moves through an electric field.

Analyze

Alpha particles are much more massive than beta particles and they have a positive charge, whereas beta particles have a negative charge.

Solve

Alpha particles, with their positive charges, will be deflected toward the negative side of the electric field. Their behavior is represented by the red arrow. Beta particles, however, are deflected toward the positive side of the electric field because they carry a negative charge; they are represented by the green arrow. Notice, too, that the alpha particles travel farther and with less arc in their deflection than the lighter beta particles. Because alpha particles have greater mass, their momentum through the electric field is higher and it would take a stronger field to deflect them to the same degree.

Think About It

The blue arrow must represent a neutral particle because it is not deflected by the electric field.

2.9. **Collect and Organize**

Using Figure P2.9, we are to determine whether the mass spectrum shown is for dichloromethane or cyclohexane.

Analyze

Dichloromethane has a molar mass of 84.93 g/mol and cyclohexane has a molar mass of 84.15 g/mol. Those masses are close to each other, so we will have to use other clues from the mass spectrum to help. The figure shows two obvious mass peaks at 86 and 88 amu and another grouping around 49 amu.

Solve

The molecular ion peaks showing up at 84, 86, and 88 amu show that this compound has isotopes of significant abundance. Cyclohexane has only carbon and hydrogen atoms and the abundance of C and H isotopes are too low to show this significantly in the mass spectrum. The m/z peak at 84 amu is due to the presence of two atoms of ^{35}Cl ($CH_2{}^{35}Cl_2$), the m/z peak at 86 amu is due to the presence of one atom of ^{35}Cl and one atom of ^{37}Cl ($CH_2{}^{35}Cl^{37}Cl$), and the m/z peak at 88 amu is due to the presence of two atoms of ^{37}Cl ($CH_2{}^{37}Cl_2$). Also, the m/z peak grouping around 50 amu is consistent with CH_2Cl_2 losing one ^{35}Cl or one ^{37}Cl atom.

Think About It

^{35}Cl is about 76% naturally abundant and ^{37}Cl is about 24% naturally abundant. The relative intensities of the isotopic peaks in the mass spectrum also reflect those differences in abundance.

2.11. **Collect and Organize**

This question asks us to correlate the position of an element in the periodic table with typical charges on the ions for the groups (or families) of elements.

Analyze

Figure 2.10 in the textbook shows the common charges on the elements used in forming compounds. That figure will help us determine which elements in monatomic form give the charges named in the question.

Solve

Highlighted elements in Figure P2.11 are K, Mg, Sc, Ag, O, and I.
(a) Elements in group 1 form 1+ ions, so K will form K^+ (dark blue). Silver (green) also typically forms a 1+ cation.
(b) Elements in group 2 form 2+ ions, so Mg forms Mg^{2+} (gray).
(c) Elements in group 3 form 3+ ions, so Sc forms Sc^{3+} (yellow).
(d) Elements in group 17 (the halogens) form 1– ions, so I forms I^- (purple).
(e) Elements in group 16 form 2– ions, so O forms O^{2-} (red).

Think About It
Elements on the left-hand side of the periodic table form cations, and elements on the right-hand side tend to form anions.

2.13. Collect and Organize
In this question we are asked to explain how Rutherford's gold-foil experiment changed the plum-pudding model of the atom.

Analyze
The plum-pudding model of the atom viewed the electrons as small particles in a diffuse, positively charged "pudding." In Rutherford's experiment, most of the α particles (positively charged particles) directed at the gold foil went straight through, but a few of them bounced back toward the source of the α particles.

Solve
The plum-pudding model could not explain the infrequent large-angle deflections of alpha particles that Rutherford's students observed. However, these deflections could be explained by the particles' colliding with tiny, dense atomic nuclei that took up little of the volume of gold atoms, but that contained all the positive charge and most of the atoms' mass (Rutherford's model).

Think About It
The nucleus is about 10^{-15} m in diameter, whereas the atom is about 10^{-10} m. That size difference has often been compared to "a fly in a cathedral."

2.15. Collect and Organize
In this question we are to explain how J. J. Thomson discovered that cathode rays were not rays of pure energy but were actually charged particles.

Analyze
Thomson's experiment directed the cathode ray through a magnetic field, and he discovered that the ray was deflected. A magnetic field would not deflect pure-energy "rays."

Solve
When Thomson observed cathode rays being deflected by a magnetic field, he reasoned that the rays were streams of charged particles because only moving charged particles would interact with a magnetic field. Pure-energy rays would not.

Think About It
Thomson's discovery of the electron in cathode rays did not eliminate the use of the term *cathode ray*. CRTs (cathode-ray tubes) are the traditional (that is, not LCD) television and computer screens.

2.17. Collect and Organize
Helium is found in pitchblende, an ore of radioactive uranium oxide found on Earth. We are asked to explain why helium is present in that ore.

Analyze
Pitchblende contains uranium oxide, and uranium is a naturally occurring radioactive element.

Solve
The helium is present because uranium (and some of its products of further decays) decays by α emission. Alpha particles, composed of two protons and two neutrons, easily pick up electrons from their environment to become helium.

Think About It
All the helium on Earth is generated in this fashion and trapped. Helium, though, once in the atmosphere, escapes into space because it is so light.

2.19. Collect and Organize

In this "thought experiment" we are asked to predict how the result would have been different if Rutherford's gold atoms in the foil had absorbed the α particles.

Analyze

If α particles were absorbed by the gold, that would mean that the α particles were somehow "reacting" with the nuclei of the gold atoms.

Solve

We would see fewer α particles either passing directly through the foil or being deflected at large angles. Absorption of an α particle by a gold atom would have transmuted the gold atom into the heavier element, thallium.

Think About It

The nuclear reaction that would describe the transmutation of gold-197 into thallium is

$$^{197}\text{Au} + \alpha \rightarrow {}^{201}\text{Tl}$$

2.21. Collect and Organize

This question asks us to consider the ratio of neutrons to protons in an element where we are given the fact that the mass number is more than twice the atomic number.

Analyze

We can find the number of neutrons for an isotope by relating the number of protons to the mass number. From that result we can then determine the neutron-to-proton ratio.

Solve

We are given an isotope in which the mass number is more than twice the number of protons. With m being the mass number and p the number of protons, we can express this relationship as

$$m > 2p$$

The mass number is also equal to the number of protons plus the number of neutrons (n),

$$m = p + n$$

Combining these expressions

$$p + n > 2p$$

and solving for n gives

$$n > 2p - p$$
$$n > p$$

Therefore, the number of neutrons in this isotope is greater than the number of protons, and the neutron-to-proton ratio is greater than 1.

Think About It

We wouldn't have had to express the relationships between the nuclear particles mathematically if the isotope had a mass number equal to twice the number of protons. Then the number of neutrons would have to be the same as the number of protons, giving a neutron-to-proton ratio of 1:1.

2.23. Collect and Organize

Given that most stable nuclides have at least the same numbers of neutrons in their nuclei as protons (and often more), we are to identify which element is an exception to that rule.

Analyze

Neutrons help stabilize the nucleus by counteracting the repulsive forces between protons.

Solve

Hydrogen, with only one proton, does not need neutrons to be stable and so is the exception.

Think About It

Hydrogen, however, can have one (for deuterium) and even two (for tritium) neutrons in its nucleus.

2.25. Collect and Organize

For each element in this question, we must look at the relationship of the neutrons, protons, and electrons. We need to determine the element's atomic number from the periodic table and, from the mass number given for the isotope, compute the number of neutrons to give the indicated isotope.

Analyze

An isotope is given by the symbol $^A_Z X$, where X is the element symbol from the periodic table, Z is the atomic number (the number of protons in the nucleus), and A is the mass number (the number of protons and neutrons in the nucleus). Often, Z is omitted because the element symbol gives us the same information about the identity of the element. To determine the number of neutrons in the nucleus for each named isotope, we subtract Z (number of protons) from A (mass number). If the elements are neutral (no charge), the number of electrons equals the number of protons in the nucleus.

Solve

	Atom	Mass Number	Atomic Number = Number of Protons	Number of Neutrons = Mass Number – Atomic Number	Number of Electrons = Number of Protons
(a)	^{14}C	14	6	8	6
(b)	^{59}Fe	59	26	33	26
(c)	^{90}Sr	90	38	52	38
(d)	^{210}Pb	210	82	128	82

Think About It

Isotopes of an element contain the same number of protons but a different number of neutrons. Thus, isotopes have different masses.

2.27. Collect and Organize

After calculating the neutron-to-proton ratio for ^4He, ^{23}Na, ^{59}Co, and ^{197}Au, we are to comment on how the ratio changes as Z increases.

Analyze

To calculate the neutron-to-proton ratios we need to determine, from the mass number and the atomic number, the number of protons and neutrons for each element:

^4He, atomic number 2, has 2 protons and 2 neutrons
^{23}Na, atomic number 11, has 11 protons and 12 neutrons
^{59}Co, atomic number 27, has 27 protons and 32 neutrons
^{197}Au, atomic number 79, has 79 protons and 118 neutrons

Solve

The neutron-to-proton ratio for each element is
(a) ^4He, 2/2 = 1.00
(b) ^{23}Na, 12/11 = 1.09
(c) ^{59}Co, 32/27 = 1.19
(d) ^{197}Au, 118/79 = 1.49
As the atomic number (Z) increases, the neutron-to-proton ratio increases.

Think About It

More neutrons are required to stabilize nuclei with more protons because of the strong repulsive forces between the positively charged protons. Neutrons bring added strong nuclear force to the nucleus to stabilize it.

2.29. Collect and Organize

To fill in the table, we have to consider how the numbers of nuclear particles relate to one another. We also need to recall how the symbols for the isotopes are written. From the table, it is apparent that we have to work backward in some cases from the number of electrons or protons and mass number for the element symbol.

Analyze

An isotope is given by the symbol $^A_Z X$, where X is the element symbol from the periodic table, Z is the atomic number (the number of protons in the nucleus), and A is the mass number (the number of protons and neutrons in the nucleus). We can determine the number of neutrons in the nucleus for the isotopes by subtracting Z (number of protons) from A (mass number). If the elements are neutral (no charge), the number of electrons equals the number of protons in the nucleus. Except for the first element, we will assume that the rest of the elements are neutral, otherwise there would be different possibilities for the numbers of proton and electrons.

Solve

Symbol	^{23}Na	^{89}Y	^{118}Sn	^{197}Au
Number of Protons	11	39	50	79
Number of Neutrons	12	50	68	118
Number of Electrons	11	39	50	79
Mass Number	23	89	118	197

Think About It

Because the nuclear particles are all related to one another, either we can work from the isotope symbol to find the number of protons, neutrons, and electrons for a particular isotope or we can work from the mass number and the number of electrons or protons to determine the number of neutrons and write the element symbol.

2.31. Collect and Organize

An isotope is given by the symbol $^A_Z X^n$, where X is the element symbol from the periodic table, Z is the atomic number (the number of protons in the nucleus), A is the mass number (the number of protons and neutrons in the nucleus), and n is the charge on the species.

Analyze

If we are given the number of protons in the nucleus, the element can be identified from the periodic table. We can determine the mass number by adding the protons to the neutrons in the nucleus for the isotope. We can determine the number of neutrons or protons in the nucleus for the isotopes by subtracting Z (number of protons) or the number of neutrons from A (mass number), respectively. We can account for the charge on the species by adding electrons (to form a negatively charged ion) or by subtracting electrons (to form a positively charged ion).

Solve

Symbol	$^{37}Cl^-$	$^{23}Na^+$	$^{81}Br^-$	$^{226}Ra^{2+}$
Number of Protons	17	11	35	88
Number of Neutrons	20	12	46	138
Number of Electrons	18	10	36	86
Mass Number	37	23	81	226

Think About It

To form a singly charged ion, there has to be one more electron (for a negative charge) or one fewer electron (for a positive charge) than the number of protons in the nucleus. For a doubly charged ion, we add or take away two electrons.

2.33. Collect and Organize

Knowing that Mendeleev labeled his groups on the left of the periodic table on the basis of the formulas of the compounds they formed with oxygen, we are to assign his labels to groups 2, 3, and 4 on the modern periodic table.

Analyze

Groups 2, 3, and 4 have cations with charges of 2+, 3+, and 4+, respectively. Oxygen forms an anion of 2– charge. The oxygen compounds that would form would balance the positive cation charge with the negative anion charge by combining the elements in that group with oxygen in whole-number ratios.

Solve

Group 2, with a 2+ charge, would form RO with oxygen; group 3, with a 3+ charge, would form an R_2O_3 compound with oxygen; and group 4, with a 4+ charge, would form an RO_2 compound with oxygen.

Think About It

This classification was based on the chemical behavior of the elements, which was Mendeleev's brilliant insight.

2.35. **Collect and Organize**

We are asked why Mendeleev did not leave spaces for the noble gases in his periodic system.

Analyze

The noble gases are characterized by their remarkable unreactivity. Unreactive elements can be quite unnoticeable because they do not form compounds with other elements.

Solve

The noble gases were not discovered until after Mendeleev put together his periodic table. He also could not have predicted the existence of the noble gases at the time since (a) none of them was isolated and characterized on the basis of their reactivity (or lack thereof) to indicate their presence in nature and (b) he arranged the elements in order of increasing mass, not atomic number. If he had been aware of atomic numbers as characteristic of the elements, he would have noticed that the atomic numbers for the noble gases were missing as a column in his table.

Think About It

The noble gases are monatomic, are colorless and odorless, and have a remarkably narrow liquid range (their boiling points and melting points are close together).

2.37. **Collect and Organize**

Knowing that the explosive TNT contains second-row elements in groups 14, 15, and 16 as well as hydrogen, we are to name those particular elements.

Analyze

The second-row elements start with lithium and end at neon.

Solve

The elements in TNT besides hydrogen are carbon (group 14), nitrogen (group 15), and oxygen (group 16).

Think About It

Many explosives have the same elements. The powerful C4 explosive is composed mainly of the explosive RDX_2, which has the chemical formula $C_3H_6N_6O_6$.

2.39. **Collect and Organize**

Given information about the elements (their group numbers and relationship to one another in the periodic table) used in catalytic converters, we are to name the elements.

Analyze

In examining the clues in parts a–c, we can guess that these will all be transition metals.

Solve
(a) The element in the fifth row of the periodic table (Rb–Xe) in group 10 is palladium (Pd).
(b) The element to the left of Pd in group 9 is rhodium (Rh).
(c) The element below Pd in group 10 is platinum (Pt).

Think About It
These metals are generally as expensive as gold. Platinum is about $1400/oz and gold is about $1200/oz. Rhodium and palladium are less expensive, at about $1000/oz and $700/oz, respectively.

2.41. Collect and Organize
We are to count the metallic elements in the third row of the periodic table.

Analyze
The third row in the periodic table begins at sodium and ends at argon. The semimetal elements begin at silicon and move into the nonmetals with phosphorus.

Solve
Three metallic elements are in the third row in the periodic table: sodium (Na), magnesium (Mg), and aluminum (Al).

Think About It
As a semimetal, silicon has some, but not all, the properties of a metal but has a structure more like that of a nonmetal.

2.43. Collect and Organize
We define *weighted average* for this question.

Analyze
An average is a number that expresses the middle of the data (here, for various masses of atoms or isotopes).

Solve
A weighted average takes into account the proportion of each value in the group of values to be averaged. For example, the average of 2, 2, 2, and 5 would be computed as $(2 + 2 + 2 + 5)/4 = 2.75$. That average shows the heavier weighting toward the values of 2.

Think About It
Because isotopes for any element are not equally present but have a range of natural abundances, all the masses in the periodic table for the elements are calculated weighted averages.

2.45. Collect and Organize
Given that the abundance of the two isotopes of an element are both 50%, we are to express the average atomic mass of the element if the mass of isotope X is m_X and the mass of isotope Y is m_Y.

Analyze
Because each isotope is present in exactly 50% abundance, the average atomic mass will be the simple average of the two masses of isotopes X and Y.

Solve

$$\text{average atomic mass} = \frac{m_X + m_Y}{2}$$

Think About It
Some elements in nature have just one naturally occurring isotope. Can you find some examples in Appendix 3?

2.47. **Collect and Organize**

In this question we are asked to explain the observation that the average atomic mass of platinum is 195.08 amu, whereas the natural abundance of ^{195}Pt is only 33.8%.

Analyze

The average atomic mass on the periodic table is the weighted average of all the masses of the naturally occurring isotopes for that element. Only if an element has one naturally occurring element will the average atomic mass match the mass of that isotope.

Solve

We are given that the ^{195}Pt isotope is not 100% abundant. Therefore, we must conclude that the other isotopes with masses greater than 195 amu have natural abundances in equal proportion to those isotopes with masses lower than 195, so that the weighted average atomic mass calculates to 195.08 amu.

Think About It

For example, platinum might have three isotopes, each in 33% abundance: ^{194}Pt, ^{195}Pt, and ^{196}Pt.

2.49. **Collect and Organize**

We are to determine which isotope of argon is most abundant, ^{36}Ar, ^{38}Ar, or ^{40}Ar.

Analyze

To help here, it is useful to know from the periodic table that the average atomic mass of argon is 39.948 amu.

Solve

Only if the highest mass isotope were most abundant would the average mass of argon be 39.948 amu. Therefore, ^{40}Ar is the most abundant isotope of argon.

Think About It

Indeed, ^{40}Ar is 99.6% abundant.

2.51. **Collect and Organize**

We have to consider the concept of weighted average atomic mass to answer this question.

Analyze

We are asked to compare two isotopes and their weighted average mass. If the lighter isotope is more abundant, the average atomic mass will be less than the average if both isotopes are equally abundant. If the heavier isotope is more abundant, the average atomic mass will be greater than the simple average of the two isotopes. We are given the mass number for the isotopes as part of the isotope symbol, and we will take that as the mass of that isotope in atomic mass units.

Solve

(a) The simple average atomic mass for ^{10}B and ^{11}B would be 10.5 amu. The actual average mass (10.811 amu) is greater than this; therefore, ^{11}B is more abundant.
(b) The simple average atomic mass for ^{6}Li and ^{7}Li would be 6.5 amu. The actual average mass (6.941 amu) is greater than this; therefore, ^{7}Li is more abundant.
(c) The simple average atomic mass for ^{14}N and ^{15}N would be 14.5 amu. The actual average mass (14.007 amu) is less than this; therefore, ^{14}N is more abundant.

Think About It

This is a quick question to answer for elements such as boron, lithium, and nitrogen that have the dominance of only two isotopes in terms of their abundance. Answering the same question for elements with more than two stable isotopes in relatively high abundances is a little harder.

2.53. **Collect and Organize**

In this question we are given the masses and abundances of the naturally occurring isotopes of copper. From that information, we can calculate the average atomic mass of copper.

Analyze

To calculate the average atomic mass, we have to consider the relative abundances according to the following formula:

$$m_x = a_1m_1 + a_2m_2 + a_3m_3 + \cdots$$

where a_n refers to the abundance of isotope n and m_n refers to the mass of isotope n. If the relative abundances are given as percentages, the value we use for a_n in the formula is the percentage divided by 100.

Solve

For the average atomic mass of copper

$$m_{Cu} = (0.6917 \times 62.9296 \text{ amu}) + (0.3083 \times 64.9278 \text{ amu}) = 63.55 \text{ amu}$$

Think About It

Because copper-63 is more abundant than copper-65, we expect that the average atomic mass for copper would be below the simple average of 64.

2.55. **Collect and Organize**

Here we are asked to find out whether the mass of magnesium on Mars is the same as here on Earth. We are given the masses of each of the three isotopes of Mg in the Martian sample. Once we calculate the weighted average for Mg for the Martian sample, we can compare it with the average mass for Mg found on Earth.

Analyze

To calculate the average atomic mass, we have to consider the relative abundances according to the following formula:

$$m_x = a_1m_1 + a_2m_2 + a_3m_3 + \cdots$$

where a_n refers to the abundance of isotope n and m_n refers to the mass of isotope n. If the relative abundances are given as percentages, the value we use for a_n in the formula is the percentage divided by 100.

Solve

For the average atomic mass of magnesium in the Martian sample

$$m_{Mg} = (0.7870 \times 23.9850 \text{ amu}) + (0.1013 \times 24.9858 \text{ amu}) + (0.1117 \times 25.9826 \text{ amu})$$
$$= 24.31 \text{ amu}$$

The average mass of Mg on Mars is the same as here on Earth.

Think About It

The mass of Mg on Mars should be close to the same value as on Earth; the magnesium on both planets arrived in the solar system via the same ancient stardust.

2.57. **Collect and Organize**

In this problem, we again use the concept of weighted average atomic mass, but here we are asked to work backward from the average mass to find the exact mass of the ^{48}Ti isotope.

Analyze

We can use the formula for finding the weighted average atomic mass, but this time our unknown quantity is one of the isotope masses. Here,

$$m_{Ti} = a_{^{46}Ti}m_{^{46}Ti} + a_{^{47}Ti}m_{^{47}Ti} + a_{^{48}Ti}m_{^{48}Ti} + a_{^{49}Ti}m_{^{49}Ti} + a_{^{50}Ti}m_{^{50}Ti}$$

Solve

$$47.867 \text{ amu} = (0.0825 \times 45.9526) + (0.0744 \times 46.9518) + (0.7372 \times m_{^{48}\text{Ti}})$$
$$+ (0.0541 \times 48.94787) + (0.0518 \times 49.94479)$$
$$m_{^{48}\text{Ti}} = 47.948 \text{ amu}$$

Think About It

That answer makes sense because the exact mass of ^{48}Ti should be close to 48 amu.

2.59. Collect and Organize

From the formulas for three ionic compounds, CaF_2, Na_2S, and Cr_2O_3, we are to calculate the masses of the formula units.

Analyze

For each formula we will sum the masses of the elements, making sure that we also account for the number of a particular element present in the formula.

Solve

(a) CaF_2: 40.078 amu + 2(18.998 amu) = 78.074 amu
(b) Na_2S: 2(22.990 amu) + 32.065 amu = 78.045 amu
(c) Cr_2O_3: 2(51.996 amu) + 3(15.999 amu) = 151.989 amu

Think About It

In determining the formula mass, use as many significant figures in your calculation as listed on the periodic table. Resist the temptation to round up or down, which would make the calculation for the mass less accurate.

2.61. Collect and Organize

For each given molecular formula, we are to determine the number of carbon atoms in each.

Analyze

The subscript in each formula after the C atom is the number of carbons in the molecule.

Solve

(a) 1
(b) 3
(c) 6
(d) 6

Think About It

Those molecules are all organic molecules, and writing their formulas as $C_aH_bN_cO_d$ followed by other elements, if present, is customary.

2.63. Collect and Organize

For a list of five compounds, we are to determine their molecular masses and then rank them in order of increasing mass.

Analyze

When we use the masses on the periodic table to calculate the molecular masses, we obtain the following:
(a) CO = 28.01 amu
(b) Cl_2 = 70.91 amu
(c) CO_2 = 44.01 amu
(d) NH_3 = 17.03 amu
(e) CH_4 = 16.04 amu

Solve

In order of increasing molecular mass: (e) CH_4 < (d) NH_3 < (a) CO < (c) CO_2 < (b) Cl_2.

Think About It

In this problem fewer significant figures were necessary for the molecular masses because we were going to compare masses, and the masses were not likely to be too close together to warrant more than four significant digits.

2.65. Collect and Organize

For describing a collection of atoms or molecules, we are asked why using the unit *dozen* to express the number of atoms or molecules we have might not be a good idea.

Analyze

A dozen is 12 objects and therefore a relatively small group.

Solve

Although a dozen is a convenient and recognizable unit for donuts and eggs, it is too small a unit to express the very large number of atoms, ions, or molecules present in a mole.

$$\frac{6.022 \times 10^{23} \text{ atoms}}{\text{mole}} \times \frac{1 \text{ dozen}}{12 \text{ atoms}} = 5.02 \times 10^{22} \text{ dozen/mole}$$

Think About It

The mole (6.022×10^{23}) is a much more convenient unit to express the number of atoms or molecules in a sample.

2.67. Collect and Organize

We are asked whether equal masses of two isotopes of an element contain the same number of atoms.

Analyze

The number of atoms present for a given mass is dependent on the molar mass of the substance. Substances (including isotopes) with higher molar masses will have fewer atoms than those with lower molar masses.

Solve

No, the isotope with the higher molar mass will contain fewer atoms than the isotope of lower molar mass.

Think About It

That difference may not be significant, however, since many isotopic molar masses are close to each other.

2.69. Collect and Organize

In this exercise, we convert the given number of atoms or molecules of each gas to moles.

Analyze

To convert the number of atoms or molecules to moles, we divide by Avogadro's number.

Solve

(a) $\dfrac{4.4 \times 10^{14} \text{ atoms of Ne}}{6.022 \times 10^{23} \text{ atoms/mol}} = 7.3 \times 10^{-10} \text{ mol Ne}$

(c) $\dfrac{2.5 \times 10^{12} \text{ molecules of O}_3}{6.022 \times 10^{23} \text{ molecules/mol}} = 4.2 \times 10^{-12} \text{ mol O}_3$

(b) $\dfrac{4.2 \times 10^{13} \text{ molecules of CH}_4}{6.022 \times 10^{23} \text{ molecules/mol}} = 7.0 \times 10^{-11} \text{ mol CH}_4$

(d) $\dfrac{4.9 \times 10^{9} \text{ molecules of NO}_2}{6.022 \times 10^{23} \text{ molecules/mol}} = 8.1 \times 10^{-15} \text{ mol NO}_2$

Think About It

The trace gas with the most atoms or molecules present also has the most moles present. In that sample of air, the amount of the trace gases decreases in the order Ne > CH_4 > O_3 > NO_2.

2.71. Collect and Organize

From the chemical formulas for various iron compounds with oxygen, we are asked to determine how many moles of iron are in 1 mol of each substance.

Analyze

The chemical formula reflects the molar ratios of the elements in the compound. If one atom of iron is in the compound's chemical formula, then 1 mol of iron is in 1 mol of the compound. Likewise, if three atoms of iron are in the chemical formula, 3 mol of iron is present in 1 mol of the substance.

Solve

(a) One atom of iron is in FeO; therefore, 1 mol of FeO contains 1 mol of iron.
(b) Two atoms of iron are in Fe_2O_3; therefore, 1 mol of Fe_2O_3 contains 2 mol of iron.
(c) One atom of iron is in $Fe(OH)_3$; therefore, 1 mol of $Fe(OH)_3$ contains 1 mol of iron.
(d) Three atoms of iron are in Fe_3O_4; therefore, 1 mol of Fe_3O_4 contains 3 mol of iron.

Think About It

The parentheses used in $Fe(OH)_3$ show that three OH units are in that compound. If the question had asked how many moles of oxygen were present in 1 mol of that substance, the answer would be 3 mol of oxygen.

2.73. Collect and Organize

We are to calculate the mass of a given number of moles of magnesium carbonate.

Analyze

To convert from moles to mass, multiply the number of moles by the molar mass of the substance. The molar mass of $MgCO_3$ is $24.30 + 12.01 + 3(16.00) = 84.31$ g/mol.

Solve

$$0.122 \text{ mol MgCO}_3 \times \frac{84.31 \text{ g}}{\text{mol}} = 10.3 \text{ g}$$

Think About It

Moles in chemistry are like a common currency in exchanging money. From moles we can calculate mass; from mass we can calculate moles.

2.75. Collect and Organize

In this exercise we convert from the moles of titanium contained in a substance to the number of atoms present.

Analyze

For each substance, we need to take into account the number of moles of titanium *atoms* present in *1 mol* of the substance. For 0.125 mol of substance, then, a substance that contains two atoms of titanium in its formula contains $0.125 \times 2 = 0.250$ mol of titanium. We can then use Avogadro's number to convert the moles of titanium to the number of atoms present in the sample.

Solve

(a) Ilmenite, $FeTiO_3$, contains one atom of Ti per formula unit, so 0.125 mol of ilmenite contains 0.125 mol of Ti.

$$0.125 \text{ mol Ti} \times \frac{6.022 \times 10^{23} \text{ Ti atoms}}{1 \text{ mol}} = 7.53 \times 10^{22} \text{ Ti atoms}$$

(b) The formula for titanium(IV) chloride is $TiCl_4$. This formula contains only one Ti atom per formula unit as well, so the answer is identical to that calculated in (a).

$$0.125 \text{ mol Ti} \times \frac{6.022 \times 10^{23} \text{ Ti atoms}}{1 \text{ mol}} = 7.53 \times 10^{22} \text{ Ti atoms}$$

(c) Ti_2O_3 contains two titanium atoms in its formula, so 0.125 mol of Ti_2O_3 contains $0.125 \times 2 = 0.250$ mol of titanium.

$$0.250 \text{ mol Ti} \times \frac{6.022 \times 10^{23} \text{ Ti atoms}}{1 \text{ mol}} = 1.51 \times 10^{23} \text{ Ti atoms}$$

(d) Ti_3O_5 contains three titanium atoms in its formula, so 0.125 mol of Ti_3O_5 contains $0.125 \times 3 = 0.375$ mol of titanium.

$$0.375 \text{ mol Ti} \times \frac{6.022 \times 10^{23} \text{ Ti atoms}}{1 \text{ mol}} = 2.26 \times 10^{23} \text{ Ti atoms}$$

Think About It

The number of atoms of titanium in 0.125 mol of each compound reflects the number of atoms of Ti in the chemical formula. Ti_2O_3 has twice the number of Ti atoms, and Ti_3O_5 has three times the number of Ti atoms, compared with the number of Ti atoms in the same number of moles of $FeTiO_3$ and $TiCl_4$.

2.77. **Collect and Organize**

Given the formulas and the moles of each substance in a pair, we are asked to decide which compound contains more moles of oxygen.

Analyze

To answer this question, we have to take into account the moles of oxygen present in the substance formulas as well as the initial number of moles specified for each substance.

Solve

(a) One mole of Al_2O_3 contains 3 mol of oxygen, and 1 mol of Fe_2O_3 also contains 3 mol of oxygen. Those compounds contain the same number of moles of oxygen.
(b) One mole of SiO_2 contains 2 mol of oxygen, and 1 mol of N_2O_4 contains 4 mol of oxygen. Therefore, N_2O_4 contains more moles of oxygen (twice as much).
(c) Three moles of CO contains 3 mol of oxygen, and 2 mol of CO_2 contains 4 mol of oxygen. Therefore, the 2 mol of CO_2 contains more oxygen.

Think About It

We cannot decide which substance has more moles of oxygen by comparing only the amounts of the substances present. If that were the case, we would have concluded wrongly that 3 mol of CO contains more moles of oxygen than 2 mol of CO_2.

2.79. **Collect and Organize**

For each aluminosilicate, we are given the chemical formula. From that formula we are asked to deduce the number of moles of aluminum in 1.50 mol of each substance.

Analyze

The number of moles of aluminum in 1 mol of each substance is reflected in its chemical formula. We need next to take into account that we are starting with 1.50 mol of each substance.

Solve

(a) Each mole of pyrophyllite, $Al_2Si_4O_{10}(OH)_2$, contains 2 mol of Al atoms. Therefore, 1.50 mol of pyrophyllite contains $1.50 \text{ mol} \times 2 = 3.00$ mol of Al.
(b) Each mole of mica, $KAl_3Si_3O_{10}(OH)_2$, contains 3 mol of Al. Therefore, 1.50 mol of mica contains 1.50 mol $\times 3 = 4.50$ mol of Al.
(c) Each mole of albite, $NaAlSi_3O_8$, contains 1 mol of Al. Therefore, 1.50 mol of albite contains 1.50 mol of Al.

Think About It

Those minerals could all be distinguished by analyzing the amount of aluminum present in the same number of moles of each substance.

2.81. **Collect and Organize**

This exercise has to compute the molar mass of various molecular compounds of oxygen.

Analyze

We can find the molar mass of each compound by adding the molar mass of each element from the periodic table, taking into account the number of moles of each atom present in 1 mol of the substance.

Solve

(a) SO_2: $32.065 + 2(15.999) = 64.063$ g/mol

(b) O_3: $3(15.999) = 47.997$ g/mol

(c) CO_2: $12.011 + 2(15.999) = 44.009$ g/mol

(d) N_2O_5: $2(14.007) + 5(15.999) = 108.009$ g/mol

Think About It

The three compounds SO_2, O_3, and CO_2 have three atoms in their chemical formula, but each has a different molar mass.

2.83. Collect and Organize

This exercise has to compute the molar mass of various flavorings.

Analyze

We can find the molar mass of each flavoring by adding the molar mass of each element from the periodic table, taking into account the number of moles of each atom present in 1 mol of the flavoring. Each flavoring contains only carbon (12.01 g/mol), hydrogen (1.01 g/mol), and oxygen (16.00 g/mol).

Solve

(a) Vanillin, $C_8H_8O_3$: $8(12.011) + 8(1.0079) + 3(15.999) = 152.148$ g/mol

(b) Oil of cloves, $C_{10}H_{12}O_2$: $10(12.011) + 12(1.0079) + 2(15.999) = 164.203$ g/mol

(c) Anise oil, $C_{10}H_{12}O$: $10(12.011) + 12(1.0079) + 15.999 = 148.204$ g/mol

(d) Oil of cinnamon, C_9H_8O: $9(12.011) + 8(1.0079) + 15.999 = 132.161$ g/mol

Think About It

Each flavoring has a distinctive odor and flavor due in part to its different chemical formula. Another factor, however, in differentiating these flavorings is their chemical structure, or the arrangement in which the atoms are attached, as shown by the structures of those flavorings:

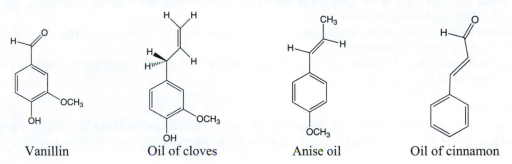

| Vanillin | Oil of cloves | Anise oil | Oil of cinnamon |

2.85. Collect and Organize

We are asked to convert a mass of carbon in grams to moles.

Analyze

We need the mass of 1 mol of carbon to compute the number of moles of carbon in the 500.0 g sample. From the periodic table, we see that the molar mass of carbon is 12.011 g/mol.

Solve

$$500.0 \text{ g C} \times \frac{1 \text{ mol}}{12.011 \text{ g}} = 41.63 \text{ mol C}$$

Think About It

Because carbon's molar mass is relatively low at 12 g/mol, 500 g of this substance contains a fairly substantial number of moles.

2.87. **Collect and Organize**

Given a molar amount of calcium titanate, we are asked to determine the number of moles and mass of Ca^{2+} ions in the substance.

Analyze

The formula for calcium titanate gives us the number of moles of Ca in the compound. Because the mass of the two missing electrons in the Ca^{2+} cation is negligible, the molar mass of the Ca^{2+} ion is taken to be the same as the molar mass of Ca. We can use that value to determine the mass of Ca^{2+} in the 0.25 mol of calcium titanate.

Solve

Because calcium titanate contains one atom of Ca in its formula, 0.25 mol of $CaTiO_3$ contains 0.25 mol of Ca^{2+} ions. The mass of Ca^{2+} ions in the sample, therefore, is

$$0.25 \text{ mol } Ca^{2+} \times \frac{40.078 \text{ g}}{1 \text{ mol}} = 10 \text{ g } Ca^{2+}$$

Think About It

Our answer makes sense. One-quarter of a mole of Ca^{2+} should give us 40/4, or about 10 g, of Ca in the 0.25 mol of calcium titanate.

2.89. **Collect and Organize**

Between two balloons filled with 10.0 g of different gases, we are to choose which balloon has more particles.

Analyze

The balloon with more particles has more moles. The greater number of moles contained in 10.0 g of a gas is for the gas with the lowest molar mass. A gas with a lower molar mass contains more moles in a 10.0 g mass and, therefore, has more moles than a 10.0 g mass of a higher molar mass gas.

Solve

(a) The molar mass of CO_2 is 44 g/mol, and the molar mass of NO is 30 g/mol. Therefore, the balloon containing NO has more particles.
(b) The molar mass of CO_2 is 44 g/mol, and the molar mass of SO_2 is 64 g/mol. Therefore, the balloon containing CO_2 has more particles.
(c) The molar mass of O_2 is 32 g/mol, and the molar mass of Ar is 40 g/mol. Therefore, the balloon containing O_2 has more particles.

Think About It

Although we could numerically determine the number of moles of gas in each balloon to make the comparisons in this problem, doing so is unnecessary because we know the relationship between moles and molar mass.

2.91. **Collect and Organize**

Given a mass of quartz, we are to determine the moles of SiO_2 present.

Analyze

To convert from mass to moles, we divide the mass given by the molar mass of SiO_2 [28.086 + 2(15.999) = 60.084 g/mol].

Solve

$$\frac{45.2 \text{ g } SiO_2}{60.084 \text{ g/mol}} = 0.752 \text{ mol } SiO_2$$

Think About It

Because the initial mass is less than the molar mass, we would expect less than 1 mol of SiO_2 to be in the quartz sample.

2.93. Collect and Organize

This exercise asks us to compute the moles of uranium and carbon (diamond) atoms in a 1 cm³ block of each element and then to compare them.

Analyze

Starting with the 1 cm³ block of each element, we can obtain the mass of the block by multiplying the density of the element. Dividing that result by the molar mass of the element gives us the moles of atoms in that block. The element block with more moles of atoms must have more atoms. We can compute the actual number of atoms by multiplying the moles of atoms for each element by Avogadro's number.

Solve

$$1 \text{ cm}^3 \text{ C} \times \frac{3.514 \text{ g}}{\text{cm}^3} \times \frac{1 \text{ mol}}{12.011 \text{ g}} \times \frac{6.022 \times 10^{23} \text{ C atoms}}{\text{mol}} = 1.762 \times 10^{23} \text{ atoms of C}$$

$$1 \text{ cm}^3 \text{ U} \times \frac{19.05 \text{ g}}{\text{cm}^3} \times \frac{1 \text{ mol}}{238.03 \text{ g}} \times \frac{6.022 \times 10^{23} \text{ U atoms}}{\text{mol}} = 4.820 \times 10^{22} \text{ atoms of U}$$

Therefore, the 1 cm³ block of diamond contains more atoms.

Think About It

We might expect that, because the block of uranium weighs so much more than the diamond block (more than five times as much), the uranium block would contain more atoms. However, we also have to take into account the very large molar mass of uranium. The result is that the diamond block has about 3.7 times more atoms in it than the same-sized block of uranium.

2.95. Collect and Organize

In this question we are asked how mass spectrometry provides information about a molecule.

Analyze

Mass spectrometry plots the mass-to-charge ratio (*m/z*) versus relative intensity or relative abundance of the peaks.

Solve

In mass spectrometry molecules of the compound are ionized with high-energy electrons to form +1 cations. When a molecule loses one electron, the ion that forms is call the *molecular ion*, M^+. This and other cations produces when the molecule fragments are separated base on their mass-to-charge (m/z) ratios and counted. The m/z ratio of a molecular ion with a charge of 1+ corresponds to the molecular mass of the compound. Often, but not always, this is the peak that appears at the highest *m/z* ratio.

Think About It

Remember that it is the *m/z* ratio that is determined in mass spectrometry. If a cation has a +2 charge—that is, it loses two electrons in the ionization step—it will appear at an *m/z* value that is half of its mass.

2.97. Collect and Organize

We are asked to consider whether the mass spectrum of CO_2 and that of C_3H_8 would show the same molecular ion peak.

Analyze

Mass spectrometry plots the mass-to-charge ratio (*m/z*) versus relative intensity or relative abundance of the peaks. Molecules with the same molecular mass will show the same *m/z* ratio for their molecular ions.

Solve

To the nearest amu, the molecular mass of CO_2 is 44 amu and for C_3H_8 the molecular mass also is 44 amu. Therefore, both molecules will show the same molecular ion peak at *m/z* 44.

Think About It
That result does not mean that the mass spectra of these two compounds will be the same, however. The two molecules will probably show different fragmentation patterns in their mass spectra.

2.99. Collect and Organize
For the explosive materials given we are to calculate the masses of their molecular ion peaks in the mass spectrum.

Analyze
The molecular ion is formed through the loss of one electron from the molecule. The mass of an electron is negligible, so the mass of the molecular ion peak is that of the molecule. We can compute the mass of a molecule using masses of the elements from the periodic table and their ratios in the molecular formula.

Solve
(a) $C_3H_6N_6O_6$: $(3 \times 12 \text{ amu}) + (6 \times 1 \text{ amu}) + (6 \times 14 \text{ amu}) + (6 \times 16 \text{ amu}) = 222$ amu
(b) $C_4H_8N_8O_8$: $(4 \times 12 \text{ amu}) + (8 \times 1 \text{ amu}) + (8 \times 14 \text{ amu}) + (8 \times 16 \text{ amu}) = 296$ amu
(c) $C_5H_8N_4O_{12}$: $(5 \times 12 \text{ amu}) + (8 \times 1 \text{ amu}) + (4 \times 14 \text{ amu}) + (12 \times 16 \text{ amu}) = 316$ amu
(d) $C_{14}H_6N_6O_{12}$: $(14 \times 12 \text{ amu}) + (6 \times 1 \text{ amu}) + (6 \times 14 \text{ amu}) + (12 \times 16 \text{ amu}) = 450$ amu

Think About It
Those explosives have the following structures. What common features do you see?

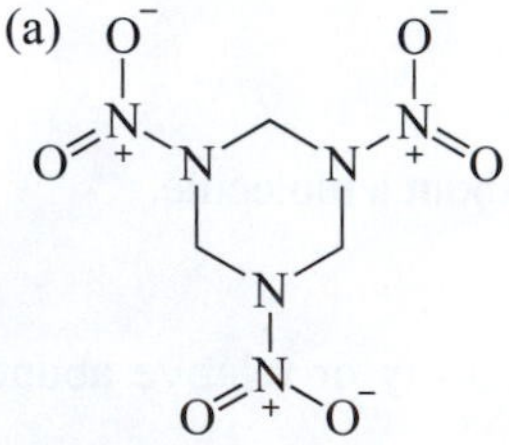

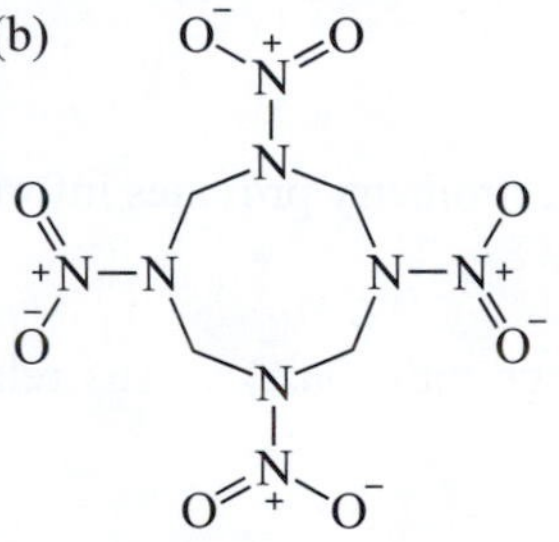

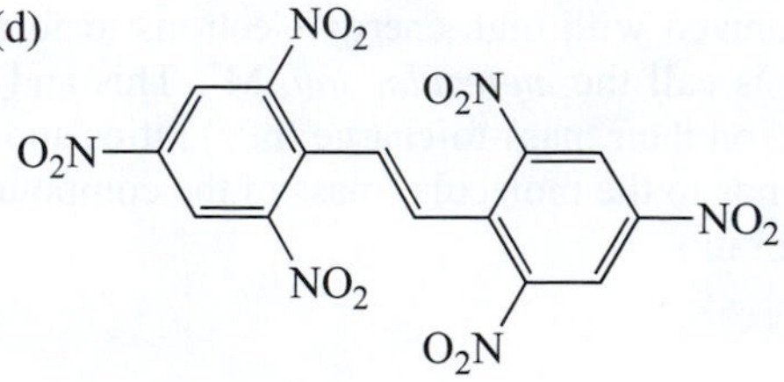

2.101. Collect and Organize
From the mass spectrum for Cl_2 shown in Figure P2.101 and given the natural abundances of ^{35}Cl and ^{37}Cl, we are asked to explain why three peaks occur around m/z 72 amu and why the peak at 70 amu is taller than the one at 74 amu.

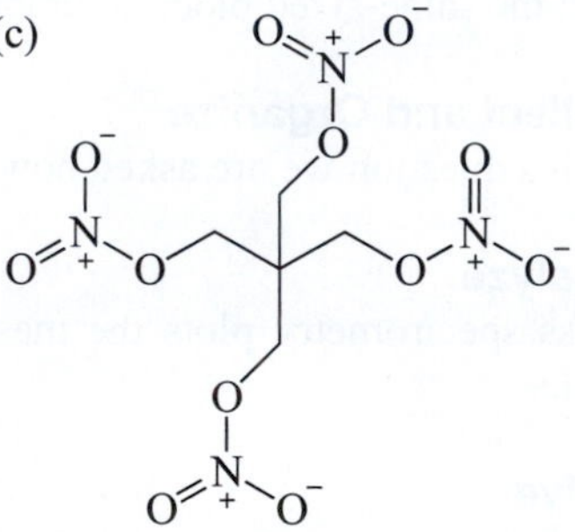

Analyze
The possible combinations of the chlorine isotopes in Cl_2 are $^{35}Cl_2$ (70 amu), $^{35}Cl^{37}Cl$ (72 amu), and $^{37}Cl_2$ (74 amu).

Solve
(a) The three peaks are due to the different combinations of isotopes in Cl_2: $^{35}Cl_2$ is at 70 amu, $^{35}Cl^{37}Cl$ is at 72 amu, and $^{37}Cl_2$ is at 74 amu.
(b) The peak at 70 amu is much more intense than the peak at 74 amu because ^{35}Cl is much more abundant than ^{37}Cl, so it is more likely that a molecule of Cl_2 will contain two ^{35}Cl isotopes than contain two ^{37}Cl isotopes.

Think About It

The mixed isotope can be made up in two ways: $^{35}Cl^{37}Cl$ and $^{37}Cl^{35}Cl$.

2.103. Collect and Organize

From the mass spectrum for H_2S shown in Figure P2.103 and given the natural abundances of four isotopes of sulfur, we are asked to explain the relative intensities as reflecting the abundance of the sulfur isotopes and the sequential loss of H from the molecule.

Analyze

The possible combinations of the sulfur isotopes in H_2S with their relative abundances are $H_2{}^{32}S$ (34 amu, 94.93%), $H_2{}^{33}S$ (35 amu, 0.76%), $H_2{}^{34}S$ (36 amu, 4.29%), and $H_2{}^{35}S$ (37 amu, 0.02%).

Solve

The biggest peak at 34 amu is the molecular ion of $H_2{}^{32}S$ (nearly 95% of all S atoms are ^{32}S). The peaks at 33 ns 32 amu are produced when electron bombardment of $H_2{}^{32}S$ molecules fragments them, forming ions $H^{32}S$ and ^{32}S atoms. The source of the small peak at 36 amu is probably the molecular ion of $H_2{}^{34}S$ (the second most abundant S isotope). The even smaller peak at 35 amu is probably a combination of mostly $H^{34}S$ and a very little $H_2{}^{33}S$ (this isotope of sulfur has a natural abundance of less than 1%). Molecular and fragment ions of the even less abundant ^{35}S isotope were probably not detected.

Think About It

Isotopic patterns such as these shown for H_2S are important in using mass spectrometry to identify compounds.

2.105. Collect and Organize

From the mass spectrum of cocaine shown in Figure P2.105, we are to determine the molar mass.

Analyze

The molar mass can be determined from a mass spectrum from the molecular ion peak. The molecular ion peak is (usually) the peak at the largest *m/z* value.

Solve

The largest *m/z* value is at 303 amu. Therefore, the molar mass of cocaine is 303 g/mol.

Think About It

The molecular ion peak is not the most intense peak in this mass spectrum.

2.107. Collect and Organize

J. J. Thomson's experiment revealed the electron and its behavior in magnetic and electric fields. In this question, we examine his experiment.

Analyze

Thomson showed that a cathode ray was deflected by a magnetic field in one direction and by an electric field in the other direction. He saw the deflection of the cathode ray when the ray hit a fluorescent plate at the end of his experimental apparatus, as shown in the textbook in Figure 2.2. The cathode ray was deflected by the electrically charged plates as shown. We can imagine the experiment proceeding from no voltage across the charged plates to low voltages and then to higher voltages. From this thought experiment, the ray must be deflected more by an increase in the voltage across the charged plates. Thomson reasoned that the cathode ray was composed of tiny charged particles, which were later called electrons.

Solve

(a) Today we call cathode rays electrons.
(b) The beam of electrons was deflected between the charged plates because they were attracted to the oppositely charged plate and repelled by the negatively-charged plate as the beam passed through the electric field. Indeed, in Figure 2.2 we see the beam deflected up towards the (+) plate.
(c) If the polarities of the plates were switched, the electron would still be deflected towards the positively charged plate, which would now be at the bottom of the tube.

Think About It

That experiment was key to the discovery of subatomic particles, which until then in the atomic theory were not known to exist. It was believed before that time that the atom was the smallest indivisible component of matter.

2.109. Collect and Organize

In this problem we are given the masses of the three isotopes of magnesium (^{24}Mg = 23.9850 amu, ^{25}Mg = 24.9858 amu, and ^{26}Mg = 25.9826 amu) and given that the abundance of ^{24}Mg is 78.99%. From that information and the average (weighted) atomic mass units of magnesium (24.3050 amu), we must calculate the abundances of the other two isotopes, ^{25}Mg and ^{26}Mg.

Analyze

The average atomic mass is derived from a weighted average of the isotopes' atomic masses. If x = abundance of ^{25}Mg and y = abundance of ^{26}Mg, the weighted average of magnesium is

$$(0.7899 \times 23.9850) + 24.9858x + 25.9826y = 24.3050$$

Because the sum of the abundances of the isotopes must add up to 1.00,

$$0.7899 + x + y = 1.00$$

So

$$x = 1.00 - 0.7899 - y = 0.2101 - y$$

Substituting this expression for x in the weighted average mass equation gives

$$(0.7899 \times 23.9850) + 24.9858(0.2101 - y) + 25.9826y = 24.3050$$

Solve

$$18.94575 + 5.249517 - 24.9858y + 25.9826y = 24.3050$$
$$0.9968y = 0.1097$$
$$y = 0.1101$$

So

$$x = 0.2101 - 0.1101 = 0.1000$$

The abundance of ^{25}Mg is $x \times 100 = 10.00\%$, and the abundance of ^{26}Mg is $y \times 100 = 11.01\%$.

Think About It

Although the abundances of ^{25}Mg and ^{26}Mg are nearly equal to each other at the end of that calculation, we cannot assume that in setting up the equation. We have to solve the problem algebraically by setting up two equations with two unknowns.

2.111. Collect and Organize

Given the diameter of silver nanoparticles and the number of atoms in one nanoparticle, we are to calculate the number of nanoparticles in 1.00 g.

Analyze

If we divide the mass (1.00 g) of silver by the molar mass of silver and then multiply the result by Avogadro's number, we will obtain the number of silver atoms in 1.00 g of nanoparticles. If we then divide that result by the number of atoms of silver in one nanoparticle, we will obtain the number of nanoparticles in the 1.00 g sample.

Solve

$$1.00 \text{ g} \times \frac{1 \text{ mol}}{107.87 \text{ g}} \times \frac{6.022 \times 10^{23} \text{ atoms}}{\text{mol}} \times \frac{1 \text{ nanoparticle}}{4.8 \times 10^7 \text{ atoms}} = 1.2 \times 10^{14} \text{ nanoparticles}$$

Think About It

We did not need the information given in this problem pertaining to the diameter of the nanoparticles.

2.113. **Collect and Organize**

Given that the average molar mass of air is 28.8 g/mol and that each mole of air had 402.5×10^{-6} mol of CO_2, we are to calculate how many micrograms of CO_2 that sample of air contains.

Analyze

This is a problem in unit analysis. If we multiply the moles of CO_2 per mole of air by the molar mass of air and then multiply that result by the molar mass of CO_2, we will obtain the mass of CO_2 per gram of air.

Solve

$$\frac{402.5 \times 10^{-6} \text{ mol } CO_2}{\text{mol air}} \times \frac{\text{mol air}}{28.8 \text{ g air}} \times \frac{44.0 \text{ g } CO_2}{\text{mol } CO_2} \times \frac{1 \text{ µg } CO_2}{1 \times 10^{-6} \text{ g } CO_2} = 615 \text{ µg } CO_2 / \text{g air}$$

Think About It

Always label your values with units and make sure that they cancel correctly to arrive at your final answer.

2.115. **Collect and Organize**

To determine the number of moles and atoms of carbon in the Hope Diamond, we have to first convert its given mass of 45.52 carats to mass in grams.

Analyze

To convert the mass of the diamond to grams, we use the relationship 1 carat = 200 mg. The moles of carbon atoms is equal to the mass of the diamond divided by the molar mass of carbon (12.011 g/mol); the number of carbon atoms in the diamond is the number of moles of carbon multiplied by Avogadro's number.

Solve

(a) The mass of the diamond in grams is

$$45.52 \text{ carats } \times \frac{200 \text{ mg}}{1 \text{ carat}} \times \frac{1 \text{ g}}{1000 \text{ mg}} = 9.104 \text{ g}$$

The number of moles of carbon atoms in the diamond is

$$9.104 \text{ g} \times \frac{1 \text{ mol}}{12.011 \text{ g}} = 0.7580 \text{ mol of C}$$

(b) The number of carbon atoms in the diamond is

$$0.7580 \text{ mol} \times \frac{6.022 \times 10^{23} \text{ atoms}}{1 \text{ mol}} = 4.565 \times 10^{23} \text{ atoms of C}$$

Think About It

Diamond is the hardest natural substance and is an electrical insulator while being an excellent conductor of heat.

CHAPTER 3 | Atomic Structure: Explaining the Properties of Elements

3.1. Collect and Organize

From the highlighted elements in Figure P3.1, we can correlate position on the periodic table with orbital (s, p, d) filling.

Analyze

The periodic table consists of the s-block elements, which are the first two columns (groups 1–2); the d-block elements, which are the next 10 columns (groups 3–12); and the p-block elements, which are the rightmost six columns (groups 13–18). The f-block elements are in the two rows at the bottom of the table. From their positions in these blocks, we can ascertain the elements' electron configurations.

Solve

(a) Group 1 elements have an ns^1 configuration, including the element shaded purple (Na). Because half-filled and filled d orbitals are predicted for transition metals, the red-shaded ($3d^5 4s^1$, Cr) and orange-shaded ($5d^{10} 6s^1$, Au) elements also have a single s electron in their outermost shells.

(b) Filled sets of s and p orbitals ($ns^2 np^6$ configuration) occur for the noble gases (group 18 elements), including the element shaded blue (Ne).

(c) Filled sets of d orbitals would occur for elements in period 4 and below, having in their electron configuration nd^{10}. Because a filled d orbital is stable for transition elements, the orange-shaded element (Au) is predicted to have a $5d^{10} 6s^1$ configuration.

(d) A half-filled d-orbital set would be predicted for the element shaded red (Cr); its electron configuration would be $3d^5 4s^1$.

(e) Filled s orbitals would occur in the outermost shells of the blue element ($2s^2 2p^6$, Ne) and the green-shaded element ($3s^2 3p^5$, Cl).

Think About It

Remember that the outermost shell of an atom includes all those electrons above the previous noble gas core.

3.3. Collect and Organize

Of the highlighted elements in Figure P3.1, we are to find the elements that form common monatomic ions (cations or anions) that are larger and smaller than the element itself.

Analyze

Nonmetals tend to form anions and metals tend to form cations. Anions are larger than the parent atom, so for part a, we are looking for elements that will probably form an anion (X^{n-}) and we are looking for an element located on the right-hand side of the periodic table. Cations are smaller than the parent atom, and so for part b we are looking for elements that will form Y^{n+} and for an element on the left-hand side of the periodic table.

Solve

(a) Both the green and blue elements are nonmetals and potentially form anions. Green (Cl) would form Cl^-, picking up an electron to fill its outermost shell. Blue (Ne), however, would not form an anion; its outermost shell is already full.

(b) Any of the metals—purple (Na), red (Cr), and yellow (Au)—will form cations smaller than their parent ions.

Think About It

Most elements in the periodic table are classified as metals, so most elements tend to form cations, not anions.

3.5. **Collect and Organize**

Of the highlighted elements in Figure P3.4, we are to find the elements that form cations or anions that are smaller than the parent atom.

Analyze

Cations are always smaller than the parent atom, so we look for elements that are likely to form cations (X^{n+}). Metals tend to form cations, so we look for elements on the left side of the periodic table.

Solve

Blue (Rb), green (Sr), and yellow (Y) are all metals and potentially lose electrons to form cations that are smaller than their corresponding atoms.

Think About It

As an atom loses more electrons, the size continues to decrease. Therefore, $X^+ > X^{2+} > X^{3+}$ in size.

3.7. **Collect and Organize**

Pictured in Figure P3.7 are three waves. Wave A is shone on a metal surface and from it an electron is emitted. We are to choose the statement that reflects the effect that waves B and C would have when they are shown on the metal surface in place of wave A.

Analyze

Wave B has a longer wavelength and thus lower energy than wave A. Wave B might be as energetic as wave A in emitting electrons, but we can't be sure because we do not know the threshold value for the photoemission for this metal. Wave C has shorter wavelength and thus higher energy than wave A, and so it will definitely also emit electrons from the metal surface.

Solve

Statement (c) is correct.

Think About It

The difference in the effect of wave A and wave C will be in the speed of the electron emitted from the metal. Electrons ejected from the metal by wave C, with a greater energy, will have greater kinetic energy than those of wave A, with a lower energy.

3.9. **Collect and Organize**

Of the four choices of the group 2 cation (M^{2+}) and the group 17 anion (X^-) from Problem 3.8, we are to choose the one that best reflects their relative ion sizes.

Analyze

The M (alkaline earth) atom would decrease its size significantly upon forming the M^{2+} cation, and the X (halogen) atom would increase greatly in size upon forming the X^- anion.

Solve

Representation (a) best shows their relative sizes. The metal will form a 2+ cation, which will very much decrease its size, whereas the halogen will form a 1– anion, which will increase the size, albeit not substantially (by adding only one electron).

Think About It

Periodic trends apply to ions as well, but we have to be careful to consider the number of electrons lost or gained. In order of increasing cation size: $Al^{3+} < Mg^{2+} < Na^+$. In order of increasing anion size: $Cl^- < S^{2-} < P^{3-}$.

3.11. **Collect and Organize**

From the emission colors of quantum dot solutions shown in Figure P3.11, we are to determine which emits the highest energy photons; the longest wavelength photons; and using Figure 3.14 in the textbook, decide which container has the largest quantum dots.

Analyze

Because wavelength and energy are inversely related, the energy of emitted light increases in energy and decreases in wavelength in the order red < orange < yellow < green < blue < violet. Figure 3.14 shows us that the larger the quantum dot, the longer the wavelength of the light emitted.

Solve

(a) The container with the quantum dots emitting the greatest energy photons is (e), violet.
(b) The container with the quantum dots emitting with the longest wavelength is (a), red.
(c) The quantum dots are the largest in (a), red.

Think About It

The color of quantum dots of a given size can also be tuned by "doping" them with other elements—for example, doping ZnS nanoparticles with copper or aluminum cations.

3.13. **Collect, Organize, and Analyze**

All forms of radiant energy (light) from gamma rays to low-energy radio waves are called *electromagnetic radiation*. Why?

Solve

All these forms of light have perpendicular, oscillating electric and magnetic fields that travel together through space, as Maxwell described.

Think About It

All forms of electromagnetic radiation travel at the speed of light (3.00×10^8 m/s in a vacuum).

3.15. **Collect and Organize**

We are asked why a lead shield is used at the dentist's office when X-ray images are taken.

Analyze

Light interacts with matter and X-rays are high-energy light, which can damage living cells.

Solve

The lead shield must protect the part of our bodies that might be exposed to X-rays but are not being imaged. Lead is a dense metal with many electrons, which interact with X-rays and absorb nearly all the X-rays before they can reach our bodies.

Think About It

Exposure to high-energy radiation (γ rays and X-rays in particular) may cause genetic damage in cells, which may lead to cancers.

3.17. **Collect and Organize**

We consider whether as molten lava cools and no longer glows, it still emits radiation.

Analyze

The electromagnetic spectrum covers a wide range of wavelengths, from low-energy radio waves to high-energy γ rays. The visible part of the spectrum is confined to 400–700 nm and is only a small portion.

Solve

As the lava cools the energy given off does not cease, but it no longer gives off energy in the visible range (so it is no longer glowing). The lava is still hot and it radiates heat, which is in the infrared region of the electromagnetic spectrum.

Think About It
Night-vision goggles detect infrared radiation from objects warmer than ambient temperature and convert it into visible (usually green) light.

3.19. Collect and Organize
We are to calculate the frequency of light of a wavelength given in nanometers ($\lambda = 546.1$ nm) and then determine whether it is that wavelength that contributes to the greenish glow of the mercury lamp.

Analyze
The wavelength of light is related to the frequency through the equation $v = c/\lambda$. Wavelength must be expressed in meters for this calculation (1 nm = 10^{-9} m). From Figure 3.1 in the textbook we can see that if the emission occurs between 520 and 550 nm, the emission is in the green region of the electromagnetic spectrum.

Solve

$$v = \frac{2.998\times10^8 \text{ m/s}}{\left(546.1 \text{ nm}\times\dfrac{1\times10^{-9} \text{ m}}{1 \text{ nm}}\right)} = 5.490\times10^{14} \text{ s}^{-1}$$

Because the wavelength of light emitted here is 546.1 nm, it is in the green region of the electromagnetic spectrum, so yes, that emission does contribute to the greenish glow of the mercury light.

Think About It
Be sure to convert nanometers to meters when using this equation.

3.21. Collect and Organize
Given the frequencies of several radio stations, we are to calculate the corresponding wavelengths.

Analyze
The equation to calculate wavelength from frequency is
$$\lambda = c/v$$
where λ is in meters, $c = 2.998 \times 10^8$ m/s, and v is in hertz (per second). We need to convert megahertz to hertz for our calculation (1 MHz = 10^6 Hz).

Solve

(a) $\lambda_{\text{KRNU}} = \dfrac{2.998\times10^8 \text{ m/s}}{90.3\times10^6 \text{ s}^{-1}} = 3.32$ m

(b) $\lambda_{\text{WBRU}} = \dfrac{2.998\times10^8 \text{ m/s}}{95.5\times10^6 \text{ s}^{-1}} = 3.14$ m

(c) $\lambda_{\text{WYLD}} = \dfrac{2.998\times10^8 \text{ m/s}}{98.5\times10^6 \text{ s}^{-1}} = 3.04$ m

(d) $\lambda_{\text{WAAF}} = \dfrac{2.998\times10^8 \text{ m/s}}{107.3\times10^6 \text{ s}^{-1}} = 2.79$ m

Think About It
Remember that the speed of electromagnetic radiation in air is approximately the same as in a vacuum (2.998×10^8 m/s).

3.23. Collect and Organize
We are to compare the frequency of 0.154 nm X-rays emitted from copper versus that of iron with a frequency of 194 pm.

Analyze
We can convert the wavelength of light to frequency by using $v = c/\lambda$. Wavelength has to be expressed in meters for this calculation (1 nm = 10^{-9} m). We will want to compare wavelengths of comparable units; let's

choose meters. Therefore, we will have to convert nanometers to meters (1 nm = 10^{-9} m) and picometers to meters (1 pm = 10^{-12} m).

Solve

The frequency of the 0.154 nm X-rays from copper is

$$v = \frac{2.998 \times 10^8 \text{ m/s}}{0.154 \times 10^{-9} \text{ m}} = 1.95 \times 10^{18} \text{ s}^{-1}$$

The frequency of the 194 pm X-rays from iron is

$$v = \frac{2.998 \times 10^8 \text{ m/s}}{194 \times 10^{-12} \text{ m}} = 1.55 \times 10^{18} \text{ s}^{-1}$$

The frequency of the X-rays emitted from copper have a higher frequency than those emitted from iron.

Think About It

You may have answered this question with no calculations. Converting the X-ray frequency of copper to picometers gives 154 nm. That is a shorter wavelength and therefore higher frequency than that of iron X-rays.

3.25. Collect and Organize

Earth is 149.6 million km from the sun. We are to calculate how long the sun's light takes to reach Earth.

Analyze

We can find the time light takes to travel from the sun to Earth by dividing the distance (149.6 million km) by the speed at which light travels (2.998×10^8 m/s).

Solve

$$\text{time} = \frac{\left(149.6 \times 10^6 \text{ km} \times \dfrac{1000 \text{ m}}{\text{km}}\right)}{2.998 \times 10^8 \text{ m/s}} = 499.0 \text{ s, or } 8.317 \text{ min}$$

Think About It

Even though light travels very fast, the large distance between Earth and the sun means that events (for example, solar flares) we witness on the sun actually happened 8 min ago.

3.27. Collect and Organize

We are asked to compare the atomic emission and absorption spectra of an element.

Analyze

Emission of light from an atom occurs when the atoms are heated to a high temperature and an electron at a high energy level falls to a lower energy level. Absorption of light occurs when the atoms absorb energy from an external source of energy to promote an electron from a lower energy level to a higher energy level. We are asked to compare the two resulting spectra for an atomic element.

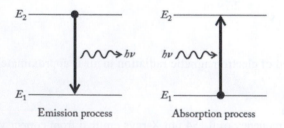

Solve

The atomic absorption spectrum consists of dark lines at wavelengths specific to the element where it absorbs energy to enter an excited state. The emission spectrum has bright lines on a dark background, with the lines appearing at the same wavelengths as the dark lines in the absorption spectrum. The bright lines are where an excited-state electron falls to a lower energy state. The wavelengths of the emission and absorption lines are the same.

Think About It

The atom absorbs only certain amounts of energy (as seen in the absorption spectrum) and, once excited to those higher energy states, emits that same energy (in the emission spectrum) to go back to its ground (lowest) energy state.

3.29. Collect and Organize

We are asked how study of the emission spectra of the elements led to identification of the dark Fraunhofer lines in the sun's spectrum.

Analyze

The emission lines for elements match the absorption lines in energy because those processes are the reverse of each other.

Solve

Because each element shows distinctive and unique absorption and emission lines, the bright emission lines observed for the pure elements could be matched to the many dark absorption lines in the spectrum of sunlight. That approach can be used to deduce the sun's elemental composition.

Think About It

The elemental composition of distant stars can be determined in that way as well.

3.31. Collect, Organize, and Analyze

We are to define the term *quantum*.

Solve

Named by Max Planck, the quantum is the smallest indivisible quantity of radiant energy that can be absorbed or emitted.

Think About It

Planck also defined the relationship between the energy of a quantum particle and its frequency ($E = h\nu$).

3.33. Collect and Organize

We are to determine whether and what color a piece of tungsten would glow if heated to 1000 K.

Analyze

We can treat the tungsten wire as a blackbody emitter. Figure 3.10 shows the change of the intensity of emitted radiation as a function of wavelength for different temperatures of a blackbody.

Solve

No, the tungsten wire would not glow because no intensity of light exists in the visible spectrum at 1000 K.

Think About It

At 2000 K, however, the tungsten wire would give off visible light. The red region of the visible spectrum has a higher intensity, so we might expect it to appear reddish at that temperature.

3.35. Collect and Organize

We are asked to calculate the energy of one UV photon of wavelength 3.00×10^{-7} m.

Analyze

The pertinent equation to consider here is $E = hc/\lambda$, where h is the Planck constant, c is the speed of light, and λ is the wavelength of the UV light.

Solve

$$E = \frac{(6.626 \times 10^{-34}\, \text{J} \cdot \text{s}) \times (2.998 \times 10^8\, \text{m/s})}{3.00 \times 10^{-7}\, \text{m}} = 6.62 \times 10^{-19}\, \text{J}$$

Think About It

That is the energy for only one photon. For an entire mole of photons, the energy would be 6.62×10^{-19} J multiplied by Avogadro's number to give 3.99×10^5 J, or 399 kJ.

3.37. Collect and Organize

From a list, we are to choose which have quantized values.

Analyze

Something is quantized if it is present only in discrete amounts and can have only whole-number multiples of the smallest amount.

Solve

(a) The elevation of a step on a moving escalator continuously changes, so that is not a quantized value.
(b) Because the doors open only *at* the floors and not *between* the floors, that value is quantized.
(c) The speed of an automobile can change smoothly, so that is not a quantized value.

Think About It

Any quantity that is quantized has to have changes occurring in discrete steps.

3.39. Collect and Organize

The kinetic energy of an ejected photon is given by the equation

$$\text{KE}_{\text{electron}} = h\nu - \Phi$$

where Φ is the work function, or the energy threshold required to eject the electron from the metal.

Analyze

In this problem, we solve for Φ:

$$\Phi = h\nu - \text{KE}_{\text{electron}}$$

where $h = 6.626 \times 10^{-34}$ J · s, $\nu =$ frequency of the light used to eject the electron, and $\text{KE}_{\text{electron}} = 7.34 \times 10^{-19}$ J. We can find ν of the irradiating light from $\nu = c/\lambda$, where $c = 2.998 \times 10^8$ m/s, and we are given $\lambda = 132$ nm $(1.32 \times 10^{-7}$ m).

Solve

$$\Phi = \left(6.626 \times 10^{-34}\, \text{J} \cdot \text{s} \times \frac{2.998 \times 10^8\, \text{m/s}}{1.32 \times 10^{-7}\, \text{m}} \right) - 7.34 \times 10^{-19}\, \text{J} = 7.71 \times 10^{-19}\, \text{J}$$

Think About It

That does not seem to be a large amount of energy to emit one electron, but remember if we were to consider a mole of electrons to be ejected, we would need $(7.71 \times 10^{-19}$ J$) \times (6.0221 \times 10^{23}/\text{mol}) = 4.64 \times 10^5$ J/mol, or 464 kJ per mol. That is the ionization energy in kilojoules per mole for this metal.

3.41. Collect and Organize

We can use the equation for the photovoltaic effect to determine whether electrons could be ejected from germanium by using visible light with a wavelengths less than 600 nm.

Analyze

As long as the energy of the light that shines on the metal is greater than the work function for germanium ($\Phi = 7.21 \times 10^{-19}$ J), the light will eject electrons and germanium would therefore be useful in a voltaic cell. The energy of the light can be calculated from $E = hc/\lambda$.

Solve

The wavelength of light that would be needed to eject electrons from germanium would be

$$\lambda = \frac{hc}{\Phi} = \frac{6.626 \times 10^{-34} \text{ J} \cdot \text{s} \times 2.998 \times 10^{8} \text{ m/s}}{7.21 \times 10^{-19} \text{ J}} = 2.76 \times 10^{-7} \text{ m, or 276 nm}$$

This is in the UV part of the electromagnetic spectrum, not in the visible region, so germanium could not be used to convert solar energy to electricity with visible light.

Think About It

The energy of light of 600 nm wavelength can be found as

$$E = \frac{6.626 \times 10^{-34} \text{ J} \cdot \text{s} \times 2.998 \times 10^{8} \text{ m/s}}{6.00 \times 10^{-7} \text{ m}} = 3.31 \times 10^{-19} \text{ J}$$

That energy is less than the work function for germanium ($\Phi = 7.21 \times 10^{-19}$ J), so germanium could not be used to convert solar energy at 600 nm to electricity.

3.43. Collect and Organize

We are to calculate the velocity (speed) of electrons ejected from Na and K when irradiated by a 300 nm light source to determine which has the higher speed. Because the work function for potassium is lower than that of sodium, we expect that an electron ejected from potassium has a higher kinetic energy.

Analyze

For each metal, $KE_{electron} = h\nu - \Phi$, where the frequency of the light irradiating the metals would be found from $\nu = c/\lambda$, where $\lambda = 300$ nm (3.00×10^{-7} m). The speed of the electron can then be found as

$$KE = \frac{1}{2} m_e u^2, \text{ or } u = \sqrt{\frac{2KE}{m_e}}$$

where m_e is the mass of an electron, 9.11×10^{-31} kg.

Solve

Potassium:

$$KE = \left(6.626 \times 10^{-34} \text{ J} \cdot \text{s} \times \frac{2.998 \times 10^{8} \text{ m/s}}{3.00 \times 10^{-7} \text{ m}} \right) - 3.68 \times 10^{-19} \text{ J} = 2.942 \times 10^{-19} \text{ J}$$

$$u = \sqrt{\frac{2 \times 2.942 \times 10^{-19} \text{ kg} \cdot \text{m}^2/\text{s}^2}{9.11 \times 10^{-31} \text{ kg}}} = 8.04 \times 10^{5} \text{ m/s}$$

Sodium:

$$KE = \left(6.626 \times 10^{-34} \text{ J} \cdot \text{s} \times \frac{2.998 \times 10^{8} \text{ m/s}}{3.00 \times 10^{-7} \text{ m}} \right) - 4.41 \times 10^{-19} \text{ J} = 2.212 \times 10^{-19} \text{ J}$$

$$u = \sqrt{\frac{2 \times 2.212 \times 10^{-19} \text{ kg} \cdot \text{m}^2/\text{s}^2}{9.11 \times 10^{-31} \text{ kg}}} = 6.97 \times 10^{5} \text{ m/s}$$

Potassium's ejected electrons have a greater speed.

Think About It

The units of 1 J = 1 kg·m^2/s^2 are useful in ensuring that the speeds are in meters per second.

3.45. **Collect and Organize**

Using the information that a red laser's power is 1 J/s (1 W), we are to calculate the number of photons emitted by the laser per second.

Analyze

The energy of the photons of red laser light ($\lambda = 630$ nm) is given by

$$E = hc/\lambda$$

That corresponds to the number of photons in a second by

$$\text{number of photons} = (1 \text{ J}) \times \left(\frac{1 \text{ photon}}{\text{energy of photon}} \right)$$

Solve

Energy of one photon with $\lambda = 630$ nm:

$$E = \frac{6.626 \times 10^{-34} \text{ J} \cdot \text{s} \times 2.998 \times 10^{8} \text{ m/s}}{6.30 \times 10^{-7} \text{ m}} = 3.153 \times 10^{-19} \text{ J/photon}$$

Number of photons per second:

$$1.00 \text{ watt} \times \frac{1 \text{ J/s}}{\text{watt}} \times \frac{1 \text{ photon}}{3.153 \times 10^{-19} \text{ J}} = 3.17 \times 10^{18} \text{ photons/s}$$

Think About It

Greater power (watts) leads to more photons being emitted each second. Also, the shorter the wavelength, the greater the energy per photon and the higher the power (watts) for the same number of photons emitted.

3.47. **Collect and Organize**

We are asked to explain how the Balmer equation is just a special case of the Rydberg equation.

Analyze

The Balmer equation, where $m > 2$ and $n = 2$, is

$$\lambda \text{ (in nm)} = \left(\frac{364.56 m^2}{m^2 - n^2} \right)$$

The Rydberg equation, where $n_2 > n_1$ (and they may be any integers), is

$$\frac{1}{\lambda} = R_{\text{H}} \left(\frac{1}{n_1^2} - \frac{1}{n_2^2} \right)$$

Solve

Although their forms appear fairly different, both predict the spectrum of the hydrogen atom. The Balmer equation is equivalent to the Rydberg equation is $n_1 = 2$. Specifically, the Balmer equation is equivalent to the Rydberg equation when $n_1 = 2$ with $n_2 = 1$ and $m = 3$.

Think About It

The Rydberg equation is more general, with variable integers for n_1 and n_2, and so can predict lines outside the visible spectrum.

3.49. **Collect and Organize**

We consider here the process of emission of light from a hydrogen atom in an excited state. Does the energy emitted depend on the values of n_1 and n_2 or only on the difference between them ($n_1 - n_2$)?

Analyze

The emission of light from an H atom in the excited state is the result of the electron dropping from a higher energy orbit to a lower-lying orbit.

Solve

Because the difference in the orbits' energy correlates with the light energy emitted by the hydrogen atom in the excited state, the difference between n levels determines emission energy as the excited H atom relaxes to its ground state.

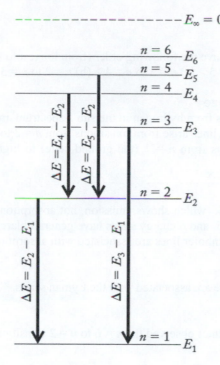

Think About It

The n levels get closer in energy as n increases so that $n_2 - n_1 > n_3 - n_2 > n_4 - n_3$, and so on.

3.51. Collect and Organize

In considering the absorption of energy for an electron to be promoted from a lower energy orbit to a higher energy orbit ($n_1 < n_2$), we can predict which transition requires the shortest wavelength of light.

Analyze

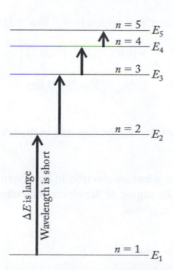

The transition with the shortest wavelength has the highest change in energy. The energy levels (n) in hydrogen are not evenly spaced; as n increases, the differences in energy between adjacent energy levels decrease.

Solve

All the transitions given involve changes between adjacent levels ($\Delta n = 1$). Because the energy levels become more closely spaced as n increases, ΔE values for energy levels for lower n values are greater and therefore have shorter wavelengths associated with them. Therefore, choice (a), where the electron is "promoted" from $n = 1$ to $n = 2$, has the shortest wavelength.

Think About It

The transition wavelengths will have the following order (from longest to shortest wavelength):
$$(d)\ n = 4\ \text{to}\ n = 5 > (c)\ n = 3\ \text{to}\ n = 4 > (b)\ n = 2\ \text{to}\ n = 3 > (a)\ n = 1\ \text{to}\ n = 2$$

3.53. Collect, Organize, and Analyze

The lines of the Fraunhofer series from hydrogen in the sun's spectrum are observed in the visible region of the electromagnetic spectrum. Those lines arise from transitions from $n = 2$ to higher energy levels ($n = 3$ to 6). We are asked whether any transitions from $n = 1$ (the ground state) to higher energy levels would be observed among the Fraunhofer lines.

Solve

From Figure 3.18 in the textbook (which shows emission, not absorption, but the two are complementary) we see that the $n = 1$ to $n = 2$, 3, 4, 5, and 6 energy states have generally larger energies associated with them than transitions from $n = 2$, so no Fraunhofer lines are associated with absorption from the ground state.

Think About It

Absorptions from the ground state are associated with the Lyman series.

3.55. Collect and Organize

We are asked to explain why Balmer observed the $n = 6$ to $n = 2$ transition in hydrogen but not that for $n = 7$ to $n = 2$.

Analyze

As seen in Figure 3.17, the energy levels become more closely spaced as the value of n increases. We can use the equation below to calculate the wavelength of the transition.

$$\frac{1}{\lambda} = [1.097 \times 10^{-2}\ (\text{nm})^{-1}] \left(\frac{1}{n_1^2} - \frac{1}{n_2^2} \right)$$

Solve

At $n = 7$, the wavelength of the electron's transition ($n = 7$ to $n = 2$) has moved out of the visible region.

$$\frac{1}{\lambda} = [1.097 \times 10^{-2}\ (\text{nm})^{-1}] \left(\frac{1}{2^2} - \frac{1}{7^2} \right)$$

$$\lambda = 397.0\ \text{nm}$$

Think About It

The transition from the $n = 7$ energy level can be detected in the UV region of the electromagnetic spectrum.

3.57. Collect and Organize

To calculate the wavelength emitted when an electron in hydrogen undergoes a transition from $n = 4$ to $n = 3$, we can use the Rydberg equation. The region of the electromagnetic spectrum corresponding to that wavelength can be found from Figure 3.1 in the textbook.

Analyze

In the Rydberg equation, $n_1 = 3$ and $n_2 = 4$.

$$\frac{1}{\lambda} = [1.097 \times 10^{-2}\ (\text{nm})^{-1}] \left(\frac{1}{n_1^2} - \frac{1}{n_2^2} \right)$$

Solve

$$\frac{1}{\lambda} = [1.097 \times 10^{-2} \text{ (nm)}^{-1}]\left(\frac{1}{3^2} - \frac{1}{4^2}\right) = 5.333 \times 10^{-4} \text{(nm)}^{-1}$$

$$\lambda = 1875 \text{ nm}$$

That wavelength occurs in the infrared region of the electromagnetic spectrum.

Think About It

In the Rydberg equation, n_2 is a higher orbit number (n) than n_1. In this way we don't get the nonsensical result of a negative wavelength.

3.59. Collect and Organize

The equation given relates the photon's energy to atomic number and the energy-level transition.

Analyze

The equation shows a direct relationship between the energy and the atomic number and between the energy and the transition of the electron between n_1 and n_2 as

$$\frac{1}{n_1^2} - \frac{1}{n_2^2}$$

Solve

(a) Because the energy of the photons is directly related to the atomic number in the equation, as Z increases, the energy increases. Because energy is inversely related to wavelength, the wavelength decreases as Z increases.

(b) The energy of the photon for $Z = 1$ (hydrogen) is

$$E = (2.18 \times 10^{-18} \text{ J})(1)^2\left(\frac{1}{1^2} - \frac{1}{2^2}\right) = 1.635 \times 10^{-18} \text{ J}$$

The wavelength of that photon is

$$\lambda = \frac{6.626 \times 10^{-34} \text{ J} \cdot \text{s} \times 2.998 \times 10^8 \text{ m/s}}{1.635 \times 10^{-18} \text{ J}} = 1.21 \times 10^{-7} \text{ m, or } 121 \text{ nm}$$

That wavelength is in the UV range. As Z increases, the energy of the photon increases and the wavelength decreases, so the transition will never be observed in the visible range.

Think About It

The $n = 1$ and $n = 2$ energy levels are too far apart in energy to give an emitted photon in the visible range.

3.61. Collect and Organize

In considering the absorption of light by an electron in a hydrogen atom, we are to consider whether absorptions in terms of wavelength and energy are additive.

Analyze

Energy and wavelength are related to each other through the equation $\lambda = hc/E$. To determine the wavelength of the absorptions in hydrogen, we can use the Rydberg equation,

$$\frac{1}{\lambda} = R_H\left(\frac{1}{n_1^2} - \frac{1}{n_2^2}\right)$$

where n_1 and n_2 are positive integers (where $n_2 > n_1$) and $R_H = 1.097 \times 10^{-2}$ nm. We use that to calculate n_1 and n_2 equal to 2 and 3 and 3 and 4, and then add the results to see whether they are equal to the calculation for n_1 and n_2 equal to 2 and 4. To determine whether the energies are additive, we can use the Bohr equation in the same manner:

$$\Delta E = -2.178 \times 10^{-18} \text{ J}\left(\frac{1}{n_{\text{final}}^2} - \frac{1}{n_{\text{intial}}^2}\right)$$

(a) $\lambda_{2\rightarrow4} = \lambda_{2\rightarrow3} + \lambda_{3\rightarrow4}$

(b) $E_{2\rightarrow4} = E_{2\rightarrow3} + E_{3\rightarrow4}$

Solve

(a) To determine whether $\lambda_{2\rightarrow4} = \lambda_{2\rightarrow3} + \lambda_{3\rightarrow4}$:

$$\frac{1}{\lambda_{2\rightarrow3}} = 1.097\times10^{-2}\,\text{nm}^{-1}\left(\frac{1}{2^2} - \frac{1}{3^2}\right) = 1.524\times10^{-3}\,\text{nm}^{-1}$$

$$\frac{1}{\lambda_{3\rightarrow4}} = 1.097\times10^{-2}\,\text{nm}^{-1}\left(\frac{1}{3^2} - \frac{1}{4^2}\right) = 5.333\times10^{-4}\,\text{nm}^{-1}$$

$$\frac{1}{\lambda_{2\rightarrow4}} = 1.097\times10^{-2}\,\text{nm}^{-1}\left(\frac{1}{2^2} - \frac{1}{4^2}\right) = 2.057\times10^{-3}\,\text{nm}^{-1}$$

Taking the inverse of those values to obtain the wavelength gives

$$\lambda_{2\rightarrow3} = \frac{1}{1.524\times10^{-3}\,\text{nm}^{-1}} = 656.2\text{ nm}$$

$$\lambda_{3\rightarrow4} = \frac{1}{5.333\times10^{-4}\,\text{nm}^{-1}} = 1875\text{ nm}$$

$$\lambda_{2\rightarrow4} = \frac{1}{2.057\times10^{-3}\,\text{nm}^{-1}} = 486.2\text{ nm}$$

Checking the equality, we see that $\lambda_{2\rightarrow4} \neq \lambda_{2\rightarrow3} + \lambda_{3\rightarrow4}$

$$\lambda_{2\rightarrow3} + \lambda_{3\rightarrow4} = 656.3\text{ nm} + 1875\text{ nm} = 2531\text{ nm}$$

$$2531\text{ nm} \neq \lambda_{2\rightarrow4}(486.2\text{ nm})$$

(b) Now we can use the Bohr equation to check whether energies are additive for those transitions:

$$\Delta E_{2\rightarrow3} = -2.178\times10^{-18}\,\text{J}\left(\frac{1}{3^2} - \frac{1}{2^2}\right) = 3.025\times10^{-19}\,\text{J}$$

$$\Delta E_{3\rightarrow4} = -2.178\times10^{-18}\,\text{J}\left(\frac{1}{4^2} - \frac{1}{3^2}\right) = 1.059\times10^{-19}\,\text{J}$$

$$\Delta E_{2\rightarrow4} = -2.178\times10^{-18}\,\text{J}\left(\frac{1}{4^2} - \frac{1}{2^2}\right) = 4.084\times10^{-19}\,\text{J}$$

Checking the equality, we see that $E_{2\rightarrow4} = E_{2\rightarrow3} + E_{3\rightarrow4}$

$$E_{2\rightarrow3} + E_{3\rightarrow4} = 3.025\times10^{-19}\,\text{J} + 1.059\times10^{-19}\,\text{J}$$

$$4.084\times10^{-19}\,\text{J} = E_{2\rightarrow4}(4.084\times10^{-19}\,\text{J})$$

Think About It

Do you think that the frequencies of those transitions would be additive?

3.63. Collect, Organize, and Analyze

The de Broglie equation is

$$\lambda = \frac{h}{mu}$$

We are to define the symbols in the equation and explain how that equation shows the wavelike properties of a particle.

Solve

In the de Broglie equation, λ is the wavelength that the particle of mass m exhibits as it travels at velocity u, with h being the Planck constant. That equation states that any moving particle has wavelike properties because a wavelength can be calculated through the equation and that the wavelength of the particle is inversely related to its momentum (mass × velocity).

Think About It

From the equation, we see that as mass or velocity increases, the wavelength of a particle decreases.

3.65. Collect, Organize, and Analyze

We consider whether the density or shape of an object affects its de Broglie wavelength.

Solve

No, the de Broglie equation relates only the mass and the speed to the wavelength, so neither the density nor the shape would affect the de Broglie wavelength.

Think About It

From the de Broglie equation we see that as the mass or the speed of a particle increases, the wavelength associated with that particle decreases.

3.67. Collect and Organize

When two objects of different masses move at the same speed, we can use the de Broglie relationship to compare their wavelengths and determine whether the given statements are true.

Analyze

From the de Broglie equation,

$$\lambda = \frac{h}{mu}$$

we see that wavelength is inversely proportional to the mass of the particle.

Solve

(a) False. Heavier particles have a *shorter* wavelength than lighter particles.

(b) True. When $m_2 = 2m_1$, then

$$\frac{\lambda_2}{\lambda_1} = \frac{h/2m_1 u}{h/m_1 u} = \frac{1}{2}$$

(c) True. Doubling the speed gives $u_2 = 2u_1$:

$$\frac{\lambda_2}{\lambda_1} = \frac{h/m2u_1}{h/mu_1} = \frac{1}{2}$$

That result is the same as doubling the mass in (b).

Think About It

Small, fast-moving particles exhibit the longest wavelengths.

3.69. Collect and Organize

Given the mass of objects (from a muon to Earth), we are to use the de Broglie equation to calculate the wavelength of the objects moving at given speeds.

Analyze

The de Broglie equation is

$$\lambda = \frac{h}{mu}$$

where the Planck constant is $h = 6.626 \times 10^{-34}$ J·s, m is the mass of the particle in kilograms, and u is the speed of the particle in meters per second. Recall that 1 J $= 1$ kg·m^2/s^2.

Solve

(a) Muon:

$$\lambda = \frac{6.626 \times 10^{-34} \text{ kg} \cdot \text{m}^2/\text{s}}{1.884 \times 10^{-28} \text{ kg} \times 325 \text{ m/s}} = 1.08 \times 10^{-8} \text{ m, or 10.8 nm}$$

(b) Electron:

$$\lambda = \frac{6.626 \times 10^{-34} \text{ kg} \cdot \text{m}^2/\text{s}}{9.10939 \times 10^{-31} \text{ kg} \times 4.05 \times 10^6 \text{ m/s}} = 1.80 \times 10^{-10} \text{ m, or 0.180 nm}$$

(c) Sprinter:

$$\lambda = \frac{6.626 \times 10^{-34} \text{ kg} \cdot \text{m}^2/\text{s}}{82 \text{ kg} \times 9.9 \text{ m/s}} = 8.2 \times 10^{-37} \text{ m, or } 8.2 \times 10^{-28} \text{ nm}$$

(d) Earth:

$$\lambda = \frac{6.626 \times 10^{-34} \text{ kg} \cdot \text{m}^2/\text{s}}{6.0 \times 10^{24} \text{ kg} \times 3.0 \times 10^4 \text{ m/s}} = 3.7 \times 10^{-63} \text{ m, or } 3.7 \times 10^{-54} \text{ nm}$$

Think About It

Only small particles with low mass generally show wavelike behavior. We can detect λ only on the order of the size of atoms, or 10^{-10} m. We therefore do not observe the sprinter's or Earth's waves.

3.71. Collect and Organize

Heisenberg's uncertainty principle states that the uncertainty in the position (Δx) multiplied by $m\Delta u$, where m is the particle's mass and Δu is the uncertainty in the particle's speed, must be equal to or greater than $h/4\pi$; that is,

$$\Delta x \cdot m\Delta u \geq \frac{h}{4\pi}$$

Analyze

For the H_2^+ particle in the cyclotron, the 3% uncertainty in velocity would be

$$\Delta u = 4 \times 10^6 \text{ m/s} \times 0.03 = 1.2 \times 10^5 \text{ m/s}$$

The mass of the H_2^+ particle would be

$$\frac{2.016 \text{ g}}{1 \text{ mol}} \times \frac{1 \text{ mol}}{6.022 \times 10^{23} \text{ particles}} \times \frac{1 \text{ kg}}{1000 \text{ g}} = 3.35 \times 10^{-27} \text{ kg}$$

Solve

Rearranging Heisenberg's equation to solve for Δx:

$$\Delta x \geq \frac{h}{4\pi m\Delta u} = \frac{6.626 \times 10^{-34} \text{ kg} \cdot \text{m}^2/\text{s}}{4\pi \times 3.35 \times 10^{-27} \text{ kg} \times 1.2 \times 10^5 \text{ m/s}} = 1.3 \times 10^{-13} \text{ m}$$

Think About It

The minimum uncertainty in position is very small, smaller than what we can measure, so in principle we can accurately determine the position of the H_2^+ particle moving at that speed.

3.73. Collect, Organize, and Analyze

We are to differentiate between a Bohr orbit and a quantum theory orbital.

Solve

The Bohr model orbit showed the quantized nature of the electron in the atom as a particle moving around the nucleus in concentric orbits, much like planets moving around the sun.

In quantum theory, an orbital is a region of space where the probability of finding the electron is high. The electron is not viewed as a particle but as a wave, and it is not confined to a clearly defined orbit; rather, we refer to the probability of the electron being at various locations around the nucleus.

Think About It
Bohr's model helped explain atomic spectra. The quantum theory of the atom helped to explain much more, including how atoms bond and the probability of an electronic transition in an atom.

3.75. Collect, Organize, and Analyze
To identify the orbital for an electron, we are asked how many quantum numbers we would need.

Solve
We need to describe the shell, the subshell, and the orbital's orientation to define a particular orbital. Therefore, we need three quantum numbers: n, ℓ, and m_ℓ.

Think About It
We could not use fewer quantum numbers to describe a particular orbital because confusion would arise as to what shell or subshell an electron belonged to or what its orientation was.

3.77. Collect and Organize
As the principal quantum number, n, increases, so does the number of orbitals available at the n level.

Analyze
The number of orbitals at each level n is n^2.

Solve
(a) For $n = 1$, only 1 orbital is available (an s orbital).
(b) For $n = 2$, 4 orbitals are available (one s and three p orbitals).
(c) For $n = 3$, 9 orbitals are available (one s, three p, and five d orbitals).
(d) For $n = 4$, 16 orbitals are available (one s, three p, five d, and seven f orbitals).
(e) For $n = 5$, 25 orbitals are available (one s, three p, five d, seven f, and nine g orbitals).

That totals 55 orbitals in the atom.

Think About It
Each subshell has an odd number of orbitals, and the number of orbitals in a particular subshell is $2\ell + 1$, where $\ell = 0$ for s orbitals, 1 for p orbitals, 2 for d orbitals, and 3 for f orbitals. Also note that the g orbitals that are when $n \geq 5$ are not occupied by any (yet) known element in the ground state, although those theoretical orbitals can be accessed in excited states.

3.79. Collect and Organize
We are to list all the possible ℓ values when $n = 4$.

Analyze
The angular momentum quantum number is related to n as $\ell = n - 1, n - 2, n - 3, \ldots, 0$.

Solve
When $n = 4$, $\ell = 3, 2, 1, 0$.

Think About It
Those ℓ values correspond to the f, d, p, and s subshells, respectively.

3.81. Collect and Organize

For each set of quantum numbers of n and ℓ, we are to write the orbital designation.

Analyze

The principal quantum number gives the shell number, which is just expressed as the number. The angular momentum quantum number, however, is given a letter designation ($\ell = 0$ is an s orbital, $\ell = 1$ is a p orbital, $\ell = 2$ is a d orbital, $\ell = 3$ is an f orbital, and $\ell = 4$ is a g orbital).

Solve

(a) $n = 6$, $\ell = 0$ represents $6s$
(b) $n = 3$, $\ell = 2$ represents $3d$
(c) $n = 2$, $\ell = 1$ represents $2p$
(d) $n = 5$, $\ell = 4$ represents $5g$

Think About It

The letter designation for the shape of the orbital provides a shorthand designation of the orbital as a number plus a letter. That system is easier than describing the orbital with two numbers (n and ℓ).

3.83. Collect and Organize

Given values for the quantum numbers n, ℓ, and m_ℓ, we are to determine the number of electrons that could occupy the orbitals described by those quantum numbers.

Analyze

The principal quantum number gives us the shell of the orbitals. That then gives the allowed values of ℓ ($n - 1$), which in turn describes the type of orbital (s, p, d, or f). The m_ℓ quantum number gives us the orientation of the orbital and its allowed values ($-\ell, -\ell + 1, \ldots, \ell - 1, \ell$), which gives us the number of orbitals available for that subshell. Each orbital can accommodate two electrons.

Solve

(a) The set of quantum numbers $n = 2$, $\ell = 0$ describes a $2s$ orbital, which two electrons can occupy.
(b) The set of quantum numbers $n = 3$, $\ell = 1$, $m_\ell = 0$ describes one of the $3p$ orbitals, which two electrons can occupy.
(c) The set of quantum numbers $n = 4$, $\ell = 2$ describes the set of $4d$ orbitals. Five d orbitals are in the subshell, so 10 electrons can occupy that orbital set.
(d) The set of quantum numbers $n = 1$, $\ell = 0$, $m_\ell = 0$ describes the $1s$ orbital, which two electrons can occupy.

Think About It

Remember that one s, three p, five d, seven f, and nine g orbitals exist in shells for which those orbitals are allowed.

3.85. Collect and Organize

Given values for the quantum numbers n, ℓ, m_ℓ, and m_s, we are to determine which combinations are allowed.

Analyze

The principal quantum number (n) can take on whole numbers starting with 1 ($n = 1, 2, 3, 4, \ldots$). The angular momentum quantum numbers (ℓ) possible for a given n value are $n - 1, n - 2, \ldots, 0$. The magnetic quantum numbers (m_ℓ) allowed for a given ℓ are $-\ell, -\ell + 1, \ldots, \ell - 1, \ell$. Allowed values for m_s are $+\frac{1}{2}$ and $-\frac{1}{2}$.

Solve

(a) For $n = 1$, the only allowed value of ℓ and m_ℓ is 0; the combination $n = 1$, $\ell = 1$, $m_\ell = 0$, $m_s = +\frac{1}{2}$ is not allowed because $\ell \neq 1$ when $n = 1$.
(b) For $n = 3$, the allowed values of ℓ are 0, 1, and 2 and when $\ell = 0$ the allowed value of m_ℓ is 0; the combination of $n = 3$, $\ell = 0$, $m_\ell = 0$, $m_s = -\frac{1}{2}$ is allowed.

(c) For $n = 1$, the only allowed value for ℓ and m_ℓ is 0; the combination $n = 1$, $\ell = 0$, $m_\ell = 1$, $m_s = -\frac{1}{2}$ is not allowed because $m_\ell \neq 1$ when $\ell = 0$.

(d) For $n = 2$, the allowed values of ℓ are 0 and 1 and when $\ell = 1$ the allowed value of $m_\ell = -1, 0, 1$; the combination of $n = 2$, $\ell = 1$, $m_\ell = 2$, $m_s = +\frac{1}{2}$ is not allowed because $m_\ell \neq 2$ when $\ell = 1$.

Think About It
For the allowed combination of quantum numbers, part b describes a $3s$ orbital.

3.87. Collect, Organize, and Analyze
We are asked what is meant by *degenerate orbitals*.

Solve
Degenerate orbitals have the same energy and are indistinguishable from one another.

Think About It
In the hydrogen atom, all the orbitals in a given n level are degenerate. Therefore, in hydrogen the $3s$, $3p$, and $3d$ orbitals, for example, all have the same energy. In multielectron atoms, however, those orbitals split in energy and are no longer degenerate.

3.89. Collect and Organize
In the filling of atomic orbitals, the $4s$ level fills before the $3d$. We are asked how that is evident in the periodic table.

Analyze
The two leftmost columns in the periodic table correspond to the s block, whereas columns 3–12 starting in period 4 correspond to the d block.

Solve
Starting with the fourth row elements, the outermost ($n =$ row number) s orbitals fill before the $(n - 1)$ d orbitals do. For example, K and Ca atoms contain $4s$ electrons, but no $3d$ electrons.

Think About It
The $6s$ orbitals fill (Cs and Ba), followed by the $4f$ orbitals (Ce–Yb) and then the $5d$ orbitals (La–Hg).

3.91. Collect and Organize
For multielectron atoms, we are to identify the subshells defined by their n and ℓ quantum numbers and then arrange them in order of increasing energy.

Analyze
The higher the energy of an orbital, the farther the electron is from the nucleus. Therefore, for different n values the order of energies is $1 < 2 < 3$, and so on. For orbitals in the same n shell, the orbitals increase in energy; that is, $s < p < d < f$ for multielectron atoms.

Solve
(a) $3d$, for $n = 3$, $\ell = 2$
(b) $7f$, for $n = 7$, $\ell = 3$
(c) $3s$, for $n = 3$, $\ell = 0$
(d) $4p$ for $n = 4$, $\ell = 1$.
In increasing order of energy: (c) $3s <$ (a) $3d <$ (d) $4p <$ (b) $7f$.

Think About It
To determine the energy of an orbital, first look to the n quantum number and then to the ℓ.

3.93. Collect and Organize

We can use the periodic table and Figure 3.33 in the textbook to write the electron configurations for several elemental species, including anions and cations.

Analyze

When a cation is formed, electrons are removed from the highest energy orbital. None of the species are transition metals, so we remove the electrons from the orbitals last filled in building the electron configuration of the element. To form an anion, we need to add electrons to the highest energy orbital or the next orbital up in energy. We use the previous noble gas configuration as the "core" to write the condensed form of the configurations.

Solve

Li^+: [He] or $1s^2$ $\qquad\qquad$ Mg^{2+}: [He]$2s^2 2p^6$ or [Ne]

Ca: [Ar]$4s^2$ $\qquad\qquad\qquad$ Al^{3+}: He]$2s^2 2p^6$ or [Ne]

F^-: [He]$2s^2 2p^6$ or [Ne]

Think About It

Because F^-, Mg^{2+}, and Al^{3+} all have the same electron configurations and thus the same number of electrons, they are isoelectronic with each other.

3.95. Collect and Organize

We are to write the condensed electron configurations (using the noble gas core configuration in brackets) for several species, including cationic and anionic species.

Analyze

When a cation is formed, electrons are removed from the highest energy orbital. To form an anion, we need to add electrons to the highest energy orbital or the next orbital up in energy.

Solve

K: [Ar]$4s^1$ $\qquad$ Ti^{4+}: [Ar] or [Ne]$3s^2 3p^6$

K^+: [Ar] $\qquad\;$ Ni: [Ar]$4s^2 3d^8$

Ba: [Xe]$6s^2$

Think About It

K^+ and Ti^{4+} are isoelectronic with each other and with Ar.

3.97. Collect and Organize

Using Figure P3.97, we are to identify which orbital diagram describes the ground-state electron configuration of Mn and Mn^{2+}.

Analyze

We can use the periodic table and Figure 3.33 in the textbook to identify the electron configuration for the ground-state configuration for neutral Mn. When a cation is formed, electrons are removed from the highest energy orbital. Mn is a transition metal, so we have to remember to remove electrons from its *s* orbital in forming the 2+ cation.

Solve

Neutral manganese atoms have two electrons in the 4*s* orbital and five electrons in the 3*d* orbital. The electrons in the 4*s* orbital will be paired, but the electrons in the 3*d* orbitals will not be paired until necessary according to Hund's rule. Therefore, orbital diagram (b) represents Mn. Furthermore, when two electrons are removed from Mn to give Mn^{2+}, the electrons are removed from the highest energy orbitals, 4*s*; therefore, orbital diagram (d) represents Mn^{2+}.

Think About It

Orbital diagrams (a) and (e) represent excited states for manganese—Mn and Mn^{2+}, respectively. Orbital diagram c represents the ground-state electron configuration of Mn^-.

3.99. Collect and Organize

To determine the number of unpaired electrons in the ground-state atoms and ions, we have to first write the electron configuration for each species and then detail how the electrons are distributed among the highest energy orbitals.

Analyze

If the highest energy orbital (s, p, d, or f) is either empty or full, the species have no unpaired electrons. If the highest energy orbital is partially full, electrons singly occupy the degenerate orbitals at that level before pairing in those orbitals (by Hund's rule).

Solve

(a) N: $[He]2s^2 2p^3$ 3 unpaired e^-
(b) O: $[He]2s^2 2p^4$ 2 unpaired e^-
(c) P^{3-}: $[Ne]3s^2 3p^6$ 0 unpaired e^-
(d) Na^+: $[Ne]$ or $[He]2s^2 2p^6$ 0 unpaired e^-

Think About It

The ground-state configuration of those elements fills the s orbital first and then places electrons into the p orbitals. That is because for a multielectron atom, $s < p$ in terms of energy for a given principal quantum level.

3.101. Collect and Organize

An atom with the electron configuration $[Ar]3d^2 4s^2$ is in the fourth period in the periodic table and is among the transition metals.

Analyze

That atom has no charge, so we do not have to account for additional or lost electrons.

Solve

The $4s$ orbital is filled for the element Ca. Two additional electrons are present in the $3d$ orbitals for the second transition metal of the fourth period: titanium, Ti. The electron-filling orbital box diagram shows two unpaired electrons.

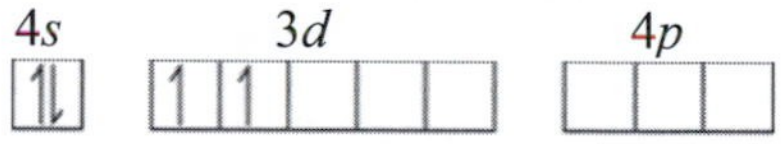

Think About It

Although we write the electron configuration so that $3d$ comes before $4s$, remember that the $4s$ orbital fills before the $3d$ in building up electron configurations.

3.103. Collect and Organize

We are to name the monatomic anion that has a filled-shell configuration of $[Ne]3s^2 3p^6$ or $[Ar]$ and determine the number of unpaired electrons in the ion in its ground state.

Analyze

Because the atom has an extra electron, to form the monatomic anion, the neutral atom would have an electron configuration of one fewer electron.

Solve

Ion's electron configuration: $[Ne]3s^2 3p^6 = X^-$
Atom's electron configuration: $[Ne]3s^2 3p^5 = X$
The atom is chlorine and the monatomic anion is chloride, Cl^-. Because electrons fill the s and p orbitals, Cl^- has no unpaired electrons in its ground state.

Think About It

When identifying elements with the electron configurations of anions, remove the electrons associated with the anionic charge to obtain the electron configuration of the neutral atom.

3.105. **Collect and Organize**

An electronic excited state exists when an electron has been placed into a higher energy orbital than would be predicted by using the filling rules shown by the periodic table.

Analyze

The order of filling for the orbitals is as follows:

$1s < 2s < 2p < 3s < 3p < 4s < 3d < 4p < 5s < 4d < 5p < 6s < 4f < 5d < 6p < 7s < 5f < 6d < 7p$

Solve

(a) Because the $2s$ orbital is lower in energy than the $2p$ orbital, the lowest energy configuration for that atom is $[He]2s^22p^4$, so the configuration $[He]2s^12p^5$ represents an excited state.

(b) The order of filling of orbitals for atoms after krypton is $5s < 4d < 5p$. This atom has 13 electrons in its outer shell: two fill the $5s$ orbital, 10 fill the $4d$ orbitals, and one is placed in a $5p$ orbital. That configuration, $[Kr]4d^{10}5s^25p^1$, does not represent an excited state.

(c) The order of filling of orbitals for atoms after argon is $4s < 3d < 4p$. This atom has 17 electrons in its outer shell: two fill the $4s$ orbital, 10 fill the $3d$ orbitals, and five are placed in the $4p$ orbitals. That configuration, $[Ar]3d^{10}4s^24p^5$, does not represent an excited state.

(d) Because the $3p$ orbital is lower in energy than the $4s$ orbital, the lowest energy configuration for that atom is $[Ne]3s^23p^3$, so the configuration $[Ne]3s^23p^24s^1$ represents an excited state.

Think About It

If each of those configurations is for neutral atoms, we can assign the elements as follows: (a) excited-state O, (b) ground-state In, (c) ground-state Br, and (d) excited-state P.

3.107. **Collect and Organize**

Iodine-131 has 53 protons, 78 neutrons, and 53 electrons as a neutral atom. We are to identify the subshell containing the highest energy electrons and compare the electron configurations of ^{131}I and ^{127}I.

Analyze

The electron configuration for iodine is $[Kr]4d^{10}5s^25p^5$. The difference between ^{131}I and ^{127}I is that ^{131}I has four more neutrons in its nucleus.

Solve

The electron configuration of iodine shows that the highest energy electrons are in the $5p$ subshell. Because the difference in isotopes is the number of neutrons present, the electron configurations of ^{131}I and ^{127}I (which are based on total number of electrons in the atom) are the same.

Think About It

Electron configurations, however, do change if the atom gains or loses electrons to become either anionic or cationic, respectively.

3.109. **Collect and Organize**

Sodium and chlorine atoms are neutral in charge, but the sodium atom in NaCl has a charge of 1+ and the chlorine atom has a charge of 1−. Those changes in charge also come with a change in size. Why?

Analyze

When we remove an electron from an atom, we reduce the repulsion for the remaining electrons in the atom. When we add electrons, we increase $e^-–e^-$ repulsion.

Solve

Sodium atoms are larger than chlorine atoms because the latter have six more positive charges in each of their nuclei, which pull the atom's electrons inward more strongly. However, Na$^+$ ions are much smaller than Na because their $n = 3$ shell is gone. In addition, Cl$^-$ ions are larger than Cl atoms because each ion has one more electron, which means greater electron-electron repulsion and an expanded cloud of electrons around the ion's nucleus.

Think About It
The change in size upon forming a cation or anion can be dramatic, as seen in Figure 3.35.

3.111. **Collect and Organize**
Using periodic trends, we are to place the atoms/ions in order of decreasing size.

Analyze
The sizes of atoms increase down a group because electrons have been added to higher n levels; the sizes of atoms decrease across a period because of increasing effective nuclear charge. When an atom loses electrons to become a cation, size decreases due to reduced electron–electron repulsion; when an atom gains electrons to become an anion, size increases due to increased electron–electron repulsion.

Solve
(a) $Al > P > Cl > Ar$
(b) $Sn > Ge > Si > C$
(c) $K > Na > Li > Li^+$
(d) $Cl^- > Cl > F > Ne$

Think About It
The largest atoms are those situated in the lower left of the periodic table; the smallest atoms are those situated at the upper right in the periodic table.

3.113. **Collect and Organize**
Ionization energy is the energy required to remove an electron from a gaseous atom.
$$X(g) \rightarrow X^+(g) + e^-$$
We are to state the trends in ionization energies down and across the periodic table.

Analyze
The ionization energy will change with effective nuclear charge (the higher the Z_{eff}, the greater the ionization energy) and with size (an electron farther away from the nucleus requires less energy to remove).

Solve
(a) As the atomic number increases down a group, electrons are added to higher n levels, leading to a decrease in effective nuclear charge and ionization energy.
(b) As the atomic number increases across a row, the effective nuclear charge increases. Therefore, the ionization energy increases across a period of elements.

Think About It
Ionization energy trends follow atomic size trends; smaller atoms require more energy to ionize than larger atoms.

3.115. **Collect and Organize**
For this question we are asked to relate the wavelength of light needed to ionize gaseous atoms as we increase atomic number down a group in the periodic table.

Analyze
As we proceed down a group in the periodic table, the atoms get larger and the electron being removed in the ionization process is farther from the nucleus and, therefore, higher in energy. Wavelength of light is inversely related to the energy through the equation $E = hc/\lambda$.

Solve
Because less energy would be required to ionize a larger atom (because the electron being removed is farther from the nucleus), the wavelength of light required would be longer for higher atomic number atoms than for lower atomic number atoms in the same group on the periodic table. In other words, in progressing down the group, the wavelength necessary for ionization would increase.

Think About It

The opposite would be true as we progress across a period; the wavelength necessary for ionization would decrease because higher energies would be required owing to increased effective nuclear charge across the elements in a period.

3.117. Collect and Organize

For this question we are asked to place elements in order of increasing first ionization energy.

Analyze

As we proceed down a group in the periodic table, the atoms get larger and the electron being removed in the ionization process is farther from the nucleus and, therefore, ionization energy decreases down a group. As we proceed across a period in the periodic table, the atoms get smaller and the electron being removed in the ionization process is closer to the nucleus and, therefore, ionization energy increases across a period.

Solve

(a) $I < Br < Cl < F$

(b) $Na < Li < Mg < Be$

(c) $O < N < F < Ne$

Think About It

The second ionization energy is much greater for an atom than the first ionization energy; in the second ionization energy, you are removing an electron from an already positively charged cation.

3.119. Collect and Organize

For some elements, the electron affinity is negative and favorable to form the ions compared with that of the neutral atom. We are to decide whether that behavior means that those elements all are present in nature as anions.

Analyze

Figure 3.38 shows electron affinity values for the representative elements. Nearly all those elements have a negative electron affinity, which indicates that the anion is lower in energy than the neutral atom.

Solve

No, a negative electron affinity does not mean that the element is present in nature as an anion. For example, sodium has an electron affinity of −52.9 kJ/mol, but it is always present in nature as Na^+, not even as Na metal. That is also true of the other metals in the table.

Think About It

However, the nonmetals such as the halogens often do have negative charges in their compounds found in nature, such as NaCl and KI.

3.121. Collect and Organize

As we descend the halogens (group 17), the electron affinity values increase (become more positive or less negative). We are to explain that trend.

Analyze

Less negative (or more positive) values of electron affinity mean that the anion formed is getting less stable than the neutral atom as the atoms in a group get larger.

Solve

Electrons are added to shells that are farther away from the nucleus with increasing Z. This means electron affinities become weaker, that is, less negative, which means their arithmetic values increase.

Think About It
The trend in descending a group often shows discrepancies for the first element in the group in period 2. The values of the electron affinity for those elements (Be, B, C, N, O, and F) are lower than expected because the electron is being added to a small atom where electron–electron repulsions are more noticeable.

3.123. **Collect and Organize**
In this question we consider the electronic properties of barium, which is responsible for the green color in many fireworks.

Analyze
We can use Figure 3.33 in the textbook to help us write the electron configuration for barium, and from the electron configuration of the lowest-energy excited state we will be able to write the quantum numbers for a $5d$ electron. The wavelength of light associated with a particular energy is $\lambda = hc/E$; we can use that formula to determine whether the energies of light emitted correspond to wavelengths in the green region of the electromagnetic spectrum (~520–560 nm; see Figure 3.1 in the textbook).

Solve
(a) The ground-state electron configuration for barium is $[Xe]6s^2$.
(b) The possible quantum numbers for a $5d$ electron are $n = 5$, $\ell = 2$, $m_\ell = -2, -1, 0, 1, 2$.
(c) The wavelength associated with an emission energy of 1.79×10^{-19} J is

$$\lambda = \frac{6.626 \times 10^{-34} \text{ J} \cdot \text{s} \times 2.998 \times 10^8 \text{ m/s}}{1.79 \times 10^{-19} \text{ J}} = 1.11 \times 10^{-6} \text{ m, or } 1110 \text{ nm}$$

That emission is in the infrared region, so it cannot account for the green color of barium-containing fireworks.
(d) The wavelength associated with an emission energy of 3.59×10^{-19} J is

$$\lambda = \frac{6.626 \times 10^{-34} \text{ J} \cdot \text{s} \times 2.998 \times 10^8 \text{ m/s}}{3.59 \times 10^{-19} \text{ J}} = 5.53 \times 10^7 \text{ m, or } 553 \text{ nm}$$

That emission is in the visible region and corresponds to the wavelength for green, so it can account for the green color of barium-containing fireworks.

Think About It
Other elements are used to generate different colors for fireworks: for example, red comes from either lithium or strontium, and yellow comes from sodium.

3.125. **Collect and Organize**
We consider the emission of energy from an He$^+$ ion from $n = 3$ to $n = 1$ compared with a stepwise relaxation of a He$^+$ ion to the ground state ($n = 3$ to $n = 2$ and then $n = 2$ to $n = 1$). We are to determine which of the statements given are true.

Analyze
Both He$^+$ ions have the same nuclear charge (2+) and the same energies for $n = 3$, $n = 2$, and $n = 1$.

Solve
(a) True. Because the energies of $n = 1, 2,$ and 3 do not depend on how the electron relaxes to the ground state, the total energy of $n = 3$ to $n = 1$ is equal to the sum of the energy of $n = 3$ to $n = 2$ and the energy of $n = 2$ to $n = 1$.

(b) False. Although the energies are additive, the wavelengths are not:

$$E_{3\to1} = hc/\lambda_{3\to1}$$
$$E_{3\to1,\,2\to1} = hc/\lambda_{3\to2} + hc/\lambda_{2\to1}$$

Those energies are equal, so

$$\frac{hc}{\lambda_{3\to1}} = \frac{hc}{\lambda_{3\to2}} + \frac{hc}{\lambda_{2\to1}}$$

$$\frac{1}{\lambda_{3\to1}} = \frac{1}{\lambda_{3\to2}} + \frac{1}{\lambda_{2\to1}} = \frac{\lambda_{2\to1}}{\lambda_{3\to2}\lambda_{2\to1}} + \frac{\lambda_{3\to2}}{\lambda_{2\to1}\lambda_{3\to2}}$$

Multiplying both sides by $\lambda_{3\to1}$ gives

$$1 = \frac{\lambda_{2\to1}\lambda_{3\to1}}{\lambda_{3\to2}\lambda_{2\to1}} + \frac{\lambda_{3\to2}\lambda_{3\to1}}{\lambda_{2\to1}\lambda_{3\to2}} = \frac{\lambda_{2\to1}\lambda_{3\to1} + \lambda_{3\to2}\lambda_{3\to1}}{\lambda_{3\to2}\lambda_{2\to1}}$$

$$\lambda_{3\to2}\lambda_{2\to1} = \lambda_{2\to1}\lambda_{3\to1} + \lambda_{3\to2}\lambda_{3\to1} = \lambda_{3\to1}\left(\lambda_{2\to1} + \lambda_{3\to2}\right)$$

$$\frac{\lambda_{3\to2}\lambda_{2\to1}}{\lambda_{2\to1} + \lambda_{3\to2}} = \lambda_{3\to1}$$

(c) True. Because the energies are additive, the frequencies also are additive:

$$E_{3\to1} = h\nu_{3\to1}$$
$$E_{3\to2,\,2\to1} = h\nu_{3\to2} + h\nu_{2\to1}$$

Because those energies are equal,

$$h\nu_{3\to1} = h\nu_{3\to2} + h\nu_{2\to1}$$
$$\nu_{3\to1} = \nu_{3\to2} + \nu_{2\to1}$$

(d) True. Using Equation 3.11 in the textbook,

For He^+: $\Delta E = -(2.178\times10^{-18}\text{ J})(2)^2\left(\frac{1}{1^2} - \frac{1}{3^2}\right) = -7.744\times10^{-18}\text{ J}$

$$\lambda = \frac{hc}{E} = \frac{6.626\times10^{-34}\text{ J}\cdot\text{s}\times2.998\times10^8\text{ m/s}}{7.744\times10^{-18}\text{ J}} = 2.565\times10^{-8}\text{ m, or }25.65\text{ nm}$$

For H^+: $\Delta E = -(2.178\times10^{-18}\text{ J})(1)^2\left(\frac{1}{1^2} - \frac{1}{3^2}\right) = -1.936\times10^{-18}\text{ J}$

$$\lambda = \frac{hc}{E} = \frac{6.626\times10^{-34}\text{ J}\cdot\text{s}\times2.998\times10^8\text{ m/s}}{1.936\times10^{-18}\text{ J}} = 1.026\times10^{-7}\text{ m, or }102.6\text{ nm}$$

Think About It

Be careful in abruptly concluding that the wavelengths of transitions are additive. Only their energies and frequencies can be added in steps to get to the overall energy.

3.127. **Collect and Organize**

Ionization energy (IE_1) is correlated with electronic structure. In this problem we examine the trends in first and second ionization energies for elements 31–36 (Ga–Kr).

Analyze

The general trend is for increasing IE_1 as atomic number (Z) increases across a period. However, electronic structure (configuration) plays a role. In particular, the IE_1 and IE_2 for Ga–Kr depend on whether the electron is being removed from an s or a p orbital.

Solve

The electron configurations for Ga–Kr for both neutral atoms (X) and singly charged cations (X^+) are as follows:

Element	Electron Configuration X	Electron Configuration X^+
Ga	$[Ar]3d^{10}4s^24p^1$	$[Ar]3d^{10}4s^2$
Ge	$[Ar]3d^{10}4s^24p^2$	$[Ar]3d^{10}4s^24p^1$
As	$[Ar]3d^{10}4s^24p^3$	$[Ar]3d^{10}4s^24p^2$
Se	$[Ar]3d^{10}4s^24p^4$	$[Ar]3d^{10}4s^24p^3$
Br	$[Ar]3d^{10}4s^24p^5$	$[Ar]3d^{10}4s^24p^4$
Kr	$[Ar]3d^{10}4s^24p^6$	$[Ar]3d^{10}4s^24p^5$

For the first ionization energy, the IEs increase in the following order:

$$Ga < Ge < Se < As < Br < Kr$$

In that series, as Z increases, IE_1 generally increases. The IE_1 of Se is less than that of As because the electron pairing ($4p^4$) in one of the p orbitals for Se lowers Se's IE_1 slightly.

For the second ionization, the IE_2 values increase in the following order:

$$Ge < Ga < As < Br < Se < Kr$$

Again, it is generally observed that as Z increases, so does the IE_2. However, Ge's second IE_2 is lower than Ga's because to ionize the second electron in Ga, we need to remove an electron from a lower energy $4s$ orbital. Also, Br's IE_2 is lower than Se's because the electron pairing ($4p^4$) in one of the p orbitals for the Br^+ ion lowers its IE_2 slightly.

Think About It

In comparing the first and second ionization energies for Ga–Kr, notice that the reversal of the general trend at As–Se in IE_1 occurs one pair to the right (Se–Br) in IE_2.

3.129. Collect and Organize

We are to determine which neutral atoms are isoelectronic with Sn^{2+} and Mg^{2+} and which 2+ ion is isoelectronic with Sn^{4+}.

Analyze

(a) The ground-state electron configurations for the neutral atoms Sn and Mg are Sn = $[Kr]4d^{10}5s^25p^2$ and Mg = $[Ne]3s^2$. To form Sn^{2+}, remove the two $5p$ electrons; to form Sn^{4+}, remove the two $5p$ electrons and the two $5s$ electrons. To form Mg^{2+}, remove the two $3s$ electrons.

(b) The neutral atom that has the same electron configuration as Sn^{2+} would have to have two $5s$ electrons and a filled $4d$ shell. The neutral atom that has the same electron configuration as Mg^{2+} would have to have a filled $n = 2$ shell (two $2s$ electrons and six $2p$ electrons).

(c) Isoelectronic species have the same number of electrons. The 2+ cation that would be isoelectronic with Sn^{4+} would have to have no $5s$ or $5p$ electrons but would have a filled $4d$ shell.

Solve

(a) Sn^{2+}: $[Kr]4d^{10}5s^2$
Sn^{4+}: $[Kr]4d^{10}$
Mg^{2+}: $[Ne]$ or $[He]2s^22p^6$

(b) Cadmium has the same electron configuration as Sn^{2+}, and neon has the same electron configuration as Mg^{2+}.

(c) Cd^{2+} is isoelectronic with Sn^{4+}.

Think About It

When writing electron configurations for ionic species, start with the neutral atom and add or remove electrons to form the ions.

3.131. **Collect and Organize**

Using the equation $Z_{eff} = Z - \sigma$, where Z is the atomic number and σ is the shielding parameter, we are to compare the Z_{eff} (effective nuclear charge) for the outermost electron in neon and argon.

Analyze

(a) In the effective nuclear charge equation given, use $Z = 10$ and $\sigma = 4.24$ for Ne and $Z = 18$ and $\sigma = 11.24$ for Ar.

(b) Shielding depends on the number of electrons lower in energy than the electron of interest.

Solve

(a) Ne: $Z_{eff} = 10 - 4.24 = 5.76$
    Ar: $Z_{eff} = 18 - 11.24 = 6.76$

(b) The outermost electron in argon is a $3p$ electron, which is mostly shielded by the electrons in the $n = 2$ level (10 electrons) and the $n = 1$ level (2 electrons), whereas the outermost electron in neon is a $2p$ electron, which is shielded only by the electrons in the $n = 1$ level (2 electrons).

Think About It

Z_{eff} is greater for the outermost electron in Ar than for Ne. The ionization energy of Ar, however, is lower than the ionization energy for Ne. The effective nuclear charge equation, therefore, doesn't seem to predict the trend in decreasing ionization energy as we descend a group in the periodic table. The effective nuclear charge equation here does not take into account the n level from which the electron is removed (ionized) to form the cation. Remember that the farther the electron is from the nucleus, the lower the energy required to remove it.

3.133. **Collect and Organize**

The p orbital has two lobes of different phases with a node between the lobes. We are asked how an electron gets from one lobe to the other without going through the node between them.

Analyze

When we think of an orbital, we should think of the electron not as a particle (which here would have to move through the node, a region of zero probability) but as a wave.

Solve

When we think of the electron as a wave, we can envision the node between the two lobes as a wave of zero amplitude and the p orbital as a standing wave.

Think About It

Remember that an orbital describes the wave function for the electron and does not specifically locate the electron as a particle.

CHAPTER 4 | Chemical Bonding: Understanding Climate Change

4.1. Collect and Organize

Aluminum is in group 13 of the periodic table, and the neutral atom has an electron configuration of $[Ne]3s^23p^1$. In this question we are to choose the correct Lewis symbol for the most stable ion of aluminum from Figure P4.1.

Analyze

Aluminum loses its three outermost electrons to form the Al^{3+} cation with an electron configuration of $[Ne]$, which leaves no electrons in the valence shell of Al^{3+}.

Solve

The Lewis symbol must correctly show both the charge and the number of valence electrons on the species. Here the charge is +3 and no valence electrons are present, so the correct Lewis structure is choice e:

$$Al^{3+}$$

Think About It

The only other correct charge–valence electron choice in this problem for Al is choice c:

$$\left[:\ddot{\underset{\cdot\cdot}{Al}}: \right]^{5-}$$

but a charge of −5 is not the most stable for the Al anion.

4.3. Collect and Organize

Given three Lewis structures for bonding between N, C, and S in thiocyanate, SCN^-, shown in Figure P4.3, we are asked which depict resonance structures and to explain our choices.

Analyze

Resonance structures show more than one valid Lewis structure for a compound. They have the same arrangement of atoms but different arrangements of electrons.

Solve

In the three structures shown, the arrangement of atoms is different; therefore, they are not resonance structures of each other. The structure that is one of the possible resonance forms for the thiocyanate ion, SCN^-, is the middle structure which has carbon as its middle atom.

Think About It

Valid resonance structures of NCS^- would be

$$\left[:N\equiv C-\ddot{\underset{\cdot\cdot}{S}}: \right]^- \longleftrightarrow \left[:\ddot{\underset{\cdot\cdot}{N}}-C\equiv S: \right]^- \longleftrightarrow \left[:\ddot{\underset{\cdot\cdot}{N}}=C=\ddot{\underset{\cdot\cdot}{S}}: \right]^-$$

4.5. Collect and Organize

From the four drawings given in Figure P4.5, we are to determine which best describes the electron density in BrCl.

Analyze

The distribution of electron density in the molecule depends on the electronegativities of the atoms in the bond. Chlorine has an electronegativity of 3.0 and bromine has an electronegativity of 2.8. The higher the negative charge, the more red the atom; the higher the positive charge, the more blue the atom; if the charge is zero on the atom, the atom is yellow-green, as shown in Figure 4.12 in the textbook.

Solve

Because chlorine has a greater electronegativity, higher electron density resides on the Cl atom, making Cl partially negative in BrCl. The difference in electronegativity, however, is small; chlorine will be slightly redder and bromine will be bluer from the neutral color of yellow-green. Therefore, choice (d), which shows the Cl as orange and Br as yellow, is the correct representation of the electron distribution in BrCl.

Think About It

For this problem, we need not necessarily know the exact values for the electronegativities of Cl and Br. From the periodic trends for electronegativity (which decreases as we go down a group), we know that the electronegativity of Cl is greater than that of Br, but with both being halogens, they are probably not very different from each other.

4.7. **Collect and Organize**

Of the three drawings of bent triatomic molecules in Figure P4.7, we are to choose the one that represents the electron density distribution in sulfur dioxide, SO_2, and explain our choice.

Analyze

Differences in electron density within a molecule depend on the different pulling powers of the atoms in the molecule for electrons (electronegativity). Because sulfur is first in the molecular formula, we can assume that it is the central atom in the molecule. In the drawings, the higher the negative charge, the more red the atom; the higher the positive charge, the more blue the atom, as shown in Figure 4.12 in the textbook.

Solve

Sulfur has a lower electronegativity (2.5) than oxygen (3.5) because as we descend a group in the periodic table, electronegativity decreases. The oxygen atoms have the higher electron density, but the difference in electronegativity is only 1.0, so partial negative charges reside on the oxygen atoms and a partial positive charge resides on the sulfur atom. Drawing (a) best represents SO_2.

Think About It

Drawing (c) shows the reverse polarity in which a higher electron density is on the sulfur atom. In that configuration, the electronegativity of sulfur is greater than that of oxygen, which we know not to be the case.

4.9. **Collect and Organize**

Of the three drawings showing the vibrations of the bent water molecule in Figure P4.9, we are to choose which will be infrared active.

Analyze

For a vibration to be infrared active, it must be asymmetric to produce the oscillating electric field that will allow it to absorb IR radiation. If the molecule bends in a vibration it will also absorb IR radiation.

Solve

Both the asymmetric stretching and bending vibrational modes in water will be infrared active; all three modes, therefore are IR-active.

Think About It

Compare those modes of water vibrations to those of carbon dioxide in Figure 4.15 in the textbook. What is similar and what is different?

4.11. **Collect and Organize**

From Figure P4.11, showing the electrostatic potential energy versus the distance between nuclei, we are to determine which curve is for KCl and which is for KF.

Analyze

The differences in the plots reflect the differences in ionic bond distances due to ionic sizes, as indicated by the position of the minimum energy along the *x*-axis, and the strength of their ionic interaction, as indicated by the depth of the energy at the minimum in the curve along the *y*-axis. The fluoride ion is smaller than the chloride ion and, because of its smaller size, will more strongly attract the K^+ ion.

Solve

The red curve represents the interaction between potassium and fluoride ions with a shorter internuclear distance and more a more negative minimum energy. The blue curve represents the interaction between potassium and chloride ions with a longer internuclear distance and a less negative minimum energy.

Think About It

The curve for potassium iodide would continue that trend for longer internuclear axis and weaker ionic bond interaction.

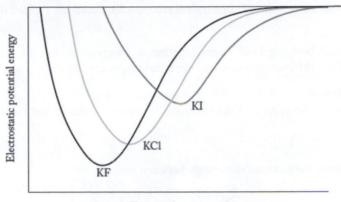

Distance between nuclei

4.13. Collect and Organize

We are asked to determine how the group number relates to the number of valence electrons in a neutral atom.

Analyze

The number of valence electrons in a neutral atom is the number of electrons in the outermost occupied shell as shown by the atoms' electron configurations. The group numbers in the periodic table range from 1 to 18, with each column numbered sequentially.

Solve

Group Number	Electron Configuration	Number of Valence e^-
1	ns^1	1
2	ns^2	2
3	$ns^2(n-1)d^1$	3
4	$ns^2(n-1)d^2$	4
5	$ns^2(n-1)d^3$	5
6	$ns^1(n-1)d^5$	6
7	$ns^2(n-1)d^5$	7
8	$ns^2(n-1)d^6$	8
9	$ns^2(n-1)d^7$	9
10	$ns^2(n-1)d^8$	10
11	$ns^1(n-1)d^{10}$	11
12	$ns^2(n-1)d^{10}$	12
13	$ns^2(n-1)d^{10}np^1$	3
14	$ns^2(n-1)d^{10}np^2$	4
15	$ns^2(n-1)d^{10}np^3$	5
16	$ns^2(n-1)d^{10}np^4$	6
17	$ns^2(n-1)d^{10}np^5$	7
18	$ns^2(n-1)d^{10}np^6$	8

The number of valence electrons equals the group number for groups 1–12. Once the *d* orbitals are filled, they are considered "core" electrons.

Think About It
Groups 13–18 have

$$\text{number of valence electrons} = \text{group number} - 10$$

4.15. Collect, Organize, and Analyze
By defining each type of bonding, we can describe the differences between covalent and ionic compounds.

Solve
Ionic compounds have attractive forces between oppositely charged ions (Coulombic forces). In ionic bonding, the electrons of one atom are not shared. In covalent compounds, bonds are formed through electron sharing.

Think About It
Both ionic and covalent bonding hold atoms together in compounds, but we will see later that ionic and covalent compounds have different physical and chemical properties (for example, melting points, solubility).

4.17. Collect and Organize
We are asked to calculate the energy of the electrostatic attraction, the strength of the ionic bond, between K^+ and Br^-.

Analyze
The equation to determine the electrostatic energy between ions is

$$E_{el} = 2.31 \times 10^{-19} \, J \cdot nm \left(\frac{Q_1 \times Q_2}{d} \right)$$

To use that equation, we need the charges on the ions (1+ and 1–) for Q_1 and Q_2 and the sum of the radii of the ions (from Figure 3.36), which is $138 + 195 = 333$ pm for d.

Solve

$$E_{el} = 2.31 \times 10^{-19} \, J \cdot nm \left(\frac{+1 \times -1}{333 \, pm \times \dfrac{1 \, nm}{1000 \, pm}} \right) = -6.94 \times 10^{-19} \, J$$

Think About It
The energy is negative, showing that bringing the two ions together gives a lower energy than if the ions were separated.

4.19. Collect and Organize
From among KCl, TiO_2, $BaCl_2$, and KI we are to determine which has the most negative lattice energy.

Analyze
We could do those calculations by using the equation

$$E_{el} = 2.31 \times 10^{-19} \, J \cdot nm \left(\frac{Q_1 \times Q_2}{d} \right)$$

but we can first see whether obvious differences exist that will make the correct prediction. The charges on the ions have a very large effect on the strength of the electrostatic interaction, more so than the sizes, so we will look there for substantial differences. KCl and KI both have 1+/1– ions, but $BaCl_2$ has a 2+/1– pairing, which will be substantially stronger, and TiO_2 has a 4+/2– pairing, which is even stronger. The more negative the lattice energy, the stronger the electrostatic interaction between ions.

Solve
TiO_2 (b) will have the most negative lattice energy because of the high charges on titanium and oxygen ions compared with the other ionic compounds on the list.

Think About It
Of the two salts, KCl and KI, potassium chloride will have the most negative lattice energy because Cl^- is smaller than I^-.

4.21. Collect and Organize

We can use Coulomb's law to rank KBr, $SrBr_2$, and $CsBr$ in order of increasing electrostatic potential energy.

Analyze

Coulomb's law states that the attraction between oppositely charged ions is directly proportional to the product of their charges and is inversely proportional to their separation distance:

$$E \propto \frac{Q_1 \times Q_2}{d}$$

The distance between the ions is taken as the sum of the ions' radii. The ionic radii for K^+ and Br^- are shown in Figure 3.36. You can look up the radii of Sr^{2+} and Cs^+ on the Internet or in some other resource. The values used in this calculation are 118 pm for Sr^{2+} and 170 pm for Cs^+.

Solve

For KBr, $E \propto \dfrac{(+1)(-1)}{(138+196)} = -0.00299$

For $SrBr_2$, $E \propto \dfrac{(+2)(-1)}{(118+196)} = -0.00637$

For $CsBr$, $E \propto \dfrac{(+1)(-1)}{(170+196)} = -0.00273$

In order of increasing ionic attraction: $CsBr < KBr < SrBr_2$.

Think About It

Because the ions in $CsBr$ are large compared with KBr, $CsBr$ has a lower ion–ion attraction. $SrBr_2$ has the highest ion–ion attraction because of the +2 charge on Sr.

4.23. Collect and Organize

When writing names for transition metal compounds, we include Roman numerals in the name. We are asked in this question what purpose Roman numerals serve.

Analyze

Transition metals often take on more than one oxidation state (or charge).

Solve

Roman numerals are used in the names of compounds of transition elements to indicate the charge on the transition metal cation.

Think About It

The indication of charge on the transition metal cation in the name of the compounds makes writing the formula for a compound easy. For example, without the *(III)* in iron(III) chloride, we would not be sure whether to write $FeCl_3$ or $FeCl_2$ because both compounds exist and exhibit different chemical and physical properties.

4.25. Collect and Organize

For the oxoanions of element X, XO_2^{2-}, and XO_3^{2-}, we are to assign one as *-ite*.

Analyze

Compound names that end in *-ite* represent oxoanions that have one fewer oxygen atom than compounds ending in *-ate*.

Solve

Between XO_2^{2-} and XO_3^{2-}, the oxoanion XO_2^{2-} would have a name ending in *-ite*.

Think About It

In Table 4.5, we can see that pattern for oxoanions of chlorine. ClO_3^- is *chlorate* and ClO_2^- is *chlorite*.

4.27. Collect and Organize

All the compounds here are oxides of nitrogen. Those are all molecular compounds composed of two nonmetallic elements. We name them by using the rules for binary compounds.

Analyze

We will use prefixes (Table 4.4) to indicate the number of oxygen atoms in those compounds. The nitrogen atom is always first in the formula, so it is named first. If only one nitrogen atom is in the formula, we do not need to use the prefix *mono-* for the nitrogen. If more than one nitrogen atom is present, however, we will indicate the number with the appropriate prefix. Also, because *oxide* begins with a vowel, saying *pentaoxide* would be awkward, we shorten the double vowel in that part of the chemical name to *pentoxide*.

Solve

(a) NO_3, nitrogen trioxide
(b) N_2O_5, dinitrogen pentoxide
(c) N_2O_4, dinitrogen tetroxide
(d) NO_2, nitrogen dioxide

(e) N_2O_3, dinitrogen trioxide
(f) NO, nitrogen monoxide
(g) N_2O, dinitrogen monoxide
(h) N_4O, tetranitrogen monoxide

Think About It

All those binary compounds of nitrogen and oxygen are uniquely named.

4.29. Collect and Organize

To predict the formula for the binary ionic compounds formed from the elements listed in the problem, we first have to decide what charges the metal and nonmetal typically have in ionic compounds. To name the compounds, we use the naming rules for ionic compounds.

Analyze

Metallic elements in group 1 of the periodic table (Na and Li) have a 1+ charge in ionic compounds, those in group 2 (Sr) have a 2+ charge, and those in group 13 (Al) have a 3+ charge. Nonmetals in group 16 (S and O) have a 2– charge, and those in group 17 (Cl) have a 1– charge. Hydrogen here has a 1– charge because it is combining with a metal and is thus a hydride. To write the formulas of the neutral salts, we must balance the charges of the anion with the charges of the cation. When we name binary ionic compounds, the cation is named first as the element, and the anion is named second with the ending *-ide* added.

Solve

(a) sodium (Na^+) and sulfur (S^{2-}): Na_2S, sodium sulfide
(b) strontium (Sr^{2+}) and chlorine (Cl^-): $SrCl_2$, strontium chloride
(c) aluminum (Al^{3+}) and oxygen (O^{2-}): Al_2O_3, aluminum oxide
(d) lithium (Li^+) and hydrogen (H^-): LiH, lithium hydride

Think About It

In naming binary ionic salts of the main group elements, we do not need to indicate the numbers of anions or cations in the formula with prefixes, making the naming of these compounds very direct.

4.31. Collect and Organize

Those compounds are all binary ionic compounds of cobalt and oxygen. Because cobalt is a transition metal and thus has more than one available oxidation state, we use the naming rules that incorporate Roman numerals to indicate the charge on the cobalt in the compound.

Analyze

In those compounds oxygen has a charge of 2–. The charge on the cobalt atoms must balance the charge on the oxygen atoms to give the neutral species listed. When we name those compounds, the metal is named first, followed by the charge in Roman numerals in parentheses. The anion is named as a separate word with the ending *-ide*.

Solve
(a) CoO: cobalt has a 2+ charge, cobalt(II) oxide.
(b) Co_2O_3: cobalt has a 3+ charge, cobalt(III) oxide.
(c) CoO_2: cobalt has a 4+ charge, cobalt(IV) oxide.

Think About It
Because the charges of the cations and anions must balance to give a neutral species, we do not have to indicate the number of oxide anions in these compounds; the cation charge dictates the number of oxides that must be present in the formula.

4.33. Collect and Organize
We are asked to identify the oxoanion from the name of a salt and write its formula with associated charge.

Analyze
The oxoanions are polyatomic ions. The element other than oxygen appears first in the name, and the ending depends on the number of oxygen atoms in the anion. Oxoanions with *-ate* as an ending have one more oxygen in their structure than those ending in *-ite*. Prefixes such as *per-* and *hypo-* can indicate the largest and smallest number of oxygens, respectively. We can use those rules and the examples in the text for chlorine (Table 4.5) as well as the polyatomic ions listed in Table 4.3 to help us write the formulas for the oxoanions in this question.

Solve
(a) hypobromite, BrO^- in analogy with hypochlorite
(b) sulfate, SO_4^{2-}
(c) iodate, IO_3^-
(d) nitrite, NO_2^-

Think About It
The names here do not really help us write the formulas; we have to just remember them. Learning them well for chlorine can help because we can name the other halogen oxoanions by analogy with chlorine oxoanions.

4.35. Collect and Organize
Each of those compounds contains a metal in combination with a polyatomic anion. Those compounds are ionic and follow the naming rules for ionic compounds.

Analyze
For those compounds, we name the metal cation first as the element name and then the anion.

Solve
(a) $NiCO_3$, nickel(II) carbonate
(b) NaCN, sodium cyanide
(c) $LiHCO_3$, lithium bicarbonate or lithium hydrogen carbonate
(d) $Ca(ClO)_2$, calcium hypochlorite

Think About It
Those compounds are named much like the binary ionic compounds. The anion name often ends in *-ide* but can end in *-ate* or *-ite*, depending on the name of the polyatomic anion.

4.37. Collect and Organize
We need to name or write the formula for each compound according to the rules for naming acids.

Analyze
Binary acids are named by placing *hydro-* in front of the element name other than hydrogen along with replacing the last syllable with *-ic* and adding *acid*. For acids containing oxoanions that end in *-ite*, the acid name becomes *-ous acid*.

Solve

(a) HF, a binary acid, hydrofluoric acid
(b) H_2SO_3, an acid of the sulfite anion, sulfurous acid
(c) phosphoric acid, acid of the phosphate anion, H_3PO_4
(d) nitrous acid, acid of the nitrite anion, HNO_2

Think About It
The rules are somewhat systematic but have to be learned and practiced.

4.39. Collect and Organize
We are asked to write the formula of an ionic salt from the name given.

Analyze
We use the rules for naming ionic salts. The first element in the name is the cation in the formula. In binary ionic salts, the anion is the element name with the ending *-ide*. If the ion is a polyatomic ion, the name of that polyatomic anion follows the name of the metal. When writing the formula, we must always balance the charges of the anion and the cation to give a neutral salt.

Solve

(a) potassium sulfide, K^+ with S^{2-} gives K_2S
(b) potassium selenide, K^+ with Se^{2-} gives K_2Se
(c) rubidium sulfate, Rb^+ with SO_4^{2-} gives Rb_2SO_4
(d) rubidium nitrite, Rb^+ with NO_2^- gives $RbNO_2$
(e) magnesium sulfate, Mg^{2+} with SO_4^{2-} gives $MgSO_4$

Think About It
Most of the anions in those salts are dianions with a 2− charge. When combining with 1+ cations, we have to balance the charge by having two cations for every anion in the formula.

4.41. Collect and Organize
The formulas for the compounds to be named all contain transition metals with variable charges. In the name, therefore, we must be sure to indicate the charge of the cation.

Analyze
The naming of those compounds follows the same system as naming other ionic compounds. We need to only add Roman numerals to indicate the charge on the transition metal cation.

Solve

(a) MnS, manganese(II) sulfide
(b) V_3N_2, vanadium(II) nitride
(c) $Cr_2(SO_4)_3$, chromium(III) sulfate
(d) $Co(NO_3)_2$, cobalt(II) nitrate
(e) Fe_2O_3, iron(III) oxide

Think About It
Adding Roman numerals to those names clearly indicates the charge of the cation. If the charges were not indicated, the way to write formulas from the names of those compounds would not be clear.

4.43. Collect and Organize
We are asked to identify a compound by name from a list of formulas.

Analyze
Sodium sulfite would have the sodium cation (Na^+) and the sulfite anion (SO_3^{2-}) in its formula. To balance the charge, two Na^+ cations would be present for every SO_3^{2-} anion.

Solve
The answer is (b), Na_2SO_3.

Think About It
To write a formula, we have to be very familiar with the names of the cations and anions. Confusing sulfite with sulfate (SO_4^{2-}) or sulfide (S^{2-}) would be easy here.

4.45. Collect and Organize
We are asked to consider how Lewis electron counting might be considered double counting.

Analyze
Lewis counts all electrons surrounding the atom in a bond, including all the electrons in shared pairs as well as electrons in lone pairs on the atom.

Solve
In the diatomic molecule XY shown here

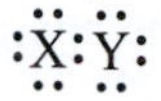

Lewis counts 6 e^- in three lone pairs on both X and Y. He also counts the two electrons shared between X and Y separately (2 e^- for X and 2 e^- for Y). However, 4 e^- are not being shared, only 2 e^-. The Lewis counting scheme seems to count the shared electrons twice.

Think About It
The octet rule uses double counting to surround each nonhydrogen atom in a Lewis structure with eight electrons.

4.47. Collect and Organize
We are to consider why water has a bonding pattern of H—O—H instead of H—H—O.

Analyze
We consider the Lewis structures of each compound. For both structures, the total number of valence electrons is 1 e^- (H) + 1 e^- (H) + 6 e^- (O) = 8 e^-. Each oxygen atom wants 8 e^- and each hydrogen atom wants 2 e^-, for a total of 12 e^-. The difference in the number of valence electrons and the number the molecule wants is 12 − 8 e^- = 4 e^-. Therefore, water has two covalent bonds.

Solve
For the H—O—H bonding pattern, the oxygen of the central atom forms bonds to the two hydrogen atoms. That uses 4 of the 8 e^-, leaving 4 e^- for the two lone pairs. Each hydrogen atom has a duet of electrons, so the lone pairs reside on oxygen and form an octet on oxygen.

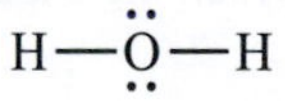

$$\text{H}-\overset{..}{\underset{..}{\text{O}}}-\text{H}$$

For H—H—O bonding, the two covalent bonds again use 4 of the 8 e^-, leaving 4 e^- for two lone pairs. If those are placed on the oxygen atom as shown here,

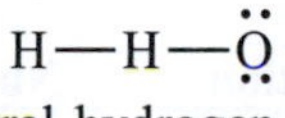

$$\text{H}-\text{H}-\overset{..}{\underset{..}{\text{O}}}$$

oxygen does not complete its octet and the central hydrogen atom has 4 e^-, not a duet. That structure would violate the Lewis structure formalism.

Think About It
Because hydrogen does not expand its duet in covalent bonding, the H atom is always terminal and never a central atom in a Lewis structure.

4.49. Collect and Organize
We are to draw the Lewis symbols for the neutral atoms K, Mg, and P.

Analyze
Lewis symbols show the number of valence electrons as dots around the element symbol. K has one valence electron, Mg has two valence electrons, and P has five valence electrons.

Solve

$$K\cdot \quad \cdot Mg\cdot \quad \cdot \overset{\displaystyle .}{\underset{\displaystyle ..}{P}}\cdot$$

Think About It

The particular placement of the electrons around the element symbol is not crucial to correct Lewis symbols. The electron dots are placed around the four sides of the element symbol and, generally, the electrons are not paired on a side until each other side also has an electron dot.

4.51. Collect and Organize

We are to draw the Lewis symbols for the ions K^+, Al^{3+}, N^{3-}, and I^-.

Analyze

To form the ions, we have to remove or add the appropriate number of electrons on the basis of each neutral element atom.

Element/Ion	Number of Valence e⁻ in Neutral Atom	Number of e⁻ in Ion
K/K⁺	1	0
Al/Al³⁺	3	0
N/N³⁻	5	8
I/I⁻	7	8

Solve

$$K^+ \quad Al^{3+} \quad \left[:\overset{\displaystyle ..}{\underset{\displaystyle ..}{N}}: \right]^{3-} \quad \left[:\overset{\displaystyle ..}{\underset{\displaystyle ..}{I}}: \right]^{-}$$

Think About It

The metals sodium and calcium lose electrons to form cations with a noble gas configuration, whereas the nonmetal sulfur gains electrons to form an anion with a noble gas configuration. The metal indium can exist in either the +3 oxidation state (to have a noble gas configuration) or the +1 oxidation state (in which two electrons remain in the $5s$ orbital).

4.53. Collect and Organize

Of B^{3+}, I^-, Ca^{2+}, and Pb^{2+} we are to identify which have a complete valence-shell octet.

Analyze

A valence-shell octet is also a noble gas configuration.

Ion	Electron Configuration	Number Valence e⁻
B³⁺	[He]	0
I⁻	[Kr]$4d^{10}5s^25p^6$	8
Ca²⁺	[Ar]	0
Pb²⁺	[Xe]$4f^{14}5d^{10}6s^2$	2

Solve

The ions I^- and Ca^{2+} have complete valence-shell octets. B^{3+} does not have an octet but rather the duet of the He atom.

Think About It

Cations are formed by loss of electrons to achieve a core noble gas configuration, which leaves no electrons in the valence shell, and anions are formed by gain of electrons to give eight electrons in the valence shell.

4.55. **Collect and Organize**

For the species N_2, HCl, NH_4^+, and CN^-, we are to determine the total number of valence electrons.

Analyze

For each species we need to add the valence electrons for each atom. If the species is charged, we need to reduce or increase the number of electrons as necessary to form cations or anions, respectively.

Solve

(a) 5 valence e^- (N) + 5 valence e^- (N) = 10 valence e^-

(b) 1 valence e^- (H) + 7 valence e^- (Cl) = 8 valence e^-

(c) 5 valence e^- (N) + 4 valence e^- (4H) – 1 e^- (negative charge) = 8 valence e^-

(d) 4 valence e^- (C) + 5 valence e^- (N) + 1 e^- (negative charge) = 10 valence e^-

Think About It

For each of those species, we can predict the number of covalent bonds between the atoms by finding the difference between what each species needs (to fill a duet for H and an octet for all other atoms) and the number of valence electrons for the molecule.

Molecule or Ion	Number of e^- Species Needs	Number of Valence e^-	Number of Covalent Bonds
(a) N_2	16	10	3
(b) HCl	10	8	1
(c) NH_4^+	16	8	4
(d) CN^-	16	10	3

4.57. **Collect and Organize**

We are to draw correct Lewis structures satisfying the octet rule for all atoms in the diatomic molecules and ions CO, O_2, ClO^-, and CN^-.

Analyze

To draw the Lewis structures, we first must determine the number of valence electrons in each structure. Then we arrange the atoms to show the bonding in the molecule by connecting the atoms with single covalent bonds. Finally, we complete the octets of the atoms bonded to the central atoms and then complete the octet of the central atom.

Solve

(a) For CO

(Step 1) The number of valence electrons in CO is

Element	C		O
Valence electrons per atom	4	+	6 = 10

(Step 2) Only two atoms are bonded, so neither is the central atom.

$$C\!-\!O$$

(Step 3) We complete the octet on the oxygen atom by adding three lone pairs.

$$C\!-\!\ddot{\underset{\cdot\cdot}{O}}:$$

(Step 4) That structure has eight electrons from three lone pairs and one bond pair. We need two more electrons (one pair) to match the valence electrons determined in step 1. We add the lone pair to the carbon atom.

$$:C\!-\!\ddot{\underset{\cdot\cdot}{O}}:$$

(Step 5) To complete the octet on the carbon atom we convert two lone pairs on the O atom to give a triple bond between the oxygen atom and the carbon atom.

$$:C\!-\!\ddot{\underset{\cdot\cdot}{O}}:$$

$$:C\!\equiv\!\underset{\cdot\cdot}{O}:$$

The Lewis structure is now complete.

(b) For O_2

(Step 1) The number of valence electrons in O_2 is

Element	2O
Valence electrons per atom	$(2 \times 6) = 12$

(Step 2) Only two atoms are bonded, so neither is the central atom.

O——O

(Step 3) We complete the octet on one oxygen atom by adding three lone pairs.

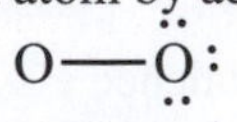

O——Ö:

(Step 4) That structure has eight electrons from three lone pairs and one bond pair. We need four more electrons (two pairs) to match the valence electrons determined in step 1. We add the lone pairs to the other oxygen atom.

:Ö——Ö:

(Step 5) To complete the octet on the left oxygen atom in the structure we convert a lone pair on the right O atom to give a double bond between the oxygen atoms.

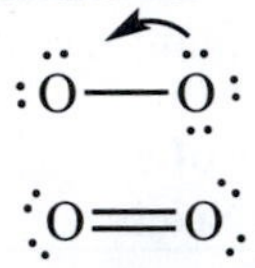

:Ö——Ö:

:O=O:

The Lewis structure is now complete.

(c) For ClO^-

(Step 1) The number of valence electrons in ClO^- is

Element	Cl		O	
Valence electrons per atom	7	+	6	= 13
Gain of electron due to charge				+1
Total valence electrons				14

(Step 2) Only two atoms are bonded, so neither is the central atom.

Cl——O

(Step 3) We complete the octet on the oxygen atoms by adding three lone pairs.

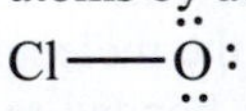

Cl——Ö:

(Step 4) That structure has eight electrons from three lone pairs and one bond pair. We need six more electrons (three pairs) to match the valence electrons determined in step 1. We add the lone pairs to the chlorine atom.

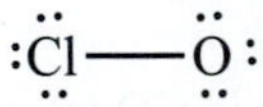

:Cl̈——Ö:

(Step 5) The Lewis structure is complete. To indicate the charge on the ion we add brackets for the structure and the charge.

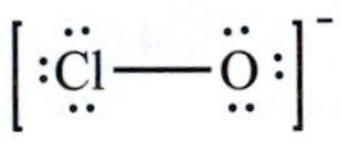

$$\left[:\ddot{C}\ddot{l}——\ddot{O}: \right]^-$$

(d) For CN^-

(Step 1) The number of valence electrons in CN^- is

Element	C		N	
Valence electrons per atom	4	+	5	= 9
Gain of electron due to charge				+1
Total valence electrons				10

(Step 2) Only two atoms are bonded, so neither is the central atom.

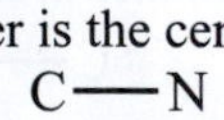

C——N

(Step 3) We complete the octet on the nitrogen atom by adding three lone pairs.

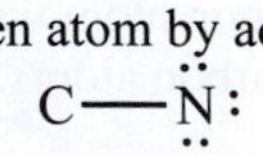

C——N̈:

(Step 4) That structure has eight electrons from three lone pairs and one bond pair. We need two more electrons (one pair) to match the valence electrons determined in step 1. We add the lone pair to the carbon atom.

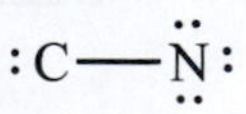

:C——N̈:

(Step 5) To complete the octet on the carbon atom we convert two lone pairs on the N atom to give a triple bond between the nitrogen atom and the carbon atom. Finally, we add brackets to the structure and indicate the charge on the anion.

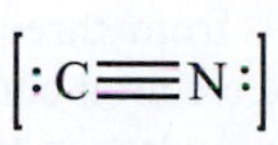

$$:C \!-\! \ddot{N}:$$

$$\left[:C \!\equiv\! N: \right]^{-}$$

The Lewis structure is now complete.

Think About It

When writing Lewis structures for ionic species, don't forget to enclose the structure in brackets and indicate the charge on the ion, as shown in this problem for ClO^- and CN^-.

4.59. Collect and Organize

We are to draw correct Lewis structures satisfying the octet rule for all atoms in the molecules and ions CCl_4, BH_3, SiF_4, BH_4^-, and PH_4^+.

Analyze

To draw the Lewis structures, we first must determine the number of valence electrons in each structure. Then we arrange the atoms to show the bonding in the molecule by connecting the atoms with single covalent bonds. Finally, we complete the octets of the atoms bonded to the central atoms and then complete the octet of the central atom.

Solve

(a) For CCl_4

(Step 1) The number of valence electrons in CCl_4 is

Element	C		4Cl	
Valence electrons per atom	4	+	(4×7)	= 28
Total valence electrons				32

(Step 2) Carbon is the atom with the fewest valence electrons (and therefore the greatest bonding capacity), so it is the central atom with the four Cl atoms bonded to it.

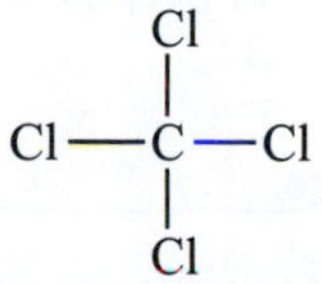

$$
\begin{array}{c}
\text{Cl} \\
| \\
\text{Cl} \!-\! \text{C} \!-\! \text{Cl} \\
| \\
\text{Cl}
\end{array}
$$

(Step 3) We complete the octet on the chlorine atoms by adding three lone pairs.

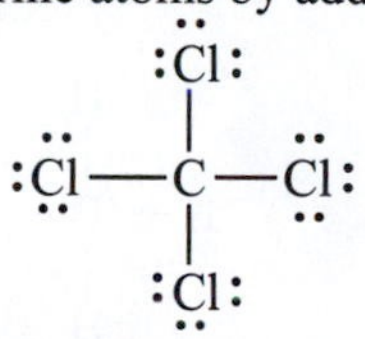

$$
\begin{array}{c}
:\!\ddot{\text{Cl}}\!: \\
| \\
:\!\ddot{\text{Cl}} \!-\! \text{C} \!-\! \ddot{\text{Cl}}\!: \\
| \\
:\!\ddot{\text{Cl}}\!:
\end{array}
$$

(Step 4) That structure has 32 electrons from 12 lone pairs and four bond pairs. That uses all the valence electrons we calculated in step 1.

(Step 5) All octets are satisfied for C and Cl, and so the structure in step 3 is the complete Lewis structure.

(b) For BH_3

(Step 1) The number of valence electrons in BH_3 is

Element	B		3H	
Valence electrons per atom	3	+	$(3 \times 1) = 3$	
Total valence electrons			6	

(Step 2) The central atom is boron with the three hydrogen atoms attached to it.

$$
\begin{array}{c}
H \\
\backslash \\
B\!-\!H \\
/ \\
H
\end{array}
$$

(Step 3) The duplet on all the H atoms is satisfied.

(Step 4) That structure has six electrons from three bond pairs. We cannot add more electrons to the structure because that matches the valence electron count determined in step 1.

(Step 5) Because we have no lone pairs on hydrogen to complete the octet for boron, this Lewis structure is complete as shown in step 2.

(c) For SiF_4

(Step 1) The number of valence electrons in SiF_4 is

Element	Si		4F	
Valence electrons per atom	4	+	(4×7)	= 32
Total valence electrons				32

(Step 2) Silicon is the atom with the fewest valence electrons and the greatest bonding capacity, so it is the central atom with the four Cl atoms bonded to it.

$$
\begin{array}{c}
F \\
| \\
F\!-\!Si\!-\!F \\
| \\
F
\end{array}
$$

(Step 3) We complete the octet on the fluorine atoms by adding three lone pairs.

$$
\begin{array}{c}
:\!\ddot{F}\!: \\
| \\
:\!\ddot{F}\!-\!Si\!-\!\ddot{F}\!: \\
| \\
:\!\ddot{F}\!:
\end{array}
$$

(Step 4) That structure has 32 electrons from 12 lone pairs and four bond pairs. That uses all the valence electrons we calculated in step 1.

(Step 5) All octets are satisfied for Si and F, and so the structure in step 3 is the complete Lewis structure.

(d) For BH_4^-

(Step 1) The number of valence electrons in BH_4^- is

Element	B		4H	
Valence electrons per atom	3	+	(4×1) = 7	
Gain of electron due to charge			+1	
Total valence electrons			8	

(Step 2) The central atom is boron with the four hydrogen atoms attached to it because B has the greatest bonding capacity.

$$
\begin{array}{c}
H \\
| \\
H\!-\!B\!-\!H \\
| \\
H
\end{array}
$$

(Step 3) The duplet on all the H atoms is satisfied.

(Step 4) That structure has eight electrons from four bond pairs. We cannot add more electrons to the structure because that matches the valence electron count determined in step 1.

(Step 5) This Lewis structure is complete since the octet is satisfied for boron. Because this species has a charge, we add brackets and indicate the charge.

$$
\left[
\begin{array}{c}
H \\
| \\
H\!-\!B\!-\!H \\
| \\
H
\end{array}
\right]^{-}
$$

(e) For PH_4^+

(Step 1) The number of valence electrons in PH_4^+ is

Element	P		4H	
Valence electrons per atom	5	+	$(4 \times 1) = 9$	
Loss of electron due to charge				-1
Total valence electrons				8

(Step 2) The central atom is phosphorus because it has the greatest bonding capacity with the four hydrogen atoms attached.

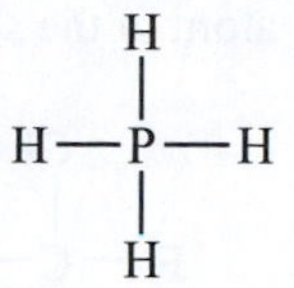

$$\begin{array}{c} H \\ | \\ H - P - H \\ | \\ H \end{array}$$

(Step 3) The duplet on all the H atoms is satisfied.

(Step 4) That structure has eight electrons from four bond pairs. We cannot add more electrons to the structure because that matches the valence electron count determined in step 1.

(Step 5) This Lewis structure is complete since the octet is satisfied for phosphorus. Because this species has a charge, we add brackets and indicate the charge.

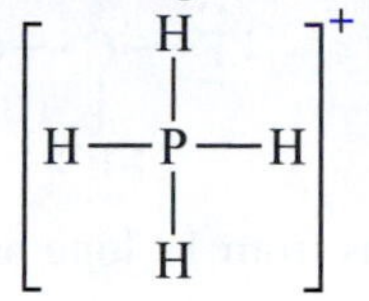

$$\left[\begin{array}{c} H \\ | \\ H - P - H \\ | \\ H \end{array}\right]^+$$

Think About It

In this problem, SiF_4 and CCl_4 are isoelectronic (have the same number of valence electrons), as are BH_4^- and PH_4^+.

4.61. Collect and Organize

Using the method described in the textbook, we are to draw Lewis structures for five chlorofluorocarbon greenhouse gases.

Analyze

To draw the Lewis structures, we first must determine the number of valence electrons in each structure. Then we arrange the atoms to show the bonding in the molecule by connecting the atoms with single covalent bonds. Finally, we complete the octets of the atoms bonded to the central atoms and then complete the octet of the central atom.

Solve

(a) CCl_3F

(Step 1) The number of valence electrons in CCl_3F is

Element	C		F		3Cl	
Valence electrons per atom	4	+	(1×7)	+	$(3 \times 7) = 32$	

(Step 2) Carbon has the most unpaired electrons (four) in its Lewis symbol and therefore has the highest bonding capacity and will be the central atom in the structure. The fluorine and chlorine atoms will each be bonded to the carbon.

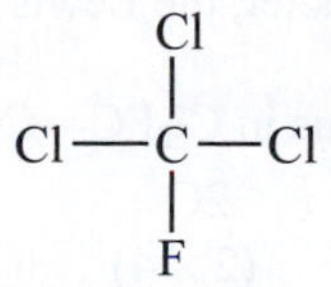

$$\begin{array}{c} Cl \\ | \\ Cl - C - Cl \\ | \\ F \end{array}$$

(Step 3) We complete the octets on the chlorine and fluorine atoms by adding three lone pairs to each.

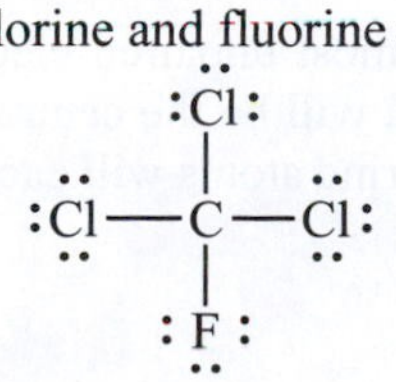

$$\begin{array}{c} :\ddot{Cl}: \\ | \\ :\ddot{Cl} - C - \ddot{Cl}: \\ | \\ :\ddot{F}: \end{array}$$

(Step 4) That structure has 32 electrons from 12 lone pairs and four bond pairs. We do not need any more valence electrons.

(Step 5) With carbon satisfied with its octet, the Lewis structure is complete.

(b) CCl_2F_2

(Step 1) The number of valence electrons in CCl_2F_2 is

Element	C	2Cl	2F
Valence electrons per atom	4 +	(2×7) +	(2×7) = 32

(Step 2) Carbon has the most unpaired electrons (four) in its Lewis symbol and therefore has the highest bonding capacity and will be the central atom in the structure. The fluorine and chlorine atoms will each be bonded to the carbon.

(Step 3) We complete the octets on the chlorine and fluorine atoms by adding three lone pairs to each.

(Step 4) That structure has 32 electrons from 12 lone pairs and four bond pairs. We do not need any more valence electrons.

(Step 5) With carbon satisfied with its octet, the Lewis structure is complete.

(c) $CClF_3$

(Step 1) The number of valence electrons in $CClF_3$ is

Element	C	Cl	3F
Valence electrons per atom	4 +	7 +	(3×7) = 32

(Step 2) Carbon has the most unpaired electrons (four) in its Lewis symbol and therefore has the highest bonding capacity and will be the central atom in the structure. The fluorine and chlorine atoms will each be bonded to the carbon.

Cl
|
F—C—F
|
F

(Step 3) We complete the octets on the chlorine and fluorine atoms by adding three lone pairs to each.

(Step 4) That structure has 32 electrons from 12 lone pairs and four bond pairs. We do not need any more valence electrons.

(Step 5) With carbon satisfied with its octet, the Lewis structure is complete.

(d) $Cl_2FC—CClF_2$

(Step 1) The number of valence electrons in $Cl_2FC—CClF_2$ is

Element	2C	3Cl	3F
Valence electrons per atom	(2×4) +	(3×7) +	(3×7) = 50

(Step 2) The carbon atoms have the most unpaired electrons (four) in their Lewis symbols and therefore have the highest bonding capacity and will be the central atoms in the structure. We are given that a C—C bond is present. The fluorine and chlorine atoms will each be bonded to the carbon atoms.

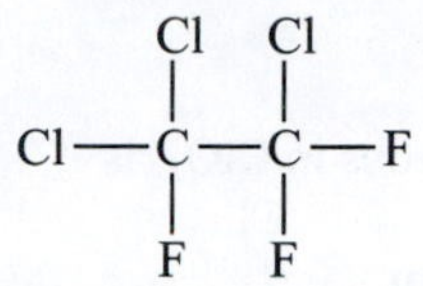

$$\begin{array}{cc} Cl & Cl \\ | & | \\ Cl-C-C-F \\ | & | \\ F & F \end{array}$$

(Step 3) We complete the octets on the chlorine and fluorine atoms by adding three lone pairs to each.

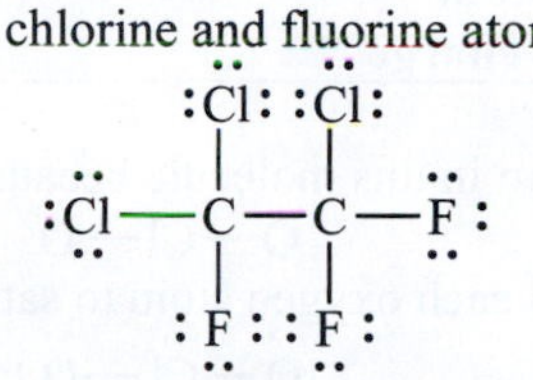

$$\begin{array}{cc} :\ddot{C}l: & :\ddot{C}l: \\ | & | \\ :\ddot{C}l-C-C-\ddot{F}: \\ | & | \\ :\ddot{F}: & :\ddot{F}: \end{array}$$

(Step 4) That structure has 50 electrons from 18 lone pairs and seven bond pairs. We do not need any more valence electrons.

(Step 5) With the carbon atoms satisfied with their octets, the Lewis structure is complete.

(e) $ClF_2C-CClF_2$

(Step 1) The number of valence electrons in $ClF_2C-CClF_2$ is

Element	2C	2Cl	2F
Valence electrons per atom	(2×4) +	(2×7) +	(4×7) = 50

(Step 2) The carbon atoms have the most unpaired electrons (four) in their Lewis symbols and therefore have the highest bonding capacity and will be the central atoms in the structure. We are given that a C—C bond is present. The fluorine and chlorine atoms will each be bonded to the carbon atoms.

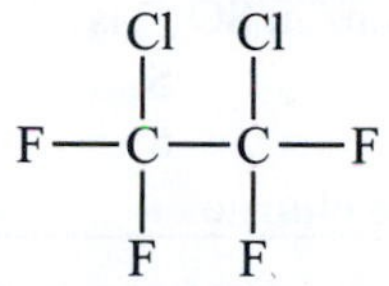

$$\begin{array}{cc} Cl & Cl \\ | & | \\ F-C-C-F \\ | & | \\ F & F \end{array}$$

(Step 3) We complete the octets on the chlorine and fluorine atoms by adding three lone pairs to each.

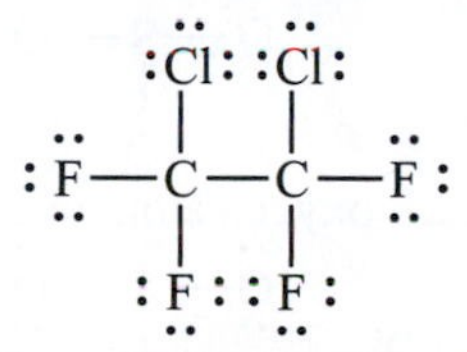

$$\begin{array}{cc} :\ddot{C}l: & :\ddot{C}l: \\ | & | \\ :\ddot{F}-C-C-\ddot{F}: \\ | & | \\ :\ddot{F}: & :\ddot{F}: \end{array}$$

(Step 4) That structure has 50 electrons from 18 lone pairs and seven bond pairs. We do not need any more valence electrons.

(Step 5) With the carbon atoms satisfied with their octets, the Lewis structure is complete.

Think About It

In determining the skeletal structure for such molecules, knowing that carbon commonly bonds to four atoms and that the halogens are usually terminal atoms is helpful.

4.63. Collect and Organize

We are to draw correct Lewis structures satisfying the octet rule for all atoms in the oxoanions ClO_2^-, SO_3^{2-}, and HCO_3^-.

Analyze

To draw the Lewis structures, we first must determine the number of valence electrons in each structure. Then we arrange the atoms to show the bonding in the molecule by connecting the atoms with single covalent bonds. Finally, we complete the octets of the atoms bonded to the central atoms and then complete the octet of the central atom. All those oxoanions have negative charges, so we must make sure to add the appropriate number of electrons to the structure.

Solve

(a) For ClO_2^-

(Step 1) The number of valence electrons in ClO_2^- is

Element	Cl		2O
Valence electrons per atom	7	+	(2×6) = 19
Gain of electron due to charge			+1
Total valence electrons			20

(Step 2) Chlorine is the central atom in this molecule because it is less electronegative than oxygen.

$$O - Cl - O$$

(Step 3) We add three lone pairs to each oxygen atom to satisfy the octet.

$$:\!\overset{..}{\underset{..}{O}} - Cl - \overset{..}{\underset{..}{O}}\!:$$

(Step 4) That structure has 16 electrons from two bond pairs and six lone pairs. We need two more electrons (one pair) to match the valence electrons determined in step 1. We place that lone pair on the chlorine atom.

$$:\!\overset{..}{\underset{..}{O}} - \overset{..}{Cl} - \overset{..}{\underset{..}{O}}\!:$$

(Step 5) The octet on chlorine in that structure is satisfied. We complete the Lewis structure by adding the brackets and charge for the ion.

$$\left[:\!\overset{..}{\underset{..}{O}} - \overset{..}{Cl} - \overset{..}{\underset{..}{O}}\!: \right]^-$$

(b) For SO_3^{2-}

(Step 1) The number of valence electrons in SO_3^{2-} is

Element	S		3O
Valence electrons per atom	6	+	(3×6) = 24
Gain of two electrons due to charge			+2
Total valence electrons			26

(Step 2) Sulfur is the central atom in this molecule because it is less electronegative than oxygen.

$$O - \underset{\displaystyle |}{\overset{\displaystyle S}{}} - O$$
$$O$$

(Step 3) We add three lone pairs to each oxygen atom to satisfy the octet.

$$:\!\overset{..}{\underset{..}{O}} - \underset{\displaystyle |}{S} - \overset{..}{\underset{..}{O}}\!:$$
$$:\!\overset{..}{\underset{..}{O}}\!:$$

(Step 4) That structure has 24 electrons from three bond pairs and nine lone pairs. We need two more electrons (one pair) to match the valence electrons determined in step 1. We place that lone pair on the sulfur atom.

$$:\!\overset{..}{\underset{..}{O}} - \overset{..}{\underset{\displaystyle |}{S}} - \overset{..}{\underset{..}{O}}\!:$$
$$:\!\overset{..}{\underset{..}{O}}\!:$$

(Step 5) The octet on sulfur in that structure is satisfied. We complete the Lewis structure by adding the brackets and charge for the ion.

$$\left[:\!\overset{..}{\underset{..}{O}} - \overset{..}{\underset{\displaystyle |}{S}} - \overset{..}{\underset{..}{O}}\!: \atop :\!\overset{..}{\underset{..}{O}}\!: \right]^{2-}$$

(c) For HCO_3^-

(Step 1) The number of valence electrons in HCO_3^- is

Element	H		C		3O
Valence electrons per atom	1	+	4	+	(3×6) = 23
Gain of electron due to charge					+1
Total valence electrons					24

(Step 2) Carbon is the central atom in this molecule because it has the highest bonding capacity. Hydrogen will be a terminal atom on one oxygen atom.

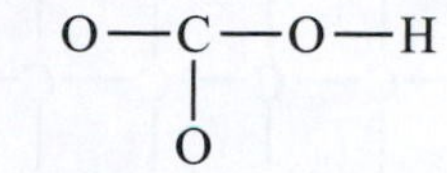

$$O—C—O—H$$
$$|$$
$$O$$

(Step 3) We add lone pairs to each oxygen atom to satisfy the octet on each.

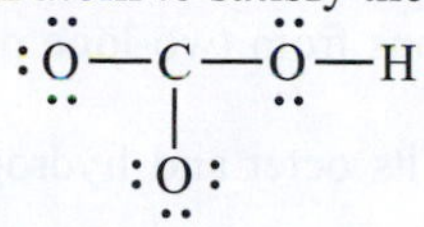

$$:\!\ddot{O}—C—\ddot{O}—H$$
$$|$$
$$:\!\ddot{O}\!:$$

(Step 4) That structure has 24 electrons from four bond pairs and eight lone pairs. We do not need any more valence electrons.

(Step 5) We can satisfy the octet on the carbon atom by forming a double bond with one of the oxygen atoms.

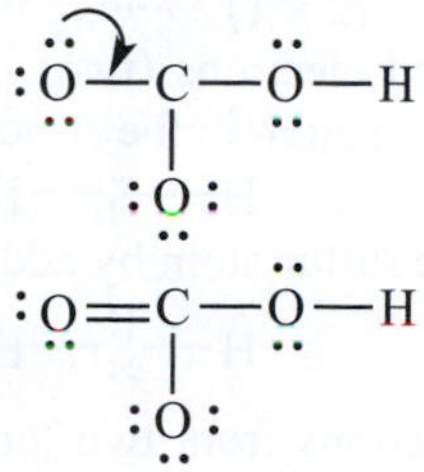

$$:\!\ddot{O}\!\!\curvearrowright\!\!C—\ddot{O}—H$$
$$|$$
$$:\!\ddot{O}\!:$$

$$:O\!\!=\!\!C—\ddot{O}—H$$
$$|$$
$$:\!\ddot{O}\!:$$

We complete the Lewis structure by adding the brackets and charge for the ion.

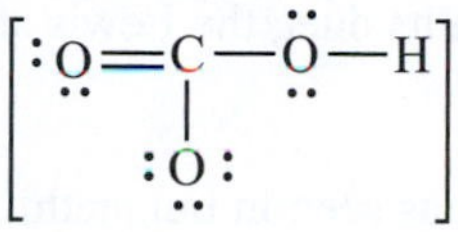

$$\left[:O\!\!=\!\!C—\ddot{O}—H\right]^{-}$$
$$|$$
$$:\!\ddot{O}\!:$$

Think About It

Hydrogen cannot be a central atom because its bonding capacity is only 1.

4.65. Collect and Organize

Using the method described in the textbook, we are to draw Lewis structures for $CH_3CH_2CH_2CH_2SH$ and H_2S.

Analyze

To draw the Lewis structures, we first must determine the number of valence electrons in each structure and then arrange the atoms to show the bonding in the molecule by connecting the atoms with single covalent bonds. Finally, we complete the octets of the atoms bonded to the central atoms and then complete the octet of the central atom. Considering that hydrogen is always terminal, we must ensure that the carbon atoms are bonded together in a chain, as indicated in the formula given in the problem.

Solve

Butanethiol

(Step 1) The number of valence electrons in $CH_3CH_2CH_2CH_2SH$ is

Element	4C	10H	S
Valence electrons per atom	(4×4) +	(10×1) +	6 = 32

(Step 2) Carbon has the most unpaired electrons (four) in its Lewis symbol and therefore has the highest bonding capacity and will be the central atom in the structure. The hydrogen and sulfur atoms will be bonded to the carbon, as indicated in the chemical formula. Also, as indicated in the formula, one H atom is bonded to the sulfur atom.

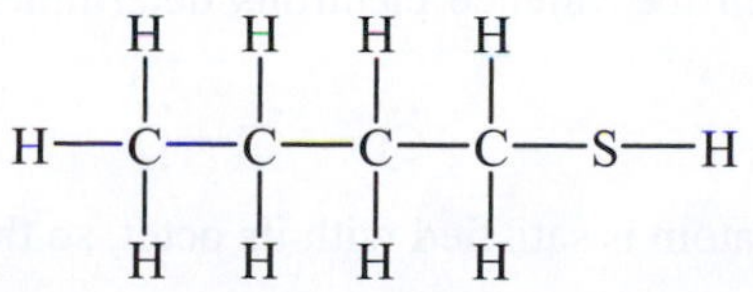

$$
\begin{array}{ccccccc}
 & H & H & H & H & & \\
 & | & | & | & | & & \\
H— & C— & C— & C— & C— & S—H \\
 & | & | & | & | & & \\
 & H & H & H & H & &
\end{array}
$$

(Step 3) We complete the octet on the sulfur atom by adding two lone pairs to it.

$$\text{H}-\overset{\displaystyle \underset{|}{\overset{|}{\text{H}}}}{\text{C}}-\overset{\displaystyle \underset{|}{\overset{|}{\text{H}}}}{\text{C}}-\overset{\displaystyle \underset{|}{\overset{|}{\text{H}}}}{\text{C}}-\overset{\displaystyle \underset{|}{\overset{|}{\text{H}}}}{\text{C}}-\overset{..}{\underset{..}{\text{S}}}-\text{H}$$

(Step 4) That structure has 32 electrons from two lone pairs and 14 bond pairs. We do not need any more valence electrons.

(Step 5) With carbon satisfied with its octet and hydrogen satisfied with its duet, the Lewis structure is complete.

Hydrogen sulfide

(Step 1) The number of valence electrons in H_2S is

Element	2H	S
Valence electrons per atom	(2×1) +	$6 = 8$

(Step 2) Sulfur has the most unpaired electrons (two) in its Lewis symbol and therefore has the highest bonding capacity and will be the central atom in the structure.

$$\text{H}-\text{S}-\text{H}$$

(Step 3) We complete the octet on the sulfur atom by adding two lone pairs to it.

$$\text{H}-\overset{..}{\underset{..}{\text{S}}}-\text{H}$$

(Step 4) That structure has eight electrons from two lone pairs and two bond pairs. We do not need any more valence electrons.

(Step 5) With hydrogen satisfied with its duet, the Lewis structure is complete.

Think About It

Carbon atoms are often bonded in chains, as seen in butanethiol. The bonding of an elemental atom to the same elemental atom to form chains is called *catenation*.

4.67. Collect and Organize

Using the method in the textbook, we are to draw Lewis structures for Cl_2O and ClO_3^-.

Analyze

To draw the Lewis structures, we first must determine the number of valence electrons. Then we arrange the atoms to show the bonding in the molecules by connecting the atoms with single covalent bonds. Finally, we complete the octets of the atoms bonded to the central atoms and then complete the octet of the central atom.

Solve

Cl_2O

(Step 1) The number of valence electrons in Cl_2O is

Element	2Cl	O
Valence electrons per atom	(2×7) +	6 = 20

(Step 2) We are given that one chlorine atom is the central atom in this structure.

$$\text{Cl}-\text{Cl}-\text{O}$$

(Step 3) We complete the octets on the oxygen and terminal chlorine atoms by adding three lone pairs to each.

$$:\overset{..}{\underset{..}{\text{Cl}}}-\text{Cl}-\overset{..}{\underset{..}{\text{O}}}:$$

(Step 4) That structure has 16 electrons from six lone pairs and two bond pairs. We need four more electrons (two pairs) to match the valence electrons determined in step 1. We add the lone pairs to the central chlorine atom.

$$:\overset{..}{\underset{..}{\text{Cl}}}-\overset{..}{\underset{..}{\text{Cl}}}-\overset{..}{\underset{..}{\text{O}}}:$$

(Step 5) The central chlorine atom is satisfied with its octet, so this Lewis structure is complete.

ClO_3^-

(Step 1) The number of valence electrons in ClO_3^- is

Element	Cl	3O	
Valence electrons per atom	7	+ (3 × 6)	= 25
Gain of electron due to charge		+1	
Total valence electrons		26	

(Step 2) We are given that the chlorine atom is the central atom in this structure.

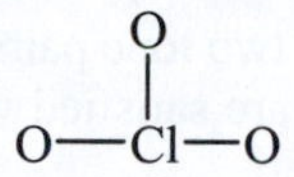

$$\begin{array}{c} O \\ | \\ O-Cl-O \end{array}$$

(Step 3) We complete the octets on the oxygen atoms by adding three lone pairs to each.

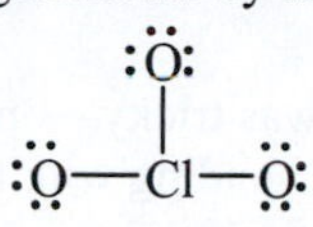

$$\begin{array}{c} :\ddot{O}: \\ | \\ :\ddot{O}-Cl-\ddot{O}: \end{array}$$

(Step 4) That structure has 24 electrons from nine lone pairs and three bond pairs. We need two more electrons (one pair) to match the valence electrons determined in step 1. We add the lone pair to the central chlorine atom.

$$\begin{array}{c} :\ddot{O}: \\ | \\ :\ddot{O}-\ddot{Cl}-\ddot{O}: \end{array}$$

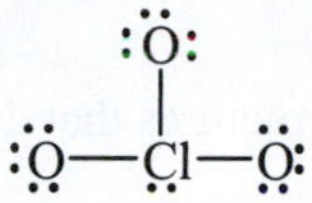

(Step 5) The central chlorine atom is satisfied with its octet, so this Lewis structure is complete. We add brackets and the charge to indicate the ion.

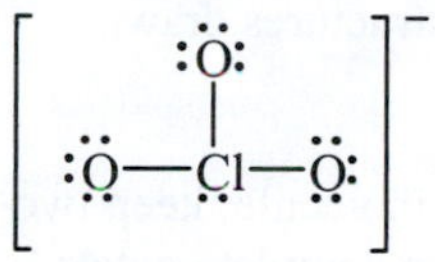

$$\left[\begin{array}{c} :\ddot{O}: \\ | \\ :\ddot{O}-\ddot{Cl}-\ddot{O}: \end{array} \right]^-$$

Think About It

In ClO_3^-, Cl is the central atom as we would guess from the formula, but in Cl_2O, oxygen could be the central atom. As shown below, the arrangement of the molecule with Cl as the central atom gives nonzero formal charges for the atoms (see Section 4.7). In Section 4.7, we also will learn that to reduce the formal charge on Cl in ClO_3^-, we can form a double bond between Cl and one O atom.

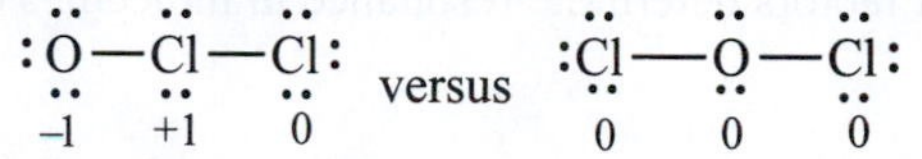

$$\begin{array}{ccc} :\ddot{O}-\ddot{Cl}-\ddot{Cl}: & & :\ddot{Cl}-\ddot{O}-\ddot{Cl}: \\ -1 \quad +1 \quad 0 & \text{versus} & 0 \quad 0 \quad 0 \end{array}$$

4.69. Collect and Organize

Using the method in the textbook, we are to draw a Lewis structure for methanol, CH_4O.

Analyze

To draw the Lewis structure, we first must determine the number of valence electrons. Then we arrange the atoms to show the bonding in the molecules by connecting the atoms with single covalent bonds. Finally, we complete the octets of the atoms bonded to the central atoms and then complete the octet of the central atom.

Solve

(Step 1) The number of valence electrons in CH_4O is

Element	C	4H	O	
Valence electrons per atom	4	+ (4 × 1)	+ 6	= 14

(Step 2) Carbon has the most unpaired electrons (four) in its Lewis symbol and therefore has the highest bonding capacity and will be the central atom in the structure. Three of the hydrogen atoms must also bond to the central carbon atom, with the last hydrogen atom bonding to the oxygen atom.

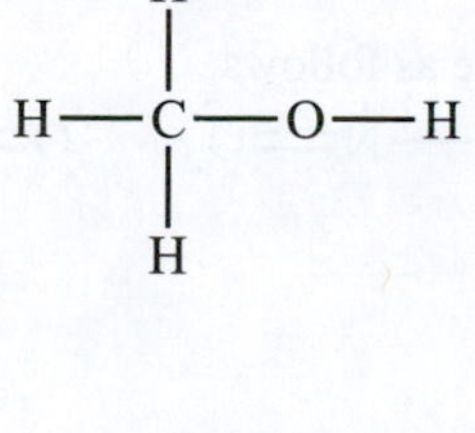

$$\begin{array}{c} H \\ | \\ H-C-O-H \\ | \\ H \end{array}$$

(Step 3) The octet on the carbon atom and the duplets on the hydrogen atoms are complete; we complete the octet on the oxygen atom by adding two lone pairs.

$$H—C—\ddot{\underset{..}{O}}—H$$
(with H above and below the C)

(Step 4) That structure has 14 electrons from two lone pairs and five bond pairs.
(Step 5) Both the carbon and oxygen atoms are satisfied with octets and the hydrogen atoms are satisfied with duets, so this Lewis structure is complete.

Think About It
Determining where to put the last hydrogen was tricky—once the three H atoms were placed on the carbon with the O atom, then we have to look to possible bonding with the O atom.

4.71. Collect and Organize
We are to explain the concept of resonance.

Analyze
Resonance structures are equivalent Lewis structures that differ only in the placement of electrons.

Solve
Resonance occurs when two or more valid Lewis structures may be drawn for a molecular species. The true structure of the species is a hybrid of the structures drawn.

Think About It
When drawing resonance structures for a molecule, keep two important items in mind: (1) each Lewis structure must be valid (that is, the atoms must have complete octets [or duets, if hydrogen atoms]), and (2) the positions of the atoms must not change—only the distribution of the electrons in bonding pairs and lone pairs will differ between resonance structures.

4.73. Collect and Organize
We are asked to describe what factors determine resonance in molecules or ions.

Analyze
We find out whether a molecule has resonance structures by drawing their Lewis structures.

Solve
A molecule or ion shows resonance when more than one correct Lewis structure exists, that is, when the electrons in the correct Lewis structure may be distributed in more than one way. Resonance is often the result of a single and a double bond to the same atom switching places within a structure or two adjacent double bonds becoming a single bond and a triple bond.

Think About It
Remember that resonance structures differ from each other only in the arrangement of the electrons, not in the atoms of the structure.

4.75. Collect and Organize
By drawing the Lewis structures of NO_2 and CO_2, we are to explain why NO_2 is more likely to exhibit resonance.

Analyze
The Lewis structures of NO_2 and CO_2 are as follows:

$$:\ddot{\underset{..}{O}}—\dot{N}=\ddot{O}: \qquad :\ddot{O}=C=\ddot{O}:$$

Solve

Either N—O bond in the NO_2 structure could be double-bonded, and the formal charges for each structure are identical, so more than one correct Lewis structure exists and NO_2 will exhibit resonance.

$$:\overset{-1}{\underset{\cdot\cdot}{O}}-\overset{+1}{N}=\overset{0}{\underset{\cdot\cdot}{O}}: \quad\longleftrightarrow\quad :\overset{0}{\underset{\cdot\cdot}{O}}=\overset{+1}{N}-\overset{-1}{\underset{\cdot\cdot}{O}}:$$

The resonance forms of CO_2 show that one is dominant (the one in which all formal charges are zero), and so the other forms contribute little to the true structure of CO_2.

$$:\overset{0}{\underset{\cdot\cdot}{O}}=\overset{0}{C}=\overset{0}{\underset{\cdot\cdot}{O}}: \quad\longleftrightarrow\quad :\overset{+1}{O}\equiv\overset{0}{C}-\overset{-1}{\underset{\cdot\cdot}{O}}: \quad\longleftrightarrow\quad :\overset{-1}{\underset{\cdot\cdot}{O}}-\overset{0}{C}\equiv\overset{+1}{O}:$$

Think About It

The following is *not* a correct Lewis structure for CO_2.

$$:O\equiv C=\overset{\cdot\cdot}{\underset{\cdot}{O}}$$

4.77. Collect and Organize

For cyclic C_4H_4, we are to draw two Lewis structures showing resonance.

Analyze

To draw the Lewis structures, we first must determine the number of valence electrons in each structure. Then we arrange the atoms to show the bonding in the molecule by connecting the atoms with single covalent bonds. Finally, we complete the octets of the atoms bonded to the central atoms and then complete the octet of the central atom. Once one Lewis structure is drawn, we can then consider alternative structures in resonance with the first.

Solve

(Step 1) The number of valence electrons in C_4H_4 is

Element	4C	4H
Valence electrons per atom	(4×4) +	(4×1) = 20

(Step 2) Carbon has the most unpaired electrons (four) in its Lewis symbol and therefore has the highest bonding capacity and will be the central atom in the structure. We are told that the carbon atoms form a ring, and so they are each bonded to two other carbon atoms and one hydrogen atom.

(Step 3) The duets on the terminal H atoms are complete.

(Step 4) That structure has 16 electrons from eight bond pairs. We need four more electrons (two pairs) to match the valence electrons determined in step 1. The carbon atoms do not have octets, so we add the lone pairs to two carbon atoms.

(Step 5) The other two carbon atoms do not have an octet. We can complete the octet for each carbon by forming two double bonds between C atoms in the ring.

The two double bonds could have been drawn for the other two carbons as well, so the two resonance forms of C_4H_4 are

Think About It
Cyclic compounds that have alternating single and double bonds often show resonance.

4.79. Collect and Organize
For N_2O_2 and N_2O_3 we are to draw Lewis structures and show all possible resonance forms.

Analyze
To draw the Lewis structures we must first determine the number of covalent bonds in each structure and then complete the octets (duets for hydrogen) as necessary and check the structure with electron bookkeeping. Once one Lewis structure is drawn, we can then consider alternative structures in resonance with the first.

Solve
For N_2O_2
  (Step 1) The number of valence electrons is

Element	2N	2O
Valence electrons per atom	(2×5) +	(2×6) = 22

  (Step 2) Nitrogen has more unpaired electrons (three) than oxygen and is less electronegative, so the two nitrogen atoms are the central atoms in the structure.

$$O—N—N—O$$

  (Step 3) We complete the octets on the oxygen atoms by adding three lone pairs to each.

$$:\ddot{O}—N—N—\ddot{O}:$$

  (Step 4) That structure has 18 electrons from six lone pairs and three bond pairs. We need four more electrons (two pairs) to match the valence electrons determined in step 1. The nitrogen atoms do not have octets yet, so we will add one lone pair to each N atom.

$$:\ddot{O}—\ddot{N}—\ddot{N}—\ddot{O}:$$

  (Step 5) We can complete the octet for each nitrogen by forming double bonds between the oxygen and nitrogen atoms.

$$:\ddot{O}{\overset{\frown}{—}}\ddot{N}—\ddot{N}{\overset{\frown}{—}}\ddot{O}: \qquad :\ddot{O}{=}\ddot{N}—\ddot{N}{=}\ddot{O}:$$

The electrons could also be distributed in five more resonance forms that also complete the octet on all the atoms. All resonance forms are shown below.

$$\ddot{O}{=}N—N{=}\ddot{O} \longleftrightarrow :O{\equiv}N—\ddot{N}—\ddot{O}: \longleftrightarrow \ddot{O}{=}N{=}N—\ddot{O}: \longleftrightarrow$$

$$:\ddot{O}—N{\equiv}N—\ddot{O}: \longleftrightarrow :\ddot{O}—\ddot{N}—N{\equiv}O: \longleftrightarrow :\ddot{O}—N{=}N{=}\ddot{O}$$

For N_2O_3
  (Step 1) The number of valence electrons is

Element	2N	3O
Valence electrons per atom	(2×5) +	(3×6) = 28

  (Step 2) Nitrogen has more unpaired electrons (three) than oxygen and is less electronegative, so the two nitrogen atoms are the central atoms in the structure.

$$\begin{array}{c} O \\ | \\ O—N—N—O \end{array}$$

(Step 3) We complete the octets on the oxygen atoms by adding three lone pairs to each.

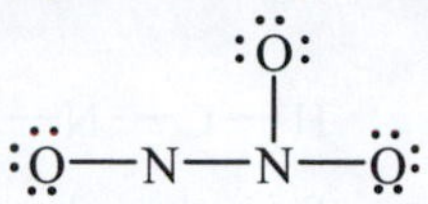

$$:\ddot{O}:$$
$$|$$
$$:\ddot{O}—N—N—\ddot{O}:$$

(Step 4) That structure has 26 electrons from nine lone pairs and four bond pairs. We need two more electrons (one pair) to match the valence electrons determined in step 1. The nitrogen atoms do not have octets yet, so we will add one lone pair to a N atom.

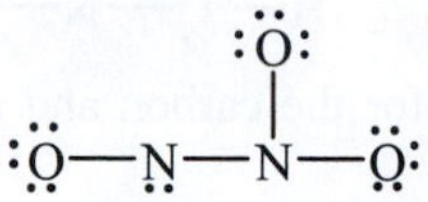

$$:\ddot{O}:$$
$$|$$
$$:\ddot{O}—N—\ddot{N}—\ddot{O}:$$

(Step 5) We can complete the octet for each nitrogen by forming double bonds between the oxygen and nitrogen atoms.

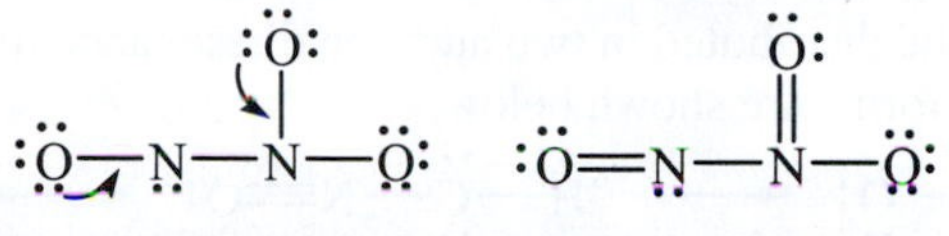

The electrons could also be distributed in three more resonance forms that also complete the octet on all the atoms. All resonance forms are shown below.

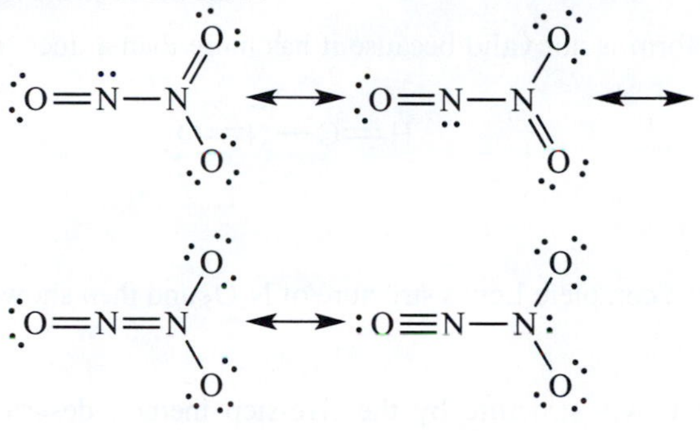

Think About It
In N_2O_3 a resonance structure that has a triple bond between the N atoms would violate the octet rule for the N bound to two O atoms and for one O atom.

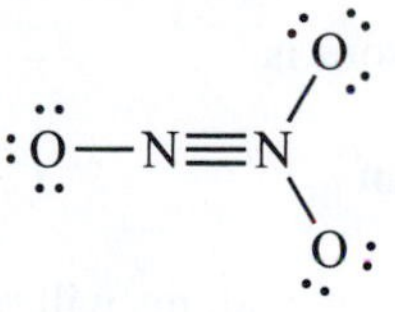

4.81. Collect and Organize
Fulminic acid has a linear structure with atom connectivity as described by the molecular formula given. From that framework we are to draw valid resonance structures for HCNO.

Analyze
We first draw one valid Lewis structure by the method described in the textbook; then we redistribute the bonding pairs and lone pairs in the structure to draw resonance forms.

Solve
For HCNO, fulminic acid
(Step 1) The number of valence electrons is

Element	H		C		N		O	
Valence electrons per atom	1	+	4	+	5	+	6	= 16

(Step 2) We are given that fulminic acid is a linear molecule with the connectivity of the atoms as

$$H—C—N—O$$

(Step 3) We complete the octets on the oxygen atom by adding three lone pairs. The duet on the terminal H atom is already satisfied.

$$H\text{---}C\text{---}N\text{---}\ddot{\underset{..}{O}}:$$

(Step 4) That structure has 12 electrons from three lone pairs and three bond pairs. We need four more electrons (two pairs) to match the valence electrons determined in step 1. The nitrogen and carbon atoms do not have octets yet, so we will add lone pairs.

$$H\text{---}\underset{..}{C}\text{---}\ddot{N}\text{---}\ddot{\underset{..}{O}}:$$

(Step 5) We can complete the octet for the carbon and nitrogen atoms by forming a triple bond between them.

$$H\text{---}\underset{..}{C}\text{---}\ddot{N}\text{---}\ddot{\underset{..}{O}}: \qquad H\text{---}C\equiv N\text{---}\ddot{\underset{..}{O}}:$$

The electrons could also be distributed in two additional resonance forms that also complete the octet on all the atoms. All resonance forms are shown below.

$$H\text{---}C\equiv N\text{---}\ddot{\underset{..}{O}}: \longleftrightarrow H\text{---}\ddot{\underset{..}{C}}\text{---}N\equiv O: \longleftrightarrow H\text{---}\ddot{C}\equiv\ddot{N}\equiv O:$$

Think About It
The following resonance form is not valid because it has more than a duet for the H atom and has less than an octet for the C atom.

$$H\equiv C\text{---}\ddot{N}\equiv\ddot{\underset{.}{O}}:$$

4.83. Collect and Organize
We are asked to first draw a complete Lewis structure of N_2O_5 and then show all its resonance forms.

Analyze
We first draw one valid Lewis structure by the five-step method described in the textbook, and then we redistribute the bonding pairs and lone pairs in the structure to draw resonance forms.

Solve
For N_2O_5
(Step 1) The number of valence electrons is

Element	2N	5O	
Valence electrons per atom	(2×5) +	(5×6) =	40

(Step 2) In this molecule one of the oxygen atoms will be a bridging atom between two "NO_2" units.

$$\underset{O}{\overset{O}{\diagdown}}N\text{---}O\text{---}N\underset{O}{\overset{O}{\diagup}}$$

(Step 3) We complete the octets on the outside oxygen atoms by adding three lone pairs to each.

$$\underset{:\ddot{O}}{\overset{:\ddot{O}}{\diagdown}}N\text{---}\ddot{\underset{..}{O}}\text{---}N\underset{\ddot{O}:}{\overset{\ddot{O}:}{\diagup}}$$

(Step 4) That structure has 40 electrons from 14 lone pairs and six bond pairs. No more valence electrons are needed.

(Step 5) We can complete the octet for the nitrogen atoms by forming a double bond between each of them and one of the oxygen atoms bonded to that nitrogen atom.

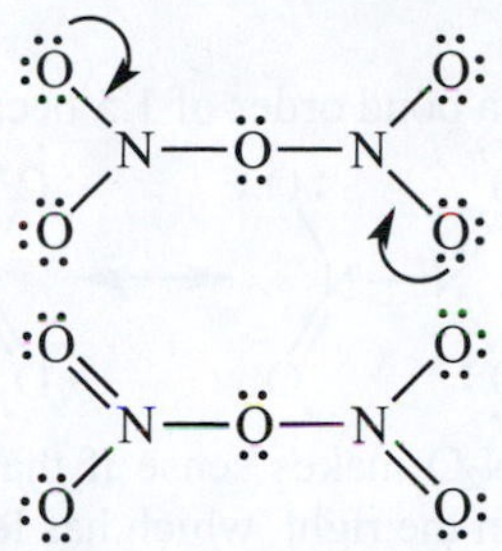

The electrons could also be distributed in three additional resonance forms that also complete the octet on all the atoms. All resonance forms are shown below.

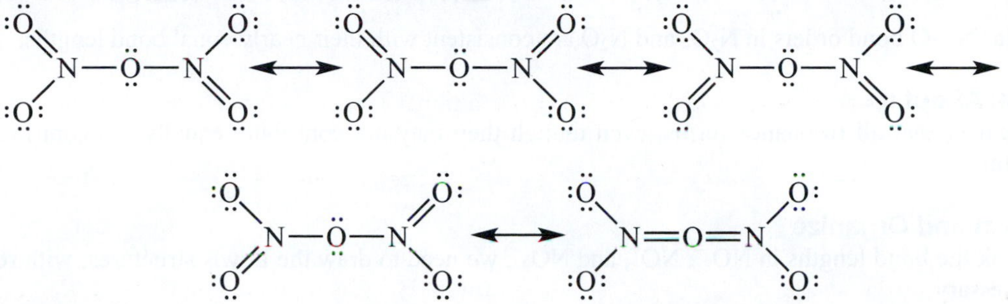

Think About It

Because the real structure of dinitrogen pentoxide is a hybrid of those four resonance structures, we would expect each nitrogen–oxygen bond to be the same length and strength, with a bond order of 1.5.

4.85. ### Collect and Organize

Using Lewis structures for NO_3^- and NO_2^-, we can compare the N—O bond lengths in those molecules.

Analyze

In Lewis structures, single bonds are longer than double bonds, which are longer than triple bonds. We must also consider any resonance forms that the molecules might have.

Solve

The Lewis structures of the nitrate ion are as follows. These show that the bond order in NO_3^- is 1.33 because of resonance.

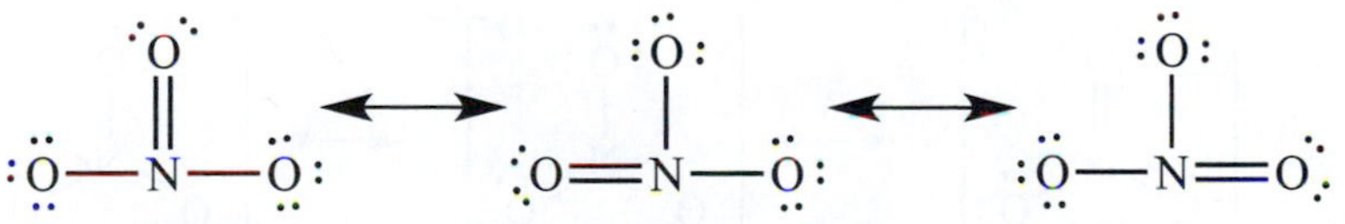

The Lewis structures of the nitrite ion are as follows. These show that the bond order in NO_2^- is 1.5 because of resonance.

The lower the bond order, the longer the bond. Therefore, the nitrogen–oxygen bond in nitrate (NO_3^-) is longer than the bond in nitrite (NO_2^-).

Think About It

Bond strength, however, is directly proportional to bond order; the higher the bond order, the stronger the bond. Here, NO_2^- has a stronger N—O bond than NO_3^-.

4.87. ### Collect and Organize

Using the Lewis structures (with resonance forms, if necessary), we can explain why the nitrogen–oxygen bonds in N_2O_4 and N_2O are nearly identical in length.

Analyze

In Lewis structures, single bonds are longer than double bonds, which are longer than triple bonds. We must also consider any resonance forms that the molecules might have.

Solve

The nitrogen–oxygen bond in N_2O_4 has a bond order of 1.5 because of four equivalent resonance forms.

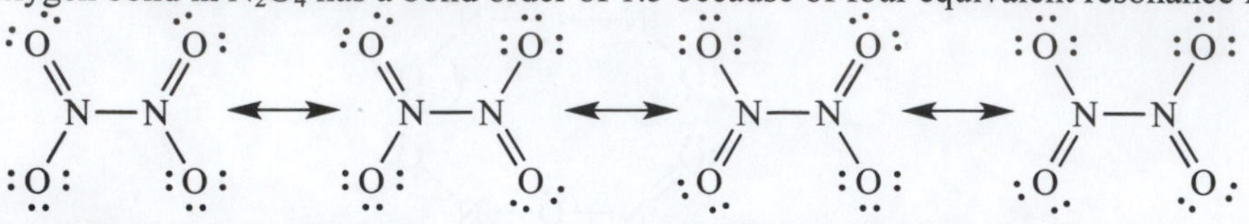

An N—O bond order of about 1.5 in N_2O makes sense if the left and middle resonance structures contribute more to the bonding than the structure on the right, which has less favorable formal charges.

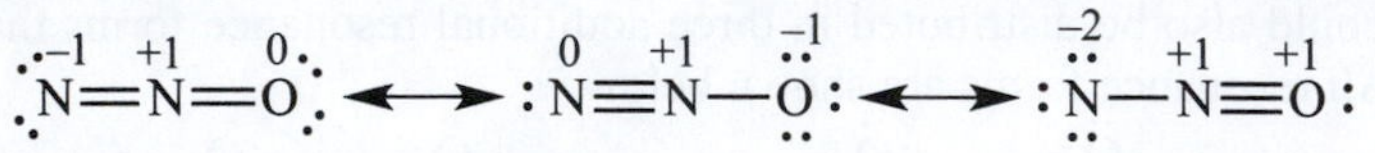

Similar N—O bond orders in N_2O_4 and N_2O are consistent with their nearly equal bond lengths.

Think About It

Remember that all resonance forms, even though they may not contribute equally, do contribute some to the structure.

4.89. **Collect and Organize**

To rank the bond lengths in NO_2^-, NO^+, and NO_3^-, we need to draw the Lewis structures, with resonance forms if necessary.

Analyze

In Lewis structures, single bonds are longer than double bonds, which are longer than triple bonds. We must also consider any resonance forms that the molecules might have.

Solve

For NO_2^-, the bond order for the N—O bond is 1.5 because of resonance.

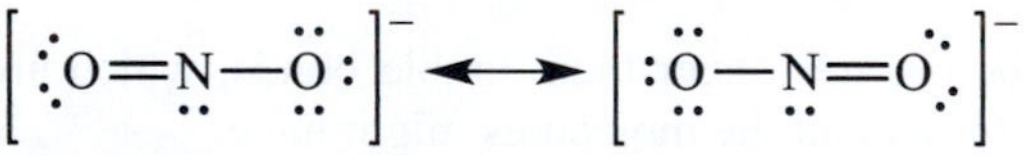

For NO^+, the bond order for the N—O bond is 3.0.

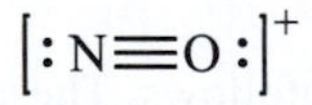

For NO_3^-, the bond order for the N—O bond is 1.33 because of resonance.

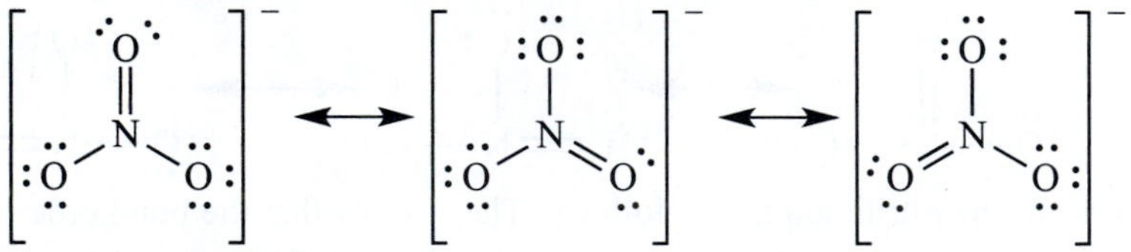

In order of increasing bond length: $NO^+ < NO_2^- < NO_3^-$.

Think About It

Resonance has a marked effect on bond order and length.

4.91. **Collect and Organize**

To rank the bond energies for NO_2^-, NO^+, and NO_3^-, we need to draw the Lewis structures, with resonance structures if necessary.

Analyze

In Lewis structures, single bonds have the lowest bond energy. Double bonds are stronger (have higher bond energies) than single bonds, and triple bonds are stronger than double bonds. The higher the bond order, the higher the bond energy.

Solve

For NO_2^-, the bond order for the N—O bond is 1.5 because of resonance.

$$\left[\ddot{\underset{..}{O}}{=}N{-}\ddot{\underset{..}{O}}{:}\right]^- \longleftrightarrow \left[{:}\ddot{\underset{..}{O}}{-}N{=}\ddot{\underset{..}{O}}{:}\right]^-$$

For NO^+, the bond order for the N—O bond is 3.0.

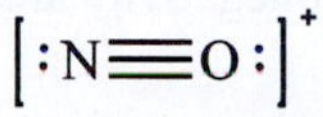

$$\left[{:}N{\equiv}O{:}\right]^+$$

For NO_3^-, the bond order for the N—O bond is 1.33 because of resonance.

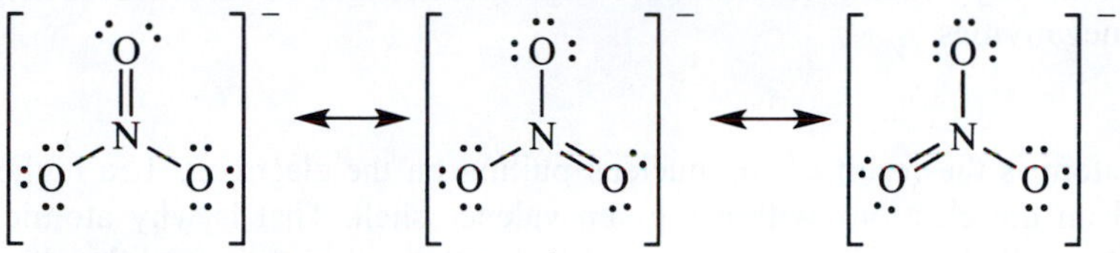

In order of increasing bond energy: $NO_3^- < NO_2^- < NO^+$.

Think About It

That ranking is the reverse of the ranking for increasing bond length in Problem 4.89.

4.93. Collect and Organize

To compare the C—C bond length between acetylene and ethane, we need to draw the Lewis structures, with resonance structures if necessary.

Analyze

In Lewis structures, single bonds have the longest bonds. Double bonds are shorter (have higher bond energies) than single bonds, and triple bonds are shorter than double bonds. The higher the bond order, the shorter the bond.

Solve

For acetylene, C_2H_2, the bond order for the C—C bond is 3.0.

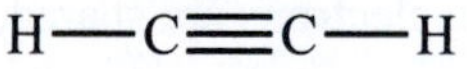

$$H{-}C{\equiv}C{-}H$$

For ethane, C_2H_6, the bond order for the C—C bond is 1.0.

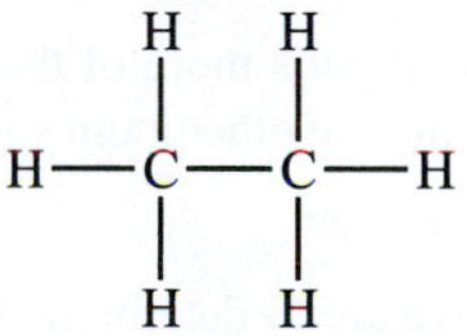

Therefore, the C—C bonds in ethane are longer than those in acetylene.

Think About It

Neither compound has any resonance structures.

4.95. Collect and Organize

We are asked how we can use electronegativity to define whether a bond is ionic or covalent.

Analyze

When a large difference in electronegativity exists, the transfer of an electron from one atom to another is likely and an ionic bond will form. When the electronegativities of two atoms are similar, the electrons will be shared in a covalent bond.

Solve

The general rule is that if an electronegativity difference of 2.0 or greater is present, the bond between the atoms is ionic. Below 2.0, the bond is covalent.

Think About It
Large differences in electronegativity, and thus the occurrence of ionic bonding, are likely between a metallic atom (low electronegativity) and a nonmetallic atom (high electronegativity).

4.97. Collect and Organize
We are asked to explain how trends in electronegativity are related to trends in atomic size.

Analyze
Small atoms such as oxygen and fluorine have high electronegativities, whereas large atoms such as cesium have low electronegativities.

Solve
The size of the atom is the result of the nucleus pulling on the electrons. The higher the nuclear charge, the stronger the pull on the electrons within a given valence shell. That is why atomic size generally decreases across a period. A small atom will form a shorter bond with another atom, and the electrons in the bond will feel a strong pull from the nucleus of a smaller atom since the bonding electrons will be "closer" to the nucleus. That stronger pull results in a higher electronegativity for smaller atoms.

Think About It
Both size and electronegativity trends are more fundamentally a reflection of the trends in effective nuclear charge and the *n* level of the valence electrons of an atom.

4.99. Collect and Organize
We are to define *polar covalent bond*.

Analyze
A covalent bond forms between atoms of close electronegativities and involves the sharing, not the transfer, of electrons between the atoms. When a bond is polar, electron density is unevenly distributed.

Solve
A polar covalent bond is one in which the electrons are shared, but not equally, by the atoms.

Think About It
The more electronegative atom in the bond pulls more of the electron density toward itself, making that atom slightly rich in electron density ($\delta-$), leaving the other atom slightly deficient in electron density ($\delta+$).

4.101. Collect and Organize
Of the bonds listed between two atoms, we are to determine which are polar and, in those bonds that are polar, which atom has the greater electronegativity.

Analyze
Polar bonds form between any two dissimilar atoms that have different electronegativity values. To determine the more electronegative atom, we can use knowledge of the periodic trends of electronegativity values in Figure 4.13 in the textbook.

Solve
The polar bonds and the atoms with the greater electronegativity (underlined) are <u>C</u>—Se, C—<u>O</u>, <u>N</u>—H, and <u>C</u>—H.

Think About It
The Cl—Cl and O $=$ O bonds cannot be polar because the bonded atoms are identical.

4.103. Collect and Organize
From the list of pairs of atoms, we are to discriminate between binary compounds with polar covalent bonds and ionic bonds.

Analyze

When the difference in electronegativity between the atoms is zero, the bond is nonpolar. If the electronegativity difference is below 2.0, the bond is polar covalent. If the electronegativity difference is 2.0 or greater, the bond is ionic.

Solve

Bond	Electronegativity Difference	Bond Type
(a) C and S	$2.5 - 2.5 = 0$	Nonpolar covalent
(b) Al and Cl	$3.0 - 1.5 = 1.5$	Polar covalent
(c) C and O	$3.5 - 2.5 = 1.0$	Polar covalent
(d) Ca and O	$3.5 - 1.0 = 2.5$	Ionic

Binary compounds of (b) Al and Cl and (c) C and O have polar covalent bonds. The binary compound of (d) Ca and O has ionic bonds.

Think About It

Although both Al—Cl and C—O bonds are polar covalent, the Al—Cl bond has greater ionic character (is more polar) than the C—O bond.

4.105. **Collect and Organize**

We are to explain how we can use formal charges to choose the best molecular structure for a given chemical formula.

Analyze

Formal charge (FC) is not a real charge on the atoms but rather a method to assign the apparent charges on atoms in covalently bonded compounds. FC is determined as follows:

$$FC = \text{(number of valence } e^- \text{ for the atom)}$$
$$- [\text{(number of } e^- \text{ in lone pairs)} + (\tfrac{1}{2} \times \text{number of } e^- \text{ in bonding pairs)}]$$

Solve

The best possible structure for a molecule, judging by formal charges, is the structure in which the formal charges are minimized and the negative formal charges are on the most electronegative atoms in the structure.

Think About It

The sum of the formal charges on the atoms in a structure must equal the charge on the molecule.

4.107. **Collect and Organize**

In a sulfur–oxygen bond, we can use their electronegativity values to predict which atom would carry the negative formal charge for the structure most likely to contribute to the bonding. We are asked whether a structure with a negative formal charge on S rather than O is more likely to contribute to bonding in a molecule containing S and O atoms.

Analyze

The more electronegative atom is more likely to carry the negative formal charge.

Solve

No. The electronegativity of oxygen (3.5) is higher than that of sulfur (2.5), so the negative formal charge must be on the O atom in the structure that contributes most to the bonding.

Think About It

The most electronegative elements are F, O, N, and Cl, and those are the elements most likely to carry negative formal charges, if they must, in Lewis structures that contribute significantly to the bonding.

4.109. **Collect and Organize**

After drawing the Lewis structures for HNC and HCN and assigning formal charges to the atoms, we are asked to analyze the differences in their formal charges (and choose the best, most stable arrangement for the atoms).

Analyze

After drawing the Lewis structures for both HNC and HCN, we assign the formal charge (FC) for each atom from the formula

$$FC = \text{(number of valence } e^- \text{ for the atom)}$$
$$- [\text{(number of } e^- \text{ in lone pairs)} + (\tfrac{1}{2} \times \text{number of } e^- \text{ in bonding pairs)}]$$

Solve

Both HNC and HCN have 10 valence electrons, and the Lewis structures with formal charges are

$$\overset{0}{H}-\overset{+1}{N}\equiv\overset{-1}{C}: \qquad \overset{0}{H}-\overset{0}{C}\equiv\overset{0}{N}:$$

The formal charges are zero for all the atoms in HCN, whereas in HNC the carbon atom, with a lower electronegativity than N, has a −1 formal charge.

Think About It

The HCN arrangement, being more stable, is the preferred structure.

4.111. **Collect and Organize**

We are to draw Lewis structures for cyanamide, H_2NCN, and assign formal charges to each atom.

Analyze

Because we are asked to draw *structures* for the compound, we suspect that the compound may show resonance. After drawing the possible resonance structures, we assign formal charges (FCs) to all atoms in each structure by using

$$FC = \text{(number of valence } e^- \text{ for the atom)}$$
$$- [\text{(number of } e^- \text{ in lone pairs)} + (\tfrac{1}{2} \times \text{number of } e^- \text{ in bonding pairs)}]$$

Solve

Cyanamide, H_2NCN, has 16 valence electrons, and the possible structures, with formal charges assigned for the atoms, are

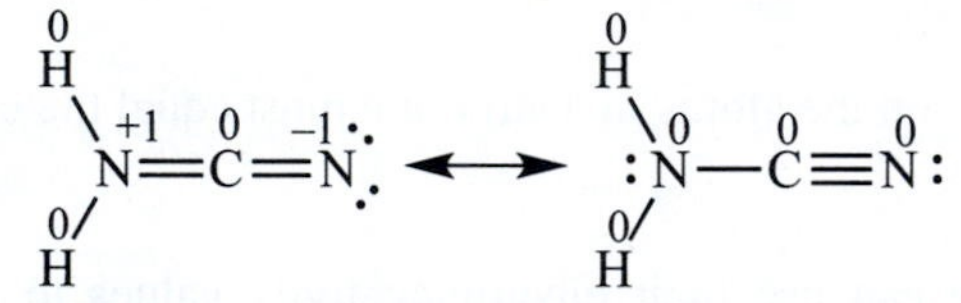

$$\begin{array}{c} \overset{0}{H} \\ \diagdown {\scriptstyle +1} \quad 0 \quad -1 \\ N = C = N \\ {\scriptstyle 0}\diagup \\ H \end{array} \longleftrightarrow \begin{array}{c} \overset{0}{H} \\ \diagdown {\scriptstyle 0} \quad 0 \quad 0 \\ :N - C \equiv N: \\ {\scriptstyle 0}\diagup \\ H \end{array}$$

The preferred structure is the one with the C triple-bonded to N because all the formal charges on the structure are zero.

Think About It

Be careful in drawing the resonance structures. The structure below is not valid because it violates the octet rule for both nitrogen atoms.

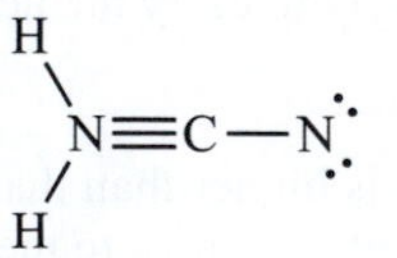

$$\begin{array}{c} H \\ \diagdown \\ N \equiv C - N: \\ \diagup \\ H \end{array}$$

4.113. **Collect and Organize**

For the arrangement of atoms in nitrous oxide (N_2O), where oxygen is the central atom, we are to assign formal charges and suggest why that structure is not stable.

Analyze

After drawing the possible resonance structures, we assign formal charges (FCs) to all atoms in each structure by using

$$FC = \text{(number of valence } e^- \text{ for the atom)}$$
$$- [\text{(number of } e^- \text{ in lone pairs)} + (\tfrac{1}{2} \times \text{number of } e^- \text{ in bonding pairs)}]$$

Solve

Nitrous oxide, N_2O, has 16 valence electrons, and the Lewis structures with formal charges assigned to the atoms are

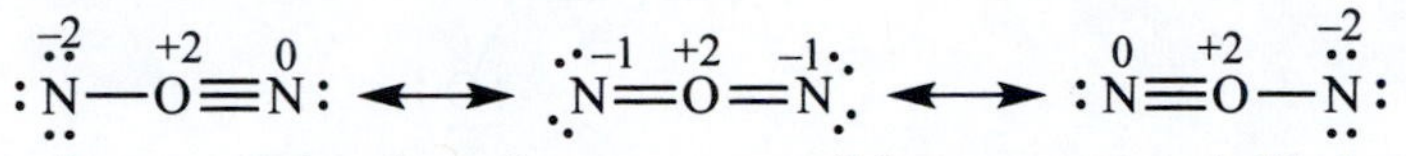

Because oxygen is more electronegative than nitrogen, none of those structures is likely to be stable because the formal charge on O is positive, when it would be predicted from electronegativity to be negative.

Think About It

The Lewis structure for the arrangement N—N—O is far better by formal charge, particularly with the resonance structure that has a N-to-N triple bond.

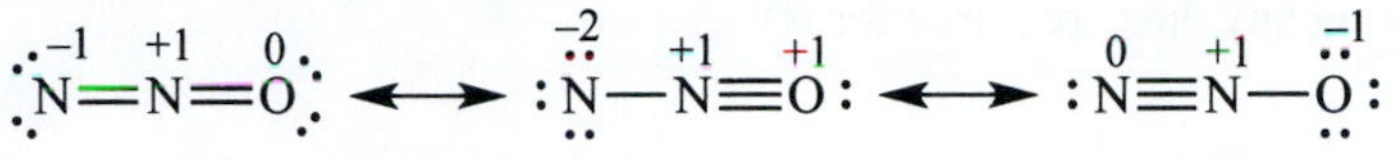

4.115. Collect and Organize

We are to draw the Lewis structures (with resonance forms) for nitromethane (CH_3NO_2) and $CNNO_2$ (which has two possible skeletal structures), and we are to determine whether the two molecules are resonance structures of each other.

Analyze

For each Lewis structure, we have to be sure that in drawing the resonance structures, we redistribute only electrons and do not move atoms.

Solve

(a) CH_3NO_2 has 24 valence electrons. Completing the octets for all the atoms (duet for hydrogen), drawing an alternative resonance structure, and assigning formal charges to the atoms gives

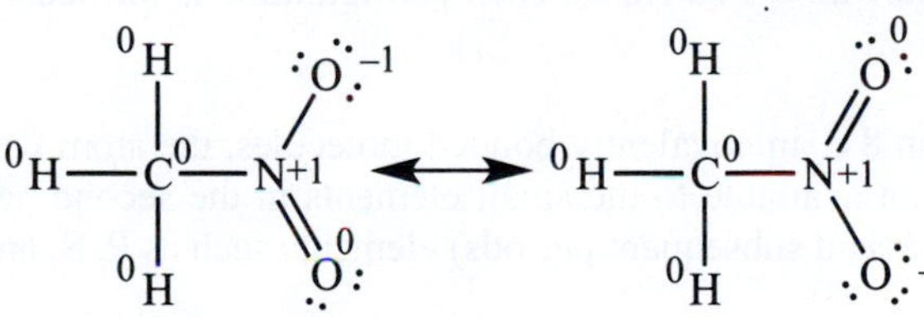

(b) $CNNO_2$ has 26 valence electrons. Completing the octets for all the atoms, drawing the alternative resonance structures, and assigning formal charges to the atoms gives

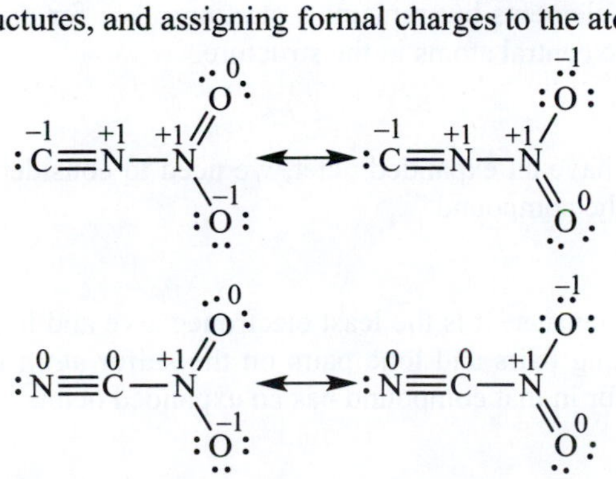

Formal charges are minimized in the two bottom structures, with the negative formal charge on the most electronegative atom (oxygen), so these are the preferred structures.

(c) The two structures of $CNNO_2$ shown in the text are not resonance structures because their atoms differ in connectivity. When two molecules have the same number of atoms, but in a different arrangement, they are called *isomers*.

Think About It

Be careful to make sure that all atoms in a resonance form have complete octets. The following Lewis structure for $NCNO_2$ is not valid because the terminal nitrogen atom has fewer than 8 e⁻.

$$:N=C=N$$

with :O: groups attached above and below the terminal N.

4.117. **Collect and Organize**

We are asked to consider whether all odd-electron molecules are inconsistent with the octet rule.

Analyze

When the octet rule is applied to the drawing of any Lewis structure, the number of electrons needed will be a multiple of 8 (ignoring any duets required for H).

Solve

The number of electrons (multiple of 8) needed to follow the octet rule is always even; therefore, yes, odd-electron molecules are always exceptions to the octet rule.

Think About It

When an odd-electron molecule (radical) either gains or loses an electron, it may then satisfy the octet rule for the atoms in the molecule.

4.119. **Collect and Organize**

We are asked why C, N, O, and F always obey the octet rule in Lewis structures.

Analyze

C, N, O, and F are all second-period elements with electron configurations of $[He]2s^2 2p^x$, where $x = 2, 3, 4,$ or 5. Once the $2p$ shell is filled with 6 e⁻, a closed-shell configuration is formed.

Solve

To accommodate more than 8 e⁻ in covalently bonded molecules, the atom would have to use orbitals beyond s and p. The d orbitals are not available to the small elements in the second period but do become available for elements in the third-period (and subsequent periods) elements such as P, S, and Cl.

Think About It

The octet rule strictly applies for only second-period elements but remains a starting place for drawing Lewis structures for compounds where larger elements are the central atoms in the structure.

4.121. **Collect and Organize**

To determine which of the sulfur–fluorine molecules have an expanded octet, we need to consider the number of electrons around the central atom required to form the compound.

Analyze

In each of those compounds sulfur is the central atom because it is the least electronegative and has the highest bonding capacity. If the number of electrons in bonding pairs and lone pairs on the sulfur atom in the Lewis structure of each compound is greater than 8, then sulfur in that compound has an expanded octet.

Solve

Molecule	Lewis Structure	Number of Electrons around S
(a) SF₆	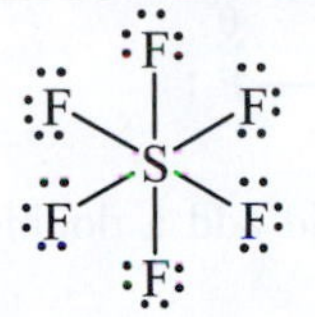	12
(b) SF₅		11
(c) SF₄	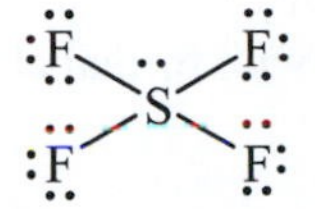	10
(d) SF₂	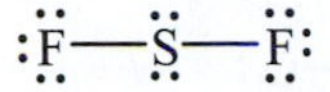	8

SF₆, SF₅, and SF₄ (a, b, and c) require sulfur to expand its octet.

Think About It

SF₅ is an odd-electron (radical) species.

4.123. **Collect and Organize**

To determine the number of electrons in the covalent bonds around each sulfur atom in the molecules, we first draw the Lewis structures for each.

Analyze

For these Lewis structures, we might have to expand octets for the sulfur atom (sulfur bonded to more than two atoms has to have more than 8 e⁻ to form the compound). We also have to consider whether the expansion of the octet on sulfur through double bonding, for example, reduces the formal charges on the atoms in the structure.

Solve

(a) SF₄O has 40 valence electrons, and its Lewis structure with formal charges assigned to its atoms is

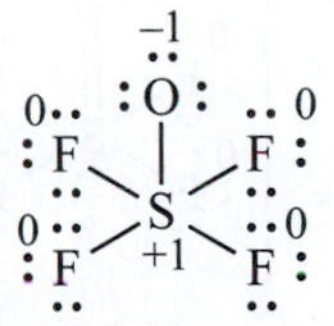

To reduce the formal charges on S and O, the oxygen may form a double bond with S.

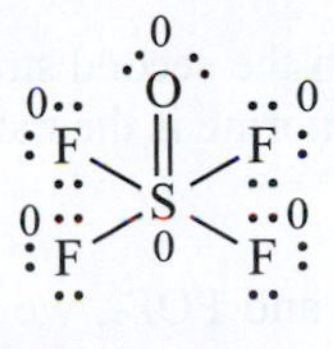

SF₄O has 12 e⁻ in six covalent bonds around sulfur.

(b) SOF_2 has 26 valence electrons, and its Lewis structure with formal charges assigned to its atoms is

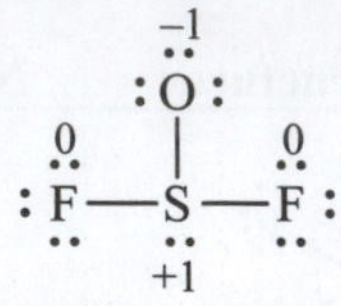

$$
\begin{array}{c}
\overset{-1}{\ddot{:}\!\ddot{O}\!:} \\
| \\
\overset{0}{:\!\ddot{F}}\!-\!\underset{+1}{S}\!-\!\overset{0}{\ddot{F}\!:}
\end{array}
$$

To reduce the formal charges on S and O, we could add a double bond between the oxygen and sulfur atoms.

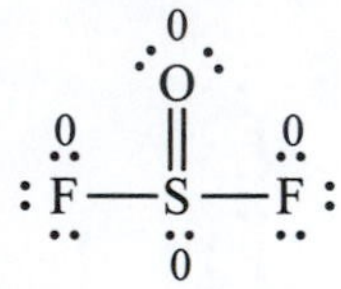

$$
\begin{array}{c}
\overset{0}{\cdot\ddot{O}\cdot} \\
\| \\
\overset{0}{:\!\ddot{F}}\!-\!\underset{0}{S}\!-\!\overset{0}{\ddot{F}\!:}
\end{array}
$$

That gives 8 e^- in four covalent bonds on sulfur in SOF_2.

(c) SO_3 has 24 valence electrons, and its Lewis structure with formal charges assigned to its atoms is

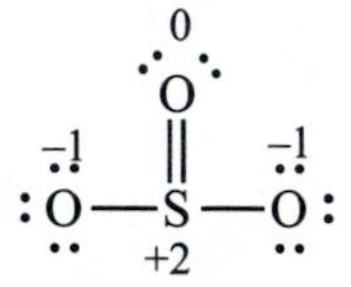

$$
\begin{array}{c}
\overset{0}{\cdot\ddot{O}\cdot} \\
\| \\
\overset{-1}{:\ddot{O}}\!-\!\underset{+2}{S}\!-\!\overset{-1}{\ddot{O}\!:}
\end{array}
$$

To reduce the formal charges on S and O, we could add double bonds between the other oxygen atoms and sulfur.

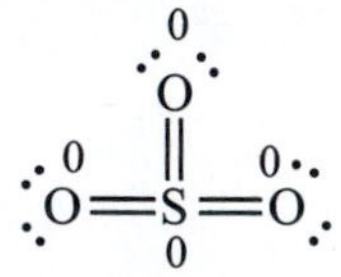

$$
\begin{array}{c}
\overset{0}{\cdot\ddot{O}\cdot} \\
\| \\
\overset{0}{\cdot\ddot{O}}\!=\!\underset{0}{S}\!=\!\overset{0}{\ddot{O}\cdot}
\end{array}
$$

That gives 12 e^- in six covalent bonds on sulfur in SO_3.

(d) SF_5^- has 42 valence electrons, and its Lewis structure with formal charges assigned to its atoms is

$$
\left[
\begin{array}{c}
\overset{0}{:\!\ddot{F}\!:} \\
\overset{0}{:\!\ddot{F}}\diagup\!\!\!\!\mid\!\!\!\!\diagdown\overset{0}{\ddot{F}\!:} \\
\overset{0}{:\!\ddot{F}}\diagdown_{\underset{-1}{S}}\diagup\overset{0}{\ddot{F}\!:}
\end{array}
\right]^-
$$

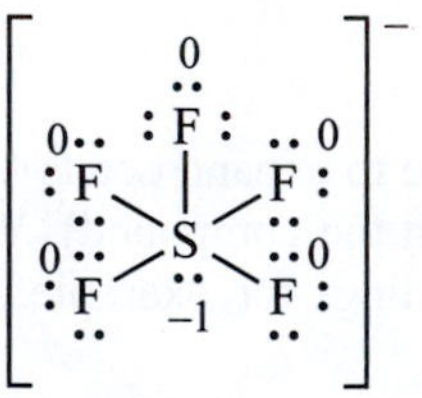

If we were to add a double bond between a fluorine atom and the sulfur atom, the formal charge buildup would *not* be preferred over the previous structure.

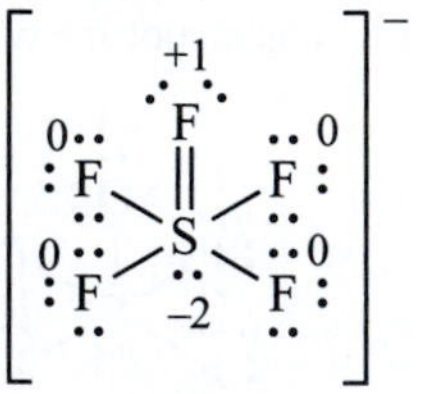

$$
\left[
\begin{array}{c}
\overset{+1}{\cdot\ddot{F}\cdot} \\
\overset{0}{:\!\ddot{F}}\diagup\!\!\!\!\|\!\!\!\!\diagdown\overset{0}{\ddot{F}\!:} \\
\overset{0}{:\!\ddot{F}}\diagdown_{\underset{-2}{S}}\diagup\overset{0}{\ddot{F}\!:}
\end{array}
\right]^-
$$

SF_5^- has 10 e^- in five covalent bonds.

Think About It

Double bonding of an atom to fluorine, as in the second structure in part d, gives a positive formal charge on F. That arrangement is never preferred since fluorine is the most electronegative element.

4.125. Collect and Organize

By drawing the Lewis structures of NOF_3 and POF_3, we are to describe the differences in bonding between those molecules.

Analyze

Nitrogen is a second-period element that cannot expand its octet, but phosphorus, as a third-period element, can expand its octet.

Solve

Both molecules have 32 valence electrons. The Lewis structures with formal charges are as follows:

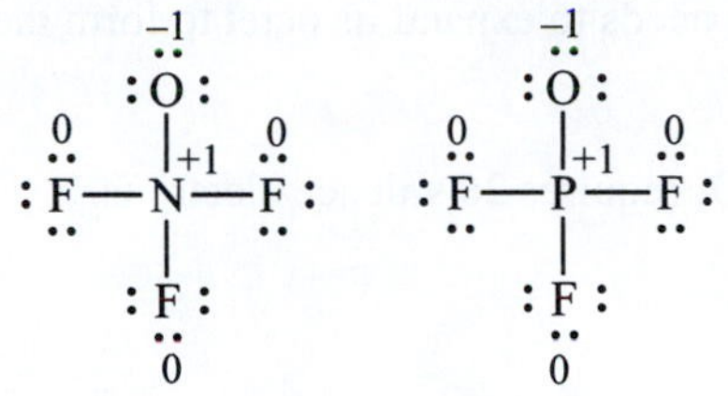

Because oxygen is more electronegative than N, those structures where the O atom has a –1 formal charge seem reasonable. However, the formal charges on P and O in POF_3 can be reduced to zero because P can expand its octet to form a double bond with O.

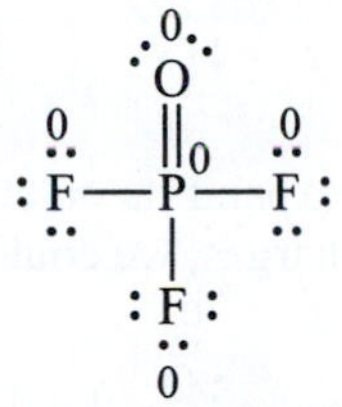

POF_3 contains a double bond and no formal charges; NOF_3 has only single bonds, and formal charges are present on the N and O atoms.

Think About It

In Lewis structures, always try to minimize formal charges. For elements in the third or higher period, you can reduce formal charges by expanding the octets to form double bonds.

4.127. **Collect and Organize**

By drawing the Lewis structures of SeF_4 and SeF_5^-, we can determine in which structure the Se atom has expanded its octet.

Analyze

Selenium, in the fourth period, expands its octet by making use of its $4d$ orbitals. For each Lewis structure, we use the method in the textbook. If the number of electrons around the central Se atom in the structures is greater than 8, selenium expands its octet to form the compound.

Solve

SeF_4 has 34 valence electrons, and its Lewis structure shows 10 electrons around the central selenium atom.

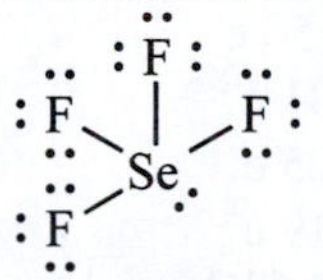

SeF_5^- has 42 valence electrons, and its Lewis structure shows 12 electrons around the central selenium atom.

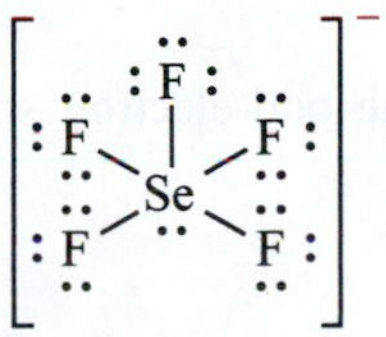

In both SeF_4 and SeF_5^-, Se has more than 8 valence electrons.

Think About It

In both SeF_4 and SeF_5^-, a lone pair of electrons is present on the selenium atom to give 10 and 12 electrons around Se, respectively.

4.129. Collect and Organize

From the arrangement of atoms given in Figure P4.129 for Cl_2O_2, we are to draw the Lewis structure and determine whether either Cl atom needs to expand its octet to form the molecule.

Analyze

The arrangement of atoms in Cl_2O_2 requires 26 valence electrons.

Solve

The Lewis structure for Cl_2O_2 is

$$
\begin{array}{c}
\overset{-1}{:\!\ddot{O}} \\
\diagdown \overset{+2}{\underset{}{\ddot{C}l}} \quad \overset{0}{Cl} \!-\! \ddot{C}l\!: \\
\diagup \\
:\!\ddot{O} \\
-1
\end{array}
$$

In that structure, neither Cl atom needs to expand its octet. However, the formal charges on the atoms of that structure are fairly high. To reduce formal charges, we could form double bonds between the Cl and O atoms.

$$
\begin{array}{c}
\overset{0}{:\!\ddot{O}} \\
\diagdown\!\!\!\diagdown \overset{0}{\underset{}{\ddot{C}l}} \quad \overset{0}{Cl} \!-\! \ddot{C}l\!: \\
\diagup\!\!\!\diagup \\
:\!O \\
0
\end{array}
$$

In that structure, all formal charges of all atoms are zero, and the central chlorine atom has an expanded octet.

Think About It

Use formal charges to seek out when an atom will expand its octet. At first glance Cl_2O_2 did not appear to have any more than 8 e^- surrounding the central Cl atom.

4.131. Collect and Organize

For each molecule combining Cl with O, we are to determine which are odd-electron molecules.

Analyze

To answer that we need only add up the valence electrons for each molecule.

Solve

(a) Cl_2O_7 has $(2\ Cl \times 7\ e^-) + (7\ O \times 6\ e^-) = 56\ e^-$
(b) Cl_2O_6 has $(2\ Cl \times 7\ e^-) + (6\ O \times 6\ e^-) = 50\ e^-$
(c) ClO_4 has $(1\ Cl \times 7\ e^-) + (4\ O \times 6\ e^-) = 31\ e^-$
(d) ClO_3 has $(1\ Cl \times 7\ e^-) + (3\ O \times 6\ e^-) = 25\ e^-$
(e) ClO_2 has $(1\ Cl \times 7\ e^-) + (2\ O \times 6\ e^-) = 19\ e^-$
The odd-electron molecules are (c) ClO_4, (d) ClO_3, and (e) ClO_2.

Think About It

For those chlorine–oxygen molecules, the odd-electron species have an odd number of Cl atoms in their formulas.

4.133. Collect and Organize

For the species named, we are to decide which atom in the molecule most likely has an unpaired electron.

Analyze

An unpaired electron in a molecule is a lone electron. This occurs when a molecule has an odd number of valence electrons. The atom that bears the odd electron is the least electronegative. The electronegativities of the atoms are S = 2.5, O = 3.5, N = 3.0, C = 2.5, H = 2.1.

Solve

(a) For SO^+, since S is less electronegative than O, the unpaired electron is on S.

$$\left[\ddot{\cdot}\ddot{S}\equiv O\ddot{\cdot}\right]^+$$

(b) For NO, since N is less electronegative than O, the unpaired electron is on N.

$$\dot{\cdot}N\equiv O\ddot{\cdot}$$

(c) For CN, since C is less electronegative than N, the unpaired electron is on C.

$$\cdot C\equiv N\ddot{\cdot}$$

(d) For OH, since H is less electronegative than O, we expect the unpaired electron to be on H. Hydrogen, however, must be bonded to O and obey the duet rule, so the unpaired electron must be on O.

$$\cdot\ddot{O}-H$$

Think About It

The unpaired electron is where a lone pair normally is shown in a Lewis structure, not where a bonding pair is drawn. That could not be done for the H in OH because the H atom would not have a lone pair on it; H cannot have more than its duet surrounding it.

4.135. **Collect and Organize**

From the Lewis structures given we can use formal charge arguments to determine which structure contributes most to the bonding in CNO.

Analyze

The resonance structure that contributes the most to the bonding has the lowest possible formal charges on the atoms and, if formal charges are present, then the negative formal charges should be on the most electronegative atoms in the structure.

Solve

The formal charge assignments on each structure are as follows:

(a) $\overset{-2}{\ddot{\cdot}\ddot{C}}-\overset{+1}{N}\equiv\overset{+1}{O}\ddot{\cdot}$ 　　　(b) $\overset{-1}{\cdot\ddot{C}}=\overset{+1}{N}=\overset{0}{\ddot{O}}\dot{\cdot}$

(c) $\overset{-1}{\ddot{\cdot}C}\equiv\overset{+1}{N}-\overset{0}{\ddot{O}}\cdot$ 　　　(d) $\overset{0}{\cdot C}\equiv\overset{+1}{N}-\overset{-1}{\ddot{O}}\ddot{\cdot}$

The structure that contributes the most to the bonding in CNO is (d).

Think About It

Because all those structures are in resonance, each resonance form contributes to the bonding. However, because structure (d) is favored with the –1 formal charge placed on the oxygen atom, the form with a $C\equiv N$ bond and with the unpaired electron on C contributes the most to the bonding.

4.137. **Collect and Organize**

We are asked to compare the atmospheric greenhouse gases to the panes of a greenhouse.

Analyze

The atmospheric gases considered greenhouse gases include CO_2 and CH_4. By acting as greenhouse gases they trap radiation, acting as a blanket to warm Earth.

Solve

Like the panes of glass in a greenhouse, the greenhouse gases in the atmosphere are transparent to visible light. Once the visible light warms Earth's surface and is reemitted as infrared (lower energy) light, the greenhouse gases absorb the infrared light, just as the panes of glass do not allow heat inside the greenhouse to escape.

Think About It

For a molecule to absorb infrared radiation and act as a greenhouse gas, it must have a molecular vibration that has a change in dipole moment upon stretching or bending. Nitrogen (N_2) and oxygen (O_2), with no dipole moment change upon stretching, are not greenhouse gases.

4.139. Collect and Organize

We are asked to consider which bond stretching in N_2O is responsible for the absorption of infrared radiation.

Analyze

Infrared radiation is absorbed by molecules with polar bonds when the fluctuating electric fields in the molecule do not cancel each other out (asymmetric stretching). The Lewis structure of N_2O shows that the molecule is linear.

$$:N\equiv N\text{—}\ddot{\underset{..}{O}}: \quad \text{or} \quad \ddot{\underset{..}{:N}}\text{=}N\text{=}\ddot{\underset{..}{O}}:$$

Solve

The stretching of either the N—N or the N—O bond gives a change in the dipole moment for this molecule; both of those stretching modes absorb IR radiation.

Think About It

N_2O is estimated to be about 300 times more potent a greenhouse gas than CO_2. But N_2O is present in much lower concentrations (320 ppb vs. 385 ppm) than CO_2.

4.141. Collect and Organize

We are asked to consider whether carbon monoxide is infrared active (can absorb IR radiation).

Analyze

Infrared radiation is absorbed by a molecular vibration (bond stretch or bend) when the bond is polar and when the fluctuating electric fields do not cancel each other out. The Lewis structure of carbon monoxide is

$$:C\equiv O:$$

Solve

The C—O bond is polar because of the difference in electronegativity of carbon and oxygen. Stretching the linear C—O bond in carbon monoxide would give a fluctuating electric field, and therefore CO does absorb IR radiation.

Think About It

Carbon monoxide is a weak greenhouse gas, but it reacts with hydroxyl (OH) radicals, which then cannot react with other greenhouse gases such as methane to "neutralize" them and reduce their effects.

4.143. Collect and Organize

For this question we consider why infrared radiation causes vibrations, but not breakage, of chemical bonds.

Analyze

Infrared radiation has wavelengths between 10^{-6} and 10^{-4} m, whereas ultraviolet radiation has wavelengths between 10^{-6} and 10^{-8} m.

Solve

The shorter the wavelength, the higher the energy of the radiation. Thus, infrared radiation, with its longer wavelengths and lower energy, causes chemical bonds only to stretch and bend. Higher energy ultraviolet radiation can cause chemical bonds to break.

Think About It
Ultraviolet radiation, along with X-ray and gamma radiation, can cause bonds to break and are classified as ionizing radiation.

4.145. **Collect and Organize**
For this question we compare the bonds in CH_2O to CO to determine whether the frequency to vibrate the C—O bond in CO is greater or less than the frequency to vibrate the C—O bond in CH_2O.

Analyze
The Lewis structure of CO shows that the C—O bond is a triple bond, whereas the C—O bond in CH_2O is a double bond.

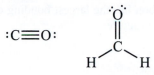

Solve
Because the triple bond in CO is stronger than the double bond in CH_2O, the energy required to vibrate the C—O bond in CO is higher and therefore, the stretching frequency for CO is higher than in CH_2O.

Think About It
The bond strength is also related to bond length. The stronger the bond, the shorter the bond distance. The triple bond in CO is 113 pm, whereas the double bond in CH_2O is 123 pm.

4.147. **Collect and Organize**
To reflect the valence electrons available for bonding, the electron dot placement around the element symbol in a Lewis symbol should show the lone electrons that may form covalent bonds with other atoms. We are to select the preferred symbol in each pair listed.

Analyze
The valence electrons available for bonding in the atom are Be = 2, Al = 3, C = 4, He = 0.

Solve
(a)   ·Be·   (b)   ·A̤l·   (c)   ·Ċ̤·   (d)   :He

Think About It
That placement method can be extended to other atoms. For example, oxygen, which tends to form two bonds to other atoms, has the Lewis dot symbol

4.149. **Collect and Organize**
Carbon disulfide could have either carbon or sulfur as the central atom in its structure. Using the formal charge assignments in the two possible Lewis structures, we are to determine which is the preferred structure.

Analyze
A structure is preferred when the formal charges on the atoms are minimized and when any negative formal charges are located on the most electronegative elements. The electronegativities of carbon and sulfur are the same, namely, 2.5.

Solve
Carbon disulfide has 16 valence electrons.

$$ \underset{:S=C=S:}{^0 \quad ^0 \quad ^0} \qquad \underset{:S=S=C:}{^0 \quad ^{+2} \quad ^{-2}} $$

When C is the central atom, the atoms all carry zero formal charge; that structure is preferred for carbon disulfide.

Think About It

Because sulfur can expand its octet, the formal charges on CSS may be reduced:

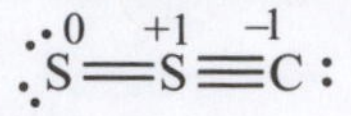

$$\ddot{S}\!\!=\!\!\overset{+1}{S}\!\!\equiv\!\!\overset{-1}{C}:$$

That still leaves formal charges on the atoms. For reasons you will see when you study valence bond theory, carbon cannot quadruple bond to reduce the formal charges on the atoms in that structure to zero.

4.151. Collect and Organize

For the poisonous gas phosgene, $COCl_2$, we are to draw the Lewis structure.

Analyze

$COCl_2$ has 24 valence electrons. Carbon has the largest bonding capacity and is the least electronegative of the atoms, so it is the central atom in the structure.

Solve

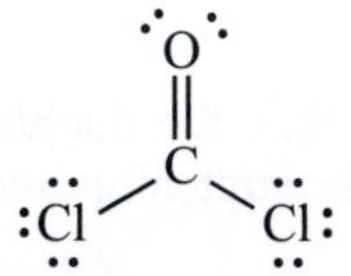

Think About It

Symptoms of human phosgene inhalation include choking, painful breathing, severe eye irritation, and skin burns. Death may result from lack of oxygen.

4.153. Collect and Organize

Neutral OCN reacts with itself to form OCNNCO, and OCN^- reacts with BrNO to give OCNNO and with Br_2 and NO_2 to give OCN(CO)NCO. For those three products we are to draw Lewis structures with any appropriate resonance forms.

Analyze

(a) OCNNCO has 30 valence electrons.
(b) BrNO has 18 valence electrons.
(c) OCN(CO)NCO has 40 valence electrons.

Solve

(a)

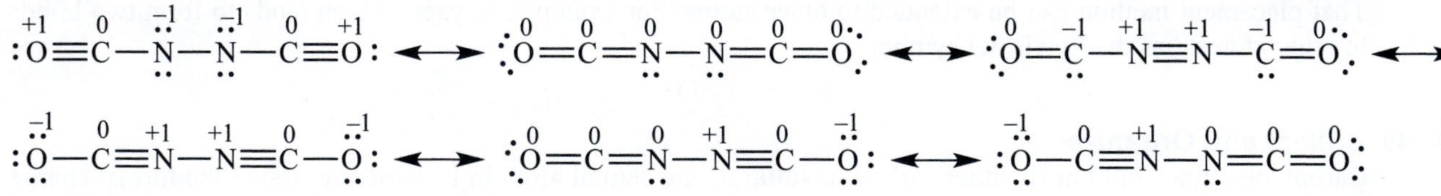

(b)

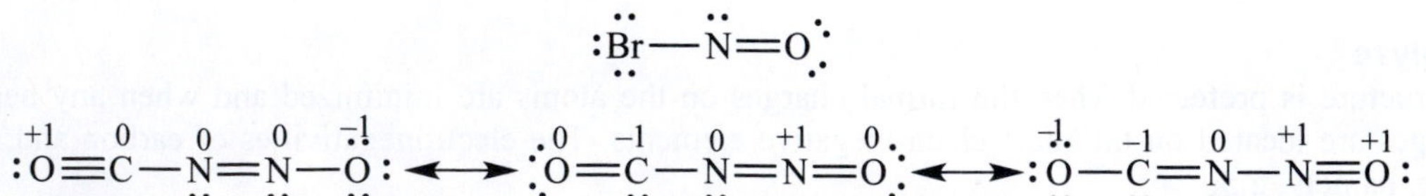

(c)

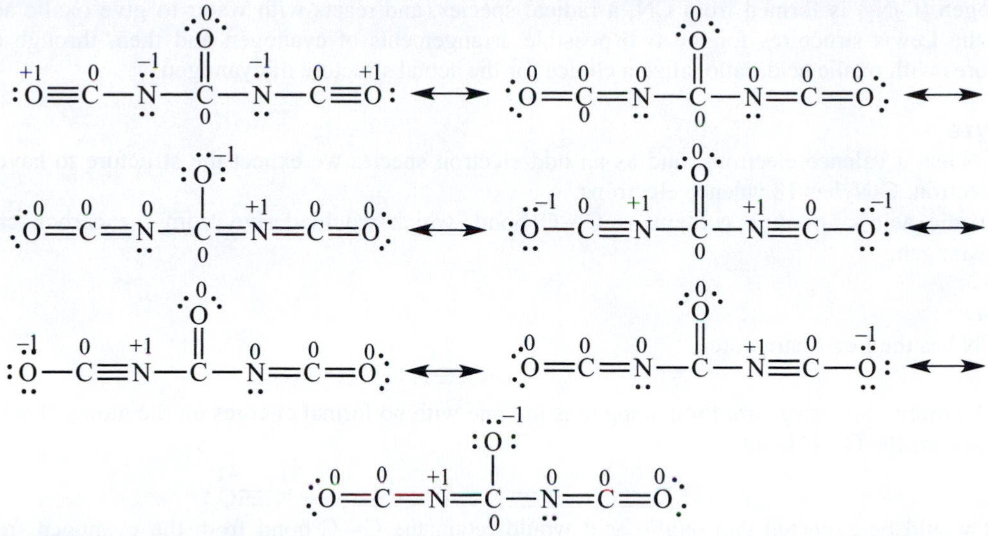

Think About It
The resonance structure that contributes the most to the bonding in each of those compounds is the one that minimizes formal charges on the atoms.

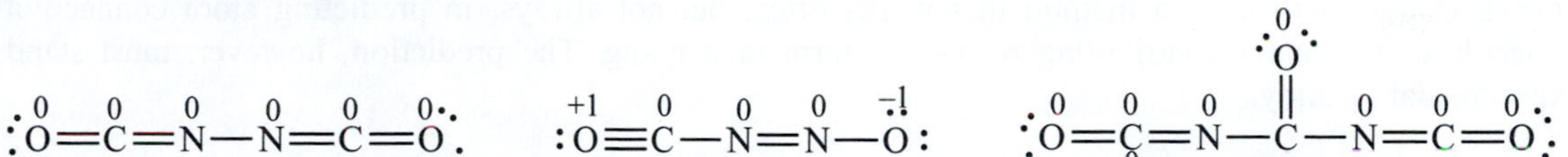

4.155. Collect and Organize
We are to draw two Lewis structures of Cl_2O_6: one structure with a chlorine–chlorine bond and the other with a Cl—O—Cl arrangement of atoms.

Analyze
Cl_2O_6 has 50 valence electrons. We will draw the Lewis structures with the minimum formal charge on all atoms.

Solve

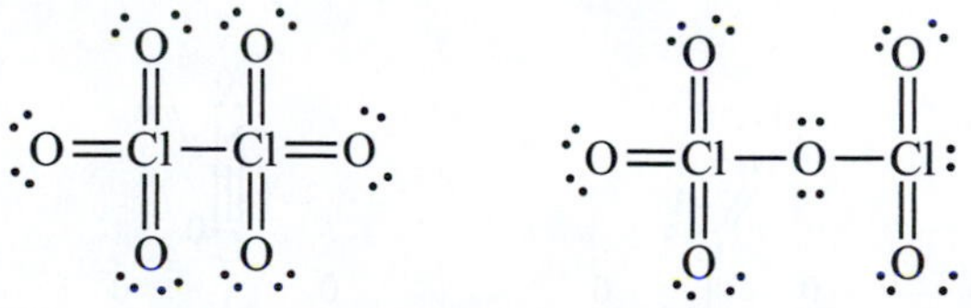

Think About It
The formal charges on all the atoms for both arrangements of atoms in Cl_2O_6 are zero. From formal charges alone, therefore, we would not be able to predict the actual atom arrangement in Cl_2O_6 because we were able to draw structures for both arrangements in which all atoms have zero formal charge. The actual structure is thought to manifest as a perchlorate salt, $ClO_2^+ClO_4^-$.

4.157. **Collect and Organize**

Cyanogen (C_2N_2) is formed from CN, a radical species, and reacts with water to give oxalic acid. We are to draw the Lewis structures for the two possible arrangements of cyanogen and then, through comparing the structures with oxalic acid, rationalize a choice for the actual structure of cyanogen.

Analyze

(a) CN has 9 valence electrons, and as an odd-electron species we expect the structure to have one unpaired electron. C_2N_2 has 18 valence electrons.

(b) Oxalic acid's structure contains a C—C bond, which would derive from the carbon–carbon bond in cyanogen.

Solve

(a) CN has the Lewis structure

$$\cdot C \equiv N :$$

The more likely structure for cyanogen is the one with no formal charges on the atoms. That is the one that contains the C—C bond:

(b) It would be expected that oxalic acid would retain the C—C bond from the cyanogen from which it is formed in the reaction of cyanogen with water. That outcome is consistent with the structure for cyanogen predicted from formal charge analysis.

Think About It

Formal charge analysis is a method that works often, but not always, in predicting atom connectivity in a molecule or the major contributing resonance form in bonding. The prediction, however, must stand up to experimental scrutiny.

4.159. **Collect and Organize**

The structure of SF_3CN shows bond lengths of 116 pm (C—N), 174 pm (S—C), and 160 pm (F—S). We can compare those bond lengths to those in Table 4.6 to determine the type of bond each might be (single, double, or triple). That information can help us draw the Lewis structure for the molecule and assign formal charges.

Analyze

From Table 4.6, the only bond length listed that corresponds to any of the bond lengths in SF_3CN is C≡N (116 pm). To draw the Lewis structure, we need 36 valence electrons.

Solve

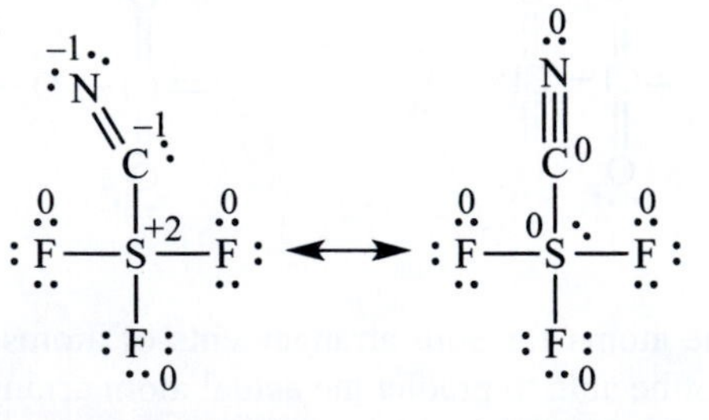

| Formal charges are nonzero and the carbon–nitrogen bond is not triple. | All formal charges are zero and the carbon–nitrogen bond is triple. |

Think About It

In the preferred structure, sulfur has expanded its octet.

4.161. **Collect and Organize**

Tellurium will expand its octet to form $TeOF_6^{2-}$. Using this information and formal charges, we can draw the best Lewis structure for this ion.

Analyze

$TeOF_6^{2-}$ has 56 valence electrons, and tellurium is the central atom, since it is the least electronegative atom in the structure.

Solve

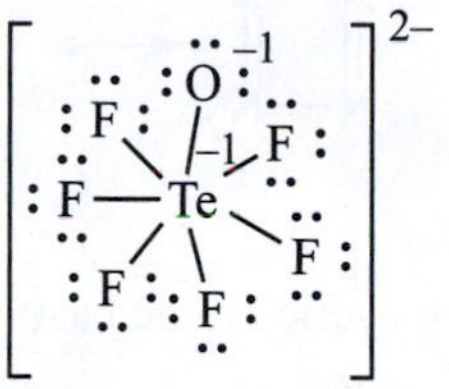

That structure has the lowest formal charges. Oxygen is highly electronegative, so it will carry a negative formal charge.

Think About It

Another resonance structure of $TeOF_6^{2-}$ would be

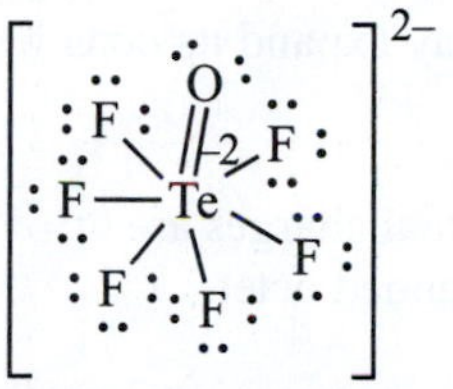

but that structure places all the negative formal charge on the least electronegative atom, so it is not preferred to the structure shown above.

4.163. **Collect and Organize**

Both calcium carbonate and magnesium hydroxide are ionic compounds. To draw their Lewis structures, we have to draw the Lewis diagrams separately for the cation and anion.

Analyze

Calcium carbonate is $CaCO_3$, where the cation is Ca^{2+} and the anion is CO_3^{2-}. Magnesium hydroxide is $Mg(OH)_2$, where the cation is Mg^{2+} and the anion is OH^-.

Solve

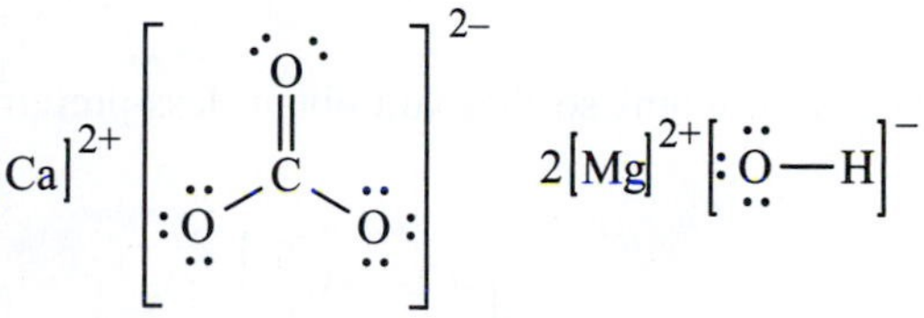

Think About It

The carbonate anion in calcium carbonate has two additional resonance structures that we could draw (see Problem 4.92).

4.165. **Collect and Organize**

In considering the Lewis structures of linear N_4 and cyclic N_4, by assigning formal charges we may be able to choose which form might be preferred by this short-lived allotrope of nitrogen.

Analyze

N_4 has 20 valence electrons.

Solve

(a, b)

$$\overset{0}{:}N\overset{+1}{\equiv}N\overset{0}{-}N\overset{-1}{=}N\overset{..}{.} \longleftrightarrow \overset{-1}{:}N\overset{+1}{=}N\overset{+1}{=}N\overset{-1}{=}N\overset{..}{.} \longleftrightarrow \overset{-1}{:}N\overset{0}{=}N\overset{+1}{-}N\overset{0}{\equiv}N:$$

All those resonance forms have atoms with nonzero formal charges. The middle structure has the most nonzero formal charges separated over three bond lengths, so that one is least preferred. The first and last resonance structures are preferred and are indistinguishable from each other.

(c) The resonance structures for cyclic N_4 have no formal charges on any of the nitrogen atoms.

$$\begin{array}{c}
\overset{..}{N}\text{---}\overset{..}{N} \\
\parallel \quad \parallel \\
\overset{..}{N}\text{---}\overset{..}{N}
\end{array}
\longleftrightarrow
\begin{array}{c}
\overset{..}{N}\text{=}\overset{..}{N} \\
\mid \quad \mid \\
\overset{..}{N}\text{=}\overset{..}{N}
\end{array}$$

Think About It

We would predict the structure of N_4 to be cyclic on the basis of formal charge considerations.

4.167. Collect and Organize

By drawing the Lewis structures for each molecule to minimize formal charges on the atoms, we can determine which molecules contain an atom with an expanded octet.

Analyze

All those molecules contain Cl, which may expand its octet because it is a third-period element.

Solve

(a) $:\overset{..}{Cl}\text{---}\overset{..}{Cl}:$ All formal charges are 0; no expanded octet

(b)
$$:\overset{..}{F}:$$
$$\mid$$
$$:\overset{..}{F}\text{---}\overset{..}{Cl}\text{---}\overset{..}{F}:$$
All formal charges are 0; expanded octet on Cl

(c)
$$:\overset{..}{I}:$$
$$\mid$$
$$:\overset{..}{I}\text{---}\overset{..}{Cl}\text{---}\overset{..}{I}:$$
All formal charges are 0; expanded octet on Cl

(d) $\left[:\overset{..}{Cl}\text{---}\overset{-1}{\overset{..}{O}}:\right]^{-}$ Formal charges are minimized and negative formal charge is on the more electronegative element; no expanded octet

ClF_3 (b) and ClI_3 (c) have an atom with an expanded octet.

Think About It

Remember to not expand octets on atoms so that you obtain less-preferred formal charges on the atoms. For example,

$$\left[:\overset{-1}{\overset{..}{Cl}}\text{=}\overset{..}{O}:\right]^{-}$$

is not preferred over the structure in part d above.

4.169. Collect and Organize

After drawing the Lewis structure (with resonance forms) for N_5^-, we can determine, using formal charges, which resonance forms contribute most to the bonding in the molecule and compare the bonding of N_5^- with that of N_3^- in terms of average bond order.

Analyze

N_5^- has 26 valence electrons.

Solve

(a, b) All possible resonance structures for N_5^- with formal charge assignment are as follows:

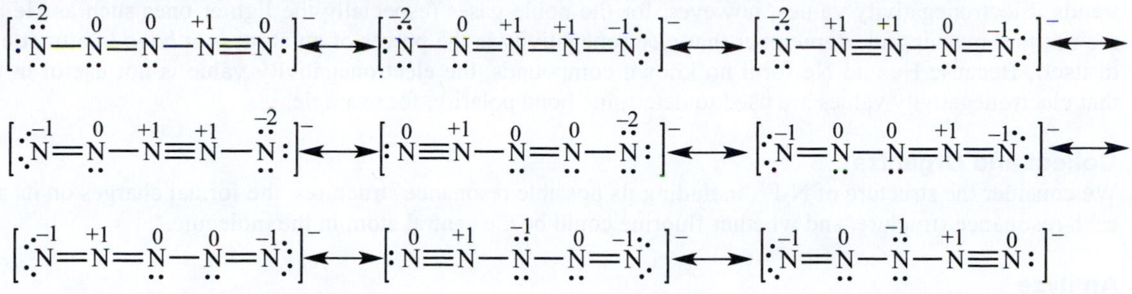

$$\begin{bmatrix} \overset{-2}{:\!\ddot{N}}\!-\!\overset{0}{N}\!\!=\!\!\overset{0}{N}\!-\!\overset{+1}{N}\!\!\equiv\!\!\overset{0}{N:} \end{bmatrix}^- \longleftrightarrow \begin{bmatrix} \overset{-2}{:\!\ddot{N}}\!-\!\overset{0}{N}\!\!=\!\!\overset{+1}{N}\!\!=\!\!\overset{+1}{N}\!\!=\!\!\overset{-1}{\ddot{N}:} \end{bmatrix}^- \longleftrightarrow \begin{bmatrix} \overset{-2}{:\!\ddot{N}}\!-\!\overset{+1}{N}\!\!\equiv\!\!\overset{+1}{N}\!-\!\overset{0}{N}\!\!=\!\!\overset{-1}{\ddot{N}:} \end{bmatrix}^- \longleftrightarrow$$

$$\begin{bmatrix} \overset{-1}{:\!\ddot{N}}\!\!=\!\!\overset{0}{N}\!-\!\overset{+1}{N}\!\!\equiv\!\!\overset{+1}{N}\!-\!\overset{-2}{\ddot{N}:} \end{bmatrix}^- \longleftrightarrow \begin{bmatrix} \overset{0}{:N}\!\!\equiv\!\!\overset{+1}{N}\!-\!\overset{0}{N}\!\!=\!\!\overset{0}{N}\!-\!\overset{-2}{\ddot{N}:} \end{bmatrix}^- \longleftrightarrow \begin{bmatrix} \overset{-1}{:\!\ddot{N}}\!\!=\!\!\overset{0}{N}\!-\!\overset{0}{N}\!\!=\!\!\overset{+1}{N}\!\!=\!\!\overset{-1}{\ddot{N}:} \end{bmatrix}^- \longleftrightarrow$$

$$\begin{bmatrix} \overset{-1}{:\!\ddot{N}}\!\!=\!\!\overset{+1}{N}\!\!=\!\!\overset{0}{N}\!-\!\overset{0}{N}\!\!=\!\!\overset{-1}{\ddot{N}:} \end{bmatrix}^- \longleftrightarrow \begin{bmatrix} \overset{0}{:N}\!\!\equiv\!\!\overset{+1}{N}\!-\!\overset{-1}{\ddot{N}}\!-\!\overset{0}{N}\!\!=\!\!\overset{-1}{\ddot{N}:} \end{bmatrix}^- \longleftrightarrow \begin{bmatrix} \overset{-1}{:\!\ddot{N}}\!\!=\!\!\overset{0}{N}\!-\!\overset{-1}{\ddot{N}}\!-\!\overset{+1}{N}\!\!\equiv\!\!\overset{0}{N:} \end{bmatrix}^-$$

The structures that contribute most have the lowest formal charges (last four structures shown). Therefore, the terminal nitrogen–nitrogen bonds will be close to the length of a double bond, and the middle nitrogen–nitrogen bonds in the structure will be close to a bond order of 1.5 (between a single and double bond).

(c) N_3^- has the Lewis structures

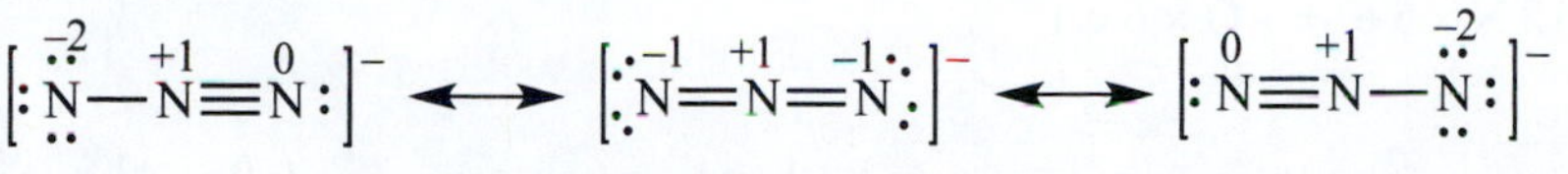

$$\begin{bmatrix} \overset{-2}{:\!\ddot{N}}\!-\!\overset{+1}{N}\!\!\equiv\!\!\overset{0}{N:} \end{bmatrix}^- \longleftrightarrow \begin{bmatrix} \overset{-1}{:\!\ddot{N}}\!\!=\!\!\overset{+1}{N}\!\!=\!\!\overset{-1}{\ddot{N}:} \end{bmatrix}^- \longleftrightarrow \begin{bmatrix} \overset{0}{:N}\!\!\equiv\!\!\overset{+1}{N}\!-\!\overset{-2}{\ddot{N}:} \end{bmatrix}^-$$

From those resonance structures we see that each bond is predicted to be of double-bond character in N_3^-. Therefore, N_5^- has two longer N—N bonds than in N_3^-. N_3^- has the higher average bond order.

Think About It

The longer (and weaker) N—N bonds in N_5^- might be the ones broken in a chemical reaction.

4.171. Collect and Organize

By plotting the electronegativity (y-axis) versus the ionization energy (x-axis) for the elements in the second period ($Z = 3$ to 9), we can determine whether the trend is linear and then estimate the electronegativity for neon, knowing that its first ionization energy is 2081 kJ/mol.

Analyze

Figure 4.13 shows the electronegativities of the second-period main group elements, and Appendix 3 lists the ionization energies. Using Excel, we can plot the values for Li, Be, B, C, N, O, and F. We can estimate the electronegativity of Ne with the equation for the line for the trend seen for the other second-period elements.

Solve

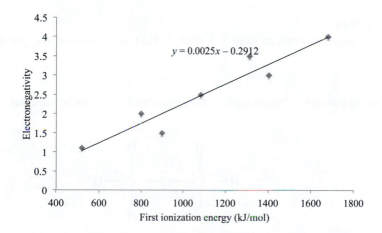

Using the equation for the best-fit line where x = the ionization energy of neon (2081 kJ/mol) gives a value of y (electronegativity) of neon:

$$y = 0.002(2081) - 0.2912 = 4.9$$

Think About It

The calculated electronegativity for neon is higher than that of fluorine, as we would expect from periodic trends. Electronegativity values, however, for the noble gases (especially the lighter ones such as He and Ne) are not that meaningful. Remember that electronegativity is the power of an atom *in a bond* to attract electrons to itself. Because He and Ne form no known compounds, the electronegativity value is not useful in the way that electronegativity values are used to determine bond polarity, for example.

4.173. Collect and Organize

We consider the structure of N_2F^+, including its possible resonance structures, the formal charges on its atoms in each resonance structure, and whether fluorine could be the central atom in the molecule.

Analyze

N_2F^+ has 16 valence electrons.

Solve

(a) Isoelectronic means that the two species have the same number of electrons. N_2O also has 16 valence electrons ($2\,N \times 5\,e^- + 1\,O \times 6\,e^-$).

(b–d)

$$\left[:\overset{-2}{\underset{..}{\overset{..}{N}}} - \overset{+1}{N} \equiv \overset{+2}{F} : \right]^+ \longleftrightarrow \left[:\overset{-1}{\overset{..}{N}} = \overset{+1}{N} = \overset{+1}{\underset{..}{F}} : \right]^+ \longleftrightarrow \left[:\overset{0}{N} \equiv \overset{+1}{N} - \overset{0}{\underset{..}{\overset{..}{F}}} : \right]^+$$

The central nitrogen atom in all the resonance structures always carries a +1 formal charge. The first and second resonance forms shown are unacceptable because they have greater than the minimal formal charges on the atoms.

(e) Yes, fluorine could be the central atom in the molecule, but that would place significant positive formal charge on the fluorine atom (the most electronegative element). These structures are unlikely:

$$\left[:\overset{-2}{\underset{..}{\overset{..}{N}}} - \overset{+3}{F} \equiv \overset{0}{N} : \right]^+ \longleftrightarrow \left[:\overset{-1}{\overset{..}{N}} = \overset{+3}{F} = \overset{-1}{\underset{..}{N}} : \right]^+ \longleftrightarrow \left[:\overset{0}{N} \equiv \overset{+3}{F} - \overset{-2}{\underset{..}{\overset{..}{N}}} : \right]^+$$

Think About It

We could try to reduce the formal charge on fluorine in part (e) by drawing

$$\left[:\overset{0}{N} \equiv \overset{+1}{F} \equiv \overset{0}{N} : \right]^+$$

As a second-period element, however, fluorine cannot expand its octet.

4.175. Collect and Organize

We are to draw the Lewis structure of dimethyl ether, C_2H_6O, given that the oxygen atom is bonded to the two carbon atoms.

Analyze

Dimethyl ether has 20 valence electrons and the hydrogen bonds all must be terminal.

Solve

Think About It

As you will learn in the next chapter, that molecule is bent and, as a result, will be a polar molecule.

4.177. **Collect and Organize**

Given that butane has a chemical formula of C_4H_{10} and that the structure contains four carbons atoms bonded in a row, we are to draw the Lewis structure of butane.

Analyze

Butane, C_4H_{10}, has 26 valence electrons. Each carbon must satisfy its octet while the hydrogen atoms have a duplet.

Solve

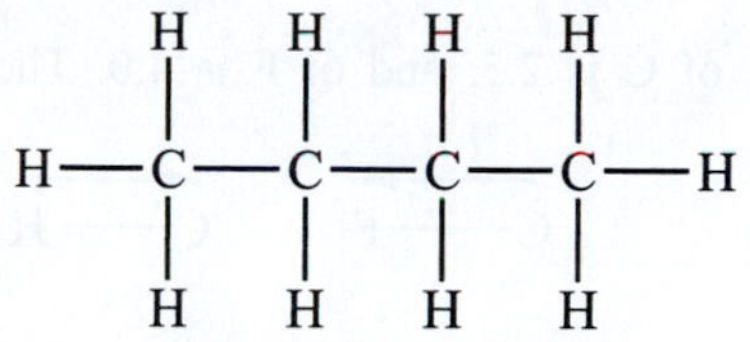

$$
\begin{array}{ccccccc}
 & H & & H & & H & & H \\
 & | & & | & & | & & | \\
H - & C & - & C & - & C & - & C & - H \\
 & | & & | & & | & & | \\
 & H & & H & & H & & H
\end{array}
$$

Think About It

That molecule belongs to the family of hydrocarbons that are *alkanes*.

CHAPTER 5 | Bonding Theories: Explaining Molecular Geometry

5.1. Collect and Organize

Dipole moments are a result of permanent partial charge separation in a molecule due to differences in atom electronegativities and molecular geometry. We can use the given structures for $C_2H_3F_3$ to determine whether we can distinguish them by their dipole moments.

Analyze

The electronegativity of H is 2.1, of C is 2.5, and of F is 4.0. The polarities of the C—F and C—H bonds, therefore, are

$$\overset{+\longrightarrow}{\text{C——F}} \qquad \overset{\longleftarrow +}{\text{C——H}}$$

In the C—F bond, the C atom carries a partial positive charge ($\delta+$) and the F atom carries a partial negative charge ($\delta-$). In the C—H bond, the C atom carries a partial negative charge ($\delta-$) and the H atom carries a partial positive charge ($\delta+$).

Solve

The geometry of structure (a) is such that all the fluorine atoms are bonded to the same carbon. When adding up the bond dipole vectors for that compound, we can see that this placement of the fluorine atoms in (a) gives the more polar molecule and, therefore, the molecule with the greater dipole moment.

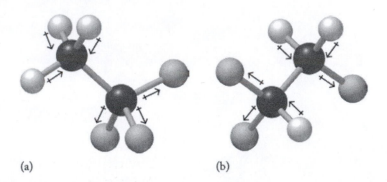

(a)   (b)

Think About It

Both the difference in electronegativities of bonded elements and the molecular geometry are important in establishing molecular polarity.

5.3. Collect and Organize

For the molecules shown, N_2F_2, H_2NNH_2, and NCCN, we are to determine whether the molecules are planar and whether any of the molecules have delocalized π electrons through resonance structures.

Analyze

To determine the planarity of the molecules, we need to draw the Lewis structures for each and determine the molecular geometry.

Solve

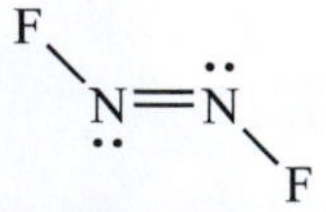

$$\overset{\text{F}}{\underset{}{\diagdown}}\text{N}\!=\!\text{N}\overset{}{\underset{\diagdown}{}}\text{F}$$

Electron pair geometry on
N = trigonal planar

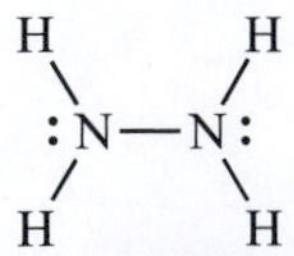

$$\overset{\text{H}\quad\text{H}}{\underset{\text{H}\quad\text{H}}{:\text{N}-\text{N}:}}$$

Electron pair geometry on
N = tetrahedral

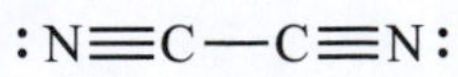

$$:N\!\equiv\!C\!-\!C\!\equiv\!N:$$

Electron pair geometry on
N = linear

Molecular geometry = bent	Molecular geometry = trigonal pyramidal	Molecular geometry = linear

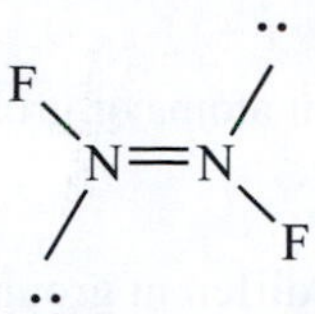

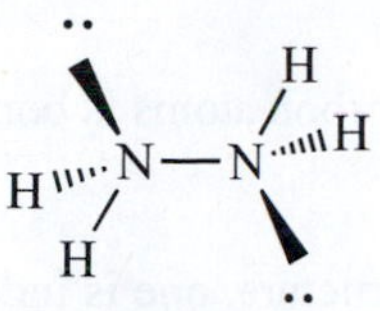

$$:\!-\!N\!\equiv\!C\!-\!C\!\equiv\!N\!-\!:$$

Planar	Not planar	Planar
No delocalized π electrons	No delocalized π electrons	No delocalized π electrons
		No resonance forms

N_2F_2 and NCCN are planar molecules. No delocalized π electrons are present in any of those molecules.

Think About It
For a molecule with two central atoms to be planar, its molecular geometry must be linear, trigonal planar, or square planar.

5.5. ### Collect and Organize
For each species, O_2^+ and O_2^{2+}, we are to fill the molecular orbital (MO) diagram for homonuclear diatomic oxygen. From the filled diagram, we can determine whether O_2^+ has more or fewer electrons in antibonding molecular orbitals than O_2^{2+}.

Analyze
The number of valence electrons in O_2^+ is $(2\ O \times 6\ e^-) - 1\ e^- = 11\ e^-$, and in O_2^{2+} the number of electrons is $(2\ O \times 6\ e^-) - 2\ e^- = 10\ e^-$. Antibonding orbitals in the MO diagram are designated with an asterisk (*).

Solve
For O_2^+, filling of the MO diagram gives

$$(\sigma_{2s})^2(\sigma_{2s}^*)^2(\sigma_{2p})^2(\pi_{2p})^4(\pi_{2p}^*)^1$$

3 electrons in antibonding orbitals

For O_2^{2+}, filling of the MO diagram gives

$$(\sigma_{2s})^2(\sigma_{2s}^*)^2(\sigma_{2p})^2(\pi_{2p})^4$$

2 electrons in antibonding orbitals

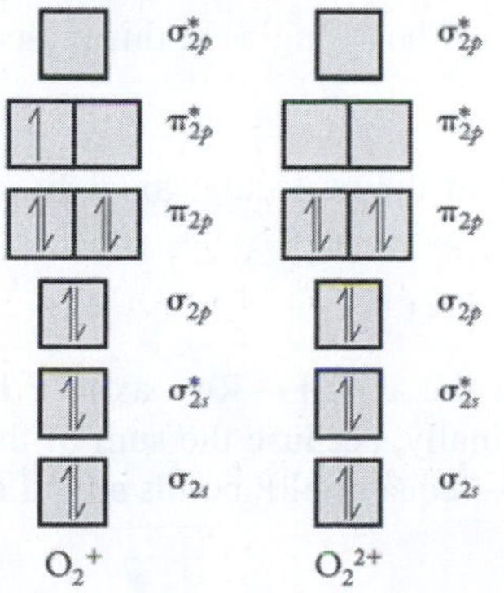

Therefore, O_2^+ has more electrons populating antibonding orbitals than O_2^{2+}.

Think About It
For those species, the bond orders are as follows:

$$O_2^+ \text{ bond order} = \tfrac{1}{2}(8-3) = 2.5$$

$$O_2^{2+} \text{ bond order} = \tfrac{1}{2}(8-2) = 3$$

5.7. **Collect and Organize**
Given the structure of a constituent of pine oil, we are to determine whether the compound is chiral.

Analyze
A compound is chiral if any of its sp^3 carbon atoms is bonded to four different atoms or groups of atoms.

Solve
Of the eight sp^3 carbon atoms in that structure, one is indeed bonded to four different groups, so the molecule is chiral.

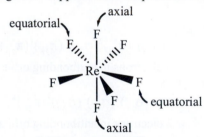

Think About It
The way to tell experimentally whether a molecule is chiral is to determine whether it rotates plane-polarized light when pure.

5.9. **Collect and Organize**
ReF_7 has a geometry in which a pentagon is "capped" on the top and bottom by atoms.

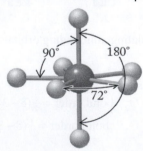

We are asked to calculate the F—Re—F bond angles in this molecule.

Analyze
Three F—Re—F angles are present in the molecule: axial F—Re—axial F, axial F—Re—equatorial F, and equatorial F—Re—equatorial F.

Solve
Looking at the diagram, we see that the axial F—Re—axial F bond is linear, so the angle is 180°. The axial F—Re—equatorial F angle is 90°. Finally, because the sum of the internal angles of the regular pentagon must add up to 360°, the equatorial F—Re—equatorial F bonds are all equal: 360°/5 = 72°.

Think About It

The equatorial fluorine atoms are quite crowded together around the Re metal center with small bond angles.

5.11. Collect and Organize

The shape of a molecule depends on the repulsions around the central atom(s) between electrons in lone pairs and bonding pairs. We are asked why repulsions between electron pairs, not between nuclei, determine molecular geometry.

Analyze

Atoms contain very small, positively charged nuclei surrounded by relatively large electron clouds.

Solve

Because the electrons take up most of the space in the atom and because the nucleus is located in the center of the electron cloud, the electron clouds repel each other before the nuclei get close enough to each other.

Think About It

All of chemistry, in essence, is due to the behavior of electrons (are they lost, gained, or shared?); that behavior, of course, is influenced by the attraction of electrons to the nucleus.

5.13. Collect and Organize

Considering that they differ by one oxygen atom, we are to explain why NO_3^- and NO_2^- have similar O–N–O bond angles.

Analyze

Bond angles are determined by molecular geometry, which is influenced by the steric number (number of lone pairs and bond pairs) around the central atom. To determine steric number, we will need to look at the Lewis structures for NO_3^- and NO_2^-.

Solve

Because the Lewis structures of both NO_3^- and NO_2^- show a steric number of 3 for the central N atoms, both ions have a trigonal planar electron-group geometry with bond angles of 120°.

Think About It

Both NO_3^- and NO_2^- have resonance structures, but resonance will not change the steric number for those molecules, so the molecular structures and bond angles will not change between resonance forms.

5.15. Collect and Organize

We are asked in this question to explain why the bond angles are different in BH_3 and NH_3.

Analyze

Molecular geometries are determined by the number of electron pairs (both in bonds and as lone pairs) around the central atom. We can draw the Lewis structures of each of these molecules to see whether they differ in the number of electron pairs around the B and N atoms.

Solve

From the Lewis structures we see that the central atoms in those two molecules have a different number steric numbers. Boron in BH_3 has three bond pairs for a steric number of 3, making its molecular geometry trigonal planar, and nitrogen has three bond pairs plus one lone pair for a steric number of 4, making its geometry trigonal pyramidal.

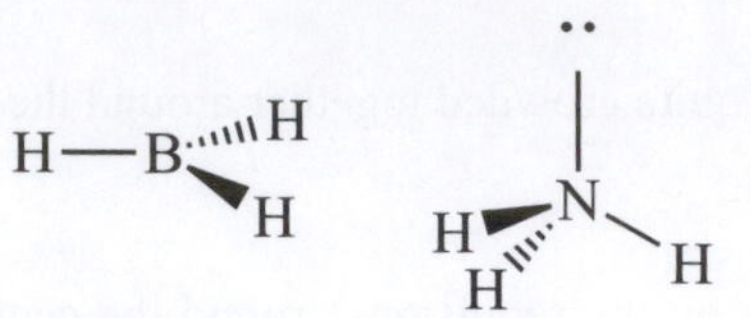

Think About It
Because of the presence of a lone pair (which takes up more space than a bond pair) on NH_3, we expect the bond angles to be slightly distorted from the ideal angle of 109.5° for the trigonal bipyramidal structure. Indeed, the H–N–H bond angle in ammonia is 107°.

5.17. Collect and Organize
We are asked to explain why the H—C—H bond angle in CH_4 is smaller than in CH_2O.

Analyze
Bond angles are determined by molecular geometry, which is influenced by the steric number (number of lone pairs and bond pairs) around the central atom. To determine the differences between CH_4 and CH_2O we will need to look at their Lewis structures.

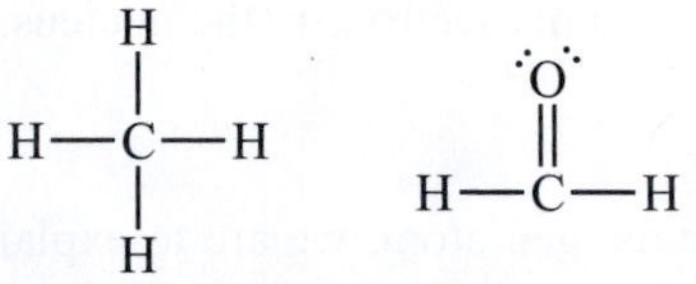

Solve
From the Lewis structures of CH_4 and CH_2O we see that the steric number of the carbon atoms in CH_4 is 4, which means a tetrahedral molecular geometry and bond angles of 109.5°. The steric number of the carbon atoms in CH_2O is 3, which means trigonal planar molecular geometry and bond angles near 120°.

Think About It
Later on in the chapter we will associate these geometries with orbital hybridizations on the carbon atom: sp^3 (tetrahedral) and sp^2 (trigonal planar).

5.19. Collect and Organize
In considering the two structures for AB_4, seesaw and trigonal pyramidal, we are to explain why the seesaw geometry has the lower energy.

Analyze
In VSEPR theory, the repulsive interactions between bonding pairs (bp) and lone pairs (lp) at 90° decrease in order as follows:

$$lp–lp \gg lp–bp \gg bp–bp$$

The lowest energy geometry has the fewest lp–lp interactions. If no lp–lp interactions are present or if they are equal for two geometries, then the geometry with the fewest lp–bp interactions has the lowest energy.

Solve

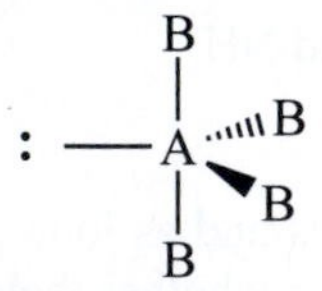

Seesaw geometry
0 lp–lp, 2 lp–bp, 4 bp–bp

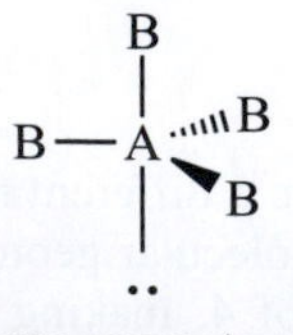

Trigonal pyramidal geometry
0 lp–lp, 3 lp–bp, 3 bp–bp

The seesaw geometry has only two lp–bp interactions at 90° (compared with three for trigonal pyramidal), so it has the lower energy.

Think About It
Remember that in comparing possible geometries by VSEPR theory, we need to only look at interactions at 90° or less.

5.21. **Collect and Organize**
We are to rank the molecules NH_2Cl, CCl_4, and H_2S in order of increasing bond angle.

Analyze
To assess the bond angles for those compounds, we need to examine their Lewis structures and molecular geometries. Each central atom in those molecules has a steric number of 4, so their electron-group geometries are all tetrahedral.

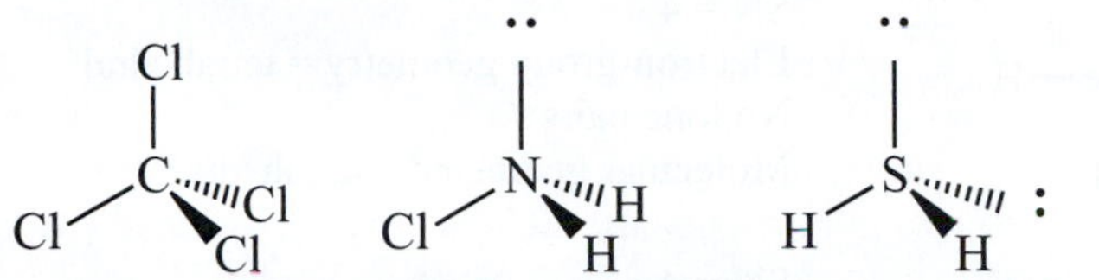

They differ, however, in the number of lone pairs on the central atom. Lone pairs take up more space than bond pairs and cause greater repulsion on neighboring bond pairs. Therefore, the more lone pairs on the central atoms, the more the bond angle between the bond pairs will be reduced.

Solve
In order of increasing bond angle, (c) H_2S < (a) NH_2Cl planar < (b) CCl_4.

Think About It
The measured bond angles are H_2S, 92.1°; NH_2Cl, 102.4°; CCl_4, 109.5°.

5.23. **Collect and Organize**
We are to determine for which electron-group geometries a linear molecular geometry is inconsistent.

Analyze
For each geometry, we will take atoms away to obtain the VSEPR geometry until we obtain a triatomic molecule and then see whether that molecular geometry is linear.

Solve
(a)

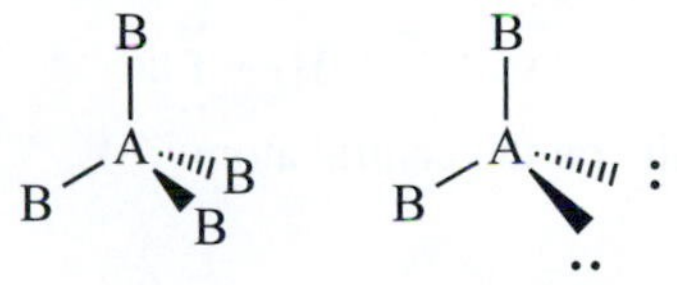

Tetrahedral Bent

(b)

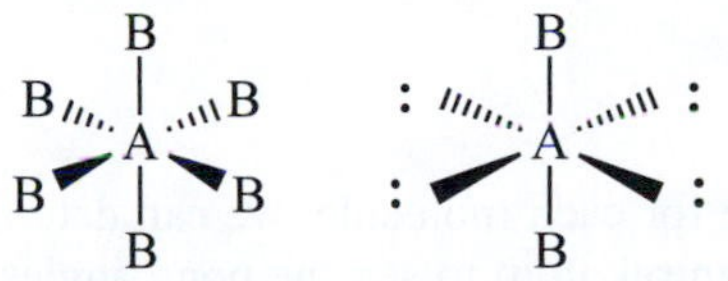

Octahedral Linear

(c)

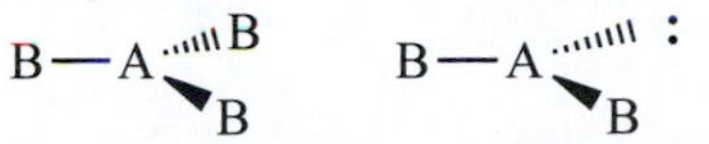

Trigonal planar Bent

Neither tetrahedral (a) nor trigonal planar (c) triatomic molecules will be linear.

Think About It
A linear triatomic molecule would also be possible for the trigonal bipyramidal electron-group geometry.

5.25. **Collect and Organize**
Using Lewis structures and VSEPR theory, we can determine the molecular geometries of GeH_4, PH_3, H_2S, and $CHCl_3$.

Analyze
After drawing the Lewis structure for each molecule, we can determine the steric number for the central atom, then locate the atoms about the central atom to see the bond angles, and finally determine the molecular shape from the location of the atoms.

Solve

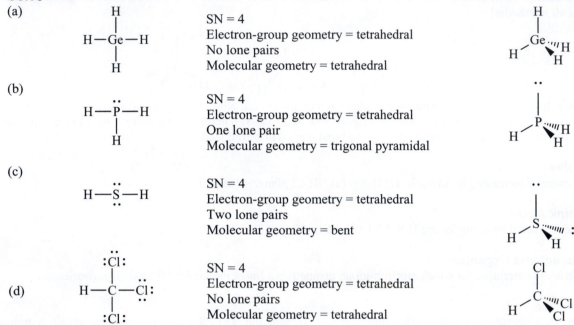

(a)

SN = 4
Electron-group geometry = tetrahedral
No lone pairs
Molecular geometry = tetrahedral

(b)

SN = 4
Electron-group geometry = tetrahedral
One lone pair
Molecular geometry = trigonal pyramidal

(c)

SN = 4
Electron-group geometry = tetrahedral
Two lone pairs
Molecular geometry = bent

(d)

SN = 4
Electron-group geometry = tetrahedral
No lone pairs
Molecular geometry = tetrahedral

Think About It
Molecules with SN = 4 may have molecular geometries of tetrahedral, trigonal pyramidal, bent, or linear—for example,

$$H—\overset{\cdot\cdot}{\underset{\cdot\cdot}{F}}:$$

depending on the number of lone pairs on the central atom.

5.27. **Collect and Organize**
Using Lewis structures and VSEPR theory, we can determine the molecular geometries of NH_4^+, CO_3^{2-}, NO_2^-, and XeF_5^+.

Analyze
After drawing the Lewis structure for each molecule, we can determine the steric number for the central atom, then locate the atoms about the central atom to see the bond angles, and finally determine the molecular shape from the location of the atoms.

Solve
(a)

SN = 4
Electron-group geometry = tetrahedral
No lone pairs
Molecular geometry = tetrahedral

(b)

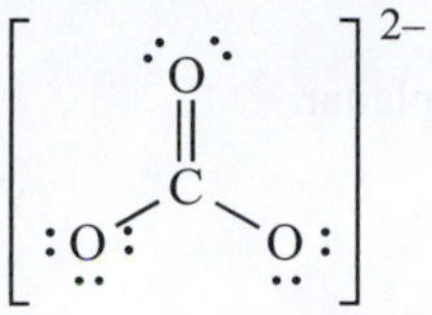

SN = 3
Electron-group geometry = trigonal planar
No lone pairs
Molecular geometry = trigonal planar

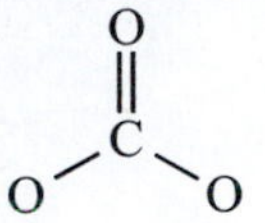

(c)

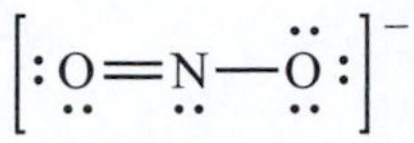

SN = 3
Electron-group geometry = trigonal planar
One lone pair
Molecular geometry = bent

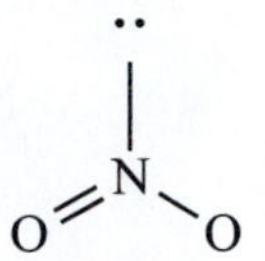

(d)

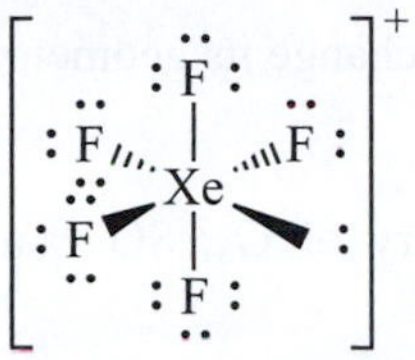

SN = 6
Electron-group geometry = octahedral
One lone pair
Molecular geometry = square pyramidal

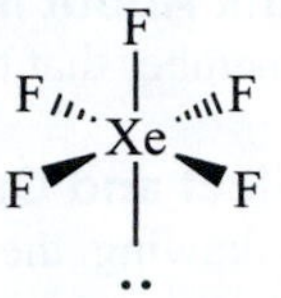

Think About It

Notice how the other resonance structure of NO_2^- also has a bent geometry because the steric number remains at 3 at the central N atom.

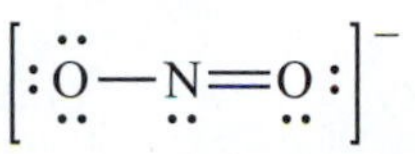

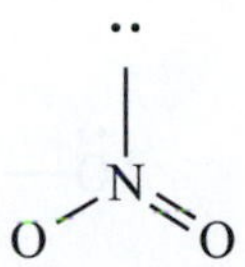

5.29. Collect and Organize

Using Lewis structures and VSEPR theory, we can determine the molecular geometries of $S_2O_3^{2-}$, PO_4^{3-}, NO_3, and NCO.

Analyze

After drawing the Lewis structure for each molecule, we can determine the steric number for the central atom, then locate the atoms about the central atom to see the bond angles, and finally determine the molecular shape from the location of the atoms.

Solve

(a)

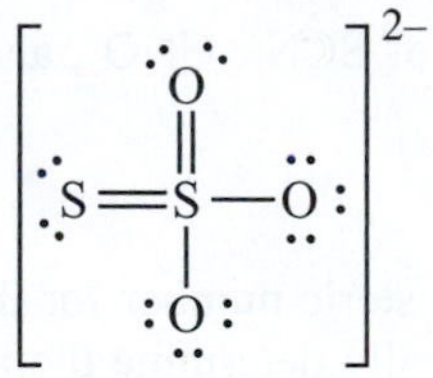

SN = 4
Electron-group geometry = tetrahedral
No lone pairs
Molecular geometry = tetrahedral

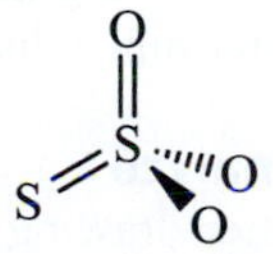

(b)

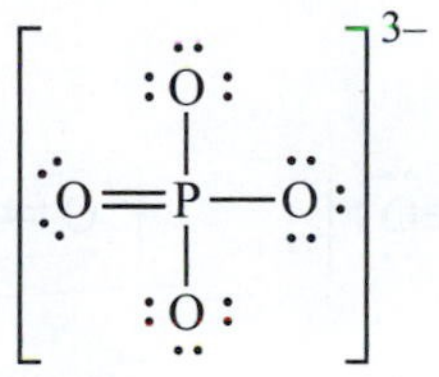

SN = 4
Electron-group geometry = tetrahedral
No lone pairs
Molecular geometry = tetrahedral

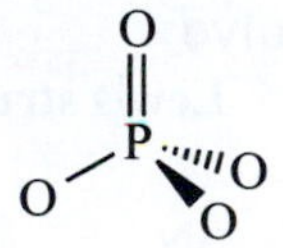

(c)

$$:\overset{..}{\underset{..}{O}}:$$

(structure with N center, double bond to O above, single bonds to two O below)

SN = 3
Electron-group geometry = trigonal planar
No lone pairs
Molecular geometry = trigonal planar

(resonance structure O=N with two O)

(d)

$$\cdot N\equiv C-\overset{..}{\underset{..}{O}}:$$

SN = 2
Electron-group geometry = linear
No lone pairs
Molecular geometry = linear

$$N-C\equiv O$$

Think About It
Remember that the presence of resonance structures for a molecule does not change its geometry.

5.31. **Collect and Organize**

By drawing the Lewis structures and determining the molecular geometry of O_3, SO_2, and CO_2, we can determine which two of these molecules have the same molecular geometry.

Analyze
After drawing the Lewis structure for each molecule, we can determine the steric number for the central atom, then locate the atoms about the central atom to see the bond angles, and finally determine the molecular shape from the location of the atoms.

Solve

Lewis structure	$:\overset{..}{\underset{..}{O}}-\overset{..}{O}=\overset{..}{O}:$	$:\overset{..}{O}=\overset{..}{S}=\overset{..}{O}:$	$:\overset{..}{O}=C=\overset{..}{O}:$
SN	3	3	2
Electron-group geometry	Trigonal planar	Trigonal planar	Linear
Number of lone pairs	1	1	0
Molecular geometry	Bent	Bent	Linear
Bond angle	<120°	<120°	180°

Both O_3 and SO_2 are bent with ~120° angles and have the same molecular geometry.

Think About It
At first those triatomic molecules may all appear to have the same geometry from their formulas. Be careful to draw correct Lewis structures because the presence of lone pairs on the central atom is important in determining overall geometry.

5.33. **Collect and Organize**

By drawing the Lewis structures and determining the molecular geometry of SCN^-, CNO^-, and NO_2^-, we can determine which two of these ions have the same molecular geometry.

Analyze
After drawing the Lewis structure for each molecule, we can determine the steric number for the central atom, then locate the atoms about the central atom to see the bond angles, and finally determine the molecular shape from the location of the atoms.

Solve

Lewis structure	$\left[:\overset{..}{\underset{..}{S}}=C=\overset{..}{\underset{..}{N}}:\right]^-$	$\left[:C\equiv N-\overset{..}{\underset{..}{O}}:\right]^-$	$\left[:\overset{..}{\underset{..}{O}}=N-\overset{..}{\underset{..}{O}}:\right]^-$
SN	2	2	3
Electron-group geometry	Linear	Linear	Trigonal planar
Number of lone pairs	0	0	1
Molecular geometry	Linear	Linear	Bent
Bond angle	180°	180°	<120°

Both SCN^- and CNO^- are linear and therefore have the same molecular geometry.

Think About It
The Lewis structures shown are the resonance structures that, by formal charge arguments, contribute most to the bonding. The other resonance structures for each molecule have the same molecular structure as for the resonance structure shown.

5.35. Collect and Organize
From the Lewis structures of S_2O and S_2O_2, we can use VSEPR theory to determine the molecular geometry of those two compounds detected in the atmosphere of Venus.

Analyze
After drawing the Lewis structure for each molecule, we can determine the steric number for the central atom, then locate the atoms about the central atom to see the bond angles, and finally determine the molecular shape from the location of the atoms. S_2O_2 has two possible geometries: one with two central S atoms and one with one central S atom.

Solve

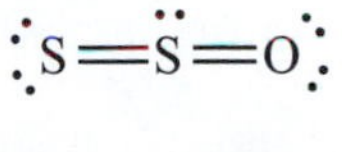

SN = 3
Electron-group geometry = trigonal planar
One lone pair
Molecular geometry = bent

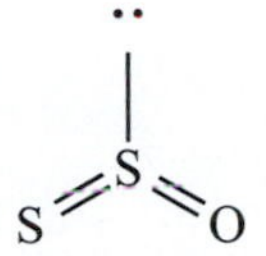

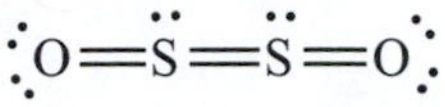

On each S, SN = 3
Electron-group geometry = trigonal planar
One lone pair
Molecular geometry = bent at each S atom
or S_2O_2 may have only one central S atom

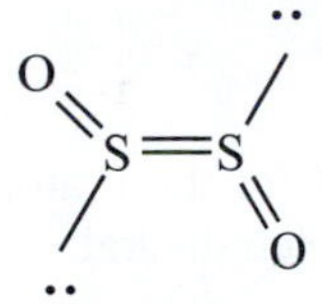

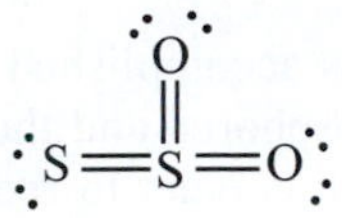

SN = 3
Electron-group geometry = trigonal planar
No lone pairs
Molecular geometry = trigonal planar

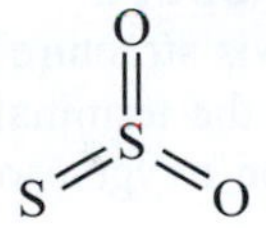

Think About It
S_2O_2 could also have the geometry

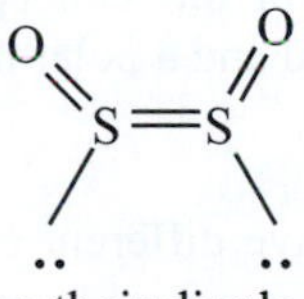

We could differentiate those isomers by measuring their dipole moments.

5.37. Collect and Organize
After drawing the Lewis structure for XeF_5^- we can use VSEPR theory to predict its molecular structure.

Analyze
Xenon brings eight electrons, a closed-shell configuration, so it must expand its octet (as we saw in Chapter 4) to bond with F atoms in XeF_5^-.

Solve

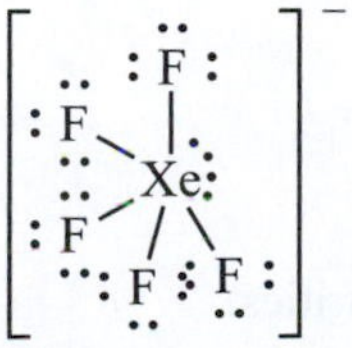

SN = 7
Electron-group geometry = pentagonal bipyramidal
Two lone pairs
Molecular geometry = pentagonal planar

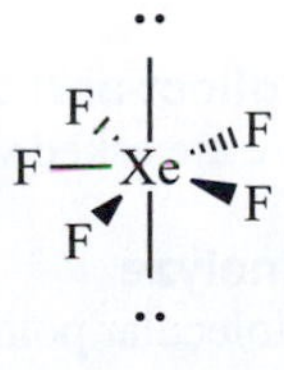

Think About It
Placing the lone pairs in the axial positions of the pentagonal bipyramid gives the lowest energy geometry because no lp–lp interactions occur for that structure.

5.39. **Collect and Organize**
Using the given skeletal structure of sarin (Figure P5.39), we can complete the Lewis structure, assign formal charges to the P and O atoms, and then predict the geometry around P by using VSEPR theory.

Analyze
The molecular formula of sarin is $C_4H_{10}FO_2P$, which has 50 e$^-$ and needs 84 e$^-$ to complete the octets (and duets) on all the atoms. That gives a difference of 34 e$^-$ for 17 covalent bonds and leaves 16 e$^-$ in eight lone pairs to complete the Lewis structure with one lone pair forming a P—O double bond.

Solve

SN = 4 for the P atom in this molecule, and no lone pairs are present, so the geometry around the P atom in sarin is tetrahedral.

Think About It
The Lewis structure drawn is for all atoms with the lowest formal charge. That was accomplished by double-bonding the terminal oxygen to phosphorus to reduce the +1 formal charge on phosphorus and the −1 formal charge on oxygen to zero. Remember from Chapter 4 that phosphorus may expand its octet to reduce formal charge.

5.41. **Collect and Organize**
Both molecules and bonds may be polar. For this concept review question, the definitions of the two terms allow us to differentiate between a polar bond and a polar molecule.

Analyze
A bond is polar when two bonded atoms have different electronegativities. Molecular polarity is the result of bond polarity and molecular geometry.

Solve
A polar bond occurs only between two atoms in a molecule. The more electronegative atom in the bond carries a partial negative charge and the least electronegative atom in the bond carries a partial positive charge. Molecular polarity takes into account all the individual bond polarities and the geometry of the molecule. A polar molecule has a permanent, measurable dipole moment.

Think About It
To determine molecular polarity we have to first determine the individual bond polarities.

5.43. **Collect and Organize**
We are asked whether a nonpolar molecule may contain polar covalent bonds.

Analyze
Molecular polarity is determined by adding the vectors of the individual bond polarities.

Solve

Yes. As long as the individual bond polarities are equal in magnitude and opposite in direction (as vectors), a molecule may be nonpolar overall even if the bonds themselves are polar.

Think About It

If the bond polarities (as vectors) do not cancel, then the molecule will be polar.

5.45. Collect and Organize

Given the structure of vinyl chloride in Figure P5.45, we are to identify the most polar bond.

Analyze

The most polar bond will be the one with the largest difference in electronegativity between the bonded atoms. The electronegativity of the atoms in vinyl chloride are: C, 2.5; H, 2.1; Cl, 3.0.

Solve

The differences in electronegativities between the bonded atoms are C–H, 0.4; C–C, 0.0; C–Cl, 0.5. Therefore, the most polar bond in vinyl chloride is C–Cl.

Think About It

That the C–H bond is nearly as polar as the C–Cl bond might have surprised you.

5.47. Collect and Organize

We can look at the bond polarities and the molecular structure by VSEPR theory to determine which molecules (CCl_4, $CHCl_3$, CO_2, H_2S, and SO_2) are polar and which are nonpolar.

Analyze

All the individual bonds in those molecules are polar, so the molecular geometry of each compound will determine the overall molecular polarity. We can represent each bond polarity with a vector, with the head of the arrow pointed towards the more electronegative atom, which carries a partial negative charge. We then visually inspect the molecule to see whether the individual bond dipoles add up or cancel out.

Solve

(a)

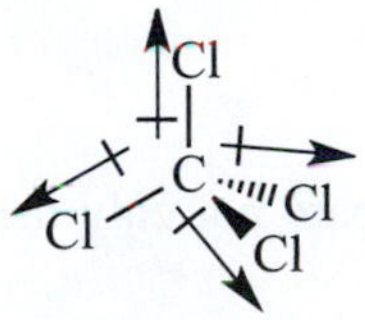

All bond polarities are equal in magnitude and cancel each other, so CCl_4 is nonpolar.

(b)

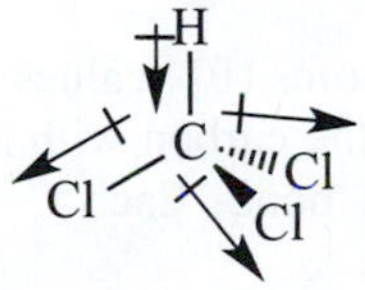

The electronegativity of the atoms in $CHCl_3$ are in order Cl > C > H. Because the bond polarities do not cancel, $CHCl_3$ is polar.

(c)

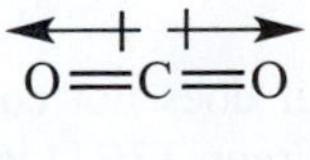

The bond polarities in CO_2 cancel, so it is nonpolar.

(d)

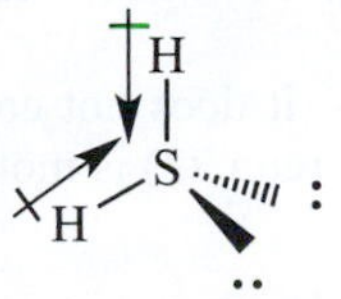

The molecular geometry of H_2S is bent, so it is polar.

(e)

The molecular geometry of SO$_2$ is bent, so it is polar.

Polar molecules are (b) CHCl$_3$, (d) H$_2$S, and (e) SO$_2$. Nonpolar molecules are (a) CCl$_4$ and (c) CO$_2$.

Think About It
Molecules with polar bonds are nonpolar only for highly symmetrical geometries (linear, trigonal planar, tetrahedral, trigonal bipyramidal, and octahedral).

5.49. Collect and Organize
By looking at the molecular structures and individual bond polarities present in CFCl$_3$, CF$_2$Cl$_2$, and Cl$_2$FCCF$_2$Cl, we can determine which of these CFCs are polar and which are nonpolar.

Analyze
To determine bond polarities, we need the electronegativity values for C (2.5), F (4.0), and Cl (3.0). In each of those molecules the halogens are bonded to the carbon atoms. Because the electronegativities of the halogens are higher than that of carbon, the bonds are polarized so that the halogen carries a partial negative charge.

Solve

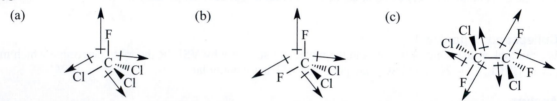

All those molecules have a tetrahedral geometry around the carbon atoms. Because of the different bond polarities of C—F and C—Cl, however, none of the molecules have bond dipoles that cancel. All those molecules (a–c) are polar.

Think About It
Only molecules with completely symmetric geometries with all the same atoms attached to the central atoms are nonpolar.

5.51. Collect and Organize
To determine which molecule is more polar in each pair given, we need to compare not only the molecules' geometries but also the magnitude of the individual bond dipoles.

Analyze
We predict the bond polarity by the difference in electronegativity of the bonded atoms (EN values are C = 2.5, Cl = 3.0, F = 4.0, H = 2.1, Br = 2.8). All the molecules have halogens bonded to the carbon with a tetrahedral geometry. The more electronegative the atom on carbon, the more polar is that bond. Each C—X bond is polarized so that the halogen carries a partial negative charge.

Solve
(a) Because the Br in CBrF$_3$ has a lower electronegativity than the Cl in CClF$_3$, it does not counteract the electron pull from the three F atoms on the C as well as a Cl atom, so CBrF$_3$ (Freon 13B1) is more polar than CClF$_3$.
(b) Because the H in CHF$_2$Cl has a lower electronegativity than the Cl in CF$_2$Cl$_2$, it does not counteract the electron pull from the two F atoms on the C as well as a Cl atom, so CHF$_2$Cl (Freon 22) is more polar than CF$_2$Cl$_2$.

Think About It
In molecules where all the bonds are polarized in the same direction (X is partially negative), the replacement of one atom by a less electronegative atom results in a more polar molecule.

5.53. **Collect and Organize**
For each of the COX_2 molecules (X = I, Br, Cl), we can use the different electronegativities of the atoms in the molecule along with the molecular geometry to place the molecules in order of increasing molecular polarity.

Analyze
The carbonyl dihalides have a trigonal planar geometry. The electronegativities of the elements in those compounds are C = 2.5, O = 3.5, Cl = 3.0, Br = 2.8, I = 2.5.

Solve
The greatest electronegativity difference in the C—X bond is between C and Cl. The least electronegativity difference in the C—X bond is between C and I. Therefore those compounds in order of increasing polarity of the C—X bond are COI_2 (ΔEN C—I = 0) < $COBr_2$ (ΔEN C—Br = 0.3) < $COCl_2$ (ΔEN C—Cl = 0.5).

Think About It
To compare the overall molecular polarity for those compounds we would have to examine how the C—X bond polarity pulls opposite the polarity of C=O bond (ΔEN = 1.0). As the electronegativity of the halogen atom decreases (Cl > Br > I), the overall molecular polarity of the COX_2 molecule increases because the C—X bonds pull less to balance the C=O bond dipole. In order of increasing molecular polarity, $COCl_2$ < $COBr_2$ < COI_2.

5.55. **Collect and Organize**
In this question we consider how atomic orbitals mix to form hybrid orbitals. Specifically, we consider what atomic orbitals need to have in common in order to mix.

Analyze
The types of hybrid orbitals we have seen in the textbook include mixing of the s and p orbitals to form sp, sp^2, and sp^3 orbitals. In making hybrid orbitals we can consider requirements of size (energy) and orientation.

Solve
Atomic orbitals will hybridize if they are of similar size (energy), which means that they should be of the same principal quantum number. Also, they must be oriented to overlap and hybridize; if the two atomic orbitals are pointed away from each other in space, they could not hybridize.

For hybridization, atomic orbitals should be of similar size:

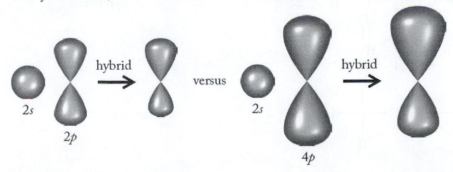

For hybridization, atomic orbitals should have the correct orientation for overlap:

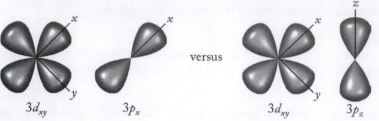

These can overlap to form hybrid orbital.	These cannot overlap to form hybrid orbital.

Think About It
Hybridization of orbitals can also be thought of as adding or overlapping the wave functions of two atomic orbitals, and hybridization often results in a new shape for the orbital to use in bonding.

5.57. Collect and Organize
From the steric number obtained from the Lewis structures of the hydrocarbons listed we can determine the hybridization of each of the underlined carbon atoms.

Analyze
Hybridization is directly obtained from the steric number (SN). When SN = 2, the hybridization is sp; when SN = 3, the hybridization is sp^2; when SN = 4, the hybridization is sp^3; when SN = 5, the hybridization is sp^3d; when SN = 6, the hybridization is sp^3d^2.

Solve

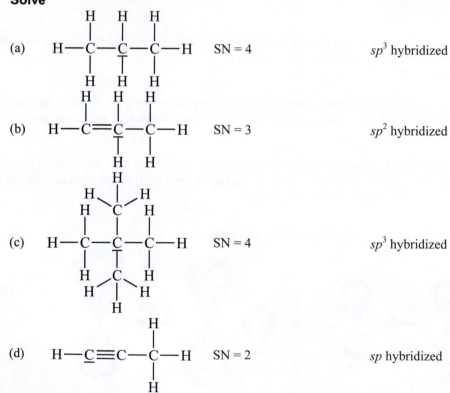

(a)   SN = 4 sp^3 hybridized

(b)   SN = 3 sp^2 hybridized

(c)   SN = 4 sp^3 hybridized

(d)   SN = 2 sp hybridized

(e) 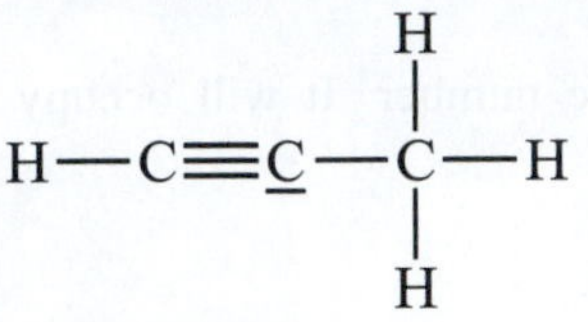 $\quad$ H$-$C$\equiv$C$-$C$-$H $\quad$ SN = 2 $\qquad$ *sp* hybridized

(with H atoms above and below the third C)

Think About It
The molecular geometries for each of these hybridized orbitals are tetrahedral (sp^3), trigonal planar (sp^2), and linear (sp).

5.59. Collect and Organize
From the two Lewis structures of N_2F_2 we are shown in Figure P5.59, we can determine whether differences in structure are due to the hybridization on the N atoms in each structure. Second, we will compare the hybridization of the N in that molecule with the hybridization of carbon in C_2H_2.

Analyze
The steric number around the N and C atoms in the Lewis structures indicates the hybridization. When SN = 2, the hybridization is *sp*; when SN = 3, the hybridization is sp^2; when SN = 4, the hybridization is sp^3.

Solve
Both Lewis structures of N_2F_2 show a steric number of 3 for the N atoms, so both structures have sp^2 hybridized orbitals on N. The difference in the structures lies in the relative placement of the F atoms in space. The steric number on each F atom is 4, so each F atom is sp^3 hybridized. Acetylene, C_2H_2, shows SN = 2 for the carbon atoms in its Lewis structure, so they are *sp* hybridized and not sp^2 hybridized as for N in N_2F_2.

$$\text{H}-\text{C}\equiv\text{C}-\text{H}$$

Think About It
The structures shown for N_2F_2 are not resonance forms, but they are isomers. They differ in the arrangement of the atoms in space, and the two isomers can be separated on the basis of their difference in polarity.

5.61. Collect and Organize
Using Lewis structures and the steric number around the central atom, we can determine how orbital hybridization changes in CO_2, NO_2, O_3, and ClO_2.

Analyze
We draw the Lewis structures in the usual way. Hybridization is directly obtained from the steric number (SN). When SN = 2, the hybridization is *sp*; when SN = 3, the hybridization is sp^2; when SN = 4, the hybridization is sp^3; when SN = 5, the hybridization is sp^3d; when SN = 6, the hybridization is sp^3d^2.

Solve

$\ddot{\text{O}}=\text{C}=\ddot{\text{O}}$	$\ddot{\text{O}}-\dot{\text{N}}=\ddot{\text{O}}$	$\ddot{\text{O}}-\ddot{\text{O}}=\ddot{\text{O}}$	$\ddot{\text{O}}=\ddot{\text{Cl}}=\ddot{\text{O}}$
			SN = 4
SN = 2	SN = 3	SN = 3	"sp^3" according to
sp	sp^2	sp^2	steric number alone

From our study of chemistry so far, the hybridization of Cl in ClO_2 is assigned as sp^3. We have to think here, though. We need to form a π bond from the Cl to both oxygen atoms, which requires that two of the *p* orbitals on Cl should not be involved in the hybridization, so that they can form parallel π bonds with the oxygens. Therefore, Cl must use low-lying *d* orbitals in place of the *p* orbitals for spd^2 hybridization to form the two σ bonds to the oxygens and to "hold" the lone pair and the electron.

Think About It

An unpaired electron in NO_2 and ClO_2 "counts" as one for the steric number. It will occupy one of the hybridized orbitals.

5.63. Collect and Organize

After drawing the Lewis structures for ClO_4^- for which formal charges are minimized, we can determine the molecular shape and the hybridization on the central Cl atom.

Analyze

We draw the Lewis structures in the usual way. The structure contributing the most to the actual geometry of the molecule is that with the lowest formal charges and with the most electronegative atom (here, oxygen) carrying a negative formal charge if necessary.

Solve

The first resonance structure has the best formal charge arrangement. The steric number for Cl in that structure is 4, which gives a tetrahedral molecular geometry. At first glance that would also mean that the hybridization would be assigned as sp^3. However, Cl forms three π bonds to three of the oxygen atoms, which requires that three of the p orbitals on Cl should not be involved in the hybridization, so that it can form parallel π bonds. Therefore, Cl must use low-lying d orbitals in place of the p orbitals for sd^3 hybridization to form the four σ bonds to oxygen.

Think About It

On central atoms with expanded octets, we need to use d orbitals in place of p orbitals for the σ-bonded hybrid orbitals so as to leave the unhybridized p orbitals available for π bonding.

5.65. Collect and Organize

From the Lewis structure of HArF, we can use the steric number to determine the molecular geometry and the hybridization of the central Ar atom.

Analyze

We draw the Lewis structure in the usual way. HArF has 16 e^- and needs 4 e^- to form the two covalent bonds from Ar to F and H. That leaves 12 e^- in six lone pairs on Ar and F to complete the structure. Because HArF is a compound of a noble gas, we expect Ar to have to an expanded octet. Hybridization is directly obtained from the steric number (SN). When SN = 2, the hybridization is sp; when SN = 3, the hybridization is sp^2; when SN = 4, the hybridization is sp^3; when SN = 5, the hybridization is sp^3d; when SN = 6, the hybridization is sp^3d^2.

Solve

$H-\ddot{A}r-\ddot{F}:$

SN = 5
sp^3d hybridized
Electron-pair geometry = trigonal bipyramidal
Molecular geometry = linear
H—Ar—F bond angle = 180°

Think About It

This question has placed many of the components of structure and bonding theories together—Lewis structure, VSEPR theory, and valence bond theory—to fully describe the molecular structure of HArF.

5.67. Collect and Organize

After drawing resonance structures for N₂O we can compare the hybridization of the central N atom among the structures.

Analyze

N_2O has 16 e⁻ and needs 24 e⁻ to complete the octets on all the atoms. The difference of 8 e⁻ corresponds to four covalent bonds. That leaves 8 e⁻ in four lone pairs to complete the structure.

Solve

$$:\overset{-1}{N}=\overset{+1}{N}=\overset{0}{O}: \longleftrightarrow :\overset{-2}{N}-\overset{+1}{N}\equiv\overset{+1}{O}: \longleftrightarrow :\overset{0}{N}\equiv\overset{+1}{N}-\overset{-1}{O}:$$

For each resonance structure the central N atom has a steric number of 2, so yes, all those resonance structures have the central N atom as *sp* hybridized.

Think About It

If we had a change in hybridization, we would also have a change in structure (how the atoms are arranged in space).

5.69. Collect and Organize

Using the Lewis structure of SOF_3^-, we can determine the geometry of the anion and describe the bonding by valence bond theory.

Analyze

SOF_3^- has 34 e⁻ and needs 40 e⁻ to complete the octets on all the atoms. To connect the atoms we need four bonds that use 8 e⁻, leaving 26 e⁻ in 13 lone pairs.

Solve

The Lewis structure of the SOF_3^- anion is below, where the sulfur atom has expanded its octet to hold 10 electrons.

Because the sulfur atom has expanded its octet to 10 electrons, the hybridization of the S atom in that anion is sp^3d.

Think About It

The geometry of the SOF_3^- anion is see-saw.

5.71. Collect and Organize

We consider in this question whether a molecule with more than one central atom can have resonance forms.

Analyze

Resonance forms are Lewis structures that show alternative (yet still valid) electron distributions in a molecule.

Solve

We have already seen several examples of molecules that have several central atoms and that have multiple resonance forms in Chapter 4. In that chapter, a good example of resonance in a molecule with more than one central atom is benzene. So, yes, molecules with more than one central atom can indeed have resonance forms.

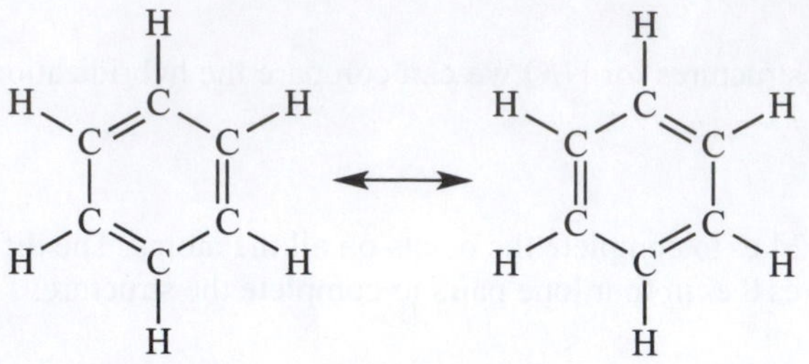

Think About It

As long as another way exists to distribute the electrons in a molecule, we have resonance.

5.73. Collect and Organize

We are asked to explain whether resonance structures are examples of delocalized electrons in a molecule.

Analyze

Resonance structures show different possible electron distributions over the atoms in the molecule. Each one contributes to the actual structure of the molecule.

Solve

To obtain the actual molecular structure, we mix the resonance forms. The molecule does not exist in one form at one instant and another form the next. The electron distribution is blurred across all the resonance forms, which in essence defines the delocalization of electrons.

Think About It

Resonance forms help us see which atoms and bonds are involved in sharing the delocalized electrons.

5.75. Collect and Organize

We are given the skeletal arrangement for the nitramide molecule. We first need to complete the Lewis structure for the molecule to describe the geometry and hybridization around both N atoms to see whether they are the same.

Analyze

H_2NNO_2 has 24 e^- and needs 36 e^-, giving a difference of 12 e^- in six covalent bonds and leaving 12 e^- in six lone pairs to complete the structure.

Solve

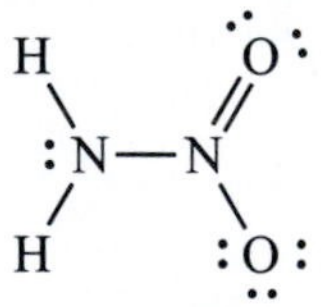

One N atom has SN = 4 with one lone pair, so it has trigonal pyramidal geometry and is sp^3 hybridized. The other N atom has SN = 3 with no lone pairs, so it has trigonal planar geometry and is sp^2 hybridized. No, the hybridization of both N atoms is not the same.

Think About It

Only the resonance structure with the lowest formal charges on the atoms is shown above.

5.77. Collect and Organize

We are given the skeletal arrangement for the sulfamate ion. We first need to complete the Lewis structure for the molecule to describe the geometry and hybridization around the S and N atoms. We are asked which atomic or hybrid orbitals overlap to form the S—O and S—N bonds.

Analyze

$SO_3NH_2^-$ has 32 e^- and needs 44 e^-, giving a difference of 12 e^- in six covalent bonds and leaving 20 e^- in 10 lone pairs to complete the structure. To reduce formal charges on the atoms, sulfur may expand its octet.

Solve

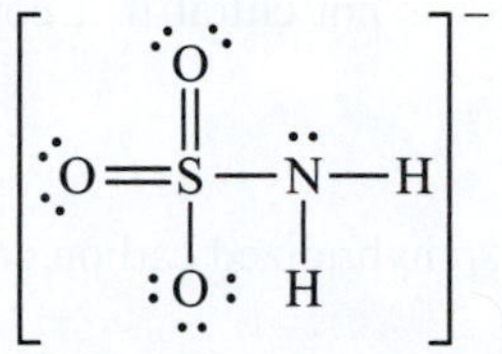

$$\left[\; \begin{array}{c} \ddot{O} \\ \parallel \\ :\!\ddot{O}=\!S\!-\!\ddot{N}\!-\!H \\ \mid \quad \mid \\ :\!\ddot{O}: \quad H \end{array} \;\right]^{-}$$

Both the S and N atoms have SN = 4 for an electron-pair geometry of tetrahedral. The presence of a lone pair on N gives that atom trigonal pyramidal geometry and the nitrogen atom is sp^3 hybridized. The steric number for S is also 4, which at first glance would also mean that the hybridization would be assigned as sp^3. However, the S forms two π bonds to two of the oxygen atoms, which requires that two p orbitals on S not be involved in the hybridization so that it can form parallel π bonds. Therefore, S must use two low-lying d orbitals in place of two of the p orbitals for spd^2 hybridization to form the four σ bonds to oxygen and nitrogen.

Think About It

On central atoms with expanded octets we need to use d orbitals in place of p orbitals for the σ-bonded hybrid orbitals so as to leave the unhybridized p orbitals available for π bonding.

5.79. **Collect and Organize**

Given the structure shown in Figure P5.79, we are to determine the molecular geometry and angle around each carbon atom.

Analyze

That structure has two types of carbon atoms: a carbon bonded to three hydrogen atoms and another carbon atom and a carbon atom doubly bonded to an oxygen atom and two other carbon atoms. The full Lewis structure and analysis of the electron pair geometry will help us assign the geometries and bond angles for those carbons.

Solve

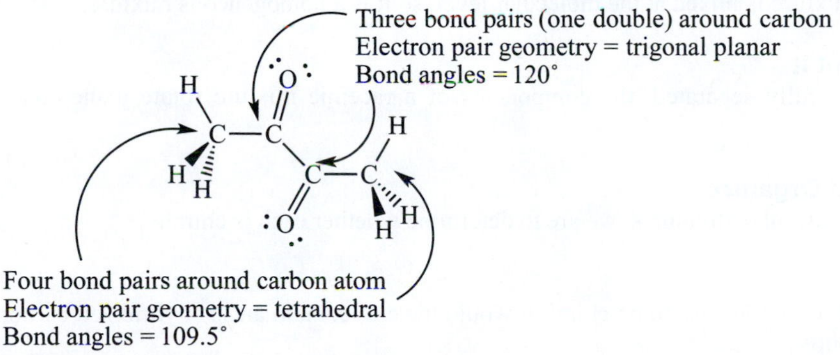

Three bond pairs (one double) around carbon
Electron pair geometry = trigonal planar
Bond angles = 120°

Four bond pairs around carbon atom
Electron pair geometry = tetrahedral
Bond angles = 109.5°

Think About It

All four atoms in trigonal planar geometry lie in the same plane.

5.81. **Collect and Organize**

Given some ordinary objects, we are to determine which are chiral.

Analyze

A chiral object is not superimposable on its mirror image.

Solve

A spoon (b) (if we ignore any design on the handle) is superimposable on its mirror image, but (a) a golf club, (c) a glove, and (d) a shoe are not, and so they are chiral objects.

Think About It

You might also learn later that an object is not chiral if it contains a plane of symmetry or has an inversion center.

5.83. **Collect and Organize**

In this question we consider whether an *sp* hybridized carbon center could be chiral.

Analyze

A chiral object is not superimposable on its mirror image.

Solve

When a carbon atom is *sp* hybridized, the SN = 2 and therefore the geometry around that carbon is linear. A carbon center is chiral when it has four different groups bonded to it. An *sp* hybridized carbon, however, has only two atoms or groups bonded to it, so it cannot be a chiral center.

Think About It

You might also learn later that an object is not chiral if it contains a plane of symmetry or has an inversion center, but the general rule of chirality for carbon atoms as having four different groups bonded to it is a reliable way to determine whether a particular carbon atom in a molecule is a chiral center.

5.85. **Collect and Organize**

We are to determine whether a racemic mixture is a homogeneous or heterogeneous mixture.

Analyze

For a mixture to be heterogeneous, we must be able to discern by eye (or with a microscope) the different components of the mixture. A racemic mixture is a mixture of two enantiomers.

Solve

A racemic mixture is mixed at the molecular level, so it is a homogeneous mixture.

Think About It

When successfully separated, the components of a racemic mixture rotate plane-polarized light in opposite directions.

5.87. **Collect and Organize**

Given four molecular structures, we are to determine whether each is chiral.

Analyze

For any of those molecules to be chiral, it would have to contain an sp^3 hybridized carbon atom bonded to four different groups.

Solve

(a) Chiral

(b) Not chiral   The carbon atoms either are not sp^3 hybridized (the doubly bonded C atoms) or do not have four different substituents (the –CH_3 groups).

(c) Chiral

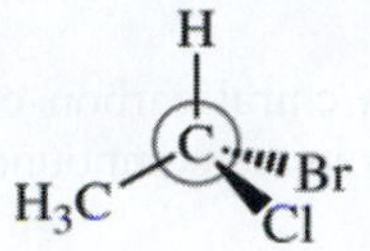

(d) Not chiral — One carbon atom is not sp^3 hybridized (the doubly bonded C═O), and the other carbon atom does not have four different substituents (the –CH$_3$ group).

Think About It
Remember to look carefully for four different substituents on the sp^3 hybridized carbons in the structures.

5.89. Collect and Organize
From the line drawings of three carboxylic acids (Figure P5.89), we are to determine which are chiral.

Analyze
A molecule is chiral if it has at least one carbon atom bonded to four different groups.

Solve
Only molecule (a) has a chiral carbon center and so is the only molecule shown that is chiral:

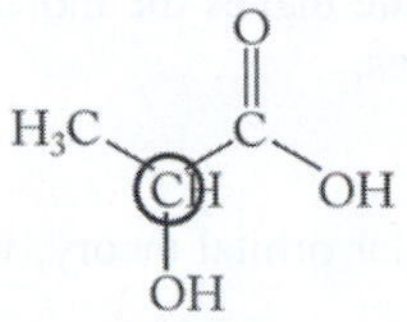

Think About It
Even though one of the carbon atoms in this molecule is bonded to three different groups (═O, –OH, and the –CH(CH$_3$)OH group, this carbon is not a chiral carbon center because it is not sp^3 hybridized, but rather sp^2 hybridized.

5.91. Collect and Organize
In each structure in Figure P5.91, we are to circle the chiral centers.

Analyze
Wherever in the molecule a carbon is bonded to four different groups, a chiral center exists.

Solve

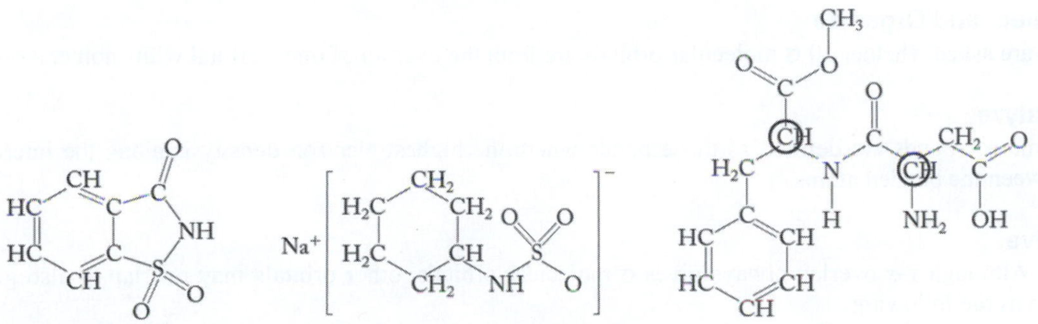

Saccharin   Sodium cyclamate   Aspartame

Think About It
Because the ring of carbon atoms in sodium cyclamate is symmetrical, the carbon to which the –NHSO$_3^-$ group is bound is not chiral.

5.93. Collect and Organize
Optical isomers are nonsuperimposable mirror images that contain a chiral carbon center with four different groups bonded to the carbon. We are asked to identify the chiral center in the compound in Figure P5.93.

Analyze
Wherever in the molecule a carbon is bonded to four different groups, a chiral center exists.

Solve

Think About It
Having even one chiral center in a molecule makes the molecule chiral, and each enantiomer will rotate plane-polarized light in equal but opposite degrees.

5.95. Collect and Organize
Between valence bond theory and molecular orbital theory, we are asked which better explains emission in the visible range.

Analyze
Emission involves the relaxation of an electron from a higher energy orbital (atomic or molecular) to a lower energy orbital.

Solve
Molecular orbital theory better explains the emission of light from a molecule because it describes electronic energy levels in its theory.

Think About It
Valence bond theory is better than molecular orbital theory in describing bond angles and molecular geometry.

5.97. Collect and Organize
We are asked whether all σ molecular orbitals are from the overlap of one *s* orbital with another *s* orbital.

Analyze
Sigma (σ) bonds are defined as those bonds where the highest electron density is along the internuclear axis between the bonded atoms.

Solve
No. Although *s–s* overlap always gives σ molecular orbitals, other orbitals may overlap to also give σ bonds such as the following:

s + p orbital   *p + p* orbital $d_{z^2} + p$ orbital

Think About It
When those atomic orbitals mix to form molecular orbitals, remember that two molecular orbitals form: the sigma bonding (σ) and the sigma antibonding (σ*) orbitals.

5.99. Collect and Organize

When an *s* orbital overlaps with another *s* orbital, a sigma (σ) bond forms. We are to consider the effectiveness of the *s*–*s* orbital overlap between orbitals of different *n* values.

Analyze

A difference between a 1*s* orbital and a 2*s* orbital, for example, is the volume that the electrons occupy. A 2*s* orbital is larger than a 1*s* orbital.

Solve

No. The overlap of 1*s* and 2*s* orbitals is not as efficient as 1*s*–1*s* or 2*s*–2*s* overlaps. The mismatch in size and energy is poor.

Think About It

One guideline for molecular orbital diagrams for homonuclear diatomic molecules states that the better mixing of orbitals from the same *n* level leads to greater bond stabilization.

5.101. Collect and Organize

We are to make a sketch to show the overlap of two 1*s* orbitals to form σ_{1s} and σ_{1s}^* molecular orbitals.

Analyze

When mixing atomic orbitals (AOs) to give molecular orbitals (MOs), the number of MOs equals the number of AOs. Here we obtain two MOs because we are mixing two 1*s* orbitals. One MO is bonding (lower in energy) and the other is antibonding (higher in energy).

Solve

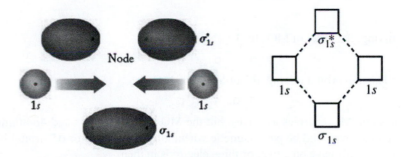

Think About It

That molecular orbital diagram is appropriate for neutral and ionic species of H_2 and He_2.

5.103. Collect and Organize

For the species N_2^+, O_2^+, C_2^+, and Br_2^{2-}, we are to place electrons into the appropriate molecular orbital (MO) energy levels to predict the bond order for each diatomic molecule.

Analyze

Because of *s*–*p* orbital mixing, the order of MOs for Li_2–N_2 is

$$\sigma_{2s}\sigma_{2s}^*\pi_{2p}\sigma_{2p}\pi_{2p}^*\sigma_{2p}^*$$

For O_2–Ne_2, which have less *s*–*p* orbital mixing, the order of the MOs is

$$\sigma_{2s}\sigma_{2s}^*\sigma_{2p}\pi_{2p}\pi_{2p}^*\sigma_{2p}^*$$

For each species we fill the MO energy levels from lowest to highest energy with the total number of electrons. The bond order (BO) is calculated from

$$\text{BO} = \tfrac{1}{2}(\text{number of e}^- \text{ in bonding MOs} - \text{number of e}^- \text{ in antibonding MOs})$$

For Br_2^{2-} we assume the same MO energies as for F_2, but the MOs involve the 4*s* and 4*p* atomic orbitals.

Solve

N_2^+ Total number of electrons = 9 e⁻

$$(\sigma_{2s})^2(\sigma_{2s}^*)^2(\pi_{2p})^4(\sigma_{2p})^1$$

BO $= \frac{1}{2}(7-2) = 2.5$

O_2^+ Total number of electrons = 11 e⁻

$$(\sigma_{2s})^2(\sigma_{2s}^*)^2(\sigma_{2p})^2(\pi_{2p})^4(\pi_{2p}^*)^1$$

BO $= \frac{1}{2}(8-3) = 2.5$

C_2^+ Total number of electrons = 7 e⁻

$$(\sigma_{2s})^2(\sigma_{2s}^*)^2(\pi_{2p})^3$$

BO $= \frac{1}{2}(5-2) = 1.5$

Br_2^{2-} Total number of electrons = 16 e⁻

$$(\sigma_{4s})^2(\sigma_{4s}^*)^2(\sigma_{4p})^2(\pi_{4p})^4(\pi_{4p}^*)^4(\sigma_{4p}^*)^2$$

BO $= \frac{1}{2}(8-8) = 0$

All species with nonzero bond order (N_2^+, O_2^+, and C_2^+) are expected to exist.

Think About It

N_2^+ and O_2^+ have the same bond order but very different MO filling. N_2^+ has two fewer electrons than O_2^+.

5.105. **Collect and Organize**

For the species N_2^+, O_2^+, C_2^{2+}, and Br_2^{2-}, we can place electrons into the appropriate molecular orbital (MO) energy levels to predict which species have one or more unpaired electrons.

Analyze

Because of s–p orbital mixing, the order of MOs for Li_2–N_2 is

$$\sigma_{2s}\,\sigma_{2s}^*\,\pi_{2p}\,\sigma_{2p}\,\pi_{2p}^*\,\sigma_{2p}^*$$

For O_2–Ne_2, which have less s–p orbital mixing, the order of the MOs is

$$\sigma_{2s}\,\sigma_{2s}^*\,\sigma_{2p}\,\pi_{2p}\,\pi_{2p}^*\,\sigma_{2p}^*$$

For Br_2^{2-} we assume the same MO energies as for F_2, but the MOs involve the $4s$ and $4p$ atomic orbitals. The species will have unpaired electrons and be paramagnetic when, after filling, a σ or σ^* orbital has one electron in it or when the π or π^* orbitals have one, two, or three electrons in them.

Solve

(a) N_2^+ Total number of electrons = 9 e⁻

$$(\sigma_{2s})^2(\sigma_{2s}^*)^2(\pi_{2p})^4(\sigma_{2p})^1$$

One unpaired electron

(b) O_2^+ Total number of electrons = 11 e⁻

$$(\sigma_{2s})^2(\sigma_{2s}^*)^2(\sigma_{2p})^2(\pi_{2p})^4(\pi_{2p}^*)^1$$

One unpaired electron

(c) C_2^{2+} Total number of electrons = 6 e⁻

$$(\sigma_{2s})^2(\sigma_{2s}^*)^2(\pi_{2p})^2$$

Two unpaired electrons

(d) Br_2^{2-} Total number of electrons = 16 e⁻

$$(\sigma_{4s})^2(\sigma_{4s}^*)^2(\sigma_{4p})^2(\pi_{4p})^4(\pi_{4p}^*)^4(\sigma_{4p}^*)^2$$

No unpaired electrons

The paramagnetic species are (a) N_2^+, (b) O_2^+, and (c) C_2^{2+}.

Think About It
The π orbital filling according to Hund's rule shows how the π orbitals have unpaired electrons when one, two, or three electrons occupy them, but not with four electrons.

$$\boxed{\uparrow}\;\boxed{}\qquad \boxed{\uparrow}\;\boxed{\uparrow}\qquad \boxed{\uparrow\downarrow}\;\boxed{\uparrow}\qquad \boxed{\uparrow\downarrow}\;\boxed{\uparrow\downarrow}$$
$$\pi\qquad\qquad\pi\qquad\qquad\pi\qquad\qquad\pi$$

5.107. **Collect and Organize**
For the species C_2^{2-}, N_2^{2-}, O_2^{2-}, and Br_2^{2-}, we can place electrons into the appropriate molecular orbital (MO) energy levels to predict which species have electrons in π^* orbitals.

Analyze
Because of $s–p$ orbital mixing, the order of MOs for $Li_2–N_2$ is

$$\sigma_{2s}\sigma_{2s}^*\pi_{2p}\sigma_{2p}\pi_{2p}^*\sigma_{2p}^*$$

For $O_2–Ne_2$, which have less $s–p$ orbital mixing, the order of the MOs is

$$\sigma_{2s}\sigma_{2s}^*\sigma_{2p}\pi_{2p}\pi_{2p}^*\sigma_{2p}^*$$

For each dianionic species, we fill the MO energy levels from lowest to highest energy with the total number of electrons. For Br_2^{2-} we assume the same MO energies as for F_2, but the MOs involve the $4s$ and $4p$ atomic orbitals.

Solve
(a) C_2^{2-} Total number of electrons = 10 e$^-$

$$(\sigma_{2s})^2(\sigma_{2s}^*)^2(\pi_{2p})^4(\sigma_{2p})^2$$

No electrons in π^* molecular orbitals

(b) N_2^{2-} Total number of electrons = 12 e$^-$

$$(\sigma_{2s})^2(\sigma_{2s}^*)^2(\pi_{2p})^4(\sigma_{2p})^2(\pi_{2p}^*)^2$$

Two electrons in π^* molecular orbitals

(c) O_2^{2-} Total number of electrons = 14 e$^-$

$$(\sigma_{2s})^2(\sigma_{2s}^*)^2(\sigma_{2p})^2(\pi_{2p})^4(\pi_{2p}^*)^4$$

Four electrons in π^* molecular orbitals

(d) Br_2^{2-} Total number of electrons = 16 e$^-$

$$(\sigma_{4s})^2(\sigma_{4s}^*)^2(\sigma_{4p})^2(\pi_{4p})^4(\pi_{4p}^*)^4(\sigma_{4p}^*)^2$$

Four electrons in π^* molecular orbitals

The species with electrons in π^* orbitals are (b) N_2^{2-}, (c) O_2^{2-}, and (d) Br_2^{2-}.

Think About It
The bond orders for each species are 3 for C_2^{2-}, 2 for N_2^{2-}, 1 for O_2^{2-}, and 0 for Br_2^{2-}.

5.109. **Collect and Organize**
For B_2, C_2, N_2, and O_2, we are to determine which increases its bond order on acquiring two electrons to become a dianion.

Analyze
Bond order increases when the two extra electrons are placed into bonding molecular orbitals (MOs). If the two extra electrons are placed into antibonding MOs, the bond order decreases.

Solve
(a) B_2 has the MO configuration of

$$(\sigma_{2s})^2(\sigma_{2s}^*)^2(\pi_{2p})^2$$

The two electrons added to form the dianion are placed into the π_{2p} orbital, so the bond order increases.

(b) C_2 has the MO configuration of

$$(\sigma_{2s})^2(\sigma_{2s}^*)^2(\pi_{2p})^4$$

The two electrons added to form the dianion are placed into the σ_{2p} orbital, so the bond order increases.

(c) N_2 has the MO configuration of

$$(\sigma_{2s})^2(\sigma_{2s}^*)^2(\pi_{2p})^4(\sigma_{2p})^2$$

The two electrons added to form the dianion are placed into the π_{2p}^* orbital, so the bond order decreases.

(d) O_2 has the MO configuration of

$$(\sigma_{2s})^2(\sigma_{2s}^*)^2(\sigma_{2p})^2(\pi_{2p})^4(\pi_{2p}^*)^2$$

The two electrons added to form the dianion are placed into the π_{2p}^* orbital, so the bond order decreases.

Bond order increases with a gain of two electrons for (a) B_2 and (b) C_2.

Think About It
The species above that have unpaired electrons (and thus are paramagnetic) are B_2, N_2^{2-}, and O_2.

5.111. Collect and Organize
For the diatomic 1+ cations of Li_2, Be_2, B_2, C_2, N_2, O_2, F_2, and Ne_2, we are to consider whether those cations always have shorter bonds than the neutral molecules.

Analyze
Shorter bonds have higher bond orders. Longer bonds have lower bond orders. Bond order decreases when the electron is removed from a bonding molecular orbital (MO). If the electron is removed from an antibonding MO, the bond order increases.

Solve
Li_2 has the MO configuration of

$$(\sigma_{2s})^2$$

Removing one electron decreases the bond order, and the bond is lengthened.
Be_2 has the MO configuration of

$$(\sigma_{2s})^2(\sigma_{2s}^*)^2$$

Removing one electron increases the bond order, and the bond is shortened.

B_2 has the MO configuration of

$$(\sigma_{2s})^2(\sigma_{2s}^*)^2(\pi_{2p})^2$$

Removing one electron decreases the bond order, and the bond is lengthened.
C_2 has the MO configuration of

$$(\sigma_{2s})^2(\sigma_{2s}^*)^2(\pi_{2p})^4$$

Removing one electron decreases the bond order, and the bond is lengthened.
N_2 has the MO configuration of

$$(\sigma_{2s})^2(\sigma_{2s}^*)^2(\pi_{2p})^4(\sigma_{2p})^2$$

Removing one electron decreases the bond order, and the bond is lengthened.
O_2 has the MO configuration of

$$(\sigma_{2s})^2(\sigma_{2s}^*)^2(\sigma_{2p})^2(\pi_{2p})^4(\pi_{2p}^*)^2$$

Removing one electron increases the bond order, and the bond is shortened.
F_2 has the MO configuration of

$$(\sigma_{2s})^2(\sigma_{2s}^*)^2(\sigma_{2p})^2(\pi_{2p})^4(\pi_{2p}^*)^4$$

Removing one electron increases the bond order, and the bond is shortened.

Ne_2 has the MO configuration of

$$(\sigma_{2s})^2(\sigma_{2s}^*)^2(\sigma_{2p})^2(\pi_{2p})^4(\pi_{2p}^*)^4(\sigma_{2p}^*)^2$$

Removing one electron increases the bond order, and the bond is shortened.

No. The cations N_2^+, C_2^+, B_2^+, and Li_2^+, which lose an electron from bonding orbitals in the corresponding neutral molecules, decrease their bond order and have longer bond lengths.

Think About It

All the 1+ cations will be paramagnetic.

5.113. Collect and Organize

In this question we are asked to draw the Lewis structure for urea given the space-filling structure in Figure P5.113 and determine whether that molecule is planar.

Analyze

First, we must draw the Lewis structure of CH_4N_2O. Then, through the electron pair geometry we can determine the molecular geometry. Molecular geometries that are planar are trigonal planar and square planar.

Solve

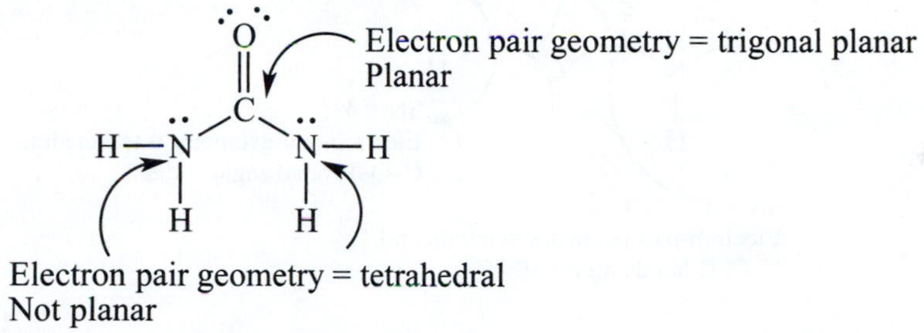

Electron pair geometry = trigonal planar
Planar

Electron pair geometry = tetrahedral
Not planar

Although the geometry around the carbon atom is planar, the geometry around the nitrogen atoms is not; that molecule is not planar overall.

Think About It

The hybridization on the carbon atom in urea is sp^2 and the hybridization on the nitrogen atoms is sp^3.

5.115. Collect and Organize

For NH_4^+ and ClO_4^-, we are asked to determine the molecular geometries of the two ions.

Analyze

First, we must draw the Lewis structures of each ion. Then, through the steric number (SN), we can determine the electron-pair geometry. If lone electron pairs are on the central atom, we have to take that into account to translate the electron-pair geometry into the molecular geometry.

Solve

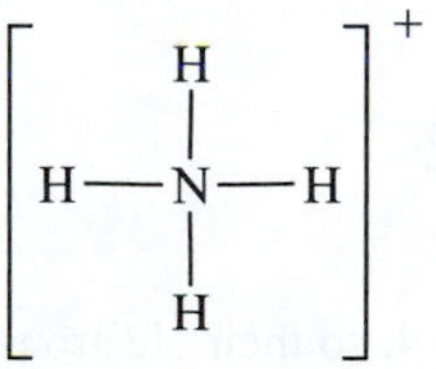

SN = 4
Electron-pair geometry = tetrahedral
Molecular geometry = tetrahedral

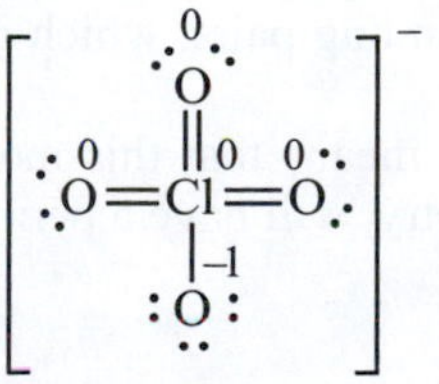

SN = 4
Electron-pair geometry = tetrahedral
Molecular geometry = tetrahedral

Think About It
The structure drawn for ClO_4^- has the lowest formal charges on the atoms. The expanded octet on Cl is possible because it is a third-period element.

5.117. Collect and Organize
We are given a skeletal structure in Figure 5.117. By completing the Lewis structure and applying VSEPR theory, we can determine the N—C—C, O═C—O, and C—O—H bond angles.

Analyze
From the Lewis structure and steric number (SN) we can determine the electron-pair geometry around the carbon and oxygen atoms. If the electron-pair geometry is linear, the bond angles are 180°; if it is trigonal planar, the bond angles are 120°; if it is tetrahedral, the bond angles are 109.5°.

Solve

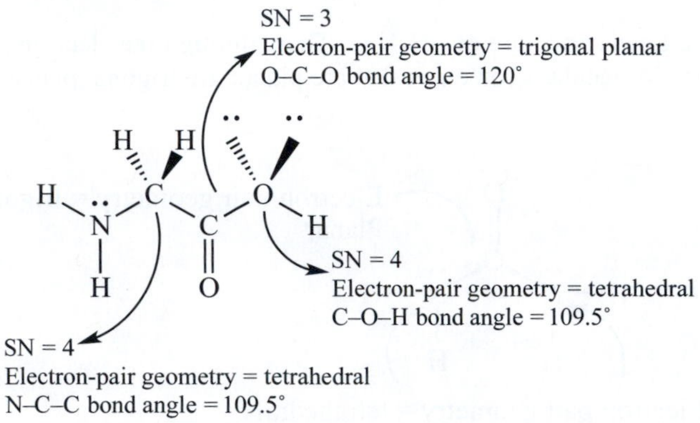

SN = 3
Electron-pair geometry = trigonal planar
O–C–O bond angle = 120°

SN = 4
Electron-pair geometry = tetrahedral
C–O–H bond angle = 109.5°

SN = 4
Electron-pair geometry = tetrahedral
N–C–C bond angle = 109.5°

Think About It
Remember, those are idealized bond angles. Because lone pairs take up more space than bonding pairs, the C—O—H bond angle is probably <109.5°.

5.119. Collect and Organize
We are given two alternate skeletal structures for Cl_2O_2. We are to complete the Lewis structures, and from those we can find the molecular geometry to determine whether either isomer is linear and whether either or both have a permanent dipole.

Analyze
Cl_2O_2 has 26 e^- and needs 32 e^-, giving a difference of 6 e^- in three covalent bonds. That leaves 20 e^- in 10 lone pairs to complete the octets on the atoms. Chlorine may expand its octet to minimize formal charges on the atoms in the Lewis structure.

Solve
The Lewis structures for the two skeletal arrangements of Cl_2O_2 are

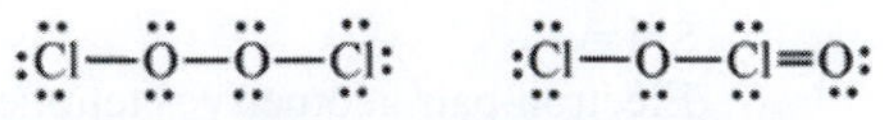

:C̈l—Ö—Ö—C̈l: :C̈l—Ö—C̈l═O:
Structure 1   Structure 2

(a) All the central atoms in both structures (O—O and O—Cl) have SN = 4, so their electron-group geometries are tetrahedral. Each central atom also has two lone pairs and two bonding pairs, which gives them a bent molecular geometry. Therefore, neither molecule is linear.
(b) Free rotation about the O—O bond in Structure 1 and its symmetry means that this molecule would not have a permanent dipole. Structure 2, however, because of its asymmetry, will have a permanent dipole.

Think About It
Those compounds may be drawn three-dimensionally as

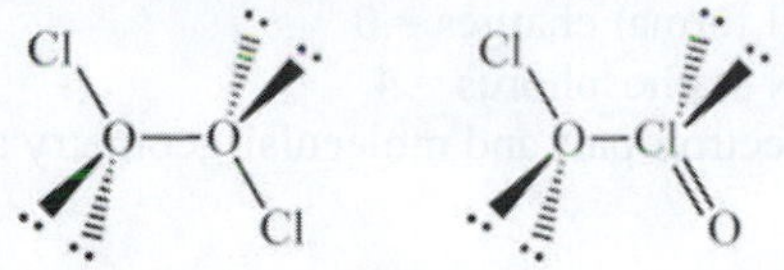

5.121. Collect and Organize
For the diatomic ion ClO^+ we are to draw the Lewis structure and complete the molecular orbital (MO) diagram (see Figure P5.121) to determine the Cl—O bond order.

Analyze
ClO^+ has 12 e^-. To complete the octets on the atoms for the Lewis structure, ClO^+ would need 16 e^-, giving a difference of 4 e^- in two covalent bonds. That leaves 8 e^- in four lone pairs to complete the structure. Chlorine may expand its octet to reduce the formal charges on the atom in the Lewis structure.

Solve

(a) 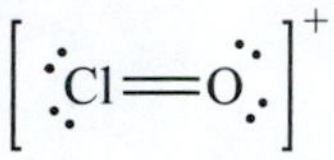 $\left[:\!\ddot{C}l\!=\!\ddot{O}: \right]^+$

(b) The MO diagram would fill as

$$(\sigma_{3s})^2(\sigma_{3s}^*)^2(\sigma_{3p})^2(\pi_{3p})^4(\pi_{3p}^*)^2$$
$$BO = \tfrac{1}{2}(8-4) = 2$$

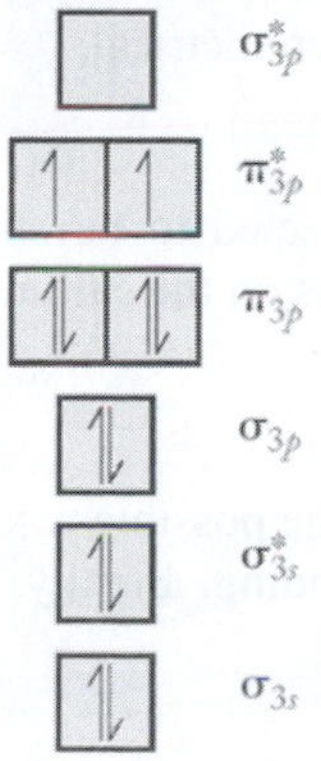

Think About It
For ClO^+ the bond order drawn in the Lewis structure matches the bond order calculated with MO theory.

5.123. Collect and Organize
Given the skeletal structure of phosphoric acid, we are to complete its Lewis structure and use VSEPR theory to determine the molecular geometry around the phosphorus atom.

Analyze
H_3PO_4 has 32 e^- and needs 46 e^-, giving a difference of 14 e^- in seven covalent bonds. That leaves 18 e^- in nine lone pairs to complete the octets on the atoms in the structure. Phosphorus may expand its octet to reduce the formal charges on the atoms.

Solve

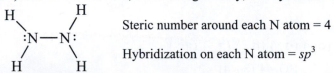

All formal charges = 0
SN at phosphorus = 4
Electron-pair and molecular geometry around P = tetrahedral

Think About It

In that structure phosphorus forms three σ bonds to OH, and a σ plus a π bond to oxygen.

5.125. Collect and Organize

After drawing the Lewis structure of NH_2NH_2 and determining the hybridization of the nitrogen atoms in that molecule, we are asked to explain whether the compound is polar.

Analyze

The hybridization and bond angles around the nitrogen atom in each of those compounds is determined by the number of electron pairs on nitrogen and will help us decide the molecule's polarity.

Solve

(a and b) The Lewis structure, molecular geometry, and hybridization for NH_2NH_2 is

Steric number around each N atom = 4

Hybridization on each N atom = sp^3

(c) Hydrazine, NH_2NH_2 is polar because the presence of a lone pair of electrons and two N—H bonds on each N atom assures that the molecule is asymmetrical.

Think About It

In reality, the bond angles might be expected to be slightly different for those compounds because of the different sizes of the CH_3 and H substituents on the sulfur atom.

5.127. Collect and Organize

For the molecule BNO we are to draw all the possible resonance structures and use formal charges to predict the structure that contributes most to the bonding. Finally, using VSEPR theory we can predict the molecular geometry at the carbon atoms.

Analyze

BNO has 14 e^- and needs 24 e^-, giving a difference of 10 e^- in five covalent bonds. That leaves 4 e^- in two lone pairs to complete the octets on the atoms. Boron in compounds may be electron deficient (for example, in BF_3), so if necessary we can arrange the electrons to satisfy the octets on the atoms other than B and leave B with an incomplete octet.

Solve

(a) and (b) The Lewis structure with resonance forms with formal charge assignments for BNO are

$$:\overset{-2}{B}-\overset{+1}{N}\equiv\overset{+1}{O}: \longleftrightarrow :\overset{-1}{B}=\overset{+1}{N}=\overset{0}{O}: \longleftrightarrow \overset{0}{B}\equiv\overset{+1}{N}-\overset{-1}{O}:$$

The structure in which boron is triply bonded to nitrogen is the best description; the −1 formal charge is on the most electronegative atom. All those structures have boron with an incomplete octet.

(c) Because SN = 2 around the central nitrogen atom, the molecular geometry is linear.

Think About It

We might expect BNO to pick up two electrons to give BNO^{2-}. That dianion would have enough electrons to complete all the atoms' octets.

$$\left[\overset{-2}{:}\text{B}\overset{+1}{\equiv}\text{N}\overset{-1}{-}\overset{..}{\underset{..}{\text{O}}}\text{:} \right]^{2-}$$

5.129. Collect and Organize

We are to draw the resonance structures of methyl isothiocyanate (CH_3NCS) and use formal charges to identify the resonance form that contributes most to the bonding. From the Lewis structure, we can determine the steric number (SN) at the carbon atoms in the molecule and predict the molecular geometry at each carbon atom.

Analyze

CH_3NCS has 22 e^- and needs 38 e^-, giving a difference of 16 e^- in eight covalent bonds. That leaves 6 e^- in three lone pairs on the molecule.

Solve

(a) and (b)

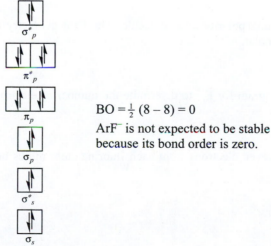

The resonance structure (at the left) in which all the formal charges equal zero contributes the most to the bonding.

(a) At the methyl (CH_3) carbon, SN = 4, so that carbon is tetrahedral. At the isothiocyanate (NCS) carbon, SN = 2, so the molecular geometry at that carbon is linear.

Think About It

The molecular geometries at the carbon atoms stay the same in all the resonance structures.

5.131. Collect and Organize

If the molecule HArF contains ArF^-, we can use molecular orbital (MO) theory to determine the bond order in ArF^-.

Analyze

ArF^- would have 16 valence electrons and bonding would involve overlap of the 3s and 3p orbitals on Ar with the 2s and 2p orbitals on F. Although the overlap would not be as effective as the overlap of orbitals of the same n level, we can assume that the overlap gives similar MOs. The MO diagram, then, would look similar to that for F_2.

Solve

$$BO = \tfrac{1}{2}(8-8) = 0$$

ArF^- is not expected to be stable because its bond order is zero.

Think About It
The argon–fluorine bond would be stable, however, as a neutral species (ArF) or as a cation (ArF$^+$).

5.133. Collect and Organize
To determine the polarity of N_2O_2, N_2O_5, and N_2O_3, we must draw Lewis structures and then consider the direction and magnitude of the individual bond dipoles.

Analyze
We draw the Lewis structures in the usual way and then use VSEPR theory to draw the structures' geometries on the basis of the steric number of the central N and O atoms. The electronegativity for O is greater than that for N, so each N—O bond is polarized so that partial negative charge is on the oxygen atom. Once we have assigned the individual bond dipoles, we can then see whether the vectors representing those bond dipoles cancel to give a nonpolar molecule or add to give a polar molecule.

Solve

The N_2O_2 molecule is nonpolar as drawn, but we can imagine the N—N bond rotating to give a polar molecule.

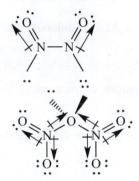

Because of free rotation around the N—N bond, N_2O_2 is nonpolar.

The N_2O_5 molecule is polar.

The N_2O_3 molecule is polar.

Think About It
Be careful to consider geometry in assigning polarity to a molecule. The bent geometry around the central oxygen atom in N_2O_5 makes that molecule polar.

5.135. Collect and Organize
We can use the molecular orbital (MO) diagram for F_2^+ to determine the number of electrons that occupy the antibonding MOs.

Analyze
F_2^+ has 13 e$^-$ to fill up the MO diagram, seven electrons from each fluorine atom minus one electron for the positive charge.

Solve

The molecular orbital diagram for F_2^+ is shown here.

$BO = \frac{1}{2}(8-5) = 1.5$

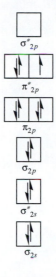

Think About It

Removing an electron from the π^* orbitals in F_2 (with a bond order of 1) increases the bond order to 1.5 in F_2^+.

5.137. **Collect and Organize**

We are asked to draw Lewis structures for trimethylamine, $(CH_3)_3N$, and for trisilylamine, $(SiH_3)_3N$, which are consistent with their respective trigonal pyramidal and trigonal planar geometries.

Analyze

The only difference in those molecules is a Si for a C atom. Both silicon and C have the same number of valence electrons, so the number of total valence electrons for each molecule is the same.

Solve

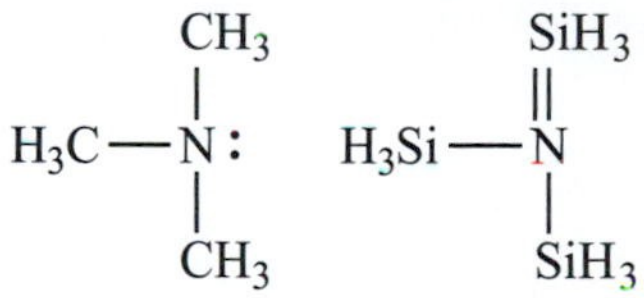

$$
\begin{array}{cc}
CH_3 & SiH_3 \\
| & \| \\
H_3C-N: & H_3Si-N \\
| & | \\
CH_3 & SiH_3
\end{array}
$$

With a lone pair on N in trimethylamine, the molecular geometry is trigonal pyramidal (from SN = 4 with one lone pair). If the lone pair on nitrogen double-bonds with the Si atom, the SN = 3 and the molecular geometry of trisilylamine is trigonal planar. Silicon can form that double bond by expanding its octet since it has available $3d$ orbitals to do so.

Think About It

Some textbooks describe the double bonding here between N and Si as "$d\pi-p\pi$ bonding."

5.139. **Collect and Organize**

From the molecular orbital diagram for S_2^{2-} we show that the bond order is 1 and determine whether that species is paramagnetic or diamagnetic.

Analyze

We can assume that the MO diagram for S_2 is similar to that of O_2 except that $3s$ and $3p$ atomic orbitals are used to make the molecular orbitals. S_2^{2-} has 14 total valence electrons.

Solve

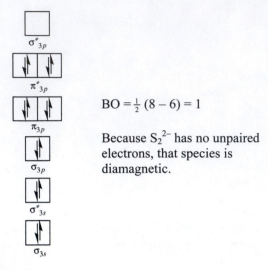

$$BO = \tfrac{1}{2}(8 - 6) = 1$$

Because S_2^{2-} has no unpaired electrons, that species is diamagnetic.

Think About It

Like O_2, neutral S_2 is expected to have a bond order of 2, which is consistent with its Lewis diagram:

The Lewis structure, however, does not predict that we also expect S_2 to be paramagnetic like O_2.

CHAPTER 6 | Intermolecular Forces: Attractions between Particles

6.1. Collect and Organize

Of the four constitutional isomers of heptane shown in Figure P6.1, we are to choose the one with the highest boiling point.

Analyze

As constitutional isomers, all those isomers of heptane have the same molecular formula, C_7H_{16}, with the same molar masses, so their boiling points cannot be distinguished by their relative masses. However, they do have structures that are different from each other in terms of branching, with (d) more branched than (c), which in turn is more branched than (b), with (a) being an unbranched, linear hydrocarbon.

Solve

The more branching, the lower the dispersion forces between the molecules and the lower the boiling point. Therefore, the linear *n*-heptane (a), with no branching of the hydrocarbon chain, is the isomer with the highest boiling point.

Think About It

In the series *n*-pentane, *n*-hexane, *n*-heptane, and *n*-octane, the molar mass difference leads us to predict that *n*-octane will have the highest boiling point.

6.3. Collect and Organize

Given in Figure P6.3 the trigonal pyramidal structures of NH_3 and PH_3 with their boiling points of 185 K and 200 K, we are asked to determine which molecule boils at the higher temperature.

Analyze

Both molecules have the same trigonal pyramidal structure and, therefore, are polar molecules. Both ammonia and phosphine have dispersion forces with those for PH_3 being greater than those for NH_3 because phosphorus has more electrons. However, nitrogen–hydrogen bonds are more polar than phosphine–hydrogen bonds because nitrogen's electronegativity is greater than that of phosphine; indeed, ammonia can form hydrogen bonds. More polar molecules will have greater dipole–dipole forces and are expected to have higher boiling points.

Solve

Ammonia (NH_3) should have a higher boiling point than phosphine (PH_3) even though its molecules experience weaker London Dispersion forces; Ammonia's molecules can form hydrogen bonds because ammonia is more polar and forms hydrogen bonds with other ammonia molecules and phosphine molecules do not.

Think About It

Hydrogen bonding is a special and strong form of dipole–dipole interaction.

6.5. Collect and Organize

Given the ball-and-stick structures in Figure P6.5, we are to identify in each the functional group.

Analyze

In those ball-and-stick structures carbon is black, hydrogen is white, and oxygen is red. They all involve an oxygen atom bond. If the oxygen is doubly bonded to a carbon atom it is a ketone, if the oxygen atom is singly bonded to two carbon atoms it is an ether, and if the oxygen atom is bonded to a carbon atom and a hydrogen atom it is an alcohol.

Solve

The line structures for those ball-and-stick representations and their classification based on the bonding of oxygen in each molecule are:

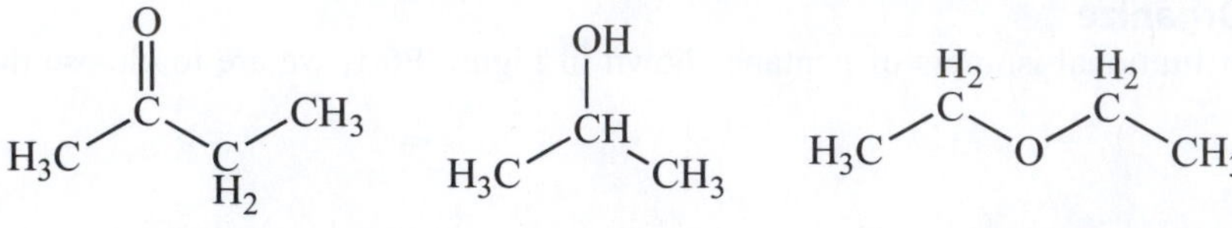

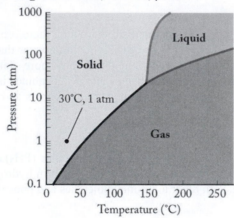

(a) ketone   (b) alcohol   (c) ether

Think About It

Alcohols can hydrogen-bond to each other and we would predict then that alcohol (b) would have a lower boiling point than the ether with the same chemical formula and molecular mass ($CH_3OCH_2CH_3$).

6.7. **Collect and Organize**

From the phase diagram for compound X shown in Figure P6.7, we are to determine whether on a hot summer day the substance will be in the solid, liquid, or gas form.

Analyze

The conditions of a hot summer day are pressure at 1 atm and temperature at about 30°C. We find that point on the phase diagram. The green-shaded area on the phase diagram in Figure P6.7 is for the solid phase, the blue-shaded area is for the liquid phase, and the pink-shaded area is for the gas phase.

Solve

The point for 1 atm and 30°C is in the green-shaded, or solid, phase.

Think About It

Heating compound X further in the closed container to above 55°C would sublime the solid into the gas phase.

6.9. **Collect and Organize**

For compound X in Problems 6.7, we are to predict the phase change, if any, that will occur when we take the compound at 100 atm from 0°C to 250°C at the same pressure.

Analyze

We can plot both points on the phase diagram. If the change from one condition to the other crosses any phase equilibrium line, a phase change will occur. The green-shaded area on the phase diagram in Figure P6.7 is for the solid phase, the blue-shaded area is for the liquid phase, and the pink-shaded area is for the gas phase.

Solve

The point for 100 atm and 0°C is in the green-shaded, or solid, phase. The point for 100 atm and 250°C is in the liquid phase (but very close to the liquid-gas equilibrium line). When we connect those two states, we see that the compound undergoes melting (crossing the solid–liquid equilibrium line). The compound, therefore, undergoes the solid-to-liquid phase change.

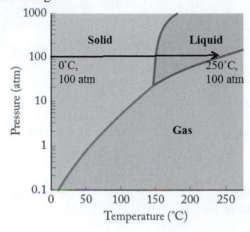

Think About It

Below the triple point, an increase in temperature at the same pressure (for this compound at a pressure less than 20 atm) results only in sublimation, not melting.

6.11. **Collect and Organize**

Using the phase diagram shown in Figure P6.7 for compound X, we are to predict whether the solid form will float on the liquid form as the liquid freezes at 300 atm.

Analyze

In the phase diagram we see that as pressure is applied at a particular temperature, say 160°C, compound X will form a solid from the liquid form. That change indicates that the solid is the denser form because it is favored at higher pressures.

Solve

Because the solid is the denser form, it will not float on the liquid form.

Think About It

The right-sloping solid-to-liquid equilibrium line shows that the solid is the denser phase. That is unlike water, which has a left-sloping solid-to-liquid line, which means that ice does float on water.

6.13. **Collect and Organize**

Using Figure P6.12 we are to determine what, if any, phase change occurs when the water at –25°C and 2500 atm undergoes heating and a pressure reduction to –15°C and 1000 atm.

Analyze

We can plot both points on the phase diagram. If the change from one condition to the other crosses the phase equilibrium line, a phase change will occur. The pink-shaded area on the phase diagram in Figure P6.12 is for the solid phase, and the blue-shaded area is for the liquid phase.

Solve

In decreasing the pressure and increasing the temperature, we see on the phase diagram that the liquid will freeze to become a solid.

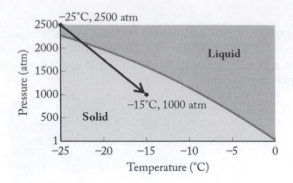

Think About It
That result for lowering pressure and increasing temperature is counterintuitive because water is quite different from most other liquids.

6.15. Collect and Organize
From Figure P6.15 showing xenon atoms, we are to identify the attractive forces between the xenon atoms and the physical state (solid, liquid, or gas).

Analyze
Xenon atoms and single atoms cannot be polar. Solids have atoms close together and have a definite shape, liquids have atoms close together but spread out to take on the shape of the container filling the bottom, and gases have atoms far apart from each other that fill the container.

Solve
The xenon atoms depicted are all close to each other and form a definite structure, so the physical state represented is a solid. Xenon atoms, being nonpolar, have only dispersion forces for the attractive forces between them.

Think About It
Dispersion forces are greater for atoms with more electrons; therefore, we expect the dispersion forces of xenon to be greater than that of neon.

6.17. Collect and Organize
We are to explain why a branched alkane (hydrocarbon) has a lower boiling point than a normal (linear) alkane if they have the same molar mass.

Analyze
A higher boiling point indicates greater intermolecular forces between the molecules in the liquid phase.

Solve
The intermolecular forces between molecules of a branched alkane are less than those of a linear, unbranched alkane because a branched molecule has less available surface area for the intermolecular forces than the linear form. Therefore, with lower intermolecular forces between the molecules, the branched alkane has a lower boiling point than its linear analogue.

Think About It
Compare the boiling points of the constitutional isomers of pentane:

36°C   27.7°C   9.5°C

6.19. **Collect and Organize**

For each pair of substances we are to determine which one has the stronger London dispersion forces.

Analyze

The more atoms and electrons in a compound, the greater the dispersion forces between the molecules.

Solve

(a) C_2Cl_6 has stronger dispersion forces than C_2F_6 because Cl is a larger atom with more electrons than F.

(b) C_3H_8 has stronger dispersion forces than CH_4 because C_3H_8 has more atoms in its molecular structure.

(c) CS_2 has stronger dispersion forces than CO_2 because S has more electrons than O.

Think About It

The greater number of electrons and atoms in a compound gives rise to stronger dispersion forces because more polarizable electrons are present on the atoms that make up the compound.

6.21. **Collect and Organize**

From the order of the boiling points of gasoline, jet fuel, kerosene, diesel oil, and fuel oil, we are to predict which fuel has hydrocarbons of the greatest average molar mass.

Analyze

Molecules with higher molar masses generally have greater intermolecular forces.

Solve

Fuel oil has hydrocarbons of the greatest molar mass among those fuels because it has the highest boiling point.

Think About It

We might also expect diesel fuel to be more viscous than gasoline owing to greater dispersion forces present because of the increased molar mass of diesel fuel.

6.23. **Collect and Organize**

Given that the dipole moment of CH_2F_2 is greater than that of CH_2Cl_2, we are to explain why the boiling point of CH_2Cl_2 is higher.

Analyze

The greater the intermolecular forces, the higher the boiling point of the substance. Intermolecular forces are a combination of all the forces and all substances have dispersion forces present as well as other forces if the molecule is polar.

Solve

Both CH_2Cl_2 and CH_2F_2 are polar, and on the basis of dipole–dipole interactions alone we would expect that the more polar molecule, CH_2F_2, would have the higher boiling point. However, because Cl is a much larger atom than F, the dispersion forces adding to the dipole–dipole forces in CH_2Cl_2 are much greater than in CH_2F_2. The lower dipole moment in CH_2Cl_2 is overtaken by the much greater dispersion forces. Therefore, the total intermolecular forces for CH_2Cl_2 are greater than the stronger dipole moment but weaker than the dispersion forces in CH_2F_2.

Think About It

That example shows how dispersion forces, which we tend to think of as generally weak, can have an important effect on the properties of substances.

6.25. **Collect and Organize**

For each pair of substances we are to determine which one has the stronger dispersion forces.

Analyze

The more atoms and electrons in a compound, the greater the dispersion forces between the molecules.

Solve

(a) CCl_4 has stronger dispersion forces than CF_4 because Cl is a larger atom with more electrons than F.

(b) C_3H_8 has stronger dispersion forces than CH_4 because C_3H_8 has more atoms in its molecular structure.

Think About It

The greater number of electrons and atoms in a compound gives rise to stronger dispersion forces because more polarizable electrons are present on the atoms that make up the compound.

6.27. Collect and Organize

Given the two structures in Figure P6.27, one trigonal pyramidal and one trigonal bipyramidal with the trigonal bipyramidal structure melting at a higher temperature, we are asked why.

Analyze

As depicted, the central atoms and the bonded atoms in each structure are the same. We must consider here the polarity of the molecular geometries and the strength of the dispersion forces between the molecules of each geometry.

Solve

The molecule that is trigonal pyramidal will be polar and have stronger dipole–dipole forces than the nonpolar molecule that is trigonal bipyramidal that has only dispersion forces. Therefore, we might expect, based only on molecular polarity, that the trigonal bipyramidal substance would have a lower melting point than the trigonal pyramidal molecule. However, the dispersion forces for the trigonal bipyramidal substance must be greater, because of much higher molar mass, than the dipole–dipole forces of the trigonal pyramidal structure, making the melting point for the trigonal bipyramidal molecule higher.

Think About It

You might have, using the color scheme for atoms in the textbook, determined that those two substances are PCl_3 and PCl_5, respectively. In that case, PCl_5 does indeed have a higher melting point ($161°C$) than PCl_3 ($-94°C$).

6.29. Collect and Organize

We are to describe how individual water molecules are oriented around dissolved anions.

Analyze

The water molecule is polar because of its bent geometry, with a partial negative charge on the oxygen atom and a partial positive charge on the hydrogen ends.

Solve

The water molecule is oriented around an anion so as to point the partially positive hydrogen atoms towards the anion. That arrangement results in attractive forces between the water molecules and the anion.

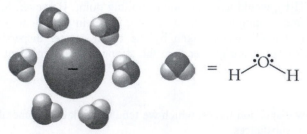

Think About It

That interaction between the anion and water molecules is an ion–dipole interaction.

6.31. Collect and Organize

We are to explain the differences in strength between dipole–dipole interactions (weaker) and ion–dipole interactions (stronger).

Analyze

The dipole–dipole interaction involves attractions between two polar molecules with slight charge separation (partial positive and negative charges) on the molecule. An ion–dipole interaction involves attractions between an ion with a full positive or negative charge and a polar molecule.

Solve

Coulomb's law states that as charge increases, the attraction of two oppositely charged species for each other increases. Because of the full positive or negative charge on the ion, the ion–dipole interaction is stronger than the dipole–dipole interaction.

Think About It

Ion–ion is the strongest of all interactions between molecules.

6.33. **Collect and Organize**

We are to explain why hydrogen bonds are considered a special class of dipole–dipole interactions.

Analyze

Hydrogen bonds can form when hydrogen is bonded to a very electronegative element (F, O, N). The hydrogen bond is very polar.

Solve

The partial positive charge on an H atom bonded to an electronegative O, N, or F atom means that H atoms and O, N, or F atoms on adjacent molecules experience dipole–dipole interactions. These particular interactions are called hydrogen bonds and are unusually strong because the small electron cloud around each of the H atoms allows neighboring O, N, or F atoms to come very close to the center of the H atom, which leads to very strong attractions predicted by Coulomb's law.

Think About It

Hydrogen bonds also are important in explaining why ice floats and how proteins fold.

6.35. **Collect and Organize**

We are to explain why CH_3F (melting point, $-142°C$) has a higher melting point than CH_4 (mp, $-182°C$).

Analyze

The higher melting point of CH_3F indicates stronger intermolecular forces between CH_3F molecules than between CH_4 molecules.

Solve

CH_3F is a polar molecule and therefore has stronger intermolecular forces than those of the nonpolar molecules of CH_4, which have only the weak dispersion forces. Because overcoming strong intermolecular forces takes more energy, CH_3F has a higher melting point than CH_4. Second, the dispersion forces between CH_3F molecules will be greater than those between CH_4 molecules because F has more electrons than H. That is then added to the dipole–dipole forces between CH_3F molecules.

Think About It

Molecular polarity and the degree of charge separation are important considerations for comparing some physical properties of compounds (such as boiling point or vapor pressure).

6.37. **Collect and Organize**

We are asked to explain why CH_4 does not form hydrogen bonds but CH_3OH does.

Analyze

Hydrogen bonds can form only when hydrogen is bonded to a very electronegative element (F, O, N).

Solve

The H atoms in methane singly bond to the relatively low-electronegativity C atom, and therefore the carbon–hydrogen bond is not polar enough to exhibit hydrogen bonding. In methanol, however, one H atom is bonded to oxygen, which is second to fluorine in electronegativity. That H shows hydrogen bonding in methanol.

Think About It

The other H atoms bonded to C in methanol, CH_3OH, are not capable of hydrogen bonding.

6.39. **Collect and Organize**

For the covalent molecules CF_4, CF_2Cl_2, CCl_4, and $CFCl_3$, we are to determine which we would expect to have dipole–dipole interactions.

Analyze

We need to first determine whether the molecules are polar or nonpolar. Polar molecules have permanent dipoles that attract each other (δ^- to δ^+); nonpolar molecules have only weak dispersion forces between them. If all the molecules are polar, then the one with the smallest dipole moment (as determined from differences in electronegativities between the atoms) would be the molecule with the weakest intermolecular forces.

Solve

From the Lewis structures of those molecules, we know that both CF_4 (a) and CCl_4 (c) are nonpolar tetrahedral molecules. Those have only dispersion forces as the intermolecular force between the molecules. CF_2Cl_2 (b) and $CFCl_3$ (d) are both polar tetrahedral molecules and therefore have dipole–dipole interactions between their molecules.

Think About It

Between CF_4 and CCl_4 we would expect CCl_4 to have the stronger dispersion forces.

6.41. **Collect and Organize**

For the covalent molecules methanol, ethane, dimethyl ether, and acetic acid, we are to determine which we would expect to hydrogen-bond to themselves.

Analyze

Hydrogen bonding can occur when hydrogen is bonded to a very electronegative atom (F, O, N) in a molecule.

Solve

Of the four molecules, only (a) methanol and (d) acetic acid can hydrogen-bond to themselves.

Think About It

Although dimethyl ether is polar, it does not have the –OH group and therefore does not hydrogen-bond to itself.

6.43. **Collect and Organize**

We are to explain which ion, Cl^-, Br^-, or I^-, has the strongest ion–dipole interactions with water.

Analyze

Ions with smaller size or higher charge will attract the dipoles of the water molecules more strongly than those of larger size or smaller charge. All the charges on those halide ions are the same, so we need consider only the size of the halide ions.

Solve

Because Cl^- has the smallest ionic radius compared to either Br^- or I^-, Cl^- and can get closer to the H atoms of water molecules, Cl^- will exhibit the strongest ion–dipole interaction with water.

Think About It

The F^- ion would be predicted to have an even stronger ion–dipole interaction with water than the Cl^- anion.

6.45. Collect and Organize
We are asked in this question how to determine which component of a solution is the solvent.

Analyze
Solutions are composed of a solvent and a solute.

Solve
The solvent is the component of the solution that is present in the largest amount on the basis of the number of moles of the substances mixed in the solution.

Think About It
The solute dissolves in the solvent and may be a solid, liquid, or gas. The solvent itself may be a solid, as in the example of an alloy.

6.47. Collect and Organize
We can define *miscible* and *soluble* to distinguish between those terms.

Analyze
Two liquids are miscible when they dissolve completely in all proportions into each other. A substance (liquid, solid, or gas) is soluble when it dissolves in a solvent.

Solve
Miscible and *soluble* are nearly the same in that both describe one substance dissolving into another. Miscibility, however, refers to two liquids that dissolve in each other. The solubility of a substance in a solvent may be limited and so may form a saturated solution or precipitate from the solution if the concentration is too high, but two liquids that are miscible are soluble in each other in all proportions.

Think About It
The classification between soluble and insoluble also is indistinct. In general, a solute that dissolves at less than 0.1 g in 1.00 L of a solvent is considered insoluble.

6.49. Collect and Organize
We are to relate the solubility of substances in water with the terms *hydrophilic* and *hydrophobic*.

Analyze
Hydrophilic means "water-loving" and *hydrophobic* means "water-fearing."

Solve
Hydrophilic substances dissolve in water. Hydrophobic substances do not dissolve, or are immiscible, in water.

Think About It
Ethanol is hydrophilic because it is miscible with water to give a homogeneous solution, but olive oil is hydrophobic because it forms a heterogeneous mixture with water that separates into oil and water layers.

6.51. Collect and Organize
From the molecules in Figure P6.28 we are to choose which would be more soluble in water and explain our reasoning.

Analyze
The more polar a molecule, the more soluble it will be in water.

Solve
Of the two chlorohydrocarbons in Figure P6.28 ($ClCH_2CH_2Cl$ and $CHCl_2CH_3$), $ClCH_2CH_2Cl$, the (a) compound, is more polar and therefore more soluble in water.

Think About It
We might expect, though, that $CHCl_2CH_3$, being somewhat polar, would still have some solubility in water.

6.53. **Collect and Organize**

For each pair of compounds, we are to determine which is more soluble in H_2O.

Analyze

Water is a polar solvent that can form hydrogen bonds to dissolved substances with X—H bonds (X = F, O, N). In each pair of compounds, the more soluble is the more polar molecule or the one that forms hydrogen bonds. In considering whether a salt is soluble in water, we have to consider the relative strengths of the ionic bonds as well as the relative strengths of the ion–dipole interactions formed on dissolution.

Solve

(a) $CHCl_3$ is polar, whereas CCl_4 is not. $CHCl_3$ is more soluble in water.
(b) CH_3OH is more polar because it has a smaller hydrocarbon chain than $C_6H_{11}OH$. CH_3OH is more soluble in water.
(c) NaF has a weaker ionic bond than MgO because its ions have lower charges. NaF is more soluble in water.
(d) BaF_2 has a weaker ionic bond than CaF_2 because Ba^{2+} is larger than Ca^{2+}. BaF_2 is more soluble in water.

Think About It

Solubility is determined by many factors: polarity, ability to hydrogen-bond, and strength of the intermolecular forces between molecules of the solute.

6.55. **Collect and Organize**

For each pair of compounds, we are to determine whether they are miscible.

Analyze

Two substances are miscible when they dissolve completely in all proportions into each other. Here we can use the general rule that "like dissolves like."

Solve

(a) Both Br_2 and benzene are nonpolar; we expect that Br_2 and benzene will be miscible.
(b) Diethyl ether ($CH_3CH_2OCH_2CH_3$) is slightly polar, whereas acetic acid (CH_3COOH) is polar; we expect that diethyl ether and acetic acid will be miscible.
(c) Both cyclohexane (C_6H_{12}) and hexane ($CH_3CH_2CH_2CH_2CH_2CH_3$) are nonpolar; we expect that cyclohexane and hexane will be miscible.
(d) Both carbon disulfide (CS_2) and carbon tetrachloride (CCl_4) are nonpolar; we expect that carbon disulfide and carbon tetrachloride will be miscible.

Think About It

Acetic acid has a high miscibility with other substances that can form hydrogen bonds.

6.57. **Collect and Organize**

Of the ionic compounds listed, NaCl, KI, $Ca(OH)_2$, and CaO, we are to determine which would be most soluble in water.

Analyze

The weaker the ionic bond, the more easily the bond breaks for the cation and the easier for the anion to dissolve in water. Ionic bonds are weakest for large ions of low charge.

Solve

KI (b) has the largest ions of lowest (1+ and 1–) charge, so it is the most soluble in water because it has the weakest ion–ion bond.

Think About It

CaO, with a 2+ cation and 2– anion, would be expected to be the least soluble in water.

6.59. Collect and Organize

From among the four compounds listed we are to choose the one most soluble in water.

Analyze

Water is a polar solvent that can form hydrogen bonds and dipole–dipole interactions with other polar dissolved substances. All the compounds listed have some degree of polarity as a result of the bent geometry around the oxygen in the middle of the compounds' structure. The compounds, though, have different hydrocarbon ($-CH_2-$) chain lengths.

Solve

The longer the $-CH_2-$ chain, the more hydrophobic (nonpolar) the molecule; therefore, (c) CH_3OCH_3, with the shortest hydrocarbon chain, will be the most soluble in water.

Think About It

Those compounds, however, are all less soluble than their alcohol counterparts [$CH_3(CH_2)_nOH$] because the alcohol $-OH$ can hydrogen-bond to water.

6.61. Collect and Organize

We are asked to differentiate between *sublimation* and *evaporation*.

Analyze

In sublimation, a substance goes from the solid to the gas phase. In evaporation, a substance goes from the liquid to the gas phase.

Solve

Although both processes end with the substance in the gas phase, sublimation "skips" a step in that the solid does not first liquefy before evaporating.

Think About It

A familiar substance that sublimes at room temperature and pressure is dry ice (solid CO_2).

6.63. Collect and Organize

We are asked to define *equilibrium line*.

Analyze

That term applies to the lines in a phase diagram.

Solve

If you are along the equilibrium line in a phase diagram, the two phases that border that line are stable and coexist at that pressure–temperature combination.

Think About It

Where the equilibrium lines meet is called the *triple point*. That is the temperature–pressure combination at which all three phases (gas, liquid, and solid) are present and stable.

6.65. Collect and Organize

We are to predict the phase most likely to be present at two different temperature–pressure combinations.

Analyze

At high temperatures, the atoms or molecules of a substance have high kinetic energies and can partially or fully break the intermolecular forces between them. At high pressures the atoms or molecules are close to each other and therefore are attracted to one another through intermolecular forces.

Solve

(a) For low temperatures and high pressures, we would expect a solid phase to be present.
(b) For high temperatures and low pressures, we would expect a gas phase to be present.

Think About It

As we decrease the pressure at low temperatures, we could melt the solid and even perhaps vaporize it, or sublime the solid.

6.67. Collect and Organize

Freeze-drying food involves subliming the ice in frozen food into the gas phase. We are asked whether the pressure for that process must be below the pressure at the triple point of water.

Analyze

The triple point is where the gas–solid, solid–liquid, and liquid–gas phase boundaries meet in a phase diagram.

Solve

From the phase diagram for water we see that above the triple point, the solid phase must change to the liquid phase to enter the gaseous phase. Below the triple point, changing the temperature at a given pressure will sublime solid water into the gas phase. Yes, the pressure used for the sublimation process for freeze-drying must be below the pressure at the triple point.

Think About It

The triple point is characteristic of a particular substance. The triple point for ethanol is different from that for water.

6.69. Collect and Organize

Using the phase diagram for water shown in Figure 6.20, we are to determine the boiling point of water at 200 atm pressure.

Analyze

In the phase diagram we need to read the temperature at the liquid–gas equilibrium line for a pressure of 200 atm.

Solve

Reading from the phase diagram the boiling point is approximately at 310°C.

Think About It

At a slightly higher pressure (>218 atm) and temperature (>374°C) the water will transform into a supercritical fluid.

6.71. Collect and Organize

We consider the phase changes that water, initially at 5.0 atm and 100°C, undergoes when the pressure is reduced to 0.5 atm while maintaining temperature at 100°C.

Analyze

In the phase diagram for water, the phase of the water at 5.0 atm and 100°C is liquid. The phase of water at 0.5 atm and 100°C is gas.

Solve

Water at 100°C vaporizes from liquid to gas when the pressure is reduced from 5.0 atm to 0.5 atm.

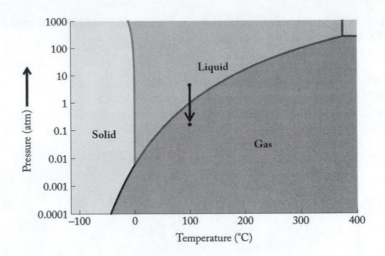

Think About It

At 100°C, water boils at 1.0 atm. At pressures lower than 1.0 atm, water at 100°C is entirely in the gaseous state.

6.73. Collect and Organize

From the phase diagram for CO_2 (Figure 6.22), we can determine the temperature below which $CO_2(s)$ sublimes to $CO_2(g)$ simply through lowering the pressure.

Analyze

The direct solid-to-gas conversion occurs below the triple point (–57°C, 5.1 atm).

Solve

The triple point of CO_2 is at –57°C. At any temperature below the triple point, $CO_2(s)$ sublimes directly to $CO_2(g)$ through lowering the pressure.

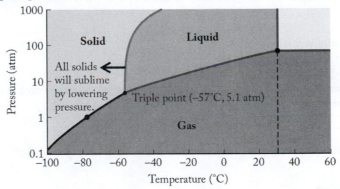

Think About It

Because the triple point of CO_2 is at a low temperature, we do not observe dry ice melting under ambient conditions (25°C, 1 atm).

6.75. Collect and Organize

We can use the phase diagram for water shown in Figure 6.20 to determine which phases of water are present at different temperature and pressure combinations.

Analyze

We use the phase diagram just like a map, locating each pressure and temperature combination and "reading" the phase at that location on the diagram. We are shown locations on the map to orient us: the normal pressure

(1 atm), melting point (0°C), and boiling point (100°C) lines are indicated as well as the conditions for the triple point (0.01°C, 0.006 atm).

Solve
(a) 2 atm and 110°C: liquid
(b) 0.5 atm and 80°C: liquid
(c) 7×10^{-3} atm and 3°C: gas

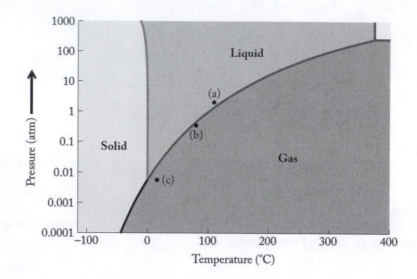

Think About It
Water at high pressures has a higher boiling point than at lower pressures, as seen in (a).

6.77. Collect and Organize
We are to consider the conversion of water at 25°C and 1 atm to water at its triple point.

Analyze
The triple point of water is 0.01°C, 0.006 atm.

Solve
To convert water from 25°C and 1 atm to its triple point, we need to both decrease the pressure to 0.006 atm and decrease the temperature to 0.01°C.

Think About It
At the triple point, all three phases (gas, liquid, and solid) coexist in equilibrium.

6.79. Collect and Organize
Water and methanol are both polar liquids capable of hydrogen bonding. We are asked why a needle floats on water but not on methanol.

Analyze
Surface tension is the resistance of a liquid to increase its surface area by moving the molecules of the liquid apart. The greater the intermolecular forces between the molecules in the liquid, the greater the surface tension.

Solve
A needle floats on water but not on methanol because of the high surface tension of water. Water has high surface tension because it can hydrogen-bond through two O—H bonds with other water molecules, whereas methanol has only one O—H bond through which to form strong hydrogen bonds. The intermolecular forces between the C—H groups on the CH_3OH molecules are only weak dispersion forces.

Think About It
The high surface tension of water also allows some insects to walk on water.

6.81. **Collect and Organize**
We are to explain why water pipes are in danger of bursting when the temperature is below the freezing point of water.

Analyze
At temperatures below freezing, the water in the pipes freezes. Because the density of ice is less than that of liquid water, the water expands as it freezes.

Solve
The expansion of water in the pipes upon freezing may create enough pressure on the wall of the pipes to cause them to burst.

Think About It
To prevent pipes from freezing during the winter months, we must drain the water from the portion of the pipe exposed to freezing temperatures.

6.83. **Collect and Organize**
We are to explain why the meniscus of liquid Hg is convex, rather than concave as it is for most liquids.

Analyze
The shape of the meniscus is due to the competing adhesive forces (liquid to glass surface that has Si—O—H bonds) and cohesive forces (liquid to liquid).

Solve
The cohesive forces in mercury are stronger than the adhesive forces of the mercury to the glass. That effect yields a convex meniscus.

Think About It
The strong metallic bonding between mercury atoms is not balanced by the adhesive Hg to Si—O—H bonds.

6.85. **Collect and Organize**
We are to describe the origin of surface tension in terms of intermolecular forces.

Analyze
Surface tension is the resistance of a liquid to increase its surface area by moving the molecules of the liquid apart. The greater the intermolecular forces between the molecules in the liquid, the greater the surface tension.

Solve
Molecules in the bulk liquid are "pulled" by all the other liquid molecules surrounding them, and they are therefore "suspended" in the bulk liquid; their intermolecular forces suspend them by pulling on all sides and directions. Molecules on the surface of a liquid, however, are pulled only by molecules under and beside them by the intermolecular forces between them, creating a tight film of molecules on the surface that we call surface tension.

Think About It
When the surface tension is greater than the force of gravity on a small object placed on top of water, the object floats.

6.87. **Collect and Organize**
We are asked which liquid, water or ethanol, will rise higher in a capillary tube.

Analyze
A liquid rises in a capillary tube until the force of gravity balances the cohesive forces between the molecules of the liquid with the adhesive forces between the liquid and the inside surface of the capillary tube. Different

liquids will have different cohesive and adhesive forces. Stronger cohesive than adhesive forces reduce the height of the liquid in the capillary tube, whereas stronger adhesive than cohesive forces increase the height of the liquid. In addition, the density of the liquid affects the height of the liquid in the capillary tube. The higher the density, the greater the force of gravity that acts on it and the lower the height in the capillary tube.

Solve
Both water and ethanol will rise in the capillary tube because of hydrogen bonding with the silica, but water will rise higher because of its stronger adhesive forces with the capillary walls as a result of its higher capacity for hydrogen bonding.

Think About It
Capillary action is a delicate balance between the cohesive and adhesive forces and the pull of gravity.

6.89. Collect and Organize
For this question we are to name the intermolecular force that exists in all substances.

Analyze
The intermolecular forces include ion–dipole, dipole–dipole (including hydrogen bonding), ion–induced dipole, dipole–induced dipole, and dispersion forces.

Solve
All molecules have electrons that can form instantaneous dipoles, so all substances have dispersion intermolecular forces.

Think About It
In general, dispersion forces are weak, but they can be substantial, especially for large molecules with many atoms.

6.91. Collect and Organize
We are to suggest a reason why methanol boils at a lower temperature than water (64.7°C vs. 100°C) even though methanol has a larger molar mass than water (32.04 g/mol vs. 18.02 g/mol).

Analyze
Both methanol and water are held together in the condensed phases (liquid and solid) by weak van der Waals forces and hydrogen bonds.

Solve
Although the dispersion forces between methanol molecules are greater than those between water molecules because methanol has more electrons and greater molar mass, water can form two hydrogen bonds compared with methanol's one hydrogen bond. That greater number of stronger interactions between water molecules raises the boiling point of water above that of methanol.

Think About It
As we add carbons to the alcohol chain for the series R—OH, the boiling point (bp) increases because of increases in the van der Waals forces between the molecules.

$$CH_3OH < CH_3CH_2OH < CH_3CH_2CH_2OH < CH_3CH_2CH_2CH_2OH$$
$$\text{bp } 64.7°C \quad \text{bp } 78.4°C \quad \text{bp } 97.2°C \quad \text{bp } 117.7°C$$

6.93. Collect and Organize
To determine which compound has the lowest boiling point of CH_4, CH_3Cl, CH_2Cl_2, $CHCl_3$, and CCl_4 we need to consider all the intermolecular forces that act between the molecules in each substance.

Analyze

$CHCl_3$, CH_2Cl_2, and $CHCl_3$ are polar molecules, so dipole–dipole interactions are present in each substance. Weak dispersion forces are also present between the molecules, but they are the only intermolecular forces present in CH_4 and CCl_4.

Solve

The substance with the lower boiling point is that with the lowest intermolecular forces. This will be between the molecules with only dispersion forces. Of CH_4 and CCl_4, CH_4 has the lowest strength dispersion forces because it has the lowest molecular mass.

Think About It

Usually, dispersion forces are so much weaker that they do not significantly add to the strength of the dominant intermolecular force between molecules.

6.95. **Collect and Organize**

We are asked whether the sublimation point of ice increases or decreases as the pressure is increased.

Analyze

Using a phase diagram for water, we can determine the behavior of the sublimation point in relation to pressure. In the phase diagram, the solid–vapor phase boundary slopes up to the right.

Solve

From the slope of the solid–vapor phase boundary, we see that as pressure increases, the temperature at which ice sublimes increases.

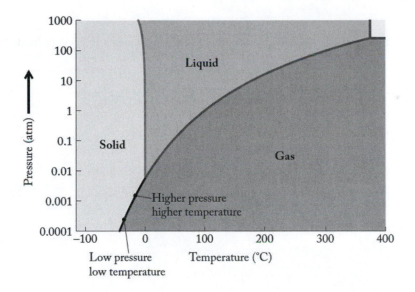

Think About It

With increased pressure, fewer water molecules enter the gas phase from the solid phase. That is also true for the liquid–vapor transition: as pressure increases, the boiling point increases.

6.97. **Collect and Organize**

Given that the melting point of hydrogen is at a higher temperature than its triple point, we are to determine whether liquid H_2 expands or contracts upon freezing.

Analyze

If the triple point is at a lower temperature than the melting point, the solid–liquid phase boundary must slope up and to the right in the phase diagram. That positive slope means that the solid phase has a higher density than the liquid phase.

Solve

Hydrogen contracts as it freezes because the phase diagram tells us that the solid phase is denser than the liquid phase.

Think About It

At very high pressures, solid hydrogen forms, in which the H—H bond of the diatomic molecule no longer exists and the solid hydrogen behaves like a metal.

6.99. Collect and Organize

From among the four molecules shown, we are to choose the one that would be soluble in both water and octanol.

Analyze

For a substance to dissolve well in both water and octanol, it should have both hydrophilic and hydrophobic groups. Hydrophilic groups are groups that might form hydrogen bonds with water or have strong bond dipoles. Hydrophobic groups are groups that are nonpolar.

Solve

(a) That molecule has few polar groups and mostly is dominated by its nonpolar carbon–hydrogen regions.
(b) That molecule has a balance of polar (–COOH and –NH$_2$) groups with nonpolar carbon–hydrogen bonds.
(c) That molecule is overall nonpolar; it is very symmetrical despite its polar C—F bonds.
(d) That molecule is dominated by polar OH groups with fewer nonpolar regions.
Because (b) has a balance of hydrophilic and hydrophobic groups, we expect that molecule to have similar solubility in water and in octanol.

Think About It

Molecule (c) is similar to Teflon, which is used to coat cooking utensils to prevent foods from sticking.

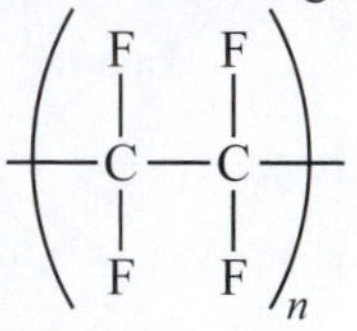

CHAPTER 7 | Stoichiometry: Mass Relationships and Chemical Reactions

7.1. Collect and Organize

For each molecule shown in Figure P7.1, we are to write empirical and molecular formulas to determine for which molecule(s) the empirical formula differs from the molecular formula.

Analyze

A molecular formula gives the number of each kind of atom in a molecule of the substance. An empirical formula is the lowest whole-number ratio of the atoms present in the substance.

Solve

(a) Molecular formula: N_2O_5; empirical formula: N_2O_5
(b) Molecular formula: N_2O_4; empirical formula: NO_2
(c) Molecular formula: NO; empirical formula: NO
(d) Molecular formula: N_2O_3; empirical formula: N_2O_3
(e) Molecular formula: NO_2; empirical formula: NO_2
Therefore, the formula for which the molecular formula differs from the empirical formula is (b) N_2O_4.

Think About It

For NO_2 and N_2O_4, the empirical formula is the same.

7.3. Collect and Organize

This exercise has to interpret diagrams (Figure P7.3) drawn from a molecular perspective in order to write a chemical reaction that includes an indication of the state of the substances (solid, liquid, or gas).

Analyze

When the atoms are isolated, they are written as atomic species. If the atoms are bound to each other, they are in the form of molecular species. Any representation of the substances that has a high degree of order (all molecules lined up) and a shape independent of the container represents the solid phase. Representations with less order that conform to the shape of the container show the liquid phase (molecules lined up with some order). Representations with randomly distributed substances that fill the container indicate the gas phase.

Solve

(a) Four atoms of X (red spheres) and four atoms of Y (blue spheres), both in the gas phase, are on the reactant side of the equation (left of the arrow). On the product side (right of the arrow), four gaseous molecules of XY (red–blue) are present. Therefore, the chemical equation reads

$$4\,X(g) + 4\,Y(g) \rightarrow 4\,XY(g)$$
$$X(g) + Y(g) \rightarrow XY(g)$$

(b) Four atoms of X (red spheres) and four atoms of Y (blue spheres), both in the gas phase, are on the reactant side of the equation. On the product side, four solid molecules of XY (red–blue) are present. Therefore, the chemical equation reads

$$4\,X(g) + 4\,Y(g) \rightarrow 4\,XY(s)$$
$$X(g) + Y(g) \rightarrow XY(s)$$

(c) Four atoms of X (red spheres) and four atoms of Y (blue spheres), both in the gas phase, are on the reactant side of the equation. On the product side, four gaseous substances are present: two molecules of XY_2 and two atoms of X. Therefore, the chemical equation reads

$$4\,X(g) + 4\,Y(g) \rightarrow 2\,XY_2(g) + 2\,X(g)$$
$$X(g) + 2\,Y(g) \rightarrow XY_2(g)$$

(d) Four molecules of X_2 (red spheres bonded together) and four molecules of Y_2 (bonded blue spheres), both in the gas phase, are on the reactant side of the equation. On the product side, eight gaseous molecules of XY (red–blue) are present. Therefore, the chemical equation reads

$$4\,X_2(g) + 4\,Y_2(g) \rightarrow 8\,XY(g)$$
$$X_2(g) + Y_2(g) \rightarrow 2\,XY(g)$$

Think About It

In parts a, b, and d when the reactants react, no leftover atoms of the reactants are present; however, in part c two atoms of X are left over in the reaction on the product side to make XY_2. Here, Y is the limiting reactant, and the number of Y atoms ultimately determines how many molecules of XY_2 will form from the reactant mixture.

7.5. Collect and Organize

Of the four hydrocarbons shown, we are to identify those that have the same percent composition.

Analyze

Molecules that have the same empirical formula will have the same percent composition.

Solve

(a) Empirical formula of C_6H_6 is CH.
(b) Empirical formula of C_2H_4 is CH_2.
(c) Empirical formula of C_3H_8 is C_3H_8.
(d) Empirical formula of C_6H_{12} is CH_2.
Therefore, (b) and (d) have the same percent composition of C and H.

Think About It

Another molecule that would have the same percent composition as C_2H_4 and C_6H_{12} would be C_4H_8.

7.7. Collect and Organize

Considering a reaction with the balanced equation

$$A + 2\,B \rightarrow C$$

in which 1.00 g of A is reacted with 4.00 g of B, we are to determine how many grams of C is formed.

Analyze

For this reaction we can assume that all the reactants form the product C.

Solve

By the law of conservation of mass, we can neither gain nor lose mass in a chemical reaction; therefore, when all of A reacts with all of B, we will have $1.00 + 4.00 = 5.00$ g of product C.

Think About It

If we did not have complete reaction between A and B to form C, we would still have a total mass of A, B, and C of 5.00 g.

7.9. Collect and Organize

Considering a reaction with the balanced equation from Question 7.8,

$$A + B \rightarrow C + D$$

we are to identify the reactant or product that would have the largest molar mass.

Analyze

The molar mass is the mass of the substance divided by the number of moles of that substance. Although we do not know the mole amount of each substance present in their masses used in the equation, we can use the fact that each one reacts in the equation in equimolar quantities.

Solve

If all are used or produced in the equation in equimolar amounts, then the reactant or product with the highest mass is the one with the highest molar mass. That is reactant B, for which the reaction uses 4.00 g.

Think About It
From the balanced chemical equation we get the information of how many moles of reactant reacts to form how many moles of product, not the number of grams.

7.11. Collect and Organize
If we alter the reaction conditions we can obtain either NO or NO_2 from the reaction of N_2 with O_2. In this problem we are asked to determine how many grams of O_2 would be produced from y grams of N_2 to form NO_2 if we know that x grams of O_2 reacts with y grams of N_2 to form the other product, NO.

Analyze
The ratios of the mass of reactants for the formation of those two products are

$$y \text{ grams } N_2 + x \text{ grams } O_2 \rightarrow NO$$
$$y \text{ grams } N_2 + ? \text{ grams } O_2 \rightarrow NO_2$$

Solve
Because twice as many oxygen atoms are present in NO_2 than in NO, we would need $2x$ grams of O_2 to react with y grams of N_2 to give NO_2.

Think About It
Here we applied the law of multiple proportions.

7.13. Collect and Organize
Given two formulas for oxides of iron, FeO and Fe_2O_3, we are to determine how much more oxygen would be required to combine with the same mass of iron used to form FeO.

Analyze
We have to be careful here: the ratio of iron to oxygen in FeO is 1:1, whereas the same ratio is 2:3 in Fe_2O_3.

Solve
We would need 3/2, or 1.5 times, as much oxygen for the reaction to form Fe_2O_3.

Think About It
You could also say that we need 50% more oxygen for the reaction.

7.15. Collect and Organize
We are asked whether the number of atoms in the reactants in a balanced chemical equation must equal the number of atoms in the products.

Analyze
A balanced chemical reaction follows the law of conservation of mass. Therefore, for each and every atom present in a reaction, we balance the number of a particular kind of atom (element) in reactants and products.

Solve
Yes, the number of atoms of reactants in a balanced chemical equation *must always* equal the number of atoms of the products.

Think About It
Elements in compounds are rearranged in chemical equations, and we may have fewer or more moles present after a reaction. For example, in the balanced equation for the production of ammonia,

$$N_2(g) + 3\ H_2(g) \rightarrow 2\ NH_3(g)$$

4 mol of reactants produces 2 mol of products. Therefore, the number of moles of reactants may not equal the number of moles of products.

7.17. Collect and Organize

For the combustion reaction of methane

$$CH_4(g) + 2\,O_2(g) \rightarrow CO_2(g) + 2\,H_2O(g)$$

we are to determine the number of moles of water vapor produced when 1 mol of methane is used in the reaction.

Analyze

From the balanced equation we see that the ratio of methane consumed to water produced is 1:2.

Solve

For every mole of methane combusted, 2 mol of water vapor will be produced.

Think About It

Likewise from the balanced equation, for every mole of methane burned, 1 mol of carbon dioxide is produced.

7.19. Collect and Organize

To balance these chemical reactions we use the three steps described in the textbook.

Analyze

To balance each equation we first write the unbalanced equation by using the chemical formulas of the reactants and products and take inventory of the atoms in the reactants and products. Next, we balance an element that is present in only one reactant and product, again taking the inventory of atoms. Finally, we balance the other elements present by placing coefficients in front of the species in the reaction so that the number of the atoms for each element is equal on both sides of the equation. If any fractional coefficients are present, we multiply the entire equation through to eliminate all fractions.

Solve

(a) The unbalanced reaction is

$$N_2(g) + O_2(g) \rightarrow NO(g)$$

Atoms: $2\,N + 2\,O \rightarrow 1\,N + 1\,O$

We can best start by balancing the N atoms by placing a 2 in front of NO on the right-hand side.

$$N_2(g) + O_2(g) \rightarrow 2\,NO(g)$$

Atoms: $2\,N + 2\,O \rightarrow 2\,N + 2\,O$

The equation is now balanced.

(b) The unbalanced reaction is

$$N_2(g) + O_2(g) \rightarrow N_2O(g)$$

Atoms: $2\,N + 2\,O \rightarrow 2\,N + 1\,O$

We start by balancing the O atoms by placing a 2 in front of N_2O on the right-hand side.

$$N_2(g) + O_2(g) \rightarrow 2\,N_2O(g)$$

Atoms: $2\,N + 2\,O \rightarrow 4\,N + 2\,O$

We next can balance the N atoms by placing a 2 in front of N_2 on the left-hand side.

$$2\,N_2(g) + O_2(g) \rightarrow 2\,N_2O(g)$$

Atoms: $4\,N + 2\,O \rightarrow 4\,N + 2\,O$

The equation is now balanced.

(c) The unbalanced reaction is

$$NO(g) + NO_3(g) \rightarrow NO_2(g)$$

Atoms: $2\,N + 4\,O \rightarrow 1\,N + 2\,O$

We can start by balancing the N atoms by placing a 2 in front of NO_2 on the right-hand side.

$$NO(g) + NO_3(g) \rightarrow 2\,NO_2(g)$$

Atoms: $2\,N + 4\,O \rightarrow 2\,N + 4\,O$

The equation is now balanced.

(d) The unbalanced reaction is
$$NO(g) + O_2(g) + H_2O(\ell) \rightarrow HNO_2(\ell)$$

Atoms: $1\,N + 4\,O + 2\,H \rightarrow 1\,N + 2\,O + 1\,H$

We can start by balancing the H atoms by placing a coefficient of 2 in front of HNO_2 on the right-hand side.
$$NO(g) + O_2(g) + H_2O(\ell) \rightarrow 2\,HNO_2(\ell)$$

Atoms: $1\,N + 4\,O + 2\,H \rightarrow 2\,N + 4\,O + 2\,H$

We can then balance the N atoms by placing a coefficient of 2 in front of NO on the left-hand side.
$$2\,NO(g) + O_2(g) + H_2O(\ell) \rightarrow 2\,HNO_2(\ell)$$

Atoms: $2\,N + 5\,O + 2\,H \rightarrow 2\,N + 4\,O + 2\,H$

We can then balance the O atoms by placing a coefficient of $\frac{1}{2}$ in front of O_2 on the left-hand side.
$$2\,NO(g) + 1/2\,O_2(g) + H_2O(\ell) \rightarrow 2\,HNO_2(\ell)$$

Atoms: $2\,N + 4\,O + 2\,H \rightarrow 2\,N + 4\,O + 2\,H$

To eliminate the fractional coefficients we multiply all the coefficients by 2.
$$4\,NO(g) + O_2(g) + 2\,H_2O(\ell) \rightarrow 4\,HNO_2(\ell)$$

Atoms: $4\,N + 8\,O + 4\,H \rightarrow 4\,N + 8\,O + 4\,H$

The equation is now balanced.

Think About It
Many gaseous oxides of nitrogen exist because nitrogen can occur in compounds with several charges. In those reactions we see nitrogen's charges as 0 (N_2), 1+ (N_2O), 2+ (NO), 4+ (NO_2), and 6+ (NO_3).

7.21. Collect and Organize
We are asked to write balanced chemical equations for the reactions described. Because we are given only names, not chemical formulas for the reactants and products, we have to be sure to correctly write formulas to balance the equations. To balance those chemical reactions we use the three steps described in the textbook.

Analyze
To balance each equation we first write the unbalanced equation by using the chemical formulas of the reactants and products and take inventory of the atoms in the reactants and products. Next, we balance an element that is present in only one reactant and product, again taking the inventory of atoms. Finally, we balance the other elements present by placing coefficients in front of the species in the reaction so that the number of the atoms for each element is equal on both sides of the equation. If any fractional coefficients are present, we multiply the entire equation through to eliminate all fractions.

Solve
(a) The unbalanced reaction is
$$N_2O_5(g) + Na(s) \rightarrow NaNO_3(s) + NO_2(g)$$

Atoms: $2\,N + 5\,O + 1\,Na \rightarrow 2\,N + 5\,O + 1\,Na$

The numbers of Na, N, and O atoms on the reactants and products side are all equal. That reaction is already balanced.

(b) The unbalanced reaction is
$$N_2O_4(g) + H_2O\,(\ell) \rightarrow HNO_3(aq) + HNO_2(aq)$$

Atoms: $2\,N + 5\,O + 2\,H \rightarrow 2\,N + 5\,O + 2\,H$

The numbers of N, O, and H atoms on the reactants and products side are all equal. That reaction is already balanced.

(c) The unbalanced reaction is
$$NO(g) \rightarrow N_2O(g) + NO_2(g)$$

Atoms: $1\,N + 1\,O \rightarrow 3\,N + 3\,O$

We can start by balancing the N atoms. We can do that by placing a 3 in front of NO on the left-hand side.

$$3\,NO(g) \rightarrow N_2O(g) + NO_2(g)$$

Atoms: $3\,N + 3\,O \rightarrow 3\,N + 3\,O$

That step also balanced the O atoms. The equation is now balanced.

Think About It
The first two chemical reactions were balanced as written, and we did not need to change the coefficients. When writing chemical equations, however, making sure that the equation is balanced is always best.

7.23. Collect and Organize
For the combustion of several hydrocarbons, we are to complete the equations and balance the equations.

Analyze
In the combustion of hydrocarbons, the only products are carbon dioxide and water. To balance the reactions we will use the three steps described in the textbook.

Solve
(a) The unbalanced reaction is

$$C_3H_8(g) + O_2(g) \rightarrow CO_2(g) + H_2O(g)$$

Atoms: $3\,C + 8\,H + 2\,O \rightarrow 1\,C + 2\,H + 3\,O$

We can start by balancing the C atoms, since they appear in only one reactant and product, by placing a 3 in front of CO_2 on the right-hand side.

$$C_3H_8(g) + O_2(g) \rightarrow 3\,CO_2(g) + H_2O(g)$$

Atoms: $3\,C + 8\,H + 2\,O \rightarrow 3\,C + 2\,H + 7\,O$

Next we can balance the H atoms by placing a 4 in front of H_2O on the right-hand side.

$$C_3H_8(g) + O_2(g) \rightarrow 3\,CO_2(g) + 4\,H_2O(g)$$

Atoms: $3\,C + 8\,H + 2\,O \rightarrow 3\,C + 8\,H + 10\,O$

Finally we can balance the O atoms by placing a 5 in front of O_2 on the left-hand side.

$$C_3H_8(g) + 5\,O_2(g) \rightarrow 3\,CO_2(g) + 4\,H_2O(g)$$

Atoms: $3\,C + 8\,H + 10\,O \rightarrow 3\,C + 8\,H + 10\,O$

The equation is now balanced.
(b) The unbalanced reaction is

$$C_4H_{10}(g) + O_2(g) \rightarrow CO_2(g) + H_2O(g)$$

Atoms: $4\,C + 10\,H + 2\,O \rightarrow 1\,C + 2\,H + 3\,O$

We can start by balancing the C atoms since they appear in only one reactant and product, by placing a 4 in front of CO_2 on the right-hand side.

$$C_4H_{10}(g) + O_2(g) \rightarrow 4\,CO_2(g) + H_2O(g)$$

Atoms: $4\,C + 10\,H + 2\,O \rightarrow 4\,C + 2\,H + 9\,O$

Next we can balance the H atoms by placing a 5 in front of H_2O on the right-hand side.

$$C_4H_{10}(g) + O_2(g) \rightarrow 4\,CO_2(g) + 5\,H_2O(g)$$

Atoms: $4\,C + 10\,H + 2\,O \rightarrow 4\,C + 10\,H + 13\,O$

Finally we can balance the O atoms by placing 13/2 in front of O_2 on the left-hand side.

$$C_4H_{10}(g) + 13/2\,O_2(g) \rightarrow 4\,CO_2(g) + 5\,H_2O(g)$$

Atoms: $4\,C + 10\,H + 13\,O \rightarrow 4\,C + 10\,H + 13\,O$

To eliminate the fractional coefficients we multiply all the coefficients by 2.

$$2\,C_4H_{10}(g) + 13\,O_2(g) \rightarrow 8\,CO_2(g) + 10\,H_2O(g)$$

Atoms: $8\,C + 20\,H + 26\,O \rightarrow 8\,C + 20\,H + 26\,O$

The equation is now balanced.

(c) The unbalanced reaction is

$$C_6H_6(\ell) + O_2(g) \rightarrow CO_2(g) + H_2O(g)$$

Atoms: $6\,C + 6\,H + 2\,O \rightarrow 1\,C + 2\,H + 3\,O$

We can start by balancing the C atoms since they appear in only one reactant and product, by placing a 6 in front of CO_2 on the right-hand side.

$$C_6H_6(\ell) + O_2(g) \rightarrow 6\,CO_2(g) + H_2O(g)$$

Atoms: $6\,C + 6\,H + 2\,O \rightarrow 6\,C + 2\,H + 13\,O$

Next we can balance the H atoms by placing a 3 in front of H_2O on the right-hand side.

$$C_6H_6(\ell) + O_2(g) \rightarrow 6\,CO_2(g) + 3\,H_2O(g)$$

Atoms: $6\,C + 6\,H + 2\,O \rightarrow 6\,C + 6\,H + 15\,O$

Finally we can balance the O atoms by placing 15/2 in front of O_2 on the left-hand side.

$$C_6H_6(\ell) + 15/2\,O_2(g) \rightarrow 6\,CO_2(g) + 3\,H_2O(g)$$

Atoms: $6\,C + 6\,H + 15\,O \rightarrow 6\,C + 6\,H + 15\,O$

To eliminate the fractional coefficients we multiply all the coefficients by 2.

$$2\,C_6H_6(\ell) + 15\,O_2(g) \rightarrow 12\,CO_2(g) + 6\,H_2O(g)$$

Atoms: $12\,C + 12\,H + 30\,O \rightarrow 12\,C + 12\,H + 30\,O$

The equation is now balanced.

(d) The unbalanced reaction is

$$C_8H_{18}(\ell) + O_2(g) \rightarrow CO_2(g) + H_2O(g)$$

Atoms: $8\,C + 18\,H + 2\,O \rightarrow 1\,C + 2\,H + 3\,O$

We can start by balancing the C atoms since they appear in only one reactant and product, by placing an 8 in front of CO_2 on the right-hand side.

$$C_8H_{18}(\ell) + O_2(g) \rightarrow 8\,CO_2(g) + H_2O(g)$$

Atoms: $8\,C + 18\,H + 2\,O \rightarrow 8\,C + 2\,H + 17\,O$

Next we can balance the H atoms by placing a 9 in front of H_2O on the right-hand side.

$$C_8H_{18}(\ell) + O_2(g) \rightarrow 8\,CO_2(g) + 9\,H_2O(g)$$

Atoms: $8\,C + 18\,H + 2\,O \rightarrow 8\,C + 18\,H + 25\,O$

Finally we can balance the O atoms by placing 25/2 in front of O_2 on the left-hand side.

$$C_8H_{18}(\ell) + 25/2\,O_2(g) \rightarrow 8\,CO_2(g) + 9\,H_2O(g)$$

Atoms: $8\,C + 18\,H + 25\,O \rightarrow 8\,C + 18\,H + 25\,O$

To eliminate the fractional coefficients we multiply all the coefficients by 2.

$$2\,C_8H_{18}(\ell) + 25\,O_2(g) \rightarrow 16\,CO_2(g) + 18\,H_2O(g)$$

Atoms: $16\,C + 36\,H + 50\,O \rightarrow 16\,C + 36\,H + 50\,O$

The equation is now balanced.

Think About It

The higher the carbon content of the hydrocarbon, the more moles of CO_2 released per mole of the hydrocarbon.

7.25. Collect and Organize

For the combustion of several gaseous hydrocarbons for which we are given the structures, we are to complete the equations and balance the equations.

Analyze

In the combustion of hydrocarbons, the only products are carbon dioxide and water. To balance the reactions we will use the three steps described in the textbook.

Solve

(a) The unbalanced reaction is

$$C_2H_4(g) + O_2(g) \rightarrow CO_2(g) + H_2O(g)$$

Atoms: $2\,C + 4\,H + 2\,O \rightarrow 1\,C + 2\,H + 3\,O$

We can start by balancing the C atoms, since they appear in only one reactant and product, by placing a 2 in front of CO_2 on the right-hand side.

$$C_2H_4(g) + O_2(g) \rightarrow 2\,CO_2(g) + H_2O(g)$$

Atoms: $2\,C + 4\,H + 2\,O \rightarrow 2\,C + 2\,H + 5\,O$

Next we can balance the H atoms by placing a 2 in front of H_2O on the right-hand side.

$$C_2H_4(g) + O_2(g) \rightarrow 2\,CO_2(g) + 2\,H_2O(g)$$

Atoms: $2\,C + 4\,H + 2\,O \rightarrow 2\,C + 4\,H + 6\,O$

Finally we can balance the O atoms by placing a 3 in front of O_2 on the left-hand side.

$$C_2H_4(g) + 3\,O_2(g) \rightarrow 2\,CO_2(g) + 2\,H_2O(g)$$

Atoms: $2\,C + 4\,H + 6\,O \rightarrow 2\,C + 4\,H + 6\,O$

The equation is now balanced.

(b) The unbalanced reaction is

$$C_3H_6(g) + O_2(g) \rightarrow CO_2(g) + H_2O(g)$$

Atoms: $3\,C + 6\,H + 2\,O \rightarrow 1\,C + 2\,H + 3\,O$

We can start by balancing the C atoms, since they appear in only one reactant and product, by placing a 3 in front of CO_2 on the right-hand side.

$$C_3H_6(g) + O_2(g) \rightarrow 3\,CO_2(g) + H_2O(g)$$

Atoms: $3\,C + 6\,H + 2\,O \rightarrow 3\,C + 2\,H + 7\,O$

Next we can balance the H atoms by placing a 3 in front of H_2O on the right-hand side.

$$C_3H_6(g) + O_2(g) \rightarrow 3\,CO_2(g) + 3\,H_2O(g)$$

Atoms: $3\,C + 6\,H + 2\,O \rightarrow 3\,C + 6\,H + 9\,O$

Finally we can balance the O atoms by placing 9/2 in front of O_2 on the left-hand side.

$$C_3H_6(g) + 9/2\,O_2(g) \rightarrow 3\,CO_2(g) + 3\,H_2O(g)$$

Atoms: $3\,C + 6\,H + 9\,O \rightarrow 3\,C + 6\,H + 9\,O$

To eliminate the fractional coefficients we multiply all the coefficients by 2.

$$2\,C_3H_6(g) + 9\,O_2(g) \rightarrow 6\,CO_2(g) + 6\,H_2O(g)$$

Atoms: $6\,C + 12\,H + 18\,O \rightarrow 6\,C + 12\,H + 18\,O$

The equation is now balanced.

(c) The unbalanced reaction is

$$C_4H_{10}(g) + O_2(g) \rightarrow CO_2(g) + H_2O(g)$$

Atoms: $4\,C + 10\,H + 2\,O \rightarrow 1\,C + 2\,H + 3\,O$

We can start by balancing the C atoms, since they appear in only one reactant and product, by placing a 4 in front of CO_2 on the right-hand side.

$$C_4H_{10}(g) + O_2(g) \rightarrow 4\,CO_2(g) + H_2O(g)$$

Atoms: $4\,C + 10\,H + 2\,O \rightarrow 4\,C + 2\,H + 9\,O$

Next we can balance the H atoms by placing a 5 in front of H_2O on the right-hand side.

$$C_4H_{10}(g) + O_2(g) \rightarrow 4\,CO_2(g) + 5\,H_2O(g)$$

Atoms: $4\,C + 10\,H + 2\,O \rightarrow 4\,C + 10\,H + 13\,O$

Finally we can balance the O atoms by placing 13/2 in front of O_2 on the left-hand side.

$$C_4H_{10}(g) + 13/2\,O_2(g) \rightarrow 4\,CO_2(g) + 5\,H_2O(g)$$

Atoms: $4\,C + 10\,H + 13\,O \rightarrow 4\,C + 10\,H + 13\,O$

To eliminate the fractional coefficients we multiply all the coefficients by 2.

$$2 \, C_4H_{10}(g) + 13 \, O_2(g) \rightarrow 8 \, CO_2(g) + 10 \, H_2O(g)$$

Atoms: $8 \, C + 20 \, H + 26 \, O \rightarrow 8 \, C + 20 \, H + 26 \, O$

The equation is now balanced.

(d) The unbalanced reaction is

$$C_4H_8(g) + O_2(g) \rightarrow CO_2(g) + H_2O(g)$$

Atoms: $4 \, C + 8 \, H + 2 \, O \rightarrow 1 \, C + 2 \, H + 3 \, O$

We can start by balancing the C atoms, since they appear in only one reactant and product, by placing a 4 in front of CO_2 on the right-hand side.

$$C_4H_8(g) + O_2(g) \rightarrow 4 \, CO_2(g) + H_2O(g)$$

Atoms: $4 \, C + 8 \, H + 2 \, O \rightarrow 4 \, C + 2 \, H + 9 \, O$

Next we can balance the H atoms by placing a 4 in front of H_2O on the right-hand side.

$$C_4H_8(g) + O_2(g) \rightarrow 4 \, CO_2(g) + 4 \, H_2O(g)$$

Atoms: $4 \, C + 8 \, H + 2 \, O \rightarrow 4 \, C + 8 \, H + 12 \, O$

Finally we can balance the O atoms by placing a 6 in front of O_2 on the left-hand side.

$$C_4H_8(g) + 6 \, O_2(g) \rightarrow 4 \, CO_2(g) + 4 \, H_2O(g)$$

Atoms: $4 \, C + 8 \, H + 12 \, O \rightarrow 4 \, C + 8 \, H + 12 \, O$

The equation is now balanced.

Think About It
Be careful in counting the atoms from a ball-and-stick structure. It is easy to miss atoms and arrive at the wrong molecular formula.

7.27. Collect and Organize
To balance the chemical equations, we use the three steps described in the textbook.

Analyze
To balance each equation we first write the unbalanced equation by using the chemical formulas of the reactants and products and take inventory of the atoms in the reactants and products. Next, we balance an element that is present in only one reactant and product, again taking the inventory of atoms. Finally, we balance the other elements present by placing coefficients in front of the species in the reaction so that the number of the atoms for each element is equal on both sides of the equation. If any fractional coefficients are present, we multiply the entire equation through to eliminate all fractions.

Solve
(a) For the reaction of sulfur dioxide with oxygen to form sulfur trioxide, the unbalanced reaction is

$$SO_2(g) + O_2(g) \rightarrow SO_3(g)$$

Atoms: $1 \, S + 4 \, O \rightarrow 1 \, S + 3 \, O$

The sulfur atoms are already balanced. To balance the O atoms, therefore, we place $\frac{1}{2}$ as the coefficient before O_2 on the left-hand side of the equation.

$$SO_2(g) + \frac{1}{2} \, O_2(g) \rightarrow SO_3(g)$$

Atoms: $1 \, S + 3 \, O \rightarrow 1 \, S + 3 \, O$

To eliminate the fractional coefficients we multiply all the coefficients by 2.

$$2 \, SO_2(g) + O_2(g) \rightarrow 2 \, SO_3(g)$$

Atoms: $1 \, S + 6 \, O \rightarrow 2 \, S + 6 \, O$

The equation is now balanced.

(b) For the reaction of hydrogen sulfide with oxygen to form sulfur dioxide and water, the unbalanced reaction is

$$H_2S(g) + O_2(g) \rightarrow SO_2(g) + H_2O(g)$$

Atoms: $2 \, H + 1 \, S + 2 \, O \rightarrow 2 \, H + 1 \, S + 3 \, O$

To balance the O atoms, therefore, we place $\frac{3}{2}$ as the coefficient before O_2 on the left-hand side of the equation.

$$H_2S(g) + 3/2\ O_2(g) \rightarrow SO_2(g) + H_2O(g)$$

Atoms: $\quad 2\,H + 1\,S + 3\,O \rightarrow 2\,H + 1\,S + 3\,O$

To eliminate the fractional coefficients we multiply all the coefficients by 2.

$$2\ H_2S(g) + 3\ O_2(g) \rightarrow 2\ SO_2(g) + 2\ H_2O(g)$$

Atoms: $\quad 4\,H + 2\,S + 6\,O \rightarrow 4\,H + 2\,S + 6\,O$

The equation is now balanced.

(c) For the reaction of hydrogen sulfide with sulfur dioxide to form sulfur and water, the unbalanced reaction is

$$H_2S(g) + SO_2(g) \rightarrow S_8(s) + H_2O(g)$$

Atoms: $\quad 2\,H + 2\,S + 2\,O \rightarrow 2\,H + 8\,S + 1\,O$

We can best start by balancing the oxygen atoms, since they appear in only one reactant and product, by placing a 2 in front of H_2O on the product side.

$$H_2S(g) + SO_2(g) \rightarrow S_8(s) + 2\ H_2O(g)$$

Atoms: $\quad 2\,H + 2\,S + 2\,O \rightarrow 4\,H + 8\,S + 2\,O$

To balance the H atoms, therefore, we place 2 as the coefficient before H_2S on the left-hand side of the equation.

$$2\ H_2S(g) + SO_2(g) \rightarrow S_8(s) + 2\ H_2O(g)$$

Atoms: $\quad 4\,H + 3\,S + 2\,O \rightarrow 4\,H + 8\,S + 2\,O$

To balance the sulfur atoms, we place a coefficient of $\frac{3}{8}$ in front of S_8 on the product side.

$$2\ H_2S(g) + SO_2(g) \rightarrow 3/8\ S_8(s) + 2\ H_2O(g)$$

Atoms: $\quad 4\,H + 3\,S + 2\,O \rightarrow 4\,H + 3\,S + 2\,O$

To eliminate the fractional coefficients we multiply all the coefficients by 8.

$$16\ H_2S(g) + 8\ SO_2(g) \rightarrow 3\ S_8(s) + 16\ H_2O(g)$$

Atoms: $\quad 32\,H + 24\,S + 16\,O \rightarrow 32\,H + 24\,S + 16\,O$

The equation is now balanced.

Think About It
Part c has large coefficients because of the formation of S_8, the most stable elemental form of sulfur.

7.29. Collect and Organize
We compare two valid ways to write a balanced equation for the combustion of ethane, and we decide whether the form of the balanced equation matters when calculating the amount of product formed from a given amount of a reactant.

Analyze
A balanced chemical reaction has the same number of a particular kind of atom (element) in the reactants as in the products.

Solve
Both equations give the same numerical answer for the amount of CO_2 produced in the reaction for a given amount of C_2H_6 because in both the molar ratio of CO_2 to C_2H_6 is 2:1.

Think About It
When we use balanced equations for stoichiometric calculations, the ratio of reactants and products to each other is important, not the actual numerical value of the coefficients.

7.31. Collect and Organize
We need to convert the given mass of carbon by which emissions would be reduced first to moles of carbon and then to the mass of carbon dioxide.

Analyze

We can convert the mass of carbon to moles by dividing by the average molar mass of carbon after converting the mass of carbon to grams from kilograms (1000 g = 1 kg). Because 1 mol of carbon dioxide contains 1 mol of carbon, the moles of carbon are equal to the moles of carbon dioxide. To find the mass of CO_2 in grams, we need only multiply the moles of CO_2 by the molar mass of CO_2 and then convert that mass (which will be in grams) to kilograms of CO_2.

Solve

(a) 5.4×10^9 kg C $\times \dfrac{1000 \text{ g}}{1 \text{ kg}} \times \dfrac{1 \text{ mol C}}{12.01 \text{ g}} = 4.5 \times 10^{11}$ mol C = mol CO_2

(b) 4.5×10^{11} mol $CO_2 \times \dfrac{44.01 \text{ g}}{1 \text{ mol } CO_2} \times \dfrac{1 \text{ kg}}{1000 \text{ g}} = 2.0 \times 10^{10}$ kg CO_2

Think About It

The mass of CO_2 emissions that would be reduced is greater than the mass of carbon burned. The molar mass of carbon dioxide is greater, so the mass of a certain molar amount of carbon dioxide is greater than that of the same molar amount of carbon.

7.33. **Collect and Organize**

To calculate the amount of CO_2 produced from the decomposition of 25.0 g of $NaHCO_3$, we need the balanced chemical equation to use in calculating the molar ratio of the CO_2 produced from a given mass of $NaHCO_3$.

Analyze

We know that the reactant for the balanced equation is $NaHCO_3$ and that the products are Na_2CO_3, H_2O, and CO_2. After we balance the equation, we can use the ratio of $NaHCO_3$ to CO_2 to find the moles of CO_2 from the moles of $NaHCO_3$ (found from the mass by dividing by the molar mass, 84.01 g/mol). From the moles of CO_2, we can find the mass of CO_2 produced by using 44.01 g/mol for the molar mass of CO_2.

Solve

(a) The unbalanced reaction is

$$NaHCO_3(s) \rightarrow CO_2(g) + H_2O(g) + Na_2CO_3(s)$$

Atoms: $1 \text{ Na} + 1 \text{ H} + 1 \text{ C} + 3 \text{ O} \rightarrow 2 \text{ Na} + 2 \text{ H} + 2 \text{ C} + 6 \text{ O}$

We notice here that the number of atoms in the products is always twice that in the reactants.
We can best start balancing the atoms by placing a 2 in front of $NaHCO_3$ on the left-hand side.

$$2 \, NaHCO_3(s) \rightarrow CO_2(g) + H_2O(g) + Na_2CO_3(s)$$

Atoms: $2 \text{ Na} + 2 \text{ H} + 2 \text{ C} + 6 \text{ O} \rightarrow 2 \text{ Na} + 2 \text{ H} + 2 \text{ C} + 6 \text{ O}$

The equation is now balanced.

(b) 25.0 g $NaHCO_3 \times \dfrac{1 \text{ mol } NaHCO_3}{84.01 \text{ g}} \times \dfrac{1 \text{ mol } CO_2}{2 \text{ mol } NaHCO_3} \times \dfrac{44.01 \text{ g } CO_2}{1 \text{ mol}} = 6.55$ g CO_2

Think About It

The mass of CO_2 produced is quite a bit less than the 25 g of $NaHCO_3$ decomposed not only because the molar mass of CO_2 is lower than that of $NaHCO_3$ but also because for every 1 mol of $NaHCO_3$ decomposed, only ½ mol of CO_2 is produced.

7.35. **Collect and Organize**

We use stoichiometric relationships to calculate the amount of a reactant, $NaAlO_2$, required to produce a given amount of cryolite, Na_3AlF_6.

Analyze

First we need to calculate the moles of Na_3AlF_6 present in 1.00 kg by using 1000 g = 1 kg and the molar mass of Na_3AlF_6 (209.94 g/mol). From that and the 3:1 ratio of $NaAlO_2$ to Na_3AlF_6 in the balanced equation, we can calculate the moles of $NaAlO_2$ required. Finally, we use the molar mass of $NaAlO_2$ (81.97 g/mol) to convert the moles into mass.

Solve

$$1.00 \text{ kg Na}_3\text{AlF}_6 \times \frac{1000 \text{ g}}{1 \text{ kg}} \times \frac{1 \text{ mol Na}_3\text{AlF}_6}{209.94 \text{ g}} \times \frac{3 \text{ mol NaAlO}_2}{1 \text{ mol Na}_3\text{AlF}_6} \times \frac{81.97 \text{ g NaAlO}_2}{1 \text{ mol}} = 1170 \text{ g, or } 1.17 \text{ kg}$$

Think About It

In that reaction we need three times more moles of the reactant $NaAlO_2$ to yield 1 mol of product, Na_3AlF_6. Be careful here. The mass of $NaAlO_2$ required is not three times the mass of the product; the molar relationship is the important factor.

7.37. Collect and Organize

We have to determine first the moles of each reactant we have and then calculate the moles and mass of oxygen that 85 g of KO_2 would theoretically produce.

Analyze

We are given the balanced chemical equation for the reaction. For every 4 mol of KO_2 used, 3 mol of O_2 is produced. The molar mass of KO_2, computed from the molar masses of the elements in the periodic table, is 71.10 g/mol, and for O_2, the molar mass is 32.00 g/mol.

Solve

The theoretical yield of oxygen from KO_2 is

$$85 \text{ g KO}_2 \times \frac{1 \text{ mol KO}_2}{71.10 \text{ g}} \times \frac{3 \text{ mol O}_2}{4 \text{ mol KO}_2} \times \frac{32.00 \text{ g O}_2}{1 \text{ mol O}_2} = 29 \text{ g O}_2$$

Think About It

Not only was KO_2 present in lower gram masses in the reaction, but it also has a significantly larger molar mass. Therefore, it is present in the least amount for that reaction. Be careful, though; you must also consider the molar ratio of the reactants to products in the balanced equation.

7.39. Collect and Organize

We are given balanced chemical reactions for converting UO_2 into UF_6 to enrich the uranium to use as a nuclear fuel. We are asked to find the mass of HF in the first reaction that will convert 5.00 kg of UO_2 to UF_4. In the second part, we calculate how much final product (UF_6) can be produced from a given mass of starting material, UO_2.

Analyze

For part a we need to find the moles of UO_2 used (from the molar mass of UO_2, 270.03 g/mol) and then use the stoichiometric ratio of UO_2 to HF (1:4) in the balanced equation to find the moles of HF required. From that we can determine the mass of HF needed by using the molar mass of HF (20.01 g/mol). For part b we first find the moles of UO_2 in 850.0 g by using the molar mass. Because 1 mol of UF_4 is produced for every mole of UO_2 reacted and 1 mol of UF_4 is used for every mole of UF_6 produced (ratio of 1:1:1), the moles of UO_2 equals the moles of UF_6 produced. From that result, we can then find the mass of UF_6 produced by using the molar mass of UF_6 (352.02 g/mol).

Solve

(a) $5.00 \text{ kg UO}_2 \times \frac{1000 \text{ g}}{1 \text{ kg}} \times \frac{1 \text{ mol UO}_2}{270.03 \text{ g}} \times \frac{4 \text{ mol HF}}{1 \text{ mol UO}_2} \times \frac{20.01 \text{ g}}{1 \text{ mol HF}} = 1480 \text{ g, or } 1.48 \text{ kg HF}$

(b) $850.0 \text{ g UO}_2 \times \frac{1 \text{ mol UO}_2}{270.03 \text{ g}} \times \frac{1 \text{ mol UF}_6}{1 \text{ mol UO}_2} \times \frac{352.02 \text{ g}}{1 \text{ mol UF}_6} = 1110 \text{ g, or } 1.11 \text{ kg UF}_6$

Think About It

Because UF_6 has a higher molar mass, the reaction produces more uranium product (in terms of mass) than the mass of the original reactant. Even though the molar mass of UF_6 is high, it is a volatile liquid and can be "distilled" in large columns to separate the isotopes. The lighter isotope, U-235, is not as heavy as the U-238 isotope and therefore can be enriched at the top of the column.

7.41. Collect and Organize

In converting chalcopyrite ($CuFeS_2$) to copper, we have to take into account that 1 mol of copper atoms is in the formula for that mineral.

Analyze

The problem asks how much copper could be produced from 1.00 kg of the mineral and looks like many other stoichiometry problems. We will have to use the molar mass of the mineral (183.52 g/mol) to find the moles of the mineral. From there, knowing that 1 mol of copper atoms is in 1 mol of the mineral, we can use the molar mass of copper (63.55 g/mol) to calculate the amount of copper that would be produced.

Solve

$$1.00 \text{ kg CuFeS}_2 \times \frac{1000 \text{ g}}{1 \text{ kg}} \times \frac{1 \text{ mol}}{183.52 \text{ g}} \times \frac{1 \text{ mol Cu}}{1 \text{ mol CuFeS}_2} \times \frac{63.55 \text{ g Cu}}{1 \text{ mol}} = 346 \text{ g Cu}$$

Think About It

Our calculation tells us that the ore is 34.6% Cu by mass.

7.43. Collect and Organize

We are asked to distinguish between *empirical formula* and *molecular formula*.

Analyze

An empirical formula gives the simplest whole-number ratio of atoms of the elements in a molecule, whereas a molecular formula gives the actual number of atoms in a molecule.

Solve

An empirical formula shows the lowest whole-number ratios of atoms in a substance. A molecular formula shows the actual numbers of each kind of atom that compose one molecule of the substance.

Think About It

Sometimes the molecular formula is equivalent to the empirical formula when the atoms in the molecular formula are in their lowest whole-number ratios.

7.45. Collect and Organize

We are asked whether the atom in a molecular formula with the largest molar mass is always the element present in the highest percentage by mass.

Analyze

The percent composition of a substance is the mass of each element in the compound divided by the molar mass of the compound. In calculating the percent mass for each element, we need to take into account how many atoms of that element are present in the molecular formula.

Solve

No, lighter elements may be present in quantities large enough to be of a greater percentage of the mass than a heavier element.

Think About It

A good example is SiO_2. Silicon is the heavier element (28 g/mol), but the presence of two oxygen atoms (16 g/mol × 2) gives 53% O by mass but only 47% Si by mass.

7.47. Collect and Organize

To calculate the percent composition for the elements in each compound we divide the molar mass of each element from the periodic table by the molar mass for the compound and convert to a percentage.

Analyze

All the chemical formulas are given for the compounds. Assume that we have 1 mol of each compound. We first compute the molar mass. Then, to find the percentage of each element, divide the mass of each element present in the compound, taking into account the presence of multiple atoms of the element if appropriate, by the molar mass of the compound and multiply by 100.

Solve

(a) Molar mass of Na_2O = 61.98 g/mol.

$$\% \text{ Na} = \frac{(22.99 \times 2) \text{ g Na}}{61.98 \text{ g}} \times 100 = 74.19\% \text{ Na}$$

$$\% \text{ O} = \frac{16.00 \text{ g O}}{61.98 \text{ g}} \times 100 = 25.81\% \text{ O}$$

(b) Molar mass of NaOH = 40.00 g/mol.

$$\% \text{ Na} = \frac{22.99 \text{ g Na}}{40.00 \text{ g}} \times 100 = 57.48\% \text{ Na}$$

$$\% \text{ O} = \frac{16.00 \text{ g O}}{40.00 \text{ g}} \times 100 = 40.00\% \text{ O}$$

$$\% \text{ H} = \frac{1.01 \text{ g H}}{40.00 \text{ g}} \times 100 = 2.52\% \text{ H}$$

(c) Molar mass of $NaHCO_3$ = 84.01 g/mol.

$$\% \text{ Na} = \frac{22.99 \text{ g Na}}{84.01 \text{ g}} \times 100 = 27.37\% \text{ Na}$$

$$\% \text{ H} = \frac{1.01 \text{ g H}}{84.01 \text{ g}} \times 100 = 1.20\% \text{ H}$$

$$\% \text{ C} = \frac{12.01 \text{ g C}}{84.01 \text{ g}} \times 100 = 14.30\% \text{ C}$$

$$\% \text{ O} = \frac{(16.00 \times 3) \text{ g O}}{84.01 \text{ g}} \times 100 = 57.13\% \text{ O}$$

(d) Molar mass of Na_2CO_3 = 106.0 g/mol.

$$\% \text{ Na} = \frac{(22.99 \times 2) \text{ g Na}}{106.0 \text{ g}} \times 100 = 43.38\% \text{ Na}$$

$$\% \text{ C} = \frac{12.01 \text{ g C}}{106.0 \text{ g}} \times 100 = 11.33\% \text{ C}$$

$$\% \text{ O} = \frac{(16.00 \times 3) \text{ g O}}{106.0 \text{ g}} \times 100 = 45.28\% \text{ O}$$

Think About It

For all those common salts of sodium, the percentage of sodium is different. That result is due not only to the different atom ratios of sodium present in the compound but also to the different molar masses of the compounds.

7.49. Collect and Organize

We cannot tell simply by looking at the chemical formula which compound has the greatest percentage of carbon by mass. We have to find the percent composition of hydrogen and carbon in each and then compare the compounds' percent carbon (% C).

Analyze

For each compound, assume that we have 1 mol of the substance and then compute the molar mass. For the percentage of carbon, divide the mass of all the carbon present in 1 mol by the molar mass and multiply by 100 to find the percentage.

Solve

(a) Naphthalene, $C_{10}H_8$:

$$\% \, C = \frac{(12.01 \times 10) \text{ g C}}{128.2 \text{ g}} \times 100 = 93.69\% \text{ C}$$

(b) Chrysene, $C_{18}H_{12}$:

$$\% \, C = \frac{(12.01 \times 18) \text{ g C}}{228.3 \text{ g}} \times 100 = 94.70\% \text{ C}$$

(c) Pentacene, $C_{22}H_{14}$:

$$\% \, C = \frac{(12.01 \times 22) \text{ g C}}{278.4 \text{ g}} \times 100 = 94.91\% \text{ C}$$

(d) Pyrene, $C_{16}H_{10}$:

$$\% \, C = \frac{(12.01 \times 16) \text{ g C}}{202.3 \text{ g}} \times 100 = 95.00\% \text{ C}$$

Pyrene, $C_{16}H_{10}$, has the highest % C by mass of all those hydrocarbons.

Think About It

Those compounds all have relatively close percent compositions, which is not obvious by looking only at their chemical formulas.

7.51. **Collect and Organize**

Given that a 3.556 g sample of aluminum oxide decomposes to 1.674 g of O_2 under high heat, we are to determine the empirical formula of the aluminum oxide.

Analyze

We are to assume here that the decomposition is complete and because of the conservation of mass, we would expect $3.556 - 1.674 = 1.882$ g of aluminum metal also is produced. To determine the empirical formula, we need to convert those masses of aluminum and oxygen to moles and determine the lowest whole-number ratio for the moles of those elements.

Solve

The moles of O atoms in the aluminum oxide is

$$\text{mol O} = 1.674 \text{ g O}_2 \times \frac{1 \text{ mol}}{31.98 \text{ g}} \times \frac{2 \text{ mol O}}{1 \text{ mol O}_2} = 0.1047 \text{ mol O}$$

The moles of Al atoms in the aluminum oxide is

$$\text{mol Al} = 1.882 \text{ g Al} \times \frac{1 \text{ mol}}{26.98 \text{ g}} = 0.0696 \text{ mol Al}$$

Dividing the moles of O atoms by the mole of Al atoms give a ratio of 1 mol Al to 1.5 mol O; multiplying by 2 to give a whole-number ratio gives an empirical formula of Al_2O_3.

Think About It

That empirical formula also makes sense because the charge on aluminum is +3 and the charge on oxygen is –2 to give a formula of Al_2O_3.

7.53. **Collect and Organize**

We are asked to compare empirical formulas for a variety of hydrocarbons to see whether any are identical.

Analyze

The empirical formula is the lowest whole-number ratio of atoms in a compound. To compare those compounds, then, we need to first write the empirical formula (lowest whole-number ratio) for each. Since many of those compounds have even numbers of carbon and hydrogen atoms, we can reduce the molecular formula by dividing by 2 until we obtain a formula that can no longer be reduced to a lowest whole-number ratio.

Solve

(a) Naphthalene, $C_{10}H_8$
    Empirical formula = C_5H_4

(b) Chrysene, $C_{18}H_{12}$
    Empirical formula = C_3H_2

(c) Anthracene, $C_{14}H_{10}$
    Empirical formula = C_7H_5

(d) Pyrene, $C_{16}H_{10}$
    Empirical formula = C_8H_5

(e) Benzoperylene, $C_{22}H_{12}$
    Empirical formula = $C_{11}H_6$

(f) Coronene, $C_{24}H_{12}$
    Empirical formula = C_2H

All those have different empirical formulas.

Think About It

Even though it was not true for the compounds in that problem, some compounds can have the same empirical formula.

7.55. Collect and Organize

For surgical-grade titanium we are to determine the empirical formula for the alloy given the percentages of titanium, aluminum, and vanadium it contains.

Analyze

The empirical formula is the lowest whole-number ratio of atoms in a compound. If we assume 100 g of surgical-grade titanium, we have 64.39 g of Ti, 24.19 g of Al, and 11.42 g of V in the sample. After we calculate the moles of each element by dividing the mass by the molar mass, we divide the molar amounts obtained by the smallest molar amount to find the lowest whole-number ratio of the elements.

Solve

$$\text{mol Ti} = \frac{64.39 \text{ g}}{47.87 \text{ g/mol}} = 1.345 \text{ mol Ti}$$

$$\text{mol Al} = \frac{24.19 \text{ g}}{26.98 \text{ g/mol}} = 0.8966 \text{ mol Al}$$

$$\text{mol V} = \frac{11.42 \text{ g}}{50.94 \text{ g/mol}} = 0.2242 \text{ mol V}$$

Dividing by the smallest molar amount (0.2242), we get a titanium–aluminum–vanadium ratio of 6:4:1. Therefore, the empirical formula for zircon is Ti_6Al_4V.

Think About It

That titanium alloy is useful as a surgical alloy because it resists corrosion, is lightweight and strong, and is biocompatible.

7.57. Collect and Organize

Given that a substance is 43.64% P and 56.36% O with a molar mass of 284 g/mol, we are to determine its empirical and molecular formulas.

Analyze

The percentage of P and O add up to 100%, so no other elements are present in that substance. To determine the empirical formula we assume 100 g of the compound; we have 43.64 g of P and 56.36 g of O. After we calculate the moles of each element by dividing the mass by the molar mass, we divide the molar amounts obtained by the smallest molar amount to find the lowest whole-number ratio of the elements (the empirical formula). We can obtain the molecular formula by comparing the molar mass of the empirical formula with the known molar mass of the substance.

Solve

(a) Assuming 100 g of product,

$$43.64 \text{ g P} \times \frac{1 \text{ mol}}{30.974 \text{ g}} = 1.409 \text{ mol P}$$

$$56.36 \text{ g O} \times \frac{1 \text{ mol}}{15.999 \text{ g}} = 3.523 \text{ mol O}$$

That is a 1:2.5 molar ratio of P to O, or a 2:5 ratio, so the empirical formula is P_2O_5.

(b) The molar mass of that empirical formula is 141.95, which is half of the known molar mass, so the molecular formula is P_4O_{10}.

Think About It

The question reflects how we would experimentally determine the formula for a new compound.

7.59. **Collect and Organize**

We use the percent composition of the asbestos mineral chrysotile to determine the empirical formula.

Analyze

If we assume 100 g of chrysotile, the percent composition (26.31% Mg, 20.27% Si, 1.45% H, and the rest O) gives us the grams of each element. We can convert those values to moles of each element via the molar masses of the elements from the periodic table. The empirical formula will be the lowest whole-number ratio of the moles of the elements in the chrysotile.

Solve

Oxygen is the only element not specified with a mass percentage. Therefore, we can determine oxygen's percentage by

$$100 - (26.31 + 20.27 + 1.45) = 51.97\%$$

The following give the moles of each element:

$$26.31 \text{ g Mg} \times \frac{1 \text{ mol}}{24.31 \text{ g}} = 1.082 \text{ mol Mg}$$

$$20.27 \text{ g Si} \times \frac{1 \text{ mol}}{28.09 \text{ g}} = 0.7216 \text{ mol Si}$$

$$1.45 \text{ g H} \times \frac{1 \text{ mol}}{1.01 \text{ g}} = 1.44 \text{ mol H}$$

$$51.97 \text{ g O} \times \frac{1 \text{ mol}}{16.00} = 3.248 \text{ mol O}$$

Dividing those by the smallest molar amount (0.7216) gives a magnesium–silicon–hydrogen–oxygen ratio of 1.5:1:2:4.5. Multiplying those values by 2 gives a whole-number ratio of 3 Mg : 2 Si : 4 H : 9 O for an empirical formula of $Mg_3Si_2H_4O_9$.

Think About It

The trick here is to recognize that the mass percentage of oxygen was not given in the original statement of the problem. Be sure to determine the moles for each element present in the compound.

7.61. **Collect and Organize**

From the percent composition of a compound containing copper, chlorine, and oxygen, we are to determine the formula. We have to convert the mass to moles and then find the lowest whole-number molar ratio for the elements in the compound.

Analyze

If we assume 100 g of the compound, the percentage of each element (24.2% Cu, 27.0% Cl, 48.8% O) gives us the mass of the elements in that 100 g amount. We can convert those masses into moles by using the molar masses of the elements from the periodic table. Then we compare the moles to find the molar ratio.

Solve

$$24.2 \text{ g Cu} \times \frac{1 \text{ mol}}{63.55 \text{ g}} = 0.381 \text{ mol Cu}$$

$$27.0 \text{ g Cl} \times \frac{1 \text{ mol}}{35.45 \text{ g}} = 0.762 \text{ mol Cl}$$

$$48.8 \text{ g O} \times \frac{1 \text{ mol}}{16.00 \text{ g}} = 3.05 \text{ mol O}$$

Dividing each mole amount by the smallest mole amount (0.381) gives a ratio of 1 Cu : 2 Cl : 8 O. The empirical formula for the compound is therefore $CuCl_2O_8$.

Think About It

Here, the molar ratio, after we divide the moles in 100 g of the substance, comes out to a whole-number ratio. We did not have to multiply to obtain whole numbers for that compound.

7.63. Collect and Organize

We consider why combustion analysis must be carried out in excess amounts of oxygen.

Analyze

In combustion analysis, compounds (usually organic) are burned in oxygen and the masses of recovered CO_2 and H_2O produced in the reaction are related to the percentages of C and H in the original compound.

Solve

The excess of oxygen is required in combustion analysis to ensure the complete reaction of the hydrogen and carbon to form water and carbon dioxide.

Think About It

Combustion in an atmosphere deficient in oxygen gives CO instead of CO_2 as the main gaseous carbon product.

7.65. Collect and Organize

We are asked whether combustion analysis can ever give the true molecular formula for a compound.

Analyze

Combustion analysis gives us the percent mass of C, H, and (by calculation of the missing mass for some compounds) O in the compound, from which we can derive the empirical formula.

Solve

Yes, the combustion analysis can give the true molecular formula for a compound, but only if the empirical formula is the same as the molecular formula and only if the compound contains only C, H, and O.

Think About It

We can confirm the molar mass of a compound by other methods such as boiling point elevation, freezing point depression, or osmotic pressure.

7.67. Collect and Organize

From the data obtained from the combustion analysis for a compound containing carbon, hydrogen, and oxygen, we are asked to determine the empirical formula of the compound. We are given the mass of carbon dioxide and water produced in the analysis. From there we can determine the molar ratios of C, H, and O.

Analyze

We can calculate the moles and masses of C and H directly from the combustion analysis results. The oxygen content will be the difference in the mass of the carbon plus hydrogen in the compound and the mass of the 0.100 g sample. We can then determine the moles of oxygen by using the molar mass of oxygen from the periodic table, and we can find the ratio of carbon to hydrogen to oxygen to determine the empirical formula.

Solve

$$0.1783 \text{ g CO}_2 \times \frac{1 \text{ mol CO}_2}{44.01 \text{ g}} \times \frac{1 \text{ mol C}}{1 \text{ mol CO}_2} = 4.051 \times 10^{-3} \text{ mol C}$$

$$4.051 \times 10^{-3} \text{ mol C} \times \frac{12.011 \text{ g C}}{1 \text{ mol}} = 4.866 \times 10^{-2} \text{ g C}$$

$$0.0734 \text{ g H}_2\text{O} \times \frac{1 \text{ mol H}_2\text{O}}{18.02 \text{ g}} \times \frac{2 \text{ mol H}}{1 \text{ mol H}_2\text{O}} = 8.147 \times 10^{-3} \text{ mol H}$$

$$8.147 \times 10^{-3} \text{ mol H} \times \frac{1.008 \text{ g H}}{1 \text{ mol}} = 8.212 \times 10^{-3} \text{ g H}$$

$$\text{Total mass of C and H} = 4.866 \times 10^{-2} \text{ g} + 8.212 \times 10^{-3} \text{ g} = 5.687 \times 10^{-2} \text{ g}$$

$$\text{Mass of O present} = 0.100 \text{ g} - 0.05687 \text{ g} = 0.043 \text{ g O}$$

$$\text{Moles of O in compound} = 0.043 \text{ g O} \times \frac{1 \text{ mol}}{15.999 \text{ g}} = 2.7 \times 10^{-3} \text{ mol O}$$

Dividing the moles of C, H, and O by the smallest molar amount (2.7×10^{-3} mol) gives a ratio of 1.5 C : 3 H : 1 O. Multiplying that by 2 to obtain whole-number ratios, we get an empirical formula of $C_3H_6O_2$.

Think About It
That problem involved an additional step to determine the mass of carbon and hydrogen present, so that we could calculate the mass (and therefore the moles) of oxygen in the compound.

7.69. Collect and Organize
From the combustion data of a given mass of a compound containing only hydrogen and carbon and the molar mass of the compound, we are to determine the empirical and molecular formulas.

Analyze
The specified compound contains only hydrogen and carbon. The water (135.0 mg) resulted from the combustion of the hydrogen, and the carbon dioxide (440.0 mg) resulted from the combustion of the carbon. First, we determine the mass of hydrogen and oxygen present in the water and the carbon dioxide. From those results, we determine the mass percentage of the hydrogen and carbon in the compound (we know the original mass of the sample used in the analysis, 135.0 mg). From the mass percentage, we can find moles and the mole ratio of carbon and hydrogen in the compound and from there determine the empirical and molecular formulas (knowing that the molar mass of the compound is 270 g/mol).

Solve
The mass and percentage of carbon and hydrogen in the compound are

$$440.0 \text{ mg CO}_2 \times \frac{1 \text{ g}}{1000 \text{ mg}} \times \frac{1 \text{ mol CO}_2}{44.01 \text{ g}} \times \frac{1 \text{ mol C}}{1 \text{ mol CO}_2} = 0.009998 \text{ mol C}$$

$$135.0 \text{ mg H}_2\text{O} \times \frac{1 \text{ g}}{1000 \text{ mg}} \times \frac{1 \text{ mol H}_2\text{O}}{18.02 \text{ g}} \times \frac{2 \text{ mol H}}{1 \text{ mol H}_2\text{O}} = 0.01498 \text{ mol H}$$

Dividing 0.01498 mol H by 0.009998 mol C gives a ratio of 1 C : 1.499 H. Multiplying by 2 to obtain a whole-number ratio gives an empirical formula of C_2H_3. The molar mass of that empirical formula is 27.05 g/mol. The molar mass of the compound is 270 g/mol. That value is 10 times the molar mass of the empirical formula. Therefore, the molecular formula is $C_{20}H_{30}$.

Think About It
The empirical formula is derived from the molar ratio of the elements in the compound. Because combustion analysis gives us the amount of carbon and hydrogen in the compound, we need to only relate the moles of CO_2 and H_2O to the C and H present in the compound.

7.71. Collect and Organize
Given that a reaction starts with equal masses of iron and sulfur, we consider the mass of iron(II) sulfide that can be produced in the reaction.

Analyze
The elements react to give a molar ratio of 1 Fe : 1 S. Because those elements have different molar masses (Fe = 55.845 and S = 32.065 g/mol), an equal mass of S contains more moles of sulfur than the same mass of iron contains moles of iron.

Solve
Excess sulfur is present at the end of the reaction, so the mass of FeS produced is (c) less than the sum of the masses of Fe and S to start.

Think About It
The limiting reactant here is iron.

7.73. Collect and Organize

We are to distinguish between *theoretical yield* and *percent yield*.

Analyze

The theoretical yield is calculated on the basis of the amounts of reactants used in a chemical reaction. The percent yield takes into account the actual experimental yield.

Solve

Theoretical yield is the greatest amount of a product possible from a reaction and assumes that the reaction runs to completion. The percent yield is the actual experimental yield divided by the theoretical yield and multiplied by 100.

Think About It

The actual yields are almost always less than the theoretical yield because of side reactions, incomplete reactions, and loss during purification steps. The percent yields for most reactions are less than 100%.

7.75. Collect and Organize

Given a reaction in which fewer moles of A than B are present, we are asked to evaluate why saying that A must be the limiting reactant is wrong.

Analyze

Stoichiometry compares the moles of reactants and products in a reaction. An important aspect of stoichiometry is the ratio of moles of reactants needed and products formed in a reaction.

Solve

Although it is correct to be sure to compare the moles of reactants in a reaction (not the masses), the ratio of the reactants in the balanced chemical equation also is important. If A and B were necessary in equal molar ratios, then yes, with fewer moles of A, A would be the limiting reactant. However, if more than 1 mol of B were required for every mole of A and fewer moles of A than B were present, we could not unequivocally say that A was the limiting reactant. For example, if 2 mol of A was required for every mole of B in the balanced chemical equation and the reaction started with 1 mol of A and 1.5 mol of B, then B would be the limiting reactant.

Think About It

Always start with the balanced chemical equation in doing any stoichiometry problem.

7.77. Collect and Organize

We have to determine the maximum amount of hollandaise sauce that can be made with the ingredients on hand.

Analyze

The ingredient that would produce the least amount of the sauce will be the limiting ingredient; therefore, that amount is the largest possible amount of sauce that can be made.

Solve

Because the sauce requires $\frac{1}{4}$ c (cup) of butter, $\frac{1}{2}$ c of water, 4 egg yolks, and the juice of one lemon, we can determine how many cups of sauce could be made from the ingredients on hand.

Two cups of butter would be enough to prepare 4 c of sauce.

Unlimited amounts of hot water are enough to prepare an unlimited amount of sauce.

Twelve eggs is enough to prepare 3 c of sauce.

Four lemons is enough to prepare 4 c of sauce.

The limiting ingredient is eggs; we have enough of all the other ingredients to make 4 c of hollandaise sauce. With the limited number of eggs, we can make 3 c of sauce.

Think About It

The ingredient most limited in supply determines how much sauce we can make. That principle is true for chemical reactions as well, although we are thinking in moles, not cups of butter or number of eggs.

7.79. Collect and Organize

If 75 metric tons of coal is contaminated with 3.0% sulfur by mass, we are asked to calculate the efficiency of the power plant SO_2 capture scrubber when 3.9 metric tons of SO_2 is captured from that 75 tons burned. We are also to calculate how much SO_2 escapes.

Analyze

We first must calculate the mass of sulfur in the coal and then convert that into how much SO_2 is produced upon combustion. The balanced equation for the production of SO_2 from S is

$$S(s) + O_2(g) \rightarrow SO_2(g)$$

From the difference between the amount of SO_2 captured and that produced, we can determine the efficiency of the scrubbers and how much SO_2 escaped.

Solve

$$75 \text{ metric tons} \times \frac{1000 \text{ kg}}{1 \text{ metric ton}} \times \frac{1000 \text{ g}}{\text{kg}} \times 0.03 = 2.25 \times 10^6 \text{ g of S in the coal}$$

$$2.25 \times 10^6 \text{ g S} \times \frac{1 \text{ mol}}{32.065 \text{ g}} \times \frac{1 \text{ mol SO}_2}{1 \text{ mol S}} \times \frac{64.064 \text{ g}}{1 \text{ mol}} \times \frac{1 \text{ kg}}{1000 \text{ g}} \times \frac{1 \text{ metric ton}}{1000 \text{ kg}} = 4.5 \text{ metric tons SO}_2$$

$$\text{efficiency} = \frac{3.9 \text{ metric tons captured}}{4.5 \text{ metric tons produced}} \times 100 = 87\%$$

$$\text{amount of SO}_2 \text{ escaped} = 4.5 - 3.9 = 0.60 \text{ metric tons}$$

Think About It

The efficiency of the scrubbers is relatively high, but still a lot of SO_2 escapes into the atmosphere in that example. SO_2 is one of the gases that mix with water vapor in the atmosphere to acidify rain.

7.81. Collect and Organize

Given balanced chemical equations for the production of $CHClF_2$ and C_2F_4, we are asked to identify the limiting reactant when 775 g each of $CHCl_3$ and HF are combined to make $CHClF_2$ and then how much C_2F_4 would be produced if the first reaction was 95%. Finally, we are asked to identify and to compute the amount of the excess reactant for the production of $CHClF_2$.

Analyze

Those calculations all involve stoichiometric analysis. We will make use of the molar masses of the reactants and products as well as the stoichiometric ratios given in the balanced chemical equations.

Solve

(a) To determine the limiting reactant, we calculate the amount of product, $CHClF_2$, that 775 g of each reactant would produce:

$$775 \text{ g CHCl}_3 \times \frac{1 \text{ mol CHCl}_3}{119.38 \text{ g}} \times \frac{1 \text{ mol CHClF}_2}{1 \text{ mol CHCl}_3} \times \frac{86.47 \text{ g}}{1 \text{ mol CHClF}_2} = 561 \text{ g CHClF}_2$$

$$775 \text{ g HF} \times \frac{1 \text{ mol HF}}{20.01 \text{ g}} \times \frac{1 \text{ mol CHClF}_2}{2 \text{ mol HF}} \times \frac{86.47 \text{ g}}{1 \text{ mol CHClF}_2} = 1670 \text{ g CHClF}_2$$

$CHCl_3$ is the limiting reactant.

(b) If the yield in the reaction to produce C_2F_2 is 95%:

$$775 \text{ g CHCl}_3 \times \frac{1 \text{ mol CHCl}_3}{119.38 \text{ g}} \times \frac{1 \text{ mol CHClF}_2}{1 \text{ mol CHCl}_3} \times \frac{1 \text{ mol C}_2\text{F}_4}{2 \text{ mol CHClF}_2} \times \frac{100.02 \text{ g}}{1 \text{ mol C}_2\text{F}_4} \times 0.95 = 308 \text{ g C}_2\text{F}_4$$

(c) The amount of HF leftover in the reaction if reaction (i) is stoichiometric

$$561 \text{ g CHClF}_2 \times \frac{1 \text{ mol CHClF}_2}{86.47 \text{ g}} \times \frac{2 \text{ mol HF}}{1 \text{ mol CHClF}_2} \times \frac{20.01 \text{ g}}{1 \text{ mol HF}} = 246 \text{ g HF used}$$

$$775 \text{ g} - 260 \text{ g} = 515 \text{ g}$$

Think About It

HF is probably used as the excess reactant in that reaction because it is the least expensive reactant.

7.83. Collect and Organize

For the production of syngas, a mixture of CO and H_2, from C and H_2O, we are asked to write a balanced reaction and to determine the percent yield of a reaction that uses 66 kg of carbon to produce 6.8 kg of H_2.

Analyze

We can use the three-step method to balance the equation for the production of syngas. Next we will calculate the theoretical yield of H_2 from 66 kg of carbon by using the molar masses of C and H_2 and the ratio of C to H_2 from the balanced chemical equation. We can calculate percent yield by dividing the actual yield (6.8 kg) by the theoretical yield and multiplying by 100.

Solve

(a) The unbalanced chemical equation is

$$C(s) + H_2O(\ell) \rightarrow CO(g) + H_2(g)$$

$$\text{Atoms:} \quad 1\,C + 2\,H + 1\,O \rightarrow 1\,C + 2\,H + 1\,O$$

The equation is already balanced.

(b)

$$66\text{ kg C} \times \frac{1000\text{ g}}{\text{kg}} \times \frac{1\text{ mol C}}{12.011\text{ g}} \times \frac{1\text{ mol H}_2}{1\text{ mol C}} \times \frac{2.016\text{ g H}_2}{1\text{ mol}} \times \frac{1\text{ kg}}{1000\text{ kg}} = 11\text{ kg}$$

$$\%\text{ yield} = \frac{6.8\text{ kg}}{11\text{ kg}} \times 100 = 61\%$$

Think About It

The hydrogen in that reaction can be used as a fuel, perhaps in a hydrogen–oxygen fuel cell.

7.85. Collect and Organize

We first have to balance the equation for the conversion of glucose into ethanol. We need the molar ratio of the reactant to the ethanol product to determine the theoretical yield of ethanol for the fermentation of 100.0 g of glucose that produces 50.0 mL of ethanol.

Analyze

First, we have to write the balanced equation for the process for part a. For part b, we need the molar masses of $C_6H_{12}O_6$ (180.16 g/mol) and C_2H_5OH (46.07 g/mol). We need the density of the ethanol (0.789 g/mL) to convert the grams of ethanol produced theoretically in the reaction to milliliters to compute the percent yield for the reaction, which is given by

$$\%\text{ yield} = \frac{\text{observed experimental yield}}{\text{theoretical yield}} \times 100$$

Solve

(a) The balanced equation is

$$C_6H_{12}O_6(aq) \rightarrow 2\,C_2H_5OH(\ell) + 2\,CO_2(g)$$

(b) The theoretical yield of C_2H_5OH is

$$100.0\text{ g C}_6\text{H}_{12}\text{O}_6 \times \frac{1\text{ mol C}_6\text{H}_{12}\text{O}_6}{180.16\text{ g}} \times \frac{2\text{ mol C}_2\text{H}_5\text{OH}}{1\text{ mol C}_6\text{H}_{12}\text{O}_6} \times \frac{46.07\text{ g C}_2\text{H}_5\text{OH}}{1\text{ mol}} \times \frac{1\text{ mL C}_2\text{H}_5\text{OH}}{0.789\text{ g}} = 64.82\text{ mL C}_2\text{H}_5\text{OH}$$

The percent yield for that reaction is

$$\%\text{ yield} = \frac{50.0\text{ mL}}{64.82\text{ mL}} \times 100 = 77.1\%$$

Think About It

The conversion of glucose by fermentation into ethanol is fairly efficient.

7.87. Collect and Organize

Hydroxyapatite is composed of calcium, phosphorus, oxygen, and hydrogen. We are asked to systematically name the compound and to find its mass percentage of calcium. We are also asked to think about how the mass percentage of calcium changes when F replaces OH in the compound.

Analyze

We use the naming rules to systematically name the hydroxyapatite. We can calculate the mass percentage of calcium by dividing the mass of calcium present (assuming 1 mol of the substance) by the total mass of the compound, taking into account that five Ca atoms are present in the molecular formula.

Solve

(a) Hydroxyapatite would be named calcium triphosphate hydroxide.

(b) If we assume 1 mol of hydroxyapatite with a molar mass of 502.31 g/mol and use the mass of calcium of 40.078 g/mol, the mass percentage of calcium in hydroxyapatite is

$$\frac{(40.08 \times 5) \text{ g}}{502.31 \text{ g}} \times 100 = 39.89\% \text{ Ca}^{2+}$$

(c) The mass percentage in $Ca_5(PO_3)_4F$ is

$$\frac{(40.08 \times 5) \text{ g}}{504.30 \text{ g}} \times 100 = 39.74\% \text{ Ca}^{2+}$$

The mass percentage of calcium decreases, but only slightly.

Think About It

The strengthening of the structure of hydroxyapatite when OH^- is replaced by fluoride happens in our teeth when we drink fluoridated water and use fluoride toothpaste.

7.89. Collect and Organize

For 5.1 metric tons of bauxite ore that is 86% aluminum oxide (Al_2O_3), we are to calculate the percent yield of the recovery if 2.3 metric tons was extracted from that ore.

Analyze

First we will use the percent composition of the ore (86%) to find the mass of Al_2O_3 in the 5.1 metric tons. Then to determine the theoretical amount of Al in the bauxite, we will convert the mass of Al_2O_3 to moles. One mole of Al_2O_3 has 2 mol of Al, which we can then convert into the mass of aluminum expected from the bauxite (the theoretical "yield"). The percent yield is the actual yield of 2.3 metric tons over the calculated theoretical yield multiplied by 100.

Solve

The amount of aluminum in the ore is

$$5.1 \text{ metric tons} \times 0.86 = 4.39 \text{ metric tons}$$

The theoretical amount of aluminum in the ore is

$$4.39 \text{ metric tons} \times \frac{1000 \text{ kg}}{\text{metric ton}} \times \frac{1000 \text{ g}}{\text{kg}} \times \frac{\text{mol } Al_2O_3}{101.96 \text{ g}} \times \frac{2 \text{ mol Al}}{1 \text{ mol } Al_2O_3} \times \frac{26.982 \text{ g}}{\text{mol}} \times \frac{1 \text{ kg}}{1000 \text{ g}} \times \frac{1 \text{ metric ton}}{1000 \text{ kg}}$$

$$= 2.32 \text{ metric tons}$$

The percent yield is

$$\frac{2.3 \text{ metric tons}}{2.32 \text{ metric tons}} \times 100 = 99\%$$

Think About It

The extraction of aluminum from bauxite is near 100%, but the process is energy intensive and therefore costly.

7.91. Collect and Organize

We are given balanced chemical equations for the extraction of gold from veins in rocks. We are asked to calculate the amounts of reactants for the process and to find the size of a block made from the gold product.

Analyze

We need the molar masses of Au (196.97 g/mol), NaCN (49.01 g/mol), NaAu(CN)$_2$ (271.99 g/mol), and Zn (65.38 g/mol). We use the density of gold (19.3 g/cm^3) to calculate the size of the block of gold for part c.

Solve

(a) If the ore is 0.009% gold by mass, then the mass of gold in 1 metric ton (1000 kg) is

$$1000 \text{ kg} \times \frac{1000 \text{ g}}{1 \text{ kg}} \times 0.00009 = 90 \text{ g Au}$$

The amount of NaCN required to extract the gold is

$$90 \text{ g Au} \times \frac{1 \text{ mol Au}}{196.97} \times \frac{8 \text{ mol NaCN}}{4 \text{ mol Au}} \times \frac{49.01 \text{ g NaCN}}{1 \text{ mol}} = 45 \text{ g NaCN}$$

(b) We can solve for the amount of Zn required by starting with the amount of Au.

$$90 \text{ g Au} \times \frac{1 \text{ mol Au}}{196.97} \times \frac{4 \text{ mol NaAu(CN)}_2}{4 \text{ mol Au}} \times \frac{1 \text{ mol Zn}}{2 \text{ mol NaAu(CN)}_2} \times \frac{65.38 \text{ g Zn}}{1 \text{ mol}} = 15 \text{ g Zn}$$

(c) The amount of gold recovered from the ore is 90 g. We use the density of gold to find the size of the block of gold for this mass.

$$90 \text{ g Au} \times \frac{1 \text{ cm}^3}{19.3 \text{ g}} = 4.7 \text{ cm}^3$$

Think About It

That problem assumes that the processing steps all go to completion (100% yield).

7.93. Collect and Organize

For this problem we must use percent compositions of the compounds that form upon heating UO$_x$(NO$_3$)$_y$(H$_2$O)$_z$ to ultimately determine the values of x, y, and z.

Analyze

We need the molar masses of uranium (238.03 g/mol) and oxygen (16.00 g/mol) to determine the molar ratios of U to O in the oxides. We also are given that the charge on uranium in the oxides may range from 3+ to 6+.

Solve

(a) If U$_a$O$_b$ is 83.22% U, then it is 16.78% O. Assuming 100 g of the compound, and using the molar masses of these elements, we obtain their molar ratios:

$$83.22 \text{ g U} \times \frac{1 \text{ mol U}}{238.03 \text{ g}} = 0.3496 \text{ mol U}$$

$$16.78 \text{ g O} \times \frac{1 \text{ mol O}}{16.00 \text{ g}} = 1.049 \text{ mol O}$$

Multiplying those molar amounts by 3 gives a whole-number ratio, for an empirical formula of UO$_3$ for that oxide. In that formula $a = 1$ and $b = 3$ with a charge on the U of 6+ (since O is 2–).

(b) If U$_c$O$_d$ is 84.8% U, then it is 15.2% O. Assuming 100 g of the compounds, and using the molar masses of these elements, we obtain their molar ratios:

$$84.8 \text{ g U} \times \frac{1 \text{ mol U}}{238.03 \text{ g}} = 0.356 \text{ mol U}$$

$$15.2 \text{ g O} \times \frac{1 \text{ mol O}}{16.00 \text{ g}} = 0.950 \text{ mol O}$$

Dividing those molar amounts by the lowest number of moles (0.356) gives a ratio of 1 U to 2.67 O. To reach a whole-number ratio, we multiply by 3 to get U_3O_8, in which $c = 3$ and $d = 8$. The charge on the U atoms is 5.33+, which indicates mixed oxidation states for the uranium cations in that ore.

(c) Upon gentle heating, $UO_x(NO_3)_y(H_2O)_z$ loses water according to the following equation:
$$UO_x(NO_3)_y(H_2O)_z \rightarrow UO_x(NO_3)_y + z\ H_2O$$
More heating of $UO_x(NO_3)_y$ gives the reaction
$$UO_x(NO_3)_y \rightarrow U_nO_m + \text{nitrogen oxides}$$
Putting those equations together,
$$UO_x(NO_3)_y(H_2O)_z \rightarrow UO_x(NO_3)_y + z\ H_2O \rightarrow U_nO_m + \text{nitrogen oxides}$$
The continued heating of the compound indicates that at the end of the reaction, we have U_3O_8 (part b). The balanced reaction then is
$$3\ UO_x(NO_3)_y(H_2O)_z \rightarrow 3\ UO_x(NO_3)_y + 3z\ H_2O \rightarrow U_3O_8 + \text{nitrogen oxides}$$
The amount in moles of H_2O present on the basis of 0.742 g of U_3O_8 is

$$0.742\ \text{g U}_3O_8 \times \frac{1\ \text{mol}}{842\ \text{g}} \times \frac{3z\ \text{mol H}_2O}{1\ \text{mol U}_3O_8} \times \frac{18.02\ \text{g H}_2O}{1\ \text{mol}} = 0.0476z\ \text{g H}_2O$$

The mass of water lost in that process is $1.328 - 1.042 = 0.286$ g of H_2O. The moles of water lost is therefore
$$0.0476z = 0.286$$
$$z = 6$$
The amount in moles of $UO_x(NO_3)_y$ from the 0.742 g of U_3O_8 is

$$0.742\ \text{g U}_3O_8 \times \frac{1\ \text{mol}}{842\ \text{g}} \times \frac{3\ \text{mol UO}_x(NO_3)_y}{1\ \text{mol U}_3O_8} = 0.00264\ \text{mol UO}_x(NO_3)_y$$

Because we know the mass of $UO_x(NO_3)_y$, for that number of moles, the molar mass of $UO_x(NO_3)_y$ is

$$\frac{1.042\ \text{g}}{0.00264\ \text{mol}} = 395\ \text{g/mol}$$

That molar mass is expressed by
$$395\ \text{g/mol} = 238\ \text{g/mol} + [(x + 3y)(16.00\ \text{g/mol})] + y(14.00\ \text{g/mol})$$
$$157\ \text{g/mol} = 16x + 62y$$
We know that the total charge on $UO_x(NO_3)_y$ is 0 (it is a neutral compound). We can express that as
$$-2(x) + -1(y) + c = 0$$
where x = number of O^{2-} ions, y = number of NO_3^- ions, and c = charge on U in the molecular formula. We are given that c could be 3+, 4+, 5+, or 6+. When $c = 6$, $-2x - y = -6$, or $2x + y = 6$. Rearranging that equation gives $y = 6 - 2x$, which, when substituted into $157\ \text{g/mol} = 16x + 62y$, gives
$$157\ \text{g/mol} = 16x + 62(6 - 2x)$$
$$157\ \text{g/mol} = -108x + 372$$
$$-215\ \text{g/mol} = -108x$$
$$2 = x$$
and then
$$y = 6 - 2x = 2$$
If $c = 5+$, 4+, or 3+, the solution does not work. The formula for $UO_x(NO_3)_y(H_2O)_z$ is $UO_2(NO_3)_2(H_2O)_6$.

Think About It
The value of the charge on U in part b must mean that it is a mixed-oxidation-state compound. Here one-third of the U atoms have a 6+ charge and two-thirds of the U atoms have a 5+ charge.

7.95. Collect and Organize

We are to compare the chemical formulas of xylose ($C_5H_{10}O_5$) and methyl galacturonate ($C_7H_{12}O_7$) to determine whether they have the same empirical formula, and we are asked to write equations for their combustion reactions.

Analyze

For two compounds to have the same empirical formula, they must have the same lowest whole-number ratio of the elements that compose them. To write a combustion reaction, recall that the other reactant is oxygen (O_2) and the products are water (H_2O) and carbon dioxide (CO_2).

Solve

(a) Those two compounds do not have the same empirical formula. The empirical formulas of xylose ($C_5H_{10}O_5$) and galacturonate ($C_7H_{12}O_7$) are CH_2O and $C_7H_{12}O_7$, respectively. Therefore, these compounds do not have the same empirical formulas.

(b) The combustion of xylose is given by

$$C_5H_{10}O_5(s) + 5\ O_2(g) \rightarrow 5\ CO_2(g) + 5\ H_2O(\ell)$$

The combustion of methyl galacturonate is given by

$$2\ C_7H_{12}O_7(s) + 13\ O_2(g) \rightarrow 14\ CO_2(g) + 12\ H_2O(\ell)$$

Think About It

The higher the carbon content of the reactant undergoing combustion, the more moles of CO_2 produced.

7.97. Collect and Organize

We are given the balanced equation in which formaldehyde (HCO_2H) is prepared from CO_2 and H_2S in the presence of FeS. First, we identify the ions that make up FeS and FeS_2 and name those iron compounds. Next, we calculate the amount of HCO_2H produced when 1.00 g of FeS, 0.50 g of H_2S, and 0.50 g of CO_2 are used in the reaction that gives 50% yield. This is a limiting reactant problem, and we must identify the limiting reactant.

Analyze

To solve the limiting reactant–percent yield problem, we need the molar masses of FeS (87.91 g/mol), H_2S (34.08 g/mol), CO_2 (44.01 g/mol), and HCO_2H (46.03 g/mol). We have to determine the theoretical yield and then consider that the reaction goes only to 50% yield.

Solve

(a and b)   FeS is iron(II) sulfide with Fe^{2+} and S^{2-}; FeS_2 is iron(IV) sulfide with Fe^{4+} and S^{2-} from your knowledge so far in this course with Fe^{4+} and S^{2-}. Actually, that compound is Fe^{2+} with S_2^{2-} and is named iron(II) persulfide.

(c) The amount of HCO_2H that could be produced from each starting material is as follows:

$$1.00\ g\ FeS \times \frac{1\ mol}{87.91\ g} \times \frac{1\ mol\ HCO_2H}{1\ mol\ FeS} \times \frac{46.03\ g}{1\ mol} = 0.524\ g\ HCO_2H\ \text{could be produced from 1.00 g FeS}$$

$$0.50\ g\ H_2S \times \frac{1\ mol}{34.08\ g} \times \frac{1\ mol\ HCO_2H}{1\ mol\ H_2S} \times \frac{46.03\ g}{1\ mol} = 0.68\ g\ HCO_2H\ \text{could be produced from 0.50 g }H_2S$$

$$0.50\ g\ CO_2 \times \frac{1\ mol}{44.01\ g} \times \frac{1\ mol\ HCO_2H}{1\ mol\ CO_2} \times \frac{46.03\ g}{1\ mol} = 0.52\ g\ HCO_2H\ \text{could be produced from 0.50 g }CO_2$$

Because the smallest amount (CO_2 and FeS nearly tie as limiting reactant) is the theoretical yield if the reaction proceeds completely (100% yield), that reaction, which goes only to 50% completion, yields 0.52 g/2 = 0.26 g of HCO_2H.

Think About It

At first that problem looks difficult because we are told the reaction gives only a 50% yield. We did not need to account for that until the end, however, and the problem is primarily a limiting reactant problem.

7.99. Collect and Organize

We are asked to balance the reactions of (a) FeO with water to form Fe_3O_4 and H_2 and then (b) the same reaction except that when CO_2 is present the reaction gives CH_4 in place of H_2.

Analyze

To balance each equation we first write the unbalanced equation by using the chemical formulas of the reactants and products. Next, we balance an element that is present in only one reactant and product. Finally, we balance the other elements present by placing coefficients in front of the species in the reaction so that the number of the atoms for each element is equal on both sides of the equation. If any fractional coefficients are present, we multiply the entire equation through to eliminate all fractions.

Solve

(a) For the reaction of FeO with H_2O to produce Fe_3O_4 and H_2, the unbalanced reaction is

$$FeO(s) + H_2O(\ell) \rightarrow Fe_3O_4(s) + H_2(g)$$

Atoms: $\quad$ 1 Fe + 2 O + 2 H $\rightarrow$ 3 Fe + 4 O + 2 H

To balance the Fe atoms we place 3 as the coefficient before FeO on the left-hand side of the equation.

$$3\,FeO(s) + H_2O(\ell) \rightarrow Fe_3O_4(s) + H_2(g)$$

Atoms: $\quad$ 3 Fe + 4 O + 2 H $\rightarrow$ 3 Fe + 4 O + 2 H

The equation is now balanced.

(b) For the reaction of FeO with water in the presence of CO_2 to give Fe_3O_4 and CH_4, the unbalanced reaction is

$$FeO(s) + H_2O(\ell) + CO_2(g) \rightarrow Fe_3O_4(s) + CH_4(g)$$

Atoms: $\quad$ 1 Fe + 4 O + 2 H + 1 C $\rightarrow$ 3 Fe + 4 O + 4 H + 1 C

To balance the H atoms we place 2 as the coefficient before H_2O on the left-hand side of the equation.

$$FeO(s) + 2\,H_2O(\ell) + CO_2(g) \rightarrow Fe_3O_4(s) + CH_4(g)$$

Atoms: $\quad$ Fe + 5 O + 4 H + 1 C $\rightarrow$ 3 Fe + 4 O + 4 H + 1 C

We will need to balance the O and Fe atoms together. To do so we place 12 as the coefficient before FeO on the left-hand side of the equation and a 4 as the coefficient before Fe_3O_4 on the right-hand side of the equation.

$$12\,FeO(s) + 2\,H_2O(\ell) + CO_2(g) \rightarrow 4\,Fe_3O_4(s) + CH_4(g)$$

Atoms: $\quad$ 12 Fe + 16 O + 4 H + 1 C $\rightarrow$ 12 Fe + 16 O + 4 H + 1 C

The equation is now balanced.

Think About It

Balancing the equation in part b is tricky and requires a little insight and trial and error. Here, when we balance two species at once, the other coefficients fall into place.

7.101. Collect and Organize

The alternative fuel E-85 is 85% (by volume) ethanol, and we are to determine how many moles of ethanol is in 1 gal of the fuel and how many moles of CO_2 would be produced if that ethanol were combusted.

Analyze

(a) If we first convert 1 gal to milliliters (1 gal = 3785 mL), we can find the milliliters of ethanol in the fuel by multiplying by 0.85 (the percent ethanol by volume in the fuel). From the milliliters of ethanol, we can calculate the mass of ethanol by multiplying by the density (0.79 g/mL). From the mass and the molar mass of ethanol (46.07 g/mol), we can determine the moles of ethanol in 1 gal of the fuel.

(b) When ethanol burns completely, it produces CO_2 and H_2O. The balanced equation for the burning of ethanol is

$$C_2H_5OH(\ell) + 3\,O_2(g) \rightarrow 2\,CO_2(g) + 3\,H_2O(\ell)$$

From the stoichiometric ratio of 1 ethanol to 2 carbon dioxide, we obtain the moles of CO_2 produced in the reaction.

Solve

(a)

$$(3785 \text{ mL} \times 0.85) \times \frac{0.79 \text{ g}}{\text{mL}} \times \frac{1 \text{ mol}}{46.07 \text{ g}} = 55 \text{ mol ethanol}$$

(b)

$$55 \text{ mol ethanol} \times \frac{2 \text{ mol CO}_2}{1 \text{ mol ethanol}} = 1.1 \times 10^2 \text{ mol CO}_2$$

Think About It

We could do the calculation for (a) all in one step, without separately calculating the milliliters of ethanol, the mass, and then the moles.

7.103. Collect and Organize

From the mass of the salt before and after dehydration (0.6240 and 0.5471 g, respectively) and the known ratio of M to Cl to H$_2$O in the compound as given by the molecular formula, MCl$_2 \cdot$ 2H$_2$O, we are to determine the identity (through calculation of the molar mass) of M.

Analyze

From the difference of the masses, we can obtain the mass of water lost and from that value calculate the moles of water lost. The moles of water lost equals the moles of Cl in the compound and is twice the moles of M in the compound. From the moles of Cl, we can find the mass of Cl in the sample. The mass of M in the sample is the total mass of the sample minus the combined masses of the water and the chlorine. Once we know the mass of M, we can divide by the moles of M in the sample found earlier to determine the molar mass of M, which identifies the metal.

Solve

$$\text{moles of water in sample} = (0.6240 \text{ g} - 0.5471 \text{ g}) \times \frac{1 \text{ mol H}_2\text{O}}{18.02 \text{ g}} = 4.27 \times 10^{-3} \text{ mol}$$

From the formula we also know that mol Cl = 4.27×10^{-3} mol and mol M = 2.14×10^{-3} mol. Therefore, the molar mass of M is

$$\text{mass of Cl in sample} = 4.27 \times 10^{-3} \text{ mol} \times \frac{35.453 \text{ g}}{1 \text{ mol Cl}} = 0.151 \text{ g}$$

$$\text{total mass of H}_2\text{O and Cl} = 0.0769 \text{ g} + 0.151 \text{ g} = 0.228 \text{ g}$$

$$\text{mass of M in sample} = 0.6240 \text{ g} - 0.228 \text{ g} = 0.396 \text{ g}$$

$$\text{molar mass of M} = \frac{0.396 \text{ g}}{2.14 \times 10^{-3} \text{ mol}} = 185 \text{ g/mol}$$

The identity of M is Re.

Think About It

That problem relies heavily on our being able to relate the moles of atoms in a compound's formula to one another.

7.105. Collect and Organize

To determine the percent yield of ammonia in the reaction described, we need a balanced chemical equation for the reaction of 6.04 kg of H$_2$ with N$_2$ to give 28.0 kg of NH$_3$. We have excess N$_2$ in the reaction, so the theoretical yield is based solely on the moles of H$_2$ present at the beginning of the reaction.

Analyze

The balanced equation for the reaction is

$$\text{N}_2(g) + 3 \text{ H}_2(g) \rightarrow 2 \text{ NH}_3(g)$$

Solve

The theoretical yield of ammonia in that reaction is

$$6.04 \text{ kg H}_2 \times \frac{1000 \text{ g}}{1 \text{ kg}} \times \frac{1 \text{ mol H}_2}{2.016 \text{ g}} \times \frac{2 \text{ mol NH}_3}{3 \text{ mol H}_2} \times \frac{17.03 \text{ g NH}_3}{1 \text{ mol}} = 34,015 \text{ g, or } 34.0 \text{ kg NH}_3$$

The percent yield is

$$\frac{28.0 \text{ kg}}{34.0 \text{ kg}} \times 100 = 82.4\%$$

Think About It

In that reaction hydrogen is the limiting reactant because the problem states that the reaction is run with an excess of nitrogen.

7.107. Collect and Organize

Sulfur dioxide can be trapped to form calcium sulfate before it enters the atmosphere. We are to write the balanced equation for the process and calculate the mass of calcium sulfate produced from each ton (t) of SO_2.

Analyze

(a) The reactants and products for the "scrubbing" of SO_2 in an unbalanced equation are

$$SO_2(g) + CaO(s) + O_2(g) \rightarrow CaSO_4(s)$$

(b) We use the stoichiometric relationship of SO_2 to $CaSO_4$ in the balanced equation to calculate how much calcium sulfate is produced. A metric ton is 1000 kg.

Solve

(a) $2 SO_2(g) + 2 CaO(s) + O_2(g) \rightarrow 2 CaSO_4(s)$

(b) $1.00 \text{ t SO}_2 \times \dfrac{1000 \text{ kg}}{1 \text{ t}} \times \dfrac{1000 \text{ g}}{1 \text{ kg}} \times \dfrac{1 \text{ mol SO}_2}{64.06 \text{ g}} \times \dfrac{2 \text{ mol CaSO}_4}{2 \text{ mol SO}_2} \times \dfrac{136.14 \text{ g}}{1 \text{ mol CaSO}_4} \times \dfrac{1 \text{ kg}}{1000 \text{ g}} \times \dfrac{1 \text{ t}}{1000 \text{ kg}} = 2.13 \text{ t}$

Think About It

The calcium sulfate produced could find use as a desiccant as well as in cement, plaster, and blackboard chalk.

7.109. Collect and Organize

Using the percent composition of a mineral, we are to determine the formula.

Analyze

The percent composition is given in percentage by mass. If we assume 100 g of the mineral, 34.55 g of it is Mg, 19.96 g of it is Si, and 45.49 g of it is O. Using the molar masses of those elements, we can calculate the moles that those masses represent for each element and then find the whole-number ratio of the elements in the compound to give the formula of the mineral.

Solve

$$34.55 \text{ g Mg} \times \frac{1 \text{ mol}}{24.31 \text{ g}} = 1.421 \text{ mol Mg}$$

$$19.96 \text{ g Si} \times \frac{1 \text{ mol}}{28.09 \text{ g}} = 0.7106 \text{ mol Si}$$

$$45.49 \text{ g O} \times \frac{1 \text{ mol}}{16.00 \text{ g}} = 2.843 \text{ mol O}$$

That gives a molar ratio of Mg:Si:O of 2:1:4; the formula for the mineral is Mg_2SiO_4.

Think About It

The SiO_4^{4-} ion is the silicate ion, and that mineral's name is therefore magnesium silicate.

CHAPTER 8 | Aqueous Solutions: Chemistry of the Hydrosphere

8.1. Collect and Organize

This question asks us to differentiate between a strong binary acid and a weak binary acid through which we are specifically to identify the acid that is weak. A binary acid is an acid containing hydrogen and another element, such as HCl. A strong acid completely dissociates, meaning that all binary molecules of HX are present in solution as H^+ and X^-. A weak acid does not completely dissociate, meaning that some of the HX molecules are in the form of H^+ and X^- and some are present in the form HX.

Analyze

From Figure P8.1, the red spheres are attached to three white spheres and have a positive charge, so they must represent the H_3O^+ that results from the dissociation of the binary acids. The green, yellow, and magenta spheres carrying a negative charge must represent the Xs in the binary acids HX. If the X^- are free in solution and surrounded only by water molecules and not combined with H^+ (a white atom), then that HX must be a strong acid. If, however, HX is found in the solution as a molecule along with H^+ and X^-, then that HX must be a weak acid.

Solve

All the green and magenta spheres are present in solution as ions, not in combination with H^+ as HX. Therefore, HX (green) and HX (magenta) must both be strong acids. HX (yellow), however, is represented as two HX molecules and three X^- (yellow) ions. Because only some of the HX (yellow) has dissociated, HX (yellow) is the weak acid.

Think About It

The extent to which a weak acid may be dissociated can vary. If this were a strong weak acid, perhaps only one HX (yellow) molecule would be represented in Figure P8.1 along with four X^- (yellow) anions.

8.3. Collect and Organize

From Figure P8.3 we are asked to identify which ions are represented by the green and blue spheres at the equivalence point for a titration of sulfuric acid (H_2SO_4) with sodium hydroxide (NaOH).

Analyze

At the equivalence point the moles of acid (H_2SO_4) have reacted with the stoichiometric number of moles of base (NaOH). The reaction that describes that titration reaction is

$$H_2SO_4(aq) + 2\ NaOH(aq) \rightarrow Na_2SO_4(aq) + H_2O(\ell)$$

Solve

All sodium salts are soluble in water and therefore Na_2SO_4 will be present in solution as Na^+ and SO_4^{2-} ions. The Na^+ cations will have water molecules surrounding them, with the O atoms oriented toward them, whereas the SO_4^{2-} anions will have water molecules surrounding them, with the H atoms oriented toward them. Therefore, the blue spheres are the Na^+ cations and the green spheres are the SO_4^{2-} anions.

Think About It

The stoichiometries of the ions in solution are correct; we see twice as many Na^+ ions as SO_4^{2-} ions, as should be the case for the ions dissociated from Na_2SO_4.

8.5. Collect and Organize

Using Figure P8.5 we are to identify the oxidation change that occurs for the nitrogen atom in the reaction.

Analyze

The figure represents the following chemical equation

$$2\ NO + O_2 \rightarrow 2\ NO_2$$

Solve

The oxidation state for O in NO and NO_2 is –2. Therefore, the oxidation state of N in NO is +2 and in NO_2 the oxidation state for N is +4. Therefore, the N atom in that reaction is oxidized from +2 to +4, for an oxidation state change of 2.

Think About It

Because this is a redox reaction, something must be reduced. Here, O_2 changes its oxidation state from 0 to –2.

8.7. **Collect and Organize**

We are to rank the molecules of N and O in Figure P8.7 in order of decreasing oxidation state of the nitrogen atom.

Analyze

In each species the oxygen atom has an oxidation state of –2. From that we can determine the oxidation state of the nitrogen atoms.

Solve

(a) The formula for this species is N_2O_5. With five O atoms, each with a –2 oxidation state, both N atoms must have an oxidation state of +5.
(b) The formula for this species is N_2O_4. With four O atoms at a –2 oxidation state, both N atoms must have an oxidation state of +4.
(c) The formula for this species is NO. With the O atom at a –2 oxidation state, the N atom must have an oxidation state of +2.
(d) The formula for this species is N_2O_3. With three O atoms at a –2 oxidation state, both N atoms must have an oxidation state of +3.
(e) The formula for this species is NO_2. With two O atoms at a –2 oxidation state, the N atom must have an oxidation state of +4.

The species in order of decreasing (less positive) oxidation state:

$$(a)\ N_2O_5 > (b)\ N_2O_4 = (e)\ NO_2 > (d)\ N_2O_3 > (c)\ NO$$

Think About It

That the oxidation state of N in N_2O_4 is the same as that in NO_2 makes sense because those two compounds have the same empirical formula.

8.9. **Collect and Organize**

For the titration of sulfuric acid with barium hydroxide, we are asked to write the overall equation and then consider how the conductivity of the solution would change during the titration.

Analyze

The reaction in the titration is an acid–base neutralization reaction. Therefore, the products are water and a salt ($BaSO_4$, which is insoluble). To think about how the conductivity changes, we need the ionic equation. Conductivity will be high for the solution when many ions are present in solution. When no ions are present, the conductivity is that of pure water (zero on our graph).

Solve

(a) Overall equation for the titration is

$$H_2SO_4(aq) + Ba(OH)_2(aq) \rightarrow BaSO_4(s) + 2\ H_2O(\ell)$$

(b) The ionic equation is

$$H^+(aq) + HSO_4^-(aq) + Ba^{2+}(aq) + 2\ OH^-(aq) \rightarrow BaSO_4(s) + 2\ H_2O(\ell)$$

Before the titration begins, we have only a solution of H_2SO_4, which is ionized in solution to 2 $H^+(aq)$ and SO_4^{2-} (aq). The conductivity of this solution is high (and above that of pure water). As the titration proceeds, the conductivity will decrease because of the formation of low-solubility $BaSO_4$ and nonionized H_2O. At the equivalence point all the H_2SO_4 has reacted; only $BaSO_4$ and H_2O are present, and therefore the conductivity will be zero (that of pure water). As we titrate with $Ba(OH)_2$ beyond the equivalence point, the conductivity will increase because of the presence of excess Ba^{2+} and OH^- ions. The graph that shows the change is (b).

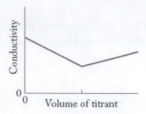

Think About It
Before the equivalence point, conductivity is due to unreacted H_2SO_4, but after the equivalence point the conductivity is due to excess $Ba(OH)_2$.

8.11. **Collect and Organize**
The molarity of a solution is the moles of solute in 1 L of solution. For each part of this problem, we are given the moles of solute in a volume (in milliliters) of solution. Molarity is abbreviated as M (for example, 2.00 M).

Analyze
To find molarity, we need to only divide the moles of solute by the volume of solution in liters. To get the volume in liters, we simply divide the milliliters of solution by 1000 or, even more simply, move the decimal three places to the left (for example, 100.0 mL = 0.1000 L).

Solve

(a) $\dfrac{0.56 \text{ mol}}{0.1000 \text{ L}} = 5.6 \ M \text{ BaCl}_2$

(b) $\dfrac{0.200 \text{ mol}}{0.2000 \text{ L}} = 1.00 \ M \text{ Na}_2\text{CO}_3$

(c) $\dfrac{0.325 \text{ mol}}{0.2500 \text{ L}} = 1.30 \ M \text{ C}_6\text{H}_{12}\text{O}_6$

(d) $\dfrac{1.48 \text{ mol}}{0.2500 \text{ L}} = 5.92 \ M \text{ KNO}_3$

Think About It
These calculations can be done quickly if you recognize, for example, that 100 mL is 1/10 of a liter. Therefore, the molarity of the solution will be 10 times the number of moles.

8.13. **Collect and Organize**
The molarity of ions in a solution is found similarly to the molarity of a solute: divide the moles of the ion present by the volume of the solution.

Analyze
In this problem, all the volumes are given in milliliters. To convert those volumes to liters, we move the decimal three places to the left (for example, 100.0 mL = 0.1000 L). The quantity of ions is given in grams. To find moles of solute we use the molar mass of each ion (recall that the mass of missing or added electrons is negligible).

Solve

(a) $\dfrac{0.29 \text{ mol NaNO}_3}{\text{L}} \times \dfrac{1 \text{ mol Na}^+}{\text{mol NaNO}_3} = 0.29 \ M \text{ Na}^+$

(b) $\dfrac{0.33 \text{ g NaCl} \times \dfrac{1 \text{ mol NaCl}}{58.44 \text{ g}} \times \dfrac{1 \text{ mol Na}^+}{1 \text{ mol NaCl}}}{0.025 \text{ L}} = 0.23 \ M \text{ Na}^+$

(c) $\dfrac{0.88 \text{ mol Na}_2\text{SO}_4}{\text{L}} \times \dfrac{2 \text{ mol Na}^+}{\text{mol Na}_2\text{SO}_4} = 1.76 \ M \text{ Na}^+$

(d)
$$\frac{0.46 \text{ g Na}_3\text{PO}_4 \times \dfrac{1 \text{ mol Na}_3\text{PO}_4}{163.94 \text{ g}} \times \dfrac{3 \text{ mol Na}^+}{1 \text{ mol Na}_3\text{PO}_4}}{0.100 \text{ L}} = 0.084 \; M \text{ Na}^+$$

Think About It
Calculating molarity given a mass of a solute involves first calculating moles of solute with the molar mass.

8.15. Collect and Organize
We are asked to calculate the grams of a solute needed to prepare a solution of a specific concentration.

Analyze
First we can find the moles of solute needed for the solution by multiplying the molarity by the volume (in liters). Once we have moles, we can use the molar mass of the solute to calculate the mass of solute needed.

Solve

(a) $1.000 \text{ L} \times \dfrac{0.200 \text{ mol}}{1 \text{ L}} \times \dfrac{58.44 \text{ g}}{1 \text{ mol}} = 11.7 \text{ g NaCl}$

(b) $0.2500 \text{ L} \times \dfrac{0.125 \text{ mol}}{1 \text{ L}} \times \dfrac{159.61 \text{ g}}{1 \text{ mol}} = 4.99 \text{ g CuSO}_4$

(c) $0.5000 \text{ L} \times \dfrac{0.400 \text{ mol}}{1 \text{ L}} \times \dfrac{32.04 \text{ g}}{1 \text{ mol}} = 6.41 \text{ g CH}_3\text{OH}$

Think About It
That is a practical calculation for preparing solutions when we know that we want to have a solution with a particular concentration.

8.17. Collect and Organize
From the concentration of ions, we are asked to calculate the total mass of the ions in 2.75 L of river water.

Analyze
We first have to convert the millimolar (mM) concentrations of each ion to molar (1000 mM = 1 M) by moving the decimal three places to the left (for example, 0.100 mM = 1.00×10^{-4} M). From the molar concentration of each ion we can find the moles (and then the mass) of each by multiplying the molarity by the volume (2.75 L). Finally, we need to add all the masses of the ions.

Solve

Mass of Ca^{2+}: $2.75 \text{ L} \times \left(\dfrac{8.20 \times 10^{-4} \text{ mol}}{\text{L}}\right) \times \left(\dfrac{40.08 \text{ g}}{1 \text{ mol}}\right) = 0.0904 \text{ g}$

Mass of Mg^{2+}: $2.75 \text{ L} \times \left(\dfrac{4.30 \times 10^{-4} \text{ mol}}{\text{L}}\right) \times \left(\dfrac{24.31 \text{ g}}{1 \text{ mol}}\right) = 0.0287 \text{ g}$

Mass of Na^+: $2.75 \text{ L} \times \left(\dfrac{3.00 \times 10^{-4} \text{ mol}}{\text{L}}\right) \times \left(\dfrac{22.99 \text{ g}}{1 \text{ mol}}\right) = 0.0190 \text{ g}$

Mass of K^+: $2.75 \text{ L} \times \left(\dfrac{2.00 \times 10^{-2} \text{ mol}}{\text{L}}\right) \times \left(\dfrac{39.10 \text{ g}}{1 \text{ mol}}\right) = 2.15 \text{ g}$

Mass of Cl^-: $2.75 \text{ L} \times \left(\dfrac{2.50 \times 10^{-4} \text{ mol}}{\text{L}}\right) \times \left(\dfrac{35.45 \text{ g}}{1 \text{ mol}}\right) = 0.0244 \text{ g}$

Mass of SO_4^{2-}: $2.75 \text{ L} \times \left(\dfrac{3.80 \times 10^{-4} \text{ mol}}{\text{L}}\right) \times \left(\dfrac{96.06 \text{ g}}{1 \text{ mol}}\right) = 0.100 \text{ g}$

Mass of HCO_3^-: $2.75 \text{ L} \times \left(\dfrac{1.82 \times 10^{-3} \text{ mol}}{\text{L}} \right) \times \left(\dfrac{61.02 \text{ g}}{1 \text{ mol}} \right) = 0.305 \text{ g}$

Total mass of ions: $0.0904 + 0.0287 + 0.0190 + 2.15 + 0.0244 + 0.100 + 0.305 = 2.72$ g

Think About It

The mass of the ions dissolved in natural water is sometimes called "total dissolved solids."

8.19. Collect and Organize

For each pharmaceutical we are given the volume and concentration of the solution. From this information we can find the moles of each solute by multiplying the volume by the concentration.

Analyze

We have to watch our units here for concentration and volume. We need to use the fact that 1000 mL = 1 L and 1000 mmol = 1 mol.

Solve

(a) $0.250 \text{ L} \times \dfrac{0.076 \text{ mol}}{\text{L}} = 0.0190$ mol acetaminophen

(b) $2.11 \text{ L} \times \dfrac{0.193 \text{ mmol}}{\text{L}} \times \dfrac{\text{mol}}{1000 \text{ mmol}} = 4.07 \times 10^{-4}$ mol chromalyn sodium

(c) $0.0475 \text{ L} \times \dfrac{5.73 \text{ mmol}}{\text{L}} \times \dfrac{1 \text{ mol}}{1000 \text{ mmol}} = 2.72 \times 10^{-4}$ mol benzocaine

(d) $14.6 \text{ L} \times \dfrac{27.4 \text{ mmol}}{\text{L}} \times \dfrac{1 \text{ mol}}{1000 \text{ mmol}} = 0.400$ mol Benadryl

Think About It

Converting milliliters to liters, liters to milliliters, millimoles to moles, and moles to millimoles by moving the decimal is convenient, but be sure that you do not mistakenly move it in the wrong direction (for example, saying that 2.56 mmol = 2560 mol would be wrong).

8.21. Collect and Organize

The table gives information on both sample size and the mass of DDT in each groundwater sample. To compare the DDT amounts among samples, we are asked to calculate the DDT in millimoles per liter.

Analyze

We have to first compute the millimoles of DDT in each sample by dividing the mass (in milligrams) by the molar mass of DDT ($C_{14}H_9Cl_5$, 354.49 mg/mmol). To find the concentration in each sample, we divide that result by the volume of the sample in liters.

Solve

Sample from the orchard: $\dfrac{0.030 \text{ mg} \times \left(\dfrac{1 \text{ mmol}}{354.49 \text{ mg}} \right)}{0.2500 \text{ L}} = 3.4 \times 10^{-4}$ mmol/L

Sample from the residentian area: $\dfrac{0.035 \text{ mg} \times \left(\dfrac{1 \text{ mmol}}{354.49 \text{ mg}} \right)}{1.750 \text{ L}} = 5.6 \times 10^{-5}$ mmol/L

Sample from the residentian area after storm: $\dfrac{0.57 \text{ mg} \times \left(\dfrac{1 \text{ mmol}}{354.49 \text{ mg}} \right)}{0.0500 \text{ L}} = 3.2 \times 10^{-2}$ mmol/L

Think About It
With all the concentrations of DDT in the samples now expressed in millimoles per liter (m*M*), we can make comparisons. The orchard is a little less contaminated than the residential area. The big surprise is that after a storm the groundwater contains nearly 600 times more DDT than before the storm.

8.23. Collect and Organize
We are given the average concentration of calcium ion in the tap water in groundwater in milligrams per liter and asked to express that in molarity, which is moles per liter.

Analyze
Because we are given milligrams of Ca^{2+} we first have to convert that mass to grams (divide by 1000) and then compute the moles from the molar mass of Ca^{2+} (40.078 g/mol). That result will be the moles of Ca^{2+} in 1 L.

Solve

$$\frac{48 \text{ mg Ca}^{2+}}{L} \times \frac{1 \text{ g}}{1000 \text{ mg}} \times \frac{1 \text{ mol}}{40.078 \text{ g}} = 1.2 \times 10^{-3} \text{ mol Ca}^{2+}$$

Think About It
That problem is made even shorter when you are comfortable in immediately stating that 10 mg = 0.010 g of Zn.

8.25. Collect and Organize
We are asked to decide whether the solubility of the substance would allow us to prepare a 1.0 *M* solution.

Analyze
If the mass of solute needed to prepare a 1.0 *M* solution is greater than the solubility limit, we will not be able to prepare the solution in 1.0 *M* concentration. We must use comparable units in making the comparisons. The solubility of those compounds could be expressed in grams per milliliter. For a 1.0 *M* solution of each, we need to weigh out 1 molar mass of the substance and dissolve it in 1000 mL. If we express that value in grams per milliliter, we will have the direct comparisons.

Solve
(a) $CuSO_4 \cdot 5 H_2O$ has a molar mass of 249.68 g/mol. For a 1.0 *M* solution, 249.68 g would be dissolved in 1 L, or 1000 mL. That is 0.25 g/mL, slightly higher than the solubility (0.231 g/mL).
(b) $AgNO_3$ has a molar mass of 169.87 g/mol. For a 1.0 *M* solution, 169.87 g would be dissolved in 1 L, or 1000 mL. That is 0.17 g/mL, lower than the solubility (1.22 g/mL).
(c) $Fe(NO_3)_2 \cdot 6 H_2O$ has a molar mass of 287.94 g/mol. For a 1.0 *M* solution, 287.94 g would be dissolved in 1 L, or 1000 mL. That is 0.29 g/mL, lower than the solubility (1.13 g/mL).
(d) $Ca(OH)_2$ has a molar mass of 74.092 g/mol. For a 1.0 *M* solution, 74.092 g would be dissolved in 1 L, or 1000 mL. That is 0.07 g/mL, higher than the solubility (0.00185 g/mL).
Therefore, we would be able to prepare 1.0 *M* solutions for substances (b) and (c) but not for (a) or (d).

Think About It
Finding the common units to make the comparison in the question is important. We could have expressed the concentration in moles per liter and converted the solubility data from grams per milliliter to moles per liter. As long as the units are comparable, we will arrive at the same answer.

8.27. Collect and Organize
We need to use the density of the coastal water given (1.02 g/mL) to calculate the volume of water for the concentration (1.09 g Mg^{2+}/kg). We also need to convert grams of Mg^{2+} into moles by using the molar mass.

Analyze
Using unit conversions, we can solve in a single step. We can use the density to find the mass in kilograms of coastal seawater directly because the density of 1.02 g/mL = 1.02 kg/L. Then we convert the grams of Mg^{2+} to moles by using the molar mass of Mg^{2+} (24.305 g/mol).

Solve

$$\frac{1.09 \text{ g Mg}^{2+}}{1 \text{ kg}} \times \frac{1.02 \text{ kg}}{1 \text{ L}} \times \frac{1 \text{ mol}}{24.305 \text{ g}} = 4.57 \times 10^{-2} \ M \text{ Mg}^{2+}$$

Think About It

Recognizing the relationships of kilograms to grams and milliliters to liters made this problem quick to solve.

8.29. **Collect and Organize**

In diluting a solution, the final concentration is less than the original concentration. Each solution is diluted to 25.0 mL.

Analyze

For each dilution, we can use Equation 8.6:

$$V_{initial} \times M_{initial} = V_{final} \times M_{final}$$

Since we are calculating the final concentration, C_{final}, we can rearrange the equation to

$$M_{final} = \frac{V_{initial} \times M_{initial}}{V_{final}}$$

Solve

(a) When 1.00 mL of 0.452 M Na$^+$ is diluted to 25.0 mL, we have $V_{initial}$ = 1.00 mL, $M_{initial}$ = 0.452 M, and V_{final} = 25.0 mL. The final concentration after diluting will be

$$M_{final} = \frac{1.00 \text{ mL} \times 0.452 \ M}{25.0 \text{ mL}} = 1.81 \times 10^{-2} \ M \text{ Na}^+$$

(b) When 2.00 mL of 3.4 mM LiCl is diluted to 25.0 mL, the final concentration will be

$$M_{final} = \frac{2.00 \text{ mL} \times 3.4 \text{ m}M}{25.0 \text{ mL}} = 2.7 \times 10^{-1} \text{ m}M \text{ LiCl}$$

(c) When 5.00 mL of 6.42×10^{-2} mM Zn^{2+} is diluted to 25.0 mL, the final concentration after diluting will be

$$M_{final} = \frac{5.00 \text{ mL} \times 6.42 \times 10^{-2} \text{ m}M}{25.0 \text{ mL}} = 1.28 \times 10^{-2} \text{ m}M \text{ Zn}^{2+}$$

Think About It

The milliliter volume units for those calculations do not need to be converted to another unit. As long as $V_{initial}$ and V_{final} are in the same units, the units will cancel in the calculations.

8.31. **Collect and Organize**

This "dilution" question is reversed: the concentration of Na$^+$ increases as the water in the puddle evaporates during the summer day. Given the initial concentration of Na$^+$ (0.449 M) and the percent evaporation, we are asked to find the final concentration of Na$^+$.

Analyze

We first have to compute volumes from the percent volume of the puddle after evaporation. If we assume a 1000 mL puddle, a reduction of the puddle volume to 23% would mean that 230 mL of the water in the puddle remains. Rearranging the dilution equation for the final concentration gives

$$M_{final} = \frac{V_{initial} \times M_{initial}}{V_{final}}$$

Solve

Here, $V_{initial}$ = 1000 mL, $M_{initial}$ = 0.449 M, and V_{final} = 230 mL. The final concentration of Na$^+$ in the puddle after evaporation is

$$M_{final} = \frac{1000 \text{ mL} \times 0.449 \ M}{230 \text{ mL}} = 1.95 \ M$$

To two siginificant digits, 2.0 M

Think About It
The concentration of the Na$^+$ in the puddle increased, as we would expect.

8.33. Collect and Organize
Given the size and amount of API in an adult dose of a cough suppressant, we are to determine the concentration and calculate the volume of the adult cough syrup needed to prepare a syrup of lower strength for children.

Analyze
The concentration of the cough suppressant in each syrup is calculated by dividing the milligrams of API by the volume of the dose. We can then use the dilution equation to determine how much of the adult syrup to dilute to obtain the children's syrup.

Solve
The adult dose concentration is

$$\frac{35 \text{ mg}}{20.0 \text{ mL}} = 1.75 \text{ mg/mL, or } 1.8 \text{ mg/mL (to two significant figures)}$$

The children's dose concentration is

$$\frac{4.00 \text{ mg}}{10.0 \text{ mL}} = 0.400 \text{ mg/mL}$$

For the child-strength cough syrup, the volume of adult syrup needed is

$$V_{adult} = \frac{V_{child} \times M_{child}}{M_{adult}} = \frac{0.400 \text{ mg/mL} \times 100.0 \text{ mL}}{1.75 \text{ mg/mL}} = 23 \text{ mL}$$

Think About It
That answer makes sense because the child-strength cough syrup is about 1/4 the strength of the adult cough syrup.

8.35. Collect, Organize, and Analyze
Electricity can conduct through a solution if it contains mobile ions.

Solve
Table salt produces Na$^+$ and Cl$^-$ ions in solution when it dissolves in polar solvents, which makes its solutions electrically conductive. Sugar does not dissociate into ions because it is not a salt. Sugar, therefore, is not a good conductor of electricity in solution, but table salt (NaCl) in solution is.

Think About It
A solution that conducts electricity is called an electrolyte, and one that does not conduct electricity is a nonelectrolyte.

8.37. Collect and Organize
We are asked to consider the composition of electrolyte versus nonelectrolyte solutions.

Analyze
Electrolyte solutions conduct electricity; nonelectrolyte solutions do not conduct electricity.

Solve
For solutions to conduct electricity, they must contain mobile charge particles. Electrolytes produce charged particles in the form of cations and anions when they dissolve in polar solvents; nonelectrolytes do not produce ions when they dissolve.

Think About It
Electrolytes may be strong or weak, depending on the degree of dissociation of the salt in the solution.

8.39. Collect and Organize
The ability to conduct electricity rises with the presence of more ions in solution.

Analyze
All the solutions contain salts, but they have different numbers of ions in their formula units and different concentrations. For each salt, we have to determine the number of ions (in terms of molarity).

Solve
(a) 1.0 M NaCl contains 1.0 M Na$^+$ + 1.0 M Cl$^-$ = 2.0 M ions.
(b) 1.2 M KCl contains 1.2 M K$^+$ + 1.2 M Cl$^-$ = 2.4 M ions.
(c) 1.0 M Na$_2$SO$_4$ contains 2.0 M Na$^+$ + 1.0 M SO$_4^{2-}$ = 3.0 M ions.
(d) 0.75 M LiCl contains 0.75 M Li$^+$ + 0.75 M Cl$^-$ = 1.5 M ions.
Therefore, the order of the solutions in decreasing ability to conduct electricity is
$$\text{(c) 1.0 } M \text{ Na}_2\text{SO}_4 > \text{(b) 1.2 } M \text{ KCl} > \text{(a) 1.0 } M \text{ NaCl} > \text{(d) 0.75 } M \text{ LiCl}$$

Think About It
Although the Na$_2$SO$_4$ and NaCl solutions are both 1 M in salt concentration, the solution of Na$_2$SO$_4$ is more conductive because it contains 3 M ions versus NaCl's 2 M ions.

8.41. Collect and Organize
The concentration of Na$^+$ ions in each solution depends on the concentration of the solution as well as the number of sodium ions in the chemical formula of the salt. The concentration of all those salt solutions is 0.025 M.

Analyze
If a substance has one Na$^+$ in its formula, then the concentration of Na$^+$ ions in a solution of that salt is equal to the concentration of the substance in solution. But if, for example, two Na$^+$ ions are in the salt's formula, the concentration of Na$^+$ ions in the solution is twice the concentration of the salt.

Solve
(a) 0.025 M NaBr is 0.025 M Na$^+$.
(b) 0.025 M Na$_2$SO$_4$ is 2 × 0.025 M = 0.050 M Na$^+$.
(c) 0.025 M Na$_3$PO$_4$ is 3 × 0.025 M = 0.075 M Na$^+$.

Think About It
Salts dissociate into their constituent ions, and so a solution may become more concentrated in a particular ion than the original concentration of salt.

8.43. Collect and Organize
We are to identify the property that makes an acid an acid.

Analyze
Acids complement bases and their reactions involve the exchange of protons (H$^+$) between the acid and the base.

Solve
Acids transfer H$^+$ to a base, and in water they increase the concentration of H$^+$ in solution.

Think About It
In aqueous solution, the base can be H$_2$O so that when the acid transfers its H$^+$ to water, H$_3$O$^+$ is produced.

8.45. Collect and Organize
In the context of acid–base reactions, we are asked to name two strong acids and two weak acids.

Analyze
A strong acid is one that completely dissociates into H$^+$ + A$^-$ in solution. A weak acid only partially dissociates and gives a mixture of H$^+$, A$^-$, and HA in solution.

Solve

Strong acids include HCl, HNO_3, $HClO_4$, H_2SO_4, HI, and HBr; weak acids include CH_3COOH, $HCOOH$, HF, and H_3PO_4.

Think About It

Strong bases include $NaOH$ and KOH; weak bases include NH_3 and anions of weak acids such as the acetate ion, CH_3COO^-.

8.47. Collect and Organize

We are to identify the property that makes a base a base.

Analyze

Bases complement acids, and their reactions involve the exchange of protons (H^+) between the acid and the base.

Solve

Bases accept H^+ from an acid, and in water they increase the concentration of OH^- in solution.

Think About It

In aqueous solution, the acid can be H_2O so that when the base accepts H^+ from water, OH^- is produced.

8.49. Collect, Organize, and Analyze

In the context of acid–base reactions, we are asked to name two strong bases and two weak bases. A strong base is one that completely dissociates into OH^- in solution. A weak base only partially ionizes when accepting H^+ and gives a mixture of B^+, OH^-, and BOH in solution.

Solve

Strong bases include $NaOH$, KOH, $CsOH$, $LiOH$, $RbOH$, $Ba(OH)_2$, $Sr(OH)_2$, and $Ca(OH)_2$; weak bases include NH_3, CH_3NH_2, and C_5H_5N.

Think About It

Strong acids include HBr and HNO_3; weak acids include acetic acid and formic acid.

8.51. Collect and Organize

In each part of this problem, we identify the acid (proton donor) and base (proton acceptor). To write the net ionic equations we have to identify the spectator ions and remove them from the ionic equation.

Analyze

For each reaction, write the species present in aqueous solution (showing dissociation). From those species, you can identify the acid and base. Then eliminate any spectator ions in the ionic equation to give the net ionic equation.

Solve

(a) Ionic equation:
$$H^+(aq) + HSO_4{}^{2-}(aq) + Ca^{2+} + 2\ OH^-(aq) \rightarrow Ca^{2+}(aq) + SO_4{}^{2-}(aq) + 2\ H_2O(\ell)$$
The acid is H_2SO_4; the base is $Ca(OH)_2$.
Net ionic equation:
$$H^+(aq) + HSO_4{}^{2-}(aq) + 2\ OH^-(aq) \rightarrow SO_4{}^{2-}(aq) + 2\ H_2O(\ell)$$

(b) Ionic and net ionic equation:
$$PbCO_3(s) + H^+(aq) + HSO_4{}^{2-}(aq) \rightarrow PbSO_4(s) + CO_2(g) + H_2O(\ell)$$
$PbCO_3$ is the base; sulfuric acid is the acid.

(c) Ionic and net equation:
$$Ca(OH)_2(s) + 2\ CH_3COOH(aq) \rightarrow Ca^{2+}(aq) + 2\ CH_3COO^-(aq) + 2\ H_2O(\ell)$$
$Ca(OH)_2$ is the base; CH_3COOH is the acid.

Think About It

Reactions (a) and (c) are neutralization reactions, whereas reaction (b) is an acid–base reaction that also forms a precipitate ($PbSO_4$) and a gas (CO_2).

8.53. **Collect and Organize**

We need to write molecular formulas for the reactants from the chemical names, determine the formulas for the products, write a balanced molecular equation, and then write the net ionic equation.

Analyze

All the reactions involve an acid–base reaction in which a proton is transferred from the acid to the base. We can use the rules of Chapter 4 to write the formulas from the chemical names. In the net ionic equation, we must eliminate all spectator ions.

Solve

(a) Molecular equation:
$$Mg(OH)_2(s) + H_2SO_4(aq) \rightarrow MgSO_4(aq) + 2\ H_2O(\ell)$$
Ionic and net ionic equations:
$$Mg(OH)_2(s) + H^+(aq) + HSO_4{}^{2-}(aq) \rightarrow Mg^{2+}(aq) + SO_4{}^{2-}(aq) + 2\ H_2O(\ell)$$
$$Mg(OH)_2(s) + H^+(aq) + HSO_4{}^{2-}(aq) \rightarrow Mg^{2+}(aq) + 2\ H_2O(\ell) + SO_4{}^{2-}(aq)$$

(b) Molecular equation:
$$MgCO_3(s) + 2\ HCl(aq) \rightarrow MgCl_2(aq) + H_2CO_3(aq)$$
The carbonic acid reacts in solution to give CO_2 and H_2O, so the ionic and net ionic equations are as follows:
$$MgCO_3(s) + 2\ H^+(aq) + 2\ Cl^-(aq) \rightarrow Mg^{2+}(aq) + 2\ Cl^-(aq) + H_2O(\ell) + CO_2(g)$$
$$MgCO_3(s) + 2\ H^+(aq) \rightarrow Mg^{2+}(aq) + H_2O(\ell) + CO_2(g)$$

(c) Molecular equation:
$$NH_3(g) + HCl(g) \rightarrow NH_4Cl(s)$$
That is also the net ionic equation because these species are not in aqueous solution and cannot form ions.

Think About It

Species that are solids or gases do not appear as ionic species. Only soluble species dissolved in water appear with the designation "(aq)" in the ionic equations.

8.55. **Collect and Organize**

We are given that lead(II) carbonate ($PbCO_3$) and lead(II) hydroxide [$Pb(OH)_2$] dissolve in acidic solutions (containing H_3O^+).

Analyze

To write the net ionic equations, we need to determine the acid–base reaction that might be occurring. Here the acid is in solution as H_3O^+, which we can write as $H^+(aq)$. That species must react with the anions of the solid salts ($CO_3{}^{2-}$ and OH^-).

Solve

Lead(II) carbonate:
$$PbCO_3(s) + 2\ H^+(aq) \rightarrow Pb^{2+}(aq) + H_2CO_3(aq)$$
Carbonic acid reacts in solution to give $H_2O(\ell)$ and $CO_2(g)$:
$$PbCO_3(s) + 2\ H^+(aq) \rightarrow Pb^{2+}(aq) + CO_2(g) + H_2O(\ell)$$
Lead(II) hydroxide:
$$Pb(OH)_2(s) + 2\ H^+(aq) \rightarrow Pb^{2+}(aq) + 2\ H_2O(\ell)$$

Think About It

Both solids, by reacting with acid to form either CO_2 with water or just water, dissolve the solid, releasing toxic Pb^{2+} ions into the water.

8.57. Collect and Organize

Solutions that contain a solute may be classified as unsaturated, saturated, or supersaturated. We are to distinguish saturated from supersaturated solutions.

Analyze

More of the solute can dissolve in an unsaturated solution, but a saturated solution contains all the solute it can hold in the solution.

Solve

A saturated solution contains the maximum concentration of a solute. A supersaturated solution *temporarily* contains more than the maximum concentration of a solute at a given temperature.

Think About It

A supersaturated solution eventually precipitates out some solute (until it reaches the point of saturation).

8.59. Collect and Organize

By defining a precipitation reaction, we are asked how snow and rain precipitating from the atmosphere are like the precipitation of a solid from a solution.

Analyze

In a chemical precipitation reaction a solid product appears when two homogeneous solutions are mixed to give a heterogeneous mixture.

Solve

Gaseous water vapor in the atmosphere is a homogeneous solution; snow and rain "fall out" of this solution in the same way that a solid precipitates from a solution.

Think About It

Hail and drizzle also would be included as atmospheric precipitation.

8.61. Collect and Organize

We are asked if a saturated solution is always also a concentrated solution.

Analyze

A saturated solution is one in which no more solute can dissolve. A concentrated solution is one that contains a large amount of solute.

Solve

A saturated solution may not be a concentrated solution if the solute is only sparingly or slightly soluble in the solution. Then the solution is a saturated dilute solution.

Think About It

Be careful when using the terms *unsaturated/saturated* and *dilute/concentrated*, which have precise meanings in chemistry.

8.63. Collect and Organize

We can use the rules in Table 8.4 to predict solubility. All the compounds listed in the problem are ionic salts and are being dissolved in water.

Analyze

Soluble salts include those of the alkali metals and the ammonium cation and those with the acetate or nitrate anion. Exceptions exist to the general solubility of halide salts (Ag^+, Cu^+, Hg_2^{2+}, and Pb^{2+} halides are insoluble) and sulfates (Ba^{2+}, Ca^{2+}, Hg_2^{2+}, Pb^{2+}, and Sr^{2+} sulfates are insoluble). All other salts are insoluble except the hydroxides of Ba^{2+}, Ca^{2+}, and Sr^{2+}.

Solve

(a) Barium sulfate is insoluble.

(b) Barium hydroxide is soluble.

(c) Lanthanum nitrate is soluble.

(d) Sodium acetate is soluble.

(e) Lead hydroxide is insoluble.

(f) Calcium phosphate is insoluble.

Think About It

Knowing well the few simple rules of solubility can help us easily predict which salts dissolve in water and which do not.

8.65. Collect and Organize

We are to write balanced molecular and net ionic equations for any precipitation reactions. For each reaction, we have to determine whether the mix of cations and anions present in solution results in an insoluble salt.

Analyze

We use the solubility rules in Table 8.4 to determine which, if any, species precipitates when the two solutions are mixed. The net ionic equation can be written from the ionic equation by eliminating any of the spectator ions, those ions not involved in forming the insoluble precipitate.

Solve

(a) The reactants in aqueous solution are Pb^{2+}, NO_3^-, Na^+, and SO_4^{2-}. If $Pb(NO_3)_2$ and Na_2SO_4 switched anionic partners they would form $PbSO_4$ and $NaNO_3$. Of those two salts, $PbSO_4$ is insoluble. The ionic equation describing that reaction is

$$Pb^{2+}(aq) + 2\ NO_3^-(aq) + 2\ Na^+(aq) + SO_4^{2-}(aq) \rightarrow PbSO_4(s) + 2\ Na^+(aq) + 2\ NO_3^-(aq)$$

The balanced reaction is

$$Pb(NO_3)_2(aq) + Na_2SO_4(aq) \rightarrow PbSO_4(s) + 2\ NaNO_3(aq)$$

The net ionic equation is

$$Pb^{2+}(aq) + SO_4^{2-}(aq) \rightarrow PbSO_4(s)$$

(b) The reactants in aqueous solution are Ni^{2+}, Cl^-, NH_4^+, and NO_3^-. If $NiCl_2$ and NH_4CO_3 switched anionic partners they would form $Ni(NO_3)_2$ and NH_4Cl. Both salts are soluble; therefore, no precipitation reaction occurs.

(c) The reactants in aqueous solution are Fe^{2+}, Cl^-, Na^+, and S^{2-}. If $FeCl_2$ and Na_2S switched anionic partners they would form FeS and $NaCl$. Of those two salts, FeS is insoluble. The ionic equation describing that reaction is

$$Fe^{2+}(aq) + 2\ Cl^-(aq) + 2\ Na^+(aq) + S^{2-}(aq) \rightarrow FeS(s) + 2\ Na^+(aq) + 2\ Cl^-(aq)$$

The balanced reaction is

$$FeCl_2(aq) + Na_2S(aq) \rightarrow FeS(s) + 2\ NaCl(aq)$$

The net ionic equation is

$$Fe^{2+}(aq) + S^{2-}(aq) \rightarrow FeS(s)$$

(d) The reactants in aqueous solution are Mg^{2+}, SO_4^{2-}, Ba^{2+}, and Cl^-. If $MgSO_4$ and $BaCl_2$ switched anionic partners they would form $MgCl_2$ and $BaSO_4$. Of those two salts, $BaSO_4$ is insoluble. The ionic equation describing that reaction is

$$Mg^{2+}(aq) + SO_4^{2-}(aq) + Ba^{2+}(aq) + 2\ Cl^-(aq) \rightarrow Mg^{2+}(aq) + 2\ Cl^-(aq) + BaSO_4(s)$$

The balanced reaction is

$$MgSO_4(aq) + BaCl_2(aq) \rightarrow MgCl_2(aq) + BaSO_4(s)$$

The net ionic equation is

$$Ba^{2+}(aq) + SO_4^{2-}(aq) \rightarrow BaSO_4(s)$$

Think About It

The net ionic equation for a precipitation reaction describes the formation of the insoluble salt from the aqueous cations and anions.

8.67. Collect and Organize

The compound that precipitates first from an evaporating solution will be the least soluble.

Analyze

The potential salts that could form are all salts of Ca^{2+}: $CaCl_2$, $CaCO_3$, and $Ca(NO_3)_2$. The solubility rules state that nitrate and chloride salts are soluble for Ca^{2+} but imply that the carbonate salt of calcium is insoluble.

Solve

The most insoluble salt, $CaCO_3$, precipitates first from the evaporating solution.

Think About It

Calcium carbonate may not have precipitated from the original more dilute solution because, $CaCO_3$, being somewhat soluble, had not yet become concentrated enough to be a saturated solution. Once the saturation point is reached through evaporation, the salt will precipitate.

8.69. Collect and Organize

To determine how much $MgCO_3$ precipitates in that reaction, we have to determine whether Na_2CO_3 or $Mg(NO_3)_2$ is the limiting reactant. The net ionic reaction for the reaction is
$$Mg^{2+}(aq) + CO_3^{2-}(aq) \rightarrow MgCO_3(s)$$

Analyze

From the given volumes of each reactant and its concentration, we first need to calculate the moles of Mg^{2+} and CO_3^{2-} present in the mixed solution. Those react in a 1:1 molar ratio to form $MgCO_3$. By comparing the moles of Mg^{2+} and CO_3^{2-} we can determine the limiting reactant. Because 1 mol of $MgCO_3$ will form from 1 mol of either Mg^{2+} or CO_3^{2-}, the moles of the limiting reactant must equal the moles of $MgCO_3$ formed. From the moles of $MgCO_3$ formed, we can calculate the mass formed by using the molar mass of $MgCO_3$ (84.31 g/mol).

Solve

$$\text{mol } CO_3^{2-} = 10.0 \text{ mL } Na_2CO_3 \times \frac{0.200 \text{ mol}}{1000 \text{ mL}} \times \frac{1 \text{ mol } CO_3^{2-}}{1 \text{ mol } Na_2CO_3} = 2.00 \times 10^{-3} \text{ mol}$$

$$\text{mol } Mg^{2+} = 5.00 \text{ mL } Mg(NO_3)_2 \times \frac{0.0500 \text{ mol}}{1000 \text{ mL}} \times \frac{1 \text{ mol } Mg^{2+}}{1 \text{ mol } Mg(NO_3)_2} = 2.50 \times 10^{-4} \text{ mol}$$

The limiting reactant is $Mg(NO_3)_2$ and 2.50×10^{-4} mol of $MgCO_3$ will form.
The mass of $MgCO_3$ produced is

$$2.50 \times 10^{-4} \text{ mol} \times \frac{84.31 \text{ g}}{1 \text{ mol}} = 2.11 \times 10^{-2} \text{ g } MgCO_3$$

Think About It

In every stoichiometric equation, the moles of the reactants are important. For species in solution, the moles can be found by multiplying the volume of the solution by the concentration, just as finding moles from a mass of substance involves dividing the mass of substance by the molar mass.

8.71. Collect and Organize

From the balanced equation, 1 mol of O_2 is required to react with 4 mol of $Fe(OH)^+$, the iron(II) species. Knowing the volume and concentration of iron(II), we are asked to find the grams of O_2 needed to form the insoluble $Fe(OH)_3$ product.

Analyze

We can find the moles of $Fe(OH)^+$ in solution by multiplying the volume of the iron(II) solution (75 mL) by its concentration (0.090 *M*). The number of moles of O_2 required in the reaction is 1/4 of the moles of $Fe(OH)^+$ present. From moles of O_2 we can use the molar mass of O_2 (32.00 g/mol) to calculate the grams of O_2 needed.

Solve

$$\text{mol Fe(OH)}^+ = 75 \text{ mL} \times \frac{0.090 \text{ mol}}{1000 \text{ mL}} = 6.75 \times 10^{-3} \text{ mol}$$

$$\text{mass of O}_2 = 6.75 \times 10^{-3} \text{ mol Fe(OH)}^+ \times \frac{1 \text{ mol O}_2}{4 \text{ mol Fe(OH)}^+} \times \frac{32.00 \text{ g O}_2}{1 \text{ mol}} = 5.4 \times 10^{-2} \text{ g}$$

Think About It

Because the molar ratio of Fe(OH)$^+$ to O$_2$ is 1:4, we require fewer moles of O$_2$ in that reaction than we have of the iron(II) species.

8.73. Collect and Organize

Given the concentration of magnesium in German and Italian mineral waters, we are asked to determine from which source (or neither) a 125.0 mL sample of water derives when 6.875 mg of MgNH$_4$PO$_4$ precipitates upon titration with a solution of ammonium hydrogen phosphate.

Analyze

We can assume that the precipitate, MgNH$_4$PO$_4$, precipitates 100% of the magnesium in the sample. Note here then that 1 mol of MgNH$_4$PO$_4$ is derived from 1 mol of magnesium ions in the sample. We need only that stoichiometric ratio to solve the problem; the details about the titration with ammonium hydrogen phosphate are not relevant.

Solve

$$\frac{6.875 \text{ mg MgNH}_4\text{PO}_4}{0.125 \text{ L}} \times \frac{1 \text{ g}}{1000 \text{ mg}} \times \frac{1 \text{ mol MgNH}_4\text{PO}_4}{137.31 \text{ g}} \times \frac{1 \text{ mol Mg}^{2+}}{1 \text{ mol MgNH}_4\text{PO}_4} \times \frac{24.31 \text{ g}}{1 \text{ mol Mg}^{2+}} \times \frac{1000 \text{ mg}}{1 \text{ g}} = 9.737 \text{ mg/L}$$

That is below the range of Mg^{2+} for both German (108–130 mg/L) and Italian (50–60 mg/L) mineral waters. Therefore, this mineral water sample derives from some other source.

Think About It

That Mg^{2+} concentration is 5–6 times lower than the Italian mineral water range and 10–12 times lower than the German mineral water range.

8.75. Collect and Organize

We are to define the connection between losses or gains of electrons and changes in oxidation numbers.

Analyze

Oxidation numbers are assigned to atoms in compounds on the basis of the number of electrons they formally bring to the species. The loss of electrons (oxidation) means that the oxidation number becomes more positive. The gain of electrons (reduction) means that the oxidation number becomes more negative.

Solve

The number of electrons gained or lost is directly related to the change in oxidation number of a species. If a species loses two electrons, the oxidation number of one of the atoms in the species will increase by 2 (for example, from +1 to +3).

Think About It

Oxidation numbers can help us decide which species is oxidized and which is reduced in a redox reaction.

8.77. Collect and Organize

The charges of all the ions are shown as superscripts for the species. We are to determine the sum of oxidation numbers for each species.

Analyze

Because the sum of the oxidation numbers of the atoms must equal the total charge on the polyatomic ion, the sum of the oxidation numbers for each species is simply the charge on the species.

Solve

(a) −1 for OH^-
(b) +1 for NH_4^+
(c) −2 for SO_4^{2-}
(d) −3 for PO_4^{3-}

Think About It

Recall that we can use the charge on a species to determine the oxidation state of an atom in the species that might have a variable oxidation state. For example, the oxidation state for S in SO_4^{2-} is +6 because the sum of the oxidation states of the four oxygen atoms is −8 (4 O^{2-}) and the overall charge is 2− on the anion.

8.79. **Collect and Organize**

Both silver and gold are placed into sulfuric acid, but only silver dissolves. We are asked which metal is the better reducing agent.

Analyze

For the metals to dissolve, they must be oxidized from their metallic state to a soluble cation (Au^{3+} or Ag^+). When a metal is oxidized, it acts as a reducing agent. Here Au does not reduce sulfuric acid, but Ag does.

Solve

Because silver dissolves (is oxidized) in sulfuric acid but gold does not, silver is more easily oxidized and is therefore the stronger of the two metals as a reducing agent.

Think About It

Oxidation and reduction reactions always occur in pairs. If a substance is reduced, it acts as an oxidizing agent; if a substance is oxidized, it acts as a reducing agent.

8.81. **Collect and Organize**

For the C_nH_{2n+2} formula of alkanes, we are to describe how the oxidation state of carbon in the compounds changes as n (or the chain length) increases.

Analyze

We can look at this by considering the oxidation state changes for a short series: CH_4, C_2H_6, C_3H_8, and C_4H_{10}. Remember that carbon is slightly more electronegative than hydrogen, so in those compounds carbon will have a negative oxidation state.

Solve

The oxidation state of carbon in CH_4 is −4, for C_2H_6 it is −3, for C_3H_8 it is −8/3 (or −2.66), and for C_4H_{10} it is −10/4 (or −2.5). From that series we see that as n increases, the oxidation state of the carbon atom in the alkanes increases (becomes less negative, or more positive).

Think About It

For C_nH_{2n+2}, the oxidation state of the carbon atoms is $-(2n+2)/n$.

8.83. **Collect and Organize**

The oxidation number for chlorine in those species varies depending on the number of oxygens to which the Cl atom is bound and the overall charge on the species.

Analyze

The oxidation number for oxygen in those species is −2. For hydrogen it is +1. The oxidation number for chlorine in each species, therefore, must be positive and can be determined by

oxidation number on Cl = charge on species − [(number of O atoms) × (−2) + (number of H atoms) × (+1)]

Solve

(a) HClO: oxidation number on Cl = 0 − [1(−2) + 1(+1)] = +1
(b) $HClO_3$: oxidation number on Cl = 0 − [3(−2) + 1(+1)] = +5
(c) $HClO_4$: oxidation number on Cl = 0 − [4(−2) + 1(+1)] = +7

Think About It
You may also determine the oxidation number by considering the charge on each ion and then adding them to get the overall charge on the species. For example, in $HClO_3$ we have H^+, O^{2-}, O^{2-}, O^{2-}, and Cl^{n+} with the sum of $+1$, -2, -2, -2, and $n+$ equaling zero, so n must be 5.

8.85. Collect and Organize
For the dehydration reaction of glucose to form C and H_2O, we are asked to determine the change of oxidation state in the reaction.

Analyze
To answer that question, we compare the oxidation state of carbon in $C_{12}H_{22}O_{11}$ to that in elemental carbon.

Solve
We start by looking at the oxidation state of C in glucose, $C_{12}H_{22}O_{11}$. The oxygen atoms will have an oxidation state of -2 (for 11 O atoms that is a total of -22), and the hydrogen atoms will have an oxidation state of $+1$ (for 22 H atoms that is a total of $+22$). The oxidation states of the O and H atoms, therefore, add to zero; therefore, the carbon atoms in glucose have an oxidation state of 0. In elemental carbon the oxidation state also is zero. Therefore, no change of oxidation state has occurred in that reaction.

Think About It
That molecule has the exactly correct stoichiometric amount of H and O to make H_2O.

8.87. Collect and Organize
To write the balanced equation to determine the moles of O_2 used in reaction with 1 mol of Fe_3O_4 to form Fe_2O_3, we have to identify the reactants and products and then balance the atoms.

Analyze
We are given the formulas for the reactant, Fe_3O_4, and the product, Fe_2O_3, so we can write the unbalanced equation for the conversion in the presence of oxygen and balance the reaction by inspection.

Solve
Write reactants and products in an unbalanced equation:
$$Fe_3O_4 + O_2 \rightarrow Fe_2O_3$$
Balance iron atoms:
$$2\ Fe_3O_4 + O_2 \rightarrow 3\ Fe_2O_3$$
Balance oxygen atoms:
$$2\ Fe_3O_4 + \tfrac{1}{2}\ O_2 \rightarrow 3\ Fe_2O_3$$
Eliminate the fractional coefficient:
$$4\ Fe_3O_4 + O_2 \rightarrow 6\ Fe_2O_3$$
For every mole of Fe_3O_4 reacted we would use 0.25 mol of O_2.

Think About It
In that reaction O_2 is the oxidizing agent and Fe_3O_4 is the reducing agent.

8.89. Collect and Organize
For every species we are asked to find the oxidation number for each atom. From those oxidation numbers, we can see which species is oxidized and which species is reduced.

Analyze
All those reactions involve species of iron, silicon, oxygen, and hydrogen. Oxygen typically has an oxidation number of -2; hydrogen typically has an oxidation number of $+1$. Oxidation numbers of pure elements are zero. Iron is the atom most likely to have a variable oxidation number. Silicon's oxidation number is usually $+4$, consistent with its position in group 14 of the periodic table. Because all the compounds are neutral, the sum of the oxidation numbers for the atoms must be zero.

Solve

(a) Reactants

  SiO_2: Si = +4, O = −2

  Fe_3O_4: Fe = +8/3, O = −2

Products

  Fe_2SiO_4: Fe = +2, Si = +4, O = −2

  O_2: O = 0

Notice that we compute an oxidation state for Fe in Fe_3O_4 as +8/3. Actually, that compound consists of FeO (Fe^{2+}) and Fe_2O_3 (Fe^{3+})—2/3 of Fe^{3+} and 1/3 of Fe^{2+}.

Oxygen is oxidized (O^{2-} to O_2) and (some) iron is reduced (Fe^{3+} to Fe^{2+}).

(b) Reactants

  SiO_2: Si = +4, O = −2

  Fe: Fe = 0

  O_2: O = 0

Products

  Fe_2SiO_4: Fe = +2, Si = +4, O = −2

Iron is oxidized (Fe^0 to Fe^{2+}) and oxygen is reduced (O_2 to O^{2-}).

(c) Reactants

  FeO: Fe = +2, O = −2

  O_2: O = 0

  H_2O: H = +1, O = −2

Products

  $Fe(OH)_3$: Fe = +3, O = −2, H = +1

Iron is oxidized (Fe^{2+} to Fe^{3+}) and oxygen is reduced (O_2 to O^{2-}).

Think About It

Molecular elemental oxygen is reduced in equations (b) and (c) and, therefore, acts as an oxidizing agent.

8.91. **Collect and Organize**

For the conversion of $FeCO_3$ to Fe_2O_3 and Fe_3O_4 in separate processes, we are to determine from balancing the redox equations how many moles of oxygen are consumed.

Analyze

We can assume that the carbon in $FeCO_3$ is converted into carbon dioxide in each process and that we are in a neutral environment for the conversions. We can balance these reactions simply by inspection.

Solve

(a) The unbalanced reaction is

$$FeCO_3(s) + O_2(g) + \rightarrow Fe_2O_3(s) + CO_2(g)$$

Atoms: 1 Fe + 1 C + 5 O → 2 Fe + 5 O + 1 C

We can start by balancing the Fe atoms by placing a coefficient of 2 in front of $FeCO_3$ on the left-hand side.

$$2 FeCO_3(s) + O_2(g) + \rightarrow Fe_2O_3(s) + CO_2(g)$$

Atoms: 2 Fe + 2 C + 8 O → 2 Fe + 5 O + 1 C

We can then balance the C atoms by placing a coefficient of 2 in front of CO_2 on the right-hand side.

$$2 FeCO_3(s) + O_2(g) + \rightarrow Fe_2O_3(s) + 2 CO_2(g)$$

Atoms: 2 Fe + 2 C + 8 O → 2 Fe + 7 O + 2 C

We can then balance the O atoms by placing a coefficient of $\frac{1}{2}$ in front of O_2 on the left-hand side.

$$2 FeCO_3(s) + 1/2 O_2(g) + \rightarrow Fe_2O_3(s) + 2 CO_2(g)$$

Atoms: 2 Fe + 2 C + 7 O → 2 Fe + 7 O + 2 C

To eliminate the fractional coefficients we multiply all the coefficients by 2.

$$4 FeCO_3(s) + O_2(g) + \rightarrow 2 Fe_2O_3(s) + 4 CO_2(g)$$

Atoms: 4 Fe + 4 C + 14 O → 4 Fe + 14 O + 4 C

The equation is now balanced. For one mole of $FeCO_3$, 0.25 mole of O_2 will be consumed.

(b) The unbalanced reaction is

$$FeCO_3(s) + O_2(g) \rightarrow Fe_3O_4(s) + CO_2(g)$$

Atoms: 1 Fe + 1 C + 5 O → 3 Fe + 6 O + 1 C

We can start by balancing the Fe atoms by placing a coefficient of 3 in front of $FeCO_3$ on the left-hand side.

$$3 FeCO_3(s) + O_2(g) \rightarrow Fe_3O_4(s) + CO_2(g)$$

Atoms: 3 Fe + 3 C + 11 O → 3 Fe + 6 O + 1 C

We can then balance the C atoms by placing a coefficient of 3 in front of CO_2 on the right-hand side.

$$3\ FeCO_3(s) + O_2(g) \rightarrow Fe_3O_4(s) + 3\ CO_2(g)$$

Atoms: $\quad 3\ Fe + 3\ C + 11\ O \rightarrow 3\ Fe + 10\ O + 3\ C$

We can then balance the O atoms by placing a coefficient of $\frac{1}{2}$ in front of O_2 on the left-hand side.

$$3\ FeCO_3(s) + 1/2\ O_2(g) \rightarrow Fe_3O_4(s) + 3\ CO_2(g)$$

Atoms: $\quad 3\ Fe + 3\ C + 10\ O \rightarrow 3\ Fe + 10\ O + 3\ C$

To eliminate the fractional coefficients we multiply all the coefficients by 2.

$$6\ FeCO_3(s) + O_2(g) \rightarrow 2\ Fe_3O_4(s) + 6\ CO_2(g)$$

Atoms: $\quad 6\ Fe + 6\ C + 20\ O \rightarrow 6\ Fe + 20\ O + 6\ C$

The equation is now balanced. For one mole of $FeCO_3$, 0.17 mole of O_2 will be consumed.

Think About It

Both redox reactions simplify to overall reactions that do not need acid as a reactant, even though the half-reactions do depend on having H^+ available for the reaction.

8.93. **Collect and Organize**

Ammonium ions (NH_4^+) are oxidized by oxygen gas to give nitrate ions (NO_3^-). We are asked to balance the reaction in acid solution.

Analyze

To balance the reaction, we use the steps outlined in Table 8.7. This reaction occurs in acidic solution so we may use H^+ and H_2O in the last two balancing steps.

Solve

Step 1. We first write the reaction expression using one mole of each of the known reactants and products.

$$NH_4^+(aq) + O_2(g) \rightarrow NO_3^-(aq)$$

Step 2. Calculate the $\Delta O.N.$ values for the elements that are reduced and oxidized.

$$NH_4^+(aq) + O_2(g) \rightarrow NO_3^-(aq)$$

N is oxidized from –3 to +5 for $\Delta O.N. = +8$

Each O in O_2 is reduced from 0 to –2 for $\Delta O.N. = 2 \times (-2) = -4$

Step 3. Insert coefficients to balance $\Delta O.N.$ values and balance the atoms that are oxidized and reduced, but not balancing O atoms. For this we add 2 as coefficient in front of O_2; the N atoms are balanced.

$$NH_4^+(aq) + 2\ O_2(g) \rightarrow NO_3^-(aq)$$

Step 4. Balance charges by adding 2 H^+ ions to the right-hand side.

$$NH_4^+(aq) + 2\ O_2(g) \rightarrow NO_3^-(aq) + 2H^+(aq)$$

Step 5. Add 2 H_2O molecules to the right-hand side and 1 O_2 to the left-hand side to balance the H and O atoms.

$$NH_4^+(aq) + 2\ O_2(g) \rightarrow NO_3^-(aq) + 2\ H^+(aq) + H_2O(\ell)$$

Think About It

That is an eight-electron oxidation in which the oxidation number of the nitrogen atom changes from –3 in NH_4^+ to +5 in NO_3^-.

8.95. **Collect and Organize**

We are to balance the redox reaction between $Fe(OH)_2^+$ and Mn^{2+} to give MnO_2 and Fe^{2+}. The problem does not specify whether the freshwater stream is acidic or basic; we assume acidic conditions here.

Analyze

To balance the reaction, we use the steps outlined in Table 8.7. If we assume that the reaction is in acidic solution, we may use H^+ and H_2O in the last two balancing steps.

Solve

Step 1. We first write the reaction expression using one mole of each of the known reactants and products.

$$Fe(OH)_2^+(aq) + Mn^{2+}(aq) \rightarrow MnO_2(s) + Fe^{2+}(aq)$$

Step 2. Calculate the ΔO.N. values for the elements that are reduced and oxidized.

$$Fe(OH)_2^+(aq) + Mn^{2+}(aq) \rightarrow MnO_2(s) + Fe^{2+}(aq)$$

Fe is reduced from +3 to +2 for ΔO.N. = −1

Mn is oxidized from +2 to +4 for ΔO.N. = +2

Step 3. Insert coefficients to balance ΔO.N. values and balance the atoms that are oxidized and reduced, but not balancing O atoms. For that we add 2 as coefficient in front of $Fe(OH)_2$ and balance the Fe atoms by adding a 2 as a coefficient in front of Fe^{2+}.

$$2 Fe(OH)_2^+(aq) + Mn^{2+}(aq) \rightarrow MnO_2(s) + 2 Fe^{2+}(aq)$$

Step 4. The charges are balanced.

$$2 Fe(OH)_2^+(aq) + Mn^{2+}(aq) \rightarrow MnO_2(s) + 2 Fe^{2+}(aq)$$

Step 5. Add 2 H_2O molecules to the right-hand side to balance the H and O atoms.

$$2 Fe(OH)_2^+(aq) + Mn^{2+}(aq) \rightarrow MnO_2(s) + 2 Fe^{2+}(aq) + 2 H_2O$$

The reaction is now balanced.

Think About It

The assumption that that reaction can be balanced under acidic conditions is a good one because dissolved CO_2 in natural waters makes them slightly acidic because of the presence of carbonic acid, H_2CO_3.

8.97. Collect and Organize

We are asked to balance the equation that describes the extraction of silver ores by cyanide in basic solution.

Analyze

To balance this reaction, we use the steps outlined in Table 8.7.

Solve

Step 1. We first write the reaction expression using one mole of each of the known reactants and products. Because the reaction occurs in aqueous solution we can use H_2O as a redox species for O_2.

$$Ag(s) + CN^-(aq) + O_2(g) \rightarrow Ag(CN)_2^-(aq) + H_2O(\ell)$$

Step 2. Calculate the ΔO.N. values for the elements that are reduced and oxidized.

$$Ag(s) + CN^-(aq) + O_2(g) \rightarrow Ag(CN)_2^-(aq) + H_2O(\ell)$$

Ag is oxidized from 0 to +1 for ΔO.N. = +1

O_2 is reduced from 0 to −2 for ΔO.N. = 2 × (−2) = −4

Step 3. Insert coefficients to balance ΔO.N. values and balance the atoms that are oxidized and reduced, but not balancing O atoms. For this we add 4 as coefficient in front of Ag and then balance the Ag, and CN^- atoms.

$$4 Ag(s) + 8 CN^-(aq) + O_2(g) \rightarrow 4 Ag(CN)_2^-(aq) + 2 H_2O(\ell)$$

Step 4. Add 4 H^+ to the left-hand-side to balance the hydrogen.

$$4 H^+(aq) + 4 Ag(s) + 8 CN^-(aq) + O_2(g) \rightarrow 4 Ag(CN)_2^-(aq) + 2 H_2O(\ell)$$

Step 5. Add 4 OH^- to both sides of the equation to make the reaction in basic solution and combine OH^- and H^+ on the left-hand side to form water and simplify

$$4 OH^-(aq) + 4 H^+(aq) + 4 Ag(s) + 8 CN^-(aq) + O_2(g) \rightarrow 4 Ag(CN)_2^-(aq) + 2 H_2O(\ell) + 4 OH^-(aq)$$

$$4 H_2O(\ell) + 4 Ag(s) + 8 CN^-(aq) + O_2(g) \rightarrow 4 Ag(CN)_2^-(aq) + 2 H_2O(\ell) + 4 OH^-(aq)$$

$$2 H_2O(\ell) + 4 Ag(s) + 8 CN^-(aq) + O_2(g) \rightarrow 4 Ag(CN)_2^-(aq) + 4 OH^-(aq)$$

The reaction is now balanced.

Think About It

Always check your final equation for both charge and atom balance.

8.99. Collect and Organize

We are asked to balance three redox reactions involving the ClO_3^- ion for the preparation of ClO_2.

Analyze

To balance this reaction, we use the steps outlined in Table 8.7. All of these reactions occur in acidic solution.

Solve

(a) Step 1. We first write the reaction expression using one mole of each of the known reactants and products.

$$ClO_3^-(aq) + SO_2(g) \rightarrow ClO_2(g) + SO_4^{2-}(aq)$$

Step 2. Calculate the ΔO.N. values for the elements that are reduced and oxidized.

$$ClO_3^-(aq) + SO_2(g) \rightarrow ClO_2(g) + SO_4^{2-}(aq)$$

Cl is reduced from +5 to +4 for ΔO.N. = −1

S is oxidized from +4 to +6 for ΔO.N. = +2

Step 3. Insert coefficients to balance ΔO.N. values and balance the atoms that are oxidized and reduced, but not balancing O atoms. For this we add a 2 in front of ClO_3^-. We then balance the Cl atoms by adding the 2 as a coefficient to ClO_2.

$$2\,ClO_3^-(aq) + SO_2(g) \rightarrow 2\,ClO_2(g) + SO_4^{2-}(aq)$$

Step 4. The charges are balanced in this reaction.

$$2\,ClO_3^-(aq) + SO_2(g) \rightarrow 2\,ClO_2(g) + SO_4^{2-}(aq)$$

Step 5. All the atoms including O are balanced in this reaction.

$$2\,ClO_3^-(aq) + SO_2(g) \rightarrow 2\,ClO_2(g) + SO_4^{2-}(aq)$$

The reaction is balanced.

(b) Step 1. We first write the reaction expression using one mole of each of the known reactants and products.

$$ClO_3^-(aq) + Cl^-(aq) \rightarrow ClO_2(g) + Cl_2(g)$$

Step 2. Calculate the ΔO.N. values for the elements that are reduced and oxidized.

$$ClO_3^-(aq) + Cl^-(aq) \rightarrow ClO_2(g) + Cl_2(g)$$

Cl is reduced from +5 to +4 for ΔO.N. = −1

Cl (in Cl^-) is oxidized from −1 to 0 for ΔO.N. = +1

Step 3. These ΔO.N. values are balanced. We can balance Cl atoms by placing 1/2 in front of Cl_2 on the right-hand side.

$$ClO_3^-(aq) + Cl^-(aq) \rightarrow ClO_2(g) + 1/2\,Cl_2(g)$$

Step 4. To balance the charge in acidic solution we add 2 H^+ to the left-hand side.

$$2\,H^+ + ClO_3^-(aq) + Cl^-(aq) \rightarrow ClO_2(g) + 1/2\,Cl_2(g)$$

Step 5. Add H_2O molecule to the left-hand side to balance the O and H atoms.

$$2\,H^+ + ClO_3^-(aq) + Cl^-(aq) \rightarrow ClO_2(g) + 1/2\,Cl_2(g) + H_2O$$

Multiplying by 2 to remove the fractional coefficient gives

$$4\,H^+(aq) + 2\,ClO_3^-(aq) + 2\,Cl^-(aq) \rightarrow 2\,ClO_2(g) + Cl_2(g) + 2\,H_2O$$

The reaction is now balanced.

(c) Step 1. We first write the reaction expression using one mole of each of the known reactants and products.

$$ClO_3^-(aq) + Cl_2(aq) \rightarrow ClO_2(g) + O_2(g)$$

Step 2. Calculate the ΔO.N. values for the elements that are reduced and oxidized.

$$ClO_3^-(aq) + Cl_2(aq) \rightarrow ClO_2(g) + O_2(g)$$

Cl in ClO_3^- is reduced from +5 to +4 for ΔO.N. = −1

Cl in Cl_2 is oxidized from 0 to +4 for ΔO.N. = +4, for two atoms of Cl this is a total of $8e^-$

O in ClO_3^- is also oxidized from −2 to 0 for ΔO.N. = +2 for two of the O atoms for a total of $4e^-$

The total number of electrons transferred in oxidation is 12.

Step 3. Insert coefficients to balance ΔO.N. values and balance the atoms that are oxidized and reduced, but not balancing O atoms. To do this we place a 12 in front of ClO_3^-

$$12\,ClO_3^-(aq) + Cl_2(aq) \rightarrow ClO_2(g) + O_2(g)$$

Then we can balance the Cl atoms by placing a 14 in front of ClO_2.

$$12\,ClO_3^-(aq) + Cl_2(aq) \rightarrow 14\,ClO_2(g) + O_2(g)$$

To balance the O atoms we add 6 H_2O to the left-hand side

$$12\,ClO_3^-(aq) + Cl_2(aq) \rightarrow 14\,ClO_2(g) + O_2(g) + 6\,H_2O(\ell)$$

Step 4. To balance the H atoms and the charge we add 12 H^+

$$12\,H^+(aq) + 12\,ClO_3^-(aq) + Cl_2(aq) \rightarrow 14\,ClO_2(g) + O_2(g) + 6\,H_2O(\ell)$$

Step 5. All charges and atoms are now balanced.

$$12\,H^+(aq) + 12\,ClO_3^-(aq) + Cl_2(aq) \rightarrow 14\,ClO_2(g) + O_2(g) + 6\,H_2O(\ell)$$

Think About It

For part c, we needed to complete the equation by recognizing that Cl_2 is reduced to Cl^- by water to produce O_2.

8.101. Collect and Organize

All the titrations involve a neutralization reaction. The moles of base (OH^-) required must equal the moles of acid (H^+) in the sample.

Analyze

First, we need to calculate the number of moles of acid from the volume and concentration of acid in the samples. Because the stoichiometry of the neutralization reaction is 1 mol OH^- : 1 mol H^+, the moles of base required is equal to the moles of H^+ in the sample. We can find the volume of base needed by dividing moles of OH^- required by the concentration of base used.

Solve

(a) $10.0 \text{ mL} \times \dfrac{0.0500 \text{ mol HCl}}{1000 \text{ mL}} \times \dfrac{1 \text{ mol H}^+}{1 \text{ mol HCl}} \times \dfrac{1 \text{ mol OH}^-}{1 \text{ mol H}^+} \times \dfrac{1000 \text{ mL}}{0.100 \text{ mol NaOH}} = 5.00 \text{ mL}$

(b) $25.0 \text{ mL} \times \dfrac{0.126 \text{ mol HNO}_3}{1000 \text{ mL}} \times \dfrac{1 \text{ mol H}^+}{1 \text{ mol HNO}_3} \times \dfrac{1 \text{ mol OH}^-}{1 \text{mol H}^+} \times \dfrac{1000 \text{ mL}}{0.100 \text{ mol NaOH}} = 31.5 \text{ mL}$

(c) $50.0 \text{ mL} \times \dfrac{0.215 \text{ mol H}_2\text{SO}_4}{1000 \text{ mL}} \times \dfrac{2 \text{ mol H}^+}{1 \text{ mol H}_2\text{SO}_4} \times \dfrac{1 \text{ mol OH}^-}{1 \text{ mol H}^+} \times \dfrac{1000 \text{ mL}}{0.100 \text{ mol NaOH}} = 215 \text{ mL}$

Think About It

Because sulfuric acid has two H^+ ions (it is a diprotic acid), we need twice as many moles of OH^- to neutralize it as for the same concentration of a monoprotic acid such as HNO_3.

8.103. Collect and Organize

Using the solubility of $Ca(OH)_2$ we first have to calculate the moles of $Ca(OH)_2$ in the solution; then we can find the volume of the $HCl(aq)$ solution to neutralize the $Ca(OH)_2$ solution.

Analyze

To find the moles of $Ca(OH)_2$ in the saturated solution, we first have to multiply the volume of the solution by the solubility of $Ca(OH)_2$. Doing so gives the grams of $Ca(OH)_2$ in the solution, which we can then convert into moles by dividing the grams of $Ca(OH)_2$ by the molar mass of $Ca(OH)_2$. Because 2 moles of OH^- is in $Ca(OH)_2$, the moles of OH^- to neutralize must be twice the moles of $Ca(OH)_2$. We can then use the 1:1 molar ratio of OH^- to H^+ in the neutralization reaction and the concentration of the HCl solution to find the volume of HCl required to neutralize the $Ca(OH)_2$ solution.

Solve

Moles of $Ca(OH)_2$ in the saturated solution:

$$10.0 \text{ mL} \times \dfrac{0.185 \text{ g}}{100.0 \text{ mL}} \times \dfrac{1 \text{ mol}}{74.09 \text{ g}} = 2.50 \times 10^{-4} \text{ mol}$$

Volume (in milliliters) of HCl required to neutralize:

$$2.50 \times 10^{-4} \text{ mol Ca(OH)}_2 \times \dfrac{2 \text{ mol OH}^-}{1 \text{ mol Ca(OH)}_2} \times \dfrac{1 \text{ mol H}^+}{1 \text{ mol OH}^-} \times \dfrac{1 \text{ mol HCl}}{1 \text{ mol H}^+} \times \dfrac{1000 \text{ mL}}{0.00100 \text{ mol HCl}} = 500 \text{ mL}$$

Think About It

Calcium hydroxide is not very soluble in water, but that neutralization requires a large volume of HCl solution because the HCl solution is fairly dilute.

8.105. **Collect and Organize**

In the titration of 25.00 mL of seawater to determine the chloride concentration, 27.80 mL of 0.5000 M AgNO$_3$ is used to reach the equivalence point. From that information we are asked to determine the concentration of Cl$^-$ in the seawater in units of millimolar and in grams per kilogram where the density of the seawater is 1.025 g/mL.

Analyze

The titration of Cl$^-$ with Ag$^+$ to form the precipitate AgCl has the following balanced equation:

$$AgNO_3(aq) + Cl^-(aq) \rightarrow AgCl(s) + NO_3^-(aq)$$

To solve for the concentration of chloride ion in the seawater we will need to first find the moles of AgNO$_3$ used to reach the equivalence point. Because that is the moles of Cl$^-$ in the 25.00 mL sample, we can calculate the concentration in moles per liter and then convert it to millimoles per liter and finally to grams of Cl$^-$ per kilogram of seawater.

Solve

$$27.80 \text{ mL AgNO}_3 \times \frac{0.5000 \text{ mol AgNO}_3}{1000 \text{ mL}} \times \frac{1 \text{ mol AgCl}}{1 \text{ mol AgNO}_3} \times \frac{1 \text{ mol Cl}^-}{1 \text{ mol AgCl}} \times \frac{1}{0.02500 \text{ L}} = 0.5560 \text{ } M$$

$$\frac{0.5560 \text{ mol}}{\text{L}} \times \frac{1000 \text{ mmol}}{1 \text{ mol}} = 556.0 \text{ m}M$$

$$\frac{0.5560 \text{ mol}}{1000 \text{ mL}} \times \frac{1 \text{ mL}}{1.025 \text{ g}} \times \frac{1000 \text{ g}}{1 \text{ kg}} \times \frac{35.453 \text{ g Cl}^-}{1 \text{ mol}} = 19.23 \text{ g/kg}$$

Think About It

Although here the stoichiometry of the titration is simple (1:1), be careful to always consider the balanced equation because that simple ratio does not always apply.

8.107. **Collect and Organize**

Anion and cation exchangers swap unwanted cations and anions in water with other ions. We are asked to explain how they can deionize water.

Analyze

To deionize water (that is, remove all the cations and anions), we would have to swap the cations and anions with ones that, when combined, produce H$_2$O.

Solve

To deionize water, cations such as Na$^+$ and Ca^{2+} are exchanged for H$^+$ at cation-exchange sites. Anions such as Cl$^-$ and SO$_4^{2-}$ are exchanged for OH$^-$ at anion-exchange sites. The released ions (H$^+$ and OH$^-$) at those sites combine to form H$_2$O.

Think About It

The electrical neutrality of the water is preserved. If we swap every Cl$^-$ and Na$^+$ for OH$^-$ and H$^+$ ions, we have kept the number of ions in the solution the same. However, because OH$^-$ and H$^+$ combine to form neutral H$_2$O, we no longer have ions in the water; it is deionized.

8.109. **Collect and Organize**

We consider why using potassium ions in place of sodium ions in resins for ion exchange would be an advantage.

Analyze

Water softeners for water use resins with Na$^+$ for the exchange with harder cations such as Mg^{2+} and Ca^{2+}.

Solve

K$^+$ is in the same group in the periodic table and so probably would function the same as Na$^+$ in the resin. K$^+$ would avoid, however, raising Na$^+$ concentrations, which may be a problem for people with hypertension. One caveat, though, is that potassium in the form of KCl is more expensive to use than the more abundant NaCl.

Think About It

Ion exchange and deionization preserves water's electrical neutrality. If we swap every Cl^- and Na^+ for OH^- and H^+ ions, we have kept the number of ions in the solution the same. However, because OH^- and H^+ combine to form neutral H_2O, we no longer have ions in the water; it is deionized.

8.111. **Collect and Organize**

In that precipitation titration, a known volume of barium nitrate is titrated into a solution of unknown sulfate concentration. We are asked to calculate how much sulfate is in the solution and express the concentration of sulfate in moles per liter (*M*).

Analyze

Using the concentration of the $Ba(NO_3)_2$ titrant and the volume of titrant used, we first find the moles of $Ba(NO_3)_2$ used in the titration. From the balanced equation

$$Ba(NO_3)_2 + SO_4^{2-} \rightarrow BaSO_4 + 2\,NO_3^-$$

we see that 1 mol of $Ba(NO_3)_2$ reacts with 1 mol of SO_4^{2-}. That 1:1 ratio means that the moles of titrant used equals the moles of SO_4^{2-} in the sample. To calculate concentration of SO_4^{2-} we divide the moles of SO_4^{2-} by the volume of the sample in liters.

Solve

$$3.19\text{ mL Ba}(NO_3)_2 \times \frac{0.0250\text{ mol Ba}(NO_3)_2}{1000\text{ mL}} \times \frac{1\text{ mol SO}_4^{2-}}{1\text{ mol Ba}(NO_3)_2} \times \frac{1}{0.1000\text{ L}} = 7.98 \times 10^{-4}\ M\text{ SO}_4^{2-}$$

Think About It

Precipitation titrations give us the concentration of a species in solution and are accurate only when the salt formed has very low solubility. If we form a salt of marginal solubility, then we leave some of the species in solution and our calculation underestimates the concentration of the species in solution.

8.113. **Collect and Organize**

From the mass percent and density of concentrated HCl we are asked to determine the molarity of HCl. Then we are to calculate the amount needed to prepare a dilute solution from the concentrated HCl and figure out how much sodium bicarbonate is needed to neutralize a spill of concentrated HCl.

Analyze

The task looks difficult, but we have all the information we need. Parts b and c depend on the answer to part a. For part a, if the solution is 36.0% by mass, then 100 g of concentrated HCl contains 36.0 g of HCl. Using the molar mass of HCl (36.46 g/mol), we can convert the grams into moles. To obtain the volume of acid we divide the 100 g of acid by the density (1.18 g/mL). We now have moles of HCl and volume of HCl to compute the molarity. To determine the volume of concentrated HCl needed to prepare a more dilute solution of HCl in part b, we use Equation 8.6:

$$V_{initial} M_{initial} = V_{final} M_{final}$$

For part c we need the chemical equation that describes the neutralization:

$$NaHCO_3 + HCl \rightarrow NaCl + H_2O + CO_2$$

Solve

(a) Molarity of concentrated HCl solution:

$$36.0\text{ g HCl solution} \times \frac{1\text{ mol HCl}}{36.46\text{ g}} = 0.987\text{ mol HCl}$$

$$100\text{ g solution} \times \frac{1\text{ mL}}{1.18\text{ g}} = 84.7\text{ mL, or }0.0847\text{ L}$$

$$\text{Molarity HCl(conc)} = \frac{0.987\text{ mol}}{0.0847\text{ L}} = 11.7\ M$$

(b) Volume of HCl(conc) required to make 0.250 L of 2.00 *M* solution:

$$V_{initial} M_{initial} = V_{final} M_{final}$$

$$V_{initial} \times 11.7\ M = 250\text{ mL} \times 2.00\ M$$

$$V_{initial} = 42.7\text{ mL}$$

(c) Mass of $NaHCO_3$ required to neutralize the spill:

$$1.75 \text{ L HCl(conc)} \times \frac{11.7 \text{ mol}}{\text{L}} \times \frac{1 \text{ mol NaHCO}_3}{1 \text{ mol HCl}} \times \frac{84.01 \text{ NaHCO}_3}{1 \text{ mol}} = 1720 \text{ g, or } 1.72 \text{ kg}$$

Think About It

That problem puts together many of the chapter's topics, including the definition of molarity, the preparation of diluted solutions, and neutralization. It also involves concepts from previous chapters, including density and reaction stoichiometry.

8.115. Collect and Organize

This is a redox reaction, so we have to use the method described in Table 8.7 to balance the reaction. From the species involved we can then identify the oxidizing and reducing agents. Finally, we are asked to calculate the amount of $Na_2S_2O_4$ needed to remove a certain amount of CrO_4^{2-}.

Analyze

The redox reaction occurs in basic solution. From the change in oxidation numbers we can identify the species that are oxidized and reduced and assign species as oxidizing or reducing agents. We need the balanced equations for the stoichiometry of reactants to calculate how much $Na_2S_2O_4$ is required to remove the CrO_4^{2-} from the wastewater. We also need the molar mass of $Na_2S_2O_4$.

Solve

(a) Step 1. We first write the reaction expression using one mole of each of the known reactants and products.

$$S_2O_4^{2-}(aq) + CrO_4^{2-}(aq) \rightarrow SO_3^{2-}(aq) + Cr(OH)_3(s)$$

Step 2. Calculate the ΔO.N. values for the elements that are reduced and oxidized.

$$S_2O_4^{2-}(aq) + CrO_4^{2-}(aq) \rightarrow SO_3^{2-}(aq) + Cr(OH)_3(s)$$

Cr is reduced from +6 to +3 for ΔO.N. $= -3$
S is oxidized from +3 to +4 for ΔO.N. $= 2 \times (+1) = +2$

Step 3. Insert coefficients to balance ΔO.N. values and balance the atoms that are oxidized and reduced, but not balancing O atoms. For this we can add a 2 in front of CrO_4^{2-} and a 3 in front of SO_3^{2-}. Then we can balance the Cr and S atoms on the right-hand side.

$$3 \text{ S}_2O_4^{2-}(aq) + 2 \text{ CrO}_4^{2-}(aq) \rightarrow 6 \text{ SO}_3^{2-}(aq) + 2 \text{ Cr(OH)}_3(s)$$

Step 4. To balance the charge in basic solution we add 2 OH$^-$ to the left-hand side.

$$2OH^-(aq) + 3 \text{ S}_2O_4^{2-}(aq) + 2 \text{ CrO}_4^{2-}(aq) \rightarrow 6 \text{ SO}_3^{2-}(aq) + 2 \text{ Cr(OH)}_3(s) +$$

Step 5. Add 2 H_2O molecules to the left-hand side to balance the H and O atoms.

$$2 \text{ H}_2O(\ell) + 2OH^-(aq) + 3 \text{ S}_2O_4^{2-}(aq) + 2 \text{ CrO}_4^{2-}(aq) \rightarrow 6 \text{ SO}_3^{2-}(aq) + 2 \text{ Cr(OH)}_3(s)$$

The reaction is now balanced.

(b) In $S_2O_4^{2-} \rightarrow SO_3^{2-}$ sulfur is oxidized from +3 to +4. In $CrO_4^{2-} \rightarrow Cr(OH)_3$ chromium is reduced from +6 to +3.

(c) The oxidizing agent is the species that is itself reduced: CrO_4^{2-}. The reducing agent is itself oxidized: $S_2O_4^{2-}$.

(d) The amount of $Na_2S_2O_4$ required is

$$100.0 \text{ L wastewater} \times \frac{0.00148 \text{ mol CrO}_4^{2-}}{1 \text{ L}} \times \frac{3 \text{ mol S}_2O_4^{2-}}{2 \text{ mol CrO}_4^{2-}} \times \frac{1 \text{ mol Na}_2S_2O_4}{1 \text{ mol S}_2O_4^{2-}} \times \frac{174.11 \text{ g}}{1 \text{ mol Na}_2S_2O_4} = 38.7 \text{ g}$$

Think About It

Be sure to always start with a balanced chemical equation before calculating the amount of reactants required or the amount of products in a reaction.

8.117. Collect and Organize

In this problem we examine the tarnishing of silver to Ag_2S and the conversion of Ag_2S back to Ag in the presence of aluminum metal. Both are redox reactions.

Analyze

We are given the chemical formulas for all the reactants and products with which to write the equations.

Solve

(a) The tarnishing of Ag occurs in the presence of H_2S and O_2 to form Ag_2S.

$$4\,Ag(s) + 2\,H_2S(g) + O_2(g) \rightarrow 2\,Ag_2S(s) + 2\,H_2O(\ell)$$

(b) The reaction of Ag_2S with Al and to remove tarnish can be balanced through the method described in Table 8.7 (note that the solution would be basic in order to produce aluminum hydroxide).

Step 1. We first write the reaction expression using one mole of each of the known reactants and products.

$$Ag_2S(s) + OH^-(aq) + Al(s) \rightarrow Al(OH)_3(s) + Ag(s) + HS^-(aq)$$

Step 2. Calculate the ΔO.N. values for the elements that are reduced and oxidized.

$$Ag_2S(s) + OH^-(aq) + Al(s) \rightarrow Al(OH)_3(s) + Ag(s) + HS^-(aq)$$

Ag is reduced from +1 to 0 for ΔO.N. $= 2 \times (-1) = -2$

Al is oxidized from 0 to +3 for ΔO.N. $= +3$

Step 3. Insert coefficients to balance ΔO.N. values and balance the atoms that are oxidized and reduced, but not balancing O atoms. For this we can add a 2 in front of Al and a 3 in front of Ag_2S. Then we can balance the Ag, Al, and S atoms on the right-hand side.

$$3\,Ag_2S(s) + OH^-(aq) + 2\,Al(s) \rightarrow 2\,Al(OH)_3(s) + 6\,Ag(s) + 3\,HS^-(aq)$$

Step 4. To balance the charges we place a coefficient of 3 in front of OH^-.

$$3\,Ag_2S(s) + 3\,OH^-(aq) + 2\,Al(s) \rightarrow 2\,Al(OH)_3(s) + 6\,Ag(s) + 3\,HS^-(aq)$$

Step 5. We can balance the H atoms by adding 3 H_2O to the left-hand side.

$$3\,H_2O(\ell) + 3\,Ag_2S(s) + 3\,OH^-(aq) + 2\,Al(s) \rightarrow 2\,Al(OH)_3(s) + 6\,Ag(s) + 3\,HS^-(aq)$$

The reaction is now balanced.

Think About It

Baking soda does not become involved in the reaction but makes the solution basic.

8.119. Collect and Organize

We are asked to predict the products of some reactions between various phosphorus, selenium, and boron oxides and water.

Analyze

When nonmetal oxides (acids and anhydrides) react with water, they give acidic solutions. For those reactions we need to simply add water to the reactant and write the formula of the acidic product. Using the naming rules from Chapter 2, we can name the acid product.

Solve

(a) $P_4O_{10} + 6\,H_2O \rightarrow 4\,H_3PO_4$ (phosphoric acid)

(b) $SeO_2 + H_2O \rightarrow H_2SeO_3$ (selenous acid)

(c) $B_2O_3 + 3\,H_2O \rightarrow 2\,H_3BO_3$ (boric acid)

Think About It

When we simply add water to the anhydride, an acid is formed from nonmetal oxides.

8.121. Collect and Organize

Both reactions described are redox reactions, and so we can use method described in Table 8.7 to write these reactions.

Analyze

In the first reaction, ClO^- (with Cl^+) is combined with I^- to give Cl^- and I_2. Here, Cl is being reduced and I is being oxidized in acidic solution. In the second reaction, I_2 is titrated with $S_2O_3^{2-}$ (with S^{2+}) to give I^- and $S_4O_6^{2-}$ (with $S^{2.5+}$). Here, sulfur is being oxidized and iodine is being reduced.

Solve

First reaction, balancing the half-reactions in acidic solution

Step 1. We first write the reaction expression using one mole of each of the known reactants and products.

$$ClO^-(aq) + I^-(aq) \rightarrow I_2(aq) + Cl^-(aq)$$

Step 2. Calculate the ΔO.N. values for the elements that are reduced and oxidized.

$$ClO^-(aq) + I^-(aq) \rightarrow I_2(aq) + Cl^-(aq)$$

Cl is reduced from +1 to –1 for ΔO.N. = –2

I$^-$ is oxidized from –1 to 0 for ΔO.N. = +1

Step 3. Insert coefficients to balance ΔO.N. values and balance the atoms that are oxidized and reduced, but not balancing O atoms. For this we can add a 2 in front of I$^-$. The Cl atoms are balanced.

$$ClO^-(aq) + 2\,I^-(aq) \rightarrow I_2(aq) + Cl^-(aq)$$

Step 4. To balance the charges we add 2 H$^+$ to the left-hand side.

$$2H^+(aq) + ClO^-(aq) + 2\,I^-(aq) \rightarrow I_2(aq) + Cl^-(aq)$$

Step 5. We can balance the H atoms by adding H_2O to the right-hand side.

$$2H^+(aq) + ClO^-(aq) + 2\,I^-(aq) \rightarrow I_2(aq) + Cl^-(aq) + H_2O(\ell)$$

The reaction is now balanced.

Second reaction, balancing the half-reactions

Step 1. We first write the reaction expression using one mole of each of the known reactants and products.

$$S_2O_3^{2-}(aq) + I_2(aq) \rightarrow I^-(aq) + S_4O_6^{2-}(aq)$$

Step 2. Calculate the ΔO.N. values for the elements that are reduced and oxidized.

$$S_2O_3^{2-}(aq) + I_2(aq) \rightarrow I^-(aq) + S_4O_6^{2-}(aq)$$

S is oxidized from +2 to +2.5 for ΔO.N. = $2 \times (+ 0.5) = +1$

I_2 is reduced from 0 to –1 for ΔO.N. = $2 \times (-1) = -2$

Step 3. Insert coefficients to balance ΔO.N. values and balance the atoms that are oxidized and reduced, but not balancing O atoms. For this we can add a 2 in front of $S_2O_3^{2-}$. Then we balance the I atoms.

$$2\,S_2O_3^{2-}(aq) + I_2(aq) \rightarrow 2\,I^-(aq) + S_4O_6^{2-}(aq)$$

Step 4. The charges for this equation are balanced.

$$2\,S_2O_3^{2-}(aq) + I_2(aq) \rightarrow 2\,I^-(aq) + S_4O_6^{2-}(aq)$$

Step 5. All the atoms are also balanced.

$$2\,S_2O_3^{2-}(aq) + I_2(aq) \rightarrow 2\,I^-(aq) + S_4O_6^{2-}(aq)$$

The reaction is now balanced.

Adding these two equations together

$$2H^+(aq) + ClO^-(aq) + 2\,I^-(aq) \rightarrow I_2(aq) + Cl^-(aq) + H_2O(\ell)$$
$$2\,S_2O_3^{2-}(aq) + I_2(aq) \rightarrow 2\,I^-(aq) + S_4O_6^{2-}(aq)$$

gives

$$2H^+(aq) + ClO^-(aq) + 2\,S_2O_3^{2-}(aq) \rightarrow Cl^-(aq) + H_2O(\ell) + + S_4O_6^{2-}(aq)$$

Think About It

The second reaction does not require H$^+$ to balance the equation. That reaction, therefore, does not depend on the presence of acid to proceed and can take place in neutral solution.

8.123. Collect and Organize

Considering the values we are given for the stream contamination by perchlorate, the flow rate of the stream, and the advisory range for perchlorate in drinking water, we are asked to determine the amount of perchlorate that flows into Lake Mead and the volume of perchlorate-free water that would be needed to reduce concentrations from 700.0 to 4 µg/L. We also compare results from three labs on replicate samples of water for perchlorate and determine which lab gave the most precise results.

Analyze
(a) To write the formulas for sodium perchlorate and ammonium perchlorate we must write neutral chemical formulas using Na^+ with ClO_4^- and NH_4^+ with ClO_4^-. The cation and anion each have a charge of 1, so the cation and anion in each salt are present in a 1:1 ratio.
(b) To calculate how much ClO_4^- enters the lake from the stream we need to multiply the flow rate (after converting to liters per day) by the concentration of ClO_4^- in the stream (and convert micrograms into kilograms).
(c) In this part we can use Equation 8.6

$$V_{initial} \times M_{initial} = V_{final} \times M_{final}$$

where $M_{initial} = 700$ µg/L, and $M_{final} = 4$ µg/L, and $V_{initial}$ is 161×10^6 gal (volume that flows in the stream each day). V_{final} is the final total volume to reduce the concentration of ClO_4^- to 4 µg/L. The volume of lake water that must be mixed with the stream water will be $V_{final} - 161 \times 10^6$ gal.
(d) When we compare the sample data for Maryland, Massachusetts, and New Mexico, the most precise data for the replicate samples must have the narrowest range of values.

Solve
(a) Sodium perchlorate, $NaClO_4$
    Ammonium perchlorate, NH_4ClO_4
(b) Perchlorate flow into Lake Mead each day:

$$\frac{161 \times 10^6 \text{ gal}}{\text{day}} \times \frac{3.785 \text{ L}}{1 \text{ gal}} \times \frac{700 \text{ µg}}{L} \times \frac{1 \times 10^{-6} \text{ g}}{1 \text{ µg}} \times \frac{1 \times 10^{-3} \text{ kg}}{1 \text{ g}} = 427 \text{ kg}$$

(c) Using the dilution equation:

$$161 \times 10^6 \text{ gal} \times 700 \text{ µg/L} = 4 \text{ µg/L} \times V_{final}$$

$$V_{final} = 2.82 \times 10^{10} \text{ gal}$$

Volume of lake water required: 2.82×10^{10} gal $- 161 \times 10^6$ gal $= 2.80 \times 10^{10}$ gal

(d) The data for the samples from Maryland range from 0.9 to 1.4 µg/L, or a range of 0.5 µg/L; the data from Massachusetts range from 0.90 to 0.95 µg/L, or a range of 0.05 µg/L; and the data from New Mexico range from 1.1 to 1.3 µg/L, or a range of 0.2 µg/L. Because the range of data is lowest for the Massachusetts sample, that lab produced the most precise results.

Think About It
Looking closely at the dilution in part c, we see that the stream water must be diluted by

$$\frac{2.82 \times 10^{10} \text{ gal}}{161 \times 10^6 \text{ gal}} = 175$$

or nearly 200 times.

8.125. Collect and Organize
From the balanced equations that we write for the two fermentation steps for the conversion of sugar to acetic acid, we are to calculate how much acetic acid could be produced (with 100% theoretical yield) from 100 g of sugar.

Analyze
The first step in the fermentation of apple juice is an anaerobic process, so only the sugars are converted to ethanol and carbon dioxide. The acid fermentation of ethanol to give acetic acid and water, however, requires oxygen as a reactant. Oxidation states for the carbon atoms in both the reactants and products of those two fermentation reactions can be deduced using the typical oxidation states of H (+1) and O (−2). Because all the carbon species are neutral, the sum of the oxidation numbers must be zero. Therefore,

oxidation number on C = 0 − [number H atoms × (+1) + number of oxygen atoms × (−2)]

To calculate the maximum acetic acid produced from the fermentation of 100 g of sugar, we first need to calculate the moles of sugar in 100 g by using the molar mass of CH_2O (30.03 g/mol). Then we use the molar ratios in the balanced equations to find the moles of acetic acid that can be produced (3:1 ratio for $CH_2O : C_2H_5OH$ and 1:1 ratio for $C_2H_3OH : HC_2H_3O_2$). Using the molar mass of acetic acid (60.05 g/mol), we can calculate the maximum mass of acetic acid that could be produced.

Solve

(a) The fermentation of the sugars is described by
$$3 \, CH_2O \rightarrow CO_2 + C_2H_5OH$$

(b) The fermentation of ethanol is described by
$$C_2H_5OH + O_2 \rightarrow CH_3COOH + H_2O$$

(c) Oxidation states for carbon in reactants and products

CH_2O: $C = 0 - [2(+1) + 1(-2)] = 0$

CO_2: $C = 0 - [2(-2)] = +4$

C_2H_5OH: $C = 0 - [6(+1) + 1(-2)] = -4$ over two carbon atoms, so oxidation number on each carbon $= -2$

CH_3COOH: $C = 0 - [1(+1) + 3(+1) + 2(-2)] = 0$

(d) The maximum amount of acetic acid that could be produced assumes that both fermentation reactions give 100% yield.

$$100 \text{ g } CH_2O \times \frac{1 \text{ mol } CH_2O}{30.03 \text{ g}} \times \frac{1 \text{ mol } C_2H_5OH}{3 \text{ mol } CH_2O} \times \frac{1 \text{ mol } HC_2H_3O_2}{1 \text{ mol } C_2H_5OH} \times \frac{60.05 \text{ g } HC_2H_3O_2}{1 \text{ mol}} = 66.7 \text{ g acetic acid}$$

Think About It

In the first step of the fermentation process the carbon of the sugar is being oxidized and reduced to CO_2 and C_2H_5OH, respectively. In the second step, ethanol is being oxidized and oxygen is being reduced.

8.127. Collect and Organize

For the two reactions that describe the formation of $CaSO_4$ (gypsum) we are asked to identify whether either is a redox reaction and to write a net ionic equation for the reaction of sulfuric acid with calcium carbonate for an alternative form of the equation.

Analyze

A redox reaction is indicated by a change in oxidation state in one of the elements in going from reactants to products in an equation. Net ionic equations are written from ionic equations by canceling any spectator ions that are present as both reactants and products.

Solve

(a) In the first reaction, the oxidation number of hydrogen stays the same (+1), but the oxidation number of S in H_2S (–2) increases to +6 in H_2SO_4. Coupled with that oxidation is the reduction of oxygen in O_2 (oxidation number = 0) to oxidation number –2 in H_2SO_4. Therefore, the first reaction is a redox reaction. In the second reaction, no oxidation numbers change (Ca = +2, C = +4, O = –2, H = +1, S = +6). That reaction is not a redox reaction and would be classified as an acid–base reaction instead.

(b) Ionic equation:
$$2 \, H^+(aq) + SO_4^{2-}(aq) + CaCO_3(s) \rightarrow CaSO_4(s) + H_2O(\ell) + CO_2(g)$$
This is also the net ionic equation.

(c) The ionic equation would change slightly on the product side:
$$2 \, H^+(aq) + SO_4^{2-}(aq) + CaCO_3(s) \rightarrow CaSO_4(s) + 2H^+(aq) + CO_3^{2-}(aq)$$
The net ionic equation would be:
$$SO_4^{2-}(aq) + CaCO_3(s) \rightarrow CaSO_4(s) + CO_3^{2-}(aq)$$

Think About It

The decomposition of H_2CO_3 to CO_2 and H_2O is not a redox reaction because no atoms change oxidation number.

8.129. **Collect and Organize**

From four reactions involving the element calcium or its compounds, we are to find the redox reactions.

Analyze

In redox reactions, an element must undergo a change in oxidation number. By assigning oxidation numbers to the elements in each compound and then comparing products versus reactants, we can find the redox reactions.

Solve

(a) Reactant oxidation numbers: Ca = +2, C = +4, O = –2
Product oxidation numbers: Ca = +2, C = +4, O = –2
That is not a redox reaction.

(b) Reactant oxidation numbers: Ca = +2, O = –2, S = +4
Product oxidation numbers: Ca = +2, O = –2, S = +4
That is not a redox reaction.

(c) Reactant oxidation numbers: Ca = +2, Cl = –1
Product oxidation numbers: Ca = 0, Cl = 0
That is a redox reaction.

(d) Reactant oxidation numbers: Ca = 0, N = 0
Product oxidation numbers: Ca = +2, N = –3
That is a redox reaction.

Think About It

In redox reactions, both reduction and oxidation must be present. For example, in part c, calcium is reduced, whereas chlorine is oxidized.

CHAPTER 9 | Thermochemistry: Energy Changes in Chemical Reactions

9.1. Collect and Organize

From the depiction of a diesel engine piston (Figure P9.1), we are to describe how the internal energy of the trapped gases changes when the piston moves up.

Analyze

Here, the system we are considering is the gases. As the piston moves up, the gases in the cylinder are compressed.

Solve

Upon compression the molecules are squeezed together and the change in volume on the system is negative. Therefore, work is done on the system and $w = -P\Delta V$, where ΔV is negative, so w is positive and the internal energy change $E = q + w$ is positive. Here, the internal energy of the gases in the cylinder increases.

Think About It

If work is done by the system, in this situation where the gases expand, the internal energy of the gases in the cylinder decreases.

9.3. Collect and Organize

All the molecules shown in Figure P9.3 are hydrocarbons with either five or six carbon atoms. On the basis of differences in their structures, we are to predict which has the highest and which has the lowest fuel value.

Analyze

For hydrocarbons, as the hydrogen-to-carbon ratio decreases the fuel value also decreases. Therefore, we will distinguish between the hydrocarbon in this group that has the most and the least fuel value on the basis of the hydrogen-to-carbon ratio.

Solve

The hydrogen-to-carbon ratios of those hydrocarbons are as follows:
(a) C_6H_{14} H:C = 2.33:1
(b) C_6H_{12} H:C = 2:1
(c) C_5H_{12} H:C = 2.4:1
(d) C_6H_{14} H:C = 2.33:1
From those ratios, (c) C_5H_{12} has the highest fuel value and (b) C_6H_{12} has the lowest fuel value.

Think About It

Although structurally different, (a) and (d) have the same H:C ratio and therefore the same fuel value.

9.5. Collect and Organize

For the reaction depicted in Figure P9.4, we are to assign signs to the values ΔE, q, and w and then determine the percent yield of the reaction if each molecule represents 1 mol.

Analyze

From our answer to Problem 9.4, we know that the reaction is exothermic and is compressed in going from reactants to products. From the diagram we have 3 mol of N_2 and 9 mol of H_2 as reactants that form 4 mol of NH_3, with 3 mol of H_2 and 1 mol of N_2 remaining at the end of the reaction.

Solve

Because the reaction volume is less at the end of the reaction, the sign of w is positive, showing that work was done on the system. Because the reaction is exothermic, the sign of q is negative, showing flow of heat from the system to the surroundings. The change in E for the reaction is negative if the heat released by the reaction is greater than the work done on the system by the surroundings by decrease of volume. For the percent yield we consider the system in comparison with the balanced equation

$$N_2(g) + 3 H_2(g) \rightarrow 2 NH_3(g)$$

The 3 mol of N_2 present in the reaction mixture would be expected to use 9 mol of hydrogen (which we have exactly) and yield 6 mol of ammonia. Only 4 mol of ammonia is produced, so the percent yield of the reaction is

$$\frac{\text{experimental yield}}{\text{theoretical yield}} \times 100 = \frac{4 \text{ mol } NH_3}{6 \text{ mol } NH_3} \times 100 = 67\%$$

Think About It
A 100% yield for this reaction would have used up all the nitrogen and the hydrogen initially present in the reaction mixture.

9.7. Collect and Organize
The diagram in Figure P9.7 shows that the volume of the cylinder on the product side is greater than that on the reactant side. We are considering that reaction at constant temperature and pressure, and we are to write a balanced equation for the reaction, calculate the enthalpy of reaction for the formation of 1 mol of CO, and determine the heat flow in the reaction in converting reactants to products.

Analyze
From the color scheme for the elements shown on the inside back cover of the textbook, the reactants are CH_4 and H_2O reacting to form CO and H_2. The reactant has four molecules of CH_4 and four molecules of H_2O. Those react to form 4 molecules of CO and 12 molecules of H_2. The enthalpy of the reaction is calculated by subtracting the sum of the enthalpies of formation of the reactants (multiplied by the number of moles in the balanced equation for each product) from the sum of enthalpies of formation of the products (again multiplied by their molar amounts from the balanced equation).

Solve
(a) The balanced equation is

$$4 CH_4(g) + 4 H_2O(g) \rightarrow 4 CO(g) + 12 H_2$$

or, more simply,

$$CH_4(g) + H_2O(g) \rightarrow CO(g) + 3 H_2$$

(b) $\Delta H^\circ_{rxn} = \left[1 \text{ mol } (\Delta H^\circ_f)CO(g) + 3 \text{ mol } (\Delta H^\circ_f)H_2(g)\right] - \left[1 \text{ mol } (\Delta H^\circ_f)CH_4(g) + 1 \text{ mol } (\Delta H^\circ_f)H_2O(g)\right]$

$\Delta H^\circ_{rxn} = \left[1 \text{ mol } CO \times -110.5 \text{ kJ/mol} + 3 \text{ mol } H_2 \times 0.0 \text{ kJ/mol}\right] -$

$\left[1 \text{ mol } CH_4 \times -74.8 \text{ kJ/mol} + 1 \text{ mol } H_2O \times -241.8 \text{ kJ/mol} +\right] = 206.1 \text{ kJ/mol}$

(c) Because that reaction is endothermic, heat flows into the reaction mixture from the surroundings.

Think About It
The volume of the cylinder decreases because the reaction yields fewer product molecules than the number of reactant molecules.

9.9. Collect and Organize
Energy and work must be related since from our everyday experience we know that doing work takes energy.

Analyze
In this context, energy is defined as the capacity to do work. Work is defined as moving an object against a force over some distance. Energy is also thought to be a fundamental component of the universe. The Big Bang theory postulates that all matter originated from a burst of energy, and Albert Einstein proposed that $m = E/c^2$ (mass equals energy divided by the speed of light squared).

Solve
Energy is needed to do work, and doing work uses energy.

Think About It
A system with high energy has the potential to do a lot of work.

9.11. Collect, Organize, and Analyze
We are to explain what is meant by a *state function*.

Solve
The value of a state function is independent of the path taken in reaching a particular state; only the initial and final values are important.

Think About It
Examples of state functions are internal energy and enthalpy, but not work and heat.

9.13. Collect and Organize
We are asked whether two particles attract or repel each other if their potential energy increases when they are moved away from each other.

Analyze
The basic definition of potential energy is the energy of position from PE = *mgh*. But potential energy is also present at the molecular level in the form of stored energy in the chemical bonds or the attraction of one molecule to another.

Solve
Because the potential energy increases as those particles are moved farther apart, those particles attract each other.

Think About It
You can think of it another way. As the molecules attract and move closer to each other, their potential energy decreases (they enter into a lower potential energy state).

9.15. Collect and Organize
Internal energy is the sum of the potential and kinetic energies of the components of a system.

Analyze
To increase kinetic energy, increase the motion of the molecules in the gas. To increase the potential energy, compress the gas to increase its pressure. It then can do work when it expands to its original volume.

Solve
We can increase the motion of the gas molecules by raising the temperature. To increase the pressure, we can compress the gas (decreasing the volume).

Think About It
Absolute internal energy is difficult to determine for a system, but changes in internal energy are easy to measure.

9.17. Collect, Organize, and Analyze
We consider how pressure–volume (*P–V*) work can have energy units (joules).

Solve
Energy is the ability to do work, so energy and work must have equivalent units. *P–V* work is equivalent to force × distance (*f* × *d*) work because pressure is force is per unit area, which has units of d^2. Volume has units of d^3. Therefore, the units on *P–V* work are equivalent

$$\frac{f}{d^2} \times d^3 = f \times d$$

Think About It

If pressure is measured using SI units of Pascals, *P–V* work gives units of Pa · L which is joules. To convert liter · atmospheres to joules, we multiply by 101.325.

9.19. Collect and Organize

For the processes listed, we are asked to identify which are exothermic and which are endothermic.

Analyze

An exothermic process transfers energy from the system to the surroundings. Thus, something feels warm or hot in an exothermic process. An endothermic process transfers energy from the surroundings to the system. Thus, something feels cool or cold in an endothermic process.

Solve

(a) When molten aluminum solidifies, energy in the form of heat is released to the surroundings to cool the metal, so that process is exothermic.

(b) When rubbing alcohol evaporates from the skin, energy in the form of heat is absorbed by the alcohol from the skin (the surroundings) to evaporate the alcohol, so that process is endothermic.

(c) When fog forms, energy in the form of heat is released by the water vapor in the air to condense, so that process is exothermic.

Think About It

The opposite processes are the reverse: melting aluminum is endothermic, condensing alcohol is exothermic, and evaporating fog is endothermic.

9.21. Collect and Organize

Internal energy is defined as

$$\Delta E = q + w$$

where q = energy transferred and w = work done on a system ($-P\Delta V$). Commonly, the energy transfer is caused by a temperature difference, so q is called "heat."

Analyze

When a process releases energy, the internal energy decreases and q is negative. The reverse is true (ΔE increases and q is positive) if the surroundings transfer energy to the system. If the volume of a system increases (ΔV is positive), then work is negative and the internal energy of the system decreases. The reverse is true (ΔE increases and w is positive) when the volume of the system decreases (for example, a gas is compressed).

Solve

When a liquid vaporizes at its boiling point, energy is absorbed from the surroundings. Thus, q is positive. The volume increases ($\Delta V > 0$), so work is done by the system on the surroundings and the sign of w is negative. More energy is transferred to the liquid than work done, so the sign of $q + w$ is positive; therefore, $\Delta E > 0$. The heat of vaporization absorbed by the water raises its internal energy.

Think About It

Remember to focus on work and energy transfer from the system's point of view. Doing so will help you be clear about the sign conventions as you learn more about thermochemistry.

9.23. Collect and Organize

The work being done is due to an expansion of the gas from 250.0 mL to 750.0 mL.

Analyze

Work is expressed as $-P\Delta V$. Here P is constant at 1.00 atm and the volume change is 500.0 mL, or 0.5000 L. We are to express work in both liter · atmospheres and joules. We can convert from one unit to another by using 101.325 J/L · atm.

Solve

$$w = -P\Delta V = (1.00 \text{ atm})(0.5000 \text{ L}) = -0.500 \text{ L} \cdot \text{atm}$$

In joules, that work is

$$-0.500 \text{ L} \cdot \text{atm} \times \frac{101.325 \text{ J}}{\text{L} \cdot \text{atm}} = -50.7 \text{ J}$$

Think About It

Because work was done by the system on the surroundings, the sign of work is negative.

9.25. Collect and Organize

For each part, we are to calculate internal energy (ΔE) from energy and work values.

Analyze

The formula for internal energy from energy transferred as heat and work is

$$\Delta E = q + w$$

We need to only add the values of q and w given.

Solve

(a) $\Delta E = 100 \text{ J} + (-50 \text{ J}) = 50.0 \text{ J}$

(b) $\Delta E = 6.2 \times 10^3 \text{ J} + 0.7 \text{ L} \cdot \text{atm} = 6200 \text{ J} + (0.7 \text{ L} \cdot \text{atm} \times 101.325 \text{ J/ L} \cdot \text{atm}) = 6300 \text{ J, or } 6.3 \text{ kJ}$

(c) $\Delta E = -615 \text{ kJ} + (-3.25 \text{ kWh}) = -615 \text{ kJ} + (-3.25 \text{ kWh} \times 3600 \text{ kJ/kWh}) = -1.23 \times 10^4 \text{ kJ}$

Think About It

When adding q and w, be sure to add values with consistent units.

9.27. Collect and Organize

The change in internal energy for a system is

$$\Delta E = q + w$$

Analyze

The system releases energy to its surroundings, so q is negative. Because the system does work on the surroundings, w is negative.

Solve

$$\Delta E = q + w = -210.0 \text{ kJ} + (-65.5 \text{ kJ}) = -275.5 \text{ kJ}$$

Think About It

Be careful to always define the signs for q and w from the point of view of the system.

9.29. Collect and Organize

Work will be done by a system on the surroundings when the volume of the system shown in Figure P9.29 increases. For each reaction we are to determine for which one work is being done by the system on the surroundings and then determine the sign of w for that reaction.

Analyze

The volume of gas is proportional to the number of moles of gas (n) at constant temperature and pressure. Because both temperature and pressure are constant for each reaction, if n increases in going from reactants to products, the volume of the system increases and work is done by the system on the surroundings.

Solve

(a) In that reaction, 3 mol of gaseous reactants forms 3 mol of gaseous products. The Δn for the reaction is 0, so the reaction does not do work on the surroundings.

(b) In that reaction, 6 mol of gaseous reactants forms 7 mol of gaseous products. The Δn for the reaction is +1, so the reaction does work on the surroundings.

(c) In that reaction, 3 mol of gaseous reactants forms 2 mol of gaseous products. The Δn for the reaction is –1, so the reaction does not do work on the surroundings.

Reaction (b) does work on the surroundings. When work is done on the surroundings, the change in volume of the system is positive. Because $w = -P\Delta V$, the sign of w is negative for ΔV that is positive.

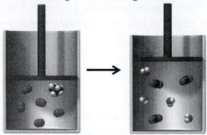

Think About It

Reaction c has $+w$ (from $-P\Delta V$, where $\Delta V = V_f - V_i$ and $V_i > V_f$, so ΔV is negative). Therefore, the surroundings did work on the system. The system was compressed.

9.31. Collect and Organize

We are asked to define a change in enthalpy.

Analyze

We can refer to the mathematical expression for the change in enthalpy (ΔH) to answer this question.

$$\Delta H = \Delta E + P\Delta V$$

Solve

By the equation above, a change in enthalpy is the sum of the change of internal energy and the product of the system's pressure and change in volume and is the heat released or absorbed by a system at constant pressure.

Think About It

Because $\Delta E = q + w$, internal energy changes may involve changes in energy or work or both.

9.33. Collect and Organize

We are asked why we assign a negative sign to ΔH for an exothermic process.

Analyze

In an exothermic process, the system transfers energy to the surroundings.

Solve

If the system transfers energy to the surroundings, its energy will be less after the process than at the start of the process, and q is negative. If the pressure is constant and only P–V work is done, then this q is ΔH, so ΔH must also be negative.

Think About It

The signs of thermodynamic quantities are always assigned from the point of view of the system. That approach can be confusing because we are often observing the process from the point of view of the surroundings.

9.35. Collect and Organize

A drainpipe gets hot when Drano is added. We are asked what the sign for ΔH is for that process.

Analyze

Enthalpy is related to energy transferred as heat (q) by the equation

$$\Delta H = q_P.$$

In this problem we will consider the drainpipe and the water as the surroundings and the Drano as the system.

Solve

Because energy is released in the reaction between the Drano and the clog in the pipe, q is released from the system to the surroundings, so q is negative and therefore ΔH is also negative.

Think About It

All exothermic reactions have a negative ΔH.

9.37. Collect and Organize

If O_2 is the stable form of oxygen under standard conditions, it must have lower energy than other forms. The reaction describes breaking the oxygen–oxygen bond in O_2. We are asked what the sign of ΔH is.

Analyze

If a species is at low energy, then it must require an input of energy to take it out of its most stable form. Breaking the oxygen–oxygen bond, therefore, must require an input of energy.

Solve

The process of breaking the oxygen–oxygen bond is endothermic, and the sign of ΔH will be positive.

Think About It

Breaking of chemical bonds is always an endothermic process.

9.39. Collect and Organize

Compression of H_2 gas will give solid H_2. To predict the sign of the enthalpy change of this transformation, we have to consider the enthalpy of the phase change from gas to solid.

Analyze

To solidify substances, we must cool them to reduce their molecular motion.

Solve

To solidify hydrogen gas, we would remove energy from the gas, so the sign of q from the point of view of the system would be negative. Because $\Delta H = q_P$, the sign of ΔH is also negative.

Think About It

The reverse reaction, in which $H_2(s)$ sublimes into $H_2(g)$, is endothermic.

9.41. Collect and Organize

We look at the definitions and units of heat capacity and specific heat to differentiate those two terms.

Analyze

Heat capacity is the amount of energy needed to raise the temperature of an object by 1°C. Specific heat is the amount of energy needed to raise the temperature of 1 g of a substance by 1°C.

Solve

The difference in the terms lies in the specificity. Specific heat is heat capacity per gram of the substance and is an intensive property of the substance. Heat capacity does not take into account how much of a substance is present; it is defined for a given object and is an extensive property.

Think About It
Specific heat for a substance is characteristic of that substance.

9.43. Collect and Organize
We consider why the heat of vaporization of water is so much greater than its heat of fusion.

Analyze
The enthalpy of fusion is the energy required to melt a solid substance. The enthalpy of vaporization is the energy required to vaporize the substance from a liquid to a gaseous state.

Solve
Melting and vaporizing are different processes, so we would not expect them to be the same. By melting ice we are giving the water molecules enough energy to temporarily break the hydrogen bonds between individual water molecules that are then re-formed with a neighboring water molecule. Only a fraction of the intermolecular hydrogen bonds must be broken when ice melts. In vaporization, however, the strong hydrogen bonds between the water molecules are completely broken. Thus, vaporizing water requires much more energy.

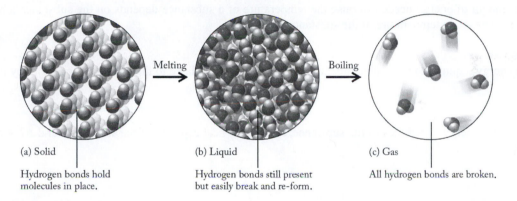

Melting → Boiling →

(a) Solid
Hydrogen bonds hold molecules in place.

(b) Liquid
Hydrogen bonds still present but easily break and re-form.

(c) Gas
All hydrogen bonds are broken.

Think About It
Usually the enthalpy of vaporization is more endothermic than the enthalpy of fusion because in the phase change from liquid to gas, individual molecules must be completely separated from one another.

9.45. Collect and Organize
Referring to Figure P9.44 we are to explain why the line segments *AB* have different slopes.

Analyze
Line segment *AB* in all three of the heating curves shows the temperature change of the liquid substance upon adding energy. In other words, it shows the heating of the liquids before the boiling point shown by line segment *BC*.

Solve
The slope of the lines *AB* correlate to the amount of heat needed to raise the temperature of the substance. They have different slopes because the liquids have different molar heat capacities.

Think About It
A steeper slope indicates that the substance has a lower heat capacity. In order of increasing molar heat capacity of the liquids, the order for these substances is blue (water), yellow (ethanol), and red (chloroform). That ranking is in agreement with the accepted molar heat capacities: water, 75.28 J/(mol·K); ethanol, 112.4 J/(mol·K); chloroform, 114.25 J/(mol·K).

9.47. **Collect and Organize**

To compare the advantage of water-cooled engines over air-cooled engines, we have to compare the heat capacity of water with that of air.

Analyze

The specific heat capacity of water (4.18 J/g·°C) is higher than that of air (1.01 J/g·°C) at typical room conditions.

Solve

Water's high heat capacity compared with that of air means that water carries away more energy from the engine for every Celsius degree rise in temperature, so water is a good choice to cool automobile engines.

Think About It

On a volume basis, water has an even higher heat capacity than air does.

9.49. **Collect and Organize**

The amount of energy needed to raise the temperature of a substance depends on the substance's heat capacity and the change in temperature of the substance.

Analyze

The energy required to raise the temperature of a substance is related to the molar heat capacity by the equation

$$q = nc_P\Delta T$$

where q = energy, n = moles of the substance, c_P = molar heat capacity of the substance, and ΔT = difference in temperature, $T_f - T_i$.

Solve

$$q \text{ required} = \left(100.0 \text{ g} \times \frac{1 \text{ mol}}{18.02 \text{ g}}\right) \times \left(\frac{75.3 \text{ J}}{\text{mol} \cdot {}^{\circ}\text{C}}\right) \times \left(100.0\,^{\circ}\text{C} - 30.0\,^{\circ}\text{C}\right) = 29,300 \text{ J, or } 29.3 \text{ kJ}$$

Think About It

If the water were cooled from 100°C to 30°C, the sign of ΔT would be negative, and therefore q would be negative, showing that the system was cooled.

9.51. **Collect and Organize**

A heating curve plots temperature as a function of energy added to a substance.

Analyze

Methanol at –100°C is a solid. As energy is added, the solid methanol increases in temperature until it reaches its melting point, at which point the added energy does not change the temperature of the methanol until all the solid methanol has melted. Only when methanol is entirely liquid will added energy increase the temperature of the liquid until the boiling point of methanol is reached. During boiling, the temperature of the methanol does not change. Once the methanol is converted to gaseous form, added energy increases the temperature of the gaseous methanol.

Relevant equations for finding the energy required for each step are as follows:

$q = nc_P\Delta T$ for energy added to the solid, liquid, and gas phases, where n = 1 mol,
c_P = heat capacity for that phase, and ΔT = temperature change for that phase
$q = n\Delta H_{fus}$ energy for melting
$q = n\Delta H_{vap}$ energy for vaporization

Solve

Step in Heating Curve	T_i, °C	T_f, °C	q	Total q
Heating methanol ice	−100	−94	$q = 1\ \text{mol} \times 48.7\ \text{J/mol} \cdot °C \times 6°C = 0.292\ \text{kJ}$	0.292 kJ
Melting methanol ice	−94	−94	$q_{fus} = 1\ \text{mol} \times 3.18\ \text{kJ/mol} = 3.18\ \text{kJ}$	3.47 kJ
Warming liquid methanol	−94	65	$q = 1\ \text{mol} \times 81.1\ \text{J/mol} \cdot °C \times 159°C = 12.9\ \text{kJ}$	16.4 kJ
Boiling liquid methanol	65	65	$q_{vap} = 1\ \text{mol} \times 35.3\ \text{kJ/mol} = 35.3\ \text{kJ}$	51.7 kJ
Heating methanol gas	65	100	$q = 1\ \text{mol} \times 43.9\ \text{J/mol} \cdot °C \times 35°C = 1.54\ \text{kJ}$	53.2 kJ

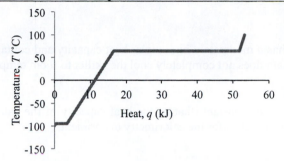

Think About It

Methanol has a wide temperature range as a liquid, making it useful as a solvent.

9.53. Collect and Organize

Sweating helps to cool athletes. During a workout, an athlete generates 233 kJ of energy. We are to calculate how much water the athlete would lose if all the energy is used to evaporate water.

Analyze

The energy generated by the athlete must vaporize the water, so the pertinent equation is

$$q = n\Delta H_{vap}$$

where ΔH_{vap} for water is 40.7 kJ/mol.

Solve

The amount of water in moles that 233 kJ of energy would vaporize is

$$233\ \text{kJ} = n \times 40.7\ \text{kJ/mol}$$

$$n = 5.72\ \text{mol H}_2\text{O}$$

Converting that to mass,

$$5.72\ \text{mol} \times \frac{18.02\ \text{g H}_2\text{O}}{1\ \text{mol}} = 103\ \text{g H}_2\text{O}$$

Think About It

Since the density of water is 1.00 g/mL, the athlete would use 886 mL of water to dissipate the energy. To rehydrate, the athlete should drink about 1 L of water.

9.55. Collect and Organize

When the water is converted into steam, the skillet must lose energy to heat and boil the water away. We are asked to calculate the change in the temperature of the pan with a mass of 1.20 kg when all 10.0 mL water initially at 25°C is converted into steam at 100.0°C.

Analyze

The equation describing the energy exchange *between* the skillet and the water is

$$q_{\text{water gained}} = -q_{\text{skillet lost}}$$

where

$$q_{\text{water gained}} = nc_P \Delta T + n\Delta H_{vap}$$

$$q_{\text{skillet lost}} = nc_P \Delta T$$

Solve

$$q_{\text{water gained}} = -q_{\text{skillet lost}}$$

$$\left[10.0 \text{ g} \times \frac{1 \text{ mol}}{18.02 \text{ g}} \times \frac{75.3 \text{ J}}{\text{mol} \cdot {}^{\circ}\text{C}} \times (100.0\,{}^{\circ}\text{C} - 25.0\,{}^{\circ}\text{C}) \right] + \left[10.0 \text{ g} \times \frac{1 \text{ mol}}{18.02 \text{ g}} \times \frac{40{,}670 \text{ J}}{\text{mol}} \right]$$

$$= -\left(1200 \text{ g} \times \frac{1 \text{ mol Fe}}{55.845} \right) \times \frac{25.19 \text{ J}}{\text{mol} \cdot {}^{\circ}\text{C}} \times \Delta T$$

$$\Delta T = -47.5\,{}^{\circ}\text{C}$$

Think About It

Because the iron skillet has a moderately high molar heat capacity and contains much iron (in terms of moles), boiling the water into steam does not completely cool the skillet to room temperature.

9.57. Collect and Organize

To know why the calorimeter constant (that is, the heat capacity of the calorimeter) is important, we need to define the system and surroundings for the calorimetry experiment.

Analyze

The system in a calorimetry experiment is defined as the substance for which, for example, the heat capacity is being measured. The calorimeter is everything but the system (that is, the calorimeter is the surroundings).

Solve

Because energy is transferred between the system and the surroundings, the heat capacity of the calorimeter (the surroundings) is important because we need to know how much energy (generated or absorbed by the system) is required to change the temperature of the surroundings (the calorimeter) to calculate the heat capacity or final temperature of the system in an experiment. Thus, it is needed because the product of heat capacity and temperature change of the calorimeter is needed to determine the quantity of heat transferred from or into the thermodynamic system inside the calorimeter.

Think About It

Calorimeter constants vary from calorimeter to calorimeter.

9.59. Collect, Organize, and Analyze

The calorimeter constant is the heat capacity of the surroundings. In replacing water in a calorimeter with another liquid, we are changing the surroundings.

Solve

The heat capacity of the new liquid is different from that of water. The liquid is part of the calorimeter and therefore part of the surroundings. Yes, the calorimeter constant must again be determined.

Think About It

If the system is expected to transfer a lot of energy to the calorimeter, then using a liquid with a higher heat capacity than water might be necessary.

9.61. Collect and Organize

Benzoic acid is often used to determine calorimeter constants. As mentioned in the text, when 1 g of it combusts, 26.38 kJ is released to the surroundings.

Analyze

The calorimeter constant is defined as

$$C_{\text{calorimeter}} = \frac{q}{\Delta T}$$

In the combustion of benzoic acid for this calorimeter, we use 5.000 g of benzoic acid and get a temperature change of 16.397°C.

Solve

$$C_{\text{calorimeter}} = \frac{\dfrac{26.38 \text{ kJ}}{\text{g benzoic acid}} \times 5.000 \text{ g}}{16.397°C} = 8.044 \ \frac{\text{kJ}}{°C}$$

Think About It

Be sure to account for how many grams of benzoic acid is used in measuring the calorimeter constant.

9.63. **Collect and Organize**

In a bomb calorimeter $q_{\text{system}} = \Delta E_{\text{comb}}$, but since the P–V work is usually small, $\Delta E_{\text{comb}} \approx \Delta H_{\text{comb}}$. So we may assume that $\Delta H_{\text{comb}} = -q_{\text{calorimeter}}$ since $q_{\text{rxn}} = -q_{\text{calorimeter}}$.

Analyze

We can find ΔH_{comb} through

$$\Delta H_{\text{comb}} = -q_{\text{calorimeter}} = -C_{\text{calorimeter}} \Delta T$$

The ΔH_{comb} we find is for the combustion of 1.200 g of cinnamaldehyde. To find ΔH_{comb} in terms of kilojoules per mole, we need to divide the calculated ΔH_{comb} by the moles of cinnamaldehyde (C_9H_8O).

Solve

$$\Delta H_{\text{comb}} = -3.640 \text{ kJ/°C} \times 12.79°C = -46.56 \text{ kJ}$$

$$\text{molar } \Delta H_{\text{comb}} = \frac{-46.56 \text{ kJ}}{\left(1.200 \text{ g} \times \dfrac{1 \text{ mol}}{132.2 \text{ g}}\right)} = -5129 \ \frac{\text{kJ}}{\text{mol}}$$

Therefore, 5129 kJ of energy are produced in that combustion.

Think About It

Expressing the enthalpies of reactions in terms of molar enthalpies allows us to compare reactions on a per-mole basis.

9.65. **Collect and Organize**

We are asked to work backward from the molar enthalpy of combustion of dimethyl phthalate to the final temperature of the calorimeter.

Analyze

First, we have to calculate ΔH_{comb} from the molar heat of combustion and the grams (which we convert into moles) of dimethyl phthalate ($C_{10}H_{10}O_4$). We can then find T_f by rearranging the equation for ΔH_{comb}.

$$\Delta H_{\text{comb}} = -C_{\text{calorimeter}} \Delta T$$

$$\Delta T = \frac{\Delta H_{\text{comb}}}{C_{\text{cal}}} = T_f - T_i$$

Solve

The ΔH_{comb} for 1.00 g of dimethyl phthalate is

$$1.00 \text{ g} \times \frac{1 \text{ mol}}{194.19 \text{ g}} \times \frac{4685 \text{ kJ}}{1 \text{ mol}} = 24.1 \text{ kJ}$$

The change in the temperature of the calorimeter is

$$\Delta T_f = \frac{24.1 \text{ kJ}}{7.854 \text{ kJ/°C}} = 3.07°C$$

Think About It
Although the molar ΔH_{comb} is large for dimethyl phthalate, that experiment combusts so little dimethyl phthalate that the change in temperature is small.

9.67. Collect and Organize
For the reaction of 0.243 g of Mg with 100.0 mL of 1.00 M HCl, we are to calculate the enthalpy of the reaction, knowing the temperature change of the solution and given the specific heat and density of the solution.

Analyze
All the energy from the reaction is transferred to the solution. We can find that heat from the equation

$$q_{solution} = mc_p\Delta T$$

Because all that heat came from the reaction of 0.243 g of Mg, the enthalpy of the reaction is

$$\Delta H_{rxn} = \frac{q_{rxn}}{\text{mol of Mg}}$$

Solve
The energy transferred in the reaction is

$$q_{solution} = \left(100.0 \text{ mL solution} \times \frac{1.01 \text{ g}}{\text{mL}}\right) \times \frac{4.18 \text{ J}}{\text{g} \cdot {}^{\circ}\text{C}} \times (33.4\,{}^{\circ}\text{C} - 22.4\,{}^{\circ}\text{C}) = 4.64 \times 10^3 \text{ J}$$

The enthalpy of the reaction per mole of Mg is

$$\Delta H_{rxn} = \frac{4.64 \times 10^3 \text{ J}}{\left(0.243 \text{ g Mg} \times \dfrac{\text{mol}}{24.305 \text{ g}}\right)} = 4.64 \times 10^5 \text{ J/mol, or } 464 \text{ kJ/mol}$$

Because that reaction is exothermic, the enthalpy of reaction is –464 kJ/mol.

Think About It
Always be sure to determine the sign of enthalpy by considering whether the reaction is exothermic or endothermic.

9.69. Collect and Organize
Compare Hess's law with the law of conservation of energy.

Analyze
The law of conservation of energy states that energy cannot be created or destroyed; it can be converted from one form into another. Hess's law states that the enthalpy change for a reaction can be obtained by summing the enthalpies of constituent reactions.

Solve
When we apply Hess's law, all the energy is accounted for in the reaction; energy is neither created nor destroyed.

Think About It
In using Hess's law, we assume the enthalpy changes in each step of the overall reaction are conserved so that their sum is equal to the enthalpy change of the overall reaction.

9.71. Collect and Organize
To calculate ΔH_f° for SO_2 from the equations given, we use Hess's law.

Analyze
The equation for ΔH_f° of SO_2 has S and O_2 as the reactants and SO_2 as the product. The sum of the other two reactions must add up to the overall ΔH_f°. In both reactions, SO_3 is produced. If we reverse the first equation and add the second equation, the overall reaction will consist of S and O_2 as the reactants and SO_2 as the product. When the first reaction is reversed, the ΔH_{rxn}° will change from exothermic to endothermic.

Solve

$$2\,SO_3(g) \rightarrow 2\,SO_2(g) + O_2(g) \qquad\qquad \Delta H^\circ_{rxn} = 196\ kJ$$

$$\tfrac{1}{4}\,S_8(s) + 3\,O_2(g) \rightarrow 2\,SO_3(g) \qquad\qquad \Delta H^\circ_{rxn} = -790\ kJ$$

$$\tfrac{1}{4}\,S_8(s) + 2\,O_2(g) \rightarrow 2\,SO_2(g) \qquad\qquad \Delta H^\circ_{rxn} = -594\ kJ$$

That will be twice that of the ΔH°_f for the formation reaction (for 1 mol of SO_2 formed):

$$\tfrac{1}{8}S_8(s) + O_2(g) \rightarrow SO_2(g)$$

Therefore, $\Delta H^\circ_f = -297\ kJ/mol$.

Think About It
Remember that enthalpy is stoichiometric. If 5 mol of SO_2 was formed, then $\Delta H^\circ_f = -1485\ kJ$.

9.73. Collect and Organize
The two equations when added in the appropriate way describe the conversion of α-spodumene into β-spodumene.

Analyze
Because α-spodumene is the reactant in the conversion, we have to reverse the first equation. That reaction will then be endothermic. Hess's law then states that we can add the equations and their corresponding ΔH values to give the overall equation and its enthalpy.

Solve

$$2\,\alpha\text{-LiAlSi}_2O_6(s) \rightarrow Li_2O(s) + 2\,Al(s) + 4\,SiO_2(s) + \tfrac{3}{2}\,O_2(g) \qquad \Delta H^\circ_{rxn} = 1870.6\ kJ$$

$$Li_2O(s) + 2\,Al(s) + 4\,SiO_2(s) + \tfrac{3}{2}\,O_2(g) \rightarrow 2\,\beta\text{-LiAlSi}_2O_6(s) \qquad \Delta H^\circ_{rxn} = -1814.6\ kJ$$

$$2\,\alpha\text{-LiAlSi}_2O_6(s) \rightarrow 2\,\beta\text{-LiAlSi}_2O_6(s) \qquad \Delta H^\circ_{rxn} = 56.0\ kJ$$

This is for the conversion of 2 mol of the α form into the β form. For 1 mol:

$$\alpha\text{-LiAlSi}_2O_6(s) \rightarrow \beta\text{-LiAlSi}_2O_6(s) \qquad\qquad \Delta H^\circ_{rxn} = 28.0\ kJ/mol$$

Think About It
Because the conversion of α-spodumene into β-spodumene is endothermic, we can say that the α form is more stable (by enthalpy) than the β form.

9.75. Collect and Organize
The two chemical reactions must add up to the overall reaction in which NOCl decomposes to nitrogen, oxygen, and chlorine. We can use Hess's law to find the enthalpy of decomposition of NOCl.

Analyze
NOCl must be on the reactant side, so the second reaction must be reversed. That reaction will then be endothermic ($\Delta H^\circ_{rxn} = 38.6\ kJ$). The first reaction also has to be reversed because we need to have N_2 and O_2 on the product side of the overall equation. That will be an exothermic reaction ($\Delta H^\circ_{rxn} = -90.3\ kJ$).

Solve

$$NO(g) \rightarrow \tfrac{1}{2}\,N_2(g) + \tfrac{1}{2}\,O_2(g) \qquad\qquad \Delta H^\circ_{rxn} = -90.3\ kJ$$

$$NOCl(g) \rightarrow NO(g) + \tfrac{1}{2}\,Cl_2(g) \qquad\qquad \Delta H^\circ_{rxn} = 38.6\ kJ$$

$$NOCl(g) \rightarrow \tfrac{1}{2}\,N_2(g) + \tfrac{1}{2}\,O_2(g) + \tfrac{1}{2}\,Cl_2(g) \qquad\qquad \Delta H^\circ_{rxn} = -51.7\ kJ$$

Multiplying that chemical reaction by 2, we obtain the ΔH°_{rxn} for the desired overall equation:

$$2\,NOCl(g) \rightarrow N_2(g) + O_2(g) + Cl_2(g) \qquad\qquad \Delta H^\circ_{rxn} = -103\ kJ$$

Think About It
That reaction, because it is exothermic, releases energy in the decomposition of NOCl into its constituent elements.

9.77. **Collect and Organize**
We are asked to explain why the standard heat of formation of carbon monoxide as a gas is difficult to measure experimentally.

Analyze
Carbon monoxide is formed from the combustion of carbon. Another product, however, also forms when carbon reacts with oxygen: CO_2.

Solve
The enthalpy of formation of CO is difficult to measure because the competing reaction to form CO_2 is favorable, so reacting C(*s*) with a limited supply of $O_2(g)$ produces a mixture of products, not just CO.

Think About It
The enthalpy of formation of carbon dioxide is more exothermic (–412.9 kJ/mol) and therefore more favorable by enthalpy than the formation of carbon monoxide (–393.5 kJ/mol).

9.79. **Collect and Organize**
The standard enthalpy of formation is the energy absorbed or evolved when 1 mol of a substance is formed from the elements, all in their standard states. Here we compare the enthalpy of formation of O_2 to that of O_3.

Analyze
Standard conditions are 1 atm and some specified temperature. Both ozone (O_3) and elemental oxygen (O_2) exist under those conditions.

Solve
No, because ozone and elemental oxygen are different forms of oxygen, their standard enthalpies of formation are different. From Appendix 4, ΔH_f° for O_2 is 0 kJ/mol (because it is an element in its most stable form under standard conditions) and ΔH_f° for O_3 is 142.7 kJ/mol.

Think About It
Because ΔH_f° for O_3 is more positive than that for O_2, ozone is less stable than O_2.

9.81. **Collect and Organize**
In calculating an estimate of the enthalpy change for a reaction by using bond energies, we are to explain why we need to know the stoichiometry of the reaction.

Analyze
Bond breaking is endothermic, and bond formation is exothermic. The stoichiometry of the reaction tells us how many bonds break and how many form.

Solve
We must account for all the bonds that break and all the bonds that form in the reaction. To do so we must start with a balanced chemical reaction. If we miss a bond that breaks, our calculated enthalpy of reaction would be too negative. If we miss a bond that forms, our calculated enthalpy of reaction would be too positive.

Think About It
Before using an equation to calculate enthalpy change, whether from enthalpies of formation or from bond energies, always start with a balanced chemical equation.

9.83. **Collect and Organize**

We are to explain why the phase of the reactants is important in enthalpy calculations, even in those involving bond energies—in particular, why the reactants and products should all be in the gas phase when we use bond energies to estimate the enthalpy of a reaction.

Analyze

The bond energy is defined as the enthalpy change required to break 1 mol of bonds in a substance in the *gas phase*.

Solve

Bond energies are based on bond formation in the gas phase from the free atoms also in the gas phase.

Think About It

Having bond energy data tabulated only for the gaseous phase ensures that the measured energies are for the bonds breaking only and do not include any intermolecular interactions. Later, you will learn more about intermolecular forces. Some can be very strong (ion–ion forces), and others are quite weak (van der Waals forces).

9.85. **Collect and Organize**

The heat of formation is reflected in a reaction when (1) 1 mol of the substance is produced, (2) the substance is produced under standard-state conditions, and (3) it is produced from the substance's constituent elements in their standard state.

Analyze

Each reaction must meet all the criteria for its ΔH°_{rxn} to be classified as a heat of formation.

Solve

(a) One mole of CO is produced from elemental carbon and oxygen in their standard states, so ΔH°_{rxn} for that reaction represents ΔH°_f.

(b) Because 2 mol of Mn_3O_4 is produced from 6 mol of MnO_4 (which is not an element) and 1 mol of O_2, that reaction does not represent ΔH°_f.

(c) Because Na_2CO_3 is produced from the reaction of 2 mol of Na, 1 mol of C, and $^3/_2$ mol of O_2 (all elements in their standard states), that reaction does represent ΔH°_f.

(d) Two moles of CH_4 is produced from 2 mol of elemental carbon and 4 mol of hydrogen each in their standard states. That at first appears to represent ΔH°_f, but the equation does not represent the product in 1 mol of product, CH_4, and therefore that reaction does not represent ΔH°_f.

Think About It

Heat of formation must involve the reaction of the *elements* to form compounds. Remember, though, that some elements, such as O_2, H_2, and N_2, are diatomic in their elemental state.

9.87. **Collect and Organize**

The heat of a reaction can be computed by finding the difference between the sum of the heats of formation of the products and the sum of the heats of formation of the reactants.

Analyze

We have to take into account the moles of products formed and the moles of reactants used as well, because enthalpy is a stoichiometric quantity.

$$\Delta H^\circ_{rxn} = \sum n_{products} \Delta H^\circ_{f,products} - \sum n_{reactants} \Delta H^\circ_{f,reactants}$$

Values for ΔH°_f for the reactants and products are found in Appendix 4.

Solve

$$\Delta H^{\circ}_{rxn} = \left[(1 \text{ mol CH}_4)(-74.8 \text{ kJ/mol}) + (2 \text{ mol H}_2\text{O})(-285.8 \text{ kJ/mol}) \right]$$
$$- \left[(4 \text{ mol H}_2)(0 \text{ kJ/mol}) + (1 \text{ mol CO}_2)(-393.5 \text{ kJ/mol}) \right]$$
$$\Delta H^{\circ}_{rxn} = -252.9 \text{ kJ}$$

Think About It

Be careful to note and find the appropriate ΔH°_f for a compound that may exist in different phases. For example, ΔH°_f of $H_2O(g) = -241.8$ kJ/mol, but ΔH°_f of $H_2O(\ell) = -285.8$ kJ/mol.

9.89. Collect and Organize

To calculate ΔH°_{rxn} for the decomposition of NH_4NO_3 to N_2O and H_2O vapor, we need the balanced equation because the enthalpy of the reaction depends on the moles of reactants consumed and moles of products formed in the reaction.

Analyze

From the balanced chemical equation and the values of ΔH°_f of the reactants and products, we use

$$\Delta H^{\circ}_{rxn} = \sum n_{products} \Delta H^{\circ}_{f,products} - \sum n_{reactants} \Delta H^{\circ}_{f,reactants}$$

Because the reaction is run at 250–300°C, the water product is in the gaseous phase.

Solve

The balanced equation for that reaction is
$$NH_4NO_3(s) \rightarrow N_2O(g) + 2 \text{ } H_2O(g)$$
We use the coefficients in the equation for ΔH°_{rxn}:
$$\Delta H^{\circ}_{rxn} = \left[(1 \text{ mol N}_2\text{O})(82.1 \text{ kJ/mol}) + (2 \text{ mol H}_2\text{O})(-241.8 \text{ kJ/mol}) \right]$$
$$- \left[(1 \text{ mol NH}_4\text{NO}_3)(-365.6 \text{ kJ/mol}) \right]$$
$$\Delta H^{\circ}_{rxn} = -35.9 \text{ kJ}$$

Think About It

The reaction is exothermic, releasing 36 kJ for every mole of NH_4NO_3 decomposed.

9.91. Collect and Organize

We are given the balanced chemical equation for the explosive reaction of fuel oil with ammonium nitrate in the presence of oxygen.

Analyze

To calculate ΔH°_{rxn} we use

$$\Delta H^{\circ}_{rxn} = \sum n_{products} \Delta H^{\circ}_{f,products} - \sum n_{reactants} \Delta H^{\circ}_{f,reactants}$$

Solve

$$\Delta H^{\circ}_{rxn} = \left[(3 \text{ mol N}_2)(0.0 \text{ kJ/mol}) + (17 \text{ mol H}_2\text{O})(-241.8 \text{ kJ/mol}) + (10 \text{ mol CO}_2)(-393.5 \text{ kJ/mol}) \right]$$
$$- \left[(3 \text{ mol NH}_4\text{NO}_3)(-365.6 \text{ kJ/mol}) + (1 \text{ mol C}_{10}\text{H}_{22})(249.7 \text{ kJ/mol}) + (14 \text{ mol O}_2)(0.0 \text{ kJ/mol}) \right]$$
$$\Delta H^{\circ}_{rxn} = -7198 \text{ kJ}$$

Think About It

That is a very exothermic reaction that occurs very fast and is therefore explosive.

9.93. Collect and Organize

We can use average bond energy data in Table A4.1 to estimate the enthalpy of three reactions.

Analyze

The enthalpy of a reaction as estimated from bond energies is given as

$$\Delta H_{rxn} = \sum \Delta H_{bond\ breaking} + \sum \Delta H_{bond\ forming}$$

where $\Delta H_{bond\ breaking}$ and $\Delta H_{bond\ forming}$ are average bond energies for the bonds in the reactants and products, respectively. In computing $\Delta H_{bond\ breaking}$ and $\Delta H_{bond\ forming}$ we have to take into account the number (or moles) of a particular type of bond that breaks or forms. We must also keep in mind that bond breaking requires energy $(+\Delta H)$, whereas bond formation releases energy $(-\Delta H)$.

Solve

(a) $\Delta H_{rxn} = [(2 \times 945\ kJ/mol) + (3 \times 436\ kJ/mol)] + [-(6 \times 391\ kJ/mol)] = 852\ kJ$

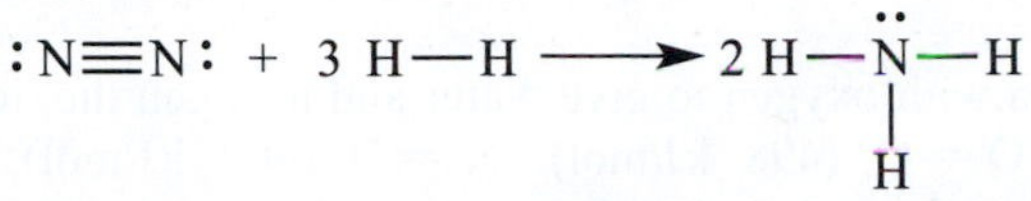

(b) $\Delta H_{rxn} = [(1 \times 945\ kJ/mol) + (2 \times 436\ kJ/mol)] + [-(1 \times 163\ kJ/mol) - (4 \times 391\ kJ/mol)] = 90\ kJ$

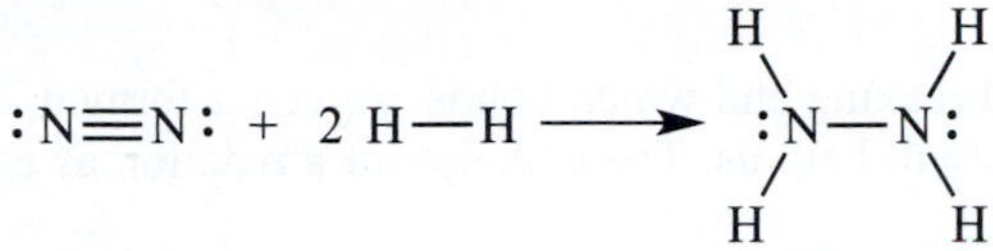

(c) $\Delta H_{rxn} = [(2 \times 945\ kJ/mol) + (1 \times 498\ kJ/mol)] + [-(2 \times 945\ kJ/mol) - (2 \times 201\ kJ/mol)] = 96\ kJ$

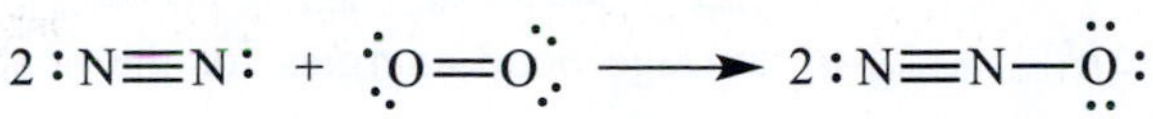

Think About It

We have to use the "best" Lewis structure for these calculations of reaction enthalpy. In part c, an alternative form (but contributing less to the bonding by formal charge arguments) for N_2O would be

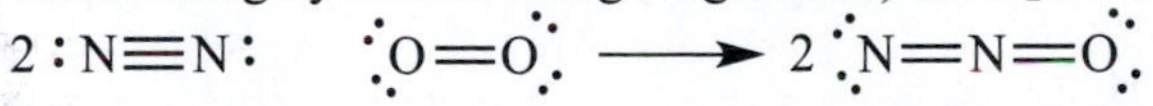

$$\Delta H_{rxn} = [(2 \times 945\ kJ/mol) + (1 \times 498\ kJ/mol)] + [-(2 \times 418\ kJ/mol) - (2 \times 607\ kJ/mol)] = 338\ kJ$$

9.95. Collect and Organize

We can use bond energies to compare the reaction enthalpies of the two reactions

$$C_2H_6 + \tfrac{5}{2} O_2 \rightarrow 2\ CO + 3\ H_2O$$
$$C_2H_6 + \tfrac{7}{2} O_2 \rightarrow 2\ CO_2 + 3\ H_2O$$

Analyze

From the Lewis structures and the average bond energy values in Table A4.1, we can estimate the reaction enthalpies. The enthalpy of a reaction as estimated from bond energies is given as

$$\Delta H_{rxn} = \sum \Delta H_{bond\ breaking} + \sum \Delta H_{bond\ forming}$$

where $\Delta H_{bond\ breaking}$ and $\Delta H_{bond\ forming}$ are average bond energies for the bonds in the reactants and products, respectively.

Solve

For the incomplete combustion of C_2H_6 to CO:

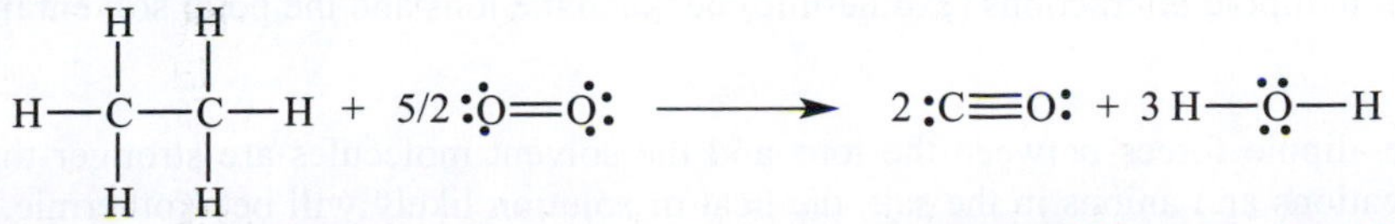

$\Delta H_{rxn} = [(6 \times 413\ kJ/mol) + (1 \times 348\ kJ/mol) + (5/2 \times 498\ kJ/mol)] + [-(2 \times 1072\ kJ/mol) - (6 \times 463\ kJ/mol)]$
$\Delta H_{rxn} = -851\ kJ/mol$

For the complete combustion of C_2H_6 to CO_2:

$$H-\overset{\overset{\displaystyle H}{|}}{\underset{\underset{\displaystyle H}{|}}{C}}-\overset{\overset{\displaystyle H}{|}}{\underset{\underset{\displaystyle H}{|}}{C}}-H + 7/2 \,\ddot{\ddot{O}}=\ddot{\ddot{O}} \longrightarrow 2\,\ddot{\ddot{O}}=C=\ddot{\ddot{O}} + 3\,H-\ddot{O}-H$$

$\Delta H_{rxn} = [(6 \times 413 \text{ kJ/mol}) + (1 \times 348 \text{ kJ/mol}) + (7/2 \times 498 \text{ kJ/mol})] + [-(4 \times 799 \text{ kJ/mol}) - (6 \times 463 \text{ kJ/mol})]$

$\Delta H_{rxn} = -1405 \text{ kJ/mol}$

The complete combustion reaction releases 554 kJ more energy than the incomplete combustion reaction.

Think About It
Although weaker $C=O$ bonds are formed in the complete combustion reaction, two such bonds are formed in CO_2, which outweighs the strong $C\equiv O$ bond in carbon monoxide in the incomplete combustion reaction.

9.97. Collect and Organize
For the reaction of ammonia with oxygen to give water and nitrogen dioxide, we can use the bond energies of the N—H (391 kJ/mol), $O=O$ (498 kJ/mol), $N=O$ (607 kJ/mol), N—O (201 kJ/mol), and O—H (463 kJ/mol) bonds to estimate the enthalpy of the reaction.

Analyze
To be sure which bonds are breaking and which bonds are being formed, drawing the Lewis structures of each of the products and reactants will help us. The enthalpy of a reaction as estimated from bond energies is given as

$$\Delta H_{rxn} = \sum \Delta H_{\text{bond breaking}} + \sum \Delta H_{\text{bond forming}}$$

where $\Delta H_{\text{bond breaking}}$ and $\Delta H_{\text{bond forming}}$ are average bond energies for the bonds in the reactants and products, respectively.

Solve

$$4\,H-\overset{\overset{\displaystyle \cdot\cdot}{}}{\underset{\underset{\displaystyle H}{|}}{N}}-H + 7\,\ddot{\cdot O}=\ddot{O}\cdot \longrightarrow 4\,\ddot{:O}-\dot{N}=\ddot{O}\cdot + 6\,H-\ddot{O}-H$$

$\Delta H_{rxn} = [(12 \times 391 \text{ kJ/mol}) + (7 \times 498 \text{ kJ/mol})] + [-(4 \times 201 \text{ kJ/mol}) - (4 \times 607 \text{ kJ/mol}) - (12 \times 463 \text{ kJ/mol})]$

$\Delta H_{rxn} = -610 \text{ kJ}$

Think About It
The bonding in NO_2 is not strictly one single bond and one double bond, as shown by the resonance structures

$$\ddot{:O}-\dot{N}=\ddot{O}\cdot \longleftrightarrow \cdot\ddot{O}=\dot{N}-\ddot{O}:$$

For calculations using bond energies, however, we can use one of the "frozen" resonance structures to assign bond energy values.

9.99. Collect and Organize
We are asked to describe how the strength of ion–dipole interactions between the solvent and the cations and anions of an ionic salt determine whether the heat of solution of that ionic salt is exothermic or endothermic.

Analyze
The heat of solution of an ionic salt involves the breaking of ionic bonds in the salt (endothermic) and the formation of ion–dipole interactions (exothermic) between the ions and the polar solvent molecules.

Solve
When the ion–dipole forces between the ions and the solvent molecules are stronger than the ion–ion forces between the cations and anions in the salt, the heat of solution likely will be exothermic. If, however, they are weaker than the ion–ion forces, the heat of solution likely will be endothermic.

Think About It
Whether the heat of solution is exothermic or endothermic does not necessarily determine whether a salt will dissolve in the solvent. For example, we might expect that an endothermic heat of solution would mean that the salt does not dissolve well. However, the dissolution of sodium chloride (table salt) is endothermic, but certainly the salt does dissolve well in water.

9.101. Collect, Organize, and Analyze
We are asked to define and describe the term *fuel value*.

Solve
We often are concerned about the energy a fuel provides per mass; fuel value refers to the thermal energy released per gram of fuel during complete combustion.

Think About It
For some fuels, such as gasoline, thinking of energy per volume (liter or gallon) might be more convenient.

9.103. Collect and Organize
For this question we compare the fuel values (without making any calculations) of CH_4 and H_2 on a per-mole and a per-gram basis.

Analyze
The balanced equations for the combustion reactions are
$$CH_4(g) + 2\,O_2(g) \rightarrow CO_2(g) + 2\,H_2O(g)$$
$$H_2(g) + \tfrac{1}{2}O_2(g) \rightarrow H_2O(g)$$
The heats of formation for these reactants and products are
$$CH_4(g) = -74.8 \text{ kJ/mol}$$
$$O_2(g) = 0.0 \text{ kJ/mol}$$
$$CO_2(g) = -393.5 \text{ kJ/mol}$$
$$H_2O(g) = -241.8 \text{ kJ/mol}$$
$$H_2(g) = 0.0 \text{ kJ/mol}$$

Solve
(a) In examining the two reactions and the associated heats of formation, we can see easily that the combustion of 1 mol of CH_4 gives off more energy than the combustion of 1 mol of H_2.
(b) However, because 1 g is 1/16 mol of CH_4, whereas 1 g is 1/2 mol of H_2 on a per-gram basis, H_2 releases more energy upon combustion per gram.

Think About It
The high-energy combustion reaction of hydrogen on a per-gram basis makes it an attractive fuel source to develop.

9.105. Collect and Organize
We are to use a Born–Haber cycle along with values for the ionization energy of K, electron affinity of Cl, sublimation energy of K, the bond energy of Cl_2, and the enthalpy of formation of KCl to determine the lattice energy for KCl.

Analyze
A Born–Haber cycle considers the formation of a compound from its elements as a sequence of steps. It uses Hess's law to calculate the value of a step with a missing thermodynamic value or the overall heat of formation for that compound. For ionic salts such as KCl, the Born–Haber cycle adds the enthalpies of sublimation of the metal, the ionization energy of the metal, the bond energy of the diatomic halogen, the electron affinity of the halogen atom, and the lattice energy of the cation and anion to give the heat of formation of the salt. In this particular case

$$\Delta H^{\circ}_{f,\,KCl} = \Delta H_{sub,\,K} + \frac{1}{2}\,BE_{Cl_2} + IE_{1,\,K} + EA_{Cl} + U_{KCl}$$

Solving for U_{KCl} gives

$$-U_{KCl} = \Delta H_{sub, K} + \frac{1}{2} BE_{Cl_2} + IE_{1, K} + EA_{Cl} - \Delta H^{\circ}_{f, KCl}$$

Solve

$$-U_{KCl} = 89 \text{ kJ/mol} + \left(\frac{1}{2} \times 243 \text{ kJ/mol}\right) + 419 \text{ kJ/mol} + (-349 \text{ kJ/mol}) - (-436.5 \text{ kJ/mol}) = 717 \text{ kJ/mol}$$

$$U_{KCl} = -717 \text{ kJ/mol}$$

Think About It

Lattice energy for an ionic solid is defined as the energy change when gaseous cations and anions form a solid. Because of the attraction of the oppositely charged ions, that is always exothermic for ionic solids.

9.107. Collect and Organize

To determine the amount of water that could be heated from 20.0°C to 45.0°C when 1.00 pound of propane is burned, we use the fuel value of propane and the heat capacity equation for water.

Analyze

First, to compute the energy (q, in kilojoules) that 1.00 pound of propane generates, we multiply the fuel value (46.35 kJ/g) by the mass of propane (after converting pounds to grams, using 1 lb = 453.6 g). Then we use the molar heat capacity equation

$$q = nc_P\Delta T$$

where $c_P = 75.3$ J/mol · °C, $\Delta T = 25.0$°C, to find the moles (n) and finally mass (m) of water that can be heated 25.0°C by 1.00 pound of propane.

Solve

Energy generated by propane:

$$1.00 \text{ lb} \times \frac{453.6 \text{ g}}{1 \text{ lb}} \times \frac{46.35 \text{ kJ}}{g} = 2.10 \times 10^4 \text{ kJ}$$

Mass of water heated 25.0°C by this q:

$$2.10 \times 10^4 \text{ kJ} \times \frac{1000 \text{ J}}{1 \text{ kJ}} = n \times 75.3 \frac{J}{\text{mol} \cdot °C} \times 25.0 °C$$

$$n = 1.11 \times 10^4 \text{ mol}$$

That is $1.11 \times 10^4 \text{ mol} \times \dfrac{18.02 \text{ g}}{\text{mol}} = 2.01 \times 10^5$ g, or 201 kg

Think About It

Propane's relatively high fuel value makes it an efficient fuel.

9.109. Collect and Organize

Once we compute the fuel value of C_5H_{12}, we use it to calculate the energy released when 1.00 kg of C_5H_{12} is burned and how much C_5H_{12} is needed to heat 1.00 kg of water by 70.0°C.

Analyze

(a) The fuel value can be calculated by dividing the absolute value of the given ΔH°_{comb} (−3535 kJ/mol) by the molar mass of C_5H_{12} (72.15 g/mol).

(b) The energy released when 1.00 kg of C_5H_{12} is burned can be found by multiplying the mass in grams by the fuel value.

(c) The molar heat equation is

$$q = nc_P\Delta T$$

where moles (n) can be determined from $m = 1.00$ kg (1000 g) water, $c_P = 75.3$ J/mol · °C, and $\Delta T = 70.0$°C. That gives us the energy (in joules) required to heat the water. The amount of C_5H_{12} needed to heat the water can then be calculated by dividing the q value by the fuel value for C_5H_{12}.

Solve

(a) Fuel value of $C_5H_{12} = \dfrac{3535 \text{ kJ}}{\text{mol}} \times \dfrac{1 \text{ mol}}{72.15 \text{ g}} = 49.00 \text{ kJ/g}$

(b) Heat released by 1.00 kg $C_5H_{12} = 1000 \text{ g} \times \dfrac{48.99 \text{ kJ}}{\text{g}} = 4.90 \times 10^4 \text{ kJ}$

(c) Energy needed to raise 1.00 kg water from 20.0°C to 90.0°C:

$$q = 1000 \text{ g} \times \dfrac{1 \text{ mol}}{18.02 \text{ g}} \times \dfrac{75.3 \text{ J}}{\text{mol} \cdot {}°\text{C}} \times 70.0°\text{C} = 2.925 \times 10^5 \text{ J, or } 293 \text{ kJ}$$

Mass of C_5H_{12} needed to generate that energy $= 292.5 \text{ kJ} \times \dfrac{1 \text{ g}}{49.00 \text{ kJ}} = 5.97 \text{ g}$

Think About It
Only about 6 g of fuel is required from the camper's stove to heat the water. The white gas has a relatively high fuel value.

9.111. Collect and Organize
For two preparations of HCl(*g*) we are to use the heat of reaction to determine whether we should cool or heat those reactions.

Analyze
If a reaction is exothermic, it would make sense to cool the reaction (not heat it) so that it does not boil or cause an explosion. If a reaction is endothermic it would make sense to gently heat the reaction to provide the necessary energy for the reaction to proceed.

Solve
The heats of those two reactions are

$H_2(g) + Cl_2(g) \rightarrow 2HCl(g)$
$\Delta H_{rxn} = [(2 \text{ mol HCl} \times -92.3 \text{ kJ/mol})] - [(1 \text{ mol } H_2 \times 0.0 \text{ kJ/mol}) + (1 \text{ mol } Cl_2 \times 0.0 \text{ kJ/mol})] = -184.6 \text{ kJ}$

$2NaCl(s) + H_2SO_4(\ell) \rightarrow 2HCl(g) + Na_2SO_4(s)$
$\Delta H_{rxn} = [(2 \text{ mol HCl} \times -92.3 \text{ kJ/mol}) + (1 \text{ mol } Na_2SO_4 \times -1387.1 \text{ kJ/mol})] -$

$[(2 \text{ mol NaCl} \times -411.2 \text{ kJ/mol}) + (1 \text{ mol } H_2SO_4 \times -814.0 \text{ kJ/mol})] = 64.7 \text{ kJ}$

The first reaction using hydrogen and chlorine gases is exothermic, so we should cool that reaction; the second reaction using sodium chloride and sulfuric acid is endothermic, so we should heat that reaction.

Think About It
This problem illustrates that whether the formation of a particular product is endothermic or exothermic depends greatly on the process.

9.113. Collect and Organize
When a sodium hydroxide solution is mixed with a sulfuric acid solution, the solution gets hot. We are asked to find the ΔH_{rxn} for the reaction.

Analyze
(a) That is a neutralization reaction that produces water and sodium sulfate.
(b) To determine the molar ratio (to see whether any reactant is in excess), we multiply the volume of reactant by its concentration.
(c) The water absorbs all the energy generated by the reaction, causing the temperature to rise. The energy then is

$$q = mc_S \Delta T$$

where m = total mass of the solution (100 g + 50.0 g), c_S is the specific heat of water (4.184 J/g · °C), and ΔT is the change in temperature ($T_f - T_i$).

Solve

(a) $2 NaOH(aq) + H_2SO_4(aq) \rightarrow 2 H_2O(\ell) + Na_2SO_4(aq)$

(b) Stoichiometry of the reaction:

$$mol\ NaOH = 100.0\ mL \times \frac{1.0\ mol}{1000\ mL} = 0.10\ mol\ NaOH$$

$$mol\ H_2SO_4 = 50.0\ mL \times \frac{1.0\ mol}{1000\ mL} = 0.050\ mol\ H_2SO_4$$

From the balanced equation, 0.050 mol H_2SO_4 would require 0.100 mol of NaOH. Therefore, neither NaOH nor H_2SO_4 is left over after the reaction and 0.10 mol of H_2O is produced in the reaction.

(c) $q = mc_s\ \Delta T = 150\ g \times \dfrac{4.184\ J}{g \cdot {}^\circ C} \times (31.4\,{}^\circ C - 22.3\,{}^\circ C) = 5.7 \times 10^3$ J, or 5.7 kJ

That is for the reaction of 0.10 mol of H_2O that is produced in that particular reaction. For 1 mol of H_2O, therefore,

$$\frac{5.71\ kJ}{0.10\ mol\ H_2O} = 57\ kJ/mol\ H_2O$$

Because the reaction transfers energy (the mixture gets hot), that reaction is exothermic and $\Delta H_{rxn} = -57$ kJ/mol H_2O.

Think About It

We can assign the sign of ΔH at the end of our calculation. If temperature rises, the reaction is exothermic and the sign of enthalpy is negative.

9.115. Collect and Organize

When a hot sample of copper is dropped into water, energy is transferred from the metal to the water until thermal equilibrium is established.

Analyze

The energy lost by the copper sample is

$$q_{lost} = m_{Cu} \times c_s \times (T_f - T_i)$$

where $m_{Cu} = 7.25$ g, $c_s = 0.385$ J/g $\cdot$ °C, and $T_i = 100.1$°C.
The energy gained by the water is

$$q_{gained} = n_{H_2O} \times c_P \times (T_f - T_i)$$

where $n_{H_2O} = 50.0$ g/(18.02 g/mol), $c_P = 75.3$ J/mol $\cdot$ °C, and $T_i = 25.0$°C.
The law of conservation of energy gives

$$-q_{lost,Cu} = q_{gained,water}$$

Solve

$$-q_{lost,Cu} = q_{gained,water}$$

$$-7.25\ g \times \frac{0.385\ J}{g \cdot {}^\circ C}\left(T_f - 100.1\,{}^\circ C\right) = 50.0\ g \times \frac{1\ mol}{18.02\ g} \times \frac{75.3\ J}{mol \cdot {}^\circ C} \times \left(T_f - 25.0\,{}^\circ C\right)$$

$$-2.79\ \frac{J}{{}^\circ C} \times T_f + 279\ J = 209\ \frac{J}{{}^\circ C} \times T_f - 5223\ J$$

$$-211.8\ \frac{J}{{}^\circ C} \times T_f = -5502\ J$$

$$T_f = 26.0\,{}^\circ C$$

Think About It

Because the heat capacity of the copper is so small and the heat capacity of water is so large, very little increase in the temperature of water occurs: only 1°C.

9.117. **Collect and Organize**
We are asked to calculate the standard heat of combustion of acetylene and determine its fuel value.

Analyze
(a) To calculate the molar enthalpy of combustion, we need to write a balanced chemical equation and then use the following equation to calculate ΔH°_{rxn}

$$\Delta H^\circ_{rxn} = \sum n_{products} \Delta H^\circ_{f,products} - \sum n_{reactants} \Delta H^\circ_{f,reactants}$$

(b) The fuel value is the energy released per gram of the fuel, so we will divide our answer in part a by the molar mass of acetylene.

Solve
(a) The balanced chemical equation for the combustion of 1 mol of C_2H_2 is

$$C_2H_2(g) + \tfrac{5}{2}O_2(g) \rightarrow 2\,CO_2(g) + H_2O(g)$$

The standard enthalpy for burning 1 mol of C_2H_2 is

$$\Delta H^\circ_{rxn} = \left[(2\text{ mol }CO_2 \times -393.5\text{ kJ/mol}) + (1\text{ mol }H_2O \times -241.8\text{ kJ/mol})\right]$$
$$-\left[(1\text{ mol }C_2H_2 \times 226.7\text{ kJ/mol}) + (\tfrac{5}{2}\text{ mol }O_2 \times 0.0\text{ kJ/mol})\right]$$
$$\Delta H^\circ_{rxn} = -1255.5\text{ kJ/mol }C_2H_2$$

(b) The fuel value of acetylene is

$$\frac{1255.5\text{ kJ}}{\text{mol}} \times \frac{1\text{ mol}}{26.037\text{ g}} = 48.22\text{ kJ/g}$$

Think About It
Some other examples of endothermic compounds:
Inorganic compounds = B_2H_6, CS_2, HI, N_2O, O_3
Organic compounds = C_2H_4, C_6H_6, $(CH_3)_2C{=}C(CH_3)_2$

9.119. **Collect and Organize**
We are asked to determine, using average bond energy value, whether the formation of polyethylene from ethylene is endothermic, is exothermic, or has no change in enthalpy.

Analyze
The enthalpy of a reaction as estimated from bond energies is given as

$$\Delta H_{rxn} = \sum \Delta H_{bond\ breaking} + \sum \Delta H_{bond\ forming}$$

where $\Delta H_{bond\ breaking}$ and $\Delta H_{bond\ forming}$ are average bond energies for the bonds in the reactants and products, respectively. In computing $\Delta H_{bond\ breaking}$ and $\Delta H_{bond\ forming}$ we have to take into account the number (or moles) of a particular type of bond that breaks or forms. We must also keep in mind that bond breaking requires energy $(+\Delta H)$, whereas bond formation releases energy $(-\Delta H)$.

Solve
The number of C–H bonds does not change, so we need not consider those in the calculation, but rather we can focus on the change in the carbon–carbon bonds.

For the formation of polyethylene from ethylene, the heat of reaction is

$$n\,H_2C{=}CH_2 \longrightarrow {-}\!\!\left[CH_2{-}CH_2\right]\!\!{}_n$$

$$\Delta H_{rxn} = [(n \times 614\text{ kJ/mol})] + [-(2n \times 348\text{ kJ/mol})]$$

where we will consider $n = 1$

$$\Delta H_{rxn} = -82\text{ kJ/mol}$$

That reaction is exothermic.

Think About It
That finding may be unexpected in that we are breaking a double bond, but remember: From that double bond we are forming two single bonds to each carbon atom. Two single bonds added together is stronger than a double bond.

9.121. Collect and Organize
From the balanced equation and the ΔH°_{rxn} we are to determine whether energy is consumed or produced in the reaction and how much energy is involved when 60.0 g of CH_3OH is used in the reaction.

Analyze
(a) We can balance that reaction by inspection. Note that $O_2(g)$ is also a product in the reaction.
(b) We are given that the reaction requires 164 kJ/mol (of methanol) of energy, so that is an endothermic reaction.
(c) To calculate the energy needed to transform 60.0 g of methanol, we need to only determine the moles of CH_3OH in 60.0 g and then multiply that result by the energy required in the reaction for 1 mol of CH_3OH (164 kJ/mol).

Solve
(a) $CH_3OH(g) + N_2(g) \rightarrow HCN(g) + NH_3(g) + \frac{1}{2} O_2(g)$
(b) Because that is an endothermic reaction, the energy is a reactant:
$$164 \text{ kJ} + CH_3OH(g) + N_2(g) \rightarrow HCN(g) + NH_3(g) + \frac{1}{2} O_2(g)$$
(c) The moles of CH_3OH used is
$$60.0 \text{ g} \times \frac{1 \text{ mol}}{32.04 \text{ g}} = 1.87 \text{ mol}$$
The enthalpy for the reaction using that amount of CH_3OH is
$$1.87 \text{ mol} \times \frac{164 \text{ kJ}}{1 \text{ mol}} = 307 \text{ kJ}$$

Think About It
The energy required or evolved in a reaction is an extensive property; it depends on the amount of reactants involved in the reaction.

9.123. Collect and Organize
We combine three reactions to give the overall reaction for the formation of $PbCO_3$.

Analyze
Because $PbCO_3$ is a product in the overall equation, reaction 3 has to be reversed. Neither reaction 1 nor reaction 2 needs to be changed.

Solve

$$
\begin{array}{ll}
Pb(s) + \frac{1}{2}O_2(g) \rightarrow PbO(s) & \Delta H^\circ_{rxn} = -219 \text{ kJ} \\
C(s) + O_2(g) \rightarrow CO_2(g) & \Delta H^\circ_{rxn} = -394 \text{ kJ} \\
\underline{PbO(s) + CO_2(g) \rightarrow PbCO_3(s)} & \underline{\Delta H^\circ_{rxn} = -86 \text{ kJ}} \\
Pb(s) + \frac{3}{2}O_2(g) + C(s) \rightarrow PbCO_3(s) & \Delta H^\circ_{rxn} = -699 \text{ kJ}
\end{array}
$$

That is for 1 mol of $C(s)$. For 2 mol of carbon in the target equation, the energy evolved in that reaction is
$$2 \text{ mol} \times \frac{-699 \text{ kJ}}{\text{mol}} = -1398 \text{ kJ}$$

Think About It
All the reactions added together are exothermic. The overall reaction, therefore, must also be exothermic. The overall enthalpy of reaction is also the enthalpy of formation of $PbCO_3$.

9.125. Collect and Organize

Using the ΔH_f° values for the reactants and products, we are asked to calculate the ΔH_{rxn}° for the decomposition of 1 mol of $NaHCO_3$.

Analyze

The overall balanced decomposition reaction is

$$2\, NaHCO_3(s) \rightarrow Na_2CO_3(s) + CO_2(g) + H_2O(g)$$

To calculate the enthalpy of the combustion, use the equation

$$\Delta H_{rxn}^\circ = \sum n_{products}\, \Delta H_{f,products}^\circ - \sum n_{reactants}\, \Delta H_{f,reactants}^\circ$$

Solve

$$\Delta H_{rxn}^\circ = \left[(1\text{ mol } CO_2 \times -393.5\text{ kJ/mol}) + (1\text{ mol } H_2O \times -241.8\text{ kJ/mol}) + (1\text{ mol } Na_2CO_3 \times -1130.7\text{ kJ/mol}) \right]$$
$$- \left[(2\text{ mol } NaHCO_3 \times -950.8\text{ kJ/mol}) \right]$$

$$\Delta H_{rxn}^\circ = 135.6\text{ kJ}$$

For the formation of 1 mol of $NaHCO_3(s)$:

$$\Delta H_f^\circ = \frac{135.6\text{ kJ}}{2\text{ mol}} = 67.8\text{ kJ/mol}$$

Think About It

Remember that enthalpy is stoichiometric. You may have to adjust the ΔH_{rxn}° according to what the problem asks for. Here the enthalpy of decomposition of 1 mol of $NaHCO_3$ was specified.

9.127. Collect and Organize

Using ΔH_f° values for urea, carbon dioxide, water, and ammonia, we calculate ΔH_{rxn}° for the conversion of urea into ammonia.

Analyze

The ΔH_{rxn}° can be found from the enthalpy of formation values by using

$$\Delta H_{rxn}^\circ = \sum n_{products}\, \Delta H_{f,products}^\circ - \sum n_{reactants}\, \Delta H_{f,reactants}^\circ$$

Solve

$$\Delta H_{rxn}^\circ = \left[(1\text{ mol } CO_2 \times -412.9\text{ kJ/mol}) + (2\text{ mol } NH_3 \times -80.3\text{ kJ/mol}) \right]$$
$$- \left[(1\text{ mol urea} \times -319.2\text{ kJ/mol}) + (1\text{ mol } H_2O \times -285.8\text{ kJ/mol}) \right]$$

$$\Delta H_{rxn}^\circ = 31.5\text{ kJ}$$

Think About It

The overall reaction is slightly endothermic.

9.129. Collect and Organize

We are given balanced chemical reactions for the combustion of two fuels, dimethylhydrazine and hydrogen, along with their enthalpy of combustion values to help us determine which fuel releases more energy per pound.

Analyze

The ΔH values are given on a per-mole-of-fuel basis. We can convert those to a per-gram basis by dividing the absolute value of the enthalpy of the combustion by the molar mass of the fuel. That energy per gram of fuel can then be converted into energy per pound by using the conversion 1 lb = 453.6 g.

Solve

For dimethylhydrazine:

$$\frac{1694 \text{ kJ}}{\text{mol}} \times \frac{1 \text{ mol}}{60.10 \text{ g}} \times \frac{453.6 \text{ g}}{1 \text{ lb}} = 12,790 \text{ kJ/lb}$$

For hydrogen:

$$\frac{286 \text{ kJ}}{\text{mol}} \times \frac{1 \text{ mol}}{2.02 \text{ g}} \times \frac{453.6 \text{ g}}{1 \text{ lb}} = 64,200 \text{ kJ/lb}$$

Pound for pound, therefore, hydrogen is a better fuel.

Think About It

Even though the combustion of hydrogen is less exothermic on a per-mole basis, its use as a fuel on a per-weight basis yields more energy than dimethylhydrazine.

CHAPTER 10 | Properties of Gases: The Air We Breathe

10.1. Collect and Organize

Four illustrations representing molecules in a balloon are shown in Figure P10.1. The spheres represent gas atoms, and we are asked to choose the drawing that best shows how the gas is distributed in the balloon.

Analyze

Gases are characterized by having no definite shape or volume. They fill the container they are in and occupy the container's entire volume. Also, the atoms or molecules in gases are far apart from each other.

Solve

Drawing (c) accurately reflects the distribution of helium atoms in a balloon. The other representations either show clusters of gas atoms (b and d) or have the gas atoms attached to the wall of the balloon (a) and therefore the helium gas does not occupy the entire volume.

Think About It

What Figure P10.1 does not show, but is good to keep in mind, is that gas molecules are in constant motion.

10.3. Collect and Organize

We are to explain how the decrease in volume of the gas by downward movement of the piston in Figure P10.2 will affect the root-mean-square speed of the gas molecules in the cylinder.

Analyze

The root-mean-square speed of a gas is described by the formula

$$u_{rms} = \sqrt{\frac{3RT}{\mathcal{M}}}$$

where R is the gas constant, T is the temperature, and $\mathcal{M}$ is the molar mass of the gas.

Solve

Because neither pressure nor volume affects the root-mean-square speed (u_{rms}) of the gas (they do not appear in the equation), and the temperature, which does affect the u_{rms}, is assumed to be constant for the change in volume in Figure P10.2, the root-mean-square speed is not affected.

Think About It

If the temperature were raised, the root-mean-square speed would increase.

10.5. Collect and Organize

When the number of gas particles in a cylinder is doubled, we are to consider how the pressure is affected, how the frequency of collisions of the particles with the walls of the pistons changes, and by how much the probable speed of the gas particles in the cylinder might change.

Analyze

Avogadro's law states that pressure of a gas is directly proportional to the number of gas particles in the container. The pressure is a measure of the frequency of collisions of the gas particles with the walls of the container; as pressure goes up, the frequency of the collisions also increases. Finally, the most probable speed of the gas particles, like the root-mean-square speed, is affected only by a change in temperature.

Solve

(a) If the volume of the cylinder does not change (that is, the piston does not move), increasing the number of gas particles in the cylinder will increase the pressure proportionately. Because the number of gas particles is doubled, the pressure will also double.

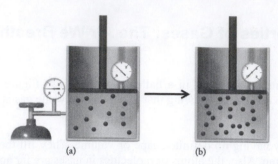

(b) The pressure is a measure of the frequency of collisions of the gas particles with the walls of the container and is directly proportional to the number of particles. Therefore, the frequency of collisions of the gas particles with the walls of the container will double because the number of particles and the pressure double.

(c) The most probable speed of the gas particles in the container does not depend on the number of gas particles present; the most probable speed is not affected.

Think About It

If the piston moves to increase the volume of the cylinder by a factor of 2 upon doubling the number of gas particles in the cylinder, the pressure and the frequency of collisions will not change.

10.7. **Collect and Organize**

For the two gas samples in which molecules of A are represented by the red spheres confined to cylinders with movable pistons in Figure P10.7 with the same volumes and at the same temperature, we are asked to determine for which sample the mole fraction of A, the partial pressure of A, and the total pressure is greater.

Analyze

The mole fraction of a component is determined by dividing the number of moles or particles of the component of interest in the mixture by the total number of moles or particles present in the mixture. Each gas in a mixture exerts its own contribution to the total pressure of the mixture. The partial pressure of a component in the gas mixture is proportional to the mole fraction of its gas particles in the mixture. Finally, the greater the total number of gas particles present in a mixture, the higher the pressure.

Solve

(a) Cylinder (a) contains 8 red spheres and 4 blue spheres. The mole fraction of A in that cylinder is 8/12 = 0.67. Cylinder (b) contains 11 red spheres and 17 blue spheres. The mole fraction in that cylinder is 11/28 = 0.39. Therefore, the red spheres have the higher mole fraction in cylinder (a).

(b) More red spheres are in cylinder (b) than in cylinder (a), so the pressure exerted by the red spheres will be higher in cylinder (b) than in cylinder (a).

(c) The total number of gas particles in cylinder (a) is 12, whereas the total number of gas particles in cylinder (b) is 28. Therefore, the pressure inside cylinder (b) will be 2.33 times as great as that in cylinder (a).

Think About It

If the pressure were to be equal for those two cylinders, the volume of cylinder (b) would be 2.33 times that of cylinder (a).

10.9. **Collect and Organize**

Effusion is the leaking of gas through a small hole in a container. We are asked which change in Figure P10.9 shows effusion of helium.

Analyze

The balloon contains fewer He atoms after effusion has taken place and, because of the presence of fewer He atoms in the balloon, the pressure inside the balloon decreases, so the balloon will shrink.

Solve

Both (a) and (b) show fewer atoms of He after effusion; however, outcome (a) is the correct choice since the balloon in (a) shrinks, whereas the balloon in (b) expands.

Think About It

We witness that effect happening to helium balloons in our everyday experience: a helium balloon made from latex usually shows noticeable shrinkage after being left overnight.

10.11. Collect and Organize

For CO_2 and SO_2 the number of molecules versus molecular speed is plotted in Figure P10.11. We are asked to identify which curve belongs to CO_2 and which belongs to SO_2 and then to determine which curve would also describe the distribution of molecular speeds for propane (C_3H_8).

Analyze

The formula that describes molecular speed (u_{rms}) is

$$u_{rms} = \sqrt{\frac{3RT}{\mathcal{M}}}$$

where R is the universal gas constant, T is the absolute temperature in kelvins, and $\mathcal{M}$ is the molar mass of the gaseous compound. The larger the molar mass, the slower the u_{rms}. Curve 1 shows more molecules at lower speeds, whereas curve 2 shows more molecules at higher speeds.

Solve

The molar masses of CO_2, SO_2, and C_3H_8 are 44, 64, and 44 g/mol, respectively. Because SO_2 has a higher molar mass than either CO_2 or C_3H_8, it is represented by curve 1. Curve 2 represents both CO_2 and C_3H_8 because they have the same molar mass.

Think About It

For heavier molecules, the distribution of speeds is narrower than for lighter molecules.

10.13. Collect and Organize

Figure P10.13 shows the plot of volume versus temperature for 1 mole of three gases. Using that plot we are to decide which gas represents the highest pressure and are asked to estimate the slope of that line to determine the pressure of that gas.

Analyze

Plotted here is Charles's law, which states that the volume of a gas is proportional to its temperature. Each of those lines represents the volume (versus temperature) of 1 mole of gas. The volume of a gas is inversely related to its pressure; the smaller the volume, the higher the pressure.

Solve

(a) At any temperature the red line shows a lowest volume of the 1 mole of gas. Therefore, the red line represents the gas at the highest pressure.
(b) The slope of the line is estimated by dividing the "rise" in the graph by the "run." For ease we will use here the (0 K, 0 cm^3) and (273 K, 10 cm^3) points

$$\text{slope} = \frac{\text{rise}}{\text{run}} = \frac{\Delta y}{\Delta x} \approx \frac{(10-0) \text{ cm}^3}{(273-0) \text{ L}} = 0.0366 \text{ cm}^3/\text{K, or } 3.66 \times 10^{-5} \text{ L/K}$$

that ratio of V/T can be used in the ideal gas law where $n = 1$ mole:

$$P = \frac{nRT}{V} = 1 \text{ mol} \times 0.08206 \frac{\text{L} \cdot \text{atm}}{\text{mol} \cdot \text{K}} \times \frac{\text{K}}{3.66 \times 10^{-5} \text{ L}} = 2240 \text{ atm}$$

Your value for pressure, depending on which points you chose on the graph and how well you were able to estimate their x and y values, may vary slightly from the value calculated above.

Think About It

The yellow line on the plot has a slope that is twice that of the red line, so the pressure of that sample of gas would be half of that calculated for the red line.

10.15. Collect and Organize

The plot in Figure P10.15 shows the volume (V) as a function of the inverse of pressure ($1/P$). Line 2 in the plot diverges from line 1 at the origin and is above line 1. We are asked which line represents the higher temperature for the same quantity of gas.

Analyze

Because V is inversely proportional to P, the plot of V versus $1/P$ will give a straight line for a sample of gas. According to Charles's law, a gas sample held at constant pressure at higher temperatures has a larger volume.

Solve

A gas sample at a higher temperature has a larger volume at a given $1/P$ value than one at a lower temperature. Therefore, line 2 represents the gas sample at higher temperature.

Think About It

The lines converge at the origin because at infinite pressure the volume of gas will be zero.

10.17. Collect and Organize

The plot in Figure P10.17 shows two lines for the relationship between volume as a function of temperature. We are asked which line is not consistent with the ideal gas law.

Analyze

By Charles's law, volume and temperature are directly proportional. Therefore, as temperature (T) increases, the volume (V) of the gas increases. That relationship is also evident in the ideal gas law,

$$PV = nRT$$

if we rearrange the equation to

$$V = \frac{nRT}{P} \quad \text{or} \quad V = cT$$

where c = constant.

Solve

Line 2 in Figure P10.17 is the line that is inconsistent with the ideal gas law because it shows that as temperature increases, the volume decreases. Line 1 is consistent with the ideal gas law because it shows the direct and linear proportionality between T and V.

Think About It

The line labeled 2 would be correct if either P were plotted on the y-axis or V were plotted on the x-axis.

10.19. Collect and Organize

Three barometers are shown in Figure P10.19. We are to match the barometers with their respective locations.

Analyze

As altitude increases, the atmospheric pressure decreases. We can combine that knowledge with the fact that the barometer will show a lower column of mercury at lower pressures.

Solve

The locations in order of increasing pressure (higher to lower altitude) are Mount Everest < San Diego < gold mine. The barometers in order of increasing pressure are (a) < (b) < (c). Therefore, barometer (b) is sensing the atmospheric pressure for San Diego, barometer (a) is on Mount Everest, and barometer (c) is for the gold mine.

Think About It

The atmospheric pressure at sea level (ignoring the effects of the local weather) would be expected to be the same worldwide.

10.21. Collect, Organize, and Analyze

We are asked to define the root-mean-square speed of gas particles.

Solve

The root-mean-square speed (u_{rms}) is the square root of the arithmetic mean of the squares of the speeds of all the particles in a gas sample. It is the speed of a particle that has the average kinetic energy of all the particles in the sample.

Think About It

The root-mean-square speed is not exactly the same as the most probable speed, which is the speed corresponding to the peak in the distribution diagram. Nor is it the same as the average speed, which is a simple average of all the particle speeds.

10.23. Collect and Organize

We are asked whether pressure changes affect the value of u_{rms} for a gas sample.

Analyze

The root-mean-square speed equation is

$$u_{rms} = \sqrt{\frac{3RT}{M}}$$

Solve

Pressure does not appear in the equation describing the root-mean-square speed; therefore, a change in pressure has no effect on u_{rms}. Only temperature affects the root-mean-square speed.

Think About It

Only temperature and molar mass affect u_{rms}.

10.25. Collect and Organize

We are to rank the various gases according to the root-mean-square speeds at 0°C.

Analyze

The equation for root-mean-square speed shows an inverse dependence of u_{rms} of the molar mass of the gas:

$$u_{rms} = \sqrt{\frac{3RT}{M}}$$

Solve

According to the u_{rms} equation, the lower the molar mass, the greater the root-mean-square speed. The molar masses of NO, NO_2, N_2O_4, and N_2O_5 are 30, 46, 92, and 108 g/mol, respectively. Therefore, N_2O_5 has the lowest u_{rms}, and NO has the highest u_{rms}. The rank order in terms of increasing root-mean-square speed is $N_2O_5 < N_2O_4 < NO_2 < NO$.

Think About It

We did not need to compute the root-mean-square speeds here. We needed only to rank the gases in order of decreasing molar mass.

10.27. Collect and Organize

Knowing that H_2 (molar mass = 2.02 g/mol) effuses four times faster than an unknown gas, X, we are to calculate the molar mass of X.

Analyze

The effusion equation needed is

$$\frac{r_{H_2}}{r_X} = \sqrt{\frac{\mathcal{M}_X}{\mathcal{M}_{H_2}}} = \frac{4}{1}$$

Rearranging to solve for $\mathcal{M}_X$ gives

$$\mathcal{M}_X = \mathcal{M}_{H_2} \times \left(\frac{r_{H_2}}{r_X}\right)^2$$

Solve

$$\mathcal{M}_X = 2.02 \text{ g/mol} \times (4)^2 = 32.3 \text{ g/mol}$$

Think About It

That gas might be oxygen, whose molar mass is 32.0 g/mol.

10.29. Collect and Organize

In this problem we are asked to graphically compare the root-mean-square speeds of methane, ethane, and propane and to predict the root mean square speed of pentane and choose the component of natural gas that effuses the fastest.

Analyze

The root-mean-square speed can be calculated knowing the molar mass determined from the formula of each hydrocarbon and given that $T = 298$ K and $R = 8.314 \text{ kg·m}^2\text{·s}^{-2}\text{·mol}^{-1}\text{·K}^{-1}$.

$$u_{rms} = \sqrt{\frac{3RT}{\mathcal{M}}}$$

Solve

(a) The root-mean-square speeds for methane, ethane, and propane are:

$$\text{For } CH_4: \quad u_{rms} = \sqrt{\frac{3 \times \dfrac{8.314 \text{ kg·m}^2}{s^2 \cdot \text{mol·K}} \times 298 \text{ K}}{\left(\dfrac{16.04 \text{ g}}{\text{mol}} \times \dfrac{\text{kg}}{1000 \text{ g}}\right)}} = 681 \text{ m/s}$$

$$\text{For } C_2H_6: \quad u_{rms} = \sqrt{\frac{3 \times \dfrac{8.314 \text{ kg·m}^2}{s^2 \cdot \text{mol·K}} \times 298 \text{ K}}{\left(\dfrac{30.07 \text{ g}}{\text{mol}} \times \dfrac{\text{kg}}{1000 \text{ g}}\right)}} = 497 \text{ m/s}$$

$$\text{For } C_3H_8: \quad u_{rms} = \sqrt{\frac{3 \times \dfrac{8.314 \text{ kg·m}^2}{s^2 \cdot \text{mol·K}} \times 298 \text{ K}}{\left(\dfrac{44.10 \text{ g}}{\text{mol}} \times \dfrac{\text{kg}}{1000 \text{ g}}\right)}} = 411 \text{ m/s}$$

(b) Plotting the root-mean-square speeds of methane, ethane, and propane gives:

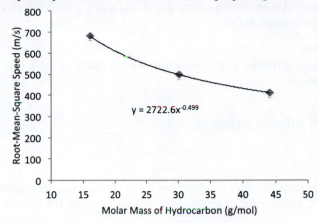

From the equation for the line we can estimate the root-mean-square speed for butane and pentane:

For C_4H_{10} with molar mass of 58 g/mol $y = 2722.6(58^{-0.499}) = 359$ m/s

For C_5H_{12} with molar mass of 84 g/mol $y = 2722.6(84^{-0.499}) = 298$ m/s

(c) The hydrocarbon with the lowest molar mass will effuse the fastest. Therefore, methane in natural gas will effuse the fastest.

Think About It
Using the formula to calculate the root-mean-square speed for butane and pentane in part b, we get values of 358 m/s and 321 m/s, respectively. How close were your estimated values?

10.31. Collect and Organize
For a sample of argon gas we are to calculate the root-mean-square speed of the atoms given that the average kinetic energy of the atoms in the sample is 5.18 kJ/mol.

Analyze
The average kinetic energy is given by

$$KE_{avg} = \frac{1}{2}mu_{rms}^2$$

The average kinetic energy is given by

$$KE_{avg} = \frac{1}{2}mu_{rms}^2$$

Rearranging that, we can calculate the root-mean-square speed:

$$u_{rms}^2 = \frac{2 \times KE_{avg}}{m}$$

$$u_{rms} = \sqrt{\frac{2 \times KE_{avg}}{m}}$$

where m is the mass of 1 mol of Ar atoms (39.948 g/mol) and 1 kJ = 1000 J, and J $= kg \cdot m^2/s^2$.

Solve

$$u_{rms} = \sqrt{\frac{2 \times KE_{avg}}{m}} = \sqrt{\frac{2 \times \left(\dfrac{5.18 \text{ kJ}}{\text{mol}} \times \dfrac{1000 \text{ J}}{\text{kJ}} \times \dfrac{kg \cdot m^2 \cdot s^2}{J}\right)}{\left(\dfrac{39.95 \text{ g}}{\text{mol}} \times \dfrac{kg}{1000 \text{ g}}\right)}} = 509 \text{ m/s}$$

Think About It

Writing the units associated with each quantity in a formula is a good habit so that you can easily see that you have converted the units appropriately.

10.33. **Collect and Organize**

By comparing the rates of diffusion of the two gases, we can calculate the molar mass of HX, which will enable us to identify the element X.

Analyze

The ratio of the rates of diffusion is given by

$$\frac{r_{NH_3}}{r_{HX}} = \sqrt{\frac{\mathcal{M}_{HX}}{\mathcal{M}_{NH_3}}}$$

The rate of diffusion of each gas is defined as how far in the tube the gas travels before it reacts with the other gas to form solid NH_4X. In this experiment, NH_3 travels 68.5 cm and HX travels $100 - 68.5$ cm $= 31.5$ cm.

Solve

Because distance is directly proportional to rate of travel, we can use the ratio of distances in place of the ratio of rates in Graham's law:

$$\frac{r_{NH_3}}{r_{HX}} = \frac{68.5 \text{ cm}}{31.5 \text{ cm}} = \sqrt{\frac{\mathcal{M}_{HX}}{17.03 \text{ g/mol}}} = 2.17$$

$$\mathcal{M}_{HX} = (2.17)^2 \times 17.03 \text{ g/mol} = 80.5 \text{ g/mol}$$

The molar mass of X will be $\mathcal{M}_{HX} - \mathcal{M}_H = 80.5 - 1.00 = 79.5$ g/mol. The element that forms HX (a halogen, in group 17 of the periodic table) and has a molar mass close to 79.5 is bromine, Br.

Think About It

Despite the simple nature of the experiment, the molar mass of an unknown gaseous compound can be fairly accurately determined.

10.35. **Collect and Organize**

Force and pressure are often used interchangeably in physics describing speed, but what is their difference in meaning in the context of gases?

Analyze

The mathematical equations describing force and pressure are

$$F = ma$$
$$P = F/A$$

Solve

Force is the product of the mass of an object and the acceleration due to gravity. Pressure uses force in its definition: it is the force an object exerts over a given area.

Think About It

A large force over a small area results in high pressure, but that same large force distributed over a very large area results in low pressure.

10.37. **Collect and Organize**

Differences in density of the liquid determine how high the liquid climbs in the tube of a Torricelli barometer. We are asked which liquid (water, ethanol, or mercury) leads to the tallest column in the barometer.

Analyze

Because the force of gravity on the liquid in the column of a barometer opposes the atmospheric pressure forcing the liquid up the column, the less dense the liquid, the less the liquid is affected by gravity, so atmospheric pressure raises the liquid higher in the column.

Solve

Among ethanol, water, and mercury liquids, ethanol has the lowest density, so it has the tallest column of liquid in a barometer.

Think About It

Mercury, having the highest density of the liquids, has the shortest column in a barometer.

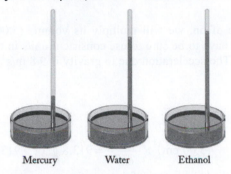

Mercury   Water   Ethanol

10.39. Collect and Organize

We are asked to explain the difference in pressure that an ice skater places on ice with a dull versus a sharp skate blade.

Analyze

Pressure is force over a unit area. The force of the ice skater on the ice is due to the skater's mass. The mass stays constant for that comparison. What must be different is the area between the skate blade and the ice.

Solve

A sharpened blade has a smaller area over which the force is distributed than that of a dull blade. Because $P = F/A$, as area (A) decreases, pressure must increase since the force due to the skater's mass is constant.

Think About It

The pressure also increases if we increase the mass of the ice skater.

10.41. Collect and Organize

We are to convert pressures expressed in kilopascals and millimeters of mercury to atmospheres.

Analyze

We need the following conversion factors:
$$1 \text{ atm} = 101.325 \text{ kPa} = 101,325 \text{ Pa}$$
$$1 \text{ atm} = 760 \text{ mmHg}$$

Solve

(a) $2.0 \text{ kPa} \times \dfrac{1 \text{ atm}}{101.325 \text{ kPa}} = 0.020 \text{ atm}$

(b) $562 \text{ mmHg} \times \dfrac{1 \text{ atm}}{760 \text{ mmHg}} = 0.739 \text{ atm}$

Think About It

A pascal also is equivalent to 1 N/m^2. That makes sense because the newton (N) is a unit of force. That force divided by area (in square meters) gives the units of pressure.

10.43. Collect and Organize

The pressure due to gravity for a cube of tin 5.00 cm on a side is given by

$$P = \frac{F}{A} = \frac{ma}{A}$$

where m = mass of the tin cube, a = acceleration due to gravity, and A = area over which the force is distributed.

Analyze

To find the mass of the cube of tin, we will multiply its volume $(5.00 \text{ cm})^3$ by the density of tin given in Appendix 3 (7.31 g/cm^3). We have to be sure to use consistent units in that calculation of the pressure exerted by the bottom face of the tin. The acceleration due to gravity is 9.8 m/s^2, so we want to express the area of the cube face in square meters.

Solve

The mass of the tin cube is

$$(5.00 \text{ cm})^3 \times \frac{7.31 \text{ g}}{\text{cm}^3} = 913.8 \text{ g, or } 0.9138 \text{ kg}$$

The pressure due to gravity of the bottom face of the cube is

$$P = \frac{0.9138 \text{ kg} \times 9.80665 \text{ m/s}^2}{(5.00 \text{ cm})^2 \times \left(\frac{1 \text{ m}}{100 \text{ cm}}\right)^2} = 3.58 \times 10^3 \ \frac{\text{kg}}{\text{m} \cdot \text{s}^2} = 3.58 \times 10^3 \text{ Pa}$$

Think About It

We can also express that pressure in atmospheres:

$$3.58 \times 10^3 \text{ Pa} \times \frac{1 \text{ atm}}{1.01325 \times 10^5 \text{ Pa}} = 3.54 \times 10^{-2} \text{ atm}$$

10.45. Collect and Organize

We are to express the highest recorded atmospheric pressure of 108.6 kPa in millimeters of mercury, atmospheres, and millibars.

Analyze

We need the following conversion factors:

1 atm = 101.325 kPa = 101,325 Pa

1 atm = 760 mmHg

1 kPa = 10 mbar

Solve

(a) $108.6 \text{ kPa} \times \dfrac{1 \text{ atm}}{101.325 \text{ kPa}} \times \dfrac{760 \text{ mmHg}}{1 \text{ atm}} = 814.6 \text{ mmHg}$

(b) $108.6 \text{ kPa} \times \dfrac{1 \text{ atm}}{101.325 \text{ kPa}} = 1.072 \text{ atm}$

(c) $108.6 \text{ kPa} \times \dfrac{10 \text{ mbar}}{1 \text{ kPa}} = 1086 \text{ mbar}$

Think About It

The highest recorded atmospheric pressure is just a little above 1 atm at 1.072 atm.

10.47. Collect and Organize

Amontons's law states that pressure and temperature are directly related: as temperature increases, so does pressure. We are asked to interpret the relationship from the kinetic molecular theory perspective.

Analyze

Pressure originates from the collision of gas particles with the walls of a container. The more collisions between gas particles and the container, the greater the pressure. Temperature is a measure of how fast molecules are moving in the gas.

Solve

The higher the temperature, the faster the gas particles move. The faster they move, the more often they collide with the walls of the container and the greater the force with which gas particles hit the walls. Both of those effects result in increased pressure as temperature is raised.

$$\frac{P}{T} = \text{constant}$$

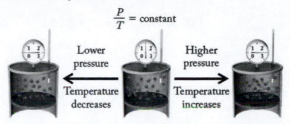

Think About It

At absolute zero, no molecular motion, and thus no pressure, would exist.

10.49. **Collect and Organize**

A balloonist needs to decrease her rate of ascent. How should she change the temperature of the gas in the balloon?

Analyze

A balloon rises because the air inside the balloon is less dense than the air outside. Therefore, fewer air molecules are inside the balloon than in an identical volume outside the balloon. Warmer gas molecules move faster and with more force, which results in a higher pressure, allowing them to escape from the bottom of the balloon.

Solve

To allow more air molecules in the balloon so as to increase the mass and therefore the density of air inside the balloon, the balloonist should decrease the temperature. Fewer "hot" gas molecules will escape, and the open design of the balloon will allow an increase in amount of cooler gas molecules that enter. That will slow the ascent of the balloon.

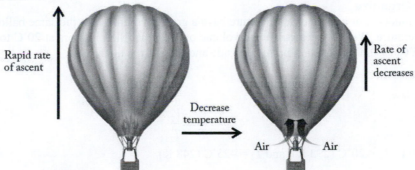

Think About It

You will see that effect if you ever have a chance to ride in a hot-air balloon. To avoid going too high, the balloonist will stop firing the burners.

10.51. Collect and Organize
Using Boyle's law, find the pressure of a gas when 1.00 mol at 1.00 atm in 3.25 L is compressed to 2.24 L.

Analyze
The equation for Boyle's law for the compression (or expansion) of a gas is

$$P_1V_1 = P_2V_2$$

For that gas compression, neither the number of moles of gas nor the temperature changes, so we do not need them in the calculation.

Solve

$$1.00 \text{ atm} \times 3.25 \text{ L} = P_2 \times 2.24 \text{ L}$$

$$P_2 = 1.45 \text{ atm}$$

Think About It
Because the volume was reduced, the pressure increased proportionately.

10.53. Collect and Organize
A 4.66 L sample of gas is heated from 273 K to 398 K without changing the pressure, and we are asked to predict the final volume of the gas. We would expect that the final volume would be greater than the initial volume according to Charles's law.

Analyze
To find V_2, use Charles's law,

$$\frac{V_1}{T_1} = \frac{V_2}{T_2}$$

where $V_1 = 4.66$ L, $T_1 = 273$ K, and $T_2 = 398$ K.

Solve

$$\frac{4.66 \text{ L}}{273 \text{ K}} = \frac{V_2}{398 \text{ K}}$$

$$V_2 = 6.79 \text{ L}$$

Think About It
As expected, the gas did expand upon heating.

10.55. Collect and Organize
Balloons filled indoors at a warmer temperature have a greater volume than the same balloons placed outside at a colder temperature. We are to calculate the volume of balloons filled indoors at 20°C to 5.0 L and then hung outside at –25°C, assuming constant pressure inside and outside the house.

Analyze
Use Charles's law,

$$\frac{V_1}{T_1} = \frac{V_2}{T_2}$$

where $V_1 = 5.0$ L, $T_1 = 20°C$ (293 K), and $T_2 = -25°C$ (248 K).

Solve

$$\frac{5.0 \text{ L}}{293 \text{ K}} = \frac{V_2}{248 \text{ K}}$$

$$V_2 = 4.2 \text{ L}$$

Think About It
The volume of the balloons decreased, as would be expected when they are placed outside in the colder temperature.

10.57. Collect and Organize

We are asked to find the pressure of an underwater site, knowing how much expansion a balloon undergoes when rising from the site to the surface. From that pressure and the fact that for every 10 m change in depth the pressure increases by 1.0 atm, we can calculate the diver's depth.

Analyze

We can use Boyle's law ($P_1V_1 = P_2V_2$) to find P_1. Once we know the pressure we can multiply it by 10 m/1 atm to arrive at the depth of the diver.

Solve

(a and b) The pressure at the site of the diver's work is

$$P_1 \times 115 \text{ L} = 1.00 \text{ atm} \times 352 \text{ L}$$

$$P_1 = 3.06 \text{ atm}$$

The pressure includes atmospheric pressure (1.00 atm), so

$$P_{additional} = 3.06 \text{ atm} - 1.00 \text{ atm} = 2.06 \text{ atm}$$

$$2.06 \text{ atm} \times \frac{10 \text{ m}}{1.0 \text{ atm}} = 21 \text{ m}$$

Think About It

If the diver had been working deeper, the balloon would have undergone a greater expansion upon reaching the surface.

10.59. Collect and Organize

We compare the effect of decreasing pressure on a gas sample (from 760 to 700 mmHg) to the effect of raising the temperature (from 10°C to 35°C) on the volume of a gas sample.

Analyze

Boyle's law describes the effect of decreasing pressure on the volume where P/V = constant. If we decrease the pressure by 50%, we have to increase the volume by 50%. Charles's law shows the direct relationship between temperature and volume. Here $\%\Delta V = \%\Delta T$.

Solve

(a) Lowering the pressure from 760 mmHg to 700 mmHg gives a percent change in V of

$$\frac{760 \text{ mmHg} - 700 \text{ mmHg}}{760 \text{ mmHg}} \times 100 = 7.9\% \text{ change in volume}$$

That will give a 7.9% increase in volume.

(b) The percent change in temperature is

$$\frac{308 \text{ K} - 273 \text{ K}}{273 \text{ K}} \times 100 = 12.8\%$$

That will give a 12.8% increase in volume.

Therefore, an increase in the temperature from 10°C to 35°C (b) gives the greatest increase in volume.

Think About It

Remember to convert temperatures to the Kelvin scale. If we had not converted to the Kelvin scale, it would have appeared that the volume in part b increased by 75%.

10.61. Collect and Organize

For various changes in temperature and external pressure, we are to predict how the volume of a gas sample changes.

Analyze

We must keep in mind the direct proportionality between V and T, as shown by Charles's law, and the inverse proportionality between V and P, as shown by Boyle's law.

Solve

(a) When the absolute temperature doubles, the volume doubles. When the pressure is doubled, the volume is halved. Combining those yields no change in the volume of the gas.

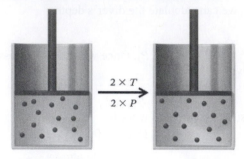

(b) When the absolute temperature is halved, the volume is halved. When the pressure is doubled, the volume is halved. Combining those gives a decrease in volume of gas sample to ¼ the original volume.

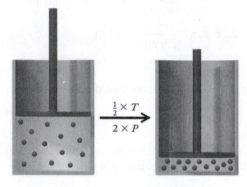

(c) Combining those effects is best looked at mathematically:

$$V_2 = \frac{PV_1 \times 1.75T_1}{T_1 \times 1.50P_1} = 1.17V_1, \text{ or an increase of } 17\%$$

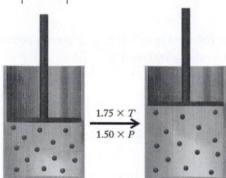

Think About It

The way to simplify this problem is to consider each change separately and then "add" the effects.

10.63. Collect and Organize

We are to calculate the volume of a weather balloon after 24 h, during which helium leaks out.

Analyze

Because volume and the amount of gas are directly proportional, the appropriate form of Avogadro's law is

$$\frac{V_1}{n_1} = \frac{V_2}{n_2}$$

We are given $V_1 = 150.0$ L and $n_1 = 6.1$ mol of He and told that the helium leaks from the balloon at the rate of 10 mmol/hr for 24 h.

Solve

The amount of gas leaked from the balloon is

$$24 \text{ h} \times \frac{10 \text{ mmol}}{1 \text{ h}} = 240 \text{ mmol} = 0.24 \text{ mol}$$

The amount of He in the balloon after 24 h is

$$6.1 \text{ mol} - 0.24 \text{ mol} = 5.86 \text{ mol}$$

The volume (V_2) of the balloon is

$$\frac{150.0 \text{ L}}{6.1 \text{ mol}} = \frac{V_2}{5.86 \text{ mol}}$$
$$V_2 = 144 \text{ L}$$

Think About It

As expected, fewer moles of gas in the balloon takes up less volume.

10.65. Collect and Organize

When the bicycle tire cools, we would expect the pressure to decrease according to Amontons's law.

Analyze

Amontons's law states that pressure and temperature are directly proportional:

$$\frac{P_1}{T_1} = \frac{P_2}{T_2}$$

We are given $P_1 = 7.1$ atm. We have to be sure to express $T_1 = 27°C$ and $T_2 = 5.0°C$ as absolute temperatures. In this problem we will assume that 7.1 atm is the actual pressure in the tire and not a guage pressure.

Solve

$$\frac{7.1 \text{ atm}}{300 \text{ K}} = \frac{P_2}{278.2 \text{ K}}$$
$$P_2 = 6.6 \text{ atm}$$

Think About It

The pressure of air in the tire did decrease as we expected when the temperature decreased.

10.67. Collect and Organize

Define standard temperature and pressure conditions and define the volume of 1 mol of an ideal gas at those conditions.

Analyze

We can use the ideal gas law, $PV = nRT$, to calculate the volume (V) of 1 mol of gas (n) at standard temperature (T) and pressure (P).

Solve

STP is defined as 1 atm and 0°C (273 K). The volume of 1 mol of gas at STP is

$$1 \text{ atm} \times V = 1.00 \text{ mol} \times 0.08206 \frac{\text{L} \cdot \text{atm}}{\text{mol} \cdot \text{K}} \times 273 \text{ K}$$
$$V = 22.4 \text{ L}$$

Think About It

The molar volume of gas at STP does not vary much depending on the identity of the gas.

10.69. Collect and Organize

Use the ideal gas law to determine the moles of air present in a bicycle tire with a volume of 2.36 L at 6.8 atm and 17.0°C.

Analyze

The ideal gas law is

$$PV = nRT$$

Rearranging to solve for n, number of moles of air, gives

$$n = \frac{PV}{RT}$$

With R in units of L·atm/mol·K, V must be in liters, pressure in atmospheres, and temperature in kelvins.

Solve

$$n = \frac{6.8 \text{ atm} \times 2.36 \text{ L}}{0.08206 \text{ L} \cdot \text{atm/mol} \cdot \text{K} \times 290 \text{ K}} = 0.67 \text{ mol}$$

Think About It

If we were to pump more moles of gas into the bicycle tire, the pressure would increase.

10.71. Collect and Organize

Given the volume of a hyperbaric chamber, the mass of oxygen in the chamber, and the temperature in the chamber, we can use the ideal gas law to calculate the pressure inside the chamber.

Analyze

Rearranging the ideal gas equation to solve for pressure, we get

$$P = \frac{nRT}{V}$$

We first have to determine the moles (n) of O_2 present in 15.00 kg of O_2.

Solve

$$\text{moles } (n) \text{ } O_2 \text{ present} = 15{,}000.00 \text{ g } O_2 \times \frac{1 \text{ mol } O_2}{31.998 \text{ g}} = 468.8 \text{ mol}$$

The pressure in the chamber due to oxygen is

$$P = \frac{468.8 \text{ mol} \times 0.08206 \text{ L} \cdot \text{atm/mol} \cdot \text{K} \times 298 \text{ K}}{4.85 \times 10^3 \text{ L}} = 2.36 \text{ atm}$$

Think About It

Hyper comes from the Greek meaning *over*. Thus, a hyperbaric chamber will have an *overpressure* of oxygen. Our answer therefore makes sense because the normal pressure of oxygen in the air is about 0.21 atm.

10.73. Collect and Organize

We are to calculate the mass of methane contained in a 250 L tank at a pressure of 255 bar and a temperature of 20°C.

Analyze

We can use the ideal gas law for this problem, but we will have to be careful to make our units consistent. Rearranging the ideal gas equation to solve for moles of methane (which we can convert to mass later by using the molar mass of methane, 16.04 g/mol), we have

$$n = \frac{PV}{RT}$$

where P is 255 bar (to be converted to atmospheres with the conversion 1 atm = 1.01325 bar). V = 250 L, and T is 20°C (293 K).

Solve

The moles of gas present in the tank is

$$n = \frac{\left(255 \text{ bar} \times \dfrac{1 \text{ atm}}{1.01325 \text{ bar}}\right) \times 250 \text{ L}}{\dfrac{0.08206 \text{ L} \cdot \text{atm}}{\text{mol} \cdot \text{K}} \times 293 \text{ K}} = 2.62 \times 10^3 \text{ mol of methane}$$

The mass of methane in the tank is

$$2.62 \times 10^3 \text{ mol CH}_4 \times \frac{16.04 \text{ g}}{\text{mol}} = 4.19 \times 10^4 \text{ g, or 41.9 kg}$$

Think About It

Remember to always convert the temperature into kelvins when using the ideal gas law.

10.75. Collect and Organize

In this problem we compare the moles of oxygen a skier breathes in at the top of a ski run with that breathed in at the bottom of the ski run. We are given that the pressure and temperature at the top of the ski run are 713 mmHg and −5°C, and at the bottom of the ski run the pressure and temperature are 734 mmHg and 0°C.

Analyze

We can use the combined gas law to solve this problem for the ratio of moles of air breathed in by the skier at the bottom of the run to that at the top of the run.

$$\frac{P_1 V_1}{n_1 T_1} = \frac{P_2 V_2}{n_2 T_2}$$

Because the volume of the skier's lungs does not change, $V_1 = V_2$, the equation reduces to

$$\frac{P_1}{n_1 T_1} = \frac{P_2}{n_2 T_2}$$

Rearranging that equation to give the ratio of moles of air the skier breathes in gives

$$\frac{n_2}{n_1} = \frac{P_2 T_1}{P_1 T_2}$$

where $P_1 = 713$ mmHg, $T_1 = -5°C$ (268 K), $P_2 = 734$ mmHg, and $T_2 = 0°C$ (273 K). We can then express the ratio as a percentage.

Solve

$$\frac{n_2}{n_1} = \frac{734 \text{ mmHg} \times 268 \text{ K}}{713 \text{ mmHg} \times 273 \text{ K}} = 1.01, \text{ or 101\%}$$

The skier breathes in 1.1% more air at the bottom of the ski run than he breathes in at the top of the run.

Think About It

The question specifically asks about how many more moles of *oxygen* the skier breathes in. We, however, calculated the increase in the percentage of *air* that the skier takes in. The answer will be the same when expressed as a percentage because the composition of air (21% oxygen) would be expected to be the same at the bottom of the run as at the top of the run.

10.77. Collect and Organize

Using data for the volume as a function of the temperatures of a 4.0 g gas sample at 1.00 atm pressure, we are asked to determine the moles of gas in the sample and then identify the gas from its molar mass.

Analyze

The relationship between temperature and volume is described by Charles's law:

$$V \propto T, \text{ or } \frac{V}{T} = \text{constant, or } V = T \times \text{constant}$$

and a plot of volume versus temperature from the rearranged ideal gas law would be

$$V = \frac{nRT}{P} = \frac{nR}{P} \times T$$

where $x = T$ and $y = V$, so the slope of the line will be nR/P. From the slope of the line we can find the number of moles of the gas in the sample

$$\text{slope} = \frac{nR}{P}, \text{ or } n = \frac{\text{slope} \times P}{R}$$

where $P = 1.00$ atm.

Solve
(a) The plot of temperature versus volume for that gas sample gives a slope of 0.0821.

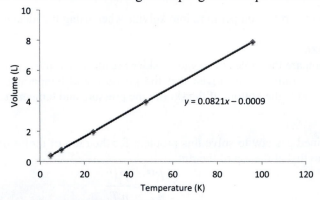

The number of moles of the gas in the sample then is

$$n = \frac{\dfrac{0.0821 \text{ L}}{\text{K}} \times 1.00 \text{ atm}}{\dfrac{0.08206 \text{ L} \cdot \text{atm}}{\text{mol} \cdot \text{K}}} = 1.00 \text{ mol}$$

(b) The molar mass of the gas is

$$\frac{4.0 \text{ g}}{1.00 \text{ mol}} = 4.0 \text{ g/mol}$$

The gas is helium.

Think About It
The graph does indeed show the direct linear relationship between the temperature and volume, as predicted by Charles's law. We see clearly on the graph that as temperature increases, the volume increases.

10.79. Collect and Organize
The density of a gas is defined as its mass per unit volume. If gases are at the same T and P, must they have the same density?

Analyze
The density of gas is derived from the ideal gas equation,

$$PV = nRT$$

where "molar density" would be given by

$$\frac{n}{V} = \frac{P}{RT}$$

The number of moles of gas n is related to its mass by

$$n = \frac{m}{\mathcal{M}}$$

where $\mathcal{M}$ = molar mass.

Solve

Substituting $n = m/\mathcal{M}$ into the molar density equation and rearranging to give $m/V = d$ gives

$$\frac{m/\mathcal{M}}{V} = \frac{P}{RT}$$

$$\frac{m}{V} = \frac{P\mathcal{M}}{RT} = d$$

Because each gas has a different molar mass, the densities of different gases are not necessarily the same for a particular temperature and pressure.

Think About It

From the equation above for density, we can see that as the molar mass of the gas increases, the density increases.

10.81. Collect and Organize

Density changes as a function of pressure and temperature. We are to predict how.

Analyze

The equation describing the density of gas is

$$d = \frac{\mathcal{M}P}{RT}$$

Solve

From the equation we see that density is directly proportional to pressure and inversely proportional to temperature. Therefore, density (a) increases with increasing pressure and (b) increases with decreasing temperature.

Think About It

Notice from the equation that density also increases as the molar mass of the gas increases.

10.83. Collect and Organize

We use the equation for the density of a gas to calculate the density of radon. Will radon concentrations be higher in the basement or on the top floor of a building?

Analyze

The density of the radon can be calculated from

$$d = \frac{\mathcal{M}P}{RT}$$

where $\mathcal{M} = 222$ g/mol, $P = 1$ atm, and $T = 298$ K. We can compare the radon density with air's density (1.2 g/L).

Solve

(a) $d = \dfrac{222 \text{ g/mol} \times 1 \text{ atm}}{0.08206 \text{ L} \cdot \text{atm/mol} \cdot \text{K} \times 298 \text{ K}} = 9.08$ g/L

(b) The density of radon is greater than the density of air (1.2 g/L), so radon is more likely to be concentrated in the basement, particularly if the source of radon is underground.

Think About It

Radon testing kits all measure radon levels in basements of homes and buildings because that is where radon is most concentrated.

10.85. **Collect and Organize**
Given the mass of an oxide of sulfur gas (0.078 g) in a known volume (30.0 mL) measured at a particular temperature (22°C) and pressure (750 mmHg), we are to determine the identity as SO_2 or SO_3.

Analyze
First we must calculate the density of the gas given the mass and the volume of the sample, and then we can use the density to calculate the molar mass of the gas to identify it as either SO_2 or SO_3. Pertinent equations are

$$d = \frac{m}{V} = \frac{\mathcal{M}P}{RT}$$

$$\mathcal{M} = \frac{dRT}{P}$$

We have to be sure to use units of volume in liters, pressure in atmospheres, and temperature in kelvins.

Solve
(a) The density of the gas is

$$d = \frac{m}{V} = \frac{0.078 \text{ g}}{30.0 \text{ mL} \times \dfrac{1 \text{ L}}{1000 \text{ mL}}} = 2.60 \text{ g/L}$$

(b) The molar mass of the gas is

$$\mathcal{M} = \frac{2.60 \text{ g/L} \times 0.08206 \text{ L} \cdot \text{atm/mol} \cdot \text{K} \times 295 \text{ K}}{\left(750 \text{ mmHg} \times \dfrac{1 \text{ atm}}{760 \text{ mmHg}} \right)} = 64 \text{ g/mol}$$

The molar mass of SO_2 is 64.1 g/mol and the molar mass of SO_3 is 80.1 g/mol. The gas in the flask, therefore, is SO_2.

Think About It
Remember that consistent units are very important when using any equation.

10.87. **Collect and Organize**
For a gas we are given the density (1.107 g/L) at a particular temperature (300 K) and pressure (740 mmHg). We are to determine by calculating the molar mass whether the gas could be CO or CO_2.

Analyze
Use the gas density equation to calculate $\mathcal{M}$:

$$d = \frac{m}{V} = \frac{\mathcal{M}P}{RT}$$

$$\mathcal{M} = \frac{mRT}{VP} = \frac{dRT}{P}$$

Solve
(a) The molar masses of the gas is

$$\mathcal{M} = \frac{1.107 \text{ g/L} \times 0.08206 \text{ L} \cdot \text{atm/mol} \cdot \text{K} \times 300 \text{ K}}{\left(740 \text{ mmHg} \times \dfrac{1 \text{ atm}}{760 \text{ mmHg}} \right)} = 28.0 \text{ g/mol}$$

(b) The molar masses of CO and CO_2 are 28.01 and 44.01 g/mol, respectively. The unknown gas could be CO.

Think About It
Be careful in this problem to convert millimeters of mercury to atmospheres for the pressure.

10.89. Collect and Organize

We are given the balanced chemical reaction for KO_2 to absorb CO_2 in a breathing device for producing O_2. We are to calculate the mass of KO_2 needed to produce 100.0 L of oxygen.

Analyze

First, we determine n_{O_2} required by using the ideal gas equation, where $P = 1.00$ atm, $V = 100.0$ L, and $T = 20°C$ (293 K). Then, using the stoichiometric relationship of KO_2 and O_2 in the balanced equation (4 KO_2:3O_2) and the molar mass of KO_2 (71.10 g/mol), we calculate the mass of KO_2 needed in the reaction.

Solve

$$n_{O_2} = \frac{1.00 \text{ atm} \times 100.0 \text{ L}}{0.08206 \text{ L} \cdot \text{atm/mol} \cdot \text{K} \times 293 \text{ K}} = 4.16 \text{ mol O}_2$$

Mass of KO_2 required:

$$4.16 \text{ mol O}_2 \times \frac{4 \text{ mol KO}_2}{3 \text{ mol O}_2} \times \frac{71.10 \text{ g KO}_2}{1 \text{ mol}} = 394 \text{ g KO}_2$$

Think About It

Because the ideal gas law allows us to compute the moles of gas, we can use the result in later stoichiometry calculations.

10.91. Collect and Organize

We are given the balanced chemical equation for the decomposition of $KClO_3$ to $O_2(g)$. We are asked to calculate the mass of $KClO_3$ needed to generate 200.0 L of $O_2(g)$ at 0.85 atm and 273 K.

Analyze

We first need to use the ideal gas law to calculate n_{O_2} generated, where $P = 0.85$ atm, $V = 200.0$ L, and $T = 273$ K. Because the stoichiometric ratio of O_2 produced to $KClO_3$ used is 3:2, we have to multiply n_{O_2} by $^2/_3$ to give n_{KClO_3}. The mass of $KClO_3$ then can be found by multiplying n_{KClO_3} by its molar mass (122.5 g/mol).

Solve

Moles of O_2 generated:

$$n_{O_2} = \frac{0.85 \text{ atm} \times 200.0 \text{ L}}{0.08206 \text{ L} \cdot \text{atm/mol} \cdot \text{K} \times 273 \text{ K}} = 7.6 \text{ mol O}_2$$

Mass of $KClO_3$ required:

$$7.6 \text{ mol O}_2 \times \frac{2 \text{ mol KClO}_3}{3 \text{ mol O}_2} \times \frac{122.5 \text{ g KClO}_3}{1 \text{ mol}} = 6.2 \times 10^2 \text{ g KClO}_3$$

Think About It

The amount of $KClO_3$ needed is large because the volume of oxygen is quite large.

10.93. Collect and Organize

We are given the balanced chemical reaction for removing CO_2 in air by reacting it with 2-aminoethanol. We are asked to calculate the volume of 4.0 M 2-aminoethanol needed on the submarine per sailor in 24 h given the rate of CO_2 exhaled by each sailor (125 mL/min) at 1.02 atm and 23°C.

Analyze

First we need to use the ideal gas equation to calculate moles of CO_2 exhaled by a sailor in 1 min, where $P = 1.02$ atm, $V = 125$ mL (0.125 L), and $T = 23°C$ (296 K). Then, we need to convert that into n_{CO_2} exhaled in 1 d (24 h) by using 60 min = 1 h. Finally, we can use the stoichiometric equation to calculate the moles of 2-aminoethanol required for each sailor, with which we can then find the volume of the solution required since we know the molarity of the 2-aminoethanol solution (4.0 M).

Solve

Moles of CO_2 exhaled in 1 min:

$$n_{CO_2} = \frac{1.02\ \text{atm} \times 0.125\ \text{L/min}}{0.08206\ \text{L} \cdot \text{atm/mol} \cdot \text{K} \times 296\ \text{K}} = 5.25 \times 10^{-3}\ \text{mol/min } CO_2$$

Moles of CO_2 exhaled in 1 d:

$$\frac{5.25 \times 10^{-3}\ \text{mol } CO_2}{1\ \text{min}} \times \frac{60\ \text{min}}{1\ \text{h}} \times 24\ \text{h} = 7.56\ \text{mol } CO_2$$

Volume of 2-aminoethanol required for 24-h period per sailor:

$$7.56\ \text{mol } CO_2 \times \frac{2\ \text{mol 2-aminoethanol}}{1\ \text{mol } CO_2} \times \frac{1\ \text{L 2-aminoethanol}}{4\ \text{mol}} = 3.78\ \text{L}$$

Think About It

Because the ideal gas law allows us to compute the moles of gas, we can use the result in later stoichiometry calculations.

10.95. **Collect, Organize, and Analyze**

We are asked to define the partial pressure of a gas.

Solve

The partial pressure of a gas is the pressure that a particular gas contributes to the total pressure.

Think About It

The sum of the partial pressures of all the gases in a mixture adds up to the total pressure.

10.97. **Collect and Organize**

The mole fraction of a gas is the ratio of moles of a gas divided by the total number of moles in the gas mixture.

Analyze

To get the mole fraction for each gas, we first sum the moles of all the gaseous components (0.70 mol of N_2 + 0.20 mol of H_2 + 0.10 mol of CH_4) to calculate the total moles of gas in the mixture. The individual mole fraction for H_2 is

$$x_{H_2} = \frac{n_{H_2}}{n_{total}}$$

where n_{H_2} is the number of moles of hydrogen and n_{total} is the sum of the moles of all components in the mixture.

Solve

Total moles of gas in the mixture equals $(0.70 + 0.20 + 0.10) = 1.00$ mol.

$$x_{H_2} = \frac{0.20\ \text{mol}}{1.00\ \text{mol}} = 0.20$$

Think About It

The mole fraction of N_2 would be 0.70 and the mole fraction of CH_4 would be 0.10. The sum of the mole fractions for all the components in the mixture adds up to 1.

10.99. **Collect and Organize**

The mole fractions of the gases in Problem 10.97 are $x(N_2) = 0.70$, $x(H_2) = 0.20$, and $x(CH_4) = 0.10$. We are to use that information to calculate the partial pressure of each gas and the total pressure of the gas mixture when the volume is 0.75 L at 10°C (283 K).

Analyze

The partial pressure of each gas is related to the total pressure: $P_x = x_x P_{total}$. We can calculate P_{total} from the ideal gas law, where $V = 0.75$ L, $n =$ total moles of gas in the mixture, and $T = 283$ K.

Solve

$$P_{total} = \frac{1.00 \text{ mol} \times 0.08206 \text{ L} \cdot \text{atm/mol} \cdot \text{K} \times 283 \text{ K}}{0.75 \text{ L}} = 31 \text{ atm}$$

$$P_{N_2} = 0.70 \times 30.96 \text{ atm} = 22 \text{ atm}$$

$$P_{H_2} = 0.20 \times 30.96 \text{ atm} = 6.2 \text{ atm}$$

$$P_{CH_4} = 0.10 \times 30.96 \text{ atm} = 3.1 \text{ atm}$$

Think About It

The sum of the partial pressures equals P_{total}.

10.101. Collect and Organize

Because the oxygen is collected over water, a portion of the volume includes gaseous water. We are given the volume of the $O_2(g)$ and $H_2O(g)$ mixture, the temperature, and the pressure. We are asked to calculate how many moles of O_2 are present in the mixture.

Analyze

At 25°C (298 K), the vapor pressure due to water is 23.8 mmHg (Table 10.4). The pressure of O_2 in the sample is

$$P_{O_2} = P_{total} - P_{H_2O} = 760 - 23.8 \text{ mmHg} = 736 \text{ mmHg}$$

Then, we can use the ideal gas equation ($PV = nRT$) to calculate the number of moles of oxygen gas where $P = 736$ mmHg (we have to convert that value to atmospheres), $V = 0.480$ L, and $T = 298$ K.

Solve

$$n_{O_2} = \frac{\left(736 \text{ mm} \times \dfrac{1 \text{ atm}}{760 \text{ mm}}\right) \times 0.480 \text{ L}}{0.08206 \text{ L} \cdot \text{atm/mol} \cdot \text{K} \times 298 \text{ K}} = 0.0190 \text{ mol } O_2$$

Think About It

Water vapor contributes little to the total pressure, so oxygen is the major gaseous component in the mixture.

10.103. Collect and Organize

We are given the balanced chemical equations for three reactions in which reactants and products are all gases. From the moles of gas consumed and the moles of gas produced, we are to determine how the pressure changes for each reaction.

Analyze

The greater the moles of gas in a sealed rigid container, the higher the pressure. If the reaction produces more moles of gas than it consumes, the pressure after the reaction is complete and it must be greater than the pressure before the reaction took place. However, if a reaction produces fewer moles of gas than it consumes, the pressure must be lower. If no change occurs in the moles of gas between reactants and products, the pressure does not change.

Solve

(a) Two moles of gas is consumed and 5 mol of gas is produced; $\Delta n = n_{products} - n_{reactants} = 3$, and the pressure is greater at the end of the reaction.

(b) Three moles of gas is consumed and 2 mol of gas is produced; $\Delta n = -1$, and the pressure is lower at the end of the reaction.

(c) Six moles of gas is consumed and 7 mol of gas is produced; $\Delta n = 1$, and the pressure is greater at the end of the reaction.

Think About It

Mathematically, if $\Delta n = n_{products} - n_{reactants}$ is positive, the pressure increases; if $\Delta n = 0$, the pressure stays the same; and if Δn is negative, the pressure decreases.

10.105. Collect and Organize

Using $PV = nRT$, we can calculate the moles of O_2 present in a human lung where the pressure is 266 mmHg with pure O_2 to compare with the oxygen content at sea level (760 mmHg) breathing in air, so as to determine how much more oxygen the high-altitude, pure-oxygen breath has.

Analyze

We will be comparing moles of oxygen, so rearranging the ideal gas equation to solve for n gives

$$n = \frac{PV}{RT}$$

At sea level, $P = 1.00$ atm and the mole fraction of oxygen in air is 0.2095. Therefore, the partial pressure of O_2 at sea level is $P_{O_2} = 1.00$ atm $\times$ 0.2095 = 0.2095 atm. At an altitude that has an atmospheric pressure of 266 mmHg, the climber is breathing 100% O_2 with $P = 266$ mmHg/760 mmHg = 0.35 atm. We are not given the values of V and T, so we can treat those as constants. As we are comparing n_{O_2} values in a ratio of $n_{O_2,8000m}/n_{O_2,\text{sea level}}$, we see that we do not need those values. In fact, we can even leave R, the gas constant, out of our final calculation.

Solve

$$n_{O_2,8000m} = \frac{0.35 \text{ atm} \times V}{RT} \qquad n_{O_2,\text{sea level}} = \frac{0.2095 \text{ atm} \times V}{RT}$$

The ratio $n_{O_2,8000m}/n_{O_2,\text{sea level}}$ is

$$\frac{\left(\dfrac{0.35 \text{ atm} \times V}{RT}\right)}{\left(\dfrac{0.2095 \text{ atm} \times V}{RT}\right)} = \frac{0.35 \text{ atm}}{0.2095 \text{ atm}} = 1.67$$

Therefore, the climber at high altitude breathing pure oxygen is breathing in 67% more oxygen than at sea level.

Think About It

Breathing pure O_2 at that great altitude more than compensates for the pressure difference. The climber actually has more oxygen in her lungs at the higher altitude than at sea level.

10.107. Collect and Organize

Given that methane has a fuel value of 55.5 kJ/g and that a water heater heats at 33,500 BTU/h (with 1 BTU = 1.055 kJ), we are asked to calculate the volume of methane used under STP conditions.

Analyze

To find the volume a gas occupies at STP (0°C, or 273 K, and 1 atm) we will use the ideal gas law. We will first, however, have to calculate the moles of gas used in 1 h by converting the BTUs consumed in 1 h to kilojoules and then calculating the mass (and then moles) of methane gas (16.01 g/mol) needed to obtain that amount of energy, using the conversion of 55.5 kJ/g.

Solve

The moles of methane required is

$$1 \text{ h} \times \frac{33,500 \text{ BTU}}{1 \text{ h}} \times \frac{1.055 \text{ kJ}}{\text{BTU}} \times \frac{1 \text{ g methane}}{55.5 \text{ kJ}} \times \frac{1 \text{ mol methane}}{16.04 \text{ g}} = 39.70 \text{ mol}$$

The volume that this methane occupies at STP is

$$V = \frac{39.70 \text{ mol} \times 0.08206 \dfrac{\text{L} \cdot \text{atm}}{\text{mol} \cdot \text{K}} \times 273\text{K}}{1 \text{ atm}} = 889 \text{ L}$$

Think About It

In answering this question, we had to first do a unit conversion and then apply the ideal gas equation. We could simplify solving this problem by recognizing that one liter of gas occupies 22.4 L at STP. From that we can multiply the moles of gas by 22.4 L/mol.

10.109. Collect and Organize

The balanced chemical equation for the production of ammonia from nitrogen and hydrogen is

$$N_2(g) + 3 H_2(g) \rightarrow 2 NH_3(g)$$

We are given the initial partial pressures of the reactants N_2 and H_2 (3.6 and 2.4 atm, respectively), and we are to calculate the percent decrease in the pressure when half the H_2 is used to produce NH_3.

Analyze

The initial pressure inside the vessel can be calculated by using the ideal gas equation, where n = total moles of H_2 and N_2. To find the pressure when half the N_2 is consumed, we first have to determine the moles of H_2, N_2, and NH_3 present in the vessel by using the reaction stoichiometry. Then we calculate the pressure from the total moles of gas in the vessel. Because we are looking for a percent decrease, we can express pressures with the unspecified "variables" R, T, and V.

Solve

The initial pressure inside the reaction vessel is all due to the presence of H_2 and N_2. So the total pressure at the start of the reaction is

$$(3.6 + 2.4) \text{ atm} = 6.0 \text{ atm}$$

Because the pressure of a gas is directly proportional to the number of moles of gas, we can think of the partial pressures like moles. So if half the hydrogen reacts, the new partial pressure of hydrogen will be

2.4 atm ÷ 2 = 1.2 atm of H_2 remaining after the reaction is complete.

Because we need only 1/3 of the moles of N_2 to react with the 1.2 atm of H_2 to form NH_3, the reaction will use 1.2 atm ÷ 3 = 0.4 atm of N_2, which will leave

3.6 atm − 0.4 = 3.2 atm of N_2 remaining after the reaction is complete.

Now we consider what partial pressure of NH_3 is produced. The stoichiometry of the reaction says that for every mole of N_2 used, 2 mol of NH_3 is produced. We have seen above that 0.4 atm of N_2 is used, which means that

$$0.4 \times 2 = 0.8 \text{ atm of } NH_3 \text{ is produced after the reaction is complete.}$$

So, the total pressure at the end of the reaction is

$$1.2 \text{ atm } H_2 + 3.2 \text{ atm } N_2 + 0.8 \text{ atm } NH_3 = 5.2 \text{ atm}$$

The difference between the final pressure and the initial pressure is

$$5.2 \text{ atm} - 6.0 \text{ atm} = -0.8 \text{ atm}$$

The percent decrease in pressure for this reaction is

$$\frac{0.8 \text{ atm}}{6.0 \text{ atm}} \times 100 = 13\%$$

Think About It

We observe a decrease in pressure as that reaction proceeds, which we expect because 4 mol of reactants produces 2 mol of products in the balanced chemical equation.

10.111. Collect, Organize, and Analyze

We are to explain why real gases deviate from ideal behavior at low temperatures and high pressures.

Solve

At low temperatures the gas particles move more slowly, and their collisions become inelastic; they stick together because of the weak attractive forces between them. The particles, therefore, do not act separately to contribute to the pressure in the container, and the pressure is lower than would be expected from the ideal gas law. Also, the gas particles take up real volume in the container and as the pressure increases, the volume of the particles takes up a greater volume of the free space in the container. That has the effect of raising the pressure–volume above what we would expect from the ideal gas law (in a plot of PV/RT vs. P).

Think About It

Gases at low pressures and not-too-cold temperatures behave most ideally.

10.113. Collect and Organize

The van der Waals constant b is associated with the volume of the gas particles. We are asked why the value of b increases as the atomic number of noble gas elements increases.

Analyze

As atomic number increases, the number of electrons and the size of the atoms increase.

Solve

Because b is a measure of the volume that the gas particles occupy, b increases as the sizes of the particles increase.

Think About It

Table 10.5 shows that the units of b are in liters per mole (L/mol). That is the volume that 1 mol of the gas particles occupies in the volume of the gas container. As the size increases from He to Ar, for example, b does indeed increase.

10.115. Collect and Organize

The van der Waals constant a is associated with the interactions between gas particles. We are asked why the value of a for Ar is greater than the value of a for He.

Analyze

As molar mass increases, the sizes of the atoms increase and have more electrons. Those electrons can have greater imbalances in their distributions around the atoms (they are more polarizable).

Solve

As the imbalance in electron distribution increases (polarizability) with greater molar mass, the weak interactions between gas particles become stronger and the value of a increases. Therefore, Ar, with 18 electrons, has a larger value of a than He, with only 4 electrons.

Think About It

Table 10.5 shows that the value of a does indeed increase as the molar mass increases in other related molecules. Compare, for example, H_2, N_2, and O_2.

10.117. Collect and Organize

Given that the plot in Figure 10.34 of PV/RT versus P for CH_4 is different from the plot for H_2 in describing how they deviate from ideal behavior, we are to consider for which gas the volume of the gas particles is more important than the attractive interactions between them at a pressure of 200 atm.

Analyze

In Figure 10.34 we see that the curve for H_2 always deviates above the ideal gas line, whereas the curve for CH_4 first deviates below the ideal gas line at lower pressures and then deviates above the line at higher pressures. A line diverging above the ideal gas line indicates that the volume of the real molecules in the gas sample is no longer approximately equal to zero, so that the free volume no longer is a good estimate of the total volume. A line that diverges below the ideal gas line indicates attractive forces between the molecules that cause them to stick together, which decreases the force of collisions with the wall of the container; therefore, the pressure decreases in relation to ideal behavior.

Solve

Figure 10.34 at $P = 200$ atm shows that CH_4 has attractive forces that cause it to deviate from ideal behavior at lower pressures but that for H_2, deviation is caused only by the real volume of the molecules. From that behavior we can say that for H_2 the effect of the volume occupied by the gas molecules is more important than the attractive forces between them.

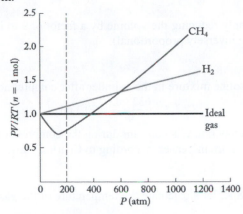

Think About It

In a plot of *PV/RT* versus *P*, we can see that CO_2 has an even larger attractive-forces effect than that of CH_4 because its negative deviation from ideal behavior (at pressures of 0–100 atm) is even greater than that for CH_4.

10.119. **Collect and Organize**

We are to calculate the pressure of 50.0 g of H_2 at 20°C (293 K) in a 1.00 L vessel by using the van der Waals equation for real gases and then using the ideal gas equation.

Analyze

The van der Waals equation is

$$\left(P + \frac{n^2 a}{V^2}\right)(V - nb) = nRT$$

where $a = 0.244$ L^2 · atm/mol^2 and $b = 0.0266$ L/mol for H_2. We first have to determine the number of moles of H_2 in 50.0 g by using the molar mass of H_2 (2.02 g/mol).

Solve

(a) Moles of H_2: 50.0 g × 1 mol/2.02 g = 24.75 mol H_2

Pressure calculation using the van der Waals equation:

$$\left(P + \frac{(24.75 \text{ mol})^2 \times 0.244 \text{ L}^2 \cdot \text{atm/mol}^2}{(1.00 \text{ L})^2}\right)(1.00 \text{ L} - 24.75 \text{ mol} \times 0.0266 \text{ L/mol})$$

$$= 24.75 \text{ mol} \times 0.08206 \text{ L} \cdot \text{atm/mol} \cdot \text{K} \times 293 \text{ K}$$

$$\left(P + 149.5 \text{ atm}\right) = 1741.8 \text{ atm}$$

$$P = 1590 \text{ atm}$$

(b) Pressure calculation using the ideal gas equation:

$$P = \frac{nRT}{V} = \frac{24.75 \text{ mol} \times 0.08206 \text{ L} \cdot \text{atm/mol} \cdot \text{K} \times 293 \text{ K}}{1.00 \text{ L}} = 595 \text{ atm}$$

Think About It

At those high pressures, hydrogen deviates a great deal from ideal behavior, as shown by the very different pressures calculated through the ideal gas equation and the van der Waals equation.

10.121. **Collect and Organize**

For a sports car engine cylinder with a volume of 633 mL in which compression achieves an increase in pressure of 9.0 times, we are asked to calculate the volume of the air and gasoline mixture after compression. We then are asked to evaluate the assumption that during compression no change in temperature occurs.

Analyze

In the compression we are simply reducing the volume by a factor of 9 to increase the pressure by a factor of 9 (since pressure and volume are inversely proportional).

Solve

(a) The volume of the air–gasoline mixture in the cylinder after compression is

$$\frac{633 \text{ mL}}{9} = 70.3 \text{ mL}$$

(b) The assumption that the temperature is constant during the compression is not realistic because when gases are compressed the temperature increases, according to Charles's law.

Think About It

If the initial temperature is 298 K, let's calculate the temperature of the gasoline–air mixture upon compression by using Charles's law:

$$\frac{V_1}{V_2} = \frac{T_1}{T_2}$$

$$\frac{633 \text{ mL}}{70.3 \text{ mL}} = \frac{298 \text{ K}}{T_2}, \text{ so } T_2 = 331 \text{ K}$$

10.123. **Collect and Organize**

We are to calculate u_{rms} for argon atoms at 0.00010 K.

Analyze

The u_{rms} is calculated through

$$u_{rms} = \sqrt{\frac{3RT}{\mathcal{M}}}$$

Solve

$$u_{rms} = \sqrt{\frac{3 \times 8.314 \text{ kg} \cdot \text{m}^2/\text{s}^2 (\text{mol} \cdot \text{K}) \times 0.00010 \text{ K}}{39.948 \text{ g/mol} \times \dfrac{1 \text{ kg}}{1000 \text{ g}}}} = 0.25 \text{ m/s}$$

Think About It

Helium atoms at the same temperature would have a greater root-mean-square speed (0.79 m/s) because they are lighter gas particles.

10.125. **Collect and Organize**

We can use the ideal gas equation to calculate the moles of air (and then the mass of air) that must be compressed into a scuba tank so that it will deliver 80 ft³ of air at 72°F and 1.00 atm pressure.

Analyze

To find the moles of air in 80 ft³ we use $PV = nRT$ rearranged to $n = PV/RT$, where $V = 80$ ft³ (which must be converted to liters by using 1 ft = 0.3048 m and 1 m³ = 1000 L), $P = 1.00$ atm, and $T = 72°F$ [which must be converted to kelvins by using °C = 5/9(°F – 32) and K = °C + 273.15]. Using the molar mass of air (which we can calculate from the mole fractions and the molar mass of each component), we can convert the moles of air into grams and then add to the weight of the 15 kg scuba tank.

Solve

Moles of air required:

$$n = \frac{1\ \text{atm} \times \left(80\ \text{ft}^3 \times \dfrac{(0.3048\ \text{m})^3}{1\ \text{ft}^3} \times \dfrac{1000\ \text{L}}{1\ \text{m}^3} \right)}{0.08206\ \text{L} \cdot \text{atm/mol} \cdot \text{K} \times \left(\dfrac{5}{9}(72 - 32) + 273.15 \right)} = 93.5\ \text{mol}$$

Molar mass of dry air (data from Table 10.1):

$$\mathcal{M}_{\text{air}} = (0.7808 \times 28.01\ \text{g/mol}) + (0.2095 \times 32.00\ \text{g/mol}) + (0.00934 \times 39.948\ \text{g/mol})$$

$$+ (0.00033 \times 44.01\ \text{g/mol}) + (0.000002 \times 16.04\ \text{g/mol}) + (0.0000005 \times 2.02\ \text{g/mol}) = 28.96\ \text{g/mol}$$

Mass of air in the tank:

$$93.5\ \text{mol} \times \frac{28.96\ \text{g}}{\text{mol}} = 2.71 \times 10^3\ \text{g, or 2.71 kg}$$

Total mass of the tank: 2.71 kg + 15 kg = 18 kg

Think About It

The air inside the tank does not add much to the mass. The tank, though, must be thick and heavy to withstand the 3000 psi of pressure needed for that amount of air.

10.127. **Collect and Organize**

We are to calculate the volume of air at 1.00 atm needed to vaporize 10.0 mL of halothane (d = 1.87 g/mL) at 20°C.

Analyze

From the volume of the liquid halothane and its density and molar mass ($C_2HBrClF_3$, 197.4 g/mol), we can calculate the moles of halothane present in the sample. Then, using $PV = nRT$, we can determine the volume that the gas occupies which, halothane being volatile, will be the volume of air needed to vaporize that 10.0 mL sample. Remember that we must convert temperature in this problem to kelvin.

Solve

Moles of halothane in 10.0 mL of liquid:

$$10.0\ \text{mL} \times \frac{1.87\ \text{g}}{\text{mL}} \times \frac{1\ \text{mol}}{197.4\ \text{g}} = 0.0947\ \text{mol}$$

Volume that 10.0 mL of liquid halothane occupies as a gas (and thus the air needed to vaporize it) at 20°C:

$$V = \frac{0.0947\ \text{mol} \times 0.08206\ \text{L} \cdot \text{atm/mol} \cdot \text{K} \times 293\ \text{K}}{1.00\ \text{atm}} = 2.28\ \text{L}$$

Think About It

Halothane, discovered in 1955, is the only inhalable anesthetic that contains bromine and is important in a basic health system.

10.129. **Collect and Organize**

We are asked to analyze the effect of molar mass on the position of the product ring formed by the reaction of either HCl or CH_3COOH with the three amines CH_3NH_2, $(CH_3)_2NH$, and $(CH_3)_3N$, which we can do by using Graham's law (refer to Figure P10.129).

Analyze

Graham's law of effusion may be used to approximately describe situations involving diffusion. It shows that the rate of diffusion (which is directly related to how far along the tube the gas travels) is inversely related to the molar mass of the gas (lighter gases travel faster).

$$\frac{r_x}{r_y} = \sqrt{\frac{\mathcal{M}_y}{\mathcal{M}_x}}$$

To answer that question, we need the molar masses of the acids and amine bases: HCl (36.46 g/mol), CH_3COOH (60.05 g/mol), CH_3NH_2 (31.06 g/mol), $(CH_3)_2NH$ (45.08 g/mol), and $(CH_3)_3N$ (59.11 g/mol).

Solve
(a) When the white ring of the product is seen exactly halfway between the two ends, it means that the gases diffused the same distance and therefore at the same rate. That would occur for gases that have nearly equal molar masses: CH_3COOH and $(CH_3)_3N$.

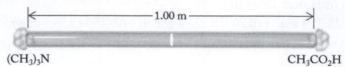

(b) The combination of gases that would produce the white ring closest to the amine end of the tube would be fast-diffusing HCl and slow-diffusing $(CH_3)_3N$.

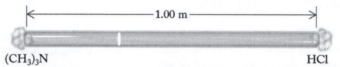

(c) Any combination that would give the product at the same location as another (within 1 cm) would have to have the same ratio of r_{acid}/r_{amine}. Those ratios from Graham's law are as follows:

Amine Base	Acid	r_{acid}/r_{amine}
CH_3NH_2	HCl	0.923
$(CH_3)_2NH$	HCl	1.11
$(CH_3)_3N$	HCl	1.27
CH_3NH_2	CH_3COOH	0.719
$(CH_3)_2NH$	CH_3COOH	0.866
$(CH_3)_3N$	CH_3COOH	0.992

To be within 1 cm of each other, the r_{acid}/r_{amine} ratio would have to be within 0.01. None of the combinations in that list would form the product at the same position as any other combination.

Think About It
That "crude" experiment can accurately identify the amine or the acid used.

10.131. **Collect and Organize**
We can calculate the partial pressure of H_2, He, and CH_4 in Uranus's atmosphere, knowing each gas's percentage by volume and the total pressure (130 kPa).

Analyze
The percent composition by volume gives us the mole fraction in percent because the volume is directly related to the number of moles of gas through the ideal gas equation. Therefore, the mole fractions of the gases are 0.83 H_2, 0.15 He, and 0.02 CH_4. The partial pressure of each gas can be calculated by using $P_X = x_X \times P_{total}$.

Solve

$$P_{H_2} = 0.83 \times 130 \text{ kPa} = 110 \text{ kPa}$$

$$P_{He} = 0.15 \times 130 \text{ kPa} = 20 \text{ kPa}$$

$$P_{CH_4} = 0.02 \times 130 \text{ kPa} = 3 \text{ kPa}$$

Think About It
The sum of the partial pressures adds up to the total pressure. That is a good way to check for any calculation errors.

10.133. **Collect and Organize**

We are given the balanced equation for ammonium nitrite decomposing to nitrogen and liquid water. We are asked to calculate the change in pressure in a 10.0 L vessel due to N_2 formation for 1.00 L of a 1.0 M NH_4NO_2 solution decomposing at 25°C.

Analyze

Because N_2 is the only gas involved in the reaction, all the pressure increase will be due to the production of N_2. From the stoichiometric equation we can calculate the moles of N_2 that will be produced (assuming 100% decomposition), and then we can use the ideal gas equation ($PV = nRT$) to calculate the pressure change.

Solve

$$\text{moles of } N_2 \text{ produced: } 1.00 \text{ L} \times \frac{1.0 \text{ mol } NH_4NO_2}{1 \text{ L}} \times \frac{1 \text{ mol } N_2}{1 \text{ mol } NH_4NO_2} = 1.0 \text{ mol}$$

$$P_{N_2} = \frac{1.0 \text{ mol} \times 0.08206 \text{ L} \cdot \text{atm/mol} \cdot \text{K} \times 298 \text{ K}}{9.0 \text{ L}} = 2.7 \text{ atm}$$

Think About It

Because the molar ratio of N_2 to NH_4NO_2 is 1:1, the moles of N_2 produced in the reaction equals the moles of NH_4NO_2 consumed.

10.135. **Collect and Organize**

From the balanced chemical equation and given the amount of nitrate that enters the swamp in one day (200.0 g), we are asked first to calculate the volume of N_2 and CO_2 that would form at 17°C (290 K) and 1.00 atm and then to calculate the density of the gas mixture at that temperature and pressure.

Analyze

Once we calculate the moles of NO_3^- (from the molar mass of $NO_3^- = 62.00$ g/mol), we can calculate the moles of N_2 and CO_2 produced by using the stoichiometric ratios from the balanced equation (2 NO_3^- : 1 N_2, 2 NO_3^- : 5 CO_2). Using the ideal gas equation, we can calculate the volume of each gas produced. We can then calculate the density of the gas mixture by determining the mass of CO_2 and N_2 produced (from the molar amounts already calculated) and dividing the total mass by the total volume the gases occupy.

Solve

(a) and (b) Volume of gases produced in the swamp:

$$\text{mol } NO_3^- = 200.0 \text{ g} \times \frac{1 \text{ mol}}{62.00 \text{ g}} = 3.226 \text{ mol}$$

$$\text{mol } N_2 \text{ produced} = 3.226 \text{ mol } NO_3^- \times \frac{1 \text{ mol } N_2}{2 \text{ mol } NO_3^-} = 1.613 \text{ mol } N_2$$

$$\text{volume of } N_2 = \frac{nRT}{P} = \frac{1.613 \text{ mol} \times 0.08206 \text{ L} \cdot \text{atm/mol} \cdot \text{K} \times 290 \text{ K}}{1.00 \text{ atm}} = 38.4 \text{ L}$$

$$\text{mol } CO_2 \text{ produced} = 3.226 \text{ mol } NO_3^- \times \frac{5 \text{ mol } CO_2}{2 \text{ mol } NO_3^-} = 8.065 \text{ mol } CO_2$$

$$\text{volume of } CO_2 = \frac{nRT}{P} = \frac{8.065 \text{ mol} \times 0.08206 \text{ L} \cdot \text{atm/mol} \cdot \text{K} \times 290 \text{ K}}{1.00 \text{ atm}} = 192 \text{ L}$$

Total volume of gases at 1.00 atm and 290 K: 38.4 L + 192 L = 230 L

(c) The density of this gas is as follows:

$$\text{mass of N}_2 \text{ produced} = 1.613 \text{ mol} \times \frac{28.01 \text{ g}}{1 \text{ mol}} = 45.18 \text{ g}$$

$$\text{mass of CO}_2 \text{ produced} = 8.065 \text{ mol} \times \frac{44.01 \text{ g}}{1 \text{ mol}} = 354.9 \text{ g}$$

$$\text{total mass of gases} = 45.18 \text{ g} + 354.9 \text{ g} = 400.1 \text{ g}$$

$$\text{density of gas mixture} = \frac{400.1 \text{ g}}{230 \text{ L}} = 1.74 \text{ g/L, or } 0.00174 \text{ g/mL}$$

Think About It

The calculation of density here drew on the simple definition of density as mass divided by volume.

10.137. Collect and Organize

From the balanced chemical equation describing the decomposition of sodium azide in the presence of silica and potassium nitrate, we are to calculate how much sodium azide we need to inflate an air bag at 20°C and compare that with how much we would need to inflate the same air bag to the same pressure on a colder day at 10°C.

Analyze

We first have to express the volume of the air bag ($40 \times 40 \times 20$ cm) in terms of liters by using the conversions 100 cm = 1 m and 1 m^3 = 1000 L. Then, we can apply the ideal gas equation to solve for the moles of N_2 gas needed, where $P = 1.25$ atm and $T = 293$ K. Once we know the moles of nitrogen required to fill the air bag, we can calculate the grams of sodium azide required by using the stoichiometric ratio of 32 mol N_2 : 20 mol NaN_3 and the molar mass of sodium azide (65.01 g/mol).

Solve

$$\text{volume of air bag in liters: } (40 \times 40 \times 20) \text{ cm}^3 \times \left(\frac{1 \text{ m}}{100 \text{ cm}}\right)^3 \times \frac{1000 \text{ L}}{1 \text{ m}^3} = 32 \text{ L}$$

$$n_{N_2} \text{ at } 20°C: n = \frac{PV}{RT} = \frac{1.25 \text{ atm} \times 32 \text{ L}}{0.08206 \dfrac{\text{L} \cdot \text{atm}}{\text{mol} \cdot \text{K}} \times 293 \text{ K}} = 1.66 \text{ mol}$$

$$\text{mass of NaN}_3 \text{ required: } 1.66 \text{ mol N}_2 \times \frac{20 \text{ mol NaN}_3}{32 \text{ mol N}_2} \times \frac{65.01 \text{ g NaN}_3}{1 \text{ mol}} = 67.6 \text{ g}$$

$$n_{N_2} \text{ at } 10°C: n = \frac{PV}{RT} = \frac{1.25 \text{ atm} \times 32 \text{ L}}{0.08206 \dfrac{\text{L} \cdot \text{atm}}{\text{mol} \cdot \text{K}} \times 283 \text{ K}} = 1.72 \text{ mol}$$

$$\text{mass of NaN}_3 \text{ required: } 1.72 \text{ mol N}_2 \times \frac{20 \text{ mol NaN}_3}{32 \text{ mol N}_2} \times \frac{65.01 \text{ g NaN}_3}{1 \text{ mol}} = 70.0 \text{ g}$$

Additional amount of NaN_3 to produce 1.25 atm at 10°C compared with 20°C: $70.0 - 67.6$ g = 2.40 g

Think About It

Manufacturers would have to make allowances for the pressure of the air bag at different ambient temperatures, but had to do so within the range of the strength of the air bag itself. We would need more NaN_3 to inflate the bag at lower temperatures, but we wouldn't want the bag to burst when deployed on a hot summer day.

10.139. **Collect and Organize**

Given that a sample of O_2 from the reaction between KO_2 and CO_2 is collected over water (with a vapor pressure of 24 torr) into a volume of 0.200 L at 25.0°C (298 K) at a pressure of 750 torr, we are to calculate the number of moles of oxygen produced in the reaction.

Analyze

First we can use the law of partial pressures to calculate the pressure in the collecting vessel due only to oxygen by subtracting the partial pressure of the water vapor from the atmospheric pressure. Then we can use the ideal gas law to solve for n the number of moles of O_2.

Solve

The pressure of oxygen in the collecting vessel is:

$$P_{O_2} = P_{atm} - P_{H_2O} = 750.0 \text{ torr} - 24.0 \text{ torr} = 726.0 \text{ torr}$$

The mol O_2 in the collecting vessel is:

$$n_{O_2} = \frac{PV}{RT} = \frac{\left(726.0 \text{ torr} \times \dfrac{1 \text{ atm}}{760 \text{ torr}}\right) \times 0.200 \text{ L}}{0.08206 \dfrac{\text{L} \cdot \text{atm}}{\text{mol} \cdot \text{K}} \times 298.16 \text{ K}} = 0.00781 \text{ mol}$$

Think About It

If we multiply this result by the molar mass of O_2 we would have produced 0.25 g of oxygen.

CHAPTER 11 | Properties of Solutions: Their Concentrations and Colligative Properties

11.1. Collect and Organize

From Figure P11.1, showing a sealed container with blue and red spheres in the solution and blue spheres in the gas phase, we are to determine which statements about the blue spheres (X) and the red spheres (Y) in the solution are true.

Analyze

No red spheres have entered the gas phase, and more X (36 blue spheres) is present in the solution phase than Y (20 red spheres). We will need to recall the definitions of *solvent*, *volatile*, and *vapor pressure* to determine the truth of each of the four statements.

Solve

(a) For X (blue spheres) to be the solvent in the solution as opposed to the solute, it would have to be present in a larger amount in the solution than Y (red spheres). More X particles than Y are present, so this statement is true.

(b) If pure Y (red spheres) were a volatile liquid, then we would expect at least a few Y to be present in the gas phase. Because none are, this statement is false.

(c) If Y spheres were absent from the solution (or in smaller amounts), then we would expect more, not less, X (blue spheres) in the gas phase above the solution because any dissolved solute lowers the vapor pressure of the solvent. Removing solute Y would increase the vapor pressure of X. This statement, therefore, is false.

(d) As in part c, if Y (red spheres) are present in the solution as a nonvolatile solute, then we expect less X (blue spheres) in the gas phase above the solution than when no Y was dissolved in the solution. That is because any dissolved solute lowers the vapor pressure of the solvent. This statement, therefore, is false.

Think About It

If Y were a volatile liquid, we would expect to see some Y in the gas phase above the solution.

11.3. Collect and Organize

From Figure P11.3, showing the particles $C_6H_{12}O_6$, NaCl, $MgCl_2$, and K_3PO_4 dissolved in water, we are to match the solution with its respective dissolved solute.

Analyze

The difference between those solutes is the number of particles that they dissociate into upon dissolution in water. Glucose ($C_6H_{12}O_6$) remains in solution as a molecule and does not dissociate, but all the other salts do break into anions and cations that are then solvated by water. NaCl dissociates into Na^+ and Cl^- ions, $MgCl_2$ dissociates into Mg^{2+} and two Cl^- ions, and K_3PO_4 dissociates into PO_4^{3-} and three K^+ ions. For the cations, the O atom of the water molecule will be oriented towards the ion; for the anions, the H atoms of the water molecules will be oriented toward the ion.

Solve

(a) In this solution, six blue spheres are present. Three have the water molecules with oxygen orientation for solvation and three have the water molecules oriented with hydrogen orientation. Therefore, this solute contains one cation and one anion and is thus NaCl.

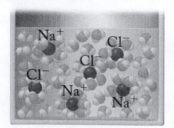

(b) In this solution, nine blue spheres are present. Three have the water molecules with oxygen orientation for solvation and six have the water molecules oriented with hydrogen orientation. Therefore, this solute contains one cation and two anions and is thus $MgCl_2$.

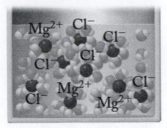

(c) In this solution, three blue spheres are present. All three have the water molecules with oxygen orientation for solvation. The absence of solvation where the hydrogen atoms are oriented toward a dissolved particle means that this solute did not dissociate and is thus $C_6H_{12}O_6$.

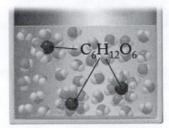

(d) In this solution, twelve blue spheres are present. Nine have the water molecules with oxygen orientation for solvation (for K^+) and three have the water molecules oriented with hydrogen orientation (for PO_4^{3-}). Therefore, this solute contains the ratio of three cations to one anion and is thus K_3PO_4.

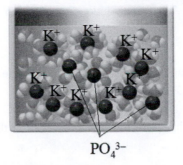

PO₄³⁻

Think About It
All three solutions represent the same molar solution of these four substances.

11.5. Collect and Organize
From the boiling points for X and Y that are read from the plot in Figure P11.5, we are to predict which substance has stronger intermolecular forces.

Analyze
The normal boiling point is defined as the temperature at which the vapor pressure of a substance is equal to 1.00 atm.

Solve
From Figure P11.5, we see that the boiling point for X is about 20°C and that the boiling point for Y is about 40°C at 1.00 atm. The substance with the stronger intermolecular forces has the higher boiling point, so Y has the stronger intermolecular forces.

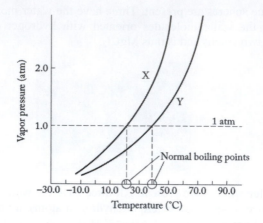

Think About It
More heat (energy) must be applied to "break" the stronger intermolecular forces in Y than in X.

11.7. Collect and Organize
From the direction of solvent flow across a semipermeable membrane indicated in Figure P11.7, we can deduce which solution (A or B) is more concentrated.

Analyze
In osmosis, solvent passes through the semipermeable membrane from low concentration of solute to high concentration of solute.

Solve
The flow of solvent in Figure P11.7 is from solution B to solution A. Therefore, solution A must be the more concentrated solution.,

Think About It
The solvent continues to flow through the membrane until the solute concentration on both sides is equal.

11.9. Collect and Organize
In considering Figure P11.9 for this problem, we are to choose the image that best describes how pressure affects the solubility of a gas.

Analyze
In the middle of the diagram the gas and the liquid contained in a cylinder are shown before the pressure is increased by moving the piston down in the cylinder. The middle diagram shows 12 molecules in the gas phase and 12 molecules in the liquid phase. As the pressure is increased, we expect some of the molecules in the gas phase (according to Henry's law) to become dissolved into the liquid phase to reduce the gas pressure in the cylinder.

Solve
Diagram (b) has more gas molecules (15) than molecules in the liquid phase (9), which shows that with increased pressure molecules evaporated from the solution. That is not correct. Diagram (a), however, has fewer molecules in the gas phase (9) and more molecules in the liquid phase (15). That is what we expect, so choice (a) is correct.

Think About It
Choice (b) would result in higher pressure in the cylinder, and we would therefore expect the piston to move up.

11.11. **Collect and Organize**
We are to define *semipermeable membrane*.

Analyze
A semipermeable membrane is a sheet-like molecular structure that acts as a boundary to allow some substances to pass but not others.

Solve
A semipermeable membrane is a boundary between two solutions through which some molecules may pass but others cannot. Usually, small molecules may pass, but large molecules are excluded.

Think About It
Semipermeable membranes are used in reverse osmosis processes and are present as the phospholipid bilayer in cell membranes. The thin film on the inside of an egg also is semipermeable.

11.13. **Collect and Organize**
We are to determine the direction of solvent flow when a semipermeable membrane separates a dilute solute from a more concentrated solution.

Analyze
Solvent flows through a semipermeable membrane from a region of more dilute solution to a region of more concentrated solution. In that way osmosis balances the concentration of solutes on both sides of the membrane.

Solve
Solvent flows across a semipermeable membrane from the more dilute solution side to the more concentrated solution side to balance the concentration of solutes on both sides of the membrane.

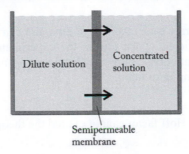

Semipermeable membrane

Think About It
The liquid level on the more dilute solution side of the membrane visibly decreases (whereas that on the more concentrated solution side increases) as osmosis proceeds in this experiment.

11.15. **Collect, Organize, and Analyze**
We are to define *reverse osmosis* and list the equipment needed to purify seawater by that process.

Solve
Reverse osmosis transfers solvent across a semipermeable membrane from a region of higher solute concentration to a region of lower solute concentration. Because reverse osmosis goes against the natural flow of solvent across the membrane, the key component needed is a pump to apply pressure to the more concentrated side of the membrane. Other components needed include a containment system, piping to introduce and remove the solutions, and a tough semipermeable membrane that can withstand the high pressures required.

Think About It
Seawater can be purified in large quantities in desalination plants by using reverse osmosis.

11.17. **Collect and Organize**

We are to explain why it is important to know whether a solute is a strong electrolyte when thinking about osmotic pressure of a solution.

Analyze

The osmotic pressure can be calculated from the equation

$$\Pi = iMRT$$

where i is the van't Hoff factor, M is the molarity of the solute in the solution, R is the gas constant ($0.08206 \text{ L} \cdot \text{atm} / \text{mol} \cdot \text{K}$), and T is the temperature (in kelvin).

Solve

A strong electrolyte completely dissociates in solution and therefore i may have a value greater than 1. If, for example, $i = 2$, as would be the case for NaCl, then the osmotic pressure would be twice that of when $i = 1$, as would be the case for glucose.

Think About It

A weak electrolyte might have values of $1 < i < 2$.

11.19. **Collect and Organize**

Measured van 't Hoff factors for electrolytes may differ from the theoretical values. We need to examine the behavior of ions in solution to explain why.

Analyze

In solution, ionic solutes are surrounded by solvent molecules but are also still attracted to other ionic solute molecules of opposite charge.

Solve

Solute ions of opposite charge attract each other in solution and can temporarily form ion pairs that reduce the total number of independent particles, which lowers the experimental value of the van 't Hoff factor. Ionic solutes, especially at high concentrations, show i values significantly less than their theoretical van 't Hoff factors because of strong ion pairing in solution.

Think About It

The measured experimental van 't Hoff factor cannot exceed the theoretical value.

11.21. **Collect and Organize**

Given the composition of two solutions separated by a semipermeable membrane, we are to determine in which direction the solvent flows.

Analyze

Solvent flows through a semipermeable membrane from a region of more dilute solution to a region of more concentrated solution. In this way, osmosis balances the concentration of solutes on both sides of the membrane. We need only to determine which side of the membrane (A or B) has the more concentrated solution, being sure to take into account the number of particles by calculating the theoretical value of i, the van 't Hoff factor. Solvent flows toward that side of the membrane.

Solve

(a) Side A has 1.25 M NaCl, which would be 2.50 M in particles ($i = 2$).
Side B has 1.50 M KCl, which would be 3.00 M in particles ($i = 2$).
Side B is more concentrated in solute. Therefore, solvent flows from side A to side B.

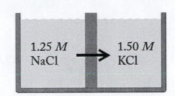

(b) Side A has 3.45 *M* CaCl$_2$, which would be 10.35 *M* in particles ($i = 3$).
Side B has 3.45 *M* NaBr, which would be 6.90 *M* in particles ($i = 2$).
Side A is more concentrated in solute. Therefore, solvent flows from side B to side A.

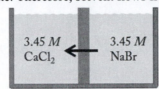

(c) Side A has 4.68 *M* glucose, which would be 4.68 *M* in particles ($i = 1$).
Side B has 3.00 *M* NaCl, which would be 6.00 *M* in particles ($i = 2$).
Side B is more concentrated in solute. Therefore, solvent flows from side A to side B.

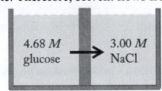

Think About It
Remember to compare the molarity of the dissolved particles, not just that of the solutions, when deciding the direction of flow across a semipermeable membrane in osmosis.

11.23. Collect and Organize
For the solutions described, we can calculate the osmotic pressure by using the formula $\Pi = iMRT$.

Analyze
For each solution we have to calculate the molarity of the solution (if not already given) and determine the theoretical value of *i* (the van 't Hoff factor) for the solute. The temperature for all solutions is $(20 + 273.15) = 293$ K.

Solve
(a) The value of *i* for 2.39 *M* CH$_3$OH is 1.
The osmotic pressure is

$$\Pi = 1 \times 2.39 \ M \times \frac{0.0821 \ \text{L} \cdot \text{atm}}{\text{mol} \cdot \text{K}} \times 293 \ \text{K} = 57.5 \ \text{atm}$$

(b) The value of *i* for 9.45 m*M* MgCl$_2$ is 3.
The osmotic pressure is

$$\Pi = 3 \times \left(\frac{9.45 \ \text{mmol}}{\text{L}} \times \frac{1 \ \text{mol}}{1000 \ \text{mmol}} \right) \times \frac{0.0821 \ \text{L} \cdot \text{atm}}{\text{mol} \cdot \text{K}} \times 293 \ \text{K} = 0.682 \ \text{atm}$$

(c) The molarity of the glycerol (C$_3$H$_8$O$_3$) solution is

$$\frac{\left(40.0 \ \text{mL} \times \dfrac{1.265 \ \text{g}}{\text{mL}} \times \dfrac{1 \ \text{mol}}{92.09 \ \text{g}} \right)}{0.250 \ \text{L}} = 2.20 \ M$$

The value of *i* for glycerol is 1.
The osmotic pressure is

$$\varPi = 1 \times 2.20 \; M \times \frac{0.0821 \; \text{L} \cdot \text{atm}}{\text{mol} \cdot \text{K}} \times 293 \; \text{K} = 52.9 \; \text{atm}$$

(d) The molarity of the $CaCl_2$ solution is

$$\frac{\left(25.0 \; \text{g} \times \dfrac{1 \; \text{mol}}{110.98 \; \text{g}} \right)}{0.350 \; \text{L}} = 0.644 \; M$$

The value of *i* for $CaCl_2$ is 3.
The osmotic pressure is

$$\varPi = 3 \times 0.644 \; M \times \frac{0.0821 \; \text{L} \cdot \text{atm}}{\text{mol} \cdot \text{K}} \times 293 \; \text{K} = 46.5 \; \text{atm}$$

Think About It
From the equation for osmotic pressure, we see that the pressure will increase as the van 't Hoff factor, the molarity of the solution, and the temperature increase.

11.25. Collect and Organize
Given the osmotic pressure of solutions at 25°C, we are to determine the molarity of each.

Analyze
Rearranging the osmotic pressure equation to solve for molarity gives

$$M = \frac{\varPi}{iRT}$$

where both ethanol and aspirin have $i = 1$. We are given $i = 2.47$ for the $CaCl_2$ solution. The temperature in kelvin is $25 + 273.15 = 298$ K.

Solve
(a) $M = \dfrac{0.674 \; \text{atm}}{1 \times 0.0821 \; \text{L} \cdot \text{atm} \,/\, \text{mol} \cdot \text{K} \times 298 \; \text{K}} = 2.75 \times 10^{-2} \; M$ ethanol

(b) $M = \dfrac{0.0271 \; \text{atm}}{1 \times 0.0821 \; \text{L} \cdot \text{atm} \,/\, \text{mol} \cdot \text{K} \times 298 \; \text{K}} = 1.11 \times 10^{-3} \; M$ aspirin

(c) $M = \dfrac{0.605 \; \text{atm}}{2.47 \times 0.0821 \; \text{L} \cdot \text{atm} \,/\, \text{mol} \cdot \text{K} \times 298 \; \text{K}} = 1.00 \times 10^{-2} \; M \; CaCl_2$

Think About It
The higher the molarity of the (particles of the) solute, the greater the osmotic pressure.

11.27. Collect and Organize
Given the plot of the osmotic pressure of three electrolyte solutions of equal concentrations varying with temperature, we are asked to determine which curve represents the solution of the strong electrolyte.

Analyze
The osmotic pressure equation

$$\varPi = iMRT$$

shows us that the higher the value of *i*, the greater the osmotic pressure.

Solve
Line A (red) represents the solution of the strong electrolyte because it has the highest osmotic pressure at all temperatures.

Think About It

All solutions show higher osmotic pressure at elevated temperatures. That behavior is reflected in the osmotic pressure equation showing a direct proportionality between osmotic pressure (Π) and temperature (T).

11.29. **Collect and Organize**

From the osmotic pressure of a solution of 188 mg of a solid nonelectrolyte dissolved in 10.0 mL of solution at 25°C, we are to calculate the molar mass of the solute.

Analyze

We can calculate the molarity from the osmotic pressure by using

$$M = \frac{\Pi}{iRT}$$

where $i = 1$, $R = 0.0821$ L · atm / mol · K, $\Pi = 4.89$ atm, and $T = 273.15 + 25°C = 298$ K. From the molarity, we can calculate the molar mass of the solute:

$$\text{molar mass of solute (g/mol)} = \frac{\text{mass of solute (g)}}{\text{mol/L of solute} \times \text{volume of solution (L)}}$$

where the mass of solute is 0.188 g and the volume of solute is 0.0100 L.

Solve

$$M = \frac{4.89 \text{ atm}}{1 \times 0.0821 \text{ L} \cdot \text{atm / mol} \cdot \text{K} \times 298 \text{ K}} = 0.1999 \text{ mol/L}$$

$$\text{molar mass} = \frac{0.188 \text{ g}}{0.1999 \text{ mol/L} \times 0.0100 \text{ L}} = 94.1 \text{ g/mol}$$

Think About It

Osmotic pressure is the best colligative property to use to measure molar mass because it is the most sensitive technique.

11.31. **Collect and Organize**

We are to calculate the osmotic pressure of a saline solution (aqueous sodium chloride) at 37°C.

Analyze

To apply the osmotic pressure equation

$$\Pi = iMRT$$

we have to convert the concentration of NaCl from percent mass (0.90%) to molarity and account for the dilution of 100.0 mL of the solution to a final volume of 350.0 mL.

Solve

We can assume that the density of the dilute saline solution is close to that of pure water (1.00 g/mL). The molarity of a 0.90% (by mass) saline solution is

$$\frac{0.90 \text{ g NaCl}}{100 \text{ g solution}} \times \frac{1.00 \text{ g solution}}{1.00 \text{ mL}} \times \frac{1000 \text{ mL}}{1 \text{ L}} \times \frac{1 \text{ mol NaCl}}{58.44 \text{ g}} = 0.154 \text{ mol/L NaCl}$$

After dilution, the molarity of the saline solution is

$$M_{initial}V_{initial} = M_{final}V_{final}$$
$$0.154 \text{ } M \times 100.0 \text{ mL} = C_{final} \times 350.0 \text{ mL}$$
$$M_{final} = 0.0440 \text{ } M$$

The osmotic pressure of that saline solution is

$$\Pi = 2 \times 0.0440 \text{ mol/L} \times \frac{0.08206 \text{ L} \cdot \text{atm}}{\text{mol} \cdot \text{K}} \times 310 \text{ K} = 2.2 \text{ atm}$$

Think About It

Physiological saline is isotonic with blood plasma.

11.33. Collect and Organize

We are to explain why the vapor pressure of a liquid increases with increasing temperature by using the ideas of kinetic molecular theory.

Analyze

As energy is applied to increase the temperature, the average kinetic energy of the molecules in a liquid increases.

Solve

When the average kinetic energy of the liquid molecules increases, more of the molecules can escape the liquid phase and enter the gas phase. More molecules in the gas phase increases the vapor pressure.

Think About It

Warm apple pie has a stronger aroma than cold apple pie.

11.35. Collect and Organize

We are to make a general statement relating vapor pressure to the strength of intermolecular forces.

Analyze

Strong intermolecular forces hold molecules or ions tightly to each other. That reduces the number of molecules or ions that enter the gas phase.

Solve

As intermolecular forces increase in strength, the vapor pressure decreases.

Think About It

We can smell some substances with weak intermolecular forces (such as acetone in nail polish remover), but we do not often smell those with strong intermolecular forces (such as table salt).

11.37. Collect and Organize

We are to rank the listed compounds in order of increasing vapor pressure.

Analyze

All the compounds (CH_3CH_2OH, CH_3OCH_3, and $CH_3CH_2CH_3$) have about the same molar mass (46 g/mol or 44 g/mol). The vapor pressure differences, then, are due to differences in the strength of the intermolecular forces between the molecules of each substance.

Solve

CH_3CH_2OH is polar and may have fairly strong hydrogen bonds. CH_3OCH_3 is polar and has dipole–dipole forces between its molecules. $CH_3CH_2CH_3$ is nonpolar and has only weak dispersion forces between its molecules. The vapor pressure is expected to increase in the order (a) CH_3CH_2OH < (b) CH_3OCH_3 < (c) $CH_3CH_2CH_3$.

Think About It

That prediction is borne out by the facts that CH_3CH_2OH is a relatively high-boiling liquid, CH_3OCH_3 has a low boiling point (is volatile), and $CH_3CH_2CH_3$ is a gas at room temperature and pressure.

11.39. Collect and Organize

Using the Clausius–Clapeyron equation, we can calculate the ΔH_{vap} of pinene given the data of vapor pressure as a function of temperature.

Analyze

We can calculate ΔH_{vap} from the relationship

$$\ln\left(P_{vap}\right) = \frac{-\Delta H_{vap}}{R}\left(\frac{1}{T}\right) + c$$

where P_{vap} is the vapor pressure measured at temperature T (in kelvin), R is the gas constant (8.315 J/mol · K), and ΔH_{vap} is the enthalpy of vaporization (in joules per mole). By plotting $\ln P_{vap}$ versus $1/T$, we find that the slope of the line will be equal to $-\Delta H_{vap}/R$.

Solve

Plotting the data as $\ln P_{vap}$ versus $1/T$ gives a straight line with a slope of -4936.37:

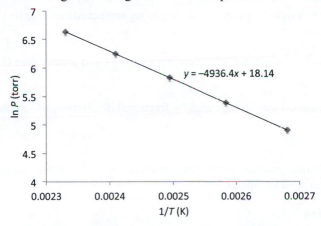

The value of ΔH_{vap}, therefore, is

$$\Delta H_{vap} = -\text{slope} \times R = 4936.4 \text{ K}^{-1} \times 8.314 \text{ J/mol} \cdot \text{K}$$

$$= 41,000 \text{ J/mol, or } 41.0 \text{ kJ/mol}$$

Think About It

We expect ΔH_{vap} to have a positive sign because the process of boiling a liquid is an endothermic process.

11.41. Collect and Organize

For two isomers of octane, isooctane and tetramethylbutane, we can use the enthalpies of vaporization (35.8 and 43.3 kJ/mol, respectively) and the boiling points (98.2°C and 106.5°C, respectively) to calculate the vapor pressure of each isomer on a summer's day (38°C).

Analyze

We can use the Clausius–Clapeyron equation to solve this problem:

$$\ln\left(\frac{P_{vap,T_1}}{P_{vap,T_2}}\right) = \frac{\Delta H_{vap}}{R}\left(\frac{1}{T_2} - \frac{1}{T_1}\right)$$

where T_1 and P_1 are the temperature and pressure on the summer day and T_2 and P_2 are the temperature and pressure at which each substance boils. Recall that the boiling point occurs when the vapor pressure equals the atmospheric pressure, which we take to be 1.00 atm for this problem. To perform the calculations, remember to convert ΔH_{vap} to joules per mole and the boiling point to temperature in kelvin.

Solve

For isooctane:

$$\ln\left(\frac{P_{vap,T_1}}{1.00 \text{ atm}}\right) = \frac{35.8\times10^3 \text{ J/mol}}{8.314 \text{ J/mol} \cdot \text{K}}\left(\frac{1}{371.4 \text{ K}} - \frac{1}{311 \text{ K}}\right)$$

$$P_{vap,T_1} = 0.105 \text{ atm, or } 80.0 \text{ torr}$$

For tetramethylbutane:

$$\ln\left(\frac{P_{vap,T_1}}{1.00 \text{ atm}}\right) = \frac{43.3\times10^3 \text{ J/mol}}{8.314 \text{ J/mol}\cdot\text{K}}\left(\frac{1}{379.7 \text{ K}} - \frac{1}{311 \text{ K}}\right)$$

$$P_{vap,T_1} = 0.0483 \text{ atm, or } 36.7 \text{ torr}$$

Think About It

That the vapor pressure of tetramethylbutane is lower than that of isooctane makes sense because its boiling point is higher.

11.43. Collect and Organize

We are to identify the physical property we can use to separate crude oil into its various useful components.

Analyze

Crude oil consists mainly of a mixture of hydrocarbons from C_1 to greater than C_{36}.

Solve

The components of crude oil can be separated by fractional distillation, which uses differences in boiling points of the compounds.

Think About It

In the fractional distillation process, components in crude oil of relatively low molar mass distill first, with the residue after distillation used in asphalt.

11.45. Collect and Organize

We can consider the relative volatilities (boiling points) of C_5H_{12} and C_7H_{16} to determine which substance is present in higher concentration in the vapor phase above an equimolar mixture of the two components.

Analyze

Boiling point is an indication of volatility and, therefore, vapor pressure of a substance. Because C_5H_{12} has fewer atoms in its structure, it has a lower boiling point than C_7H_{16}. A lower boiling point means that C_5H_{12} is more volatile.

Solve

Because C_5H_{12} is more volatile, it is present in the highest concentration in the vapor phase above a mixture of C_5H_{12} and C_7H_{16}.

Think About It

In a fractional distillation of these two compounds, C_5H_{12} is separated first from the mixture.

11.47. Collect and Organize

To calculate the vapor pressure of a mixture of 25 g of methanol and 75 g of ethanol, we can use the knowledge that the total vapor pressure of a mixture of two volatile components is the sum of the products of the individual vapor pressures and the mole fraction of that component in the mixture. We are given the vapor pressures of pure methanol (92 torr) and pure ethanol (45 torr).

Analyze

The total pressure can be calculated by

$$P_{total} = x_{methanol}P^{\circ}_{methanol} + x_{ethanol}P^{\circ}_{ethanol}$$

where P° is the vapor pressure of the pure methanol or ethanol at standard temperature and x is the mole fraction of each component.

Solve

The moles of each component are

$$25 \text{ g} \times \frac{1 \text{ mol}}{32.04 \text{ g}} = 0.780 \text{ mol methanol}$$

$$75 \text{ g} \times \frac{1 \text{ mol}}{46.07 \text{ g}} = 1.63 \text{ mol ethanol}$$

Total moles $= 1.63 + 0.780 = 2.41$ mol

The mole fraction of each component is

$$x_{\text{methanol}} = \frac{0.780 \text{ mol}}{2.41 \text{ mol}} = 0.324$$

$$x_{\text{ethanol}} = \frac{1.63 \text{ mol}}{2.41 \text{ mol}} = 0.676$$

The vapor pressure of the mixture at 20°C is

$$P_{\text{total}} = (0.324 \times 92 \text{ torr}) + (0.676 \times 45 \text{ torr}) = 60 \text{ torr}$$

Think About It

Because methanol is more volatile than ethanol, the proportion of methanol in the vapor state compared with that of ethanol is greater than the proportion of methanol to ethanol in the liquid state.

11.49. Collect and Organize

We are asked to explain how all the water ends up in one compartment in Figure 11.17.

Analyze

The figure shows two compartments, one originally containing seawater and the other originally containing pure water. After a time, the seawater compartment is filled with water, whereas the pure water compartment is empty. This is an application of Raoult's law regarding the different vapor pressure of pure water from that of seawater.

Solve

The vapor pressure of water in the compartment of pure water is greater than that of the seawater because the seawater contains dissolved solids that lower its vapor pressure. Therefore, water evaporates faster from the pure water than from the seawater. The water in the gas phase condenses at the same rate into both containers, but once in the seawater the water evaporates more slowly. Over time this leads to the transfer of water from the pure water compartment to the seawater compartment because the seawater will always contain dissolved salts even after dilution and, therefore, evaporate more slowly than pure water.

Think About It

Be careful to remember that the transfer of water in this situation is due only to different rates of evaporation, not different rates of condensation.

11.51. Collect and Organize

We are to differentiate between *molarity* and *molality*.

Analyze

Both terms describe a concentration of a solute in a solution.

Solve

Molarity is the moles of the solute in 1 L of solution. Molality is the moles of solute in 1 kg of solvent.

Think About It

Molality is used when a change in temperature would change the volume of a solution.

11.53. Collect and Organize

Seawater freezes at a lower temperature than freshwater. Our explanation of why must we relate the amounts of dissolved solutes in those two naturally occurring waters.

Analyze

Seawater is quite "salty" and therefore has more dissolved solutes than freshwater.

Solve

Because seawater has a higher concentration of dissolved salt, it has a lower freezing point compared to freshwater. The presence of nonvolatile solutes shifts the solid–liquid line on the phase diagram to a lower temperature.

Think About It

Freshwater has some dissolved solutes in it as well, just not as much as seawater.

11.55. Collect and Organize

For a solution composed of 3.5 mol of H_2O and 1.5 mol of $C_6H_{12}O_6$, we are to calculate the mole fraction of water and from that value calculate the vapor pressure of water for the solution given the vapor pressure of pure water at 25°C as 23.8 torr.

Analyze

The mole fraction of a substance is defined as

$$x = \frac{\text{mol component of interest}}{\text{total mol in mixture}}$$

and the vapor pressure of a solution is proportional to its mole fraction of solvent,

$$P_{\text{solution}} = x_{\text{solvent}} P_{\text{solvent}}$$

Solve

The mole fraction of water in the glucose solution is

$$x = \frac{0.35 \text{ mol}}{3.5 \text{ mol} + 1.5 \text{ mol}} = 0.70$$

and the vapor pressure of the solution is

$$P_{\text{solution}} = 0.70 \times 23.8 \text{ torr} = 17 \text{ torr}$$

Think About It

Because the solution is not pure water, the vapor pressure of the solution is less than that for pure water.

11.57. Collect and Organize

For each solution, we are to calculate the molality given the moles of substance and the mass of water it is dissolved into.

Analyze

Molality is the number of moles of solute present in 1 kg of solvent.

Solve

(a) $\dfrac{0.875 \text{ mol}}{1.5 \text{ kg}} = 0.58 \ m$ glucose

(b) $\dfrac{11.5 \text{ mmol}}{65 \text{ g}} \times \dfrac{1 \text{ mol}}{1000 \text{ mmol}} \times \dfrac{1000 \text{ g}}{1 \text{ kg}} = 0.18 \ m$ acetic acid

(c) $\dfrac{0.325 \text{ mol}}{290.0 \text{ g}} \times \dfrac{1000 \text{ g}}{1 \text{ kg}} = 1.12 \ m$ $NaHCO_3$

Think About It
Be sure to express the mass of the solvent in kilograms when calculating molality.

11.59. **Collect and Organize**
For each solution, we are to calculate the molality of a substance in a solution given the molarity and the density of that solution.

Analyze
Molality is the number of moles of solute present in 1 kg of solvent. Assuming we have 1 L of solution, we will first have to convert the volume of the solution to mass using the density of the solution. From that we will have to calculate the mass of the solvent by subtracting the mass of the solute obtained using the molar mass and the concentration from the mass of the solution. Then we can divide the moles of solute by the mass of the solvent in kg.

Solve
(a) For 1.30 M $CaCl_2$:

$$\text{mass of } CaCl_2 \text{ in 1 L solution} = 1 \text{ L} \times \frac{1.30 \text{ mol}}{\text{L}} \times \frac{110.98 \text{ g}}{\text{mol}} = 144.27 \text{ g}$$

$$\text{mass of 1 L of solution} = 1 \text{ L} \times \frac{1000 \text{ mL}}{\text{L}} \times \frac{1.113 \text{ g}}{\text{mL}} = 1113 \text{ g}$$

$$\text{total mass of solvent} = 1113 \text{ g} - 144.27 \text{ g} = 968.72 \text{ g, or } 0.96872 \text{ kg}$$

$$\text{molality of solution} = \frac{1.30 \text{ mol}}{0.96872 \text{ kg}} = 1.34 \text{ } m$$

(b) For 2.02 M fructose:

$$\text{mass of fructose in 1 L solution} = 1 \text{ L} \times \frac{2.02 \text{ mol}}{\text{L}} \times \frac{180.16 \text{ g}}{\text{mol}} = 363.92 \text{ g}$$

$$\text{mass of 1 L of solution} = 1 \text{ L} \times \frac{1000 \text{ mL}}{\text{L}} \times 1.139 = 1139 \text{ g}$$

$$\text{total mass of solvent} = 1139 \text{ g} - 363.92 \text{ g} = 775.08 \text{ g, or } 0.77508 \text{ kg}$$

$$\text{molality of solution} = \frac{2.02 \text{ mol}}{0.77508 \text{ kg}} = 2.61 \text{ } m$$

(c) For 8.94 M ethylene glycol

$$\text{mass of ethylene glycol in 1 L solution} = 1 \text{ L} \times \frac{8.94 \text{ mol}}{\text{L}} \times \frac{62.07 \text{ g}}{\text{mol}} = 554.91 \text{ g}$$

$$\text{mass of 1 L of solution} = 1 \text{ L} \times \frac{1000 \text{ mL}}{\text{L}} \times \frac{1.069 \text{ g}}{\text{mL}} = 1069 \text{ g}$$

$$\text{total mass of solvent} = 1069 \text{ g} - 554.91 \text{ g} = 514.09 \text{ g, or } 0.51409 \text{ kg}$$

$$\text{molality of solution} = \frac{8.94 \text{ mol}}{0.51409 \text{ kg}} = 17.4 \text{ } m$$

(d) For 1.97 M LiCl

$$\text{mass of LiCl in 1 L solution} = 1 \text{ L} \times \frac{1.97 \text{ mol}}{\text{L}} \times \frac{42.39 \text{ g}}{\text{mol}} = 83.51 \text{ g}$$

$$\text{mass of 1 L of solution} = 1 \text{ L} \times \frac{1000 \text{ mL}}{\text{L}} \times \frac{1.046 \text{ g}}{\text{mL}} = 1046 \text{ g}$$

$$\text{total mass of solvent} = 1046 \text{ g} - 83.51 \text{ g} = 962.49 \text{ g, or } 0.96249 \text{ kg}$$

$$\text{molality of solution} = \frac{1.97 \text{ mol}}{0.96249 \text{ kg}} = 2.05 \text{ } m$$

Think About It
There we have to be careful to convert our units for mass and volume.

11.61. Collect and Organize

For each solution of a given molality listed, we are to calculate the mass of the solution that contains 0.100 mol of the solute.

Analyze

Because molality is expressed as moles per kilogram, we can calculate the mass of solution required to give 0.100 mol by dividing 0.100 mol by the molality of the solution (or multiplying by the inverse of the molality). That gives the mass of solvent in the solution that also contains 0.100 mol of the solute. For the total solution mass, we have to add the mass of the solute to the mass of the solvent.

Solve

(a) $0.100 \text{ mol} \times \dfrac{1 \text{ kg}}{0.334 \text{ mol}} = 0.299 \text{ kg, or } 299 \text{ g solvent}$

$0.100 \text{ mol} \times \dfrac{80.04 \text{ g}}{1 \text{ mol}} = 8.004 \text{ g NH}_4\text{NO}_3$

Total mass of solution = $299 \text{ g} + 8.004 \text{ g} = 307 \text{ g NH}_4\text{NO}_3$ solution

(b) $0.100 \text{ mol} \times \dfrac{1 \text{ kg}}{1.24 \text{ mol}} = 0.0806 \text{ kg, or } 80.6 \text{ g solvent}$

$0.100 \text{ mol} \times \dfrac{62.07 \text{ g}}{1 \text{ mol}} = 6.207 \text{ g ethylene glycol (HOCH}_2\text{CH}_2\text{OH)}$

Total mass of solution = $80.6 \text{ g} + 6.207 \text{ g} = 86.8 \text{ g ethylene glycol solution}$

(c) $0.100 \text{ mol} \times \dfrac{1 \text{ kg}}{5.65 \text{ mol}} = 0.0177 \text{ kg, or } 17.7 \text{ g solvent}$

$0.100 \text{ mol} \times \dfrac{110.98 \text{ g}}{1 \text{ mol}} = 11.098 \text{ g CaCl}_2$

Total mass = $17.7 \text{ g} + 11.098 \text{ g} = 28.8 \text{ g CaCl}_2$ solution

Think About It

Remember that in molality, the mass of the solvent is used and that the mass of solute must be added to determine the mass of the solution needed.

11.63. Collect and Organize

We are to express lethal concentrations of ammonia, nitrite ion, and nitrate ion (1.1 mg/L, 0.40 mg/L, and 1361 mg/L, respectively) in molality units.

Analyze

Because the aqueous solutions of those species are dilute, we can assume that the densities of the solutions are 1.00 g/mL. We have to express each milligram amount in moles by using the molar masses of ammonia (17.03 g/mol), nitrite (NO_2^-, 46.01 g/mol), and nitrate (NO_3^-, 62.00 g/mol). The mass of the solvent (water) is the mass of the solution minus the mass of solute dissolved in the solution. From that mass of solvent (expressed in kilograms) and the moles of solute, we can calculate the molality. However, the mass of the solutes is insignificant compared with the mass of the solvent, so we can safely assume that the mass of the water is 1.00 kg.

Solve

For NH_3 $\dfrac{1.1 \text{ mg}}{1.00 \text{ kg}} \times \dfrac{1.00 \text{ g}}{1000 \text{ mg}} \times \dfrac{1 \text{ mol}}{17.03 \text{ g}} = 6.5 \times 10^{-5} \ m \ NH_3$

For NO_2^- $\dfrac{0.40 \text{ mg}}{1.00 \text{ kg}} \times \dfrac{1.00 \text{ g}}{1000 \text{ mg}} \times \dfrac{1 \text{ mol}}{46.01 \text{ g}} = 8.7 \times 10^{-6} \ m \ NO_2^-$

For NO_3^- $\dfrac{1361 \text{ mg}}{1.00 \text{ kg}} \times \dfrac{1.00 \text{ g}}{1000 \text{ mg}} \times \dfrac{1 \text{ mol}}{62.00 \text{ g}} = 2.195 \times 10^{-2} \ m \ NO_3^-$

Think About It
Dissolved substances are often expressed as mass/mass or mass/volume, but the numbers of particles in a mixture are best expressed as moles/mass or moles/volume.

11.65. Collect and Organize
Using the equation for freezing point depression, we can determine the freezing point of a benzene solution of cinnamaldehyde.

Analyze
The change in freezing point of a solution is

$$\Delta T_f = K_f m$$

where K_f is the freezing point depression constant for the solvent and m is the molality of the solute in the solution. We are given that 75 mg of cinnamaldehyde (C_9H_8O, 132.16 g/mol) is dissolved in 1.00 g (or 1.00×10^{-3} kg) of benzene ($K_f = 4.3°C/m$ and normal $T_f = 5.5°C$). We have to convert the mass of the cinnamaldehyde to moles by dividing by the molar mass before calculating the molality.

Solve
Molality of cinnamaldehyde solution:

$$\frac{0.075 \text{ g} \times \dfrac{1 \text{ mol}}{132.16 \text{ g}}}{1.00 \times 10^{-3} \text{ kg}} = 0.567 \ m$$

Freezing point depression due to the presence of cinnamaldehyde in the benzene solution is

$$\Delta T_f = \frac{4.3°C}{m} \times 0.567 \ m = 2.44°C$$

The freezing point of the solution is

$$T_{f,\text{solution}} = 5.5°C - 2.44°C = 3.1°C$$

Think About It
Increasing the amount of cinnamaldehyde in the solution will further decrease the freezing point.

11.67. Collect and Organize
We can use the melting point depression equation to determine what molality of a nonvolatile, nonelectrolyte solute would cause the melting point of camphor to change by 1.000°C.

Analyze
We can rearrange the melting point depression equation to solve for molality:

$$\Delta T_f = K_f m$$

$$m = \frac{\Delta T_f}{K_f}$$

For camphor, $K_f = 39.7°C/m$.

Solve

$$m = \frac{1.000°C}{39.7°C/m} = 2.52 \times 10^{-2} \ m$$

Think About It
Because of camphor's large K_f value, even a little solute dissolved in it greatly depresses the freezing point.

11.69. Collect and Organize

We are given the concentration (in milligrams per milliliter) of an aqueous solution of saccharin and the K_f for water and asked to determine the melting point of the solution.

Analyze

The melting point depression of the solution can be calculated by using

$$\Delta T_f = K_f m$$

where m is the molality of the saccharin solution, which is calculated by knowing the amount of saccharin ($C_7H_5O_3NS$, 183.19 g/mol) dissolved in 1.00 mL of solution (186 mg) and the density of the solution (1.00 g/mL). We are given water's melting point depression constant ($K_f = 1.86°C/m$). Once we know how much the melting point is depressed, we can subtract that value from 0.00°C (the melting point of pure H_2O) to arrive at the new melting point of the saccharin solution.

Solve

The molality of the solution is

$$0.186 \text{ g} \times \frac{1 \text{ mol}}{183.19 \text{ g}} = 1.015 \times 10^{-3} \text{ mol saccharin}$$

The mass of water in solution = 1.00 g, or 1.00×10^{-3} kg

$$\text{molality of solution} = \frac{1.015 \times 10^{-3} \text{ mol}}{1.00 \times 10^{-3} \text{ kg}} = 1.015 \ m$$

The freezing point depression of the solution is

$$\Delta T_f = \frac{1.86°C}{m} \times 1.015 \ m = 1.89°C$$

The freezing point of this solution is 0.00°C − 1.89°C = −1.89°C

Think About It

The freezing point of this solution is noticeably lower than that of pure water.

11.71. Collect and Organize

Of the aqueous solutions named, we are to determine which has the lowest freezing point.

Analyze

Freezing point depression is a colligative property and, as such, depends not on the identity of the solute but on the number of particles dissolved in the solution. Molecular substances dissolve as single molecules, so their particle molality is equal to the molality of the solution. Ionic substances, by contrast, dissolve to form two or more particles (ions) in solution; therefore, the molality of particles in solution is a multiple (two times, three times, etc.) of the molality of the salt in the solution.

Solve

Glucose is a molecular solid, so its particle molality in the solution is 0.5 m. Sodium chloride forms Na^+ and Cl^- in solution, so its particle molality is 0.50 m × 2 = 1.0 m. Calcium chloride forms Ca^{2+} and 2 Cl^- in solution, so its particle molality is 0.5 m × 3 = 1.5 m. The greater the particle molality, the larger the freezing point depression. Therefore, the aqueous solution that has the lowest freezing point is 0.5 m $CaCl_2$.

Think About It

In comparing colligative properties, we have to compare not only the molality of the solutions but also the behavior of the substance in the solvent (does it break into ions when dissolved, and how many?).

11.73. Collect and Organize

Of the aqueous solutions named [0.0400 m NH_4NO_3, 0.0165 m LiCl, and 0.0105 m $Cu(NO_3)_2$], we are to determine which one has the highest boiling point.

Analyze

Boiling point elevation is a colligative property and, as such, depends not on the identity of the solute but on the number of particles dissolved in the solution. Ionic substances dissolve to form two or more particles (ions) in solution; therefore, the molality of particles in solution is a multiple (two times, three times, etc.) of the molality of the salt in the solution.

Solve

Ammonium nitrate forms NH_4^+ and NO_3^- in solution, so a 0.0400 m solution of NH_4NO_3 is 0.0800 m in particles. Lithium chloride forms Li^+ and Cl^- in solution, so a 0.0165 m solution of $LiCl$ is 0.0330 m in particles. Copper(II) nitrate forms Cu^{2+} and 2 NO_3^- in solution, so a 0.0105 m solution is 0.0315 m in particles. The greater the particle molality, the higher the boiling point. Therefore, the aqueous solution with the highest boiling point is the 0.0400 m NH_4NO_3 solution.

Think About It

In comparing colligative properties, we have to compare not only the molality of the solutions but also the behavior of the substance in the solvent (does it break into ions when dissolved, and how many?).

11.75. Collect and Organize

For each aqueous solution, we are given the concentration of a species and the van 't Hoff factor with which to calculate the boiling point elevation of the solution. We are to rank the solutions in order of increasing boiling point. We need to use the boiling point elevation constant for water ($K_b = 0.52°C/m$).

Analyze

The van 't Hoff factor is a ratio that compares the measured value of a colligative property with the value expected if the dissolved substance were molecular. It gives, in essence, the number of particles in solution. Using the van 't Hoff factor, the calculation for boiling point elevation is

$$\Delta T_b = iK_b m$$

Solve

(a) For 0.06 m $FeCl_3$ ($i = 3.4$) in water:

$$\Delta T_b = 3.4 \times 0.52°C/m \times 0.06\ m = 0.11°C$$

(b) For 0.10 m $MgCl_2$ ($i = 2.7$):

$$\Delta T_b = 2.7 \times 0.52°C/m \times 0.10\ m = 0.14°C$$

(c) For 0.20 m KCl ($i = 1.9$):

$$\Delta T_b = 1.9 \times 0.52°C/m \times 0.20\ m = 0.20°C$$

Therefore, in order of increasing boiling point, (a) 0.06 m $FeCl_3(aq)$ < (b) 0.10 m $MgCl_2(aq)$ < (c) 0.20 m $KCl(aq)$.

Think About It

The van 't Hoff factors are all less than the theoretical value. That is due to ion pairing of the solute ions in solution.

11.77. Collect and Organize

From the measured freezing point depression of a solution of 0.150 g of caffeine (a nonelectrolyte molecular compound) in 10.0 g of camphor, we are to calculate the molar mass of caffeine. Then, given the mass ratios of C, H, N, and O in caffeine, we can write the molecular formula.

Analyze

We can calculate the molality of the caffeine solution by using

$$m = \frac{\Delta T_f}{iK_f}$$

where $i = 1$, $\Delta T_f = 3.07°C$, and $K_f = 39.7°C/m$. From the molality of the solution, we can calculate the molar mass of caffeine:

$$\text{molar mass of caffeine (g/mol)} = \frac{\text{mass of caffeine (g)}}{\text{molality of solution (mol/kg)} \times \text{mass of solvent (kg)}}$$

where the mass of caffeine is 0.150 g and the mass of solvent is 0.0100 kg. From the elemental analysis, we can determine the empirical formula for caffeine, which we can then relate to the molar mass derived from the freezing point depression measurement and then write the molecular formula.

Solve

$$m = \frac{3.07°C}{1 \times 39.7°C/m} = 0.07733\ m$$

$$\text{molar mass} = \frac{0.150\ g}{0.07733\ \text{mol/kg} \times 0.0100\ kg} = 194\ g/mol$$

If we assume 100 g of caffeine,

$$49.49\ g \times \frac{1\ \text{mol}}{12.011\ g} = 4.120\ \text{mol C}$$

$$5.15\ g \times \frac{1\ \text{mol}}{1.008\ g} = 5.11\ \text{mol H}$$

$$28.87\ g \times \frac{1\ \text{mol}}{14.007\ g} = 2.061\ \text{mol N}$$

$$16.49\ g \times \frac{1\ \text{mol}}{15.999\ g} = 1.031\ \text{mol O}$$

That gives a molar ratio of 4C:5H:2N:1O, for an empirical formula of $C_4H_5N_2O$ with a formula mass of 97.1 g/mol. The measured molar mass is 194 g/mol, so the molar mass is twice the formula mass. Therefore, the molecular formula for caffeine is $C_8H_{10}N_4O_2$.

Think About It
Given careful measurement of any of the colligative properties of a solution of a solute, we can accurately determine the molar mass of a substance. When that information is coupled with the results of elemental analysis, we can then write the molecular formula. That is a powerful technique to characterize new substances prepared or isolated in the laboratory.

11.79. Collect and Organize
We are to explain why Henry's law constant for CO_2 is so much larger than that for N_2 or O_2.

Analyze
We are given a hint that CO_2 reacts with water. That reaction gives carbonic acid:

$$H_2O(\ell) + CO_2(g) \rightarrow H_2CO_3(aq)$$

The higher the Henry's law constant, the greater the solubility of the gas at that temperature.

Solve
The reaction of CO_2 with water forms H_2CO_3, which partially ionizes. This decreases the concentration of CO_2 in the solution allowing more of it to dissolve.

Think About It
N_2 and O_2 simply dissolve in the water. Without reactions with water to give another species, the water becomes saturated with those gases.

11.81. Collect and Organize
We are to describe the composition of the bubbles observed when heating water.

Analyze
The water is at 60°C, well below its boiling point, so the bubbles are not composed of gaseous H_2O.

Solve
The first bubbles to form contain atmospheric gases (N_2, O_2, CO_2) that had dissolved in the water but become less soluble during heating.

Think About It
That observation is consistent with the knowledge that as temperature increases, the solubility of a gas decreases.

11.83. Collect and Organize
Given that the mole fraction of O_2 in air is 0.209 and that arterial blood has 0.25 g of O_2 per liter at 37°C and 1 atm, we can calculate k_H for O_2 in blood by using Henry's law.

Analyze
Henry's law is defined as
$$C_{gas} = k_H P_{gas}$$
where C_{gas} is the concentration of dissolved gas, k_H is Henry's law constant, and P_{gas} is the pressure of the gas. Rearranging the equation to solve for k_H gives
$$k_H = \frac{C_{gas}}{P_{gas}}$$
The units of k_H are usually expressed as moles per liter per atmosphere (mol/L · atm). The concentration of O_2 in blood in moles per liter (mol/L) is
$$\frac{0.25 \text{ g } O_2}{L} \times \frac{1 \text{ mol}}{32.00 \text{ g}} = 7.81 \times 10^{-3} \text{ mol/L}$$
The partial pressure of O_2 in the air is
$$0.209 \times 1 \text{ atm} = 0.209 \text{ atm}$$

Solve
$$k_H = \frac{7.81 \times 10^{-3} \text{ mol/L}}{0.209 \text{ atm}} = 3.7 \times 10^{-2} \text{ mol/L} \cdot \text{atm}$$

Think About It
As the temperature changes, so too does k_H. As temperature increases, less O_2 is soluble and the value of k_H decreases.

11.85. Collect and Organize
Using the k_H value of 3.7×10^{-2} mol/L · atm from Problem 11.83, we are to calculate the solubility of O_2 in the blood of a climber on Mt. Everest ($P_{atm} = 0.35$ atm) and a scuba diver ($P = 3.0$ atm).

Analyze
The concentration (solubility) of O_2 in the blood can be calculated by using Henry's law:
$$C_{O_2} = k_H P_{O_2}$$
For each case we need the pressure of O_2 in the atmosphere. If we assume that the mole fraction of O_2 stays at 0.209, the partial pressure of O_2 for each is as follows:
$$\text{For the alpine climber, } P_{O_2} = 0.209 \times 0.35 \text{ atm} = 7.32 \times 10^{-2} \text{ atm}$$
$$\text{For the scuba diver, } P_{O_2} = 0.209 \times 3 \text{ atm} = 0.627 \text{ atm}$$

Solve

(a) For the alpine climber

$$C_{O_2} = \frac{3.7 \times 10^{-2} \text{ mol}}{L \cdot atm} \times 7.32 \times 10^{-2} \text{ atm} = 2.7 \times 10^{-3} \text{ } M$$

(b) For the scuba diver

$$C_{O_2} = \frac{3.7 \times 10^{-2} \text{ mol}}{L \cdot atm} \times 0.627 \text{ atm} = 2.3 \times 10^{-2} \text{ } M$$

Think About It

The scuba diver has more than eight times the concentration of O_2 in her arterial blood than the alpine climber.

11.87. Collect and Organize

We consider whether $CaCl_2$ would melt ice at $-20°C$. We are given that 70.1 g of $CaCl_2$ ($i = 2.5$) is dissolved in 100.0 g of water at that temperature.

Analyze

From the mass of $CaCl_2$ that dissolves in 100.0 g of water, we can compute the molality of the $CaCl_2$ solution. We can then use the freezing point depression equation

$$\Delta T_f = iK_f m$$

to calculate ΔT_f. The K_f for water is $1.86°C/m$. If ΔT_f is greater than 20°C, then $CaCl_2$ would melt ice at $-20°C$.

Solve

$$\text{molality of CaCl}_2 \text{ solution} = \frac{70.1 \text{ g} \times \dfrac{1 \text{ mol}}{110.98 \text{ g}}}{0.1000 \text{ kg}} = 6.316 \text{ } m$$

Freezing point depression is $\Delta T_f = 2.5 \times 1.86°C/m \times 6.316 \text{ } m = 29°C$.

The freezing point of the $CaCl_2$ solution would be $-29°C$. That is lower than $-20°C$. Yes, the $CaCl_2$ could melt ice at $-20°C$.

Think About It

$CaCl_2$ is an excellent deicer. It gives three particles, ideally, when it dissolves and is very soluble in water, so the ΔT_f for a saturated solution of $CaCl_2$ is very large.

11.89. Collect and Organize

Given the ΔT_f for NH_4Cl and $(NH_4)_2SO_4$ in aqueous solution, we are to calculate the value of i, the van 't Hoff factor, for each solution.

Analyze

The van 't Hoff factor may be calculated by using

$$i = \frac{\Delta T_f}{K_f m}$$

where $\Delta T_f = 0.322°C$ for $m = 0.0935$ m NH_4Cl and $\Delta T_f = 0.173°C$ for $m = 0.0378$ m $(NH_4)_2SO_4$. K_f for water is $1.86°C/m$.

Solve

For 0.0935 m NH_4Cl:

$$i = \frac{0.322°C}{1.86°C/m \times 0.0935 \text{ } m} = 1.85$$

For 0.0378 m $(NH_4)_2SO_4$:

$$i = \frac{0.173°C}{1.86°C/m \times 0.0378 \text{ } m} = 2.46$$

Think About It

If no ion pairs were in either solution, i for NH_4Cl would be 2 and i for $(NH_4)_2SO_4$ would be 3. The greater difference between the expected i and the calculated i for the $(NH_4)_2SO_4$ solution ($3 - 2.46 = 0.54$, compared with $2 - 1.85 = 0.15$ for the NH_4Cl solution) means that the ions in the $(NH_4)_2SO_4$ solution form ion pairs to a greater extent.

11.91. Collect and Organize

From the boiling point elevation ($2.45°C$) of a solution prepared by dissolving 0.111 g of eugenol (a nonelectrolyte) in 1.00×10^{-3} kg of chloroform (with $K_b = 3.63°C/m$), we are to calculate the molar mass of eugenol. With the elemental analysis data provided, we can then determine the molecular formula.

Analyze

We can calculate the molality of the solution from the measured boiling point elevation:

$$m = \frac{\Delta T_b}{K_b}$$

From the molality, we can calculate the molar mass:

$$\frac{\text{g of eugenol}}{\text{kg of chloroform} \times \text{molality of solution (mol/kg)}} = \frac{\text{g}}{\text{mol}}$$

From the elemental analysis, we can determine the empirical formula of eugenol, which we can then relate to the molar mass to write the molecular formula.

Solve

$$m = \frac{2.45°C}{3.63°C/m} = 0.6749 \text{ mol/kg}$$

$$\text{molar mass} = \frac{0.111 \text{ g}}{1.00 \times 10^{-3} \text{ kg} \times 0.6749 \text{ mol/kg}} = 164 \text{ g/mol}$$

From the elemental analysis, if we assume 100 g,

$$73.17 \text{ g C} \times \frac{1 \text{ mol}}{12.011 \text{ g}} = 6.092 \text{ mol C}$$

$$7.32 \text{ g H} \times \frac{1 \text{ mol}}{1.008 \text{ g}} = 7.26 \text{ mol H}$$

$$19.51 \text{ g O} \times \frac{1 \text{ mol}}{15.999} = 1.219 \text{ mol O}$$

That gives a molar ratio of 5C:6H:1O, for an empirical formula of C_5H_6O with a formula mass of 82.10 g/mol. That is one-half the measured molar mass, so the molecular formula of eugenol is $C_{10}H_{12}O_2$.

Think About It

The formula for eugenol is indeed $C_{10}H_{12}O_2$, and that compound has the following structure:

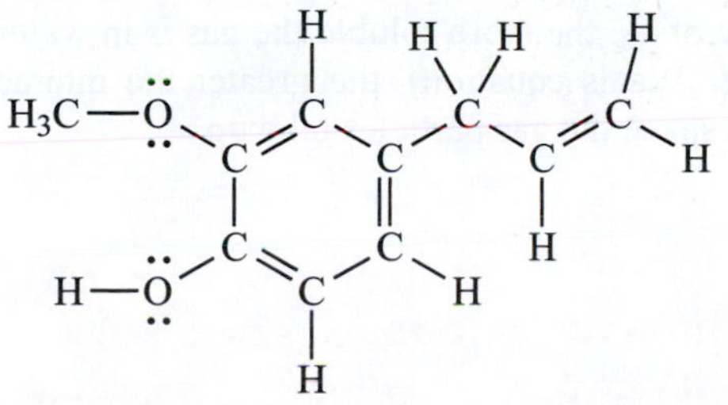

11.93. **Collect and Organize**

From the osmotic pressure of 6.50 torr at 23.1°C of a solution composed of 7.50 mg of a protein in 5.00 mL of aqueous solution, we are to determine the molar mass of the protein.

Analyze

Using the osmotic pressure equation, we can calculate the molarity of the protein solution:

$$M = \frac{\Pi}{iRT}$$

The protein is a nonelectrolyte molecular solid, so $i = 1$. The temperature in kelvin is $23.1 + 273.15 = 296.25$ K. The pressure needs to be in units of atmospheres, so

$$6.50 \text{ torr} \times \frac{1 \text{ atm}}{760 \text{ torr}} = 8.55 \times 10^{-3} \text{ atm}$$

Once we have calculated the molarity, we can determine the molar mass of the protein by using

$$\text{molar mass of protein (g/mol)} = \frac{\text{mass of protein in sample (g)}}{\text{mol/L of protein in solution} \times \text{volume of solution (L)}}$$

Solve

$$M = \frac{8.55 \times 10^{-3} \text{ atm}}{1 \times 0.0821 \text{ L} \cdot \text{atm / mol} \cdot \text{K} \times 296.25 \text{ K}} = 3.515 \times 10^{-4} M$$

$$\text{molar mass of protein} = \frac{7.50 \times 10^{-3} \text{ g}}{3.515 \times 10^{-4} \text{ mol/L} \times 0.00500 \text{ L}} = 4270 \text{ g/mol}$$

Think About It

Although that molar mass may seem large, it is small for proteins, which may have molar masses of several hundred thousand grams per mole.

11.95. **Collect and Organize**

From the data for a and b in the van der Waals equation for gases (Table 10.5) and the k_H (Table 11.1) for the gases He, O_2, N_2, CH_4, and CO_2, we are to plot the value of a and b as a function of k_H. From that we will determine which gas appears to be an outlier and then suggest a reason why that might be so.

Analyze

The data we will plot are

Gas	k_H (mol / (L · atm)	a (L^2 · atm / mol^2)	b (L/mol)
He	3.5×10^{-4}	0.0341	0.02370
O_2	1.3×10^{-3}	1.36	0.0318
N_2	6.7×10^{-4}	1.39	0.0391
CH_4	1.5×10^{-3}	2.25	0.0428
CO_2	3.5×10^{-2}	3.59	0.0427

Recall that the larger the value of k_H, the more soluble the gas is in water; the larger the value of a (the pressure correction term in the van der Waals equation), the greater the interaction between particles of the gas; the larger the value of b, the more space the gas particles take up.

Solve

(a) The graphs of a versus k_H and of b versus k_H are shown below.

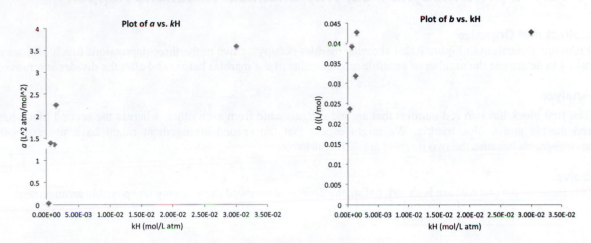

In both plots, the data points for CO_2 (the rightmost point on each) are outliers on the graph.

(b) Both graphs indicate that CO_2 dissolves to a much greater extent than what would be predicted from their values of a and b. That is because CO_2, unlike the other gases, reacts with water to form the soluble ions H^+ and HCO_3^-.

Think About It

The equation describing the reaction of carbon dioxide with water is

$$CO_2(g) + H_2O(\ell) \rightarrow H_2CO_3(aq).$$

CHAPTER 12 | Thermodynamics: Why Chemical Reactions Happen

12.1. Collect and Organize

From the illustration in Figure P12.1 showing marbles occupying two of the three depressions in a block, we are asked to determine the number of possible arrangements of the marbles before and after the divider is removed.

Analyze

The first block has two red marbles that are indistinguishable from each other, whereas the second block has a red marble and a blue marble. We might expect that the second arrangement might have more possible arrangements because the two marbles are distinguishable.

Solve

(a) Because the marbles are both red, before the divider is removed there is only one possible arrangement:

R	R	

When the divider is removed, there are three possible arrangements:

| R | R | | | R | | R | | | R | R |

(b) When there is one red marble and one blue marble before the divider is removed, there are two possible arrangements:

| R | B | | | B | R | |

When the divider is removed, there are six possible arrangements:

| R | B | | | B | R | | | | B | R |

| R | | B | | | R | B | | B | | R |

Think About It

Because there are more possible arrangements for the block with two different marbles, that system has a greater entropy both before and after the divider is removed.

12.3. Collect and Organize

From the illustration in Figure P12.3 of two tires of the same volume inflated at the same temperature but with more air in one than the other, we are to determine which has the greater internal pressure and which has the greater entropy.

Analyze

The tire on the right of Figure P12.3 has more particles of gas in the same volume. As we learned in Chapter 10 in Avogadro's law, pressure is directly proportional to the number of particles in a given volume. Entropy is directly proportional to the number of microstates for a system,

$$S = k \ln W$$

where S is entropy, k is the Boltzmann constant, and W is the number of microstates. A microstate is a unique distribution of particles among energy levels. As the number of particles in a system increases, the number of microstates increases.

Solve

According to Avogadro's law, the tire on the right (with the blue gas), with more gas particles, has the greater internal pressure. For this tire at higher pressure, more microstates are available because of the greater number of gas particles. Therefore, the tire on the right (with the blue gas) also has the greater entropy.

Think About It

When more particles are introduced into a system, the more dispersed the energy is in the system.

12.5. **Collect and Organize**

For the system in Figure P12.5, showing a random distribution of two gases, we are to assess the probability that the bottom bulb will collect gas A, which has a molar mass twice that of gas B.

Analyze

The system is already at high entropy because both gases can randomly move throughout the volume.

Solve

For gas A to collect in the bottom bulb of the apparatus, the effect of gravity on the more massive gas particles would have to overwhelm the tendency for gases to be randomly mixed because of entropy. Individual gas molecules are not really affected by gravity, and entropy is the dominant effect. Therefore, atoms of A will not eventually fill the bottom bulb and separate from atoms of B. The probability that gases A and B will separate in the apparatus to occupy separate bulbs is very low. Each gas would in that situation be confined to a smaller volume. That change would involve a decrease in entropy.

Think About It

Gases have enough energy to move anywhere in a container, so the larger the volume the gas occupies, the larger the entropy of the system.

12.7. **Collect and Organize**

For the change depicted in Figure P12.6, we are to consider whether the temperature would be more likely to occur at low or high temperature or would be unaffected by temperature entirely.

Analyze

In the process depicted in Figure P12.6, gas A_2 condenses with a decrease in entropy.

Solve

The change in which A_2 condenses is more likely to occur at low temperatures because at lower temperatures particles have less energy and may condense because of intermolecular forces that overwhelm their energy to remain in the gas phase.

Think About It

From the process shown in Figure P12.6, we can also determine that gas A_2 has a higher boiling point than gas B_2.

12.9. **Collect and Organize**

For the processes of phase changes (melting, vaporization, condensation, freezing, sublimation, and deposition), we are to determine which would have plots of energy versus temperature for ΔH and $T\Delta S$ like that shown in Figure P12.8.

Analyze

As temperature increases, the value of $T\Delta S$ decreases, and so the value of ΔS must be negative.

Solve

Phase changes processes that have $-\Delta S$ are condensation, freezing, and deposition.

Think About It

Those processes are all spontaneous at low temperature. That finding is consistent with our analysis of the plot in Problem 12.8.

12.11. Collect and Organize

We consider what happens to the sign of ΔS when we reverse a process.

Analyze

When the sign of ΔS is negative, the process is not favored by entropy, and entropy is decreasing. When the sign of ΔS is positive, the process is favored by entropy, and entropy is increasing.

Solve

If a process favored by entropy ($+\Delta S$) is reversed, the reverse process will not be favored ($-\Delta S$). Therefore, the ΔS when a process is reversed has its sign reversed.

Think About It

In terms of order and disorder in a chemical or physical process, a process that has increasing disorder ($+\Delta S$) has increasing order ($-\Delta S$) when the process is reversed.

12.13. Collect and Organize

By analyzing the possible outcomes in flipping three coins, we can determine the total number of possible microstates and which of those microstates is most likely.

Analyze

Flipping a coin has two possibilities: heads (H) or tails (T). Because two outcomes are possible for each coin, n coins will have 2^n different possible outcomes. Three coins will have $2^3 = 8$ possibilities.

Solve

The eight possible microstates and their values where $H = +1$ and $T = -1$ are as follows:

H H H	$(+1 +1 +1) = +3$	T T T	$(-1 -1 -1) = -3$
H T H	$(+1 -1 +1) = +1$	T H T	$(-1 +1 -1) = -1$
H H T	$(+1 +1 -1) = +1$	T T H	$(-1 -1 +1) = -1$
T H H	$(-1 +1 +1) = +1$	H T T	$(+1 -1 -1) = -1$

The most likely microstates have sums of $+1$ and -1.

Think About It

As the number of coins increases, the number of possible microstates increases significantly. For $n = 4$, there are 16 possibilities, for $n = 5$ there are 32, and for $n = 6$ there are 64.

12.15. Collect and Organize

From the standard molar entropy value of liquid water at 298 K, we are to calculate the number of microstates available to a single molecule of H_2O.

Analyze

The value of $S°$ for $H_2O(\ell)$ from Appendix 4 is 69.9 J/mol · K, and the equation relating the molar entropy to the number of possible microstates is

$$S = k_B \ln W$$

where S is the entropy of the system under consideration, k_B is the Boltzmann constant (1.382×10^{-23} J/K), and W is the number of microstates possible.

Solve

We must rearrange the equation to solve for W and use Avogadro's number to determine the possible microstates for just one molecule of liquid water, not an entire mole of liquid water molecules.

$$W = e^{S/k_B}$$

$$\text{where } S/k_B = \frac{\left(69.9\dfrac{J}{mol \cdot K}\right)\left(\dfrac{1 \text{ mol}}{6.0221 \times 10^{23} \text{ molecules}}\right)}{1.381 \times 10^{-23}\dfrac{J}{K}} = 8.405$$

$$W = e^{8.405} = 4.47 \times 10^3$$

Think About It
For an entire mole of water molecules,
$$W = e^{S/k_B}$$

$$\text{where } S/k_B = \frac{\left(69.9 \dfrac{J}{mol \cdot K}\right)}{1.381 \times 10^{-23} \dfrac{J}{K}} = 5.06 \times 10^{24}$$

$W = e^{5.06 \times 10^{24}}$ is beyond the capabilities of most scientific calculators.

12.17. Collect and Organize
For the three containers shown in Figure P12.17, each containing five gaseous particles, we are to rank them in order of the increasing number of microstates for the molecules in the containers.

Analyze
The higher the entropy of the system, the more possible microstates for the system. All the substances are gases, so no difference in the entropy due to difference in phase is present. What is different is the number of atoms in the five molecules in each container. Using the atomic color palette in the textbook, we can even assign those molecules (where red = oxygen and blue = nitrogen) as (a) NO_2, (b) O_2, and (c) N_2O_2.

Solve
The larger the molecules, the greater the entropy because a larger molecule with more atoms has more rotational energy levels, so they have more accessible microstates. Of those three gases, (c) is the largest and (b) is the smallest. In order of increasing number of accessible microstates: (b) < (a) < (c).

Think About It
We expect, too, that the molar entropies would increase in the same order: (b) O_2 (205.0 J/mol · K) < (a) NO_2 (240 J/mol · K) < (c) N_2O_2 (unstable, so no value found).

12.19. Collect and Organize
Given five ionic solutes, we are to predict which experiences the greatest increase in entropy when 0.0100 mol of the substance dissolves in 1 L of water.

Analyze
All those substances are ionic substances and, when dissolved in water, separate into the constituent cations and anions. Therefore, each will experience an increase in entropy upon dissolution. However, the one that produces the most dissolved ions will experience the greatest increase in entropy when dissolved into the same volume of water at the same initial concentration of the ionic salt.

Solve
(a) $CaCl_2$ dissolves into one Ca^{2+} cation and two Cl^- anions in solution, giving a total of three dissolved ions.
(b) NaBr dissolves into one Na^+ cation and one Br^- anion in solution, giving a total of two dissolved ions.
(c) KCl dissolves into one K^+ cation and one Cl^- anion in solution, giving a total of two dissolved ions.
(d) $Cr(NO_3)_3$ dissolves into one Cr^{3+} cation and three NO_3^- anions in solution, giving a total of four dissolved ions.
(e) LiOH dissolves into one Li^+ cation and one OH^- anion in solution, giving a total of two dissolved ions.
Because (d) $Cr(NO_3)_3$ yields the most dissolved ions, it is the salt that experiences the greatest increase in entropy upon dissolution.

Think About It
The total concentration of dissolved ions (both cation and anion) for the $Cr(NO_3)_3$ solution is 0.0400 M.

12.21. **Collect and Organize**

Given pairs of systems, we are asked to determine which of the pair has the greater entropy.

Analyze

The system with the most accessible microstates will have greater entropy. Systems that have larger molecules (with the particles being in the same phase) will have greater entropy. Gases have significantly greater entropy than either liquids or solids. Systems that have more particles (with the particles in the same phase) have greater entropy.

Solve

(a) Both $S_8(g)$ and $S_2(g)$ are present in the same number of particles and in the same phase. $S_8(g)$ has greater entropy because it is a larger molecule than $S_2(g)$.

(b) Both $S_8(s)$ and $S_2(g)$ are present in the same number of particles but are present in different phases. $S_2(g)$ has greater entropy because it is in the gas phase, compared with the entropy of S_8 in the solid phase.

(c) Both $O_2(g)$ and $O_3(g)$ are present in the same number of particles and in the same phase. $O_3(g)$ has greater entropy because it is a larger molecule than $O_2(g)$.

Think About It

For the same mass of gaseous S_8 and S_2, S_2 would have the greater entropy, as it would have more particles present because of its lower molar mass.

12.23. **Collect and Organize**

We are to predict which has the higher standard molar entropy, diamond or the fullerenes.

Analyze

Diamond is a three-dimensional, highly ordered network of covalently bonded carbon atoms. Fullerenes are also made up of covalently bonded carbon atoms, but they form discrete structures instead of an extended network.

Solve

Fullerenes, with less extensive bonding, have a higher standard molar entropy than diamond.

Think About It

We would expect the standard molar entropy of graphite to be between that of diamond and the fullerenes because of its intermediate structure.

12.25. **Collect and Organize**

By considering the phase and size of a compound, we can rank the compounds in each series in order of increasing standard molar entropy.

Analyze

In general, compounds in the same phase that are larger have greater $S°$.

Solve

(a) $CH_4(g) < CF_4(g) < CCl_4(g)$

(b) $CH_2O(g) < CH_3CHO(g) < CH_3CH_2CHO(g)$

(c) $HF(g) < H_2O(g) < NH_3(g)$

Think About It

Larger molecules have greater $S°$ because they have more opportunities for internal and rotational motion.

12.27. **Collect and Organize**

Given ice cubes (the system) in a glass of lemonade (the surroundings), we are to determine the signs of ΔS for the system and for the surroundings as the ice cools the lemonade from 10.0°C to 0.0°C.

Analyze

Cooling is the result of decreased molecular motion, which is correlated to a decrease in entropy.

Solve

The sign of ΔS_{surr} is negative because the lemonade is cooling, which decreases the motion of the lemonade molecules. The sign of ΔS_{sys} is positive because the ice is melting. When a phase change from solid to liquid occurs in the lemonade, disorder, and therefore entropy of the substance, increases.

Think About It

Here, as long as $\Delta S_{sys} > -\Delta S_{surr}$, the process is spontaneous.

12.29. Collect and Organize

Given three combinations of signs for changes in entropy for the system, the surroundings, and the universe, we are to determine which combinations are possible.

Analyze

For a process to occur, the second law of thermodynamics must be obeyed. That law states that the combination of the change in entropy for a system and the change in entropy for the surroundings must be equal to the change in entropy for the universe. Therefore, ΔS_{univ}, which is equal to $\Delta S_{sys} + \Delta S_{surr}$, must be greater than zero for a process to occur and

$$\Delta S_{sys} + \Delta S_{surr} > 0$$

Solve

(a) If $\Delta S_{sys} > 0$ and $\Delta S_{surr} > 0$, then $\Delta S_{univ} > 0$ for all ΔS_{surr} and ΔS_{sys}. That combination of entropy changes is always possible.

(b) If $\Delta S_{sys} > 0$ and $\Delta S_{surr} < 0$, then $\Delta S_{univ} > 0$ when $\Delta S_{sys} > -\Delta S_{surr}$. That combination of entropy changes is possible.

(c) If $\Delta S_{sys} > 0$ and $\Delta S_{surr} > 0$, then ΔS_{univ} must be > 0 for all ΔS_{surr} and ΔS_{sys}. Therefore, ΔS_{univ} cannot be <0 for those changes in entropy for the system and the surroundings, so that combination of entropy changes is not possible.

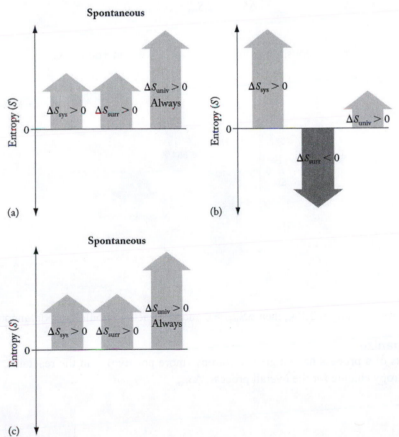

Think About It

In part c, ΔS_{univ} would be calculated as <0 if ΔS_{sys} and ΔS_{surr} were both <0 or if $\Delta S_{sys} + \Delta S_{surr} < 0$. That process, however, would not be spontaneous.

12.31. Collect and Organize

For the reaction of H_2S with O_2 to form S_8 and H_2O, we are asked to predict whether the change in entropy for the system increases or decreases.

Analyze

Both reactants and one product are gases in this reaction, but sulfur (S_8) is produced as a solid. Entropy for gases is much greater than that for solids.

Solve

Because there are 9 moles of gas as reactants and only 2 moles of gas and 3 moles of solid (with much lower entropy) as products, the number of gaseous, high-entropy species decreases for the reaction. Therefore, the entropy of the system decreases.

Think About It

That does not predict that the reaction is not spontaneous, however. We must also examine the change in entropy of the universe; if $\Delta S_{univ} > \Delta S_{sys}$, the reaction will theoretically proceed.

12.33. Collect and Organize

Given that a chemical reaction has $\Delta S_{sys} = -66.0$ J/K, we are to determine the value of ΔS_{surr} needed for the reaction to be nonspontaneous.

Analyze

For a nonspontaneous process,

$$\Delta S_{univ} = \Delta S_{sys} + \Delta S_{surr} < 0$$

Solve

For the reaction to be nonspontaneous, ΔS_{surr} must be less positive than +66.0 J/K.

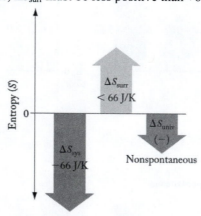

Think About It

If ΔS_{surr} is more positive than 66.0 J/K, then $\Delta S_{univ} > 0$ and the reaction would be spontaneous.

12.35. Collect and Organize

When the products of a process have a greater entropy (more positive) than the reactants, we are to determine the sign of the entropy change for the overall process, ΔS_{rxn}°.

Analyze

The entropy change for a process is calculated from

$$\Delta S^{\circ}_{rxn} = \sum n_{products} S^{\circ}_{products} - \sum n_{reactants} S^{\circ}_{reactants}$$

Solve

If $nS^{\circ}_{products} > mS^{\circ}_{reactants}$, then ΔS°_{rxn} for the process is positive.

Think About It

A positive ΔS°_{rxn} means that the process resulted in more energy dispersal in the reaction.

12.37. Collect and Organize

We are asked to predict the sign of ΔS°_{rxn} for precipitation reactions and explain our choice.

Analyze

Precipitation reactions involve the formation of a solid from dispersed ionic or molecular solutes in a solution.

Solve

The sign of ΔS°_{rxn} will be negative (less than zero) because in precipitation reactions solid products will have a lower molar entropy than the dispersed ions or molecules in the solution from which the precipitate forms.

Think About It

Precipitation reactions may be endothermic or exothermic, so the ΔS_{surr} may be either negative or positive.

12.39. Collect and Organize

Using values of S° for the reactants and products for four atmospheric reactions, we are to calculate ΔS°_{rxn}. Standard molar entropies are in Appendix 4.

Analyze

A change in entropy for a reaction is

$$\Delta S^{\circ}_{rxn} = \sum n_{products} S^{\circ}_{products} - \sum n_{reactants} S^{\circ}_{reactants}$$

Solve

(a) $\Delta S^{\circ}_{rxn} = (2 \text{ mol NO} \times 210.7 \text{ J/mol} \cdot \text{K})$

$\qquad - [(1 \text{ mol N}_2 \times 191.5 \text{ J/mol} \cdot \text{K}) + (1 \text{ mol O}_2 \times 205.0 \text{ J/mol} \cdot \text{K})]$

$\qquad = 24.9 \text{ J/K}$

(b) $\Delta S^{\circ}_{rxn} = (2 \text{ mol NO}_2 \times 240.0 \text{ J/mol} \cdot \text{K})$

$\qquad - [(2 \text{ mol NO} \times 210.7 \text{ J/mol} \cdot \text{K}) + (1 \text{ mol O}_2 \times 205.0 \text{ J/mol} \cdot \text{K})]$

$\qquad = -146.4 \text{ J/K}$

(c) $\Delta S^{\circ}_{rxn} = (1 \text{ mol NO}_2 \times 240.0 \text{ J/mol} \cdot \text{K})$

$\qquad - [(1 \text{ mol NO} \times 210.7 \text{ J/mol} \cdot \text{K}) + (\frac{1}{2} \text{ mol O}_2 \times 205.0 \text{ J/mol} \cdot \text{K})]$

$\qquad = -73.2 \text{ J/K}$

(d) $\Delta S^{\circ}_{rxn} = (1 \text{ mol N}_2\text{O}_4 \times 304.2 \text{ J/mol} \cdot \text{K}) - (2 \text{ mol NO}_2 \times 240.0 \text{ J/mol} \cdot \text{K})$

$\qquad = -175.8 \text{ J/K}$

Think About It

Because the balanced equation in part b is twice that in part c, the value of the entropy change for the reaction is doubled.

12.41. Collect and Organize

Given the ΔS°_{rxn} for the conversion of Cl and O_3 into ClO and O_2 and the S° for Cl, O_3, and O_2 from Appendix 4, we are to calculate S° for ClO.

Analyze

To calculate ΔS°_{rxn} we need $S^{\circ}_{O_3} = 238.8$ J/mol·K, $S^{\circ}_{Cl} = 165.2$ J/mol · K, and $S^{\circ}_{O_2} = 205.0$ J/mol·K along with the balanced equation

$$Cl(g) + O_3(g) \rightarrow ClO(g) + O_2(g)$$

The change in entropy for a reaction is

$$\Delta S^{\circ}_{rxn} = \sum n_{products} S^{\circ}_{products} - \sum n_{reactants} S^{\circ}_{reactants}$$

Solve

$$\Delta S^{\circ}_{rxn} = 19.9 \text{ J/K} = \left[\left(1 \text{ mol ClO} \times S^{\circ}_{ClO}\right) + \left(1 \text{ mol } O_2 \times 205.0 \text{ J/mol·K}\right)\right]$$
$$- \left[\left(1 \text{ mol Cl} \times 165.2 \text{ J/mol·K}\right) + \left(1 \text{ mol } O_3 \times 238.8 \text{ J/mol·K}\right)\right]$$
$$19.9 \text{J/K} = -199.0 \text{ J/K} + S^{\circ}_{ClO}$$
$$S^{\circ}_{ClO} = 218.9 \text{ J/mol·K}$$

Think About It

Because ΔS°_{rxn} is positive, the entropy for the system (the reaction) increases when Cl and O_3 react to form ClO and O_2.

12.43. Collect and Organize

We are to explain what the sign of ΔG tells us about the spontaneity of a reaction.

Analyze

Free energy is related to the entropy change of the universe by

$$\Delta G_{sys} = -T\Delta S_{univ}$$

Solve

When ΔS_{univ} is positive, the process is spontaneous, so for spontaneous processes according to the equation

$$\Delta G_{sys} = -T\Delta S_{univ}$$

ΔG_{sys} is negative. Likewise, for a nonspontaneous process, ΔG_{sys} is positive.

Think About It

A negative free energy change for a reaction means that the free energy at the end of the process is lower than at the beginning of the process.

12.45. Collect and Organize

We are to explain why 19th-century scientists believed that all exothermic reactions were spontaneous.

Analyze

A reaction is spontaneous when its free-energy change is negative according to the equation

$$\Delta G = \Delta H - T\Delta S$$

When H is negative, the reaction is exothermic.

Solve

The enthalpy of reactions is usually on the order of kilojoules per mole, whereas the entropy change for reactions is usually on the joules-per-mole scale. Therefore, many exothermic reactions are spontaneous because the magnitude of ΔH is greater than the magnitude of ΔS, so ΔG is usually negative, indicating a spontaneous reaction when the reaction is exothermic ($-\Delta H$).

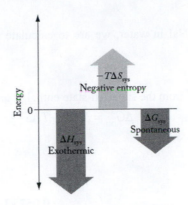

Think About It

The assumption that all exothermic reactions are spontaneous, though, is incorrect. If the value of $T\Delta S$ is more negative than the value of H, then ΔG would be positive and the reaction would be nonspontaneous.

12.47. Collect and Organize

For the sublimation of dry ice at room temperature, we are to determine the signs of ΔS, ΔH, and ΔG.

Analyze

The reaction describing the sublimation is

$$CO_2(s) \rightarrow CO_2(g)$$

Solve

ΔS is positive because the solid is subliming to a gas, which has much greater entropy.
ΔH is positive because heat is required to effect the phase change.
ΔG is negative because at room temperature (25°C) sublimation occurs spontaneously.

Think About It

The spontaneity of that process depends on temperature. At lower temperatures the value of $T\Delta S$ in the equation

$$\Delta G = \Delta H - T\Delta S$$

would not be great enough to give a negative value for ΔG.

12.49. Collect and Organize

For the processes described, we are to determine whether they are spontaneous.

Analyze

Spontaneous processes occur without outside intervention once they are started.

Solve

(a) A tornado forms spontaneously under certain weather conditions.
(b) A broken cell phone does not spontaneously fix itself.
(c) You will not spontaneously get an A in this course.
(d) Hot soup spontaneously cools before being served.
Therefore, (a) and (d) are spontaneous.

Think About It

Spontaneous processes are favored by enthalpy, entropy, or both according to the equation

$$\Delta G = \Delta H - T\Delta S$$

12.51. Collect and Organize

For the dissolution of NaBr and NaI in water, we are to calculate the value of $\Delta G°$ knowing the $\Delta H°_{sol}$ and $\Delta S°_{sol}$ for both of those soluble salts.

Analyze

We can calculate the value of $\Delta G°$ from the standard-state enthalpy and entropy of the reaction by using

$$\Delta G°_{rxn} = \Delta H°_{rxn} - T\Delta S°_{rxn}$$

where $T = 298$ K.

Solve

For NaBr:

$$\Delta G°_{rxn} = -0.60 \text{ kJ/mol} - \left(298 \text{ K} \times \frac{0.057 \text{ kJ}}{\text{mol} \cdot \text{K}}\right) = -18 \text{ kJ/mol}$$

For NaI:

$$\Delta G°_{rxn} = -7.5 \text{ kJ/mol} - \left(298 \text{ K} \times \frac{0.074 \text{ kJ}}{\text{mol} \cdot \text{K}}\right) = -30 \text{ kJ/mol}$$

Think About It

Be sure to use consistent units for $\Delta H°$ and $\Delta S°$ in calculations. In this problem, we have to change the $\Delta S°$ values given in units of J/mol · K to kJ/mol · K.

12.53. Collect and Organize

For the reaction of C(*s*) with $H_2O(g)$ to produce $H_2(g)$ and CO(*g*), we are to calculate the value of $\Delta G°_{rxn}$ from $\Delta G°_f$ in Appendix 4.

Analyze

We can calculate $\Delta G°_{rxn}$ by using the values for $\Delta G°_f$ for C(*s*), $H_2O(g)$, $H_2(g)$, and CO(*g*) in Appendix 4.

$$\Delta G°_{rxn} = \sum n_{products} \Delta G°_{f,products} - \sum n_{reactants} \Delta G°_{f,reactants}$$

Solve

$$\Delta G°_{rxn} = \left[\left(1 \text{ mol } H_2 \times 0.0 \text{ kJ/mol}\right) + \left(1 \text{ mol CO} \times -137.2 \text{ kJ/mol}\right)\right]$$
$$- \left[\left(1 \text{ mol } H_2O \times -228.6 \text{ kJ/mol}\right) + \left(1 \text{ mol C} \times 0.0 \text{ kJ/mol}\right)\right]$$
$$= 91.4 \text{ kJ}$$

Think About It

For that calculation we used carbon in the form of graphite. We would not want to use diamond as a reactant.

12.55. Collect and Organize

For the reaction of 1 mol of NO with ½ mol of oxygen given to produce 1 mol of NO_2, we are to use the appropriate values of $\Delta G°_f$ from Appendix 4 to calculate the value of $\Delta G°_{rxn}$.

Analyze

The change in free energy for that reaction is found from $\Delta G°_f$ values for the reactants and products by using

$$\Delta G°_{rxn} = \sum n_{products} \Delta G°_{f,products} - \sum n_{reactants} \Delta G°_{f,reactants}$$

Solve

$$\Delta G°_{rxn} = \left(1 \text{ mol } NO_2 \times 51.3 \text{ kJ/mol}\right) - \left[\left(1 \text{ mol NO} \times 86.6 \text{ kJ/mol}\right) + \left(0.5 \text{ mol } O_2 \times 0.0 \text{ kJ/mol}\right)\right]$$
$$= -35.3 \text{ kJ}$$

Think About It

That combustion reaction is spontaneous under standard conditions. We might also predict that the change in entropy for that reaction would be negative since 1 mol of product gases is formed from 1.5 mol of reactant gases.

12.57. Collect and Organize

For the reaction of 1 mol of gaseous sulfur trioxide with 1 mol of gaseous water to form 1 mol of liquid sulfuric acid, we are to use the appropriate values of ΔG_f° from Appendix 4 to calculate the value of ΔG_{rxn}°.

Analyze

We can use the values in Appendix 4 for ΔG_f° of the products and reactants to calculate ΔG_{rxn}°:

$$\Delta G_{rxn}^\circ = \sum n_{products} \Delta G_{f,products}^\circ - \sum n_{reactants} \Delta G_{f,reactants}^\circ$$

Solve

$$\Delta G_{rxn}^\circ = \left(1 \text{ mol } H_2SO_4 \times -690.0 \text{ kJ/mol}\right) - \left[\left(1 \text{ mol } SO_3 \times -371.1 \text{ kJ/mol}\right) + \left(1 \text{ mol } H_2O \times -228.6 \text{ kJ/mol}\right)\right]$$

$$= -90.3 \text{ kJ}$$

Think About It

Because ΔG_{rxn}° is negative, that reaction is spontaneous at standard conditions.

12.59. Collect and Organize

We consider whether exothermic reactions are spontaneous only at low temperature.

Analyze

Spontaneity is shown by a negative free-energy value (ΔG) in the equation

$$\Delta G = \Delta H - T\Delta S$$

For exothermic reactions, we also know that ΔH is negative.

Solve

No. As we can see from the equation, if ΔS is positive, the value of $T\Delta S$ would be positive. Then ΔG would always be negative, regardless of the temperature, and the exothermic reaction would be spontaneous at all temperatures.

Think About It

If ΔS is negative for an exothermic reaction, then it is true that the reaction would be spontaneous only at low temperatures.

12.61. Collect and Organize

For the reaction of $C(s)$ with $H_2O(g)$ to produce $H_2(g)$ and $CO(g)$ in Problem 12.53, we are to predict the lowest temperature at which the reaction is spontaneous.

Analyze

To calculate the lowest temperature at which CO and H_2 form from C and steam, we must first calculate ΔH_{rxn}° and ΔS_{rxn}° by using Appendix 4. For a spontaneous reaction, ΔG is negative. Therefore, if we set $\Delta G = 0$ and solve for T, that would give the temperature at which the reaction changes spontaneity.

$$\Delta G = 0 = \Delta H - T\Delta S$$

$$T = \frac{\Delta H}{\Delta S}$$

We must be sure to have consistent units for ΔH and ΔS for that calculation.

Solve

First calculating ΔH°_{rxn} and ΔS°_{rxn}:

$$\Delta H^\circ_{rxn} = \left[\left(1 \text{ mol } H_2 \times 0.0 \text{ kJ/mol}\right) + \left(1 \text{ mol } CO \times -110.5 \text{ kJ/mol}\right)\right]$$
$$- \left[\left(1 \text{ mol } H_2O \times -241.8 \text{ kJ/mol}\right) + \left(1 \text{ mol } C \times 0.0 \text{ kJ/mol}\right)\right]$$
$$= 131.3 \text{ kJ}$$

$$\Delta S^\circ_{rxn} = \left[\left(1 \text{ mol } H_2 \times 130.6 \text{ J/mol} \cdot K\right) + \left(1 \text{ mol } CO \times 197.7 \text{ J/mol} \cdot K\right)\right]$$
$$- \left[\left(1 \text{ mol } H_2O \times 188.8 \text{ J/mol} \cdot K\right) + \left(1 \text{ mol } C \times 5.7 \text{ J/mol} \cdot K\right)\right]$$
$$= 133.8 \text{ J/K}$$

Solving for T when $\Delta G^\circ_{rxn} = 0$ gives

$$T = \frac{131.3 \text{ kJ}}{0.1338 \text{ kJ/K}} = 981.3 \text{ K, or } 708.2°C$$

The lowest temperature at which this reaction is spontaneous is just above 981.3 K, or 708.2°C.

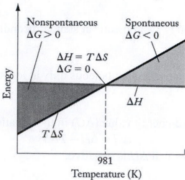

Think About It

As in Problem 12.53, we used carbon in the form of graphite to calculate the changes in enthalpy and entropy for the reaction. We would not want to use diamond as a reactant.

12.63. **Collect and Organize**

For the vaporization of hydrogen peroxide at 1.00 atm, we will first calculate $\Delta H°$ and $\Delta S°$. Assuming that calculated quantities (using data listed in Appendix 4 at 298 K) are not temperature dependent, we are to calculate the boiling point of hydrogen peroxide.

Analyze

To calculate both $\Delta H°$ and $\Delta S°$ for the phase change

$$H_2O_2(\ell) \rightarrow H_2O_2(g)$$

we need the following (from Appendix 4): ΔH°_f of $H_2O_2(\ell)$ = −187.8 kJ/mol, ΔH°_f of $H_2O_2(g)$ = −136.3 kJ/mol, $S°$ of $H_2O(\ell)$ = 109.6 J/mol · K, and $S°$ of $H_2O_2(g)$ = 232.7 J/mol · K. To calculate the boiling point, we can rearrange the Gibbs free-energy equation to solve for the temperature where ΔG°_{rxn} = 0 because at the boiling point $H_2O_2(\ell)$ and $H_2O_2(g)$ are in equilibrium:

$$\Delta G^\circ_{rxn} = 0 = \Delta H^\circ_{rxn} - T\Delta S^\circ_{rxn}$$

$$T_b = \frac{\Delta H^\circ_{rxn}}{\Delta S^\circ_{rxn}}$$

Solve

$$\Delta H^\circ_{rxn} = \left(1 \text{ mol } H_2O_2(g) \times -136.3 \text{ kJ/mol}\right) - \left(1 \text{ mol } H_2O_2(\ell) \times -187.8 \text{ kJ/mol}\right)$$

$$= 51.5 \text{ kJ}$$

$$\Delta S^\circ_{rxn} = \left(1 \text{ mol } H_2O_2(g) \times 232.7 \text{ J/mol} \cdot K\right) - \left(1 \text{ mol } H_2O_2(\ell) \times 109.6 \text{ J/mol} \cdot K\right)$$

$$= 123.1 \text{ J/K, or } 0.123 \text{ kJ/K}$$

$$T_b = \frac{51.5 \text{ kJ}}{0.123 \text{ kJ/K}} = 418 \text{ K, or } 145°C$$

Think About It

Because the calculated value of the boiling point temperature is a little different from the actual T_b (160°C), we can see that some temperature dependence exists for ΔH and ΔS.

12.65. Collect and Organize

For each reaction we will calculate ΔH°_{rxn} and ΔS°_{rxn} to determine the sign of each to predict whether each reaction is spontaneous only at low temperatures, only at high temperatures, or at all temperatures.

Analyze

The standard free energy and standard entropy of a reaction can be calculated by using Appendix 4 values for ΔH°_f and ΔS°.

$$\Delta H^\circ_{rxn} = \sum n \, \Delta H^\circ_{f,products} - \sum m \, \Delta H^\circ_{f,reactants}$$

$$\Delta S^\circ_{rxn} = \sum n \, \Delta S^\circ_{products} - \sum m \, \Delta S^\circ_{reactants}$$

Spontaneity is indicated by a negative free-energy value through the equation

$$\Delta G^\circ_{rxn} = \Delta H^\circ_{rxn} - T\Delta S^\circ_{rxn}$$

We can see that an exothermic reaction will be spontaneous at all temperatures if entropy is positive and at low temperatures if entropy is negative. An endothermic reaction will never be spontaneous at any temperature if the entropy is negative but will be spontaneous at high temperature if entropy is positive.

Solve

(a) $\Delta H^\circ_{rxn} = \left(2 \text{ mol } NO_2 \times 33.2 \text{ kJ/mol}\right)$

$$- \left[\left(2 \text{ mol } NO \times 90.3 \text{ kJ/mol}\right) + \left(1 \text{ mol } O_2 \times 0.0 \text{ kJ/mol}\right)\right]$$

$$= -114.2 \text{ kJ}$$

$$\Delta S^\circ_{rxn} = \left(2 \text{ mol } NO_2 \times 240.0 \text{ J/mol} \cdot K\right)$$

$$- \left[\left(2 \text{ mol } NO \times 210.7 \text{ J/mol} \cdot K\right) + \left(1 \text{ mol } O_2 \times 205.0 \text{ J/mol} \cdot K\right)\right]$$

$$= -146.4 \text{ J/K}$$

Because that reaction is an exothermic reaction with negative entropy, the reaction will be spontaneous (i) only at low temperature.

(b) Because that reaction is an exothermic reaction with negative entropy, the reaction will be spontaneous (i) only at low temperature.

(c) $\Delta H^\circ_{rxn} = \left[\left(2 \text{ mol } H_2O \times -241.8 \text{ kJ/mol}\right) + \left(1 \text{ mol } N_2O \times 82.1 \text{ kJ/mol}\right)\right]$

$$- \left(1 \text{ mol } NH_4NO_3 \times -365.6 \text{ kJ/mol}\right)$$

$$= -35.9 \text{ kJ}$$

$$\Delta S^{\circ}_{rxn} = \left[\left(2 \text{ mol } H_2O \times 188.8 \text{ J/mol}\cdot K \right) + \left(1 \text{ mol } N_2O \times 219.9 \text{ J/mol}\cdot K \right) \right]$$
$$- \left(1 \text{ mol } NH_4NO_3 \times 151.1 \text{ J/mol}\cdot K \right)$$
$$= 446.4 \text{ J/K}$$

Because that reaction is an exothermic reaction with positive entropy, the reaction will be spontaneous (iii) at all temperatures.

Think About It

For a reaction to be spontaneous only at high temperatures (ii), it would be not favored by enthalpy (endothermic) and would be favored by entropy (positive entropy change).

12.67. Collect and Organize

For the reaction of CO with H_2 to produce CH_3OH, we are to use data in Appendix 4 to calculate ΔG°_{rxn} and then calculate the value of ΔG_{rxn} at 475 K.

Analyze

We can calculate the standard free energy by using Appendix 4 values for ΔG°_f:

$$\Delta G^{\circ}_{rxn} = \sum n \Delta G^{\circ}_{f,products} - \sum m \Delta G^{\circ}_{f,reactants}$$

To determine the value of ΔG_{rxn} 475 K, we will need to first calculate the values of ΔH° and ΔS° for the reaction through

$$\Delta H^{\circ}_{rxn} = \sum n_{products} \Delta H^{\circ}_{f,products} - \sum n_{reactants} \Delta H^{\circ}_{f,reactants}$$

$$\Delta S^{\circ}_{rxn} = \sum n_{products} \Delta S^{\circ}_{products} - \sum n_{reactants} \Delta S^{\circ}_{reactants}$$

and then apply the equation

$$\Delta G_{rxn} = \Delta H_{rxn} - T\Delta S_{rxn}$$

If the value of ΔG at 475 K is negative, the reaction is spontaneous.

Solve

(a) The value ΔG°_{rxn} is

$$\Delta G^{\circ}_{rxn} = \left(1 \text{ mol } CH_3OH \times -166.4 \text{ kJ/mol} \right)$$
$$- \left[\left(1 \text{ mol } CO \times -137.2 \text{ kJ/mol} \right) + \left(2 \text{ mol } H_2 \times 0.0 \text{ kJ/mol} \right) \right]$$
$$= -29.2 \text{ kJ}$$

That reaction is spontaneous under standard conditions.

(b) $$\Delta H^{\circ}_{rxn} = \left(1 \text{ mol } CH_3OH \times -238.7 \text{ kJ/mol} \right)$$
$$- \left[\left(1 \text{ mol } CO \times -110.5 \text{ kJ/mol} \right) + \left(2 \text{ mol } H_2 \times 0.0 \text{ kJ/mol} \right) \right]$$
$$= -128.2 \text{ kJ}$$

$$\Delta S^{\circ}_{rxn} = \left(1 \text{ mol } CH_3OH \times 126.8 \text{ J/mol}\cdot K \right)$$
$$- \left[\left(1 \text{ mol } CO \times 197.7 \text{ J/mol}\cdot K \right) + \left(2 \text{ mol } H_2 \times 130.6 \text{ J/mol}\cdot K \right) \right]$$
$$= -332.1 \text{ J/K}$$

$$\Delta G_{rxn} = \Delta H_{rxn} - T\Delta S_{rxn} = -128.2 \text{ kJ} - 475K\left(-0.3321 \text{ kJ/K} \right) = 29.5 \text{ kJ}$$

That reaction is not spontaneous at 475 K.

Think About It

For that reaction as the temperature is raised, the $T\Delta S$ term in the equation $\Delta G = \Delta H - T\Delta S$ becomes more negative because ΔS for the reaction is negative. That negative term, subtracted from the ΔH value, makes ΔG more positive.

12.69. **Collect and Organize**

We are asked to describe how two reactions must complement each other so that a nonspontaneous reaction can be driven by a decrease in free energy of a spontaneous reaction.

Analyze

The overall free energy of the combined reactions must give an overall negative free energy value.

Solve

The spontaneous reaction must have a decrease in free energy that is greater than the increase in free energy for the nonspontaneous reaction.

Think About It

A nonspontaneous process can therefore be coupled to a spontaneous one and thus be "forced" to go.

12.71. **Collect and Organize**

We are to explain the importance of why some steps in glycolysis to form ATP from ADP are spontaneous.

Analyze

A reaction that is spontaneous has a negative value for its free energy change.

Solve

ATP has the important function of energy storage and transfer. The cell must make that important molecule and so the overall process must be spontaneous. Even if some of the glycolysis steps are nonspontaneous, at least some glycolysis steps must be spontaneous to produce ATP.

Think About It

In Figure 12.22, the only difference between ADP and ATP is the presence of an additional phosphate group in ATP.

12.73. **Collect and Organize**

Here we explore the free energy of the steps used in the steam–methane reforming process to produce hydrogen from methane.

Analyze

We will use the values for ΔG_f° in Appendix 4 for the reactant and products to determine the free energy for the reaction of methane with water and for carbon monoxide with water. The free energy of each reaction can be calculated using

$$\Delta G_{rxn}^\circ = \sum n \Delta G_{f,products}^\circ - \sum m \Delta G_{f,reactants}^\circ$$

Then we will couple the two reactions to calculate the ΔG_{rxn}° for the overall process and predict whether it is spontaneous under standard conditions.

Solve

(a) For the reaction of 1 mol of methane with 1 mol of water to form 1 mol of carbon monoxide and 3 mol of hydrogen:

$$\Delta G_{rxn}^\circ = \left[\left(1 \text{ mol CO} \times -137.2 \text{ kJ/mol} \right) + \left(3 \text{ mol } H_2 \times 0.0 \text{ kJ/mol} \right) \right]$$
$$- \left[\left(1 \text{ mol } CH_4 \times -50.8 \text{ kJ/mol} \right) + \left(1 \text{ mol } H_2O \times -228.6 \text{ kJ/mol} \right) \right]$$
$$= 142.2 \text{ kJ}$$

(b) For the reaction of 1 mol of carbon monoxide with 1 mol of water to form 1 mol of carbon dioxide and 1 mol of hydrogen:

$$\Delta G_{rxn}^\circ = \left[\left(1 \text{ mol } CO_2 \times -394.4 \text{ kJ/mol} \right) + \left(1 \text{ mol } H_2 \times 0.0 \text{ kJ/mol} \right) \right]$$
$$- \left[\left(1 \text{ mol CO} \times -137.2 \text{ kJ/mol} \right) + \left(1 \text{ mol } H_2O \times -228.6 \text{ kJ/mol} \right) \right]$$
$$= -28.6 \text{ kJ}$$

(c) Adding those reactions:

$$CH_4(g) + H_2O(g) \rightarrow CO(g) + 3\,H_2(g)$$
$$\underline{+\ CO(g) + H_2O(g) \rightarrow CO_2(g) + H_2(g)}$$
$$CH_4(g) + 2\,H_2O(g) \rightarrow CO_2(g) + 4\,H_2(g)$$

(d) $\Delta G^\circ_{overall} = 142.2 + (-28.6) = 113.6$ kJ. That reaction is not spontaneous at standard conditions.

Think About It

If the temperature is raised, however, the reaction does become spontaneous.

12.75. **Collect and Organize**

We are asked to calculate the free energy for the reaction between Fe_2O_3 and CO to form Fe and CO_2 at 1450°C.

Analyze

To determine the value of ΔG_{rxn} at 1450°C (1723 K), we will need to first calculate the values of ΔH° and ΔS° for the reaction through

$$\Delta H^\circ_{rxn} = \sum n_{producrs}\,\Delta H^\circ_{f,products} - \sum n_{reactants}\,\Delta H^\circ_{f,reactants}$$

$$\Delta S^\circ_{rxn} = \sum n_{products}\,\Delta S^\circ_{products} - \sum n_{reactants}\,\Delta S^\circ_{reactants}$$

and then apply the equation with $T = 1723$ K.

$$\Delta G_{rxn} = \Delta H_{rxn} - T\Delta S_{rxn}$$

Solve

$$\Delta H^\circ_{rxn} = \left[\left(2\text{ mol Fe} \times 0.0\text{ kJ/mol}\right) + \left(3\text{ mol CO}_2 \times -393.5\text{ kJ/mol}\right)\right]$$
$$- \left[\left(1\text{ mol Fe}_2\text{O}_3 \times -824.2\text{ kJ/mol}\right) + \left(3\text{ mol CO} \times -110.5\text{ kJ/mol}\right)\right]$$
$$= -24.8\text{ kJ}$$

$$\Delta S^\circ_{rxn} = \left[\left(2\text{ mol Fe} \times 27.3\text{ J/mol}\cdot\text{K}\right) + \left(3\text{ mol CO}_2 \times 213.8\text{ J/mol}\cdot\text{K}\right)\right]$$
$$- \left[\left(1\text{ mol Fe}_2\text{O}_3 \times 87.4\text{ J/mol}\cdot\text{K}\right) + \left(3\text{ mol CO} \times 197.7\text{ J/mol}\cdot\text{K}\right)\right]$$
$$= 15.5\text{ J/K}$$

$$\Delta G_{rxn} = \Delta H_{rxn} - T\Delta S_{rxn} = -24.8\text{ kJ} - 1723\text{K}\left(0.0155\text{ kJ/K}\right) = -51.5\text{ kJ}$$

Think About It

The reaction is spontaneous at that temperature, as indicated by the negative value of the free energy.

12.77. **Collect and Organize**

Given the boiling point (23.8°C) and molar heat of vaporization (24.8 kJ/mol) of trichlorofluoromethane, we are to calculate the molar entropy of evaporation of that liquid.

Analyze

Because phase changes are isothermal processes (meaning they take place at constant temperature),

$$\Delta S = \frac{q_{rev}}{T}$$

We also know that $q_{rev} = \Delta H_{rxn}$, so

$$\Delta S_{rxn} = \frac{\Delta H_{rxn}}{T_b}$$

where ΔH_{rxn} is the enthalpy of the phase change and T is the temperature at which the phase change occurs, expressed in kelvin.

Solve

$$\Delta S = \frac{24.8 \text{ kJ/mol}}{296.95 \text{ K}} = 8.35 \times 10^{-2} \text{ kJ/mol} \cdot \text{K, or } 83.5 \text{ J/mol} \cdot \text{K}$$

Think About It

The positive entropy change makes sense because in vaporizing the chlorofluorocarbon, we are increasing the entropy of the system.

12.79. Collect and Organize

For the decomposition of solid NH_4Cl into gaseous NH_3 and HCl, we are to calculate the temperature at which $\Delta G^\circ_{rxn} = 0$.

Analyze

To answer that, we need to first calculate ΔH°_{rxn} and ΔS°_{rxn} by using values in Appendix 4. Then we can calculate T by rearranging the free-energy equation:

$$\Delta G^\circ_{rxn} = \Delta H^\circ_{rxn} - T\Delta S^\circ_{rxn} = 0$$

$$T = \frac{\Delta H^\circ_{rxn}}{\Delta S^\circ_{rxn}}$$

Solve

$$\Delta H^\circ_{rxn} = \left[\left(1 \text{ mol } NH_3 \times -46.1 \text{ kJ/mol}\right) + \left(1 \text{ mol } HCl \times -92.3 \text{ kJ/mol}\right) \right]$$
$$- \left(1 \text{ mol } NH_4Cl \times -314.4 \text{ kJ/mol}\right)$$
$$= 176.0 \text{ kJ}$$

$$\Delta S^\circ_{rxn} = \left[\left(1 \text{ mol } NH_3 \times 192.5 \text{ J/mol} \cdot \text{K}\right) + \left(1 \text{ mol } HCl \times 186.9 \text{ J/mol} \cdot \text{K}\right) \right]$$
$$- \left(1 \text{ mol } NH_4Cl \times 94.6 \text{ J/mol} \cdot \text{K}\right)$$
$$= 284.8 \text{ J/K}$$

$$T = \frac{176.0 \text{ kJ}}{0.2848 \text{ kJ/K}} = 618.0 \text{ K, or } 344.8^\circ\text{C}$$

Think About It

The reaction is favored by entropy but not by enthalpy. It is spontaneous at high temperature.

12.81. Collect and Organize

For the gas-phase reaction of NO with H_2 to form N_2 and H_2O, we can use the values of ΔG°_f in Appendix 4 to calculate ΔG°_{rxn} and then determine whether the reaction is spontaneous.

Analyze

We can use the values for ΔG°_f of the products and reactants to calculate ΔG°_{rxn} :

$$\Delta G^\circ_{rxn} = \sum n_{products} \Delta G^\circ_{f,products} - \sum n_{reactants} \Delta G^\circ_{f,reactants}$$

If ΔG°_{rxn} is negative, then the reaction is spontaneous.

Solve

$$\Delta G^\circ_{rxn} = \left[\left(1 \text{ mol } N_2 \times 0.0 \text{ kJ/mol}\right) + \left(2 \text{ mol } H_2O \times -228.6 \text{ kJ/mol}\right) \right]$$
$$- \left[\left(2 \text{ mol } NO \times 86.6 \text{ kJ/mol}\right) + \left(2 \text{ mol } H_2 \times 0.0 \text{ kJ/mol}\right) \right]$$
$$= -630.4 \text{ kJ}$$

Yes, the reaction is spontaneous at standard temperature and pressure.

Think About It

Because 4 mol of gas combines as reactants and forms 3 mol of gas as products, we predict that ΔS°_{rxn} is negative. From the data in Appendix 4, we see that this is indeed the case:

$$\Delta S^{\circ}_{rxn} = \left[\left(1 \text{ mol N}_2 \times 191.5 \text{ J/mol} \cdot \text{K} \right) + \left(2 \text{ mol H}_2\text{O} \times 188.8 \text{ J/mol} \cdot \text{K} \right) \right]$$
$$- \left[\left(2 \text{ mol NO} \times 210.7 \text{ J/mol} \cdot \text{K} \right) + \left(2 \text{ mol H}_2 \times 130.6 \text{ J/mol} \cdot \text{K} \right) \right]$$
$$= -113.5 \text{ J/K}$$

12.83. Collect and Organize

For HCN we are to calculate the normal boiling point (T_b) given ΔH°_f and S° values for $HCN(\ell)$ and $HCN(g)$.

Analyze

At the boiling point, $\Delta G = 0$ because the system is at equilibrium. Therefore,

$$\Delta G = 0 = \Delta H^{\circ}_{vap} - T_b \Delta S^{\circ}_{vap}$$

$$T = \frac{\Delta H^{\circ}_{vap}}{\Delta S^{\circ}_{vap}}$$

Solve

For the vaporization process

$$HCN(\ell) \rightarrow HCN(g)$$
$$\Delta H_{vap} = \left(1 \text{ mol HCN}(g) \times 135.1 \text{ kJ/mol} \right) - \left(1 \text{ mol HCN}(\ell) \times 108.9 \text{ kJ/mol} \right)$$
$$= 26.2 \text{ kJ}$$
$$\Delta S_{vap} = \left(1 \text{ mol HCN}(g) \times 202 \text{ J/mol} \cdot \text{K} \right) - \left(1 \text{ mol HCN}(\ell) \times 113 \text{ J/mol} \cdot \text{K} \right)$$
$$= 89 \text{ J/K}$$

$$T_b = \frac{26.2 \text{ kJ}}{0.089 \text{ kJ/K}} = 294 \text{ K}$$

Think About It

The actual boiling point of $HCN(\ell)$ is 299 K, just about room temperature, so our calculation is approximately correct.

12.85. Collect and Organize

Given the melting point T_m (3422°C) and ΔH_{fus} (35.4 kJ/mol) of tungsten, we are to calculate ΔS_{fus}.

Analyze

The reaction for that process is

$$W(s) \rightarrow W(\ell)$$

At the melting point, $\Delta G = 0$ because the system is at equilibrium. Therefore,

$$\Delta G = 0 = \Delta H_{fus} - T_m \Delta S_{fus}$$

$$\Delta S_{fus} = \frac{\Delta H_{fus}}{T_m}$$

Solve

$$\Delta S_{fus} = \frac{35.4 \text{ kJ/mol}}{3695 \text{ K}} = 0.00958 \text{ kJ/mol} \cdot \text{K, or } 9.58 \text{ J/mol} \cdot \text{K}$$

Think About It

Tungsten has the highest melting point of all the metals, making it useful as a filament in incandescent lightbulbs.

12.87. **Collect and Organize**

Given the transition temperature (369 K) and the enthalpy change for the interconversion of two allotropes of S_8 (297 J/mol), we are to calculate the entropy change for the transition.

Analyze

At the transition temperature, $\Delta G = 0$ because the system is at equilibrium. Therefore,

$$\Delta G = 0 = \Delta H_{trans} - T\Delta S_{trans}$$

$$\Delta S_{trans} = \frac{\Delta H_{trans}}{T}$$

Solve

$$\Delta S_{trans} = \frac{297 \text{ J/mol}}{369 \text{ K}} = 0.805 \text{ J/mol} \cdot \text{K}$$

Think About It

That transition, not favored by enthalpy (ΔH positive), is favored by entropy (ΔS positive).

12.89. **Collect and Organize**

Using the values given for ΔH_f° and S° for $CaCO_3$, CaO, and CO_2, we are to explain why S° of $CaCO_3$ is higher than that of CaO and calculate the temperature at which the pressure of CO_2 over $CaCO_3$ is 1.0 atm.

Analyze

By considering the phase and size of each compound, we can rank the compounds in order of increasing standard molar entropy. The reaction involved is

$$CaCO_3(s) \rightarrow CaO(s) + CO_2(g)$$

To calculate the temperature at which the partial pressure of CO_2 is 1.0 atm, we must recognize that at that temperature the reaction will be at equilibrium, $\Delta G = 0$.

Solve

S° for $CaCO_3$ is greater than S° for CaO because more atoms are in $CaCO_3$.

To calculate the temperature at which the pressure of CO_2 is 1.0 atm, we must first calculate ΔH_{rxn}° and ΔS_{rxn}°:

$$\Delta H_{rxn}^\circ = \left[\left(1 \text{ mol CaO} \times -636 \text{ kJ/mol}\right) + \left(1 \text{ mol CO}_2 \times -394 \text{ kJ/mol}\right)\right]$$

$$-\left(1 \text{ mol CaCO}_3 \times -1207 \text{ kJ/mol}\right)$$

$$= 177 \text{ kJ}$$

$$\Delta S_{rxn}^\circ = \left[\left(1 \text{ mol CaO} \times 40 \text{ J/mol} \cdot \text{K}\right) + \left(1 \text{ mol CO}_2 \times 214 \text{ J/mol} \cdot \text{K}\right)\right]$$

$$-\left(1 \text{ mol CaCO}_3 \times 93 \text{ J/mol} \cdot \text{K}\right)$$

$$= 161 \text{ J/K}$$

$$\Delta G = 0 = 177 \text{ kJ} - T \times 0.161 \text{ kJ/K}$$

$$T = 1099 \text{ K, or } 826°C$$

Think About It

Although that reaction is endothermic, it is favored by entropy and so is spontaneous at high temperature.

12.91. **Collect and Organize**

For the process in which a linear chain dicarboxylic acid decomposes to CO_2 and an alkylcarboxylic acid, we are asked to predict the sign of ΔH and ΔS. We are also to predict the spontaneity for the reverse reaction.

Analyze

(a) To predict the signs of ΔH and ΔS, we should recall that breaking bonds takes energy, but energy is released when bonds are formed; also, the generation of gaseous products from a solid or liquid reactant increases the disorder of the system.

Solve

(a) We can estimate the enthalpy of reaction by using the bond energy values in Table 4.6 in the textbook. In the reaction, 1 mol of C—C bonds (between one carboxylic acid group and the alkyl chain), 1 mol of C—O bonds, 1 mol of C=O bonds, and 1 mol of O—H bonds (all from the carboxylic acid group) are broken, whereas 1 mol of C—H bonds and 2 mol of C=O bonds (in CO_2) are formed.

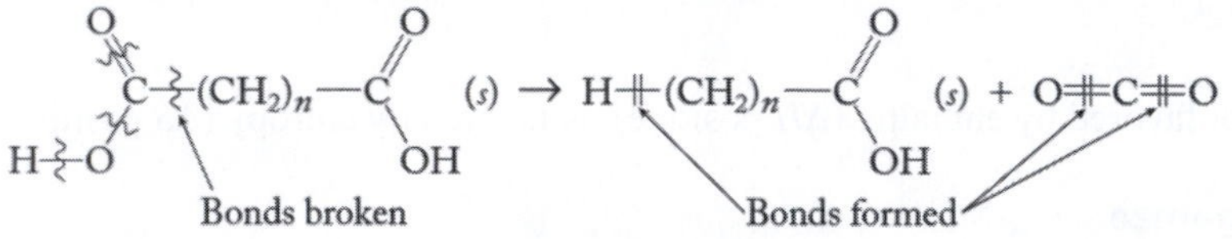

Using the bond energy values, we get an enthalpy of reaction of

$$\Delta H = \left[\left(1 \times 348 \text{ kJ/mol}\right) + \left(1 \times 358 \text{ kJ/mol}\right) + \left(1 \times 743 \text{ kJ/mol}\right) + \left(1 \times 463 \text{ kJ/mol}\right)\right]$$
$$+ \left[\left(1 \times -413 \text{ kJ/mol}\right) + \left(2 \times -799 \text{ kJ/mol}\right)\right]$$
$$= -99 \text{ kJ}$$

The sign of ΔH is negative, and the reaction is exothermic.

Because a gaseous product (CO_2) is formed in that reaction, ΔS is positive.

(b) The reverse reaction would have $+\Delta H$ and $-\Delta S$ and, therefore, would never be spontaneous.

Think About It

The forward reaction, being favored by both entropy and enthalpy, is spontaneous at all temperatures.

CHAPTER 13 | Chemical Kinetics: Clearing the Air

13.1. Collect and Organize

For the plot of Figure P13.1, we are to identify which curves represent $[N_2O]$ and $[O_2]$ over time for the conversion of N_2O to N_2 and O_2 according to the equation

$$2\,N_2O(g) \rightarrow 2\,N_2(g) + O_2(g)$$

Analyze

As the reaction proceeds, the concentration of the reactant, N_2O, decreases and the concentration of the product, O_2, increases. The rate at which N_2O is used up in the reaction is twice the rate at which O_2 is produced.

Solve

The green line represents $[N_2O]$, and the red line represents $[O_2]$.

Think About It

$[N_2]$, represented by the blue line, increases twice as fast as $[O_2]$ because two N_2 molecules are produced for every one O_2 molecule in the reaction.

13.3. Collect and Organize

For three initial concentrations of reactant A shown in Figure P13.3, we are to choose which would have the fastest rate for the conversion 2 A→B.

Analyze

We are given that the reaction is second order in A. The rate law is written as follows:

$$\text{Rate} = k[A]^2$$

As the concentration of A increases, the rate increases.

Solve

Figure P13.3(b) has the fastest reaction rate because it has the highest concentration of A.

Think About It

The higher the concentration of reactant molecules, the more often they collide, which increases the rate of the reaction.

13.5. Collect and Organize

For the reaction profile in Figure P13.5, we are to identify the parts of the energy diagram.

Analyze

The energy of the reactants is indicated at the beginning of the reaction; the energy of the products is indicated at the end of the reaction. The activation energy is the energy that the reaction must attain to form products; for the forward reaction it will be the energy change from the reactants to the energy of the transition state, and for the reverse reaction it will be the energy change from the products to the energy of the transition state. The energy change of the reaction is the energy difference between that of the products and that of the reactants.

Solve

(a) The energy of the reactants is shown as (1).
(b) The energy of the products is shown as (5).
(c) The activation energy of the forward reaction is shown as (2).
(d) The activation energy for the reverse reaction is shown as (4).
(e) The energy change of the reaction is shown as (3).

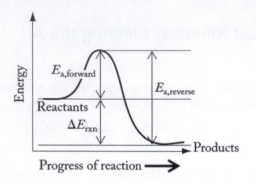

Think About It

Be careful not to assume that reaction (b) is slow because it is endothermic. The rate of a reaction does not depend on the relative energies of the reactants and the products.

13.7. **Collect and Organize**

For the three reaction profiles in Figure P13.7, we are to choose the one that has the smallest rate constant.

Analyze

The rate of a reaction (and therefore the rate constant) is determined by the activation energy (E_a) of the slowest step. All the reactions shown consist of a single step. The E_a is the energy difference between the reactants (on the left-hand side of the graph) and the transition state (the highest point on the reaction profile curve).

Solve

Reaction profile (b) has the largest E_a and therefore is the slowest reaction.

Think About It

Be careful not to assume that reaction (b) is slow because it is endothermic. The rate of a reaction does not depend on the relative energies of the reactants and the products.

13.9. **Collect and Organize**

We are to match the reaction profile in Figure P13.9 with the possible reactions given.

Analyze

The reaction profile shows a two-step reaction that has a slightly larger activation energy for its second step than for its first step.

Solve

Reaction (c) is correct because it is a two-step reaction with the first step faster than the second.

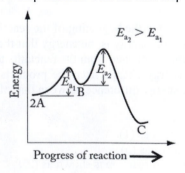

Think About It

Reaction (b), which occurs in a single step, would show only one transition state and one activation energy in its reaction profile.

13.11. **Collect and Organize**

Given the reaction profile of an uncatalyzed reaction (Figure P13.11), we are to choose the reaction profile corresponding to the catalyzed reaction.

Analyze

A catalyst increases the rate of reaction by decreasing the activation energy of the reaction through an alternative pathway to the products. That pathway usually involves more steps.

Solve

Reaction profile (b) correctly shows the catalyzed reaction.

Think About It

Reaction profiles (a) and (c) cannot be correct because the initial uncatalyzed endothermic reaction is represented as exothermic. A catalyst cannot change the relative energies of the reactants and products in a reaction.

13.13. **Collect and Organize**

Of the highlighted elements in Figure P13.13, we are to choose which forms gaseous oxides associated with photochemical smog.

Analyze

Photochemical smog is the result of sunlight interacting with NO_x produced by automobile emissions and volatile organic compounds (VOCs) released into the atmosphere.

Solve

Nitrogen (light blue) forms the volatile oxides that are components of photochemical smog.

Think About It

Sunlight causes a reaction of NO_x and VOCs to produce peroxyacyl nitrates that are very irritating to the lungs.

13.15. **Collect and Organize**

Of the highlighted elements in Figure P13.15, we are to identify which are widely used as heterogeneous catalysts.

Analyze

Heterogeneous catalysts have a different phase from the reactants. We read in Section 13.6 about the specific metals used in catalytic converters.

Solve

Both the transition metals, palladium (blue) and platinum (yellow-orange), can serve as heterogeneous catalysts and were identified in the chapter as catalysts.

Think About It

Because catalytic converters contain precious metals such as rhodium, platinum, and palladium, interest in recycling the metals from catalytic converters is great.

13.17. **Collect and Organize**

By considering the levels of O_3 during the day as seen in Figure 13.1, we are to explain why $[O_3]_{max}$ occurs later in the day than $[NO]_{max}$ and $[NO_2]_{max}$.

Analyze

Ozone in the troposphere (the lowest portion of Earth's atmosphere) is due to the reaction of O_2 with O generated from the interaction of UV light with NO_2.

Solve

The presence of NO_2 in the atmosphere and ample sunlight allows the O atoms to react with O_2 to generate O_3. The reactant NO_2 is present in the atmosphere as a result of automobile exhausts, which build up during the day. The buildup of O_3 lags until later in the day, until $[NO_2]$ increases and the sunlight becomes stronger as midday approaches.

Think About It

Ozone is a very reactive gas and is irritating to lung tissues.

13.19. Collect and Organize

We are to explain why NO concentration does not increase after the evening rush hour.

Analyze

The reaction in the troposphere (lower atmosphere) that produces NO is

$$NO_2(g) \xrightarrow{\text{sunlight}} NO(g) + O(g)$$

Solve

In the evening the sunlight (and UV radiation) is less intense, so the photochemical breakdown of NO_2 does not occur to as great an extent as after the morning rush hour.

Think About It

The use of catalytic converters to reduce the NO_x to N_2 and O_2 in automobile exhaust has greatly helped to reduce photochemical smog in large urban and suburban centers.

13.21. Collect and Organize

Using ΔH_f° for NO, O_2, and NO_2 in Appendix 4, we can calculate ΔH_{rxn}° for

$$2\,NO(g) + O_2(g) \rightarrow 2\,NO_2(g)$$

Analyze

The ΔH_{rxn}° can be calculated by using

$$\Delta H_{rxn}^\circ = \sum n_{products} \Delta H_{f,products}^\circ - \sum n_{products} \Delta H_{f,reactants}^\circ$$

Solve

$$\Delta H_{rxn}^\circ = \left(2\ \text{mol } NO_2 \times 33.2\ \text{kJ/mol}\right) - \left[\left(2\ \text{mol } NO \times 90.3\ \text{kJ/mol}\right) + \left(1\ \text{mol } O_2 \times 0.0\ \text{kJ/mol}\right)\right]$$
$$= -114.2\ \text{kJ}$$

Think About It

That reaction is exothermic and, therefore, favored by enthalpy.

13.23. Collect, Organize, and Analyze

For the reaction of N_2 with O_2 to produce N_2O and N_2O_5, we are to write balanced chemical equations.

Solve

(a) $N_2(g) + \frac{1}{2}\,O_2(g) \rightarrow N_2O(g)$

or $2\,N_2(g) + O_2(g) \rightarrow 2\,N_2O(g)$

(b) $N_2(g) + \frac{5}{2}\,O_2(g) \rightarrow N_2O_5(g)$

or $2\,N_2(g) + 5\,O_2(g) \rightarrow 2\,N_2O_5(g)$

Think About It

Balanced chemical equations usually are written with whole-number coefficients.

13.25. **Collect and Organize**
We are to explain the difference between the average rate and the instantaneous rate of a reaction.

Analyze
The rate of reaction is measured from the change in concentration of a reactant or product over time. The difference between the average and instantaneous rates is the length of time over which the change in concentration is measured.

Solve
The average rate is the rate averaged over a fairly long time, whereas the instantaneous rate is the rate at a specific moment in time (or over a very, very short time).

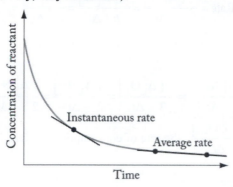

Think About It
The rate of a reaction is always positive.

13.27. **Collect and Organize**
We are to explain why the average rates of most reactions change over time.

Analyze
The forward rate of a reaction as measured from the average rate depends on the concentration of reactants.

Solve
As the reaction proceeds, the concentrations of the reactants decrease. Because most reactions depend on the availability (that is, concentration) of reactants to proceed, the decrease in reactant concentrations lowers the reaction rate.

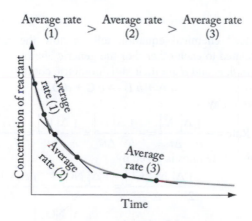

Think About It
Reactions that do not depend on the concentrations of the reactants, although rare, do not show a change in rate as the reaction proceeds.

13.29. Collect and Organize

Given the balanced equation for the reaction of ammonia with oxygen to produce H^+, NO_2^-, and H_2O, we are asked to relate the formation of products and consumption of reactants.

Analyze

The coefficients in the balanced chemical equation tell us how the rate of formation of products and consumption of reactants are related to each other. For the generic chemical equation where A, B, C, and D are the reactants and products and a, b, c, and d are their stoichiometric coefficients,

$$a\,A + b\,B \rightarrow c\,C + d\,D$$

the relationship of the rates is given by

$$\text{Rate} = -\frac{1}{a}\frac{\Delta[A]}{\Delta t} = -\frac{1}{b}\frac{\Delta[B]}{\Delta t} = \frac{1}{c}\frac{\Delta[C]}{\Delta t} = \frac{1}{d}\frac{\Delta[D]}{\Delta t}$$

Solve

For the reaction given

$$\text{Rate} = -\frac{1}{2}\frac{\Delta[NH_3]}{\Delta t} = -\frac{1}{3}\frac{\Delta[O_2]}{\Delta t} = \frac{1}{2}\frac{\Delta[H^+]}{\Delta t} = \frac{1}{2}\frac{\Delta[NO_2^-]}{\Delta t} = \frac{1}{2}\frac{\Delta[H_2O]}{\Delta t}$$

(a) $-\dfrac{\Delta[NH_3]}{\Delta t} = \dfrac{\Delta[H^+]}{\Delta t} = \dfrac{\Delta[NO_2^-]}{\Delta t}$, the rate of consumption of NH_3, is equal to the rate of formation of H^+ and NO_2^-.

(b) $\dfrac{\Delta[NO_2^-]}{\Delta t} = -\dfrac{2}{3}\dfrac{\Delta[O_2]}{\Delta t}$, the rate of formation of NO_2^-, is two-thirds the rate of consumption of O_2.

(c) $\dfrac{\Delta[NH_3]}{\Delta t} = \dfrac{2}{3}\dfrac{\Delta[O_2]}{\Delta t}$, the rate of consumption of NH_3, is two-thirds the rate of consumption of O_2.

Think About It

The negative sign is used in front of the expressions involving the consumption of a reactant to give a positive reaction rate because the change in concentration, $[X]_f - [X]_i$, or $\Delta[X]$, is negative for reactants because $[X]_f < [X]_i$.

13.31. Collect and Organize

For the balanced decomposition reaction of NO_2Cl into NO_2 and Cl_2 with the rate of formation of NO_2 measured as $5.7 \times 10^{-6}\ M \cdot s^{-1}$, we are to calculate the rate of formation of Cl_2 and the rate of disappearance of NO_2Cl.

Analyze

The coefficients in the balanced chemical equation tell us how the rate of formation of products and consumption of reactants are related to each other. For the generic chemical equation where A, B, C, and D are the reactants and products and a, b, c, and d are their stoichiometric coefficients,

$$a\,A + b\,B \rightarrow c\,C + d\,D$$

the relationship of the rates is given by

$$\text{Rate} = -\frac{1}{a}\frac{\Delta[A]}{\Delta t} = -\frac{1}{b}\frac{\Delta[B]}{\Delta t} = \frac{1}{c}\frac{\Delta[C]}{\Delta t} = \frac{1}{d}\frac{\Delta[D]}{\Delta t}$$

For that reaction the relationship of the rates is given by

$$rate = -\frac{1}{2}\frac{\Delta[NO_2Cl]}{\Delta t} = \frac{1}{2}\frac{\Delta[NO_2]}{\Delta t} = \frac{1}{1}\frac{\Delta[Cl_2]}{\Delta t}$$

and where $5.7 \times 10^{-6}\ M \cdot s^{-1} = \dfrac{\Delta[NO_2]}{\Delta t}$

Solve

(a) $\dfrac{\Delta[Cl_2]}{\Delta t} = \dfrac{1}{2}\dfrac{\Delta[NO_2]}{\Delta t} = \dfrac{1}{2}(5.7\times10^{-6}\ M\cdot s^{-1}) = 2.85\times10^{-6}\ M\cdot s^{-1}$

(b) $-\dfrac{1}{2}\dfrac{\Delta[NO_2Cl]}{\Delta t} = \dfrac{1}{2}\dfrac{\Delta[NO_2]}{\Delta t} = -5.7\times10^{-6}\ M\cdot s^{-1}$

Think About It

The negative sign is used in front of the expressions involving the consumption of a reactant to give a positive reaction rate because the change in concentration, $[X]_f - [X]_i$, or $\Delta[X]$, is negative for reactants because $[X]_f < [X]_i$.

13.33. Collect and Organize

Using the balanced equation describing the reaction of SO_2 with CO, we are to write expressions to compare the rates of formation of products and the rates of consumption of reactants.

Analyze

From the balanced equation, we see that the reaction may be expressed as

$$\text{Rate} = -\frac{\Delta[SO_2]}{\Delta t} = -\frac{1}{3}\frac{\Delta[CO]}{\Delta t} = \frac{1}{2}\frac{\Delta[CO_2]}{\Delta t} = \frac{\Delta[COS]}{\Delta t}$$

Solve

(a) $\text{Rate} = \dfrac{\Delta[CO_2]}{\Delta t} = -\dfrac{2}{3}\dfrac{\Delta[CO]}{\Delta t}$

(b) $\text{Rate} = \dfrac{\Delta[COS]}{\Delta t} = -\dfrac{\Delta[SO_2]}{\Delta t}$

(c) $\text{Rate} = \dfrac{\Delta[CO]}{\Delta t} = 3\dfrac{\Delta[SO_2]}{\Delta t}$

Think About It

Those relative rates make sense according to the stoichiometry of the reaction. For every 1 mol of SO_2 used in the reaction, 2 mol of CO_2 and 1 mol of COS are produced. So, for example, the concentration of CO_2 will increase twice as fast as the concentration of SO_2 decreases.

13.35. Collect and Organize

Using the relative rate expressions and the rate of the consumption of ClO in two reactions, we are to calculate the rate of change in the formation of the products of the two reactions.

Analyze

(a) For that reaction,

$$\text{Rate} = -\frac{1}{2}\frac{\Delta[ClO]}{\Delta t} = \frac{\Delta[Cl_2]}{\Delta t} = \frac{\Delta[O_2]}{\Delta t}$$

(b) For that reaction,

$$\text{Rate} = -\frac{\Delta[ClO]}{\Delta t} = -\frac{\Delta[O_3]}{\Delta t} = \frac{\Delta[O_2]}{\Delta t} = \frac{\Delta[ClO_2]}{\Delta t}$$

For each reaction we are given $\Delta[ClO]/\Delta t$. We can use that value in the relationships above to calculate the rate of change in (a) the concentration of Cl_2 and O_2 and (b) the concentration of O_2 and ClO_2.

Solve

(a) $\text{Rate} = \dfrac{\Delta\left[Cl_2\right]}{\Delta t} = \dfrac{\Delta\left[O_2\right]}{\Delta t} = -\dfrac{1}{2}\dfrac{\Delta\left[ClO\right]}{\Delta t} = -\dfrac{1}{2}\times -2.3\times10^7 \ M/s$

$= 1.2\times10^7 \ M/s$

(b) $\text{Rate} = \dfrac{\Delta\left[O_2\right]}{\Delta t} = \dfrac{\Delta\left[ClO_2\right]}{\Delta t} = -\dfrac{\Delta\left[ClO\right]}{\Delta t} = -\left(-2.9\times10^4 \ M/s\right)$

$= 2.9\times10^4 \ M/s$

Think About It

The rates of the formation of products are positive because $[X]_f > [X]_i$, so $[X]_f - [X]_i = \Delta[X]$ is positive.

13.37. Collect and Organize

Given the $[O_3]$ over time when it reacts with NO_2^-, we are to calculate the average reaction rate for two intervals.

Analyze

The average rate of reaction can be found according to

$$\dfrac{\Delta\left[O_3\right]}{\Delta t} = \dfrac{\left[O_3\right]_f - \left[O_3\right]_i}{t_f - t_i}$$

Solve

Between 0 and 100 μs:

$$-\dfrac{\Delta\left[O_3\right]}{\Delta t} = \dfrac{\left(9.93\times10^{-3} - 1.13\times10^{-2}\right) M}{\left(100 - 0\right) \mu s} = 1.4\times10^{-5} \ M/\mu s$$

Between 200 and 300 μs:

$$-\dfrac{\Delta\left[O_3\right]}{\Delta t} = \dfrac{\left(8.15\times10^{-3} - 8.70\times10^{-3}\right) M}{\left(300 - 200\right) \mu s} = 5.50\times10^{-6} \ M/\mu s$$

Think About It

As the reaction proceeds, the rate of consumption of ozone decreases as a result of the decreasing reactant concentrations.

13.39. Collect and Organize

After we plot $[ClO]$ versus time and $[Cl_2O_2]$ versus time, we can determine the instantaneous rate of change at 1 s for each compound.

Analyze

The instantaneous rate is the slope of the line that is tangent to the curve at the time we are interested in. We can estimate fairly well the instantaneous rate at 1 s from the plots by choosing two points that are close to 1 s to calculate the slope. Since we are given only values for $[ClO]$, we need to calculate $[Cl_2O_2]$ for each time.

$$[Cl_2O_2]_t = \dfrac{\left(\left[ClO\right]_0 - \left[ClO\right]_t\right)}{2}$$

where $[ClO]_0 = 2.60\times10^{11} M$

Time, s	$[ClO]_t$, molecules/cm^3	$[Cl_2O_2]_t$, molecules/cm^3
0	2.60×10^{11}	0.00
1	1.08×10^{11}	7.60×10^{10}
2	6.83×10^{10}	9.59×10^{10}
3	4.99×10^{10}	1.05×10^{11}
4	3.93×10^{10}	1.10×10^{11}
5	3.24×10^{10}	1.14×10^{11}
6	2.76×10^{10}	1.16×10^{11}

Initially, no Cl_2O_2 is present, so $[Cl_2O_2]_0 = 0$ molecules/cm^3.

Solve
For the change in concentration of ClO versus time, we obtain the following plot:

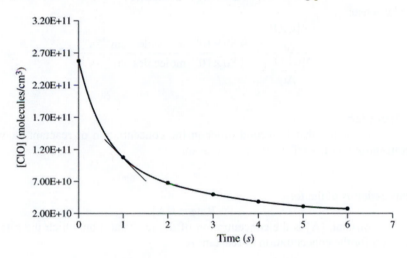

To get a fairly good estimate from the data given for the instantaneous rate, we can choose two points from the data set that surrounds the data point of interest and calculate $\Delta[ClO]/\Delta t$.
For $t = 1$ s, we can use the points $t = 0$ s and $t = 2$ s:

$$-\frac{\Delta[ClO]}{\Delta t} = \frac{(6.83 \times 10^{10} - 2.60 \times 10^{11}) \text{ molecules/cm}^3}{(2-0)\ s} = 9.59 \times 10^{10} \text{ molecules} \cdot \text{cm}^{-3} \cdot \text{s}^{-1}$$

Using a graphing program that calculates the slope of the tangent to the line at $t = 1$ s, we get an instantaneous rate of 8.28×10^{10} molecules $\cdot$ cm^{-3} $\cdot$ s.

For the change in concentration of Cl_2O_2 versus time, we obtain the following plot.

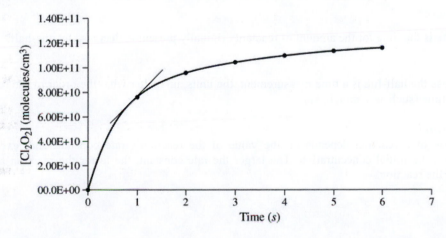

To get a fairly good estimate from the data given for the instantaneous rate, we can choose two points from the data set that surrounds the data point of interest and calculate $\Delta[Cl_2O_2]/\Delta t$.

For $t = 1$ s, we can use the points $t = 0$ s and $t = 2$ s:

$$\frac{\Delta[Cl_2O_2]}{\Delta t} = \frac{(9.59 \times 10^{10} - 0) \text{ molecules/cm}^3}{(2-0) \text{ } s} = 4.80 \times 10^{10} \text{ molecules} \cdot cm^{-3} \cdot s^{-1}$$

Using a graphing program that calculates the slope of the tangent to the line at $t = 1$ s, we get an instantaneous rate of 4.13×10^{10} molecules $\cdot cm^{-3} \cdot$ s.

Think About It

Because we expect the rate of disappearance of ClO to be twice the rate of appearance of Cl_2O_2 from the balanced equation

$$2 \text{ ClO}(g) \rightarrow Cl_2O_2(g)$$

our answers make sense:

$$\frac{\dfrac{\Delta[ClO]}{\Delta t}}{\dfrac{\Delta[Cl_2O_2]}{\Delta t}} = \frac{9.59 \times 10^{10} \text{ molecules} \cdot cm^{-3} \cdot s^{-1}}{4.80 \times 10^{10} \text{ molecules} \cdot cm^{-3} \cdot s^{-1}} = 2.0$$

13.41. Collect and Organize

For a decomposition reaction that is second order in the concentration of reactant A, we are to explain how doubling the concentration of A affects the rate constant.

Analyze

The rate law expression is of the form

$$\text{Rate} = k[A]^2$$

where k is the rate constant, [A] is the concentration of the reactant(s) on which the rate depends, and 2 is the order of the reaction for the concentration of reactant A.

Solve

Although the rate of the reaction will be squared, the value of the rate constant will not change upon changing the concentration of a reactant.

Think About It

The rate constant, however, will change for a reaction with temperature change.

13.43. Collect and Organize

We are asked whether the units of the half-life for a second-order reaction can have the same units as those of the half-life for a first-order reaction.

Analyze

The half-life is the time for the amount of reactant originally present to decrease by one-half.

Solve

Yes. Because the half-life is a time measurement, the units, no matter what the order of the reaction, are always in units of time (such as s, min, h, yr).

Think About It

The half-life of a reaction depends on the value of the reaction's rate constant and, except for first-order reactions, on the initial concentration. The larger the rate constant, the faster the reaction and the shorter the half-life of the reaction.

13.45. Collect and Organize

For a second-order reaction, we are to predict how doubling $[A]_0$ affects the half-life.

Analyze

For a second-order reaction,

$$t_{1/2} = \frac{1}{k[A]_0}$$

Solve

From the equation for the half-life of a second-order reaction, we see that doubling $[A]_0$ halves the half-life.

Think About It

The half-life of a second-order reaction, like that of a first-order reaction, is inversely related to the rate constant.

13.47. Collect and Organize

For each rate law expression, we are to determine the order of the reaction with respect to each reactant and the overall reaction order.

Analyze

The order of a reaction is the experimentally determined dependence of the rate of a reaction on the concentration of the reactants involved in the reaction. The order of a reaction in the rate law expression is shown as the power to which the concentration of a particular reactant is raised. The overall reaction order is the sum of the powers of the reactants in the rate law expression.

Solve

(a) For the rate law expression Rate = $k[A][B]$, the reaction is first order in both A and B and second order overall.
(b) For Rate = $k[A]^2[B]$, the reaction is second order in A, first order in B, and third order overall.
(c) For Rate = $k[A][B]^3$, the reaction is first order in A, third order in B, and fourth order overall.

Think About It

The higher the order of the reaction for a particular reactant, the more a change in concentration of that reactant affects the reaction rate.

13.49. Collect and Organize

For each reaction described, we are to write the rate law and determine the units for k, using the units M for concentration and s for time.

Analyze

The general form of the rate law is

$$\text{Rate} = k[A]^x[B]^y$$

where k is the rate constant, A and B are the reactants, and x and y are the orders of the reaction with respect to each reactant as determined by experiment.

Solve

(a) Rate = $k[O][NO_2]$
Because rate has units of M/s and each concentration has units of M,

$$k = \frac{M/s}{M^2} = M^{-1}s^{-1}$$

(b) Rate = $k[NO]^2[Cl_2]$
Because rate has units of M/s and each concentration has units of M,

$$k = \frac{M/s}{M^3} = M^{-2}s^{-1}$$

(c) Rate = k [CHCl$_3$][Cl$_2$]$^{\frac{1}{2}}$

Because rate has units of M/s and each concentration has units of M,

$$k = \frac{M/s}{M^{3/2}} = M^{-\frac{1}{2}}\, s^{-1}$$

(d) Rate = $k[O_3]^2[O]^{-1}$

Because rate has units of M/s and each concentration has units of M,

$$k = \frac{M/s}{M^2/M} = s^{-1}$$

Think About It

The units of the rate constant clearly depend on the overall order of the reaction.

13.51. Collect and Organize

Given the changes in rate of the decomposition of BrO to Br$_2$ and O$_2$ when [BrO] is changed, we are to predict the rate law in each case.

Analyze

The general form of the rate law for the reaction is

$$\text{Rate} = k[\text{BrO}]^x$$

where x is an experimentally determined exponent.

Solve

(a) If the rate doubles when [BrO] doubles, then $x = 1$ and the rate law is

$$\text{Rate} = k[\text{BrO}]$$

(b) If the rate quadruples when [BrO] doubles, then $x = 2$ and the rate law is

$$\text{Rate} = k[\text{BrO}]^2$$

(c) If the rate is halved when [BrO] is halved, then $x = 1$ and the rate law is

$$\text{Rate} = k[\text{BrO}]$$

(d) If the rate is unchanged when [BrO] is doubled, then $x = 0$ and the rate law is

$$\text{Rate} = k[\text{BrO}]^0 = k$$

Think About It

For that reaction, the relationship is straightforward between the change in rate when the concentration of the reactant was changed to determine x. To determine x for more complicated reactions, use

$$\frac{\text{rate}_2}{\text{rate}_1} = \left(\frac{[A]_2}{[A]_1}\right)^x$$

$$\ln\left(\frac{\text{rate}_2}{\text{rate}_1}\right) = x \ln\left(\frac{[A]_2}{[A]_1}\right)$$

$$x = \frac{\ln\left(\dfrac{\text{rate}_2}{\text{rate}_1}\right)}{\ln\left(\dfrac{[A]_2}{[A]_1}\right)}$$

13.53. Collect and Organize

Given that the rate of the reaction quadruples when both [NO] and [ClO] are doubled, we are to identify what additional information we would need to write the rate law for the reaction

$$\text{NO}(g) + \text{ClO}(g) \rightarrow \text{NO}_2(g) + \text{Cl}(g)$$

Analyze

The general form of the rate law for this reaction is

$$\text{Rate} = k[\text{NO}]^x[\text{ClO}]^y$$

Solve

If the rate quadruples when both [NO] and [ClO] are doubled, the rate law could be any of the following:

$$\text{Rate} = k[\text{NO}][\text{ClO}]$$
$$\text{Rate} = k[\text{NO}]^2$$
$$\text{Rate} = k[\text{ClO}]^2$$

To differentiate among those, we need to determine the change in the rate when only [NO] or [ClO] is changed.

Think About It

If the rate is only doubled when [NO] and [ClO] are independently changed, then the rate law is

$$\text{Rate} = k[\text{NO}][\text{ClO}]$$

If the rate is quadrupled when [NO] is doubled but remains constant if [ClO] is doubled, then the rate law is

$$\text{Rate} = k[\text{NO}]^2$$

If the rate is quadrupled when [ClO] is doubled but remains constant if [NO] is doubled, then the rate law is

$$\text{Rate} = k[\text{ClO}]^2$$

13.55. Collect and Organize

For the reaction of NO_2 with O_3 to produce NO_3 and O_2, we are to write the rate law given that the reaction is first order in both NO_2 and O_3. From the rate law and given the rate constant, we can calculate the rate of the reaction for a given [NO_2] and [O_3]. From that we can calculate the rate of appearance of NO_3 and the rate of the reaction when [O_3] is doubled.

Analyze

The general form of the rate law for that reaction is

$$\text{Rate} = k[\text{NO}_2]^x[\text{O}_3]^y$$

The rate of consumption of reactants and formation of products is

$$\text{Rate} = -\frac{\Delta[\text{NO}_2]}{\Delta t} = -\frac{\Delta[\text{O}_3]}{\Delta t} = \frac{\Delta[\text{NO}_3]}{\Delta t} = \frac{\Delta[\text{O}_2]}{\Delta t}$$

Solve

(a) $\text{Rate} = k[\text{NO}_2][\text{O}_3]$

(b) $\text{Rate} = \dfrac{1.93 \times 10^4}{M \cdot s} \times 1.8 \times 10^{-8}\,M \times 1.4 \times 10^{-7}\,M = 4.9 \times 10^{-11}\,M/s$

(c) $\text{Rate} = \dfrac{\Delta[\text{NO}_3]}{\Delta t} = 4.9 \times 10^{-11}\,M/s$

(d) When [O_3] is doubled, the rate of the reaction doubles.

Think About It

When [O_3] = 2.8×10^{-7} M (double that in part b), the rate of reaction is 9.73×10^{-11} M/s, which is twice that calculated in part b, so our prediction in part d is correct.

13.57. Collect and Organize

By comparing the rate constants for four reactions that are all second order, we can determine which reaction is the fastest if all the initial concentrations are the same.

Analyze

The reaction with the largest rate constant has the fastest reaction rate.

Solve
Reaction (c) has the largest value of k, so it proceeds the fastest.

Think About It
The slowest reaction is a reaction with the smallest value of k.

13.59. Collect and Organize
Given the information that the rate of the reaction between NO and NO_2 with water doubles when either [NO] or [NO_2] doubles and the rate does not depend on [H_2O], we can write the rate law for the reaction.

Analyze
The general form of the rate law for this reaction is
$$Rate = k[NO]^x[NO_2]^y[H_2O]^z$$
Doubling of the reaction rate with doubling of [NO] or [NO_2] means the reaction is first order in those reactants. Because the rate does not depend on [H_2O], the reaction is zero order in that reactant.

Solve
The rate law for this reaction is
$$Rate = k[NO]^1[NO_2]^1[H_2O]^0 = k[NO][NO_2]$$

Think About It
If [NO] and [NO_2] are doubled simultaneously, the rate of reaction quadruples.

13.61. Collect and Organize
In the reaction of ClO_2 with OH^- the rate of the reaction was measured for various concentrations of both reactants. From the data we are to determine the rate law and calculate the rate constant, k.

Analyze
To determine the dependence of the rate on a change in the concentration of a particular reactant, we can compare the reaction rates for two experiments in which the concentration of that reactant changes but the concentrations of the other reactants remain constant. Once we have the order of the reaction for each reactant, we can write the rate law expression. To calculate the rate constant for the reaction, we can rearrange the rate law to solve for k and use the data from any of the experiments.

Solve
Using experiments 1 and 2, we find that the order of the reaction with respect to ClO_2 is 1:

$$\frac{rate_1}{rate_2} = \left(\frac{[ClO_2]_{0,1}}{[ClO_2]_{0,2}} \right)^x$$

$$\frac{0.0248 \ M/s}{0.00827 \ M/s} = \left(\frac{0.060 \ M}{0.020 \ M} \right)^x$$

$$3.00 = 3.01^x$$

$$x = 1$$

Using experiments 2 and 3, we find that the order of the reaction with respect to [OH^-] also is 1:

$$\frac{rate_3}{rate_2} = \left(\frac{[OH^-]_{0,3}}{[OH^-]_{0,2}} \right)^x$$

$$\frac{0.0247 \ M/s}{0.00827 \ M/s} = \left(\frac{0.090 \ M}{0.030 \ M} \right)^x$$

$$2.99 = 3.0^x$$

$$x = 1$$

The rate law for this reaction is Rate = $k[ClO_2][OH^-]$.

Rearranging the rate law expression to solve for k and using the data from experiment 1 gives

$$k = \frac{\text{rate}}{[\text{ClO}_2][\text{OH}^-]} = \frac{0.0248 \ M/s}{0.060 \ M \times 0.030 \ M} = 14 \ M^{-1} \ s^{-1}$$

Think About It
We may use any of the experiments in the table to calculate k. Each experiment's data give the same value of k as long as the experiments were all run at the same temperature.

13.63. **Collect and Organize**
In the reaction of H_2 with NO, the rate of the reaction was measured for various concentrations of both reactants. From the data we are to determine the rate law and calculate the rate constant, k.

Analyze
To determine how the rate depends on a change in the concentration of a particular reactant, we can compare the reaction rates for two experiments in which the concentration of that reactant changes, but the concentrations of the other reactants remain constant. Once we have the order of the reaction for each reactant, we can write the rate law expression. To calculate the rate constant for the reaction, we can rearrange the rate law to solve for k and use the data from any of the experiments.

Solve
Using experiments 1 and 2, we find that the order of the reaction with respect to NO is 2:

$$\frac{\text{rate}_2}{\text{rate}_1} = \left(\frac{[\text{NO}]_{0,2}}{[\text{NO}_2^-]_{0,1}} \right)^x$$

$$\frac{0.0991 \ M/s}{0.0248 \ M/s} = \left(\frac{0.272 \ M}{0.136 \ M} \right)^x$$

$$4.00 = 2.00^x$$

$$x = 2$$

Using experiments 3 and 4, we find that the order of the reaction with respect to H_2 is 1:

$$\frac{\text{rate}_4}{\text{rate}_3} = \left(\frac{[\text{H}_2]_{0,4}}{[\text{H}_2]_{0,3}} \right)^x$$

$$\frac{1.59 \ M/s}{0.793 \ M/s} = \left(\frac{0.848 \ M}{0.424 \ M} \right)^x$$

$$2.01 = 2.00^x$$

$$x = 1$$

The rate law for this reaction is Rate = $k[\text{NO}]^2[\text{H}_2]$.
Rearranging the rate law expression to solve for k and using the data from experiment 1 gives

$$k = \frac{0.0248 \ M/s}{(0.136 \ M)^2 \times 0.212 \ M} = 6.32 \ M^{-2} \ s^{-1}$$

Think About It
We may use any of the experiments in the table to calculate k. Each experiment's data give the same value of k as long as the experiments were all run at the same temperature.

13.65. Collect and Organize

From the data given for the concentration of NO_3 as it decomposes to NO_2 and O_2, we are to calculate the value of k for that reaction.

Analyze

We are given a single data set and the fact that the reaction is second order. For this second-order reaction the plot of $1/[NO_3]$ versus time gives a straight line with a slope equal to k, the reaction rate constant.

Solve

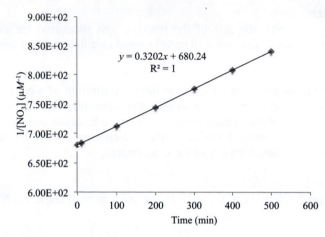

The slope $= k = 0.32\ \mu M^{-1} \cdot min^{-1}$.

Think About It

The half-life of this second-order reaction is

$$t_{1/2} = \frac{1}{k[NO_3]_0} = \frac{1}{\left(0.3202\ \mu M^{-1} \cdot min^{-1}\right) \times \left(1.470 \times 10^{-3}\ \mu M\right)}$$

$$= 2124\ min,\ or\ 35.4\ h$$

13.67. Collect and Organize

From the data given for concentration of NH_3 as it decomposes to N_2 and H_2, we are to determine the rate law and calculate the value for k for that reaction.

Analyze

We are given a single data set. If a plot of $[NH_3]$ versus time yields a straight line, the reaction is zero order. If a plot of $\ln[NH_3]$ versus time yields a straight line, the reaction is first order. If a plot of $1/[NH_3]$ versus time yields a straight line, the reaction is second order. The slope of the line on the appropriate graph gives the rate constant k for the reaction (slope $= -k$ for a zero- or first-order reaction and slope $= +k$ for a second-order reaction).

Solve

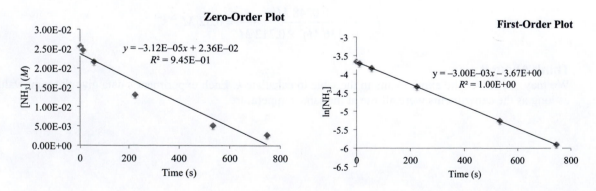

Second-Order Plot

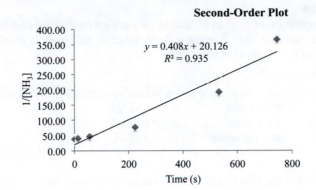

$y = 0.408x + 20.126$
$R^2 = 0.935$

The first-order plot is linear, so that the rate law is Rate = $k[NH_3]$. The slope = $-k = -0.003$ s^{-1} and thus $k = 0.003$ s^{-1}.

Think About It
The half-life of that first-order reaction is

$$t_{1/2} = \frac{0.693}{k} = \frac{0.693}{0.0030 \text{ s}^{-1}} = 231 \text{ s, or 3.9 min}$$

13.69. Collect and Organize
For the decomposition reaction of N_2O_5 to NO_2 and O_2, we are given that the reaction is first order with a rate constant, k, equal to 6.32×10^{-4} s^{-1}. We are to calculate the amount of N_2O_5 that remains after 1 h of reaction time when the initial concentration of N_2O_5 is 0.50 mol/L and determine the percentage of N_2O_5 that has reacted.

Analyze
The integrated rate law for a first-order reaction is
$$\ln[A] = -kt + \ln[A]_0$$

Solve
The concentration of N_2O_5 remaining after 1 h is
$$\ln[N_2O_5] = -kt + \ln[N_2O_5]_0$$

$$\ln[N_2O_5] = -\left[\left(\frac{6.32 \times 10^{-4}}{s}\right) \times \left(1 \text{ h} \times \frac{60 \text{ min}}{1 \text{ h}} \times \frac{60 \text{ s}}{1 \text{ min}}\right)\right] + \ln(0.50 \text{ mol/L}) = -2.9683$$

$$[N_2O_5] = e^{-2.9683} = 0.051 \text{ mol/L}$$

The percentage of N_2O_5 reacted is
$$\frac{0.50 \text{ mol/L} - 0.0514 \text{ mol/L}}{0.50 \text{ mol/L}} \times 100 = 90\%$$

Think About It
The number of half-lives needed to reduce the concentration of N_2O_5 to 10% can be calculated:

$$\frac{10}{100} = 0.50^n$$

$$0.10 = 0.50^n$$

$$\ln 0.10 = n \ln 0.50$$

$$n = 3.32$$

13.71. **Collect and Organize**

For the decomposition reaction of N_2O to N_2 and O_2, we are given that the plot of $\ln[N_2O]$ versus time is linear. We are to write the rate law and then determine the number of half-lives it would take for the concentration of N_2O to become 6.25% of its original concentration.

Analyze

The integrated rate laws for zero-, first-, and second-order reactions with their half-life equations are as follows:

$[A] = -kt + [A]_0$ zero order $t_{1/2} = [A]_0/2k$

$\ln[A] = -kt + \ln[A]_0$ first order $t_{1/2} = 0.693/k$

$1/[A] = kt + 1/[A]_0$ second order $t_{1/2} = 1/k[A]_0$

Solve

(a) Since we are given that the plot of $\ln[N_2O]$ versus time is linear, the reaction is first order and the rate law is Rate = $k[N_2O]$.

(b) The number of half-lives needed to reduce the concentration to 6.25% would be, where n = number of half-lives,

$$\frac{6.25}{100} = 0.50^n$$

$$0.0625 = 0.50^n$$

$$\ln 0.0625 = n \ln 0.50$$

$$n = 4$$

Think About It

The number of half-lives needed to reduce the concentration of a reactant to products does not depend on the order of the reaction or on the magnitude of the rate constant.

13.73. **Collect and Organize**

From the data given for the concentration of ^{32}P as it decays, we are to determine the rate law and calculate the value for k for that radioactive decay, the half-life of the reaction, and the days it takes for 90% of the sample to decay.

Analyze

Radioactive decay follows first-order kinetics. That is confirmed in part b, where we are told to determine the first-order rate constant. A plot of $\ln[^{32}P]$ versus time for that decay gives a straight line with slope = $-k$. The half-life of a first-order decay is given by

$$t_{1/2} = \frac{0.693}{k}$$

From the rate constant we can determine the time, t, it takes for 90% of the sample to decay by using the first-order integrated rate law expression:

$$\ln[A] = -kt + \ln[A]_0$$

where $[A] = 10$ and $[A]_0 = 100$ for 90% decay.

Solve

(a)

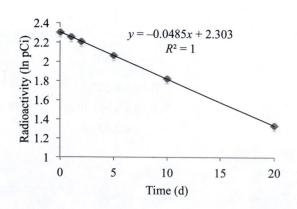

$$y = -0.0485x + 2.303$$
$$R^2 = 1$$

The rate law for that radioactive decay is Rate = $k[^{32}P]$.

(b) $k = -\text{slope} = 0.0485 \text{ day}^{-1}$

(c) $t_{1/2} = \dfrac{0.693}{0.0485 \text{ day}^{-1}} = 14.3 \text{ days}$

(d) $\ln[A] = -kt + \ln[A]_0$

$\ln[10] = -(0.0485 \text{ day}^{-1})t + \ln[100]_0$

$t = 47.5 \text{ days}$

Think About It

The number of half-lives needed to reduce the amount of ^{32}P to 10% (90% decay) would be, where n = number of half-lives,

$$\frac{10}{100} = 0.50^n$$

$$0.10 = 0.50^n$$

$$\ln 0.10 = n \ln 0.50$$

$$3.32 = n$$

13.75. Collect and Organize

Given that the dimerization of ClO to Cl_2O_2 is second order, we are to determine the value of the rate constant and to calculate the half-life of this reaction.

Analyze

We are given a single data set and knowledge that the reaction is second order. For that second-order reaction a plot of $1/[\text{ClO}]$ versus time gives a straight line with a slope equal to k, the reaction rate constant. The half-life of this second-order reaction is

$$t_{1/2} = \frac{1}{k[\text{ClO}]_0}$$

where $[\text{ClO}]_0$ is the initial concentration of ClO used in the reaction.

Solve

(a)

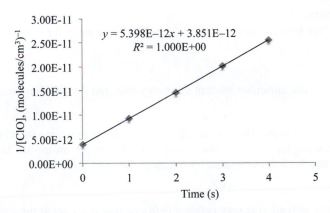

The value of k for that reaction is $k = \text{slope} = 5.40 \times 10^{-12} \text{ cm}^3 \text{ molecules}^{-1} \text{ s}^{-1}$.

(b) The half-life of the reaction is

$$t_{1/2} = \frac{1}{k[\text{ClO}]_0} = \frac{1}{\dfrac{5.40 \times 10^{-12} \text{ cm}^3}{\text{molecules} \cdot \text{s}} \times \dfrac{2.60 \times 10^{11} \text{ molecules}}{\text{cm}^3}} = 0.712 \text{ s}$$

Think About It

For a second-order reaction, as we increase the initial concentration of the reactant the half-life gets shorter.

13.77. Collect and Organize

From the data for the pseudo–first-order hydrolysis of sucrose provided, we are to write the rate law and determine the value of the pseudo–first-order rate constant, k'.

Analyze

To obtain the value of k', we plot ln[sucrose] over time. The slope of the line is equal to $-k$.

Solve

The rate law is Rate = $k[C_{12}H_{22}O_{11}][H_2O] = k'[C_{12}H_{22}O_{11}]$.

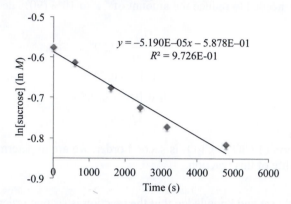

$$y = -5.190E{-}05x - 5.878E{-}01$$
$$R^2 = 9.726E{-}01$$

The pseudo–first-order plot gives $k' = -$slope $= 5.19 \times 10^{-5} \text{ s}^{-1}$.

Think About It

If we knew the concentration of water in the hydrolysis reaction, we could calculate the value of k by using

$$k = \frac{k'}{[H_2O]_0}$$

13.79. Collect and Organize

We are asked to explain how the magnitude of the activation energy for a reaction is related to the rate of the reaction.

Analyze

The activation energy is the minimum amount of energy required for colliding molecules to react.

Solve

The greater the activation energy, the more energy required for the reaction to occur and the slower the reaction. That is so because the higher the activation energy, the fewer the molecules that will possess the minimum energy required to react. With fewer molecules reacting, the reaction is slower.

Think About It

A reaction with a lower activation energy is faster than one that is higher at the same temperature.

13.81. Collect and Organize

We are asked to describe the circumstances for a reaction in which the activation energy of the forward reaction is less than that of the reverse reaction.

Analyze

For the activation energy of the forward reaction to be less than that of the reverse reaction, the energy of the reactants must be closer to the energy of the transition state than the energy of the products.

Solve

For the activation energy of the forward reaction to be less than the activation energy of the reverse reaction, the energy of the products must be lower than the energy of the reactants and therefore the reaction is spontaneous $(-\Delta G)$.

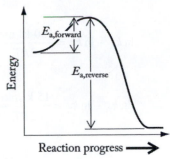

Think About It

Just because a reaction is favored thermodynamically does not necessarily mean that it is fast. That reaction could have a quite high activation energy in the forward direction and therefore be quite slow.

13.83. Collect and Organize

We are to explain why the order of a reaction is independent of temperature, yet the value of k changes with temperature.

Analyze

We need to consider how temperature affects the motion and collision of the reactants.

Solve

An increase in temperature increases the frequency and the kinetic energy at which the reactants collide. That speeds up the reaction, changing the value of k. The activation energy of the slowest step in the reaction, however, is not affected by a change in temperature and therefore the order of the reaction is unaffected.

Think About It

As a general empirical observation, heating a reaction by 10°C doubles the rate of reaction.

13.85. Collect and Organize

In comparing two first-order reactions with different activation energies, we are to decide which would show a larger increase in its rate as the reaction temperature is increased.

Analyze

We can use the Arrhenius equation to mathematically determine which reaction would be most accelerated by an increase in temperature:

$$\ln k = \ln A - \frac{E_a}{RT}$$

Solve

Let's assume that $T_2 = 2T_1$. For either reaction the difference in the rate constants is as follows:

$$\ln k_{T_1} = \ln A - \frac{E_a}{RT_1} \qquad \ln k_{T_2} = \ln A - \frac{E_a}{RT_2}$$

$$\ln k_{T_2} - \ln k_{T_1} = \frac{-E_a}{RT_2} + \frac{E_a}{RT_1} = \frac{E_a}{R}\left(\frac{1}{T_1} - \frac{1}{T_2}\right)$$

But because $T_2 = 2T_1$,

$$\ln k_{T_2} - \ln k_{T_1} = \frac{E_a}{R}\left(\frac{1}{T_1} - \frac{1}{2T_1}\right) = \frac{E_a}{2RT_1}$$

For $E_a = 150$ kJ/mol,

$$\ln k_{T_2} - \ln k_{T_1} = \frac{150 \text{ kJ/mol}}{2RT_1}$$

For $E_a = 15$ kJ/mol,

$$\ln k_{T_2} - \ln k_{T_1} = \frac{15 \text{ kJ/mol}}{2RT_1}$$

Comparing these as a ratio,

$$\frac{\ln k_{T_2} - \ln k_{T_1} \text{ for } E_a = 150 \text{ kJ/mol}}{\ln k_{T_2} - \ln k_{T_1} \text{ for } E_a = 15 \text{ kJ/mol}} = \frac{\dfrac{150 \text{ kJ/mol}}{2RT_1}}{\dfrac{15 \text{ kJ/mol}}{2RT_1}} = 10$$

Therefore, the reaction with the larger activation energy (150 kJ/mol) would be accelerated more than the reaction with the lower activation energy (15 kJ/mol) when heated.

Think About It

Our derivation shows that different reactions with different activation energies will not accelerate in the same way when they are heated.

13.87. **Collect and Organize**

We can use an Arrhenius plot of rate constant versus temperature for the reaction of O and O_3 to determine the activation energy (E_a) and the value of the frequency factor (A).

Analyze

The Arrhenius equation is

$$\ln k = \frac{-E_a}{R}\left(\frac{1}{T}\right) + \ln A$$

If we plot $\ln k$ (y-axis) versus $1/T$ (x-axis), we obtain a straight line with slope $m = -E_a/R$. The activation energy is therefore calculated from

$$E_a = -\text{slope} \times R$$

where $R = 8.314$ J/mol · K. The y-intercept (b) is equal to $\ln A$, so we can calculate the frequency factor from $b = \ln A$ or $e^b = A$.

Solve

The Arrhenius plot gives a slope of –2060 and a y-intercept of 0.00160.

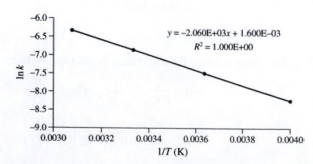

$$E_a = -(-2060 \text{ K} \times 8.314 \text{ J/mol} \cdot \text{K}) = 1.713 \times 10^4 \text{ J/mol, or } 17.1 \text{ kJ/mol}$$
$$b = 0.00160 = \ln A$$
$$A = e^{0.00160} = 1.002$$

Think About It
Once we have the values of E_a and A from the plot, we can calculate the value of k at any temperature.

13.89. Collect and Organize
We can use an Arrhenius plot of rate constant versus temperature for the reaction of N_2 with O_2 to form NO to determine the activation energy (E_a), the frequency factor (A), and the rate constant for the reaction at 300 K.

Analyze
The Arrhenius equation is

$$\ln k = \frac{-E_a}{R}\left(\frac{1}{T}\right) + \ln A$$

If we plot $\ln k$ (y-axis) versus $1/T$ (x-axis), we obtain a straight line with slope $m = -E_a/R$. The activation energy is therefore calculated from

$$E_a = -\text{slope} \times R$$

where $R = 8.314$ J/mol · K. The y-intercept (b) is equal to $\ln A$, so we can calculate the frequency factor from $b = \ln A$ or $e^b = A$. Once E_a and A are known, we may use another form of the Arrhenius equation to calculate k at any temperature:

$$k = A\exp\left(-\frac{E_a}{RT}\right)$$

Solve
The Arrhenius plot gives a slope of -37758 and a y-intercept of 24.641.

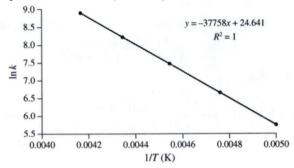

$y = -37758x + 24.641$
$R^2 = 1$

(a) $E_a = -(-37758 \text{ K} \times 8.314 \text{ J/mol} \cdot \text{K}) = 3.14 \times 10^5$ J/mol, or 314 kJ/mol
(b) $A = e^{24.641} = 5.03 \times 10^{10}$

(c) $k = 5.03 \times 10^{10} \exp\left(-\dfrac{3.14\times10^5 \text{ J/mol}}{8.314 \text{ J/mol}\cdot\text{K}\times300 \text{ K}}\right) = 1.06\times10^{-44} \text{ } M^{-1/2} \text{ s}^{-1}$

Think About It
Alternatively, k can be calculated from the original Arrhenius equation:

$$\ln k = \frac{-E_a}{RT} + \ln A = \frac{-3.14\times10^5 \text{ J/mol}}{8.314 \text{ J/mol}\cdot\text{K}\times300 \text{ K}} + 24.641$$
$$= -101.25$$
$$k = e^{-101.25} = 1.07\times10^{-44} \, M^{-1/2} \text{ s}^{-1}$$

13.91. Collect and Organize
We can use an Arrhenius plot of rate constant versus temperature for the reaction of ClO_2 and O_3 to determine the activation energy (E_a), the value of the frequency factor (A), and the value of k at 245 K.

Analyze
The Arrhenius equation is

$$\ln k = \frac{-E_a}{R}\left(\frac{1}{T}\right) + \ln A$$

If we plot $\ln k$ (y-axis) versus $1/T$ (x-axis), we obtain a straight line with slope $m = -E_a/R$. The activation energy is therefore calculated from

$$E_a = -\text{slope} \times R$$

where $R = 8.314$ J/mol · K. The y-intercept (b) is equal to $\ln A$, so we can calculate the frequency factor from $b = \ln A$ or $e^b = A$. Once E_a and A are known, we may use another form of the Arrhenius equation to calculate k at any temperature:

$$k = A \exp\left(-\frac{E_a}{RT}\right)$$

Solve

(a) The Arrhenius plot gives a slope of -4698.7 and a y-intercept of 27.872.

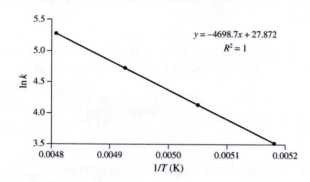

$$E_a = -(-4698.7 \text{ K} \times 8.314 \text{ J/mol} \cdot \text{K}) = 3.91 \times 10^4 \text{ J/mol, or } 39.1 \text{ kJ/mol}$$
$$b = 27.872 = \ln A$$
$$A = e^{27.872} = 1.27 \times 10^{12}$$

(b) The rate constant at 245 K is

$$k = 1.27 \times 10^{12} \exp\left(\frac{3.91 \times 10^4 \text{ J/mol}}{8.314 \text{ J/mol} \cdot \text{K} \times 245 \text{ K}}\right) = 5.85 \times 10^3 \text{ M}^{-1}\text{s}^{-1}$$

Think About It

Once we have the values of E_a and A from the plot, we can calculate the value of k at any temperature.

13.93. **Collect and Organize**

By comparing the ratio of the rate constants for the decomposition of azomethane at 600°C and 610°C, we are to assess the general rule that for every 10°C rise in temperature, the rate of a reaction doubles.

Analyze

We can use the form of the Arrhenius equation relating the rate constants for a reaction at different temperatures:

$$\ln \frac{k_1}{k_2} = \frac{E_a}{R}\left(\frac{1}{T_2} - \frac{1}{T_1}\right)$$

Solve

For the equation we use $k_2 = 2.00 \times 10^8$ s^{-1}, $T_2 = 873$ K (600°C), $T_1 = 883$ K (610°C), and $E_a = 2.14 \times 10^4$ J/mol with $R = 8.314$ J/mol · K.

$$\ln \frac{k_1}{2.00 \times 10^8 \, s^{-1}} = \frac{2.14 \times 10^4 \, \text{J/mol}}{8.314 \, \text{J/mol} \cdot \text{K}} \left(\frac{1}{873 \, \text{K}} - \frac{1}{883 \, \text{K}} \right) = 0.03391$$

$$\frac{k_1}{2.00 \times 10^8 \, s^{-1}} = e^{0.03391} = 1.034$$

$$k_1 = 1.034 \times \left(2.00 \times 10^8 \, s^{-1} \right) = 2.07 \times 10^8 \, s^{-1}$$

The ratio of those rate constants is

$$\frac{k_{610°C}}{k_{600°C}} = \frac{2.07 \times 10^8 \, s^{-1}}{2.00 \times 10^8 \, s^{-1}} = 1.04$$

That finding does not represent a doubling of the rate for the reaction. If that were the case, the ratio of the rate constant at 610°C to the rate constant at 600°C would be 2. The rule does not apply to that temperature domain. The rule applies best when the reaction temperatures are near 298 K.

Think About It

This problem illustrates well how general rules do not apply to every situation, so be cautious in using general rules to make predictions.

13.95. **Collect and Organize**

By comparing the rate laws for two reactions, we can determine whether their mechanisms are similar.

Analyze

For the reaction between NO and H_2,

$$\text{Rate} = k[\text{NO}]^2[\text{H}_2]$$

For the reaction between NO and Cl_2,

$$\text{Rate} = k[\text{NO}][\text{Cl}_2]$$

Solve

No. The different rate laws for the two reactions indicate different mechanisms.

Think About It

However, even if two rate laws were similar, that would not necessarily mean that their reaction mechanisms are similar.

13.97. **Collect, Organize, and Analyze**

We are to identify the conditions under which a bimolecular reaction shows pseudo–first-order behavior.

Solve

Pseudo–first-order kinetics occurs when one reactant is in sufficiently high concentration that it does not change appreciably during the reaction.

Think About It

We solved problems relating to pseudo–first-order reactions earlier in this chapter (Problems 13.77 and 13.78).

13.99. **Collect and Organize**

We are asked to draw reaction profiles that fit the reaction A → B (which has E_a = 50.0 kJ/mol) for three mechanisms: (a) a single elementary step, (b) a two-step reaction with $E_{a,\text{step 2}}$ = 15 kJ/mol, and (c) a two-step reaction in which the second step is rate determining.

Analyze

The reaction profile plots energy versus reaction progress and shows the presence of an activated complex at highest energy, the reaction intermediates in the "valleys" between reactant and products, and the relative activation energies in multistep reactions. The rate-determining step is the slowest step in the mechanism and has the largest activation energy in the reaction profile.

Solve

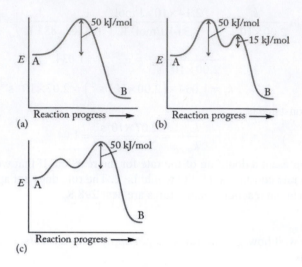

Think About It
Those reaction profiles could also be drawn for a nonspontaneous reaction. Reactant A would then be lower in energy than product B.

13.101. Collect and Organize
For each elementary step given, we are to write the rate law and determine whether the step is uni-, bi-, or termolecular.

Analyze
The rate law for an elementary step in a mechanism is written in the form
$$\text{Rate} = k[A]^x[B]^y[C]^z$$
where A, B, and C are the reactants involved in the elementary reaction and x, y, and z are the stoichiometric coefficients for the respective reactants in the elementary reaction.

Solve
(a) Rate = $k[SO_2Cl_2]$. Because that elementary step involves only a molecule of SO_2Cl_2, it is unimolecular.
(b) Rate = $k[NO_2][CO]$. Because that elementary step involves a molecule of NO_2 and a molecule of CO, it is bimolecular.
(c) Rate = $k[NO_2]^2$. Because that elementary step involves two molecules of NO_2, it is bimolecular.

Think About It
Termolecular elementary reactions are rare.

13.103. Collect and Organize
From two elementary steps that describe a reaction mechanism for the reaction of H_2O_2 with KI, we are to write the overall chemical equation and the rate law and then identify the catalysts and intermediates from the reaction mechanism.

Analyze
To write the overall chemical reaction we need to add the elementary steps, being sure to cancel the intermediates in the reaction. A catalyst is a species used in a reaction but is regenerated or not consumed. An intermediate is a species generated in a reaction and then used in a later step of the reaction.

Solve

(a) The overall reaction is

$$H_2O_2(aq) + I^-(aq) \rightarrow H_2O(aq) + IO^-(aq)$$

$$\underline{H_2O_2(aq) + IO^-(aq) \rightarrow H_2O(aq) + O_2(g) + I^-(aq)}$$

$$2\,H_2O_2(aq) \rightarrow 2\,H_2O(aq) + O_2(g)$$

(b) The rate law predicted by that mechanism with the first step being the slow step is

$$\text{Rate} = k[H_2O_2][I^-]$$

(c) I^- is the catalyst in the reaction.

(d) IO^- is the intermediate in the reaction.

Think About It

Remember that the rate of a reaction is always dependent on the slowest, or rate-limiting, step.

13.105. Collect and Organize

We are given the mechanism by which N_2 reacts with O_2 to form NO. For a given rate law of

$$\text{Rate} = k[N_2][O_2]^{1/2}$$

we are to determine which step in the mechanism is the rate-determining step.

Analyze

To determine which step in the proposed mechanism might be the slowest, we can write the rate law for the mechanism when the first, second, or third step is slow and then match the theoretical rate law to the experimental rate law.

Solve

If the first step is slow, the rate law is

$$\text{Rate} = k_1[O_2]$$

That does not match the experimental rate law, so the first step is not the slowest step in the mechanism.

If the second step is slow, the rate law is

$$\text{Rate} = k_2[O][N_2]$$

Because O is an intermediate, we use the first step to express its concentration in terms of concentrations of the reactants. For a fast step occurring before a slow step in a mechanism,

$$\text{Rate}_{\text{forward}} = \text{Rate}_{\text{reverse}}$$

$$k_1[O_2] = k_{-1}[O]^2$$

Rearranging to solve for [O],

$$[O] = \left(\frac{k_1}{k_{-1}}[O_2] \right)^{1/2}$$

Substituting that into the rate law from the second step gives

$$\text{Rate} = k_2 \left(\frac{k_1}{k_{-1}}[O_2] \right)^{1/2} [N_2] = k[O_2]^{1/2}[N_2]$$

That rate law matches the experimental rate law, so the rate-determining step is the second step.
We should check to see whether the mechanism of the third step is the slow step and might also give the experimental rate law. From the logic above,

$$\text{Rate} = k_3[N][O]$$

From the second fast step in the mechanism,

$$k_2[O][N_2] = k_{-2}[NO][N]$$

Solving for [N], an intermediate, gives

$$[N] = \frac{k_2[O][N_2]}{k_{-2}[NO]}$$

From the first fast step in the mechanism,

$$k_1[O_2] = k_{-1}[O]^2$$

solving for $[O]^2$ gives

$$[O]^2 = \frac{k_1}{k_{-1}}[O_2]$$

Substituting those expressions into the rate law from the third step in the mechanism gives

$$\text{Rate} = k_3 \frac{k_2}{k_{-2}} \frac{k_1}{k_{-1}} \frac{[O_2][N_2]}{[NO]}$$

$$= \frac{k[N_2][O_2]}{[NO]}$$

That rate law does not match the experimental rate law.

Think About It
Just because the rate law for a mechanism matches the experimental rate law does not mean that the mechanism is *the correct mechanism*. Another mechanism might also give the same experimental rate law.

13.107. Collect and Organize
We are given the mechanism by which NO reacts with Cl_2 to produce $NOCl_2$. For a given rate law of

$$\text{Rate} = k[NO][Cl_2]$$

we are to determine which step in the mechanism is the rate-determining step.

Analyze
To determine which step in the proposed mechanism might be the slowest, we can write the rate law for the mechanism when the first, second, or third step is slow and then match the rate law to the experimental rate law.

Solve
If the first step is slow, the rate law is

$$\text{Rate} = k_1[NO][Cl_2]$$

That matches the experimental rate law, so the first step is the rate-determining step.
We should check to see whether the rate law for the mechanism with the second step as the slow step,

$$\text{Rate} = k_2[NOCl_2][NO]$$

might also give the experimental rate law. If the second step is slow, then

$$\text{Rate}_1 = \text{Rate}_{-1}$$
$$k_1[NO][Cl_2] = k_{-1}[NOCl_2]$$
$$[NOCl_2] = \frac{k_1}{k_{-1}}[NO][Cl_2]$$

Substituting that into the rate law expression gives

$$\text{Rate} = k_2\left(\frac{k_1}{k_{-1}}[NO][Cl_2]\right)[NO] = k[NO]^2[Cl_2]$$

That does not match the experimental rate law, so the second step in the mechanism is not the rate-determining step.

Think About It
Just because the rate law for a mechanism matches the experimental rate law does not mean the mechanism is *the correct mechanism*. Another possible mechanism might also give the same experimental rate law.

13.109. Collect and Organize
From the mechanisms given, we are to determine which are possible for the thermal decomposition and which are possible for the photochemical decomposition of NO_2. We are given the rate laws: for the thermal decomposition reaction, $\text{Rate} = k[NO_2]^2$; for the photochemical decomposition, $\text{Rate} = k[NO_2]$.

Analyze

Using the slowest elementary step in the mechanism, we can write the rate law expression for each mechanism and then determine which is consistent with the order of the reaction given for each process.

Solve

For mechanism a, the first step in the mechanism is slow, so the rate law is

$$\text{Rate} = k[NO_2]$$

For mechanism b, the second step in the mechanism is slow, so the rate law is

$$\text{Rate} = k_2[N_2O_4]$$

Using the first step to express $[N_2O_4]$ in terms of the concentrations of the reactant NO_2 gives

$$k_1[NO_2]^2 = k_{-1}[N_2O_4]$$

$$[N_2O_4] = \frac{k_1}{k_{-1}}[NO_2]^2$$

Substituting into the rate expression from the second step,

$$\text{Rate} = \frac{k_2 k_1}{k_{-1}}\left[NO_2\right]^2 = k\left[NO_2\right]^2$$

For mechanism c, the first step in the mechanism is slow, so the rate law is

$$\text{Rate} = k[NO_2]^2$$

Therefore, mechanisms b and c are consistent with the thermal decomposition of NO_2 and mechanism a is consistent with the photochemical decomposition of NO_2.

Think About It

To distinguish between the two possible mechanisms for thermal decomposition, we might try to detect the different intermediates formed in each. Detection of the formation of N_2O_4 would support mechanism b over mechanism c.

13.111. Collect and Organize

We are asked whether a catalyst affects both the rate and the rate constant of a reaction.

Analyze

A catalyst speeds up a reaction by providing an alternative pathway (mechanism) to the products; that pathway has a lower activation energy.

Solve

Yes. Because the reaction is faster (affecting the rate) and the activation energy is lowered (affecting the value of k), a catalyst affects both the rate of the reaction and the value of the rate constant.

Think About It

A "negative" catalyst that slows down a reaction would increase E_a and decrease k for a reaction. We call those "negative catalysts" *inhibitors*.

13.113. Collect and Organize

We are asked whether a substance (catalyst) that increases the rate of a reaction also increases the rate of the reverse reaction.

Analyze

A catalyst speeds up a reaction by providing an alternative, lower-activation-energy pathway (mechanism) to the products.

Solve

Yes, both the reverse and forward reaction rates are increased when a catalyst is added to a reaction. The activation energies of both processes are lowered by the different pathway that the catalyst provides for the reaction.

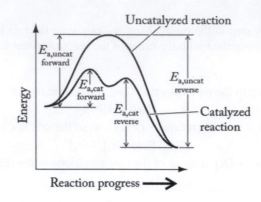

Think About It

Likewise, an inhibitor would decrease the rates of both forward and reverse reactions.

13.115. Collect and Organize

We are to explain whether the concentration of a homogeneous catalyst could appear in the rate law.

Analyze

A catalyst is used in a reaction and later regenerated.

Solve

The concentration of a homogeneous catalyst may not appear in the rate law if the catalyst is not involved in the rate-limiting step. However, if the catalyst is involved in the slowest step of the mechanism,, it can appear in the rate law.

Think About It

Catalysts speed up a reaction by lowering the activation energy, usually of the rate-limiting step in the mechanism.

13.117. Collect and Organize

Given the mechanism for the reaction of NO and N_2O to form N_2 and O_2, we are to determine whether NO or N_2O is used in the reaction as the catalyst.

Analyze

A catalyst is used in a reaction and later regenerated and provides a lower-energy pathway to the products by lowering the activation energy of the reaction, thereby speeding up the reaction.

Solve

We can assume that the presence of either NO or N_2O, if either is a catalyst, increases the rate of the reaction. In examining the mechanism we see that N_2O is a reactant, not a catalyst, but that NO is a catalyst because it is used in the reaction and then regenerated. Thus, NO is a catalyst for the decomposition of N_2O.

Think About It

If the slow step of this mechanism were the first step, the rate law would be
$$\text{Rate} = k[\text{NO}][\text{N}_2\text{O}]$$
If the second step were slow, the rate law would be

$$\text{Rate} = k \frac{\left[\text{NO}\right]^2 \left[\text{N}_2\text{O}\right]^2}{\left[\text{N}_2\right]^2}$$

13.119. Collect and Organize

Using the Arrhenius equation, we can compute and compare the rate constants for the reaction of O_3 with O versus the reaction of O_3 with Cl.

Analyze

We are given values A and E_a for each reaction at 298 K. The Arrhenius equation is

$$\ln k = \frac{-E_a}{RT} + \ln A$$

Solve

For the reaction of O_3 with O,

$$\ln k = \frac{-17.1 \times 10^3 \text{ J/mol}}{8.314 \text{ J/mol} \cdot \text{K} \times 298 \text{ K}} + \ln\left(8.0 \times 10^{-12} \text{ cm}^3/\text{molecules} \cdot \text{s}\right)$$

$$\ln k = -32.45$$

$$k = 8.05 \times 10^{-15} \text{ cm}^3/\text{molecules} \cdot \text{s}$$

For the reaction of O_3 with Cl,

$$\ln k = \frac{-2.16 \times 10^3 \text{ J/mol}}{8.314 \text{ J/mol} \cdot \text{K} \times 298 \text{ K}} + \ln\left(2.9 \times 10^{-11} \text{ cm}^3/\text{molecules} \cdot \text{s}\right)$$

$$\ln k = -25.14$$

$$k = 1.21 \times 10^{-11} \text{ cm}^3/\text{molecules} \cdot \text{s}$$

Therefore, the reaction of O_3 with Cl has the larger rate constant.

Think About It

Our answer is consistent with a qualitative look at the activation energies and frequency factors for the two reactions. The higher activation energy and lower frequency factor for the reaction of O_3 with O give a smaller reaction rate constant.

13.121. **Collect and Organize**

We are to explain why a glowing wood splint burns faster in a test tube filled with O_2 than in air.

Analyze

Air is composed of about 21% O_2.

Solve

When the concentration of a reactant (O_2 for the combustion reaction) increases, the rate of reaction also increases. As we place the glowing wood in pure O_2, the rate of combustion increases.

Think About It

If the wood splint were placed in a test tube filled with argon, the combustion reaction would stop.

13.123. **Collect and Organize**

We are to explain why a person submerged in cold water is less likely to have a lack of oxygen for a given period than a person submerged in a warm pool.

Analyze

Chemical reactions are slower at colder temperatures than at warmer temperatures.

Solve

The bodily reactions that use O_2 are slower at colder temperatures, so the person submerged in an ice-covered lake uses less of the already dissolved oxygen in his or her system than the person in a warm pool.

Think About It

Rapid-cooling technology is being investigated at Argonne National Laboratory for use in surgery patients and heart attack victims to reduce the damage done to cells by lack of oxygen in the blood.

13.125. Collect and Organize

For the case in which $rate_{reverse} \ll rate_{forward}$, we are to consider whether the method to determine the rate law (initial concentrations and initial rates) would work at other times, not just at the start of the reaction.

Analyze

The method that uses initial rates and concentrations to determine the rate law is under the condition in which no reverse reaction is occurring.

Solve

Yes, we could use that method at other times, not just $t = 0$, to determine the rate law if the rate of the reverse reaction is much slower than the forward reaction as long as [products] $\ll$ [reactants] at the time so that no appreciable reverse reaction is occurring.

Think About It

We will see in Chapter 15 that when the rate of the reverse reaction equals the rate of the forward reaction, the reaction is at equilibrium.

13.127. Collect and Organize

In the plot of $1/[X] - 1/[X]_0$ as a function of time, t, we are asked how the rate constant, k, can be determined.

Analyze

The plot of $1/[X] - 1/[X]_0$ as a function of time, t, is the plot for a second-order rate equation.

$$Rate = k[X]^2$$

$$\frac{1}{[X]} = kt + \frac{1}{[X]_0}$$

Solve

In the plot, $1/[X] - 1/[X]_0$ divided by $t - t_0$ is the slope of the line that corresponds to k, the reaction rate constant. All we need to do to determine k from that plot is to determine the slope of the line.

Think About It

The integrated form of the rate law allows us to obtain the value of k from the concentration-versus-time data from a single experiment.

13.129. Collect and Organize

We are asked to explain why an elementary step cannot have a rate law that is zero order.

Analyze

An elementary step in a reaction mechanism describes the collisions of molecular or atomic species taking place in a reaction.

Solve

For an elementary step to take place, some involvement from a molecular or atomic species must occur. Therefore, there cannot be *no* dependence (or zero order) of the reactant in an elementary step of a reaction mechanism.

Think About It

Because most elementary steps are either unimolecular or bimolecular, the rate expressions of elementary steps are usually first or second order.

13.131. Collect and Organize

Given the balanced equation for the reaction between NO_2 and O_3 to produce N_2O_5 and O_2, we are asked to relate the rates of change in $[NO_2]$, $[O_3]$, $[N_2O_5]$, and $[O_2]$.

Analyze

From the balanced equation, the rate of formation of products and consumption of reactants is

$$\text{Rate} = -\frac{1}{2}\frac{\Delta[NO_2]}{\Delta t} = -\frac{\Delta[O_3]}{\Delta t} = \frac{\Delta[N_2O_5]}{\Delta t} = \frac{\Delta[O_2]}{\Delta t}$$

Solve

The rate of consumption of O_3 is the same as the rate of formation of N_2O_5 and O_2 and one-half the rate of consumption of NO_2.

Think About It

The rate of consumption of N_2O_5 is half the rate of formation of NO_2 and twice the rate of formation of O_2.

13.133. Collect and Organize

We can write the rate law from the order of the decomposition reaction determined in Problem 13.132. From that we are to calculate the value of the rate constant at the experimental temperature and write the complete rate law expression.

Analyze

From Problem 13.132, we know that the reaction is first order in $[N_2O_5]$.

Solve

Because the reaction is a first-order decomposition reaction, the rate law expression is

$$\text{Rate} = k[N_2O_5]$$

Using the data in experiment 1 in Problem 13.132,

$$1.8 \times 10^{-5} \ M/s = k \times 0.050 \ M$$
$$k = 3.6 \times 10^{-4} \ s^{-1}$$

The complete rate law expression is then

$$\text{Rate} = (3.6 \times 10^{-4} \ s^{-1}) \times [N_2O_5]$$

Think About It

The data from experiment 2 in Problem 13.132 would give the same value of k.

13.135. Collect and Organize

For the reaction of NO with O_3 to produce NO_2 and O_2, we can use the information that the reaction is first order in both NO and O_3 along with the values of the rate constants at two temperatures to determine whether the reaction occurs in one or many steps and to calculate the activation energy, the rate of the reaction at another concentration of the reactants, and the rate constants at two other temperatures.

Analyze

To answer the questions, we need to use the Arrhenius equation,

$$\ln k = \frac{-E_a}{RT} + \ln A$$

and the rate law expression, which states that the reaction is first order in both NO and O_3:

$$\text{Rate} = k[NO][O_3]$$

Solve

(a) Because the rate law in which the reaction is first order in both NO and O_3 is consistent with that in which the reaction would occur in a single step, the reaction might indeed occur in a single step.

(b) We can calculate the activation energy for the reaction by comparing the rate constant at the two temperatures:

$$\ln k_{25°C} = \frac{-E_a}{R}\left(\frac{1}{298 \ K}\right) + \ln A$$

$$\ln k_{75°C} = \frac{-E_a}{R}\left(\frac{1}{348 \ K}\right) + \ln A$$

Subtracting $\ln k$ at 25°C from $\ln k$ at 75°C gives

$$\ln k_{75°C} - \ln k_{25°C} = \frac{-E_a}{R}\left(\frac{1}{348 \text{ K}} - \frac{1}{298 \text{ K}}\right) = \ln\left(\frac{k_{75°C}}{k_{25°C}}\right)$$

$$\ln\left(\frac{3000 \ M^{-1}s^{-1}}{80 \ M^{-1} \ s^{-1}}\right) = \frac{-E_a}{8.314 \text{ J/mol} \cdot \text{K}}\left(\frac{1}{348 \text{ K}} - \frac{1}{298 \text{ K}}\right)$$

(c) We can use the rate law expression to calculate the rate of the reaction at 25°C when $[NO] = 3 \times 10^{-6} \ M$ and $[O_3] = 5 \times 10^{-9} \ M$. We are given in the statement of the problem that k at 25°C is 80 $M^{-1} \ s^{-1}$.

$$\text{Rate} = 80 \ M^{-1} \ s^{-1} \times (3 \times 10^{-6} \ M) \times (5 \times 10^{-9} \ M) = 1.2 \times 10^{-12} \ M/s$$

(d) To use the Arrhenius equation to calculate the values of k at 10°C and 35°C, we have to first determine the value of the frequency factor A. To do so we can use the value for E_a and k for 25°C:

$$\ln\left(80 \ M^{-1} \ s^{-1}\right) = \frac{-6.25 \times 10^4 \text{ J/mol}}{8.314 \text{ J/mol} \cdot \text{K}}\left(\frac{1}{298 \text{ K}}\right) + \ln A$$

$$\ln A = 29.608$$

$$A = 7.22 \times 10^{12}$$

So the value of k at 10°C (283 K) is

$$\ln k = \frac{-6.25 \times 10^4 \text{ J/mol}}{8.314 \text{ J/mol} \cdot \text{K}}\left(\frac{1}{283 \text{ K}}\right) + \ln(7.22 \times 10^{12})$$

$$\ln k = 3.045$$

$$k = 21 \ M^{-1} \ s^{-1}$$

The value of k at 35°C (308 K) is

$$\ln k = \frac{-6.25 \times 10^4 \text{ J/mol}}{8.314 \text{ J/mol} \cdot \text{K}}\left(\frac{1}{308 \text{ K}}\right) + \ln(7.22 \times 10^{12})$$

$$\ln k = 5.20$$

$$k = 1.8 \times 10^2 \ M^{-1} \ s^{-1}$$

Think About It

The values of k at 10°C and 35°C calculated above make sense because they are a little lower and a little higher, respectively, than the value of k at 25°C. Also, an Arrhenius plot using the two k values at 25°C and 75°C could be used to determine the activation energy and the frequency factor values in that problem.

13.137. Collect and Organize

We consider the mechanism for the exchange of $H_2^{16}O$ around a Na^+ cation for $H_2^{18}O$ to write the rate law for the rate-determining step. We also need to think about the relative energies of the reactants and products if we are to sketch the reaction energy profile.

Analyze

(a) We are given that the first step of the reaction is rate determining, so that is the step from which we write the rate law expression.

(b) In deciding which has the higher energy, the reactants or products, for the reaction profile we need to consider the relative strength of the Na^+–$^{16}OH_2$ interaction versus that of the Na^+–$^{18}OH_2$ interaction.

Solve

(a) Rate = $k[Na(H_2O)_6^+]$

(b) Neither. The ion–dipole interaction should not be significantly different for $H_2^{18}O$ versus $H_2^{16}O$, so the energy of the reactants and the products in the reaction profile will be about the same.

Think About It

In reality, an isotope effect exists, in which the $H_2^{18}O$–Na^+ interaction is slightly stronger than the $H_2^{16}O$–Na^+ interaction, so the energy of the product in that reaction is slightly lower than the energy of the reactants.

13.139. Collect and Organize

For the reaction of NO with ONOO⁻, we are to use the provided data to determine the rate law for the reaction. We are also to draw the Lewis structure with resonance forms to describe the bonding in the ONOO⁻ anions and then use bond energies to estimate ΔH°_{rxn}.

Analyze

(a) To determine the rate law, we can compare how changing the concentrations of NO and ONOO⁻ affect the rate of the reaction.

(b) After drawing the resonance forms for ONOO⁻, we can determine the preferred structure by assigning formal charges to the atoms in each resonance form.

(c) The ΔH°_{rxn} may be estimated from bond energies by using

$$\Delta H^{\circ}_{rxn} = \sum \Delta H_{bond\ breaking} + \sum \Delta H_{bond\ forming}$$

Solve

(a) For the order of reaction with respect to ONOO⁻, we can compare experiments 1 and 2:

$$\frac{rate_2}{rate_1} = \left(\frac{[ONOO^-]_{0,2}}{[ONOO^-]_{0,1}}\right)^x$$

$$\frac{1.02 \times 10^{-11}\ M/s}{2.03 \times 10^{-11}\ M/s} = \left(\frac{0.625 \times 10^{-4}\ M}{1.25 \times 10^{-4}\ M}\right)^x$$

$$0.502 = 0.500^x$$

$$x = 1$$

The data table does not have two experiments for which the concentration of ONOO⁻ stays the same. For the order of reaction with respect to NO, we can compare experiments 2 and 3 as long as we take into account the knowledge that the reaction is first order in ONOO⁻. Between those two experiments we see that as the ONOO⁻ concentration is quadrupled, we would expect that the rate would be quadrupled. The rate of the reaction when the NO concentration is halved simultaneously with the quadrupling of the rate on changing the ONOO⁻, however, doubles:

$$\frac{rate_3}{rate_2} = \left(\frac{2.03 \times 10^{-11}\ M/s}{1.02 \times 10^{-11}\ M/s}\right) = 1.99$$

Because the $\Delta[ONOO^-] = 4$, which would quadruple the rate of the reaction, the $\Delta[NO]$ must affect the rate by $1.99/4 = 0.498$. Therefore,

$$0.498 = \left(\frac{0.625 \times 10^{-4}\ M}{1.25 \times 10^{-4}\ M}\right)^x$$

$$0.498 = 0.500^x$$

$$x = 1$$

The rate law for this reaction is Rate = k[NO][ONOO⁻].

We can use any experiment to calculate the value of k. Using the data from experiment 1:

$$2.03 \times 10^{-11}\ M/s = k \times (1.25 \times 10^{-4}\ M) \times (1.25 \times 10^{-4}\ M)$$

$$k = 1.30 \times 10^{-3}\ M^{-1}\ s^{-1}$$

(b)

Preferred structure

(c)

Bonds broken = O—O (146 kJ/mol)

Bonds formed = N—O (201 kJ/mol)

$$\Delta H^{\circ}_{rxn} = [(1\ mol \times 146\ kJ/mol) + (1\ mol \times -201\ kJ/mol)] = -55\ kJ$$

Think About It

To solve that problem you had to draw on several concepts you have learned so far in this course. You had to review not only how to write a rate law given kinetic data but also how to draw resonance structures and how bond energies might be used to estimate the enthalpy of a reaction.

13.141. **Collect and Organize**

Using data for [HNO$_2$] over time, we can determine the order of the isotopic exchange reaction with respect to [HNO$_2$]. We are also asked whether the rate of the reaction will depend on [H$_2{}^{18}$O].

Analyze

We are given a single data set. If the plot of [HNO$_2$] versus time yields a straight line, the reaction is zero order. If the plot of ln[HNO$_2$] versus time yields a straight line, the reaction is first order. If the plot of 1/[HNO$_2$] versus time yields a straight line, the reaction is second order.

Solve

(a)

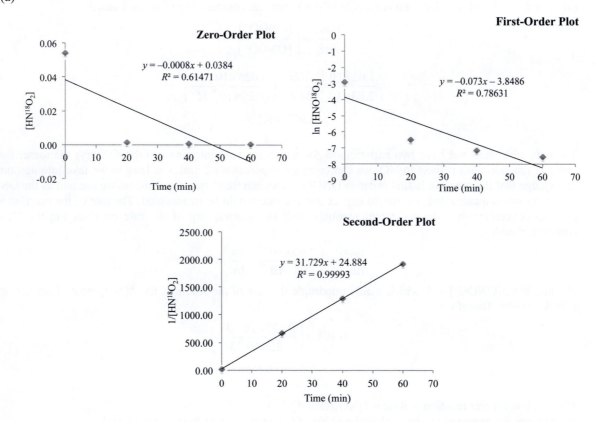

That reaction is second order in [HNO$_2$].

(b) Because the reaction mixture has a very large [H$_2{}^{18}$O], we cannot observe a change in [H$_2{}^{18}$O] over time. The rate might depend on [H$_2{}^{18}$O], but we cannot tell from the information given.

Think About It

If it were possible, we could place the reaction in a nonreactive solvent and then vary the [H$_2{}^{18}$O] over time to determine the rate's dependence on its concentration.

13.143. **Collect and Organize**

Using the raw data obtained for four experiments in which the concentrations of NH_2 and NO were varied, we are to write the rate law and determine the value of k for the reaction between NH_2 and NO at 1200 K.

Analyze

To determine the rate law, we can compare how changing the concentrations of NH_2 and NO affects the rate of the reaction. Once we have determined the order of the reaction with respect to each reactant, we can write the rate law expression and use any of the experiments to calculate the value of the rate constant, k.

Solve

(a) For the order of the reaction with respect to NH_2, we can use the data from experiments 1 and 2:

$$\frac{\text{rate}_2}{\text{rate}_1} = \left(\frac{[NH_2]_{0,2}}{[NH_2]_{0,1}}\right)^x$$

$$\frac{0.24 \ M/s}{0.12 \ M/s} = \left(\frac{2.00 \times 10^{-5} \ M}{1.00 \times 10^{-5} \ M}\right)^x$$

$$2.0 = 2.00^x$$

$$x = 1$$

For the order of the reaction with respect to NO, we can use the data from experiments 2 and 3:

$$\frac{\text{rate}_3}{\text{rate}_2} = \left(\frac{[NO]_{0,3}}{[NO]_{0,2}}\right)^x$$

$$\frac{0.36 \ M/s}{0.24 \ M/s} = \left(\frac{1.50 \times 10^{-5} \ M}{1.00 \times 10^{-5} \ M}\right)^x$$

$$1.5 = 1.50^x$$

$$x = 1$$

The rate law expression is Rate = $k[NH_2][NO]$.

(b) Using experiment 1 to calculate the value of k,

$$0.12 \ M/s = k \times (1.00 \times 10^{-5} \ M) \times (1.00 \times 10^{-5} \ M)$$

$$k = 1.2 \times 10^9 \ M^{-1} \ s^{-1}$$

Think About It

Any of the experiments listed in the data table would give us the same value of k in the calculation in part b.

14.1. Collect and Organize

For two reversible reactions, we are given the reaction profiles (Figure P14.1). The profile for the conversion of A to B shows that reactant A has a lower free energy than product B. The profile for the conversion of C to D shows that C has a higher free energy than D. From the profiles, we are to determine which reaction has the larger k_f, the smaller k_r, and the larger value of K_c.

Analyze

The magnitude of the rate constant is inversely related to the magnitude of the activation energy, E_a. The reaction with the larger k_f, the smaller k_r, and the larger K_c is that reaction with the lowest E_a for the forward reaction and where $k_f > k_r$.

Solve

The reaction C → D has the larger k_f, the smaller k_r, and the larger K_c (because it has the lowest activation energy for the forward direction).

Think About It

Remember that a large k (rate constant) means that the reaction is fast and therefore the reaction has a low activation energy.

14.3. Collect and Organize

From Figure P16.3, showing 26 blue spheres (product B) and 13 red spheres (reactant A), we are to write the chemical equation of the equilibrium reaction and calculate the value of K_c.

Analyze

(a) In this reaction, A is transformed into B. We represent a system at equilibrium by using double-headed reaction arrows between the reactants and products. We will assume that one molecule of A produces one molecule of B in the reaction.

(b) The value of K_c is the ratio of the concentration of the products (number of B spheres) and the concentration of the reactants (number of A spheres) raised to their respective stoichiometric coefficients from the balanced chemical equation.

Solve

(a) A $\rightleftharpoons$ B

(b) $K_c = \dfrac{26}{13} = 2.0$

Think About It

If the chemical equation were written as $2A \rightarrow B$, then K_c would be

$$K_c = \frac{[B]}{[A]^2} = \frac{26}{(13)^2} = 0.15$$

14.5. Collect and Organize

By comparing the relative distributions of reactants A and B with product AB at two temperatures, as shown in Figure P14.5, we can determine whether the reaction is endothermic or exothermic.

Analyze

From the equation

$$\Delta G = -RT \ln K = \Delta H - T\Delta S$$

we see that as temperature rises for an exothermic reaction, ΔG becomes less negative and therefore K decreases. If, however, the temperature is raised in an endothermic reaction, ΔG becomes more negative and K

increases. Therefore, if products increase upon raising the temperature, the reaction is endothermic; if products decrease, the reaction is exothermic.

Solve

At 300 K the equilibrium mixture is 6 A, 10 B, and 5 AB. That gives an equilibrium constant of

$$K_{300\ K} = \frac{5}{6 \times 10} = 0.083$$

At 400 K, the equilibrium mixture is 3 A, 7 B, and 8 AB. That gives an equilibrium constant of

$$K_{400\ K} = \frac{8}{3 \times 7} = 0.38$$

As temperature increases for the reaction, K increases, indicating that more products form at higher temperatures, so the reaction is endothermic.

Think About It

We assume in this problem that the difference in entropy for the two temperatures at which this reaction is run is minimal, so only ΔH contributes to the difference in ΔG at the two temperatures.

14.7. Collect and Organize

We are asked to determine from Figure P14.7 whether the reaction depicted reaches equilibrium.

Analyze

Equilibrium is a state where the composition of the reaction is not changing.

Solve

A reaction is at equilibrium when the rate of the forward reaction equals the rate of the reverse reaction. As a result, the concentration of neither the products nor the reactants will change at equilibrium. In Figure P14.7, we see that both the concentrations of the reactants and the products level off after 50 μs. That is the point of equilibrium. At 20 μs the concentrations of the reactants and the products are still changing, so at that point, the reaction is not at equilibrium.

Think About It

Chemical equilibrium is a dynamic process. At the molecular level the forward and reverse reactions are still occurring. Because they occur at the same rate, however, we observe no macroscopic changes in the concentrations of the reactants and products in the mixture.

14.9. Collect and Organize

Given two equilibrium constants (0.3 and 3) for the equilibrium between a β-diketone (A, red spheres) and the enol form (B, blue spheres), we are to draw the number of A and B molecules once the reaction has reached equilibrium.

Analyze

For each equilibrium constant we start with only molecules of A. As the equilibrium constant increases, more B will be observed for the larger equilibrium constant. Therefore, we expect more B to be produced for the equilibrium constant of 3. The equilibrium constant expression for this conversion is

$$K_c = \frac{[B]_{eq}}{[A]_{eq}}$$

where $[B]_{eq} = x$ and $[A]_{eq} = [A]_{initial} - x$.

Solve

(a) At equilibrium, which initially has 26 molecules of A, $[A]_{eq} = 26 - x$ and $[B]_{eq} = x$.

$$K_c = 0.30 = \frac{[B]_{eq}}{[A]_{eq}} = \frac{x}{26 - x}$$

$$0.30(26 - x) = x$$

$$7.80 - 0.3x = x$$

$$7.80 = 1.3x$$

$$x = 6$$

Therefore, at equilibrium, 6 blue spheres and 20 red spheres are present.

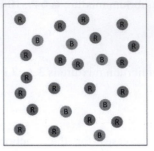

(b) At equilibrium, which initially has 28 molecules of A, $[A]_{eq} = 26 - x$ and $[B]_{eq} = x$.

$$K_c = 3.0 = \frac{[B]_{eq}}{[A]_{eq}} = \frac{x}{28 - x}$$

$$3.0(28 - x) = x$$

$$84 - 3.0x = x$$

$$84 = 4.0x$$

$$x = 21$$

Therefore, at equilibrium, 21 blue spheres and 7 red spheres are present.

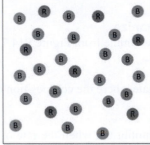

Think About It

This clearly shows the effect on the ratio of products to reactants for a reaction with an equilibrium constant greater than 1 in comparison with that of an equilibrium constant less than 1.

14.11. Collect and Organize

We are to differentiate between a reaction that goes to completion and one that is reversible.

Analyze

Both can be interpreted or defined in terms of chemical equilibrium.

Solve

A reaction that goes to completion has a very, very large equilibrium constant. A reaction that is reversible has products that can be converted back into products.

Think About It

Reversible reactions in the practical sense tend to have equilibrium constants close to 1.

14.13. Collect and Organize

For a reaction where $k_f > k_r$, we are to determine whether K is greater than, less than, or equal to 1.

Analyze

The equilibrium constant defined in terms of the rate constants for a reaction is

$$K = \frac{k_f}{k_r}$$

Solve

When $k_f > k_r$, K is greater than 1.

Think About It

When $K > 1$, more products are present at equilibrium than reactants.

14.15. **Collect and Organize**

For the decomposition of N_2O to N_2 and O_2, we are to identify the species present after one day from the given molar masses. The initial reaction mixture contains $^{15}N_2O$, N_2, and O_2.

Analyze

The only isotope in the reaction for oxygen is ^{16}O, but for nitrogen both ^{15}N and ^{14}N are present at the beginning of the reaction. After one day, the ^{14}N will be incorporated into N_2O and ^{15}N will be incorporated into N_2.

Solve

Molar Mass	Compound	How Present
28	$^{14}N_2$	Originally present
29	$^{15}N^{14}N$	From decomposition of $^{15}N^{14}NO$
30	$^{15}N_2$	From decomposition of $^{15}N_2O$
32	O_2	Originally present
44	$^{14}N_2O$	From combination of $^{14}N_2$ and O_2
45	$^{15}N^{14}NO$	From combination of $^{15}N^{14}N$ and O_2
46	$^{15}N_2O$	Originally present

Think About It

The redistribution of ^{15}N from N_2O to N_2 and of ^{14}N from N_2 to N_2O shows that both forward and reverse reactions occur in a dynamic equilibrium process.

14.17. **Collect and Organize**

From the rate laws and rate constants for the forward and reverse reactions for

$$A \rightleftharpoons B$$

we are to calculate the value of the equilibrium constant.

Analyze

At equilibrium, the rate of the forward reaction equals the rate of the reverse reaction. The equilibrium constant is the ratio of the concentrations of the products raised to their coefficients to the concentrations of the reactants raised to their respective coefficients.

Solve

For the reaction $A \rightleftharpoons B$ we are given that the forward reaction is first order in A:

$$\text{Rate}_f = k_1[A], \text{ where } k_1 = 1.50 \times 10^{-2} \text{ s}^{-1}$$

and that the reverse reaction is first order in B:

$$\text{Rate}_r = k_{-1}[B], \text{ where } k_{-1} = 4.50 \times 10^{-2} \text{ s}^{-1}$$

At equilibrium, $\text{rate}_f = \text{rate}_r$, so

$$k_1[A] = k_{-1}[B]$$

Rearranging that to give the equilibrium constant expression allows us to solve for K_c:

$$K_c = \frac{[B]}{[A]} = \frac{k_1}{k_{-1}} = \frac{1.50 \times 10^{-2} \text{ s}^{-1}}{4.50 \times 10^{-2} \text{ s}^{-1}} = 0.333$$

Think About It

For the reverse equilibrium

$$B \rightleftharpoons A$$

the equilibrium constant is 1/0.333, or 3.00.

14.19. Collect and Organize
From the equation relating K_c and K_p, we can determine under what conditions $K_c = K_p$.

Analyze
The equation relating K_c and K_p is

$$K_p = K_c(RT)^{\Delta n}$$

Solve
K_p equals K_c when $\Delta n = 0$. That is true when the number of moles of gaseous products equals the number of moles of gaseous reactants in the balanced chemical equation.

Think About It
The value of K_p may also be less than K_c (for $\Delta n < 0$) or greater than K_c (for $\Delta n > 0$).

14.21. Collect and Organize
Given three reactions involving nitrogen oxides, we are to write K_c and K_p expressions.

Analyze
The K_c expression uses concentration (molarity) units for the reactants and the products, whereas the K_p expression uses partial pressure. The K expression for a balanced chemical reaction takes the general form

$$w\text{A} + x\text{B} \rightleftharpoons y\text{C} + z\text{D}$$

$$K_c = \frac{[\text{C}]^y [\text{D}]^z}{[\text{A}]^w [\text{B}]^x} \quad \text{and} \quad K_p = \frac{(P_\text{C})^y (P_\text{D})^z}{(P_\text{A})^w (P_\text{B})^x}$$

Solve

(a) $K_c = \dfrac{[\text{N}_2\text{O}_4]}{[\text{N}_2][\text{O}_2]^2}$ and $K_p = \dfrac{\left(P_{\text{N}_2\text{O}_4}\right)}{\left(P_{\text{N}_2}\right)\left(P_{\text{O}_2}\right)^2}$

(b) $K_c = \dfrac{[\text{NO}_2][\text{N}_2\text{O}]}{[\text{NO}]^3}$ and $K_p = \dfrac{\left(P_{\text{NO}_2}\right)\left(P_{\text{N}_2\text{O}}\right)}{\left(P_{\text{NO}}\right)^3}$

(c) $K_c = \dfrac{[\text{N}_2]^2[\text{O}_2]}{[\text{N}_2\text{O}]^2}$ and $K_p = \dfrac{\left(P_{\text{N}_2}\right)^2 \left(P_{\text{O}_2}\right)}{\left(P_{\text{N}_2\text{O}}\right)^2}$

Think About It
K_c uses concentration units (usually in moles per liter), whereas K_p uses partial pressure units (usually in atmospheres).

14.23. Collect and Organize
Given a plot of the concentration versus time for the decomposition of N_2O to N_2 and O_2 (Figure P14.23), we are to estimate the value of K_c.

Analyze
The amounts of N_2O, N_2, and O_2 are given in concentration units and the form of the K_c expression is

$$K_c = \frac{[\text{N}_2][\text{O}_2]^{1/2}}{[\text{N}_2\text{O}]}$$

The concentrations of each species at equilibrium can be read from the graph as those concentrations that stop changing. That occurs after 5 s and gives $[\text{N}_2] = 0.00030\ M$, $[\text{O}_2] = 0.00014\ M$, and $[\text{N}_2\text{O}] = 7.10 \times 10^{-6}\ M$.

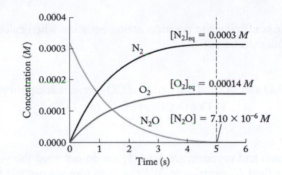

Solve

$$K_c = \frac{[N_2][O_2]^{1/2}}{[N_2O]} = \frac{(0.00030 \; M)(0.00014 \; M)^{1/2}}{(7.10 \times 10^{-6} \; M)} = 0.50$$

Think About It

Because that equilibrium constant is neither high nor low, at equilibrium the distribution of the reactants and products is expected to be roughly equal. A general rule is that if $K < 0.01$, the reactants are favored in the equilibrium, and if $K > 100$, the products are favored in the equilibrium.

14.25. Collect and Organize

Given the balanced chemical equation and the equilibrium partial pressures of all species in the decomposition reaction of H_2S to give H_2 and S, we are to calculate the value of K_p for the reaction.

Analyze

The form of the K_p expression for that reaction is

$$K_p = \frac{\left(P_{H_2}\right)\left(P_S\right)}{\left(P_{H_2S}\right)}$$

Solve

$$K_p = \frac{\left(P_{H_2}\right)\left(P_S\right)}{\left(P_{H_2S}\right)} = \frac{(0.045 \; \text{atm})(0.030 \; \text{atm})}{(0.020 \; \text{atm})} = 0.068$$

Think About It

Remember that equilibrium constants, K, have no units.

14.27. Collect and Organize

Given the equilibrium molar concentrations of N_2, O_2, and NO, we are to calculate K_c for

$$N_2(g) + O_2(g) \rightleftharpoons 2 \, NO(g)$$

Analyze

The equilibrium constant expression for that reaction is

$$K_c = \frac{[NO]^2}{[N_2][O_2]}$$

Solve

$$K_c = \frac{\left(3.1 \times 10^{-3} \; M\right)^2}{\left(3.3 \times 10^{-3} \; M\right)\left(5.8 \times 10^{-3} \; M\right)} = 0.50$$

Think About It
Be careful to account for the coefficients in the mass action equation when calculating K_c. Here we must be sure to square the equilibrium concentration of NO.

14.29. Collect and Organize
Given the initial moles of H_2O and CO and the moles of CO_2 present at equilibrium, we are to determine K_c for
$$H_2O(g) + CO(g) \rightleftharpoons H_2(g) + CO_2(g)$$

Analyze
Because the ratios of reactants and products are 1:1:1:1, we do not need the volume of the vessel because that volume would cancel in the final K_c expression. We can therefore simply use the mole amounts. Furthermore, we can calculate the amounts of reactants and products present at equilibrium from the initial amounts of the reactants, the given moles of the product CO_2 formed (8.3×10^{-3} mol), and the stoichiometry of the reaction.

Solve

Reaction	$H_2O(g)$ +	$CO(g)$ $\rightleftharpoons$	$H_2(g)$ +	$CO_2(g)$
	H_2O	CO	H_2	CO_2
Initial (mol)	0.0150	0.0150	0	0
Change (mol)	$-x$	$-x$	$+x$	$+x$
Equilibrium (mol)	$0.0150 - x$	$0.0150 - x$	x	x

At equilibrium, we know that $[CO_2] = x = 8.3 \times 10^{-3}$ mol. That gives
$$H_2O = CO = 0.0150 - x = 0.0067 \text{ mol}$$
$$H_2 = CO_2 = x = 8.3 \times 10^{-3} \text{ mol}$$
$$K_c = \frac{[H_2][CO_2]}{[H_2O][CO]} = \frac{(8.3 \times 10^{-3} \text{ mol})(8.3 \times 10^{-3} \text{ mol})}{(0.0067 \text{ mol})(0.0067 \text{ mol})} = 1.5$$

Think About It
We must have a balanced chemical reaction so that we can stoichiometrically relate the quantity of products to the reactants and correctly calculate the value of the equilibrium constant.

14.31. Collect and Organize
We are given $K_p = 32$ for the following reaction at 298 K:
$$A(g) + B(g) \rightleftharpoons AB(g)$$
We are to calculate the value of K_c, which we can do using the relationship
$$K_p = K_c (RT)^{\Delta n}$$

Analyze
The change in the number of moles of gas for that reaction (Δn) is
$$1 \text{ mol AB} - (1 \text{ mol A} + 1 \text{ mol B}) = -1$$

Solve

$$32 = K_c \times \left(0.08206 \frac{\text{L} \cdot \text{atm}}{\text{mol} \cdot \text{K}} \times 298 \text{ K} \right)^{-1}$$
$$K_c = 780$$

Think About It
When Δn is positive, K_p is greater than K_c, but when Δn is negative, K_p is less than K_c. When Δn is zero, K_p equals K_c.

14.33. **Collect and Organize**

We are given $K_p = 1.45 \times 10^{-5}$ for the following reaction at 500°C:

$$N_2(g) + 3\,H_2(g) \rightleftharpoons 2\,NH_3(g)$$

We are to calculate the value of K_c.

Analyze

Rearranging the equation relating K_p and K_c to solve for K_c gives

$$K_c = \frac{K_p}{(RT)^{\Delta n}}$$

The change in the number of moles is

$$(2\ \text{mol NH}_3) - (1\ \text{mol N}_2 + 3\ \text{mol H}_2) = -2$$

The temperature in kelvin is $500 + 273 = 773$ K.

Solve

$$K_c = \frac{1.45 \times 10^{-5}}{\left(0.08206\dfrac{\text{L} \cdot \text{atm}}{\text{mol} \cdot \text{K}} \times 773\ \text{K}\right)^{-2}} = 0.0583$$

Think About It

Here, $K_p < K_c$ because Δn is negative. When $\Delta n = 0$, $K_p = K_c$.

14.35. **Collect and Organize**

We are asked to determine in which of the three reactions the values of K_c and K_p are equal.

Analyze

K_p and K_c are equal when $\Delta n = 0$ in the relationship

$$K_p = K_c (RT)^{\Delta n}$$

where Δn is the change in the number of moles of gas in the reaction:

$$\Delta n = (\text{moles of gaseous products}) - (\text{moles of gaseous reactants})$$

Solve

(a) $\Delta n = 2 - 3 = -1$
(b) $\Delta n = 1 - 1 = 0$
(c) $\Delta n = 2 - 2 = 0$

Reactions (b) and (c) have $K_p = K_c$. Only reaction (a) has $K_p \neq K_c$.

Think About It

Remember that Δn is the difference in the moles of *gaseous* products and *gaseous* reactants. In reaction (b), Fe(*s*) and FeO(*s*) are not considered.

14.37. **Collect and Organize**

Given $K_c = 5.0$ for the following reaction at 327°C (600 K)

$$Cl_2(g) + CO(g) \rightleftharpoons COCl_2(g)$$

we are to calculate the value of K_p at 327°C. We can use the relationship

$$K_p = K_c (RT)^{\Delta n}$$

Analyze

For that reaction

$$\Delta n = (1\ \text{mol COCl}_2) - (1\ \text{mol Cl}_2 + 1\ \text{mol CO}) = -1$$

Solve

$$K_p = 5.0 \times \left(0.08206 \frac{\text{L} \cdot \text{atm}}{\text{mol} \cdot \text{K}} \times 600 \text{ K} \right)^{-1} = 0.10$$

Think About It

Because Δn is negative, the value of K_p is less than that of K_c.

14.39. Collect and Organize

We consider how K changes when the coefficients of a balanced reaction are scaled up or down.

Analyze

To answer the question, let's consider the following reactions:

$$(1) \quad 2\,A + B \rightleftharpoons C$$

$$(2) \quad 4\,A + 2\,B \rightleftharpoons 2\,C$$

The equilibrium constant expressions for the reactions are

$$K_1 = \frac{[C]}{[A]^2[B]} \quad \text{and} \quad K_2 = \frac{[C]^2}{[A]^4[B]^2}$$

Those are related by

$$K_2 = (K_1)^2$$

Solve

When we scale the coefficients of a reaction up or down, the new value of the equilibrium constant is the first K raised to the power of the scaling constant.

Think About It

Scaling reaction 1 by $^1/_3$ gives

$$\frac{2}{3}\,A + \frac{1}{3}\,B \rightleftharpoons \frac{1}{3}\,C$$

$$K = \frac{[C]^{1/3}}{[A]^{2/3}\,[B]^{1/3}}, \text{ or } (K_1)^{1/3}$$

14.41. Collect and Organize

Given the equilibrium constant for the reaction of 1 mol of $I_2(g)$ with 1 mol of $Br_2(g)$ to give 2 mol of $IBr(g)$, we are to calculate the value of the equilibrium constant for the reaction of ½ mol of I_2 and Br_2 to give 1 mol of IBr.

Analyze

The K_c expressions for those reactions are

$$K_{c_1} = \frac{\left[IBr \right]^2}{\left[I_2 \right]\left[Br_2 \right]} \quad \text{and} \quad K_{c_2} = \frac{\left[IBr \right]}{\left[I_2 \right]^{1/2}\left[Br_2 \right]^{1/2}}$$

K_{c_2} for the second reaction is, therefore, related to that of the first by

$$K_{c_2} = \left(K_{c_1} \right)^{1/2}$$

Solve

$$K_{c_2} = \left(K_{c_1} \right)^{1/2} = (120)^{1/2} = 11.0$$

Think About It

When we multiply a chemical reaction by a number, the new value of the equilibrium constant is the first equilibrium constant raised to that number.

14.43. **Collect and Organize**

For the reaction of NO with NO_3, we are to explain how K_c for the reverse reaction relates to K_c of the forward reaction by writing their equilibrium constant expressions.

Analyze

The form of the equilibrium constant expression for the forward reaction is the ratio of the product of the concentration of the products raised to their stoichiometric coefficients to the product of the concentration of the reactants raised to their stoichiometric coefficients:

$$K_c = \frac{[\text{products}]^y}{[\text{reactants}]^x}$$

Solve

The form of the equilibrium constant expression for the forward reaction is

$$K_{c,\text{forward}} = \frac{[NO_2]^2}{[NO][NO_3]}$$

For the reverse reaction, the equilibrium constant expression is

$$K_{c,\text{reverse}} = \frac{[NO][NO_3]}{[NO_2]^2}$$

Examining those two expressions, we see that

$$K_{c,\text{reverse}} = \frac{1}{K_{c,\text{forward}}}$$

Think About It

Another way to think about that relationship is

$$K_{c,\text{forward}} \times K_{c,\text{reverse}} = 1$$

14.45. **Collect and Organize**

For two reactions of different stoichiometry for the reaction of SO_2 with O_2 to produce SO_3, we are to write the equilibrium constant expressions and explain how they are related.

Analyze

Equilibrium constant expressions take the form

$$w\text{A} + x\text{B} \rightleftharpoons y\text{C} + z\text{D}$$

$$K_c = \frac{[\text{C}]^y[\text{D}]^z}{[\text{A}]^w[\text{B}]^x}$$

Solve

The equilibrium expressions for the reactions are

$$K_c = \frac{[SO_3]}{[SO_2][O_2]^{1/2}} \qquad \text{and} \qquad K_c' = \frac{[SO_3]^2}{[SO_2]^2[O_2]}$$

Those expressions are related by

$$K_c' = \left(K_c\right)^2$$

Think About It

If the second reaction were

$$\tfrac{1}{2}SO_2(g) + \tfrac{1}{4}O_2(g) \rightleftharpoons \tfrac{1}{2}SO_3(g)$$

then

$$K_c'' = \frac{[SO_3]^{1/2}}{[SO_2]^{1/2}[O_2]^{1/4}}$$

14.47. Collect and Organize

Given the value of K_c for the reaction

$$2\,SO_2(g) + O_2(g) \rightleftharpoons 2\,SO_3(g)$$

as 2.4×10^{-3}, we are to calculate K_c for three other forms of that reaction at the same temperature.

Analyze

When a reaction is multiplied by a factor x, the new equilibrium constant is $(K_c)^x$. When a reaction is reversed, the new equilibrium constant is $1/K_c$.

Solve

(a) That reaction is the original equation multiplied by $\frac{1}{2}$. The new K_c is

$$\left(2.4 \times 10^{-3}\right)^{1/2} = 4.9 \times 10^{-2}$$

(b) That reaction equation is the reverse of the original equation. The new K_c is

$$1/\left(2.4 \times 10^{-3}\right) = 420$$

(c) That reaction equation is the reverse of the original equation multiplied by $\frac{1}{2}$. The new K_c is

$$1/\left(2.4 \times 10^{-3}\right)^{1/2} = 20$$

Think About It

Because we can relate the equilibrium constants for different forms of a reaction, we need to tabulate only one of the equilibrium constants. All others can be calculated from that value.

14.49. Collect and Organize

We are asked to calculate K for a reaction given K_c values for two other reactions. We can add the two equations and reverse the resulting equation and then calculate the equilibrium constant for this new chemical equation.

Analyze

The K_c expressions for the reactions and the overall reactions are

$$A + 2\,B \rightleftharpoons C \qquad K_1 = \frac{[C]}{[A][B]^2} = 3.3$$

$$C \rightleftharpoons 2\,D \qquad K_2 = \frac{[D]^2}{[C]} = 0.041$$

$$\rule{6cm}{0.4pt}$$

$$A + 2\,B \rightleftharpoons 2\,D \qquad K_3 = \frac{[D]^2}{[A][B]^2}$$

where $K_3 = K_1 \times K_2$. The equilibrium constant for the reaction

$$2\,D \rightleftharpoons A + 2\,B$$

is the inverse of K_3.

Solve

$$K = \frac{1}{0.041 \times 3.3} = 7.4$$

Think About It

When we add reactions, the new equilibrium constant is the product of the equilibrium constants of the reactions that were combined.

14.51. **Collect, Organize, and Analyze**
We are to define the term *reaction quotient*.

Solve
The reaction quotient is the ratio of the concentrations of the products raised to their stoichiometric coefficients to the concentrations of reactants raised to their stoichiometric coefficients. The reaction quotient has the same form as the equilibrium constant expression, but the reaction concentrations (or partial pressures) are not necessarily at their equilibrium values.

Think About It
If the reaction quotient, Q, is greater than K, the reaction mixture must reduce its concentration of products to attain equilibrium. When Q is less than K, the reaction mixture must increase its concentration of products to attain equilibrium.

14.53. **Collect and Organize**
We are asked what it means when $Q = K$.

Analyze
Both K and Q take the form of the ratio of the concentrations of products raised to their stoichiometric coefficients to the concentrations of reactants raised to their stoichiometric coefficients.

Solve
When $Q = K$ the system is at equilibrium.

Think About It
Whenever $Q \neq K$, the reaction is not at equilibrium and it adjusts its relative concentrations of reactants and products so that $Q = K$.

14.55. **Collect and Organize**
We are asked whether the reaction

$$A(aq) \rightleftharpoons B(aq)$$

where $[A(aq)] = 0.10\ M$, $[B(aq)] = 2.0\ M$, and $K = 22$ is at equilibrium. If the reaction is not at equilibrium, we are to state in which direction the reaction will proceed to reach equilibrium.

Analyze
The reaction quotient (Q) is

$$\frac{[B]}{[A]} = Q$$

where [B] and [A] are the concentrations of A and B in the reaction mixture. If $Q > K$, the reaction will proceed to the left to reach equilibrium. If $Q < K$, the reaction will proceed to the right. If $Q = K$, the reaction is at equilibrium.

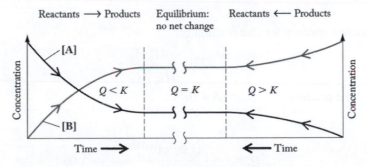

Solve

$$Q = \frac{2.0\ M}{0.10\ M} = 20$$

No, the reaction is not at equilibrium. $Q < K$, so that reaction proceeds to the right to reach equilibrium.

Think About It

In that reaction, more B forms as the reaction proceeds to equilibrium.

14.57. Collect and Organize

Given two sets of reactant and product concentrations for the reaction of A and B to form C (where $K_p = 1.00$), we are to determine whether either reaction mixture is at equilibrium. The temperature is 300 K.

Analyze

Those systems are at equilibrium when $Q = K_p = 1.00$. For the reaction where A, B, and C are expressed in terms of molarity, we must convert K_p to K_c by using

$$K_p = K_c (RT)^{\Delta n}$$

where for that reaction $\Delta n = -1$.

Solve

(a) $Q_p = \dfrac{1.0\ \text{atm}}{(1.0\ \text{atm})(1.0\ \text{atm})} = 1.0$

That reaction mixture is at equilibrium.

(b) Rearranging the equation to solve for K_c gives

$$K_c = \frac{K_p}{(RT)^{\Delta n}} = \frac{1.00}{\left[\left(0.08206\ \dfrac{\text{L}\cdot\text{atm}}{\text{mol}\cdot\text{K}}\right)\times 300\ \text{K}\right]^{-1}} = 24.6$$

$$Q_c = \frac{1.0\ M}{(1.0\ M)(1.0\ M)} = 1.0$$

Because $Q_c < K_c$, that reaction mixture is not at equilibrium and shifts to the right to attain equilibrium.

Think About It

Both K_p and K_c are close to 1, and so those reaction mixtures have roughly equal proportions of reactants and products when they reach equilibrium.

14.59. Collect and Organize

By comparing Q versus K for the reaction of N_2 with O_2 to form NO, we can determine in which direction the reaction will proceed to attain equilibrium.

Analyze

The form of the reaction quotient for that reaction is

$$Q_p = \frac{(P_{NO})^2}{(P_{N_2})(P_{O_2})}$$

Because $\Delta n = 0$ for that reaction, $K_p = K_c = 1.5 \times 10^{-3}$.

Solve

$$Q_p = \frac{(1.00\times 10^{-3})^2}{(1.00\times 10^{-3})(1.00\times 10^{-3})} = 1.00$$

Because $Q > K$, the system is not at equilibrium and proceeds to the left.

Think About It

Here $K_p = K_c$. That is not always the case, so be careful to notice which value is provided for K and in what units the amounts of the reactants and products are expressed.

14.61. Collect and Organize

Given initial concentrations of the reactants X and Y and product Z and the value of the equilibrium constant ($K_c = 1.00$ at 350 K), we can use the reaction quotient to determine in which direction the reaction will shift to reach equilibrium.

Analyze

The reaction quotient for that reaction has the form

$$Q_c = \frac{[Z]}{[X][Y]}$$

Solve

$$Q_c = \frac{[Z]}{[X][Y]} = \frac{0.2\ M}{0.2\ M \times 0.2\ M} = 5$$

Because $Q_c > K_c$, the reaction proceeds to the left (a), producing more X and Y.

Think About It

To calculate the equilibrium concentration of each reactant and product, we solve for x in the equation

$$1.00 = \frac{0.2 - x}{(0.2 + x)^2}$$

14.63. Collect and Organize

For the dissolution of CuS to give Cu^{2+} and S^{2-} in aqueous solution, we are asked to write the K_c expression.

Analyze

Because any pure solids or liquids do not change in concentration during a reaction, their "concentrations" do not appear in the mass action (equilibrium constant) expression.

Solve

$$K_c = [Cu^{2+}][S^{2-}]$$

Think About It

Because the reactant in the dissolution reaction is a pure solid, only the concentrations of the products influence the position of the equilibrium.

14.65. Collect and Organize

For the decomposition reaction of $CaCO_3$ to CO_2 and CaO, we are to explain why $[CaCO_3]$ and $[CaO]$ do not appear in the K_c expression.

Analyze

The K_c expression includes the concentrations of the reactants and products that may change during the reaction attaining equilibrium.

Solve

The concentrations (expressed as densities) of pure solids ($CaCO_3$ and CaO) do not change during the reaction to attain equilibrium, and so they do not appear in the equilibrium constant expression.

Think About It

The form of K_c for that reaction is simply $K_c = [CO_2]$.

14.67. Collect and Organize

We are asked whether the value of K increases when more reactants are added to a reaction already at equilibrium.

Analyze

The general form of the equilibrium constant expression for a reaction is

$$w\text{A} + x\text{B} \rightleftharpoons y\text{C} + z\text{D}$$

$$K_c = \frac{[\text{C}]^y[\text{D}]^z}{[\text{A}]^w[\text{B}]^x}$$

Solve

No, the equilibrium constant is not changed when the concentration of the reactants is increased. The relative concentrations of the reactants and products in that case adjust until they achieve the value of K. Only temperature affects the value of the equilibrium constant.

Think About It

The value of the reaction quotient Q decreases below the value of K when reactants are added to a system previously at equilibrium, and the reaction shifts to the right.

14.69. Collect and Organize

Given that the K for the binding of CO to hemoglobin is larger than that for the binding of O_2 to hemoglobin, we are to explain how the treatment of CO poisoning by administering pure O_2 to a patient works.

Analyze

By giving a patient pure O_2 to breathe, we increase the partial pressure of O_2 to which the hemoglobin in the patient's blood is exposed. That oxygen can displace the CO bound to the hemoglobin through Le Châtelier's principle.

Solve

Combining the two equations, we can write the expression for the displacement of CO bound to hemoglobin by O_2:

$$\text{Hb(CO)}_4 \rightleftharpoons \text{Hb} + 4\,\text{CO}(g)$$

$$\underline{\text{Hb} + 4\,O_2(g) \rightleftharpoons \text{Hb(O}_2)_4}$$

$$\text{Hb(CO)}_4 + 4\,O_2(g) \rightleftharpoons \text{Hb(O}_2)_4 + 4\,\text{CO}(g)$$

As the concentration of O_2 increases, the reaction shifts to the right and the CO on the hemoglobin is displaced.

Think About It

If we were given the values of the equilibrium constants for the reactions, we could calculate the new equilibrium constant for the overall reaction.

14.71. Collect and Organize

We are to interpret Henry's law through Le Châtelier's principle.

Analyze

The dissolution of a gas (let's use oxygen in this example) in a liquid (let's use water) can be written as a chemical equation:

$$O_2(g) \rightleftharpoons O_2(aq)$$

Solve

According to Le Châtelier's principle, an increase in the partial pressure (or concentration) of O_2 above the water shifts the equilibrium to the right so that more oxygen dissolves in the water. That result is consistent with Henry's law.

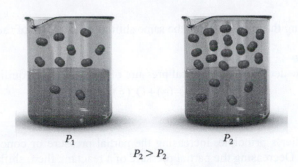

P_1 P_2

$P_2 > P_2$

Think About It
The solubilities of different gases in a liquid are different from each other, but all gases are more soluble in a liquid when present at higher partial pressures.

14.73. ### Collect and Organize
Of four reactions, we are to determine which will shift its equilibrium to form more products when the mixture is compressed to half its original volume.

Analyze
Decreasing the volume by half doubles the partial pressures of all the gaseous reactants and products in the reactions. That would cause a shift in the position of the equilibrium towards the side of the reaction with fewer moles of gas.

Solve
(a) That equilibrium shifts to the left, forming more reactants.
(b) That equilibrium shifts to the right, forming more products.
(c) That equilibrium shifts neither to the left nor to the right, as the number of moles of gas is the same on both sides of the equation.
(d) That equilibrium shifts to the right, forming more products.
Reactions (b) and (d) will form more products when the volume of the mixture is decreased by half.

Think About It
The opposite shifts occur in the position of the equilibrium when the volumes of the reaction mixtures are increased.

14.75. ### Collect and Organize
We are to predict the effect on the position of the equilibrium
$$2\,O_3(g) \rightleftharpoons 3\,O_2(g)$$
with various changes in concentration and volume.

Analyze
Increasing the concentration of a reactant shifts the equilibrium to the right, whereas increasing the concentration of a product shifts the equilibrium to the left. An increase in volume decreases the pressure and shifts the equilibrium towards the side of the reaction with more moles of gas.

Solve
(a) Increasing the concentration of the reactant, O_3, shifts the equilibrium to the right, increasing the concentration of the product, O_2.
(b) Increasing the concentration of the product, O_2, shifts the equilibrium to the left, increasing the concentration of the reactant, O_3.
(c) Decreasing the volume of the reaction to 1/10 its original volume shifts the equilibrium to the left, increasing the concentration of the reactant, O_3.

Think About It
Adding O_2 and decreasing the volume cause the same shift in the position of the equilibrium.

14.77. Collect and Organize
We are to determine how decreasing the partial pressure of O_2 affects the equilibrium

$$2\,SO_2(g) + O_2(g) \rightleftharpoons 2\,SO_3(g)$$

Analyze
According to Le Châtelier's principle, increasing the partial pressure or concentration of a reactant shifts the equilibrium to the right. Decreasing the partial pressure of a reactant, then, shifts the equilibrium to the left.

Solve
Decreasing the partial pressure of O_2 in that reaction shifts the equilibrium to the left towards the reactants.

Think About It
At equilibrium, then, we have less SO_3 product when we reduce the partial pressure of O_2 in the reaction mixture.

14.79. Collect and Organize
For three processes at equilibrium, we are to determine which shows an increased yield of product C at increasing temperatures.

Analyze
Endothermic processes have *heat* as a reactant. When the temperature is raised for those reactions, the equilibrium shifts to the right, increasing the amount of product C formed.

Solve
Reaction (a) is the only endothermic process ($\Delta H > 0$), so it is the only process for which the product yield increases with increasing temperature.

Think About It
Temperature changes do not affect the amount of product formed for reaction (b), for which $\Delta H = 0$.

14.81. Collect and Organize
We are to explain why equilibrium calculations are simpler when no product is present at the start of the reaction and when the value of the equilibrium constant, K, is very small.

Analyze
When no product is present, the reaction can only proceed to products; no reverse reaction can occur. When K is very small, the reaction does not proceed very far to the right. The concentrations of products formed once the reaction achieves equilibrium are very small.

Solve
When no products are present at the beginning of a reaction, we know that, no matter what the value of the equilibrium constant, the reaction will proceed to the right, so the equilibrium expression for the reaction is

$$X + Y \rightleftharpoons Z$$

$$K = \frac{x}{([X]-x)([Y]-x)}$$

We do not have to determine the reaction quotient for that situation because $Q = 0 < K$.

When K is very small, the amount of reactants transformed into products is small, and so at equilibrium the concentration of the reactants is approximately equal to the initial concentrations. Therefore, the approximation

$$K = \frac{x}{([X]-x)([Y]-x)} \approx \frac{x}{([X])([Y])}$$

for the reaction

$$X + Y \rightleftharpoons Z$$

is valid. That makes our calculations easier and avoids having to use the quadratic equation.

Think About It
The assumption is considered valid if $x < 5\%$ for the value of both A and B.

14.83. **Collect and Organize**
For the decomposition of PCl_5 to PCl_3 and Cl_2 ($K_p = 23.6$ at 500 K), we are to calculate the equilibrium partial pressures given the initial partial pressures of PCl_5 and PCl_3. We also are to determine how the concentrations of PCl_3 and PCl_5 change when more Cl_2 is added to the system already at equilibrium.

Analyze
(a) To calculate the partial pressures of all the species present, we set up a RICE table. Because the initial partial pressure of Cl_2 is 0.0 atm, we know that the reaction proceeds to the right to attain equilibrium.
(b) To determine how the equilibrium shifts when Cl_2 is added, we apply Le Châtelier's principle.

Solve
(a)

Reaction	$PCl_5(g)$	$\rightleftharpoons$	$PCl_3(g)$	+	$Cl_2(g)$
	P_{PCl_5}		P_{PCl_3}		P_{Cl_2}
Initial	0.560 atm		0.500 atm		0.00 atm
Change	$-x$		$+x$		$+x$
Equilibrium	$0.560 - x$		$0.500 + x$		x

After placing those values into the equilibrium constant expression, we can solve for x by using the quadratic formula:

$$23.6 = \frac{(0.500 + x)(x)}{(0.560 - x)}$$

$$x^2 + 24.1x - 13.216 = 0$$

$$x = 0.536 \text{ or } -24.6$$

Because $0.500 + x$ would be negative if $x = -24.6$, the actual root for this problem is $x = 0.536$. The equilibrium partial pressures of PCl_5, PCl_3, and Cl_2 are

$$P_{PCl_5} = 0.560 - 0.536 = 0.024 \text{ atm}$$

$$P_{PCl_3} = 0.500 + 0.536 = 1.036 \text{ atm}$$

$$P_{Cl_2} = 0.536 \text{ atm}$$

(b) When the partial pressure of Cl_2 is increased, the partial pressure (or concentration) of PCl_3 decreases and the partial pressure of PCl_5 increases.

Think About It
Because $K > 1$, the products of that reaction are favored.

14.85. **Collect and Organize**
For the initial concentrations of H_2O and Cl_2O as 0.00432 M in the equilibrium reaction

$$H_2O(g) + Cl_2O(g) \rightleftharpoons 2\,HOCl(g)$$

where $K_c = 0.0900$ at 25°C, we are to calculate the equilibrium concentrations of H_2O, Cl_2O, and HOCl.

Analyze
We first set up a RICE table to solve this problem. We assume here that the initial concentration of HOCl = 0.00 M. The equilibrium constant expression for that reaction is

$$K_c = \frac{[HOCl]^2}{[H_2O][Cl_2O]}$$

Solve

Reaction	$H_2O(g)$	$+$	$Cl_2O(g)$	$\rightleftharpoons$	$2\ HOCl(g)$
	[H_2O]		[Cl_2O]		HOCl
Initial	0.00432 M		0.00432 M		0
Change	$-x$		$-x$		$+2x$
Equilibrium	$0.00432 - x$		$0.00432 - x$		$2x$

After placing those values into the equilibrium constant expression, we can solve for x by taking the square root of both sides:

$$0.0900 = \frac{(2x)^2}{(0.00432 - x)^2}$$

$$0.300 = \frac{2x}{0.00432 - x}$$

$$1.296 \times 10^{-3} - 0.30x = 2x$$

$$x = 5.63 \times 10^{-4}$$

The concentration of all the gases at equilibrium are

$$[H_2O] = [Cl_2O] = 0.00432 - x = 3.76 \times 10^{-3}\ M$$
$$[HOCl] = 2x = 1.13 \times 10^{-3}\ M$$

Think About It

That reaction, with its equilibrium constant less than 1, favors reactants over product at equilibrium.

14.87. **Collect and Organize**

Given $K_p = 2 \times 10^6$ at 25°C for the reaction of 1 mol of NO with $^1/_2$ mol of O_2 to give 1 mol of NO_2, we are to calculate the equilibrium ratio of the partial pressure of NO_2 to that of NO in air, where the partial pressure of oxygen is 0.21 atm.

Analyze

The K_p expression for that reaction is

$$K_p = \frac{P_{NO_2}}{P_{NO} \times \left(P_{O_2}\right)^{1/2}}$$

Solve

When $P_{O_2} = 0.21$ atm,

$$K_p = \frac{P_{NO_2}}{P_{NO} \times (0.21)^{1/2}} = 2 \times 10^6$$

Rearranging that equation to solve for the ratio of the partial pressure of NO_2 to the partial pressure of NO gives

$$\frac{P_{NO_2}}{P_{NO}} = 2 \times 10^6 \times (0.21)^{1/2} = 9 \times 10^5$$

Think About It

The high value of K indicates that the product NO_2 is highly favored. That finding is consistent with the high value we calculated for the ratio of the partial pressures.

14.89. **Collect and Organize**

Given that $K_p = 1.5$ at 700°C, we are to calculate P_{CO_2} and P_{CO} at equilibrium for the reaction

$$CO_2(g) + C(s) \rightleftharpoons 2\ CO(g)$$

where the initial partial pressures of CO_2 and CO are 5.0 atm and 0.0 atm, respectively.

Analyze

Because carbon is a pure solid, it does not appear in the equilibrium constant expression:

$$K_p = \frac{\left(P_{CO}\right)^2}{P_{CO_2}}$$

Solve

Reaction	$CO_2(g) + C(s) \rightleftharpoons 2\,CO(g)$	
	P_{CO_2}	P_{CO}
Initial	5.0 atm	0.0 atm
Change	$-x$	$+2x$
Equilibrium	$5.0 - x$	$2x$

After placing those values into the equilibrium constant expression, we can solve for x by using the quadratic formula:

$$1.5 = \frac{\left(2x\right)^2}{(5.0-x)}$$

$$4x^2 + 1.5x - 7.5 = 0$$

$$x = 1.20 \text{ or } -1.57$$

The value of $x = -1.57$ would give a negative partial pressure for CO, so $x = 1.20$. The partial pressures of the gases at equilibrium are

$$P_{CO} = 2x = 2.4 \text{ atm}$$
$$P_{CO_2} = 5.0 - x = 3.8 \text{ atm}$$

Think About It

Checking our results should give a value close to 1.5, the equilibrium constant value:

$$\frac{\left(2.4\right)^2}{3.8} = 1.5$$

14.91. **Collect and Organize**

For the decomposition of NO_2 to NO and O_2 when $P_{O_2} = 0.136$ atm at equilibrium, we are asked to calculate the partial pressures of NO and NO_2 at equilibrium at 1000 K, where $K_p = 158$, and to calculate the total pressure in the flask at equilibrium.

Analyze

(a) To calculate the partial pressures of NO and NO_2, we set up a RICE table, where we start with pure NO_2 (*A*) and end with $P_{O_2} = 0.136$ atm $= x$ at equilibrium.

(b) The total pressure is

$$P_T = P_{NO_2} + P_{NO} + P_{O_2}$$

Solve

(a)

Reaction	$2\,NO_2(g)$	$\rightleftharpoons$	$2\,NO(g)$	$+$	$O_2(g)$
	P_{NO_2}		P_{NO}		P_{O_2}
Initial	*A* atm		0.00 atm		0.00 atm
Change	$-2x$		$+2x$		$+x$
Equilibrium	$A - 2x$		$2x$		x

We know that at equilibrium $x = 0.136$ atm, so

$$K_p = 158 = \frac{(P_{NO})^2 (P_{O_2})}{(P_{NO_2})^2} = \frac{(2x)^2 (x)}{(P_{NO_2})^2} = \frac{(0.272)^2 (0.136)}{(P_{NO_2})^2}$$

$$(P_{NO_2})^2 = \frac{(0.272)^2 \times (0.136)}{158} = 6.37 \times 10^{-5}$$

$$P_{NO_2} = (6.37 \times 10^{-5})^{1/2} = 7.98 \times 10^{-3} \text{ atm}$$

At equilibrium,

$$P_{NO} = 2x = 0.272 \text{ atm}$$
$$P_{O_2} = x = 0.136 \text{ atm}$$
$$P_{NO_2} = 7.98 \times 10^{-3} \text{ atm}$$

(b) $P_T = 7.98 \times 10^{-3} \text{ atm} + 0.272 \text{ atm} + 0.136 \text{ atm} = 0.416 \text{ atm}$

Think About It
The amount of NO_2 initially present can also be calculated:

$$A - 2x = 7.98 \times 10^{-3} \text{ atm}$$

Because $x = 0.136$ atm,

$$A = 7.98 \times 10^{-3} + 2 \times 0.136 = 0.280 \text{ atm}$$

14.93. **Collect and Organize**
For the equilibrium reaction

$$N_2(g) + O_2(g) \rightleftharpoons 2\,NO(g)$$

the value of K_p is 0.050 at 2200°C. We are to calculate the partial pressures of N_2, O_2, and NO at equilibrium given initial partial pressures for those gases of 0.79 atm for N_2, 0.21 atm for O_2, and 0.00 atm for NO.

Analyze
We set up a RICE table to solve this problem. The equilibrium constant expression for that reaction is

$$K_p = \frac{(P_{NO})^2}{(P_{O_2})(P_{N_2})}$$

Solve

Reaction	$N_2(g)$ +	$O_2(g)$ ⇌	2 NO(g)
	P_{N_2}	P_{O_2}	P_{NO}
Initial	0.79 atm	0.21 atm	0.00 atm
Change	$-x$	$-x$	$+2x$
Equilibrium	$0.79 - x$	$0.21 - x$	$2x$

After placing those values into the equilibrium constant expression, we can solve for x by using the quadratic formula:

$$0.050 = \frac{(2x)^2}{(0.79 - x)(0.21 - x)}$$

$$0.050 = \frac{4x^2}{(0.1659 - x + x^2)}$$

$$3.95x^2 + 0.050x - 0.008295 = 0$$

$$x = 0.03993 \quad \text{or} \quad -0.05260$$

The value of $x = -0.05260$ would give a negative partial pressure for NO, so $x = 0.03993$. The partial pressures of the gases at equilibrium are

$$P_{O_2} = 0.21 - x = 0.17 \text{ atm}$$

$$P_{N_2} = 0.79 - x = 0.75 \text{ atm}$$

$$P_{NO} = 2x = 0.080 \text{ atm}$$

Think About It

Using those equilibrium partial pressures, we can check our answers, which should give $K_p = 0.050$:

$$\frac{(0.080)^2}{(0.75)(0.17)} = 0.050$$

14.95. **Collect and Organize**

For the following reaction

$$2\,H_2S(g) \rightleftharpoons 2\,H_2(g) + S_2(g)$$

the value of K_c is 2.2×10^{-4} at 1400 K. We are to calculate the equilibrium concentration of H_2S, given that the initial $[H_2S]$ is 6.00 M.

Analyze

We are told to assume that the initial concentrations of H_2 and S_2 are 0.00 M. Because the equilibrium constant is small, we may be able to make a simplifying assumption. The equilibrium constant expression for that reaction is

$$K_c = \frac{[H_2]^2[S_2]}{[H_2S]^2}$$

Solve

Reaction	$2\,H_2S(g)$	$\rightleftharpoons$	$2\,H_2(g)$	$+$	$S_2(g)$
	$[H_2S]$		$[H_2]$		$[S_2]$
Initial	6.00 M		0.00 M		0.00 M
Change	$-2x$		$+2x$		$+x$
Equilibrium	$6.00 - 2x$		$2x$		x

Because the equilibrium constant is small (the initial concentration is more than 500 times the value of K), the reaction does not proceed very far to the right and we may be able to make a simplifying assumption. After placing those values into the equilibrium constant expression, we can solve for x:

$$2.2 \times 10^{-4} = \frac{(2x)^2(x)}{(6.00 - 2x)^2} \approx \frac{(2x)^2(x)}{(6.00)^2}$$

$$x = 0.126$$

Checking that assumption shows that it is valid:

$$\frac{2 \times 0.126}{6.00} \times 100 = 4\%$$

The equilibrium $[H_2S]$ is $6.00 - 2x = 5.75$ M.

Think About It

Without the simplifying assumption, we would have to solve a cubic equation.

14.97. **Collect and Organize**

For the following reaction, $K_c = 5.0$ at 600 K:

$$CO(g) + Cl_2(g) \rightleftharpoons COCl_2(g)$$

We are asked to calculate the partial pressures of the gases at equilibrium given that the initial partial pressures of CO and Cl_2 are 0.265 atm and that of $COCl_2$ is 0.000 atm.

Analyze

We have to first calculate K_p from K_c, where $\Delta n = -1$:

$$K_p = K_c (RT)^{\Delta n} = \dfrac{5.0}{\left(0.08206 \; \dfrac{\text{L} \cdot \text{atm}}{\text{mol} \cdot \text{K}}\right) \times 600 \text{ K}} = 0.102$$

We can set up a RICE table to solve for the equilibrium partial pressures of the gases.

Solve

Reaction	$CO(g)$	+	$Cl_2(g)$	$\rightleftharpoons$	$COCl_2(g)$
	P_{CO}		P_{Cl_2}		P_{COCl_2}
Initial	0.265 atm		0.265 atm		0.000 atm
Change	$-x$		$-x$		$+x$
Equilibrium	$0.265 - x$		$0.265 - x$		x

After placing those values into the equilibrium constant expression, we can solve for x by using the quadratic formula:

$$0.102 = \dfrac{x}{(0.265 - x)^2}$$

$$0.102 = \dfrac{x}{0.0702 - 0.530x + x^2}$$

$$0.102x^2 - 1.05406x + 0.00716 = 0$$

$$x = 10.33 \text{ or } 0.00680$$

The value of $x = 10.33$ would give negative partial pressures for CO and Cl_2, so $x = 0.00680$. The partial pressures of the gases at equilibrium are

$$P_{CO} = P_{Cl_2} = 0.265 - x = 0.258 \text{ atm}$$

$$P_{COCl_2} = x = 0.00680 \text{ atm}$$

Think About It

Checking our result, we get the value of K_p:

$$\dfrac{0.00680}{(0.258)^2} = 0.102$$

14.99. Collect and Organize

We are to calculate the concentrations of all the gases at equilibrium for the reaction

$$CO(g) + H_2O(g) \rightleftharpoons CO_2(g) + H_2(g)$$

given that the initial concentrations of all the gases are 0.050 M and that $K_c = 5.1$ at 700 K.

Analyze

The equilibrium constant expression for that reaction is

$$K_c = \dfrac{[CO_2][H_2]}{[CO][H_2O]}$$

We have to use Q, the reaction quotient, to determine the direction in which the reaction goes to reach equilibrium:

$$Q = \dfrac{(0.050)(0.050)}{(0.050)(0.050)} = 1.0$$

$Q < K$, so the reaction proceeds to the right.

Solve

Reaction	$CO(g)$	+	$H_2O(g)$	$\rightleftharpoons$	$CO_2(g)$	+	$H_2(g)$
	[CO]		[H₂O]		[CO₂]		[H₂]
Initial	0.050 M		0.050 M		0.050 M		0.050 M
Change	$-x$		$-x$		$+x$		$+x$
Equilibrium	$0.050 - x$		$0.050 - x$		$0.050 + x$		$0.050 + x$

$$5.1 = \frac{(0.050 + x)^2}{(0.050 - x)^2}$$

Taking the square root of both sides:

$$2.258 = \frac{(0.050 + x)}{(0.050 - x)}$$

$$0.1129 - 2.258x = 0.050 + x$$

$$0.0629 = 3.258x$$

$$0.0193 = x$$

The concentrations of the gases at equilibrium are

$$[CO] = [H_2O] = 0.050 - x = 0.031 \ M$$
$$[CO_2] = [H_2] = 0.050 + x = 0.069 \ M$$

Think About It

We did not need to use the quadratic formula here because with the concentrations of the reactants equal to each other and the concentrations of the products equal to each other, we could simplify the math by taking the square root of both sides of the equation to solve for x.

14.101. Collect and Organize

We are to calculate the concentrations of all the gases at equilibrium for the reaction

$$NO(g) + NO_2(g) \rightleftharpoons N_2O_3(g)$$

given that the initial amounts of NO and NO_2 are 15 g and 69 g, respectively, contained in a 4.0 L flask. For that reaction at 25°C, $K_p = 0.535$.

Analyze

The equilibrium constant expression for that reaction is

$$K_c = \frac{[N_2O_3]}{[NO][NO_2]}$$

We are given the mass of each reactant. We can convert those masses to molar concentrations by using the molar mass of the reactants and the volume of the flask to give the initial concentration of the reactants for the RICE table. Because we are given the value of K_p, we will have to calculate the value of K_c from K_p by using

$$K_c = \frac{K_p}{(RT)^{\Delta n}}$$

where $\Delta n = -1$, $R = 0.08206 \ L \cdot atm/mol \cdot K$, and $T = 298 \ K$.
Finally, the flask initially contains only reactants, so we know that the equilibrium will proceed from initial conditions to the right to form products.

Solve

The initial concentrations of the reactants are

$$15 \ g \ NO \times \frac{1 \ mol}{30.01 \ g} \times \frac{1}{4.0 \ L} = 0.125 \ M \ NO$$

$$69 \ g \ NO_2 \times \frac{1 \ mol}{46.01 \ g} \times \frac{1}{4.0 \ L} = 0.375 \ M \ NO_2$$

The value of K_c is

$$K_c = \frac{K_p}{(RT)^{\Delta n}} = \frac{0.535}{\left(0.08206\dfrac{\text{L}\cdot\text{atm}}{\text{mol}\cdot\text{K}}\times298\text{K}\right)^{-1}} = 13.08$$

Reaction	NO(g)	+	NO$_2$(g)	⇌	N$_2$O$_3$(g)
	[NO]		[NO$_2$]		[N$_2$O$_3$]
Initial	0.125 *M*		0.375 *M*		0.0 *M*
Change	$-x$		$-x$		$+x$
Equilibrium	$0.125 - x$		$0.375 - x$		x

$$13.08 = \frac{x}{(0.125-x)(0.375-x)}$$

$$13.08 = \frac{x}{(0.046875 - 0.500x + x^2)}$$

$$0.613125 - 6.54x + 13.08x^2 = x$$

$$13.08x^2 - 7.54x + 0.613125 = 0$$

$$x = 0.09796$$

The concentrations of the gases at equilibrium are

$$[\text{NO}] = 0.125 - x = 0.027 \ M$$
$$[\text{NO}_2] = 0.375 - x = 0.277 \ M$$
$$[\text{N}_2\text{O}_3] = = x = 0.098 \ M$$

Think About It

This problem is multilayered in that we must realize at the start that K_c must be calculated from K_p and that we have to compute the concentrations in moles per liter of each reactant before using the RICE table.

14.103. Collect and Organize

We are to relate the sign of $\Delta G°$ to the magnitude of K.

Analyze

The equation relating $\Delta G°$ to K is

$$\Delta G° = -RT \ln K$$

Solve

Yes, the value of $\Delta G°$ is positive when $K < 1$. Because $\ln K$ is negative for $K < 1$ and because $\Delta G° = -RT \ln K$, the value of $\Delta G°$ is positive.

Think About It

Therefore, reactions with $K < 1$ are not spontaneous.

14.105. Collect and Organize

For a reaction that starts with 100% reactants, we are to determine in which direction the reaction shifts when it is spontaneous ($\Delta G° < 0$).

Analyze

The equation relating $\Delta G°$ to K is

$$\Delta G° = -RT \ln K$$

Solve

When $\Delta G° < 0$, the $\ln K$ of the reaction is positive, giving $K > 1$. The reaction favors the formation of products, so the reaction will shift to the right.

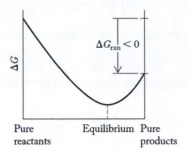

Think About It

The reaction will proceed to the right until the mixture of reactants and products as expressed in the reaction quotient Q matches the value of the equilibrium constant K.

14.107. **Collect and Organize**

Given three reactions with associated $\Delta G°$ values, we are asked to determine which reaction has the largest value of K at 25°C.

Analyze

The equilibrium constant of a reaction is related to the Gibbs free energy through the equation

$$\Delta G° = -RT \ln K$$

Rearranging that equation gives

$$\ln K = \frac{-\Delta G°}{RT}$$

From doing so we see that the more negative the value of ΔG, the larger K will be.

Solve

Two of the reactions have negative values of $\Delta G°$ and are spontaneous (b and c); they will have $K > 1$. The reaction with the more negative free energy (reaction c) has the largest equilibrium constant.

Think About It

The value of K for reaction (c) is

$$\ln K = \frac{-\Delta G°}{RT} = \frac{27.9 \text{ kJ}}{\left(8.314 \times 10^{-3} \text{ kJ/mol} \cdot \text{K}\right) \times 298 \text{ K}} = 11.26$$

$$K = e^{11.26} = 7.8 \times 10^4$$

14.109. **Collect and Organize**

For the balanced equilibrium reaction

$$2 \text{ SO}_2(g) + \text{O}_2(g) \rightleftharpoons 2 \text{ SO}_3(g)$$

for which $K_p = 3.4$ at 1000 K, we are asked to calculate the enthalpy change of the reaction under standard conditions, to calculate the value of K_p at 298 K, and finally to calculate the free energy change for the reaction under standard conditions at 298 K to compare it with the value obtained using Appendix 4.

Analyze

(a) The enthalpy of the reaction under standard conditions may be calculated using the values for the enthalpy of formation for the reactants and products in Appendix 4.

$$\Delta H°_{rxn} = \sum n\Delta H°_{f,products} - \sum n\Delta H°_{f,reactants}$$

(b) We can calculate the value of the equilibrium constant at 298 K from the value of K at 1000 K and the enthalpy of the reaction through the equation

$$\ln\left(\frac{K_2}{K_1}\right) = -\frac{\Delta H^\circ}{R}\left(\frac{1}{T_2} - \frac{1}{T_1}\right)$$

(c) The value of the equilibrium constant at 298 from part b can be used to calculate the standard free energy of the reaction through

$$\Delta G^\circ = -RT\ln K$$

and the value of the free energy from values in Appendix 4 can be calculated through the following equation for comparison

$$\Delta G^\circ_{rxn} = \sum n\Delta G^\circ_{f,products} - \sum n\Delta G^\circ_{f,reactants}$$

Solve

(a) The enthalpy of the reaction under standard conditions is

$$\Delta H^\circ_{rxn} = (2 \text{ mol SO}_3)(-395.7 \text{ kJ/mol}) - \left[(2 \text{ mol SO}_2)(-296.8 \text{ kJ/mol}) + (1 \text{ mol O}_2)(0.0 \text{ kJ/mol})\right]$$

$$\Delta H^\circ_{rxn} = -197.8 \text{ kJ}$$

(b) The value of the equilibrium constant at 298 from the calculated value of the enthalpy in part a and from the value of K at 1000 K is

$$\ln\left(\frac{K_{298}}{3.4}\right) = -\frac{(-197.8\times10^3 \text{ J/mol})}{8.314 \text{ J/mol}\cdot\text{K}}\left(\frac{1}{298 \text{ K}} - \frac{1}{1000 \text{ K}}\right) = 56.045$$

$$\left(\frac{K_{298}}{3.4}\right) = e^{56.045} = 2.188\times10^{24}$$

$$K_{298} = 7.4\times10^{24}$$

(c) From the value of K at 298 K, we calculate the free energy at standard conditions:

$$\Delta G^\circ = -(8.314 \text{ J/mol}\cdot\text{K})\times(298 \text{ K})\times\ln(7.4\times10^{24}) = -141.9 \text{ kJ/mol}$$

From the standard free energy values in Appendix 4, the free energy at 298 K is

$$\Delta G^\circ_{rxn} = (2 \text{ mol SO}_3)(-371.1 \text{ kJ/mol}) - \left[(2 \text{ mol SO}_2)(-300.1 \text{ kJ/mol}) + (1 \text{ mol O}_2)(0.0 \text{ kJ/mol})\right]$$

$$\Delta G^\circ_{rxn} = -142.0 \text{ kJ}$$

Those values are very close to each other.

Think About It

That reaction is exothermic, so it makes sense that the equilibrium constant is larger at 298 K than at 1000 K.

14.111. Collect and Organize

Here we can add two equations to calculate ΔG° and K_p at 298 K for the reaction

$$N_2(g) + 2 O_2(g) \rightleftharpoons 2 NO_2(g)$$

Analyze

To obtain the overall reaction, we simply add the two reactions given and add their ΔG° values. We can calculate the value of K_p by using

$$\Delta G^\circ = -RT\ln K$$

where $K = K_p$ because ΔG°, as stated in the text, "represents a change in free energy under standard conditions, and the standard state for a gaseous reactant or product is one in which its partial pressure is 1 bar (<1 atm). Thus, the ΔG°, of a reaction in the gas phase is linked by equation above to its K_p value."

$$K_p = K_c(RT)^{\Delta n}$$

where Δn for that reaction is -1.

Solve

Adding the equations gives the overall equation and $\Delta G°$:

$$N_2(g) + O_2(g) \rightleftharpoons 2\,NO(g) \qquad \Delta G° = 173.2 \text{ kJ}$$
$$2\,NO(g) + O_2(g) \rightleftharpoons 2\,NO_2(g) \qquad \Delta G° = -69.7 \text{ kJ}$$
$$\overline{N_2(g) + 2\,O_2(g) \rightleftharpoons 2\,NO_2(g) \qquad \Delta G° = 103.5 \text{ kJ}}$$

Rearranging to calculate K_p,

$$\ln K_p = \frac{-\Delta G°}{RT} = \frac{-103.5 \text{ kJ/mol}}{(8.314 \times 10^{-3} \text{ kJ/mol} \cdot \text{K}) \times 298 \text{ K}} = -41.77$$

$$K_p = e^{-41.77} = 7.24 \times 10^{-19}$$

Think About It

Because the free-energy change is large and positive, K is much less than 1 and the reactants are greatly favored over the product at equilibrium.

14.113. Collect and Organize

Using the relationship

$$\ln\left(\frac{K_2}{K_1}\right) = -\frac{\Delta H°}{R}\left(\frac{1}{T_2} - \frac{1}{T_1}\right)$$

we can determine whether a reaction is endothermic or exothermic if the value of K decreases with increasing T.

Analyze

In the equation, $K_2 < K_1$, so $\ln(K_2/K_1) < 0$ and $T_2 > T_1$, so $(1/T_2 - 1/T_1) < 0$.

Solve

For $\ln(K_2/K_1)$ to be negative when $(1/T_2 - 1/T_1)$ is also negative, the value of ΔH must be negative and the reaction must be exothermic.

Think About It

An endothermic reaction, by contrast, shows an increased K with increasing T because $\Delta H > 0$ and thus $\ln(K_2/K_1) > 0$, so $K_2/K_1 > 0$.

14.115. Collect and Organize

Given that K_p for the reaction of CO with H_2O increases as the temperature decreases, we can use the relationship

$$\ln\left(\frac{K_2}{K_1}\right) = -\frac{\Delta H°}{R}\left(\frac{1}{T_2} - \frac{1}{T_1}\right)$$

to determine whether the reaction is endothermic or exothermic.

Analyze

In the equation $K_2 > K_1$, so $\ln(K_2/K_1) > 0$. Because $T_2 < T_1$, $(1/T_2 - 1/T_1) > 0$.

Solve

For $\ln(K_2/K_1)$ to be positive when $(1/T_2 - 1/T_1)$ is also positive, ΔH must be negative and the reaction must be exothermic.

Think About It

An endothermic reaction, by contrast, shows a decreased K with a decrease in temperature.

14.117. Collect and Organize

Given $\Delta H° = 180.6$ kJ and $K_c = 4.10 \times 10^{-4}$ for the conversion of N_2 and O_2 to NO at 2000°C, we are to calculate the value of K_c at 25°C.

Analyze

The relationship between equilibrium constants at two temperatures is given by

$$\ln\left(\frac{K_2}{K_1}\right) = -\frac{\Delta H°}{R}\left(\frac{1}{T_2} - \frac{1}{T_1}\right)$$

where $\Delta H°$ is the enthalpy of the reaction (in joules per mole), R is the gas constant (in joules per mole per kelvin), and T_1 and T_2 are the temperatures (in kelvin).

Solve

$$\ln\left(\frac{4.1\times10^{-4}}{K_2}\right) = -\frac{1.806\times10^5 \text{ J/mol}}{8.314 \text{ J/mol}\cdot\text{K}}\left(\frac{1}{2273 \text{ K}} - \frac{1}{298 \text{ K}}\right)$$

$$\ln\left(\frac{4.1\times10^{-4}}{K_2}\right) = 63.337$$

$$\frac{4.1\times10^{-4}}{K_2} = e^{63.337} = 3.214\times10^{27}$$

$$K_2 = 1.3\times10^{-31}$$

Think About It

As we would expect, the equilibrium constant for that endothermic reaction decreases as the reaction is cooled.

14.119. Collect and Organize

Given that $K_1 = 1.5 \times 10^5$ at 430°C and $K_2 = 23$ at 1000°C, we are to calculate the standard enthalpy of the reaction between NO and O_2 to produce NO_2.

Analyze

We can use the following equation to solve that problem:

$$\ln\left(\frac{K_2}{K_1}\right) = -\frac{\Delta H°}{R}\left(\frac{1}{T_2} - \frac{1}{T_1}\right)$$

Solve

$$\ln\left(\frac{23}{1.5\times10^5}\right) = -\frac{\Delta H°}{8.314 \text{ J/mol}\cdot\text{K}}\left(\frac{1}{1273 \text{ K}} - \frac{1}{703 \text{ K}}\right)$$

$$\Delta H° = -1.15\times10^5 \text{ J/mol, or } -115 \text{ kJ/mol}$$

Think About It

The result that this reaction, where K decreases with increasing T, is exothermic and consistent with our answer to Problem 14.114.

14.121. Collect and Organize

Given the percentage of decomposition of CO_2 at three temperatures and that each equilibrium mixture begins with 1 atm of CO_2, we are to determine whether the reaction is endothermic and then calculate K_p at each temperature. We are also to comment on the decomposition reaction as a remedy for global warming.

Analyze

The amount of CO_2 decomposed increases with increasing temperature. Therefore, $K_2 > K_1$ in the equation

$$\ln\left(\frac{K_2}{K_1}\right) = -\frac{\Delta H°}{R}\left(\frac{1}{T_2} - \frac{1}{T_1}\right)$$

where $T_2 > T_1$, so $(1/T_2 - 1/T_1) < 0$. To calculate K_p for each temperature, we set up a RICE table:

Reaction	2 CO$_2$(g)	$\rightleftharpoons$	2 CO(g)	+	O$_2$(g)
	P_{CO_2}		P_{CO}		P_{O_2}
Initial	1 atm		0 atm		0 atm
Change	$-2x$		$+2x$		$+x$
Equilibrium	$1 - 2x$		$2x$		x

We know that $2x$ is equal to the percent decomposition divided by 100 and that the form of the equilibrium expression is

$$K_p = \frac{\left(P_{CO}\right)^2\left(P_{O_2}\right)}{\left(P_{CO_2}\right)^2}$$

Because the percent decomposition increases with temperature, the value of K_p is expected to increase as T increases.

Solve

Because the value of K increases with increasing temperature, $\ln(K_2/K_1) > 0$. With $(1/T_2 - 1/T_1) < 0$, the value of ΔH must be positive, so that reaction is endothermic.

At 1500 K, $2x = 0.00048$, $x = 0.00024$, and $1 - 2x = 0.99952$:

$$K_p = \frac{(0.00048)^2(0.00024)}{(0.99952)^2} = 5.5 \times 10^{-11}$$

At 2500 K, $2x = 0.176$, $x = 0.088$, and $1 - 2x = 0.824$:

$$K_p = \frac{(0.176)^2(0.088)}{(0.824)^2} = 4.0 \times 10^{-3}$$

At 3000 K, $2x = 0.548$, $x = 0.274$, and $1 - 2x = 0.452$:

$$K_p = \frac{(0.548)^2(0.274)}{(0.452)^2} = 0.40$$

As predicted, the value of the equilibrium constant increases with increasing temperature. That reaction, however, does not favor products even at very high temperature and so the process is not a viable source of CO and is not a remedy to decrease CO_2 as a contributor to global warming. Also, the process produces poisonous CO gas.

Think About It

By using values in Appendix 4, we can confirm that the reaction is endothermic.

14.123. **Collect and Organize**

By combining the two equations given, we can calculate the overall equilibrium constant, K, for the reaction and then calculate the concentration of X^{2-} in an equilibrium mixture where $[H_2X]_{eq} = 0.1$ M and $[HCl]_{eq} = [H_3O^+]_{eq} = 0.3$ M.

Analyze

When we add equations, the overall equilibrium constant is the product of the individual equilibrium constants. From the overall reaction we can write the equilibrium constant expression and from that solve for $[X^{2-}]$ at equilibrium.

Solve

$$H_2X(aq) + H_2O(\ell) \rightleftharpoons HX^-(aq) + H_3O^+(aq) \qquad K_1 = 8.3 \times 10^{-8}$$

$$\underline{HX^-(aq) + H_2O(\ell) \rightleftharpoons X^{2-}(aq) + H_3O^+(aq) \qquad K_2 = 1 \times 10^{-14}}$$

$$H_2X(aq) + 2H_2O(\ell) \rightleftharpoons X^{2-}(aq) + 2H_3O^+(aq) \qquad K_3 = K_1 \times K_2 = 8.3 \times 10^{-22}$$

$$K_c = \frac{\left[X^{2-}\right]\left[H_3O^+\right]^2}{\left[H_2X\right]}$$

We know that $K_c = 8.3 \times 10^{-22}$, $[H_3O^+] = 0.3$ M, and $[H_2X] = 0.1$ M for the saturated solution. Rearranging the equilibrium constant expression and solving for $[X^{2-}]$ gives

$$\left[X^{2-}\right] = \frac{K_c \times \left[H_2X\right]}{\left[H_3O^+\right]^2} = \frac{8.3 \times 10^{-22} \times 0.1}{(0.3)^2} = 9 \times 10^{-22} \, M$$

Think About It

Our answer makes sense because with a very small overall equilibrium constant, we expect a very small concentration of product, X^{2-}.

14.125. **Collect and Organize**

For the reaction of CH_4 with S_8 to form CS_2 and H_2, we are to calculate the value of the equilibrium constant by using thermodynamic data from Appendix 4 for the reaction at both 25°C and 500°C.

Analyze

We have to calculate the value of K_p from ΔG at 500°C. To do so, we must first determine $\Delta H°$ and $\Delta S°$ from tabulated values in Appendix 4:

$$\Delta H° = [(4 \times 115.3) + (8 \times 0.0)] - [(4 \times -74.8) - (1 \times 0.0)] = 760.4 \text{ kJ}$$
$$\Delta S° = [(4 \times 237.8) + (8 \times 130.6)] - [(4 \times 186.2) - (1 \times 32.1)] = 1219 \text{ J/K}$$

We can then calculate the value of $\Delta G_{25°C}$ and $\Delta G_{500°C}$ through

$$\Delta G_{25°C} = \Delta H° - T\Delta S° = 760.4 \text{ kJ} - (298 \text{ K} \times 1.219 \text{ kJ/K}) = 397.1 \text{ kJ}$$
$$\Delta G_{500°C} = \Delta H° - T\Delta S° = 760.4 \text{ kJ} - (773 \text{ K} \times 1.219 \text{ kJ/K}) = -181.9 \text{ kJ}$$

The value of K_p for each temperature is calculated though the value of $\Delta G°$:

$$\ln K_p = -\frac{\Delta G°}{RT}$$

Solve

At 25°C (298 K):

$$\ln K_p = -\frac{397.1 \text{ kJ/mol}}{\left(\dfrac{8.314 \times 10^{-3} \text{ kJ}}{\text{mol} \cdot \text{K}}\right) \times 298 \text{ K}} = -160.3$$

$$K_p = e^{-160.3} = 2.47 \times 10^{-70}$$

At 500°C (773 K):

$$\ln K_p = -\frac{-181.9 \text{ kJ/mol}}{\left(\dfrac{8.314 \times 10^{-3} \text{ kJ}}{\text{mol} \cdot \text{K}}\right) \times 773 \text{ K}} = 28.30$$

$$K_p = e^{28.30} = 1.96 \times 10^{12}$$

Think About It

The reaction changed its spontaneity upon heating because it is an endothermic reaction with a favorable entropy change.

14.127. **Collect and Organize**

Given $\Delta G° = -418.6$ kJ for the reaction in which CaO reacts with SO_2 in the presence of O_2 to form solid $CaSO_4$, we can calculate K and then determine the P_{SO_2} in that reaction when the partial pressure of oxygen is 0.21 atm.

Analyze

To calculate K from $\Delta G°$, we use

$$\ln K = -\frac{\Delta G°}{RT}$$

The equilibrium constant expression for the reaction is

$$K_p = \frac{1}{\left(P_{SO_2}\right)\left(P_{O_2}\right)^{1/2}}$$

Solve

$$\ln K = -\frac{-418.6 \text{ kJ/mol}}{8.314\times10^{-3} \text{ kJ/mol·K}\times298\text{ K}} = 168.956$$

$$K = e^{168.956} = 2.38\times10^{73}$$

$$2.38\times10^{73} = \frac{1}{P_{SO_2}\times(0.21)^{1/2}}$$

$$P_{SO_2} = 9.2\times10^{-74} \text{ atm}$$

Think About It

Essentially, all the SO_2 is "scrubbed" by the CaO in that reaction, making it an efficient method to remove SO_2 from exhaust gases.

CHAPTER 15 | Acid–Base Equilibria: Proton Transfer in Biological Systems

15.1. Collect and Organize

Figure P15.1 shows four lines to describe the possible dependence of percent ionization of acetic acid with concentration. We are to choose the one that best represents the trend for this weak acid.

Analyze

The ionization of acetic acid is described by the following chemical reaction:

$$CH_3COOH(aq) \rightleftharpoons CH_3COO^-(aq) + H^+$$

The degree of ionization is the ratio of the quantity of a substance that is ionized to the concentration of the substance before ionization.

Solve

According to Figure 15.9, the change in degree of ionization of a weak acid with concentration is not linear and is best described by the red line in Figure P15.1. The degree of ionization increases with decreasing acetic acid concentration.

Think About It

The percent ionization could be calculated for each concentration if we knew the equilibrium concentration of the acetate ion in solution and the initial concentration of acetic acid dissolved.

$$\% \text{ ionization} = \frac{\left[H^+\right]_{\text{equilibrium}}}{\left[\text{acetic acid}\right]_{\text{initial}}} \times 100$$

15.3. Collect and Organize

The bar graph in Figure P15.3 shows the percent ionization for three aqueous solutions of HOX, where X = Cl, Br, or I. From the graph, we can compare the K_a values for those three weak acids to determine which bar represents the percent ionization of HIO.

Analyze

The K_a values for the acids (from Table A5.1 in the Appendix) are as follows: $K_a = 2.9 \times 10^{-8}$ for HClO, $K_a = 2.3 \times 10^{-9}$ for HBrO, $K_a = 2.3 \times 10^{-11}$ for HOI. From that we see that HIO is the weakest acid because it has the smallest K_a value.

Solve

The weakest acid has the lowest percent ionization, so the third bar (shortest one) corresponds to HOI. Notice, too, that as the electronegativity of X in HXO decreases, so too does the acidity and percent ionization.

Think About It

Even though all three acids are weak, the percent ionization differs greatly among them. The strongest of those acids, HClO, has a percent ionization of just over 0.5%, whereas the weakest acid, HIO, has a percent ionization of well under 0.1%, a more than fivefold difference.

15.5. Collect and Organize

From the structures of piperidine and morpholine shown in Figure P15.5, we are to predict and explain which is the stronger base.

Analyze

Morpholine's structure differs from piperidine's only by having an oxygen atom replacing one of the carbon atoms in the ring.

Solve

The basicity of both those molecules depends on the availability of a lone pair of electrons on the N atom in the structure. Because oxygen is a very electronegative element, its presence in the ring of morpholine will draw electrons in the lone pair toward itself and make those electrons less available to make the N less basic. Therefore, piperidine is a stronger base than morpholine.

Think About It

The pK_b of piperidine is 2.78 and that of morpholine is 5.64. Therefore, an aqueous solution of piperidine is more basic than an aqueous solution of morpholine.

15.7. Collect and Organize

From the structure of pseudoephedrine shown in Figure P15.7, we are to predict whether it is acidic, basic, or neutral and to identify the functional group responsible for that property.

Analyze

Pseudoephedrine's structure contains both a hydroxyl (–OH) group as an alcohol functional group and an amine group. The alcohol group is not acidic, but the amine group has a lone pair of electrons on the nitrogen atom, which can act as a base.

Solve

(a) Pseudoephedrine is basic.
(b) The amine functional group in pseudoephedrine makes this substance basic.

Think About It

In over-the-counter medications, pseudoephedrine is often in the form of a hydrochloride or sulfate salt to increase solubility in aqueous media.

15.9. Collect and Organize

Given the structure of phosphorous acid in Figure P15.9, we are to identify the ionizable hydrogen atoms.

Analyze

Ionizable hydrogen atoms are bonded to electronegative atoms such as oxygen in chemical structures.

Solve

Phosphorous acid has two ionizable hydrogen atoms. The ionizable protons are the ones bonded to the oxygen atoms.

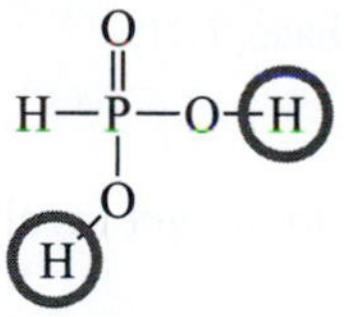

Think About It

Phosphoric acid, by contrast, has three ionizable protons.

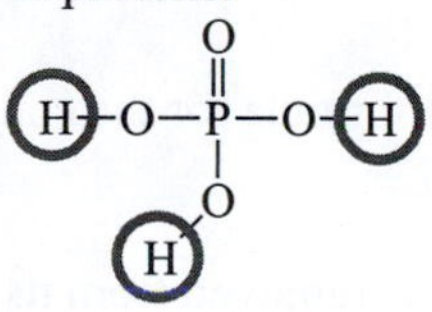

15.11. Collect and Organize

For HF(*aq*), we are to identify the Brønsted–Lowry acid and base.

Analyze

A Brønsted–Lowry acid is a proton donor. A Brønsted–Lowry base is a proton acceptor.

Solve

HF is a weak acid in water. In the equilibrium that is established, HF acts as a Brønsted–Lowry acid, donating its proton to H_2O, the Brønsted–Lowry base:

$$HF(aq) + H_2O(\ell) \rightleftharpoons H_3O^+(aq) + F^-(aq)$$

Think About It

Because hydrofluoric acid is a weak acid in water, it only partially dissociates in water to give H_3O^+ and F^-.

15.13. Collect and Organize

For $NH_3(aq)$, we are to identify the species that act as the Brønsted–Lowry acid and base.

Analyze

NH_3 is soluble in water. We need, then, to consider the behavior of NH_3 in water. A Brønsted–Lowry acid is a proton donor. A Brønsted–Lowry base is a proton acceptor.

Solve

NH_3 is a weak base in water. It acts as a Brønsted–Lowry base, removing a proton from H_2O, the Brønsted–Lowry acid:

$$NH_3(aq) + H_2O(\ell) \rightleftharpoons OH^-(aq) + NH_4^+(aq)$$

Think About It

In Problems 15.11 and 15.12, water acted as a Brønsted–Lowry base. In this problem, it acts as an acid. That dual acid–base behavior makes water *amphoteric*.

15.15. Collect and Organize

For three acid–base reactions, we are to identify which reactant is the acid and which reactant is the base.

Analyze

For all these reactions that involve the transfer of a proton between the acid and base, we can apply the Brønsted–Lowry definitions of acid and base. A Brønsted–Lowry acid is a proton donor. A Brønsted–Lowry base is a proton acceptor.

Solve

(a) HCl is the acid. It transfers H^+ to the base, NaOH.
(b) HCl is the acid. It transfers H^+ to the base, $MgCO_3$.
(c) H_2SO_4 is the acid. It transfers H^+ to the base, NH_3.

Think About It

In reactions (a) and (b), Na^+, Mg^{2+}, and Cl^- do not get involved in the reaction. They are spectator ions. The net ionic equations are

$$H^+(aq) + OH^-(aq) \rightarrow H_2O(\ell)$$
$$CO_3^{2-}(aq) + 2\,H^+(aq) \rightarrow CO_2(g) + H_2O(\ell)$$

15.17. Collect and Organize

For each species listed, we are to write the formula for the conjugate base.

Analyze

The conjugate base form of a species has H^+ removed from its formula.

Solve

The conjugate base of HNO_2 is NO_2^-.
The conjugate base of $HClO$ is ClO^-.
The conjugate base of H_3PO_4 is $H_2PO_4^-$.
The conjugate base of NH_3 is NH_2^-.

Think About It
Be sure to account for the change in charge when H^+ is removed to form the conjugate base.

15.19. Collect and Organize
For the species HSO_4^-, we are to write the formula for the conjugate acid and the conjugate base.

Analyze
The conjugate acid form of a species has an H^+ added to its formula. The conjugate base form of a species has an H^+ removed from its formula.

Solve
The conjugate acid of HSO_4^- is H_2SO_4; the conjugate base of HSO_4^- is SO_4^{2-}.

Think About It
Be sure to account for the change in charge when the base adds H^+ to form the conjugate acid and conjugate base.

15.21. Collect and Organize
Given that the concentration of a nitric acid solution is 0.65 M, we are to calculate the concentration of H^+ ions in the solution.

Analyze
Nitric acid is a strong acid that completely dissociates in water, so the concentration of H^+ ions is stoichiometrically related to the concentration of HNO_3:
$$HNO_3(aq) \rightarrow H^+(aq) + NO_3^-(aq)$$

Solve
$$[H^+] = 0.65 \ M$$

Think About It
The concentration of OH^- in strong base solutions, likewise, is the same as the concentration of the strong base dissolved into the solution.

15.23. Collect and Organize
Given that a solution is 0.0205 M in the strong base $Ba(OH)_2$, we are asked to calculate the concentration of OH^-.

Analyze
$Ba(OH)_2$, being a strong base, completely dissociates according to the equation
$$Ba(OH)_2(aq) \rightarrow Ba^{2+}(aq) + 2 \ OH^-(aq)$$
Therefore, $[OH^-] = 2 \times [Ba(OH)_2]$.

Solve
$$2 \times 0.0205 \ M = 0.0410 \ M = [OH^-]$$

Think About It
Be sure to account for both OH^- ions in this strong base.

15.25. Collect and Organize
We are asked to explain why H_2SO_4 is a stronger acid (greater K_{a_1}) than H_2SeO_4.

Analyze
The only difference in these acids is the central atom. Sulfur and selenium belong to the same group in the periodic table. These elements differ in size and electronegativity.

Solve
Sulfur is more electronegative than selenium. The higher electronegativity on the sulfur atom stabilizes the anion HSO_4^- more than the anion $HSeO_4^-$. Therefore, H_2SO_4 is a stronger acid.

Think About It
We would expect that trend to continue, so we predict that H_2TeO_4 is a weaker acid than H_2SeO_4.

15.27. Collect and Organize
We are to predict which acid of a pair is stronger.

Analyze
The more oxygen atoms bound to the central atom and the more electronegative the central atom (X) in the acid, the more acidic is the compound.

Solve
(a) H_2SO_3 is a stronger acid than H_2SeO_3.
(b) H_2SeO_4 is a stronger acid than H_2SeO_3.

Think About It
The presence of oxygen atoms bound to the central atom in an oxyacid can have a dramatic effect on acidity.

15.29. Collect and Organize
We are to explain why the pH value decreases for solutions as acidity increases.

Analyze
The pH of a solution is calculated through
$$pH = -\log[H^+]$$

Solve
Because the pH function is a $-\log$ function, as $[H^+]$ increases, the value of $-\log[H^+]$ decreases.

Think About It
The pH scale is typically 0–14 for concentrations of H^+ from $1\ M$ to $1 \times 10^{-14}\ M$, but values of pH may be negative or greater than 14.

15.31. Collect and Organize
We are asked to describe a solution in which the pH may be negative.

Analyze
The pH scale is often seen as 0–14. That occurs when $[H^+]$ is between $1\ M$ and $1 \times 10^{-14}\ M$.

Solve
When $[H^+]$ is greater than $1\ M$, the pH of the solution is negative.

Think About It
For example, a $3.00\ M$ solution of HCl has a pH of
$$pH = -\log[H^+] = -\log(3.00\ M) = -0.48$$

15.33. **Collect and Organize**
For the autoionization of ethanol, we are to draw the Lewis structures of the ions that would form.

Analyze
For ethanol, the autoionization is described by
$$2\ CH_3CH_2OH(\ell) \rightleftharpoons CH_3CH_2OH_2^+(ethanol) + CH_3CH_2O^-(ethanol)$$

Solve

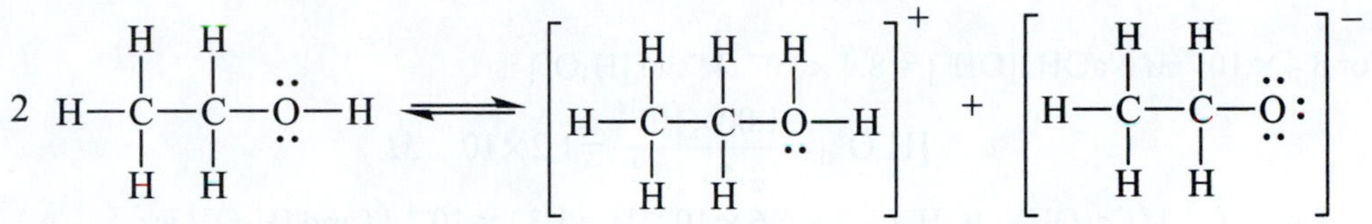

Think About It
Another species that might autoionize is acetic acid:
$$2\ CH_3COOH(\ell) \rightleftharpoons CH_3COOH_2^+(acetic\ acid) + CH_3COO^-(acetic\ acid)$$

15.35. **Collect and Organize**
Given either the $[OH^-]$ or $[H^+]$ for a solution, we are asked to calculate the pH and pOH and determine whether the solution is acidic, basic, or neutral.

Analyze
To calculate the pH or the pOH from the $[H^+]$ or $[OH^-]$, respectively, we use
$$pH = -\log[H^+]$$
$$pOH = -\log[OH^-]$$
To find the pOH from the pH, and vice versa, we use the relationship
$$pH + pOH = 14$$

If the pH of a solution is less than 7, the solution is acidic. If the pH is equal to 7, the solution is neutral. If the pH is greater than 7, the solution is basic.

Solve
(a) $pH = -\log(5.3 \times 10^{-3}) = 2.28$
  $pOH = 14 - pH = 11.72$
  This solution is acidic.
(b) $pH = -\log(3.8 \times 10^{-9}) = 8.42$
  $pOH = 14 - pH = 5.58$
  This solution is basic.
(c) $pH = -\log(7.2 \times 10^{-6}) = 5.14$
  $pOH = 14 - pH = 8.86$
  This solution is acidic.
(d) $pOH = -\log(1.0 \times 10^{-14}) = 14.00$
  $pH = 14 - pOH = 0.00$
  This solution is acidic.

Think About It
To determine how many significant figures to include in your answers when computing the pH or pOH, remember that the first number in the pH or pOH gives the location of the decimal point. The significant digits, therefore, follow the decimal point.

15.37. **Collect and Organize**

Given the concentration of a strong acid or base in solution, we are to calculate either the $[OH^-]$ or $[H_3O^+]$ for that solution.

Analyze

The $[H_3O^+]$ and $[OH^-]$ in a solution are related to each other through the autoionization constant for water, K_w:

$$[H_3O^+] \times [OH^-] = 1.00 \times 10^{-14}$$

Solve

(a) For $8.4 \times 10^{-4}\ M$ NaOH, $[OH^-] = 8.4 \times 10^{-4}\ M$ and $[H_3O^+]$ is

$$[H_3O^+] = \frac{1.00 \times 10^{-14}}{8.4 \times 10^{-4}} = 1.2 \times 10^{-11}\ M$$

(b) For $6.6 \times 10^{-5}\ M$ Ca(OH)$_2$, $[OH^-] = 2 \times 6.6 \times 10^{-5}\ M = 1.32 \times 10^{-4}\ M$ and $[H_3O^+]$ is

$$[H_3O^+] = \frac{1.00 \times 10^{-14}}{1.32 \times 10^{-4}} = 7.6 \times 10^{-11}\ M$$

(c) For $4.5 \times 10^{-3}\ M$ HCl, $[H_3O^+] = 4.5 \times 10^{-3}\ M$ and $[OH^-]$ is

$$[OH^-] = \frac{1.00 \times 10^{-14}}{4.5 \times 10^{-3}} = 2.2 \times 10^{-12}\ M$$

(d) For $2.9 \times 10^{-5}M$ HCl, $[H_3O^+] = 2.9 \times 10^{-5}\ M$ and $[OH^-]$ is

$$[OH^-] = \frac{1.00 \times 10^{-14}}{2.9 \times 10^{-5}} = 3.4 \times 10^{-10}\ M$$

Think About It

To determine how many significant figures to include in your answers when computing the pH or pOH, remember that the first number in the pH or pOH gives the location of the decimal point. The significant digits, therefore, follow the decimal point.

15.39. **Collect and Organize**

Given the concentration of solutions or mixtures of strong acids and bases, we are to calculate the pH and pOH.

Analyze

To calculate the pH or the pOH from the $[H^+]$ or $[OH^-]$, respectively, we use

$$pH = -\log[H^+]$$
$$pOH = -\log[OH^-]$$

To find the pOH from the pH, and vice versa, we use the relationship

$$pH + pOH = 14$$

For mixtures we will have to consider the neutralization reaction between the acid and base. If acid is in excess, the solution will be acidic. If base is in excess, the solution will be basic.

Solve

(a) For [HCl] = 0.155 M, $[H_3O^+]$ = 0.155 M
    pH = $-\log(0.155\ M)$ = 0.81
    pOH = 14.00 − 0.81 = 13.19

(b) For [HNO$_3$] = 0.00500 M, $[H_3O^+]$ = 0.00500 M
    pH = $-\log(0.00500\ M)$ = 2.301
    pOH = 14.00 − 2.301 = 11.699

(c) For a 2:1 mixture of 0.0125 M HCl and 0.0125 M NaOH and assuming that 1 L volumes are mixed, there will be an excess of 0.0125 mol of HCl in 3 L of total volume. The concentration of H_3O^+ is therefore 0.0125 mol/3.00 L = 0.00417 M
    pH = $-\log(0.00417\ M)$ = 2.380
    pOH = 14.00 − 2.301 = 11.612

(d) For a 3:1 mixture of 0.0125 M H_2SO_4 (a strong acid that forms HSO_4^- as conjugate base) and 0.0125 M KOH and assuming that 1 L volumes are mixed, there will be an excess of 0.0250 mol of H_2SO_4 in 4 L of total volume. The concentration of H_3O^+ is therefore 0.0250 mol/4.00 L = 0.00625 M

$$pH = -\log(0.00625\ M) = 2.204$$
$$pOH = 14.00 - 2.301 = 11.796$$

Think About It

The formula $pH + pOH = pK_w = 14.00$ is derived from the K_w expression

$$K_w = [H_3O] \times [OH^-] = 1.00 \times 10^{-14}$$
$$pK_w = -\log([H_3O] \times [OH^-]) = pH + pOH = 14.00$$

15.41. Collect and Organize

Given that the concentration of LiOH in a solution is $1.33 \times 10^{-9}\ M$, we are to calculate its pH.

Analyze

This is a very dilute solution of the strong base LiOH. To calculate the pH, we must take into account the $[OH^-]$ present in water and set up a RICE table to solve the problem.

Solve

Reaction	$2\ H_2O(aq)\ \rightleftharpoons$	$H_3O^+(aq)$	+	$OH^-(aq)$
		$[H_3O^+]$, M		$[OH^-]$, M
Initial		0		1.33×10^{-9}
Change		$+x$		$+x$
Equilibrium		x		$(1.33 \times 10^{-9}) + x$

$$K_w = [H_3O]^+[OH^-] = 1.00 \times 10^{-14}$$
$$1.00 \times 10^{-14} = (x)(1.33 \times 10^{-9} + x)$$
$$x^2 + 1.33 \times 10^{-9}x - 1.00 \times 10^{-14} = 0$$
$$x = 9.934 \times 10^{-8}$$
$$pH = -\log(9.934 \times 10^{-8}) = 7.003$$

Think About It

This solution is just slightly basic, as we would expect.

15.43. Collect and Organize

We are to rank 1 M solutions of CH_3COOH, HNO_2, $HClO$, and HCl in order of decreasing concentration of H_3O^+ and in order of increasing acid strength. To do so we need the K_a values for each acid.

Analyze

The K_a values for those acids are listed in Appendix 5. The greater the value of the K_a, the greater the concentration of H^+ in the solution of the acid. The K_a values for the acids are as follows: CH_3COOH, 1.76×10^{-5}; HNO_2, 4.0×10^{-4}; $HClO$, 2.9×10^{-8}; and HCl, >>1.

Solve

(a) In order of largest K_a (strongest acid) to smallest K_a (weakest acid), HCl > HNO_2 > CH_3COOH > $HClO$.
(b) In order of increasing acid strength, $HClO$ < CH_3COOH < HNO_2 < HCl.

Think About It

The weak acids in this series (excluding HCl) have a wide range in acidities (about 10,000-fold, by their K_a values).

15.45. Collect and Organize

We are to explain why the electrical conductivity of 1.0 M HNO_3 is much better than that of 1.0 M HNO_2.

Analyze

Solutions with a larger concentration of dissolved ions conduct electricity better than those with lower concentrations of dissolved ions.

Solve

HNO_3 is a strong acid and is therefore completely dissociated in water, separating into H^+ and NO_3^- ions, each in 1.0 M concentration, for a total ion concentration of 2.0 M. HNO_2, however, only weakly dissociates in water:

$$HNO_2(aq) \rightleftharpoons NO_2^-(aq) + H^+(aq)$$

so it produces significantly less than 2.0 M ions in solution. HNO_3, therefore, with more dissolved ions in solution, is a better conductor of electricity.

Think About It

We could use a RICE table to calculate exactly how many NO_2^- ions are present in a solution of 0.10 M HNO_2.

15.47. Collect and Organize

The formula for hydrofluoric acid is HF. From that we can write the mass action expression for this weak acid.

Analyze

The general form of the mass action expression for weak acids, on the basis of $HA + H_2O(\ell) \rightleftharpoons A^- + H_3O^+$, is

$$K_a = \frac{[H_3O^+][A^-]}{[HA]}$$

Solve

$$HF(aq) + H_2O(\ell) \rightleftharpoons F^-(aq) + H_3O^+(aq)$$

$$K_a = \frac{[H_3O^+][F^-]}{[HF]}$$

Think About It

Hydrofluoric acid is a weak acid, so the $[F^-]$ and $[H_3O^+]$ at equilibrium is less than the original $[HF]$.

15.49. Collect and Organize

Given that the K_a of alanine is less when it is dissolved in ethanol than it is when dissolved in water, we are to determine which solvent more fully ionizes alanine and which solvent is the stronger Brønsted–Lowry base.

Analyze

The larger the value of K_a, the more a substance has ionized. The solvent in which an acid ionizes the most must be the strongest Brønsted–Lowry base toward that acid.

Solve

(a) Because K_a for alanine is greater in water than in ethanol, alanine in water is ionized more than in ethanol.
(b) Water is the stronger base for alanine than ethanol because alanine is ionized more in water.

Think About It

This question shows that acid–base strengths depend on the basicity of the solvent.

15.51. Collect and Organize

Given that CH_3NH_2 is slightly basic in water, we can write the equation describing its reaction with water to identify the species in the reaction that is the Brønsted–Lowry acid and the species that is the base.

Analyze

Acting as a base, CH_3NH_2 accepts H^+ from surrounding water molecules.

Solve

The reaction describing the basicity of CH_3NH_2 is

$$CH_3NH_2(aq) + H_2O(\ell) \rightleftharpoons CH_3NH_3^+(aq) + OH^-(aq)$$

In that reaction, H_2O is the acid and CH_3NH_2 is the base.

Think About It

In another solvent, such as diethylamine, methylamine may act as an acid:

$$CH_3NH + (CH_3CH_2)_2NH(\ell) \rightleftharpoons CH_3NH^- + (CH_3CH_2)_2NH_2^+$$

That occurs because diethylamine ($K_b = 8.6 \times 10^{-4}$) is a stronger base than methylamine ($K_b = 4.4 \times 10^{-4}$).

15.53. Collect and Organize

Using the measured percent ionization of 1.00 M lactic acid of 2.94%, we are to calculate the K_a of this weak acid.

Analyze

The equilibrium equation from Appendix 5 that describes the ionization of lactic acid is

$$CH_3CHOHCOOH(aq) \rightleftharpoons H_3O^+(aq) + CH_3CHOHCOO^-(aq)$$

We can set up a RICE table to solve this problem, where $x = 1.00\ M \times 0.0294 = 0.0294$. The K_a expression is

$$K_a = \frac{[H_3O^+][CH_3CHOHCOO^-]}{[CH_3CHOHCOOH]}$$

Solve

Reaction	$CH_3CHOHCOOH(aq)$ $\rightleftharpoons$	$H_3O^+(aq)$ +	$CH_3CHOHCOO^-(aq)$
	[$CH_3CHOHCOOH$], M	[H_3O^+], M	[$CH_3CHOHCOO^-$], M
Initial	1.00	0	0
Change	$-x = -0.0294$	$+x = +0.0294$	$+x = +0.0294$
Equilibrium	0.9706	0.0294	0.0294

$$K_a = \frac{(0.0294)^2}{0.9706} = 8.91 \times 10^{-4}$$

Think About It

Compare that answer with the K_a at 25°C value of 1.4×10^{-4} listed in Appendix 5. The difference may be attributable to a temperature difference because body temperature is 37°C.

15.55. Collect and Organize

We are given the [H_3O^+] for an equilibrium solution of an unknown acid with an initial concentration of 0.125 M. From that information we can calculate the degree of ionization and K_a for this weak acid.

Analyze

The equilibrium expression for the unknown acid is

$$HA(aq) \rightleftharpoons H_3O^+(aq) + A^-(aq)$$

with a K_a expression of

$$K_a = \frac{[H_3O^+][A^-]}{[HA]}$$

We can set up a RICE table to show how this weak acid ionizes; for this acid, $[H_3O^+]$ at equilibrium is 4.07×10^{-3} M.

Solve

Reaction	HA(aq)	$\rightleftharpoons$	H_3O^+(aq)	+	A^-(aq)
	[HA], M		$[H_3O^+]$, M		$[A^-]$, M
Initial	0.125		0		0
Change	$-x = -0.00407$		$+x = +0.00407$		$+x = +0.00407$
Equilibrium	0.12093		0.00407		0.00407

$$K_a = \frac{(0.00407)^2}{0.12093} = 1.37 \times 10^{-4}$$

$$\text{Degree of ionization} = \frac{[H_3O^+]_{eq}}{[HA]_{initial}} = \frac{0.00407\ M}{0.125\ M} = 0.0326,\ \text{or}\ 3.26\%$$

Think About It

In this problem we can use the $[H_3O^+]$ as equivalent to x in the RICE table, enabling us to calculate both the K_a and the degree of ionization for the acid.

15.57. Collect and Organize

Given that formic acid has $K_a = 1.77 \times 10^{-4}$, we can calculate the pH of a 0.055 M aqueous solution of this weak acid.

Analyze

To solve this problem we set up a RICE table where the initial concentration of formic acid, HCOOH, is 0.055 M. We let x be the amount of formic acid that ionizes. The equilibrium and K_a expressions are

$$\text{HCOOH}(aq) \rightleftharpoons H_3O^+(aq) + \text{HCOO}^-(aq)$$

$$K_a = \frac{[H_3O^+][\text{HCOO}^-]}{[\text{HCOOH}]} = 1.8 \times 10^{-4}$$

The pH can be calculated through the equation pH $= -\log[H_3O^+]$.

Solve

Reaction	HCOOH(aq)	$\rightleftharpoons$	H_3O^+(aq)	+	HCOO$^-$(aq)
	[HCOOH], M		$[H_3O^+]$, M		[HCOO$^-$], M
Initial	0.055		0		0
Change	$-x$		$+x$		$+x$
Equilibrium	$0.055 - x$		x		x

$$1.77 \times 10^{-4} = \frac{x^2}{0.055 - x} \approx \frac{x^2}{0.055}$$

$$x = 3.12 \times 10^{-3}$$

We should first check the simplifying assumption we made above before calculating the pH.

$$\frac{3.12 \times 10^{-3}}{0.0055} \times 100 = 5.67\%$$

That value is a little over 5%, so technically we should solve this by the quadratic equation (which we do below). However, if we allow this simplifying assumption to be valid, the pH of the solution would be calculated by

$$\text{pH} = -\log 3.12 \times 10^{-3} = 2.51$$

Now, solving by the quadratic equation gives

$$(1.77 \times 10^{-4})(0.055 - x) = x^2$$
$$x^2 + 1.77 \times 10^{-4}x - 9.735 \times 10^{-6} = 0$$
$$x = 3.033 \times 10^{-3}$$
$$pH = -\log 3.033 \times 10^{-3} = 2.52$$

Think About It

The difference between the pH values when making the simplifying assumption and solving the equation exactly by using the quadratic equation is $2.52 - 2.51 = 0.01$, which is small.

15.59. Collect and Organize

By comparing the pH of rain in a weather system in the Midwest of 5.02 with the pH of the rain in that same system when it reached New England of 4.66, we can calculate how much more acidic the rain in New England was.

Analyze

We want to find the ratio

$$\frac{[H^+]_{\text{New England}}}{[H^+]_{\text{Midwest}}}$$

Because $pH = -\log[H^+]$, the $[H^+] = 1 \times 10^{-pH}$. Therefore,

$$\frac{[H^+]_{\text{New England}}}{[H^+]_{\text{Midwest}}} = \frac{1 \times 10^{-4.66}}{1 \times 10^{-5.02}}$$

Solve

$$\frac{[H^+]_{\text{New England}}}{[H^+]_{\text{Midwest}}} = \frac{1 \times 10^{-4.66}}{1 \times 10^{-5.02}} = \frac{2.19 \times 10^{-5}\,M}{9.55 \times 10^{-6}\,M} = 2.3$$

The rain in New England in that weather system was 2.3 times more acidic than the rain in the Midwest.

Think About It

Among the causes of the acidity of the rain in New England are the coal-burning electricity power plants in the Midwest, which expel SO_2 and SO_3 into the air that make H_2SO_3 and H_2SO_4.

15.61. Collect and Organize

Given the K_b for aminoethanol of 3.1×10^{-5}, we are to determine whether aminoethanol is a stronger or weaker base than ethylamine ($pK_b = 3.36$) and then calculate the pH of a $1.67 \times 10^{-2}\,M$ solution of aminoethanol and the $[OH^-]$ in a $4.25 \times 10^{-4}\,M$ aminoethanol solution.

Analyze

The equilibrium and K_b expressions for aminoethanol are

$$HOCH_2CH_2NH_2(aq) + H_2O(\ell) \rightleftharpoons HOCH_2CH_2NH_3^+(aq) + OH^-(aq)$$

$$K_b = \frac{[OH^-][HOCH_2CH_2NH_3^+]}{[HOCH_2CH_2NH_2]}$$

We can set up a RICE table to solve this problem, where x is the amount of aminoethanol that ionizes. By solving for x, we can calculate $[OH^-]$, from which we can determine the pOH and pH:

$$pOH = -\log[OH^-]$$
$$pH = 14 - pOH$$

Solve

(a) Because the pK_b for aminoethanol $[-\log(3.1 \times 10^{-5}) = 4.51]$ is greater than that of ethylamine, aminoethanol is a weaker base than ethylamine.

(b) We can set up a RICE table to solve this problem, where x is the amount of aminoethanol that ionizes and the concentration of OH^- produced. By solving for x, we can calculate $[OH^-]$, from which we can determine the pOH and pH:

Reaction	$HOCH_2CH_2NH_2(aq)$ $\rightleftharpoons$	$OH^-(aq)$ +	$HOCH_2CH_2NH_3^+(aq)$
	$[HOCH_2CH_2NH_2]$, M	$[OH^-]$, M	$[HOCH_2CH_2NH_3^+]$, M
Initial	1.67×10^{-2}	0	0
Change	$-x$	$+x$	$+x$
Equilibrium	$1.67 \times 10^{-2} - x$	x	x

$$3.1 \times 10^{-5} = \frac{x^2}{1.67 \times 10^{-2} - x} \approx \frac{x^2}{1.67 \times 10^{-2}}$$

$$x = 7.2 \times 10^{-4}$$

Checking the simplifying assumption shows that our simplifying assumption is valid:

$$\frac{7.2 \times 10^{-4}}{1.67 \times 10^{-2}} \times 100 = 4.3\%$$

The pH of this solution is

$$pOH = -\log 7.2 \times 10^{-4} = 3.14$$
$$pH = 14 - 3.14 = 10.86$$

(c) We can set up a RICE table to solve this problem, where x is the amount of aminoethanol that ionizes and the concentration of OH^- produced. By solving for x, we can calculate $[OH^-]$:

Reaction	$HOCH_2CH_2NH_2(aq)$ $\rightleftharpoons$	$OH^-(aq)$ +	$HOCH_2CH_2NH_3^+(aq)$
	$[HOCH_2CH_2NH_2]$, M	$[OH^-]$, M	$[HOCH_2CH_2NH_3^+]$, M
Initial	4.25×10^{-4}	0	0
Change	$-x$	$+x$	$+x$
Equilibrium	$4.25 \times 10^{-4} - x$	x	x

$$3.1 \times 10^{-5} = \frac{x^2}{4.25 \times 10^{-4} - x}$$

$$3.1 \times 10^{-5} \times (4.25 \times 10^{-4} - x) = x^2$$

$$x^2 + 3.1 \times 10^{-5} x - 1.32 \times 10^{-8} = 0$$

$$x = 1.00 \times 10^{-4} = [OH^-]$$

Think About It

Our answer in part b of a pH > 7 is consistent with aminoethanol's behavior as a base in aqueous solution.

15.63. Collect and Organize

Given the pK_b of morphine, a weak base, we are asked to calculate the pH of a 0.0018 M solution; given the pK_a of the conjugate acid of codeine, a weak base, we are asked to calculate the pH of a 0.00027 M solution.

Analyze

From the pK_b values, we can calculate the K_b for both morphine and codeine:

$$K_b(\text{morphine}) = 1 \times 10^{-pK_b} = 1 \times 10^{-5.79} = 1.62 \times 10^{-6}$$

$$pK_b = 14.00 - pK_a = 14.00 - 8.21 = 5.79$$

$$K_b(\text{codeine}) = 1 \times 10^{-pK_b} = 1 \times 10^{-5.79} = 1.62 \times 10^{-6}$$

The equilibrium and K_b expressions for the ionization of morphine and codeine are

$$\text{morphine}(aq) + H_2O(\ell) \rightleftharpoons \text{morphineH}^+(aq) + OH^-(aq)$$

$$K_b = \frac{[OH^-][\text{morphineH}^+]}{[\text{morphine}]} = 1.62\times10^{-6}$$

$$\text{codeine}(aq) + H_2O(\ell) \rightleftharpoons \text{codeineH}^+(aq) + OH^-(aq)$$

$$K_b = \frac{[OH^-][\text{codeineH}^+]}{[\text{codeine}]} = 1.62\times10^{-6}$$

We can set up a RICE table to solve these problems, where x is the amount of morphine and codeine that ionizes. By solving for x, we can calculate $[OH^-]$, from which we can determine the pOH and pH:

$$pOH = -\log[OH^-]$$
$$pH = 14 - pOH$$

Solve

(a) For a 0.0018 M morphine:

Reaction	Morphine(aq) $\rightleftharpoons$	OH$^-$(aq) +	MorphineH$^+$(aq)
	[Morphine]	[OH$^-$]	[MorphineH$^+$]
Initial	0.0018	0	0
Change	$-x$	$+x$	$+x$
Equilibrium	$0.0018 - x$	x	x

$$1.62\times10^{-6} = \frac{x^2}{0.0018 - x}$$

$$1.62\times10^{-6}\times(0.0018 - x) = x^2$$

$$x^2 + 1.62\times10^{-6}x - 2.92\times10^{-9} = 0$$

$$x = 5.32\times10^{-5}$$

Thus, we can calculate the pH:

$$pOH = -\log 5.32 \times 10^{-5} = 4.27$$
$$pH = 14 - 4.27 = 9.73$$

(b) For a 0.00027 M codeine:

Reaction	Codeine(aq) $\rightleftharpoons$	OH$^-$(aq) +	CodeineH$^+$(aq)
	[Codeine]	[OH$^-$]	[CodeineH$^+$]
Initial	0.00027	0	0
Change	$-x$	$+x$	$+x$
Equilibrium	$0.00027 - x$	x	x

$$1.62\times10^{-6} = \frac{x^2}{0.00027 - x}$$

$$1.62\times10^{-6}\times(0.00027 - x) = x^2$$

$$x^2 + 1.62\times10^{-6}x - 4.37\times10^{-10} = 0$$

$$x = 2.01\times10^{-5}$$

Thus, we can calculate the pH:

$$pOH = -\log(2.01 \times 10^{-5}) = 4.70$$
$$pH = 14 - 5.89 = 9.30$$

Think About It

Our calculation of the pH of both solutions as greater than 7 is consistent with morphine and codeine behavior as bases in aqueous solution.

15.65. Collect and Organize

We are to explain why for H_3PO_4, $K_{a_1} > K_{a_2} > K_{a_3}$.

Analyze

The equations describing these acid dissociation constants are as follows:

$$H_3PO_4(aq) \rightleftharpoons H_2PO_4^-(aq) + H^+(aq) \qquad K_{a_1}$$
$$H_2PO_4^-(aq) \rightleftharpoons HPO_4^{2-}(aq) + H^+(aq) \qquad K_{a_2}$$
$$HPO_4^{2-}(aq) \rightleftharpoons PO_4^{3-}(aq) + H^+(aq) \qquad K_{a_3}$$

Solve

With each successive ionization, it becomes more difficult to remove H^+ from a species that is negatively charged. Therefore, it is harder to remove H^+ from HPO_4^{2-} than from $H_2PO_4^-$ and from H_3PO_4. That trend is reflected in decreasing K_a values as H_3PO_4 is ionized.

Think About It

From Appendix 5, we can compare the K_a values for phosphoric acid: $K_{a_1} = 7.11 \times 10^{-3}$, $K_{a_2} = 6.32 \times 10^{-8}$, $K_{a_3} = 4.5 \times 10^{-13}$. Those values span 10 orders of magnitude.

15.67 Collect and Organize

We are asked to explain why H_2GeO_3 is weaker than the weak acid H_2CO_3.

Analyze

Germanium is in the same row as C in the periodic table, but is farther down. Therefore, differences in the electronegativity between C and Ge will result.

Solve

Germanium is much less electronegative than carbon and therefore the conjugate base $HGeO_3^-$ is less stable than HCO_3^-.

Think About It

Because silicon has about the same electronegativity as germanium, we might expect that H_2SiO_3 has about the same acidity as H_2GeO_3.

15.69. Collect and Organize

We have to use the K_{a_2} of H_2SO_4 to calculate the pH of a solution of 0.75 *M* H_2SO_4.

Analyze

The first H^+ is completely removed from the H_2SO_4, and the initial concentrations of the species in solution before the second ionization are $[H_3O^+] = 0.75$ *M*, $[H_2SO_4] = 0.0$ *M*, and $[HSO_4^-] = 0.75$ *M*. The equation describing the second ionization is

$$HSO_4^-(aq) \rightleftharpoons SO_4^{2-}(aq) + H_3O^+(aq)$$

$$K_a = \frac{[H_3O^+][SO_4^{2-}]}{[HSO_4^-]} = 1.2 \times 10^{-2}$$

Solve

Reaction	$HSO_4^-(aq)$	$\rightleftharpoons$	$H_3O^+(aq)$	+	$SO_4^{2-}(aq)$
	$[HSO_4^-]$, *M*		$[H_3O^+]$, *M*		$[SO_4^{2-}]$, *M*
Initial	0.75		0.75		0
Change	$-x$		$+x$		$+x$
Equilibrium	$0.75 - x$		$0.75 + x$		x

Plugging equilibrium concentrations into the K_a expression gives

$$1.2 \times 10^{-2} = \frac{x(0.75 + x)}{0.75 - x}$$

Solving that by the quadratic equation gives

$$x^2 + 0.762x - 0.00900 = 0$$
$$x = 0.0116$$
$$[H_3O^+] = 0.75 + 0.0116 = 0.762$$
$$pH = -\log 0.762 = 0.12$$

Think About It
Had we not considered the second ionization of H_2SO_4, we would have underestimated the acidity of the solution by 0.0067 pH units.

15.71. Collect and Organize
Given a 0.250 M solution of ascorbic acid and the fact that it is a weak triprotic acid (1.0×10^{-5} and 5×10^{-12}, respectively), we are to calculate the pH.

Analyze
Because K_{a1} (1.0×10^{-5}) for ascorbic acid is so much larger than the second or third ionization constants, we can say that the second and third ionizations contribute little to the $[H^+]$ in the solution. Therefore, we can solve this by examining only the first ionization.

Solve

Reaction	Ascorbic acid	$\rightleftharpoons$	Ascorbate$^-$	+	H$^+$
	[Ascorbic acid]		[Ascorbate$^-$]		[H$^+$]
Initial	0.250		0		0
Change	$-x$		$+x$		$+x$
Equilibrium	$0.250 - x$		x		x

$$1.0 \times 10^{-5} = \frac{x^2}{0.250 - x} \approx \frac{x^2}{0.250}$$
$$x = 1.58 \times 10^{-3}$$

Checking the simplifying assumption shows that it is valid:

$$\frac{1.58 \times 10^{-3}}{0.250} \times 100 = 0.63\%$$

The pH is found from the $[H^+]$:

$$pH = -\log 1.58 \times 10^{-3} = 2.80$$

Think About It
This solution is about as acidic as vinegar.

15.73. Collect and Organize
Given a 0.00100 M solution of nicotine and the pK_{b_1} and pK_{b_2} (from Appendix 5) for this weak dibasic compound, we are to calculate the pH.

Analyze
The pK_{b_2} (1.3×10^{-11}) is so much smaller than the pK_{b_1} (1.0×10^{-6}) for nicotine that we may ignore the contribution of the second ionization of nicotine to the $[OH^-]$ in the solution. We can therefore solve this problem by examining only the first ionization.

Solve

Reaction	Nicotine(aq) + H$_2$O(ℓ) $\rightleftharpoons$ NicotineH$^+$(aq) + OH$^-$(aq)		
	[Nicotine]	[NicotineH$^+$]	[OH$^-$]
Initial	0.00100	0	0
Change	$-x$	$+x$	$+x$
Equilibrium	$0.00100 - x$	x	x

$$1.0 \times 10^{-6} = \frac{x^2}{0.00100 - x} \approx \frac{x^2}{0.00100}$$

$$x = 3.16 \times 10^{-5}$$

Checking the simplifying assumption shows that it is valid:

$$\frac{3.16 \times 10^{-5}}{0.00100} \times 100 = 3.2\%$$

We can calculate the pOH and pH from the [OH$^-$]:

$$\text{pOH} = -\log 3.16 \times 10^{-5} = 4.500$$
$$\text{pH} = 14 - 4.500 = 9.500$$

Think About It

The relatively high pH of this dilute solution of nicotine shows that this compound is a strong weak base.

15.75. Collect and Organize

Given a 0.01050 M solution of quinine and the K_{b1} and K_{b2} for this weak dibasic compound, we are to calculate the pH.

Analyze

The K_{b2} (1.4×10^{-9}) is so much smaller than K_{b1} (3.3×10^{-6}) for quinine that we may ignore the contribution of the second ionization of quinine to the [OH$^-$] in the solution. We can therefore solve this problem by examining only the first ionization.

Solve

Reaction	Quinine(aq) + H$_2$O(ℓ) $\rightleftharpoons$ QuinineH$^+$(aq) + OH$^-$(aq)		
	[Quinine]	[QuinineH$^+$]	[OH$^-$]
Initial	0.01050	0	0
Change	$-x$	$+x$	$+x$
Equilibrium	$0.01050 - x$	x	x

$$3.3 \times 10^{-6} = \frac{x^2}{0.01050 - x} \approx \frac{x^2}{0.01050}$$

$$x = 1.86 \times 10^{-4}$$

Checking the simplifying assumption shows that it is valid:

$$\frac{1.86 \times 10^{-4}}{0.01050} \times 100 = 1.8\%$$

The pOH and pH can be calculated from the [OH$^-$]:

$$\text{pOH} = -\log 1.86 \times 10^{-4} = 3.730$$
$$\text{pH} = 14 - 3.730 = 10.270$$

Think About It

Quinine has a complicated molecular structure that includes aromatic rings, an alcohol, an amine, an alkene, and an ether as functional groups.

15.77. Collect and Organize

We are asked to explain why an aqueous solution of NaF is basic and why an aqueous solution of NaCl is neutral.

Analyze

Sodium salts are all soluble in aqueous solution. For NaF the solution will contain $Na^+(aq)$ and $F^-(aq)$; for NaCl the solution will contain $Na^+(aq)$ and $Cl^-(aq)$. If any of those ions react with water, they may make the solution acidic or basic.

Solve

In neither solution does Na^+ react with water; if it did, it would produce the strong base NaOH. An aqueous solution of NaF is basic because F^- reacts with water to produce the weak acid HF along with OH^-. An aqueous solution of Cl^- is neutral because Cl^- does not react with water; if it did, the strong acid HCl would be produced.

Think About It

Solutions in which the anion of a salt reacts with water produce basic solutions; solutions in which the cation of a salt reacts with water produce acidic solutions. If both the cation and the anion of the salt react with water, the solution may be acidic (if the cation has the larger K_a) or basic (if the anion has the larger K_b) or neutral (if the K_a and K_b are equal).

15.79. Collect and Organize

Of the three salts given, we are to determine which, when dissolved in water, gives an acidic solution.

Analyze

To give an acidic solution, the cation of the salt must donate a proton to water without the anion's reacting with water, or if the anion hydrolyzes, then the pK_a of the cation must be lower than the pK_b of the anion.

Solve

Both NH_4^+ and CH_3COO^- of ammonium acetate hydrolyze:

$$NH_4^+(aq) + H_2O(\ell) \rightleftharpoons NH_3(aq) + H_3O^+(aq) \qquad pK_a = 9.25$$
$$CH_3COO^-(aq) + H_2O(\ell) \rightleftharpoons CH_3COOH(aq) + OH^-(aq) \qquad pK_b = 9.25$$

Because $pK_a = pK_b$, this salt's solution is neutral.
Only NH_4^+ of ammonium nitrate hydrolyzes. This gives an acidic solution:

$$NH_4^+(aq) + H_2O(\ell) \rightleftharpoons NH_3(aq) + H_3O^+(aq)$$

Only $HCOO^-$ of sodium formate hydrolyzes. This gives a basic solution:

$$HCOO^-(aq) + H_2O(\ell) \rightleftharpoons HCOOH(aq) + OH^-(aq)$$

Therefore, of the three salts, only ammonium nitrate dissolves to give an acidic solution.

Think About It

Remember that neither Na^+ nor NO_3^- hydrolyzes because it would form either a strong base or a strong acid, both of which are always 100% ionized.

$$Na^+(aq) + H_2O(\ell) \not\rightleftharpoons NaOH(aq) + H^+(aq)$$
$$NO_3^-(aq) + H_2O(\ell) \not\rightleftharpoons HNO_3(aq) + OH^-(aq)$$

15.81. **Collect and Organize**

We consider why lemon juice is used to reduce the fishy odor due to the presence of $(CH_3)_3N$ in not-so-fresh seafood.

Analyze

Trimethylamine is a weak base, and lemon juice contains citric acid, which is a weak acid.

Solve

The citric acid in the lemon juice neutralizes the volatile trimethylamine to make a nonvolatile dissolved salt that neutralizes the fishy odor:

$$HOC(CH_2)_2(COOH)_3(aq) + (CH_3)_3N(aq) \rightleftharpoons HOC(CH_2)_2(COOH)_2COO^-(aq) + (CH_3)_3NH^+(aq)$$

Think About It

Because the pK_b of trimethylamine of 4.19 and the pK_a of citric acid of 3.13 are lower than the pK_a of trimethylammonium (9.81) and the pK_b of citrate (10.87), this equilibrium lies to the right, favoring the products.

15.83. **Collect and Organize**

Given $K_a = 2.1 \times 10^{-11}$ for the conjugate acid of saccharin, we are asked to calculate the value of pK_b for saccharin.

Analyze

From the K_a we can calculate the pK_a:

$$pK_a = -\log K_a$$

From the pK_a we can calculate the pK_b because $pKw = 14.00$ at 25°C:

$$pK_b = 14.00 - pK_a$$

Solve

$$pK_a = -\log(2.1 \times 10^{-11}) = 10.68$$
$$pK_b = 14.00 - 10.68 = 3.32$$

Think About It

Alternatively, we could calculate the K_b from K_a by using

$$K_b = \frac{K_w}{K_a}$$

and then calculate pK_b by using

$$pK_b = -\log K_b$$

15.85. **Collect and Organize**

From Appendix 5 we know that the K_a of HF is 6.8×10^{-4}. Using that value, we are to calculate the pH of a solution that is 0.00339 M in NaF.

Analyze

When NaF dissolves in water, the F^- ion hydrolyzes to give a basic solution:

$$F^-(aq) + H_2O(\ell) \rightleftharpoons HF(aq) + OH^-(aq)$$

The K_b for that reaction is

$$K_b = \frac{K_w}{K_a} = \frac{1 \times 10^{-14}}{6.8 \times 10^{-4}} = 1.47 \times 10^{-11}$$

We can solve for $[OH^-]$ by using a RICE table and then compute the pH.

Solve

Reaction	$F^-(aq)$ + $H_2O(\ell)$	$\rightleftharpoons$	HF(aq)	+	$OH^-(aq)$
	[F$^-$]		[HF]		[OH$^-$]
Initial	0.00339		0		0
Change	$-x$		$+x$		$+x$
Equilibrium	$0.00339 - x$		x		x

$$1.47 \times 10^{-11} = \frac{x^2}{0.00339 - x} \approx \frac{x^2}{0.00339}$$

$$x = 2.23 \times 10^{-7}$$

Checking the simplifying assumption shows that it is valid:

$$\frac{2.23 \times 10^{-7}}{0.00339} \times 100 = 0.0066\%$$

The pH is calculated from the [OH$^-$]:

$$pOH = -\log 2.23 \times 10^{-7} = 6.65$$
$$pH = 14 - 6.65 = 7.35$$

Think About It

Because HF is a moderately strong weak acid, F^-, its conjugate base, is a fairly weak base.

15.87. Collect and Organize

For the compounds CH_3NH_2, CH_3COOH, $Ca(OH)_2$, and $HClO_4$, we are asked to identify the Arrhenius acids, the Arrhenius bases, the Brønsted–Lowry acids, and the Brønsted–Lowry bases.

Analyze

An Arrhenius acid produces H^+ in water and an Arrhenius base produces OH^- in water. A Brønsted–Lowry acid donates H^+ to a base and a Brønsted–Lowry base accepts a H^+ from an acid.

Solve

(a) The Arrhenius acids are CH_3COOH and $HClO_4$.
(b) The Arrhenius base is $Ca(OH)_2$.
(c) The Brønsted–Lowry acids are CH_3COOH and $HClO_4$.
(d) The Brønsted–Lowry bases are $Ca(OH)_2$ and CH_3NH_2.

Think About It

CH_3NH_2 is not classified as an Arrhenius base because when it dissolves into aqueous solution it does not directly release OH^- ions into the solution.

15.89. Collect and Organize

We are asked to consider and explain if all Arrhenius bases are also Brønsted–Lowry bases and if all Brønsted–Lowry bases are also Arrhenius bases.

Analyze

An Arrhenius acid produces H^+ in water and an Arrhenius base produces OH^- in water. A Brønsted–Lowry acid donates H^+ to a base and a Brønsted–Lowry base accepts a H^+ from an acid.

Solve

Yes, all Arrhenius bases are Brønsted–Lowry bases because Arrhenius bases are all H^+ acceptors. However, the Arrhenius definition of bases is defined by their behavior in aqueous solution, but Brønsted–Lowry bases can accept H^+ ions in the gas phase or in nonaqueous media. An example would be NH_3, which accepts a proton from itself (autoionizes) in liquid ammonia to give NH_4^+ and NH_2^-; no OH^- is produced.

Think About It

The Brønsted–Lowry definition is generally more useful as an acid–base model in that it considers the exchange of the H^+ ion between the acid and base in any medium.

15.91. Collect and Organize

High-sulfur fuels, when burned, produce acid rain that erodes marble. We are to describe those chemical reactions.

Analyze

Sulfur burned in air produces sulfur oxides such as SO_3. That nonmetal oxide is an acid anhydride. The parent acid forms upon reaction with water in clouds to produce acid rain. The acid, then, erodes (dissolves) $CaCO_3$ in marble.

Solve

In burning S-containing fuels:

$$S(s) + O_2(g) \rightarrow SO_2(g)$$

The sulfur in SO_2 is further oxidized in the atmosphere in a series of reactions that can be summarized as:

$$2\, SO_2(g) + O_2(g) \rightarrow 2\, SO_3(g)$$

Sulfur trioxide combines with water vapor to form sulfuric acid, which dissolves rain and falls to the earth as dilute sulfuric acid.

$$SO_3(g) + H_2O(\ell) \rightarrow H_2SO_4(aq)$$

Should the sulfuric acid come into contact with marble, the reaction is

$$H_2SO_4(aq) + CaCO_3(s) \rightarrow H_2O(\ell) + CO_2(g) + CaSO_4(s)$$

The slightly soluble $CaSO_4$ is washed away by the rain and the marble structure slowly dissolves.

Think About It

H_2CO_3 is initially formed in the last reaction, but it decomposes to H_2O and CO_2 according to the equilibrium expression

$$H_2CO_3(aq) \rightleftharpoons H_2O(\ell) + CO_2(g)$$

15.93. Collect and Organize

Given the change in pH of a lake from 6.1 to 4.7 when 400 gallons of 18 M H_2SO_4 was added to the lake, we are to calculate the volume of the lake.

Analyze

To make this calculation easier, we can assume that we ionize only the first proton on H_2SO_4. First, we must calculate the moles of H_2SO_4 added to the lake and then determine the increase in the concentration of H^+ in the lake in going from pH 6.1 to pH 4.7. Knowing that we simply divide the moles of acid added by the molarity change of H^+ in the lake, we can obtain the size of the lake.

Solve

The amount of H_2SO_4 added is

$$400\ \text{gal} \times \frac{3.78\ \text{L}}{1\ \text{gal}} \times \frac{18\ \text{mol}}{\text{L}} = 27{,}216\ \text{mol}$$

The change in lake $[H^+]$ is

$$\text{Initial } [H^+] = 1 \times 10^{-6.1} = 7.94 \times 10^{-7}\ M$$
$$\text{Final } [H^+] = 1 \times 10^{-4.7} = 2.00 \times 10^{-5}\ M$$
$$\Delta[H^+] = 1.92 \times 10^{-5}\ M \text{ increase}$$

The size of the lake is

$$27{,}216\ \text{mol} \times \frac{1\ \text{L}}{1.92 \times 10^{-5}\ \text{mol}} = 1.4 \times 10^{9}\ \text{L}$$

Think About It

That volume is equivalent to 1.4×10^6 m³. If the lake were 10 m deep, it would cover an area of 1.4×10^5 m². If thought of as a square, that is 374 m on a side.

15.95. Collect and Organize

Given the structure of Prozac in Figure P15.95, we are asked about its acid–base character and the solubility of its HCl salt.

Analyze

The functional group that is likely to show acid–base character is the amine group ($-NHCH_3$) on one end of the Prozac molecule.

Solve

(a) In water, amines show basic character. They pick up a proton from water to form ammonium cations and OH^-. Therefore, dissolving Prozac in water gives a slightly basic solution.

(b) The N atom on Prozac is more likely to react with HCl than with the O atom.

(c) The HCl salt of Prozac is more soluble because it is charged and water molecules form stronger ion–dipole forces around the molecule than the dipole–induced dipole forces between the neutral molecule and water.

Think About It

Being soluble in water also helps to deliver the drug to the body.

15.97. Collect and Organize

After we draw the structure of its conjugate base, we are to explain why C_5F_5H is such a strong acid in comparison with other organic acids.

Analyze

The presence of many nearby electronegative atoms increases the acidity of a proton.

Solve

(a)

(b) C_5F_5H is very acidic because the presence of five very electronegative F atoms on the carbon ring stabilizes the anion formed when the proton is lost.

Think About It

The formation of a stable aromatic ring upon deprotonation enhances the stability of the conjugate base above the stability imparted by the presence of F atoms on the ring.

15.99. Collect and Organize

For the reaction of nitric acid with sulfuric acid to produce NO_2^+, H_3O^+, and HSO_4^-, we are to determine whether the reaction is a redox reaction and to identify the acid, base, conjugate acid, and conjugate base in the reaction by drawing the Lewis structures.

Analyze

A redox reaction is indicated by a change in the oxidation state. In each reactant and product the oxidation state of H and O will be the same (+1 and –2). We should therefore look for changes in the oxidation state in N and S. From the Lewis structures we will be able to identify the acid (the proton donor) and the base (the proton acceptor) and the conjugates.

Solve

(a) This is not a redox reaction. In HNO_2 nitrogen has an oxidation state of +5 and in NO_2^+, nitrogen also has an oxidation state of +5. Likewise, the oxidation state of S in H_2SO_4 and HSO_4^- is +6 in both species.

(b) The reaction depicted in Lewis structures is

From that we see that H_2SO_4 donates H^+ to act as the acid to the –OH group of HNO_3, which acts as the base. The conjugate acid, therefore, is H_3O^+ and the conjugate base is HSO_4^-.

Think About It

That finding indicates that sulfuric acid is a stronger acid than nitric acid. Remember that in water both acids are strong acids and are completely dissociated.

15.101. Collect and Organize

For solutions of various concentrations and acid or base strength, we are to choose which is the most acidic and which is the most basic.

Analyze

Of the solutions listed, H_2SO_3, H_2SO_4, and $NaHSO_4$ are acids—they react with water to form H_3O^+ and the conjugate bases (HSO_3^-, HSO_4^-, and SO_4^{2-}). Na_2SO_4 and Na_2SO_3 are bases—the anions of those salts react with water to form OH^- and the conjugate acids (SO_4^{2-} and SO_3^{2-}).

Solve

H_2SO_3 and HSO_4^- are much weaker acids than H_2SO_4 and despite their higher concentration, 0.10 M H_2SO_4 will be the most acidic solution. The 0.30 M solution of Na_2SO_3 will be the most basic solution because it can react with two molar equivalents of water to form the weak acid H_2SO_3 (to give nearly 2 molar equivalents of OH^-), whereas Na_2SO_4 will react with only 1 molar equivalent of water to form the weak acid HSO_4^- (to give about 1 molar equivalent of OH^-) and not react further to give the strong acid H_2SO_4.

Think About It

Both concentration and strength need to be considered simultaneously for this problem.

15.103. Collect and Organize

For each pair of acids or bases, we are to predict which will have a higher pH (be least acidic or most basic).

Analyze

In comparing the two acids or bases, we have to consider both their relative strength and their relative concentrations.

Solve

(a) 1.00×10^{-3} M NaOH
(b) 0.345 mM HBrO
(c) 45 mM Ba(OH)$_2$
(d) 1.6 mM C$_6$H$_7$N
(e) 252 mM KOH
(f) 105 mM CH$_3$NH$_2$
(g) 1.50×10^{-5} M pyridine
(h) 5 mM C$_7$H$_5$O$_2$Cl

Think About It
The stronger the acid, the lower the pH, so we would choose the weaker acid as having a higher pH.

15.105. Collect and Organize
We are to calculate the pH of a 0.125 M solution of sodium propanoate solution given that the pK_a for propionic acid pK_a is 4.85.

Analyze
To solve this problem we will set up a RICE table for the reaction of propanoate ($C_3H_5O_2^-$) with water.
$$C_3H_5O_2^-(aq) + H_2O(\ell) \rightarrow C_3H_5O_2H(aq) + OH^-(aq)$$
This equilibrium expression needs a K_b value to solve, which we can obtain from the pK_a value:
$$pK_b = 14.00 - pK_a = 14.00 - 4.85 = 9.15$$
$$K_b = 10^{-9.15} = 7.08 \times 10^{-10}$$

Solve

Reaction	$C_3H_5O_2^-(aq)$	$\rightleftharpoons$	$C_3H_5O_2H(aq)$	+	$OH^-(aq)$
	$[C_3H_5O_2^-]$, M		$[C_3H_5O_2H]$, M		$[OH^-]$, M
Initial	0.125		0		0
Change	$-x$		$+x$		$+x$
Equilibrium	$0.125 - x$		x		x

$$7.08 \times 10^{-10} = \frac{x^2}{0.125 - x} \approx \frac{x^2}{0.125}$$
$$x = 9.41 \times 10^{-6}$$

The pH is found from the $[H_3O^+]$:
$$pOH = -\log 9.41 \times 10^{-6} = 5.03$$
$$pH = 14.00 - 5.03 = 8.97$$

Think About It
This solution is basic because of the hydrolysis of the base $C_3H_5O_2^-$ to form the weak conjugate acid $C_3H_5O_2H$.

15.107. Collect and Organize
For each acid we are to identify the conjugate base and calculate the K_b for each base to determine which conjugate base is the strongest.

Analyze
The conjugate base for each acid has one H^+ removed from the formula of the acid. The K_b and K_a values for conjugates are related to each other through K_w:
$$K_w = K_a \times K_b = 1.00 \times 10^{-14}$$
To get the K_b value for a conjugate base, we need to only then divide 1.00×10^{-14} by the K_a value of the corresponding acid, which we can look up in Appendix 5.

Solve
(a) Conjugate base = $ClCH_2COO^-$; $K_b = 7.1 \times 10^{-12}$
(b) Conjugate base = NH_3, $K_b = 1.76 \times 10^{-5}$
(c) Conjugate base = CN^-; $K_b = 1.6 \times 10^{-5}$
(d) Conjugate base = $CH_3CH_2O^-$; $K_b = 77$
$CH_3CH_2O^-$ (ethoxide) is the strongest base.

Think About It
Ethoxide is a strong base in water, as evidenced from its K_b value > 1.

15.109. **Collect and Organize**

For each acid–base equation we are asked to write the net ionic equation and to identify the Brønsted–Lowry acids and bases.

Analyze

The net ionic equations will cancel out any spectator ions from soluble salts in the reactions. The Brønsted–Lowry acid is that species that donates H^+ and the Bronsted–Lowry base is that species that accepts H^+.

Solve

(a) $H^+(aq) + OH^-(aq) \rightarrow H_2O(\ell)$; acid = HNO_3, base = $Ca(OH)_2$

(b) $CO_3^{2-}(aq) + 2H^+(aq) \rightarrow CO_2(g) + H_2O(\ell)$; acid = H_2SO_4, base = Na_2CO_3

(c) $CH_3NH_2(aq) + H^+(aq) \rightarrow CH_3NH_3^+(aq)$; acid = HBr, base = CH_3NH_2

(d) $2\ CH_3COOH(aq) + Mg(OH)_2(s) \rightarrow Mg^{2+}(aq) + 2\ CH_3COO^-(aq) + H_2O(\ell)$; acid = CH_3COOH, base = $Mg(OH)_2$

(e) $CaO(s) + H_2O(\ell) \rightarrow Ca(OH)_2(s)$; acid = H_2O, base = CaO

(f) $LiH(s) + H_2O(\ell) \rightarrow Li^+(aq) + OH^-(aq) + H_2(g)$; acid = H_2O, base = LiH

(g) $Ba^{2+}(aq) + 2\ OH^-(aq) + H^+(aq) + HSO_4^-(aq) \rightarrow Ba\ SO_4(s) + 2\ H_2O(\ell)$; acid = H_2SO_4, base = $Ba(OH)_2$

(h) $SH^-(aq) + H^+(aq) \rightarrow H_2S(g)$; acid = HNO_3, base = $NaSH$

Think About It

In reaction (e) a solid reactant produces a solid product.

CHAPTER 16 | Additional Aqueous Equilibria: Chemistry and the Oceans

16.1. Collect and Organize

From Figure P16.1, we are to choose which titration curve represents a strong acid and which represents a weak acid, each of 1 M concentration at the start of the titration.

Analyze

A strong acid is completely ionized in solution and has a lower initial pH than a solution with the same concentration of a weak acid, which is only partially ionized in solution.

Solve

The blue titration curve represents the titration of a 1 M solution of strong acid. The red titration curve represents the titration of a 1 M solution of weak acid. That is because the pH of the strong acid is expected to be much lower than that of the weak acid at the start of the titration (where no base has yet been added).

Think About It

The equivalence point of the titration of the strong acid (pH 7) does not equal that of the weak acid (pH 10).

16.3. Collect and Organize

For the red titration curve in Figure P16.1, we are to choose the indicator, according to its pK_a, that would be best for the titration.

Analyze

The best indicator is the one with a pK_a nearest to the end point of the titration.

Solve

The end point for the red curve is at approximately pH 10. Therefore, the best indicator is the one with a pK_a of 9.0.

Think About It

The lower pK_a indicators would show a color change before the end point of the titration was reached. Using those would therefore underestimate the concentration of the weak acid in the original solution.

16.5. Collect and Organize

We are shown two titration curves in Figure P16.5. The blue curve has one equivalence point and the red curve has two equivalence points. We are to assign each curve to either Na_2CO_3 or $NaHCO_3$.

Analyze

Both bases being titrated are soluble sodium salts. The equation describing the titration of CO_3^{2-} (Na^+ is a spectator ion) shows CO_3^{2-} to be "dibasic"; it reacts in two steps to form H_2CO_3.

$$CO_3^{2-}(aq) + H^+(aq) \rightarrow HCO_3^-(aq)$$
$$HCO_3^-(aq) + H^+(aq) \rightarrow H_2CO_3(aq)$$

HCO_3^-, however, reacts with acid in one step; it is "monobasic":

$$HCO_3^-(aq) + H^+(aq) \rightarrow H_2CO_3(aq)$$

Solve

The red titration curve represents the titration of Na_2CO_3 because it shows two equivalence points. The blue titration curve represents the titration of $NaHCO_3$ because it shows one equivalence point.

Think About It

Both titration curves start at high pH because of the hydrolysis of CO_3^{2-} and HCO_3^- in water:

$$CO_3^{2-}(aq) + H_2O(\ell) \rightarrow HCO_3^-(aq) + OH^-(aq)$$
$$HCO_3^-(aq) + H_2O(\ell) \rightarrow H_2CO_3(aq) + OH^-(aq)$$

16.7. **Collect and Organize**

Figure P16.7 shows three beakers: one containing a yellow solution, one containing a near colorless solution, and the last one containing a blue solution. Given that the bromthymol blue indicator in each beaker is yellow in acidic solutions and blue in basic solutions, we are to match solutions of the dissolved salts NH_4Cl, $NH_4C_2H_3O_2$, and $NaC_2H_3O_2$ to the correct beaker.

Analyze

We must consider the hydrolysis of the constituent cation and anion in each salt. In NH_4Cl, the NH_4^+ ion reacts with water to give an acidic solution:

$$NH_4^+(aq) + H_2O(\ell) \rightleftharpoons NH_3(aq) + H_3O^+(aq)$$

but Cl^- does not. In $NaC_2H_3O_2$ the Na^+ ion does not hydrolyze, but the acetate ion does to give a basic solution:

$$C_2H_3O_2^-(aq) + H_2O(\ell) \rightleftharpoons HC_2H_3O_2(aq) + OH^-(aq)$$

In $NH_4C_2H_3O_2$ both cation and anion hydrolyze, giving a nearly neutral solution.

Solve

NH_4Cl is dissolved in the yellow solution, $NaC_2H_3O_2$ is dissolved in the blue solution, and $NH_4C_2H_3O_2$ is dissolved in the near colorless solution.

Think About It

The relative magnitude of the K_a of NH_4^+ (5.7×10^{-10}) in comparison with the K_b of $C_2H_3O_2^-$ (5.7×10^{-10}) shows that both salts hydrolyze to the same extent and that we should expect a solution of $NH_4C_2H_3O_2$ to be approximately neutral.

16.9. **Collect and Organize**

For the solution shown in Figure P16.9, we are to determine whether it is a weak acid, weak base, or buffer and indicate which particles will change and how they will change upon the addition of NaOH.

Analyze

In the figure the black atoms are carbons, the white atoms are hydrogens, the red atoms are oxygens, and the orange atoms are sodium. The species present in this solution therefore are HCO_2^-, HCO_2H, and Na^+.

Solve

(a) Because the acid (HCO_2H) is present with its conjugate base ($NaHCO_2$), this solution is a buffer.
(b) Upon addition of NaOH, the acid HCO_2H will react to form H_2O and the conjugate base, HCO_2^-.

Think About It

That solution was prepared by dissolving $NaCO_2H$ in a solution of HCO_2H, making a buffer. If the figure did not show Na^+ ions, we would not have a buffer solution; the presence of HCO_2^- would be due only to the hydrolysis of HCO_2H.

16.11. **Collect and Organize**

We consider why a solution composed of a weak acid with its conjugate base controls the pH better than a solution of only the weak acid.

Analyze

A buffer acts to absorb both OH^- and H^+ added to the solution.

Solve

The presence of both the acid and the conjugate base in a buffer means that the system can absorb both added H^+ and OH^-. Because the concentration of the conjugate base (present only because of a small degree of hydrolysis of the base) is too small, a buffer composed only of a weak acid cannot absorb much H_3O^+.

Think About It

A good buffer has relatively high and roughly equal concentrations of the weak acid and its conjugate base.

16.13. **Collect and Organize**
We are asked to identify a suitable buffer system for pH = 3.0.

Analyze
A good buffer system has the pK_a within one pH unit of the desired pH. We can use Appendix 5 to identify some potential acid–conjugate base systems that have pK_as around 3.0.

Solve
For an effective buffer pH = $pK_a \pm 1$. Buffer systems that have the pK_a in the range of 2.0–4.0 include ascorbic acid–ascorbate (pK_a = 4.04), bromoacetic acid–bromoacetate (pK_a = 2.70), chloroacetic acid–chloroacetate (pK_a = 2.85), citric acid–citrate (pK_a = 3.13), fluoroacetic acid–fluoroacetate (pK_a = 2.59), formic acid–formate (pK_a = 3.75), hydrofluoric acid–fluoride (pK_a = 3.17), iodoacetic acid–iodoacetate (pK_a = 3.12), lactic acid–lactate (pK_a = 3.85), malonic acid–malonate (pK_a = 2.82), nitrous acid–nitrite (pK_a = 3.40), phosphoric acid–phosphate (pK_a = 2.16), and pyruvic acid–pyruvate (pK_a = 2.55). Of those, the best choice according to the pH criteria would be HF–NaF.

Think About It
HF, however, is highly toxic, so it is more likely that another system such as citric acid–citrate would be chosen.

16.15. **Collect, Organize, and Analyze**
We are asked to define *buffer capacity*.

Solve
Buffer capacity is the amount of acid or base that a buffer can neutralize while maintaining its pH within a specific desired range.

Think About It
The more concentrated the buffer, the more acid or base it can absorb and the greater its buffer capacity.

16.17. **Collect and Organize**
For buffers made from formic acid, hydrofluoric acid, and acetic acid with the sodium salts of their conjugate bases, we are to rank them in order of decreasing pH.

Analyze
For a buffer the pH $\approx$ pK_a. The pK_a values of those acids are formic acid, 3.75; hydrofluoric acid, 3.17; and acetic acid, 4.75.

Solve
In order of decreasing pH: acetic acid–sodium acetate > formic acid–sodium formate > hydrofluoric acid–sodium fluoride.

Think About It
All those buffers, however, are in the acidic range.

16.19. **Collect and Organize**
We are asked how diluting a buffer of pH = 4.00 with an equal volume of water (to double the total volume) affects the pH of the buffer solution.

Analyze
Diluting the buffer will affect the concentration of the acid and its conjugate base equally.

Solve
No significant change will occur in the pH of the buffer.

Think About It
The pH of the buffer can be changed by having different concentrations of the acid and its conjugate base, however. It is only the case that pH = pK_a when [base] = [acid] in the Henderson–Hasselbalch equation.

$$pH = pK_a + \log \frac{[base]}{[acid]}$$

16.21. Collect and Organize
For a buffer that is 0.200 *M* chloroacetic acid and 0.100 *M* sodium chloroacetate, we can use the Henderson–Hasselbalch equation to calculate the buffer's pH (pK_a for chloroacetic acid = 2.85).

Analyze
The Henderson–Hasselbalch equation is

$$pH = pK_a + \log \frac{[base]}{[acid]}$$

For this problem, [base] = [sodium chloroacetate] = 0.100 *M* and [acid] = [chloroacetic acid] = 0.200 *M*.

Solve

$$pH = 2.85 + \log \frac{0.100}{0.200} = 2.55$$

Think About It
If we were to increase the concentration of base in relation to the acid, the pH of the solution would increase because the value of log([base]/[acid]) in the Henderson–Hasselbalch equation becomes more positive.

16.23. Collect and Organize
For a buffer that is 0.110 *M* HPO_4^{2-} and 0.220 *M* $H_2PO_4^-$, we can use the Henderson–Hasselbalch equation to calculate the buffer's pH (pK_a for $H_2PO_4^-$ = 7.19).

Analyze
The Henderson–Hasselbalch equation is

$$pH = pK_a + \log \frac{[base]}{[acid]}$$

For this problem, [base] = [HPO_4^{2-}] = 0.110 *M* and [acid] = [$H_2PO_4^-$] = 0.220 *M*.

Solve

$$pH = 7.19 + \log \frac{0.110}{0.220} = 6.89$$

Think About It
Be careful here to correctly identify the species that is the acid and which is the conjugate base; if you switch them, the pH will be higher rather than lower than 7.19.

16.25. Collect and Organize
Given the pH of an acetic acid–sodium acetate buffer solution (pH = 5.75) and where the pK_a of acetic acid is 4.75, we can use the Henderson–Hasselbalch equation to calculate the ratio of sodium acetate to acetic acid in the buffer solution.

Analyze

Rearranging the Henderson–Hasselbalch equation to solve for the ratio gives

$$pH = pK_a + \log \frac{[\text{sodium acetate}]}{[\text{acetic acid}]}$$

$$\log \frac{[\text{sodium acetate}]}{[\text{acetic acid}]} = pH - pK_a$$

$$\frac{[\text{sodium acetate}]}{[\text{acetic acid}]} = 1 \times 10^{(pH - pK_a)}$$

Solve

The pK_a = 4.75, so pH – pK_a = 5.75 – 4.75 = 1.00 and the ratio of acetate to acetic acid is

$$\frac{[\text{acetate}]}{[\text{acetic acid}]} = 1.00 \times 10^1 = 10$$

Think About It

Because [sodium acetate] > [acetic acid], the pH of that buffer is greater than the pK_a of acetic acid.

16.27. **Collect and Organize**

We are asked to compute the masses of bromoacetic acid and sodium bromoacetate needed to prepare 1.00 L of a buffer with pH = 3.00 given that the total concentration of the two components is 0.200 M.

Analyze

First we can use the Henderson–Hasselbalch equation to calculate the ratio of [sodium bromoacetate] to [bromoacetic acid]. The pK_a of bromoacetic acid is 2.70.

$$pH = pK_a + \log \frac{[\text{sodium bromoacetate}]}{[\text{bromoacetic acid}]}$$

$$\log \frac{[\text{sodium bromoacetate}]}{[\text{bromoacetic acid}]} = pH - pK_a$$

$$\frac{[\text{sodium bromoacetate}]}{[\text{bromoacetic acid}]} = 1 \times 10^{(pH - pK_a)}$$

Then we can use the fact that

$$[\text{sodium bromoacetate}] + [\text{bromoacetic acid}] = 0.200 \ M$$

to determine the concentrations of each component in the buffer solution. From there we will calculate the masses of each needed by using their molar masses and knowing that we are preparing a 1.00 L solution.

Solve

$$\frac{[\text{sodium bromoacetate}]}{[\text{bromoacetic acid}]} = 1 \times 10^{(pH - pK_a)} = 1 \times 10^{(3.00 - 2.70)} = 2.00$$

Therefore, [sodium bromoacetate] = 2 × [bromoacetic acid]

Combining that with [sodium bromoacetate] + [bromoacetic acid] = 0.200 M gives

$$2 \times [\text{bromoacetic acid}] + [\text{bromoacetic acid}] = 0.200 \ M$$

which reduces to

$$3 \times [\text{bromoacetic acid}] = 0.200 \ M$$
$$[\text{bromoacetic acid}] = 0.0667 \ M$$

The concentrations of each component are [bromoacetic acid] = 0.0667 M, [sodium bromoacetate] = 0.1333 M. For the buffer solution we need then:

$$1.00 \text{ L} \times \frac{0.0667 \text{ mol}}{\text{L}} \times \frac{138.95 \text{ g}}{\text{mol}} = 9.26 \text{ g bromoacetic acid}$$

$$1.00 \text{ L} \times \frac{0.1333 \text{ mol}}{\text{L}} \times \frac{160.93 \text{ g}}{\text{mol}} = 21.46 \text{ g sodium bromoacetate}$$

Think About It

That is a lot of each component to dissolve, but that is the result of their high molar masses.

16.29. Collect and Organize

We are asked to compute the masses of dimethylamine and dimethylammonium chloride needed to prepare 0.500 L of a buffer with pH = 12.00 given that the total concentration of the two components is 0.300 M.

Analyze

The pK_a for use in the Henderson–Hasselbalch equation of dimethylammonium chloride is

$$pK_a = 14.00 - pK_b = 14.00 - 3.23 = 10.77$$

First we can use the Henderson–Hasselbalch equation to calculate the ratio of [dimethylamine] to [dimethylammonium chloride].

$$pH = pK_a + \log \frac{[\text{dimethylamine}]}{[\text{dimethylammonium chloride}]}$$

$$\log \frac{[\text{dimethylamine}]}{[\text{dimethylammonium chloride}]} = pH - pK_a$$

$$\frac{[\text{dimethylamine}]}{[\text{dimethylammonium chloride}]} = 1 \times 10^{(pH - pK_a)}$$

Then we can use the fact that

$$[\text{dimethylamine}] + [\text{dimethylammonium chloride}] = 0.300 \text{ } M$$

to determine the concentrations of each component in the buffer solution. From there we will calculate the masses of each needed by using their molar masses and knowing that we are preparing a 0.500 L solution.

Solve

$$\frac{[\text{dimethylamine}]}{[\text{dimethylammonium chloride}]} = 1 \times 10^{(pH - pK_a)} = 1 \times 10^{(12.00 - 10.77)} = 1.70$$

Therefore, [dimethylamine] = 1.70 × [dimethylammonium chloride]
Combining that with [dimethylamine] + [dimethylammonium chloride] = 0.300 M gives

$$1.70 \times [\text{dimethylammonium chloride}] + [\text{dimethylammonium chloride}] = 0.300 \text{ } M$$

which reduces to

$$2.70 \times [\text{dimethylammonium chloride}] = 0.300 \text{ } M$$
$$[\text{dimethylammonium chloride}] = 0.111 \text{ } M$$

The concentrations of each component are [dimethylammonium chloride] = 0.111 M, [sodium acetate] = 0.189 M. For the buffer solution we need then:

$$0.500 \text{ L} \times \frac{0.111 \text{ mol}}{\text{L}} \times \frac{81.55 \text{ g}}{\text{mol}} = 4.53 \text{ g dimethylammonium chloride}$$

$$0.500 \text{ L} \times \frac{0.189 \text{ mol}}{\text{L}} \times \frac{45.08 \text{ g}}{\text{mol}} = 4.26 \text{ g dimethylamine}$$

Think About It

Dimethylamine is a liquid at normal temperatures, so we would probably measure it for that buffer in volume.

$$4.26 \text{ g} \times \frac{\text{mL}}{0.650 \text{ g}} = 6.55 \text{ mL}$$

16.31. **Collect and Organize**

We can use the Henderson–Hasselbalch equation to determine the pH of a solution prepared by mixing equal volumes of 0.05 M NH$_3$ and 0.025 M HCl.

Analyze

Mixing equal volumes of solutions dilutes them both. Therefore, after mixing and before reaction, [NH$_3$] = 0.025 M and [HCl] = 0.0125 M in the combined solution. When HCl reacts with NH$_3$, NH$_4^+$ is produced according to the equation

$$NH_3(aq) + H^+(aq) \rightarrow NH_4^+(aq)$$

Stoichiometrically, that would give [NH$_3$] = 0.0125 M and [NH$_4^+$] = 0.0125 M after complete reaction with H$^+$. Because that is a solution of an acid (NH$_4^+$) and its conjugate base (NH$_3$), we can use the Henderson–Hasselbalch equation to calculate the pH of the solution. To do so we need the pK_a of NH$_4^+$ (9.25) from Appendix 5 (Table A5.1).

Solve

$$pH = pK_a + \log\frac{[NH_3]}{[NH_4^+]} = 9.25 + \log\frac{0.0125}{0.0125} = 9.25$$

Think About It

Because the HCl added to the NH$_3$ in that solution converts exactly half of the NH$_3$ to NH$_4^+$, the ratio of the concentrations equals 1 and the pH of the solution equals the pK_a of NH$_4^+$.

16.33. **Collect and Organize**

We are to calculate the volume of 0.422 M NaOH that has to be added to 0.500 L of 0.300 M acetic acid to raise the pH of the solution to 4.00.

Analyze

We can use the Henderson–Hasselbalch equation to first calculate the ratio [acetate]/[acetic acid] needed to give a solution of pH 4.00. We then use that ratio to determine how much NaOH to add to convert acetic acid in solution to sodium acetate to give that ratio. The pK_a of acetic acid is 4.75.

Solve

$$pH = pK_a + \log\frac{[\text{sodium acetate}]}{[\text{acetic acid}]}$$

$$4.00 = 4.75 + \log\frac{[\text{sodium acetate}]}{[\text{acetic acid}]}$$

$$4.00 - 4.75 = \log\frac{[\text{sodium acetate}]}{[\text{acetic acid}]}$$

$$\frac{[\text{sodium acetate}]}{[\text{acetic acid}]} = 1 \times 10^{(4.00-4.75)} = 0.1778$$

The [sodium acetate] = 0.1778 × [acetic acid]
The total concentration of acetic acid and sodium acetate in the solution is

$$[\text{acetic acid}] + [\text{sodium acetate}] = 0.300$$

Therefore, combining those two expressions gives

$$[\text{acetic acid}] + 0.1778 \times [\text{acetic acid}] = 0.300$$
$$1.1778 \times [\text{acetic acid}] = 0.300$$
$$[\text{acetic acid}] = 0.2547 \ M$$
$$\text{so [sodium acetate]} = 0.178 \times [\text{acetic acid}] = 0.04534 \ M$$

To obtain that concentration of 0.04534 M sodium acetate in 0.500 L:

$$0.500 \text{ L} \times \frac{0.04534 \text{ mol}}{\text{L}} = 0.02267 \text{ mol sodium acetate}$$

$$0.02267 \text{ mol sodium acetate} \times \frac{1 \text{ mol NaOH}}{1 \text{ mol sodium acetate}} \times \frac{1 \text{ L}}{0.422 \text{ mol NaOH}} = 0.05372 \text{ L, or } 53.72 \text{ mL}$$

Think About It
The relatively small-volume addition of NaOH does not appreciably change the concentration of the acid and the base, so we may ignore the change of volume there.

16.35. Collect and Organize
We are to compare the pH of 1.00 L of a buffer that is 0.120 M HNO_2 and 0.150 M $NaNO_2$ before and after 1.00 mL of 11.6 M HCl is added.

Analyze
We can use the Henderson–Hasselbalch equation in both cases. After the addition of HCl, however, the amounts (calculated in moles) of HNO_2 and NO_2^- have to be adjusted before using the Henderson–Hasselbalch equation. The pK_a of HNO_2 is 3.40.

Solve
(a) Without added HCl, the pH of the buffer solution is

$$pH = 3.40 + \log \frac{0.150}{0.120} = 3.50$$

(b) Because we have 1.00 L of the buffer solution, we originally have 0.120 mol of HNO_2 and 0.150 mol of NO_2^- in the solution. Adding 1.00 mL of 11.6 M HCl adds

$$1.00 \text{ mL} \times \frac{11.6 \text{ mol}}{1000 \text{ L}} = 0.0116 \text{ mol H}^+$$

That will increase the moles of HNO_2 to 0.120 mol + 0.0116 mol = 0.132 mol and decrease the moles of NO_2^- to 0.150 mol − 0.0116 mol = 0.138 mol. Because the volume of the solution is 1.00 L, the concentrations of those species are $[HNO_2] = 0.132$ M and $[NO_2^-] = 0.138$ M. Using the Henderson–Hasselbalch equation to calculate the pH gives

$$pH = 3.40 + \log \frac{0.138}{0.132} = 3.42$$

Think About It
The pH of the buffer changed very little. The change in pH of 1.00 L of water after adding 0.0120 mol of H^+ would be from 7.00 to 1.92.

16.37. Collect and Organize
We are asked whether the pH at the equivalence point in all titrations of a strong base with a strong monoprotic acid is the same.

Analyze
A strong base such as OH^- (for example, from NaOH or KOH), when titrated with strong acid, produces water plus a dissolved salt.

Solve
Yes. The pH at the equivalence point of strong bases during titrations with strong acids is always pH 7.00, as seen by the formation of H_2O and salts that do not hydrolyze, as shown in the following examples:

$$NaOH(aq) + HCl(aq) \rightarrow H_2O(\ell) + Na^+(aq) + Cl^-(aq)$$
$$Ca(OH)_2(aq) + 2 \, HNO_3(aq) \rightarrow 2 \, H_2O(\ell) + Ca^{2+}(aq) + 2 \, NO_3^-(aq)$$
$$Na_2O(aq) + 2 \, HClO_4(aq) \rightarrow H_2O(\ell) + 2 \, Na^+(aq) + 2 \, ClO_4^-(aq)$$

Think About It

Likewise, the equivalence point is at pH 7.00 for all titrations of strong acids with strong bases.

16.39. Collect and Organize

We are to describe two properties of phenolphthalein that makes it a good acid–base indicator for the first equivalence point in an alkalinity titration.

Analyze

An indicator is a visual indicator of the pH of a solution.

Solve

Phenolphthalein changes color near the pH of the first equivalence point and it is colorless at low pH, which means that it will not obscure the color change of a second indicator added after the first equivalence point to detect the second equivalence point.

Think About It

In using an indicator in a titration we have to be careful to choose an indicator that changes color in the pH range of the expected equivalence point.

16.41. Collect and Organize

Given that the half-equivalence point for a titration is pH 4.44, we are asked to predict the pH at the half-equivalence point for a titration for a weak acid solution in which twice as much NaOH is needed to reach the equivalence point.

Analyze

The half-equivalence point is where the pH = pK_a.

Solve

Because the identity of the weak acid is not changed and the pH at the equivalence point is equal to the pK_a of the acid, the pH at the half-equivalence point for both solutions of weak acid is 4.44.

Think About It

The equivalence point for the second solution requires twice the volume of NaOH solution because the concentration of the weak acid in that solution is twice that of the first solution.

16.43. Collect and Organize

We are to calculate the pH along various points of the titration curve when 25.0 mL of 0.100 M acetic acid ($K_a = 1.76 \times 10^{-5}$) is titrated with 0.125 M NaOH.

Analyze

For each step of the titration we have to consider the moles of NaOH added that react with the moles of CH_3COOH initially present:

$$25.0 \text{ mL} \times \frac{0.100 \text{ mol}}{1000 \text{ mL}} = 0.00250 \text{ mol CH}_3\text{COOH}$$

For each point in the titration curve, the moles of OH^- (from NaOH) added are as follows:

$$10.0 \text{ mL} \times \frac{0.125 \text{ mol}}{1000 \text{ mL}} = 0.00125 \text{ mol OH}^-$$

$$20.0 \text{ mL} \times \frac{0.125 \text{ mol}}{1000 \text{ mL}} = 0.00250 \text{ mol OH}^-$$

$$30.0 \text{ mL} \times \frac{0.125 \text{ mol}}{1000 \text{ mL}} = 0.00375 \text{ mol OH}^-$$

The OH⁻ reacts with the CH_3COOH in solution to give CH_3COO^-, the conjugate base of acetic acid. Our strategy for the problem is to first react as much of the added OH⁻ with acetic acid as possible and then use the equilibrium K_a expression to calculate the pH:

$$CH_3COOH(aq) \rightleftharpoons CH_3COO^-(aq) + H^+(aq)$$

$$K_a = \frac{[CH_3COO^-][H^+]}{[CH_3COOH]}$$

Or we can use the equivalent equilibrium K_b expression:

$$CH_3COO^-(aq) + H_2O(\ell) \rightleftharpoons CH_3COOH(aq) + OH^-(aq)$$

$$K_b = \frac{[CH_3COOH][OH^-]}{[CH_3COO^-]}$$

For that calculation we have to be careful to determine the molarity of the species in solution by remembering that the volume in a titration increases through the addition of the titrant.

Solve

When 10.0 mL of OH⁻ is added, the 0.00125 mol of OH⁻ reacts with the 0.00250 mol of CH_3COOH to produce 0.00125 mol of CH_3COO^- and leaves 0.00125 mol of CH_3COOH unreacted. Because the total volume of the solution is now 25.0 + 10.0 = 35.0 mL, the molarity of those species is

$$\frac{0.00125 \text{ mol}}{0.0350 \text{ L}} = 0.0357 \text{ M}$$

Using a RICE table to calculate the pH of this solution gives the following:

Reaction	$CH_3COOH(aq)$	$\rightleftharpoons$	$CH_3COO^-(aq)$	+	$H^+(aq)$
	[CH₃COOH]		[CH₃COO⁻]		[H⁺]
Initial	0.0357		0.0357		0
Change	$-x$		$+x$		$+x$
Equilibrium	$0.0357 - x$		$0.0357 + x$		x

$$1.76 \times 10^{-5} = \frac{x(0.0357 + x)}{0.0357 - x} \approx \frac{x(0.0357)}{0.0357}$$

$$x = 1.76 \times 10^{-5}$$

The simplifying assumption is valid (0.05%), so

$$pH = -\log 1.76 \times 10^{-5} = 4.754$$

When 20.0 mL of OH⁻ is added, we have added an equal number of moles of OH⁻ as moles of CH_3COOH initially present. That reaction produces 0.00250 mol of CH_3COO^-, so using the K_b expression to calculate the pH of the solution makes sense here. Because the total volume of the solution is now 45.0 mL, the molarity of those species is

$$\frac{0.00250 \text{ mol}}{0.0450 \text{ L}} = 0.0556 \text{ M}$$

Using a RICE table to calculate the pH of this solution gives the following:

Reaction	$CH_3COO^-(aq) + H_2O(\ell) \rightleftharpoons$	$CH_3COOH(aq)$	+	$OH^-(aq)$
	[CH₃COO⁻]	[CH₃COOH]		[OH⁻]
Initial	0.0556	0		0
Change	$-x$	$+x$		$+x$
Equilibrium	$0.0556 - x$	x		x

$$K_b = \frac{K_w}{K_a} = \frac{1 \times 10^{-14}}{1.76 \times 10^{-5}} = 5.68 \times 10^{-10} = \frac{x^2}{0.0556 - x} \approx \frac{x^2}{0.0556}$$

$$x = 5.62 \times 10^{-6}$$

The simplifying assumption is valid (0.01%), so

$$pOH = -\log 5.62 \times 10^{-6} = 5.250$$
$$pH = 14 - 5.250 = 8.750$$

When 30.0 mL of OH^- is added, we convert all the CH_3COOH to 0.00250 mol of CH_3COO^- and have 0.00125 mol of OH^- remaining. Because the total volume of the solution is now 55.0 mL, the molarity of those species is

$$\frac{0.00250 \text{ mol}}{0.0550 \text{ L}} = 0.0455 \ M \ CH_3COO^-$$

$$\frac{0.00125 \text{ mol}}{0.0550 \text{ L}} = 0.0227 \ M \ OH^-$$

Using a RICE table to calculate the pH of this solution gives the following:

Reaction	$CH_3COO^-(aq) + H_2O(\ell) \rightleftharpoons CH_3COOH(aq)$		$+ \quad OH^-(aq)$
	$[CH_3COO^-]$	$[CH_3COOH]$	$[OH^-]$
Initial	0.0455	0	0.0227
Change	$-x$	$+x$	$+x$
Equilibrium	$0.0455 - x$	x	$0.0227 + x$

$$5.68 \times 10^{-10} = \frac{x(0.0227 + x)}{0.0455 - x} \approx \frac{x(0.0227)}{0.0455}$$

$$x = 1.14 \times 10^{-9}$$

The simplifying assumption is valid, so

$$pOH = -\log 0.0227 = 1.644$$
$$pH = 14 - 1.644 = 12.356$$

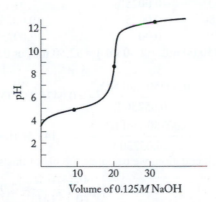

Volume of 0.125M NaOH

Think About It
When exactly half the moles of strong base are added in the titration of a weak acid (as in this problem, in which 10 mL of the titrant was added), this point is the midpoint of the titration. At this point pH = pK_a of the weak acid.

16.45. Collect and Organize
We are to calculate the pH and sketch the titration curve for the titration of 50.0 mL of a 0.200 M solution of the weak acid HNO_2 with 1.00 M NaOH for several points along the titration.

Analyze
The value of K_a for HNO_2 is 4.0×10^{-4} ($pK_a = 3.398$). First we set up a RICE table where the initial concentrations of H^+ and NO_2^- are 0.00 M. For each step of the titration we have to consider the moles of NaOH added that react with the moles of HNO_2 initially present:

$$50.0 \text{ mL} \times \frac{0.200 \text{ mol}}{1000 \text{ mL}} = 0.0100 \text{ mol } HNO_2$$

For each point in the titration curve the moles of OH^- (from NaOH) added are as follows:

$$2.50 \text{ mL} \times \frac{1.00 \text{ mol}}{1000 \text{ mL}} = 0.00250 \text{ mol } OH^-$$

$$5.00 \text{ mL} \times \frac{1.00 \text{ mol}}{1000 \text{ mL}} = 0.00500 \text{ mol } OH^-$$

$$7.50 \text{ mL} \times \frac{1.00 \text{ mol}}{1000 \text{ mL}} = 0.00750 \text{ mol } OH^-$$

$$10.00 \text{ mL} \times \frac{1.00 \text{ mol}}{1000 \text{ mL}} = 0.01000 \text{ mol } OH^-$$

The OH^- reacts with the HNO_2 in solution to NO_2^-, the conjugate acid of HNO_2. Our strategy for the problem is to first react as much of the added OH^- with HNO_2 as possible and then use the Henderson–Hasselbalch equation to calculate the pH:

$$pH = pK_a + \log \frac{[NO_2^-]}{[HNO_2]}$$

Solve

When 2.50 mL of OH^- is added, the 0.00250 mol of OH^- reacts with the 0.0100 mol of HNO_2. Using a RICE table to determine the mol of HNO_2 and NO_2^- after the reaction of HNO_2 with OH^-:

Reaction	$HNO_2(aq)$ +	$OH^-(aq)$ →	$NO_2^-(aq)$ +	$H_2O(\ell)$
	HNO_2 (moles)	OH^- (moles)	NO_2^- (moles)	
Initial	0.0100	0.00250	0	
Change	−0.00250	−0.0025	+0.00250	
Final	0.00750	0.00	0.00250	

The total volume of the sample (50.00 mL + 2.50 mL) = 52.50 mL, or 0.0520 L, and the concentrations of NO_2^- and HNO_2 are

$$\frac{0.00250 \text{ mol } NO_2^-}{0.05250 \text{ L}} = 0.0476 \text{ } M \text{ } NO_2^-$$

$$\frac{0.00750 \text{ mol } HNO_2^-}{0.05250 \text{ L}} = 0.143 \text{ } M \text{ } HNO_2$$

Using the Henderson–Hasselbalch equation, we calculate the pH of this solution:

$$pH = 3.398 + \log \frac{0.0476 \text{ } M}{0.143 \text{ } M} = 2.92$$

When 5.00 mL of OH^- is added, the 0.00500 mol of OH^- reacts with the 0.0100 mol of HNO_2. Using a RICE table to determine the mol of HNO_2 and NO_2^- after the reaction of HNO_2 with OH^-:

Reaction	$HNO_2(aq)$ +	$OH^-(aq)$ →	$NO_2^-(aq)$ +	$H_2O(\ell)$
	HNO_2 (moles)	OH^- (moles)	NO_2^- (moles)	
Initial	0.0100	0.00500	0	
Change	−0.00500	−0.00500	+0.00500	
Final	0.00500	0.00	0.00500	

The total volume of the sample (50.00 mL + 5.00 mL) = 55.00 mL, or 0.0550 L, and the concentrations of NO_2^- and HNO_2 are

$$\frac{0.00500 \text{ mol } NO_2^-}{0.05500 \text{ L}} = 0.0909 \text{ } M \text{ } NO_2^-$$

$$\frac{0.00500 \text{ mol } HNO_2^-}{0.05500 \text{ L}} = 0.0909 \text{ } M \text{ } HNO_2$$

Using the Henderson–Hasselbalch equation, we calculate the pH of this solution:

$$pH = 3.398 + \log \frac{0.0909 \text{ } M}{0.0909 \text{ } M} = 3.40$$

When 7.50 mL of OH⁻ is added, the 0.00750 mol of OH⁻ reacts with the 0.0100 mol of HNO_2. Using a RICE table to determine the mol of HNO_2 and NO_2^- after the reaction of HNO_2 with OH⁻:

Reaction	$HNO_2(aq)$ +	$OH^-(aq)$ →	$NO_2^-(aq)$ +	$H_2O(\ell)$
	HNO_2 (moles)	OH^- (moles)	NO_2^- (moles)	
Initial	0.0100	0.00750	0	
Change	−0.00750	−0.00750	+0.00750	
Final	0.00250	0.00	0.00750	

The total volume of the sample (50.00 mL + 7.50 mL) = 57.50 mL, or 0.0575 L, and the concentrations of NO_2^- and HNO_2 are

$$\frac{0.00750 \text{ mol } NO_2^-}{0.05750 \text{ L}} = 0.1304 \ M \ NO_2^-$$

$$\frac{0.00250 \text{ mol } HNO_2^-}{0.05750 \text{ L}} = 0.0435 \ M \ HNO_2$$

Using the Henderson–Hasselbalch equation, we calculate the pH of this solution:

$$pH = 3.398 + \log\frac{0.1304 \ M}{0.0435 \ M} = 3.87$$

When 10.00 mL of OH⁻ is added, the 0.0100 mol of OH⁻ reacts with the 0.0100 mol of HNO_2. That is the equivalence point in the titration. Using a RICE table to determine the mol of HNO_2 and NO_2^- after the reaction of HNO_2 with OH⁻:

Reaction	$HNO_2(aq)$ +	$OH^-(aq)$ →	$NO_2^-(aq)$ +	$H_2O(\ell)$
	HNO_2 (moles)	OH^- (moles)	NO_2^- (moles)	
Initial	0.0100	0.0100	0	
Change	−0.0100	−0.01000	+0.0100	
Final	0.000 0	0.00	0.0100	

The total volume of the sample (50.00 mL + 10.00 mL) = 60.00 mL, or 0.06000 L, and the concentration of NO_2^- is

$$\frac{0.01000 \text{ mol } NO_2^-}{0.06000 \text{ L}} = 0.167 \ M \ NO_2^-$$

The solution now contains only NO_2^-, which will partially hydrolyze according to the equilibrium expression

$$NO_2^-(aq) + H_2O(\ell) \rightleftharpoons HNO_2(aq) + OH^-(aq)$$

Using a RICE table to set up the equilibrium expression:

Reaction	$NO_2^-(aq) + H_2O(\ell)$ ⇌	$HNO_2(aq)$ +	$OH^-(aq)$
	$[NO_2^-]$	$[HNO_2]$	$[OH^-]$
Initial	0.167	0.0	0.0
Change	−x	+x	+x
Equilibrium	0.167 − x	x	x

The pH of the solution can be found through the K_b expression for NO_2^-:

$$K_b = \frac{K_w}{K_a} = \frac{1 \times 10^{-14}}{4.0 \times 10^{-4}} = 2.50 \times 10^{-11} = \frac{x^2}{0.167 - x} \approx \frac{x^2}{0.167}$$

$$x = 2.04 \times 10^{-6}$$

$$pOH = -\log(2.04 \times 10^{-6}) = 5.69$$

$$pH = 14.00 - 5.69 = 8.31$$

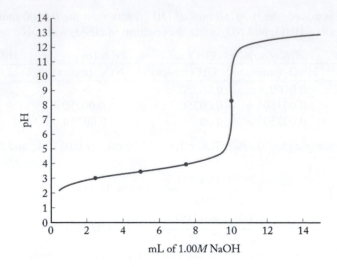

mL of 1.00*M* NaOH

Think About It
Notice that at the half-equivalence point (5.00 mL of titrant), the pH = pK_a.

16.47. Collect and Organize
We are asked to calculate the volume of 0.100 *M* HCl that would be required to titrate 250 mL of 0.0100 *M* Na$_2$CO$_3$ to the first equivalence point.

Analyze
To titrate to the first equivalence point we will need one molar equivalent of HCl.

Solve
$$250 \text{ mL Na}_2\text{CO}_3 \times \frac{0.0100 \text{ mol Na}_2\text{CO}_3}{1000 \text{ mL}} \times \frac{1 \text{ mol HCl}}{1 \text{ mol Na}_2\text{CO}_3} \times \frac{1000 \text{ mL}}{0.100 \text{ mol HCl}} = 25.0 \text{ mL}$$

Think About It
Although we showed the full calculation above, you might have seen immediately from reading the problem that because the concentrations of the titrant was ten times the concentration of the sample were equal, one-tenth of the volume would be needed to reach the first equivalence point. Likewise, you might have guessed that to reach the second equivalence point would require 50 mL of the titrant.

16.49. Collect and Organize
We are to calculate the concentration of ammonia in a window cleaner sample if 25.34 mL of 1.162 *M* HCl is required to titrate a 10.00 mL sample and then to consider whether the volume of titrant would change if we diluted the 10.00 mL sample to 50 mL before performing the titration.

Analyze
At the titration (equivalence) point, the moles of ammonia present will equal the moles of titrant added. Once we know the moles of ammonia present in the sample, we calculate the concentration of the sample by dividing the moles of ammonia by the original volume of the sample.

Solve
(a) The concentration of ammonia in the sample is
$$25.34 \text{ mL HCl} \times \frac{1.162 \text{ mol HCl}}{1000 \text{ mL}} \times \frac{1 \text{ mol NH}_3}{1 \text{ mol HCl}} \times \frac{1}{0.01000 \text{ L}} = 2.945 \text{ } M$$

(b) Because the titrant is reacting with the moles of NH_3 present in the sample and is unrelated to the concentration of the sample solution, there is no effect on the volume of the titrant needed for this titration if we dilute the sample.

Think About It
If you dilute the sample as in (b), you have to remember to divide the total number of moles of the species in the sample solution by the original volume of the sample, not the volume of the diluted solution that is titrated.

16.51. Collect and Organize
For each titration described, we are to predict whether the pH at the equivalence point is less than 7, equal to 7, or greater than 7.

Analyze
Because the equivalence point produces 100% of the conjugate acid or conjugate base in the titration, the pH of the solution is determined by the hydrolysis reactions (if any) of that conjugate acid or base. Therefore, the titration of a weak acid (HA) produces the conjugate base (A^-) that reacts with water to produce the weak acid and OH^-; the solution is basic at the equivalence point. The titration of a weak base (B) produces the conjugate acid (BH^+) that reacts with water to produce the weak base and H_3O^+; the solution is acidic at the equivalence point. If the conjugate acid or base produced is the conjugate of a strong acid or base, then no hydrolysis occurs and the solution is neutral.

Solve
(a) Quinine (a weak base) titrated with HCl will have a pH < 7 at the equivalence point.
(b) Pyruvic acid (a weak acid) titrated with $Ca(OH)_2$ will have a pH > 7 at the equivalence point.
(c) Hydrobromic acid (a strong acid) titrated with $Sr(OH)_2$ will have a pH = 7 at the equivalence point.

Think About It
The titration of HBr with a strong base produces Br^- and water. Br^- is a very, very weak base and does not react with water.

16.53. Collect and Organize
For the titration of a 100 mL sample of 0.0125 M ascorbic acid with 0.010 M NaOH, we are asked to determine how many equivalence points the titration curve would have and to choose appropriate pH indicators for the titration.

Analyze
Ascorbic acid is a diprotic acid. We need to calculate the pH of each equivalence point and then, using Figure 16.5 in the textbook, choose an indicator that changes color around the pH for each equivalence point. The moles of ascorbic acid being titrated in this sample is

$$100 \text{ mL} \times \frac{0.0125 \text{ mol}}{1000 \text{ mL}} = 1.25 \times 10^{-3} \text{ mol}$$

Solve
For the first equivalence point the volume of titrant required is

$$1.25 \times 10^{-3} \text{ mol} \times \frac{1000 \text{ mL}}{0.010 \text{ mol}} = 125 \text{ mL}$$

The total volume of the solution at the first equivalence point is 100 mL + 125 mL = 225 mL, or 0.225 L. At the equivalence point all the ascorbic acid (H_2AA) has been converted to the conjugate base HAA^- and the concentration of HAA^- will be

$$\frac{1.25 \times 10^{-3} \text{ mol}}{0.225 \text{ L}} = 5.56 \times 10^{-3} \text{ } M$$

That reacts with water to form H_2AA and OH^-, which gives the equilibrium expression

$$K_b = \frac{K_w}{K_a} = \frac{1\times10^{-14}}{9.11\times10^{-5}} = 1.10\times10^{-10} = \frac{x^2}{0.00556-x} \approx \frac{x^2}{0.00556}$$

$$x = 7.81\times10^{-7}$$

$$pOH = -\log(7.81\times10^{-7}) = 6.11$$

$$pH = 14.00 - 6.11 = 7.89$$

Phenolphthalein or phenol red would be an appropriate pH color indicator for that equivalence point.

For the second equivalence point, the volume of titrant required would be twice that of the first equivalence point, or 250 mL. The total volume of the solution at the first equivalence point is 100 mL + 250 mL = 350 mL, or 0.350 L. At the equivalence point all the deprotonated ascorbic acid (HAA^-) has been converted to the conjugate base AA^{2-} and the concentration of HAA^{2-} will be

$$\frac{1.25\times10^{-3} \text{ mol}}{0.350 \text{ L}} = 3.57\times10^{-3} \ M$$

That reacts with water to form HAA^- and OH^-, which gives the equilibrium expression

$$K_b = \frac{K_w}{K_a} = \frac{1\times10^{-14}}{5\times10^{-12}} = 2\times10^{-3} = \frac{x^2}{0.00357-x}$$

$$2\times10^{-3}\times(0.00357-x) = x^2$$

$$x^2 + 2\times10^{-3}x - 7.14\times10^{-6} = 0$$

$$x = 1.85\times10^{-3}$$

$$pOH = -\log(1.85\times10^{-3}) = 2.73$$

$$pH = 14.00 - 2.73 = 11.27$$

Alizarin yellow R would be an appropriate pH color indicator for that equivalence point.

Think About It
The color changes we might expect to see in that titration are colorless to pink and then pink to red.

16.55. Collect and Organize
We are asked whether all Lewis bases (electron-pair donators) are also Brønsted–Lowry bases (proton acceptors).

Analyze
The difference between a Lewis base and a Brønsted–Lowry base hinges on what the base donates or accepts. For Lewis bases, an electron pair is donated to the acid; for Brønsted–Lowry bases, a proton is accepted by the base.

Solve
No. A substance could be a Lewis base without being a Brønsted–Lowry base if it can donate an electron pair but does not accept a proton.

Think About It
$Cl^-(aq)$ is an example of a Lewis base that is not a Brønsted–Lowry base. It may share an electron pair, but it does not accept a proton. If it did accept a proton it would form HCl, which is 100% ionized in solution.

16.57. Collect and Organize
We are asked whether all Brønsted–Lowry acids (proton donors) are also Lewis acids (electron-pair acceptors).

Analyze

The difference between a Brønsted–Lowry acid and a Lewis acid hinges on what the acid donates or accepts. For a Brønsted–Lowry acid, a proton is donated to the base from the acid. For a Lewis acid, the acid accepts an electron pair.

Solve

Yes. Any substance that is a Brønsted–Lowry acid is a Lewis acid because H^+ is an electron-pair acceptor.

Think About It

Not all Lewis acids are Brønsted–Lowry acids, however.

16.59. **Collect and Organize**

After drawing the Lewis structures for water, we can show how the electron pairs move and bonds form in the formation of OH^- and H_3O^+, which are produced in the autoionization of water and then identify the Lewis acid and Lewis base in the reaction.

Analyze

The Lewis acid in the reaction of two water molecules will be the one that accepts H^+ from the Lewis base.

Solve

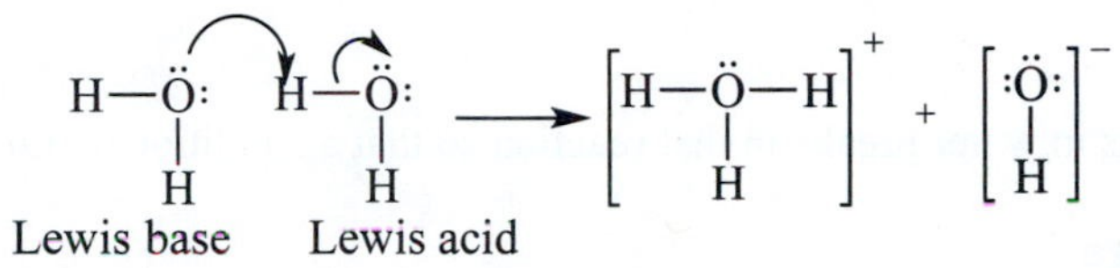

Think About It

Water is *amphoteric*; it can act as either an acid or a base.

16.61. **Collect and Organize**

After drawing the Lewis structures for SO_2, H_2O, and H_2SO_3, we can show how the electron pairs move and bonds form and break in the formation of H_2SO_3 and then identify the Lewis acid and Lewis base in the reaction.

Analyze

All the compounds have covalent bonding. A Lewis acid accepts an electron pair in an acid–base reaction, whereas a Lewis base donates an electron pair.

Solve

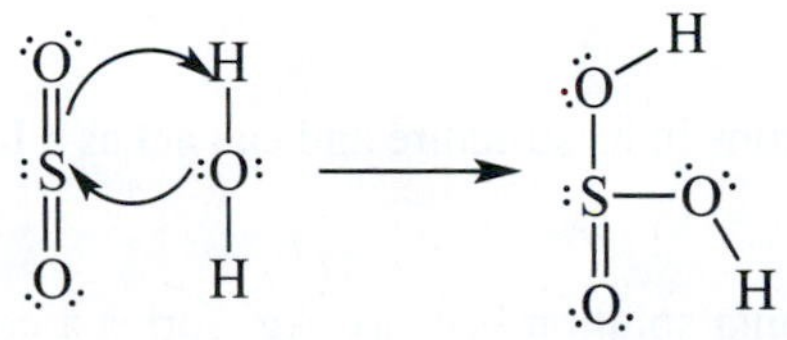

In that reaction SO_2 both accepts an electron pair (Lewis acid) and accepts a proton (Brønsted–Lowry base and Lewis base). Likewise, H_2O both donates an electron pair (Lewis base) and donates a proton (Brønsted–Lowry acid and Lewis acid). Both SO_2 and H_2O in that reaction act as both Lewis acids and Lewis bases.

Think About It

One of the O—H bonds in water breaks in that reaction so that two O—H bonds are present in the product, H_2SO_3.

16.63. Collect and Organize

After drawing the Lewis structures for $B(OH)_3$, H_2O, $B(OH)_4^-$, and H^+, we can show how the electron pairs move and bonds form and break in the formation of $B(OH)_4^-$ and then identify the Lewis acid and Lewis base in the reaction

Analyze

All the compounds have covalent bonding. A Lewis acid accepts an electron pair in an acid–base reaction, whereas a Lewis base donates an electron pair.

Solve

$B(OH)_3$ is the Lewis acid, and H_2O is the Lewis base.

Think About It

One of the O—H bonds in water breaks in that reaction so that an additional –OH group on B is formed.

16.65. Collect and Organize

We are asked which molecules or ions (H_2O or Cl^-) surround Ca^{2+} when $CaCl_2$ is dissolved in water.

Analyze

$CaCl_2$ is completely soluble in water; the Ca^{2+} and Cl^- ions are 100% dissociated.

Solve

Water molecules occupy the inner coordination sphere of Ca^{2+} ions.

Think About It

The oxygen atoms, which carry partial negative charge, are pointed towards the Ca^{2+} ion in the coordination sphere.

16.67. Collect and Organize

We are asked to explain why an ammonia solution dissolves insoluble (in water) AgCl.

Analyze

Ammonia has a lone pair of electrons in its structure and can act as a Lewis base to the Ag^+ ions.

Solve

AgCl dissolves in aqueous ammonia solution because Ag^+ forms a complex with ammonia, $Ag(NH_3)_2^+$, which is soluble.

Think About It

Rust stains, an insoluble iron oxide, will similarly dissolve when treated with sodium oxalate.

16.69. Collect and Organize

Given that a solution is prepared by dissolving 0.00100 mol of $Ni(NO_3)_2$ and 0.500 mol of NH_3 in 1.00 L, we are to calculate the concentration of $Ni(H_2O)_6^{2+}$ at equilibrium.

Analyze

When $Ni(NO_3)_2$ dissolves in water, $Ni(H_2O)_6^{2+}$ and NO_3^- ions form. The presence of NH_3 in the solution, however, forms the ammonia complex, $Ni(NH_3)_6^{2+}$, from the exchange of H_2O molecules on Ni^{2+} with NH_3 molecules as described by the equilibrium expression

$$Ni(H_2O)_6^{2+}(aq) + 6\ NH_3(aq) \rightleftharpoons Ni(NH_3)_6^{2+} + 6\ H_2O(\ell)$$

The formation constant for that reaction is $K_f = 5.5 \times 10^8$ and the formation constant expression is

$$K_f = \frac{[Ni(NH_3)_6^{2+}]}{[Ni(H_2O)_6^{2+}][NH_3]^6} = 5.5 \times 10^8$$

Because K_f is so large we can assume that 100% of the Ni^{2+} ions are converted to $Ni(NH_3)_6^{2+}$ complexes. We therefore consider the dissociation of $Ni(NH_3)_6^{2+}$ in solution according to the equilibrium equation and expression

$$Ni(NH_3)_6^{2+}(aq) \rightleftharpoons Ni(H_2O)_6^{2+} + 6\ NH_3(aq)$$

$$K_d = \frac{1}{K_f} = \frac{1}{5.5 \times 10^8} = 1.82 \times 10^{-9} = \frac{[Ni(H_2O)_6^{2+}][NH_3]^6}{[Ni(NH_3)_6^{2+}]}$$

Upon complete reaction we have 0.00100 mol of $Ni(NH_3)_6^{2+}$ and $[0.500\ \text{mol} - 6(0.00100\ \text{mol})] = 0.494$ mol of NH_3 (an excess of ammonia is present for that solution). Those are in 1.0 L volume, so the initial $[Ni(NH_3)_6^{2+}] = 0.00100\ M$ and the initial $[NH_3] = 0.494\ M$.

Solve

We set up a RICE table to solve this problem:

Reaction	$Ni(NH_3)_6^{2+}(aq)$	$\rightleftharpoons$	$Ni(H_2O)_6^{2+}(aq)$	$+$	$6\ NH_3(aq)$
	$[Ni(NH_3)_6^{2+}]$, M		$[Ni(H_2O)_6^{2+}]$, M		$[NH_3]$, M
Initial	0.00100		0.00		0.494
Change	$-x$		$+x$		$+6x$
Equilibrium	$0.00100 - x$		x		$0.494 + 6x$

$$1.82 \times 10^{-9} = \frac{[x][0.494 + 6x]^6}{[0.00100 - x]} \approx \frac{[x][0.494]^6}{[0.00100]}$$

$$x = 1.25 \times 10^{-10}$$

The $[Ni(H_2O)_6]^{2+}$ in the equilibrium solution is $1.25 \times 10^{-10}\ M$.

Think About It

In addition to the very small dissociation constant, the presence of excess ammonia in this solution further reduces the release of Ni^{2+} through a molecular version of the common ion effect.

16.71. **Collect and Organize**

For a 0.250 L solution of 0.00100 mol of $Co(NO_3)_2$, 0.100 mol of NH_3, and 0.100 mol of ethylenediamine, we are to calculate the concentration of $Co(H_2O)_6^{2+}$ in the solution at equilibrium.

Analyze

When $Co(NO_3)_2$ dissolves in water, $Co(H_2O)_6^{2+}$ and NO_3^- ions form. The presence of ammonia (NH_3) and ethylenediamine ($H_2NCH_2CH_2NH_2$, or "en," a chelating ligand) in the solution, however, means that two complexes will form, $Co(en)_3^{2+}$ and $Co(NH_3)_6^{2+}$ from the exchange of H_2O molecules on Cu^{2+} with three en molecules or six en molecules as described by the equilibrium expressions

$$Co(H_2O)_6^{2+}(aq) + 3\ en(aq) \rightleftharpoons Co(en)_3^{2+} + 6\ H_2O(\ell)$$

$$Co(H_2O)_6^{2+}(aq) + 6\ NH_3(aq) \rightleftharpoons Co(NH_3)_6^{2+} + 6\ H_2O(\ell)$$

The formation constant for those equilibria are $K_f(en) = 8.7 \times 10^{13}$ and $K_f(NH_3) = 7.7 \times 10^4$. Because K_f for the formation of the en complex is so large, we can assume that 100% of the Cu^{2+} ions are converted to $Co(en)_3^{2+}$ complexes. We therefore consider the reverse equation, the dissociation of $Co(en)_3^{2+}$ in solution, to be the overwhelming contributor to the concentration of Co^{2+} in the equilibrium solution and for which the equilibrium equation and expression are

$$Co(en)_3^{2+}(aq) \rightleftharpoons Co(H_2O)_6^{2+}(aq) + 3\ en(aq)$$

$$K_d = \frac{1}{K_f} = \frac{1}{8.7 \times 10^{13}} = 1.149 \times 10^{-13} = \frac{[Co(H_2O)_6^{2+}][en]^3}{[Co(en)_3^{2+}]}$$

Upon complete reaction of the en ligand with Co^{2+} in the mixture, we have 1.00×10^{-3} mol $Co(en)_3^{2+}$ and $[1.00 \times 10^{-1} - 3(1.00 \times 10^{-3} \text{ mol})] = 9.70 \times 10^{-2}$ mol en (this solution has an excess of ethylenediamine). Those are in 0.250 L volume, so the initial $[Co(en)_3^{2+}] = 4.00 \times 10^{-3}$ M and the initial $[en] = 3.88 \times 10^{-1}$ M.

Solve

We set up a RICE table to solve this problem:

Reaction	$Co(en)_3^{2+}(aq) \rightleftharpoons$	$Co(H_2O)_6^{2+}(aq)$	+	3 en(aq)
	$[Co(en)_3^{2+}]$, M	$[Co(H_2O)_6^{2+}]$, M		$[en]$, M
Initial	4.00×10^{-3}	0.00		3.88×10^{-1}
Change	$-x$	$+x$		$+3x$
Equilibrium	$4.00 \times 10^{-3} - x$	x		$3.88 \times 10^{-1} + 3x$

$$1.149 \times 10^{-14} = \frac{[x][3.88 \times 10^{-1} + 3x]^3}{[4.00 \times 10^{-3} - x]} \approx \frac{[x][3.88 \times 10^{-1}]^3}{[4.00 \times 10^{-3}]}$$

$$x = 7.87 \times 10^{-16}$$

The $[Co(H_2O)_6]^{2+}$ in the equilibrium solution is 7.87×10^{-16} M.

Think About It

The equilibrium concentration of $Co^{2+}(aq)$ is so small that it is, for practical purposes, zero.

16.73. Collect and Organize

For four chloride salts, we are to identify which would give an acidic solution.

Analyze

All the salts are soluble. The Cl^- does not hydrolyze, but the cations might.

Solve

(a) $Ca^{2+}(aq)$ does not hydrolyze because it would form the strong base $Ca(OH)_2$, which is 100% ionized in solution.

(b) $Cr^{3+}(aq)$ does hydrolyze to form an acidic solution according to the equation

$$Cr^{3+}(aq) + 3\ H_2O(\ell) \rightleftharpoons Cr(OH)_3(s) + 3\ H^+(aq)$$

(c) $Na^+(aq)$ does not hydrolyze because it would form the strong base NaOH, which is 100% ionized in solution.

(d) $Fe^{3+}(aq)$ does hydrolyze to form an acidic solution according to the equation

$$Fe^{3+}(aq) + 3\ H_2O(\ell) \rightleftharpoons Fe(OH)_3(s) + 3\ H^+(aq)$$

Both (b) $CrCl_3$ and (d) $FeCl_3$ produce an acidic solution.

Think About It

According to Table 16.1, Cr^{3+} is among the most acidic of the hydrated metal ions.

16.75. Collect and Organize

We are asked how the oxidation of Fe^{2+} to Fe^{3+} affects the acidity of the ozone solution. To answer this we are to compare the acidity of $Fe^{2+}(aq)$ with that of $Fe^{3+}(aq)$.

Analyze

Both $Fe^{2+}(aq)$ and $Fe^{3+}(aq)$ hydrolyze according to the equations

$$Fe^{2+}(aq) + 2\ H_2O(\ell) \rightleftharpoons Fe(OH)_2(s) + 2\ H^+(aq)$$
$$Fe^{3+}(aq) + 3\ H_2O(\ell) \rightleftharpoons Fe(OH)_3(s) + 3\ H^+(aq)$$

Solve

An aqueous solution of Fe^{3+} is more acidic and has a lower pH because Fe^{3+} hydrolyzes more than Fe^{2+}. The higher ionic charge of Fe^{3+} polarizes the O—H bonds in water bound to Fe^{3+} more than Fe^{2+} does.

Think About It

Fe^{3+} could also hydrolyze to produce 3 mol of H^+ rather than 2 mol of H^+ from Fe^{2+} hydrolysis.

16.77. Collect and Organize

For the amphiprotic $Cr(OH)_3$, we are to write chemical equations to describe this property.

Analyze

Amphiprotic compounds react with both acids and bases.

Solve

In basic solution $Cr(OH)_3$ adds OH^- to form the soluble $Cr(OH)_4^-$ ion:

$$Cr(OH)_3(s) + OH^-(aq) \rightleftharpoons Cr(OH)_4^-(aq)$$

In acidic solution $Cr(OH)_3$ reacts with H^+ to form Cr^{3+} and water:

$$Cr(OH)_3(s) + 3\,H^+(aq) \rightleftharpoons Cr^{3+}(aq) + 3\,H_2O(\ell)$$

Think About It

Other amphiprotic hydroxide compounds that behave similarly to $Cr(OH)_3$ are $Al(OH)_3$ and $Zn(OH)_2$.

16.79. Collect and Organize

We are to explain why Ca^{2+}, Mg^{2+}, and Fe^{3+} combined with a strong base (OH^-) are insoluble, but Al^{3+} is soluble.

Analyze

All those metal cations form insoluble hydroxide compounds (see Appendix 5 for K_{sp} values ranging from 1.9×10^{-33} for Al^{3+} to 4.7×10^{-6} for Ca^{2+}). $Al(OH)_3$, however, is amphoteric and reacts with OH^- to form a complex ion.

Solve

$Al(OH)_3$ reacts with OH^- in solution to form soluble $Al(OH)_4^-$. The other ions do not form that type of soluble complex ion and therefore remain insoluble as $Mg(OH)_2$, $Ca(OH)_2$, and $Fe(OH)_3$ in strongly basic solution.

Think About It

If either Cr^{3+} or Zn^{2+} contaminated the ore, each would be soluble as a complex ion, $Cr(OH)_4^-$ or $Zn(OH)_4^{2-}$.

16.81. Collect and Organize

From the K_a of $Al^{3+}(aq)$ we can calculate the pH of a solution that is 0.25 M in $Al(NO_3)_3$.

Analyze

The nitrate ions of $Al(NO_3)_3$ do not react with water. The reaction of Al^{3+} with water, however, gives an acidic solution with $K_a = 1 \times 10^{-5}$:

$$Al^{3+}(aq) + 2\,H_2O(\ell) \rightleftharpoons Al(OH)^{2+}(aq) + H_3O^+(aq)$$

Solve

Reaction	$Al^{3+}(aq) + 2\,H_2O(\ell) \rightleftharpoons Al(OH)^{2+}(aq) +$		$H_3O^+(aq)$
	$[Al^{3+}]$	$[Al(OH)^{2+}]$	$[H_3O^+]$
Initial	0.25	0	0
Change	$-x$	$+x$	$+x$
Equilibrium	$0.25 - x$	x	x

$$1 \times 10^{-5} = \frac{(x)(x)}{0.25 - x} \approx \frac{x^2}{0.25}$$

$$x = 1.58 \times 10^{-3}$$

The pH of the solution is therefore

$$pH = -\log(1.58 \times 10^{-3}) = 2.80$$

Think About It
Aluminum, with its small size and high positive charge as Al^{3+}, gives fairly acidic aqueous solutions.

16.83. Collect and Organize
From the K_a of $Fe^{3+}(aq)$ we can calculate the pH of a solution that is 0.100 M in $Fe(NO_3)_3$.

Analyze
The nitrate ions of $Fe(NO_3)_3$ do not react with water. The reaction of Fe^{3+} with water, however, gives an acidic solution with $K_a = 3 \times 10^{-3}$:

$$Fe^{3+}(aq) + 2\,H_2O(\ell) \rightleftharpoons Fe(OH)^{2+}(aq) + H_3O^+(aq)$$

Solve

Reaction	$Fe^{3+}(aq) + 2\,H_2O(\ell) \rightleftharpoons$	$Fe(OH)^{2+}(aq) +$	$H_3O^+(aq)$
	$[Fe^{3+}]$	$[Fe(OH)^{2+}]$	$[H_3O^+]$
Initial	0.100	0	0
Change	$-x$	$+x$	$+x$
Equilibrium	$0.100 - x$	x	x

$$3 \times 10^{-3} = \frac{(x)(x)}{0.100 - x}$$

$$3 \times 10^{-3}(0.100 - x) = x^2$$

$$3 \times 10^{-4} - 3 \times 10^{-3}x = x^2$$

$$x^2 + 3 \times 10^{-3}x - 3 \times 10^{-4} = 0$$

$$x = 0.0159$$

The pH of the solution is

$$pH = -\log(0.0159) = 1.80$$

Think About It
Fe^{3+} is the most acidic hydrated metal ion listed in Table 16.1, so we expect that solution to be quite acidic.

16.85. Collect and Organize
We are to sketch the titration curve for the titration of 25 mL of 0.5 M $FeCl_3$ with 0.50 M NaOH.

Analyze
The Fe^{3+} ions react with OH^- to form, in steps, $Fe(OH)_3$:

$$Fe^{3+}(aq) + OH^-(aq) \rightleftharpoons Fe(OH)^{2+}(aq)$$
$$Fe(OH)^{2+}(aq) + OH^-(aq) \rightleftharpoons Fe(OH)_2^+(aq)$$
$$Fe(OH)_2^+(aq) + OH^-(aq) \rightleftharpoons Fe(OH)_3(s)$$

The titration curve shows three equivalence points. The concentration of OH^- in the titrant equals the concentration of Fe^{3+} in the solution. The equivalence points will, therefore, occur at 25, 50, and 75 mL.

Solve

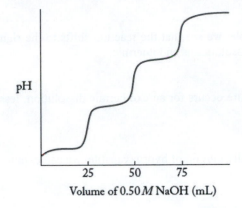

Volume of 0.50 *M* NaOH (mL)

Think About It
Remember that $Fe(OH)_3$ does not further react with OH^-.

16.87. Collect, Organize, and Analyze
We are asked to compare the terms *molar solubility* and *solubility product*.

Solve
Molar solubility is the mole of a substance that dissolves in liter of solution. The solubility product is the equilibrium constant for the dissolution of an ionic substance.

Think About It
Solubility has units of grams or moles per volume of solution but, like other equilibrium constants, the solubility product is unitless.

16.89. Collect and Organize
By comparing the K_{sp} values of $MgCO_3$, $CaCO_3$, and $SrCO_3$ we can identify which cation (Mg^{2+}, Ca^{2+}, or Sr^{2+}) precipitates first as carbonate mineral.

Analyze
From Appendix 5, the K_{sp} values are as follows:

$$MgCO_3 \quad K_{sp} = 6.8 \times 10^{-6}$$
$$CaCO_3 \quad K_{sp} = 5.0 \times 10^{-9}$$
$$SrCO_3 \quad K_{sp} = 5.6 \times 10^{-10}$$

Solve
Because $SrCO_3$ has the lowest K_{sp}, the cation Sr^{2+} precipitates first as a carbonate mineral.

Think About It
The order of solubility from least to most soluble for those carbonates is $SrCO_3 < CaCO_3 < MgCO_3$.

16.91. Collect and Organize
For $SrSO_4$, whose K_{sp} increases as the temperature increases, we are to determine whether the dissolution is exothermic or endothermic.

Analyze
We can include heat as a reactant (for an endothermic reaction) or as a product (for an exothermic reaction) and apply Le Châtelier's principle:

$$SrSO_4(s) + \text{heat} \rightleftharpoons Sr^{2+}(aq) + SO_4^{2-}(aq)$$

$$SrSO_4(s) \rightleftharpoons Sr^{2+}(aq) + SO_4^{2-}(aq) + heat$$

Solve

Applying Le Châtelier's principle, we see that the reaction shifts to the right and more $SrSO_4$ dissolves as the temperature is increased. The dissolution is endothermic.

Think About It

The opposite effect of temperature occurs for an exothermic dissolution: less solid dissolves as the temperature is increased.

16.93. **Collect and Organize**

By writing the equation for the dissolution of hydroxyapatite, we can explain why acidic substances erode tooth enamel.

Analyze

The solubility of hydroxyapatite is described by

$$Ca_5(PO_4)_3OH(s) \rightleftharpoons OH^-(aq) + 3\ PO_4^{3-}(aq) + 5\ Ca^{2+}(aq)$$

Solve

Acidic substances react with the OH^- released upon dissolution of hydroxyapatite. The equilibrium is shifted to the right, dissolving more hydroxyapatite.

Think About It

The equilibrium would be shifted in the opposite direction (to the left) in an alkaline environment.

16.95. **Collect and Organize**

Given the $[Ba^{2+}]$ in a saturated solution of $BaSO_4$ ($1.04 \times 10^{-5}\ M\ Ba^{2+}$), we are to calculate the value of K_{sp} for $BaSO_4$:

$$BaSO_4(s) \rightleftharpoons Ba^{2+}(aq) + SO_4^{2-}(aq)$$

Analyze

The K_{sp} expression is

$$K_{sp} = [Ba^{2+}][SO_4^{2-}]$$

Because for every mole of $BaSO_4$ that dissolves we get 1 mol of Ba^{2+} and 1 mol of SO_4^{2-}, the molarities of Ba^{2+} and SO_4^{2-} are the same for this saturated solution of $BaSO_4$.

Solve

$$K_{sp} = (1.04 \times 10^{-5})(1.04 \times 10^{-5}) = 1.08 \times 10^{-10}$$

Think About It

Because $BaSO_4$ is quite insoluble, we can add SO_4^{2-} to a solution of dissolved Ba^{2+} to quantitatively precipitate the barium out of solution. After weighing the dried precipitate, we can then calculate how much Ba^{2+} was present in the original solution.

16.97. **Collect and Organize**

Given that $K_{sp} = 1.0 \times 10^{-6}$, we are to calculate $[Cu^+]$ and $[Cl^-]$ for a saturated solution of CuCl.

Analyze

The solubility equation and K_{sp} expression for CuCl are

$$CuCl(s) \rightleftharpoons Cu^+(aq) + Cl^-(aq) \qquad K_{sp}[Cu^+][Cl^-]$$

The $[Cu^+] = [Cl^-]$ in that solution because for every particle of CuCl that dissolves we get one particle of Cu^+ and one particle of Cl^-.

Solve

Let $[Cu^+] = [Cl^-] = x$. The K_{sp} expression becomes

$$K_{sp} = 1.0 \times 10^{-6} = (x)(x)$$
$$x = 1.0 \times 10^{-3}$$

Therefore, $[Cu^+] = [Cl^-] = 1.0 \times 10^{-3}$ M.

Think About It

We do not need to know how much CuCl is originally placed into the solution because it does not appear in the K_{sp} expression as a pure solid.

16.99. Collect and Organize

Given the K_{sp} of $CaCO_3$ (9.9×10^{-9}), we are to calculate the solubility of that substance in units of grams per milliliter.

Analyze

The solubility equation and K_{sp} expression for $CaCO_3$ are

$$CaCO_3(s) \rightleftharpoons Ca^{2+}(aq) + CO_3^{2-}(aq) \qquad K_{sp} = [Ca^{2+}][CO_3^{2-}]$$

In that solution, $[Ca^{2+}] = [CO_3^{2-}]$ because for every $CaCO_3$ that dissolves one Ca^{2+} and one CO_3^{2-} are produced. We can then say that

$$K_{sp} = x^2$$

and we can solve for x, which is the molar solubility (moles per liter) of $CaCO_3$. To convert that to grams per milliliter, we multiply by the molar mass of $CaCO_3$ (100.09 g/mol) and divide by 1000 mL/L.

Solve

$$K_{sp} = 9.9 \times 10^{-9} = x^2$$

$$x = 9.95 \times 10^{-5}$$

$$\frac{9.95 \times 10^{-5} \text{ mol}}{L} \times \frac{100.09 \text{ g}}{1 \text{ mol}} \times \frac{1 \text{ L}}{1000 \text{ mL}} = 9.96 \times 10^{-6} \text{ g/mL}$$

Think About It

The value of x that we calculate in the K_{sp} expression is the molar solubility of the solid because it is that amount ("x") that dissolves into the solution.

16.101. Collect and Organize

Given the K_{sp} for the dissolution of AgOH (1.52×10^{-8}) in Appendix 5, we are to calculate the pH of a saturated solution.

Analyze

The solubility equation and K_{sp} expression for AgOH are

$$AgOH(s) \rightleftharpoons Ag^+(aq) + OH^-(aq) \qquad K_{sp} = [Ag^+][OH^-]$$

Letting $[Ag^+] = [OH^-] = x$ (because the stoichiometry is 1:1), we can solve for x by using the K_{sp} expression. The pH of the solution will then be

$$pH = 14 - (-\log x)$$

Solve

$$K_{sp} = 1.52 \times 10^{-8} = x^2$$

$$x = 1.233 \times 10^{-4}$$

$$pH = 14 - \left(-\log 1.233 \times 10^{-4}\right) = 10.091$$

Think About It
Even though the K_{sp} of AgOH is not high, that solution is quite basic.

16.103. Collect and Organize
Using the common-ion effect, we can determine in which 0.1 *M* solution (NaCl, Na_2CO_3, NaOH, or HCl) the most $CaCO_3$ dissolves.

Analyze
Whenever a common ion is already present in the solution, the $CaCO_3$ is less soluble. Any solution, therefore, with Ca^{2+} or CO_3^{2-} would have lower solubility of $CaCO_3$ than water. We also should look for solutions that might react with either Ca^{2+} or CO_3^{2-} and shift the solubility equilibrium to the right.

Solve
(a) NaCl(*aq*) has neither an ion common to $CaCO_3$ nor ions that react with either Ca^{2+} or CO_3^{2-}. $CaCO_3$ has the same solubility in that NaCl solution as in water.
(b) The 0.1 *M* Na_2CO_3 solution is 0.1 *M* in CO_3^{2-}. That decreases the solubility of $CaCO_3$.
(c) NaOH(*aq*) has neither an ion common to $CaCO_3$ nor reacts with either Ca^{2+} or CO_3^{2-}. $CaCO_3$ has the same solubility in that NaOH solution as water.
(d) A solution of HCl reacts with CO_3^{2-} to form H_2CO_3, which then decomposes to H_2O and CO_2. That reaction shifts the solubility equilibrium to the right, so more $CaCO_3$ dissolves.
The solution of (d) 0.1 *M* HCl dissolves the most $CaCO_3$.

Think About It
A higher concentration of acid dissolves even more $CaCO_3$ as the equilibrium shifts to the right:
$$CaCO_3(s) + 2\,H^+(aq) \rightleftharpoons Ca^{2+}(aq) + H_2CO_3(aq)$$

16.105. Collect and Organize
Given the average concentrations of SO_4^{-2} and Sr^{2+} in seawater (0.028 *M* and 9×10^{-5} *M*, respectively) and the K_{sp} of $SrSO_4$ (3.4×10^{-7}), we are to determine whether the concentration of Sr^{2+} is controlled by the relative insolubility of $SrSO_4$.

Analyze
The solubility equation and K_{sp} expression for $SrSO_4$ are
$$SrSO_4(s) \rightleftharpoons Sr^{2+}(aq) + SO_4^{2-}(aq) \qquad K_{sp} = [Sr^{2+}][SO_4^{2-}]$$

Solve
$$K_{sp} = 3.4 \times 10^{-7} = [Sr^{2+}][SO_4^{2-}]$$
$$= [Sr^{2+}](0.028\ M)$$
$$[Sr^{2+}] = 1.21 \times 10^{-5}\ M$$

That value is the expected concentration of Sr^{2+} in seawater with a known $[SO_4^{2-}]$ of 0.028 *M*. The concentration of Sr^{2+} possibly is controlled by the insolubility of the sulfate salt because $[Sr^{2+}] \times [SO_4^{2-}]$ exceeds the K_{sp} of $SrSO_4$.

Think About It
As the $[SO_4^{2-}]$ increases, the solubility of $SrSO_4$ decreases because of the common-ion effect.

16.107. Collect and Organize
Given 125 mL of solution that is 0.375 *M* in $Ca(NO_3)_2$, we are asked whether CaF_2 will precipitate when 245 mL of a 0.255 *M* NaF solution is added.

Analyze
The K_{sp} for CaF_2 from Appendix 5 is 3.9×10^{-11}. When the two solutions are mixed, the total volume is 370 mL and the $[Ca^{2+}]$ and $[F^-]$ in the mixed solution is

$$125 \text{ mL} \times 0.375 M = 370 \text{ mL} \times [\text{Ca}^{2+}]$$

$$[\text{Ca}^{2+}] = 0.127 \ M$$

$$245 \text{ mL} \times 0.255 M = 370 \text{ mL} \times [\text{F}^{-}]$$

$$[\text{F}^{-}] = 0.169 \ M$$

From the K_{sp} expression

$$K_{sp} = [\text{Ca}^{2+}][\text{F}^{-}]^2 = 3.9 \times 10^{-11}$$

If the $[\text{Ca}^{2+}]_{initial} \times [\text{F}^{-}]^2_{initial} > K_{sp}$, then CaF_2 will precipitate.

Solve

$[\text{Ca}^{2+}]_{initial} \times [\text{F}^{-}]^2_{initial} = 0.127 \times (0.169)^2 = 3.63 \times 10^{-3}$. That is greater than the value of K_{sp} for CaF_2, so it will precipitate from the mixed solution.

Think About It

Because CaF_2 has a small solubility product constant, we expect that most of the ions will precipitate as CaF_2. Here the Ca^{2+} ions are in excess (0.0469 mol) compared with F^{-} (0.0625 mol), so the final solution will have only a small amount of F^{-} in solution.

16.109. Collect and Organize

When a 0.250 M solution of $\text{Pb}^{2+}(aq)$ is added to a solution that is 0.010 M in Br^{-} and SO_4^{2-}, we can use the values of K_{sp} for PbBr_2 and PbSO_4 to determine which anion is the first to precipitate. We are then asked to calculate the concentration of the anion that precipitates first at the moment that the second ion starts to precipitate. That will be when the solution is saturated in the lead salt of the first anion to precipitate.

Analyze

The K_{sp} value shows that PbSO_4 has a smaller solubility product constant (1.8×10^{-8}) than PbBr_2 (6.6×10^{-6}).

Solve

(a) PbSO_4, with a smaller solubility product constant, will precipitate first from the solution, so the SO_4^{2-} anion will precipitate first.

(b) When Br^{-} begins to precipitate, the amount of Pb^{2+} that would be present is

$$K_{sp} = 6.6 \times 10^{-6} = [\text{Pb}^{2+}][\text{Br}^{-}]^2$$

$$[\text{Pb}^{2+}] = \frac{6.6 \times 10^{-6}}{(0.010)^2} = 6.6 \times 10^{-2} \ M$$

The $[\text{SO}_4^{2-}]$ in the solution when $[\text{Pb}^{2+}] = 6.6 \times 10^{-2} \ M$ is

$$K_{sp} = 1.8 \times 10^{-8} = [\text{Pb}^{2+}][\text{SO}_4^{2-}]$$

$$[\text{Pb}^{2+}] = \frac{1.8 \times 10^{-8}}{6.6 \times 10^{-2}} = 2.7 \times 10^{-7} \ M$$

Think About It

We do not need a RICE table to solve this problem.

16.111. Collect and Organize

We are asked whether the pH of the solution changes when a cook adds more baking soda to water used in a recipe and to explain why or why not.

Analyze

Baking soda is a soluble sodium salt with the formula $NaHCO_3$. In solution, that salt forms $Na^+(aq) + HCO_3^-$. Na^+ does not react with water, but HCO_3^- does, which changes the pH of the solution.

Solve

Yes, the pH of the solution increases because of the increase in hydrolysis of HCO_3^- according to the equation

$$HCO_3^-(aq) + H_2O(\ell) \rightleftharpoons H_2CO_3(aq) + OH^-(aq)$$

Think About It

H_2CO_3 decomposes at baking temperatures to give $CO_2(g)$ and $H_2O(\ell)$.

16.113. Collect and Organize

Given the equation describing the solubility of Ag_2O and the concentration of OH^- in a saturated solution, we are to calculate the value of the K_{sp}.

Analyze

The K_{sp} expression for the solubility of Ag_2O is

$$K_{sp} = [Ag^+]^2 \times [OH^-]^2$$

In that expression the $[OH^-] = 1.6 \times 10^{-4}$ M and $[Ag^+] = 1.6 \times 10^{-4}$ M by stoichiometry according to the solubility equation:

$$Ag_2O(s) + H_2O(\ell) \rightleftharpoons 2\,Ag^+(aq) + 2\,OH^-(aq)$$

Solve

$$K_{sp} = [Ag^+]^2 \times [OH^-]^2$$
$$K_{sp} = (1.6 \times 10^{-4})^2 \times (1.6 \times 10^{-4})^2 = 6.6 \times 10^{-16}$$

Think About It

Be sure to include the exponents in the K_{sp} expressions as appropriate for the solubility equation.

16.115. Collect and Organize

For a drop in the pH of the oceans by 0.77 pH units as a result of increased CO_2 in the atmosphere, we are asked to explain with chemical equations how CO_2 might cause that decrease. We are also to calculate the extent of the increase in acidity that change would cause. Finally, we are to explain how that decrease could affect coral reef survival.

Analyze

Carbon dioxide gas dissolves in water to form carbonic acid, a weak diprotic acid. Coral reefs are composed of $CaCO_3$, which reacts with acids.

Solve

(a) When the partial pressure of CO_2 in the atmosphere increases, the following equilibria are shifted to the right (Le Châtelier's principle):

$$CO_2(g) \rightleftharpoons CO_2(aq)$$
$$CO_2(aq) + H_2O(\ell) \rightleftharpoons H_2CO_3(aq)$$
$$H_2CO_3(aq) \rightleftharpoons H^+(aq) + HCO_3^-(aq)$$
$$HCO_3^-(aq) \rightleftharpoons H^+(aq) + CO_3^{2-}(aq)$$

The acidity of the oceans increases (lower pH) because of increased ionization of the weak acid carbonic acid.

(b) $pH_f - pH_i = -0.77$

$$-\log[H^+]_f - \left(-\log[H^+]_i\right) = -0.77$$

$$\log[H^+]_f - \log[H^+]_i = 0.77$$

$$\log\frac{[H^+]_f}{[H^+]_i} = 0.77$$

$$\frac{[H^+]_f}{[H^+]_i} = 10^{0.77} = 5.9$$

$$[H^+]_f = 5.9[H^+]_i$$

The acidity of the ocean will be about six times greater.

(c)   Because oyster shells are composed primarily of $CaCO_3$, they are likely to dissolve more as the pH decreases.

$$CaCO_3(s) + 2\,H^+(aq) \rightarrow Ca^{2+}(aq) + H_2CO_3(aq)$$

Think About It
Our answer to part b is consistent with our knowledge that one pH unit represents a 10-fold difference in $[H^+]$.

16.117. Collect and Organize
By estimating the equivalence points on the titration curve shown in Figure P16.117, we are to choose the statement that is true about the alkalinity of the sample that contains carbonate, bicarbonate, or both.

Analyze
The titration curve shows that the initial pH is high (the solution is basic) with two equivalence points estimated to be at pH = 8.0 and 4.0. The first equivalence point is after 10 mL of the titrant is added and the second after 30 mL of titrant is added.

Solve
For the curve to show two equivalence points, the sample must contain some CO_3^{2-}. If the sample was only CO_3^{2-} initially, the second equivalence point would be at twice the volume of HCl for the first equivalence point; we would expect the equivalence point to be reached when 20 mL HCl were added. The fact that the second equivalence point requires more HCl (30 mL) indicates that the sample must also have HCO_3^-. The sample, then, must be a mixture of CO_3^{2-} and HCO_3^-. For a 50:50 mixture of CO_3^{2-} and HCO_3^-, an additional 10 mL of HCl titrant would be needed to reach the second titration point (where HCO_3^- in the sample is in an equimolar amount with HCl) in addition to that needed if the sample were 100% CO_3^{2-} (where we expect the equivalence point to be at 20 mL). Therefore, the sample has a 50:50 ratio of HCO_3^- to CO_3^{2-} (e).

Think About It
How do you think the titration curve would change if the ratio of HCO_3^{2-} to CO_3^{2-} were 2:1?

16.119. Collect and Organize
Given sets of three acids, we are to determine which would be best to use to prepare a buffer of a specific pH.

Analyze
The best buffers have the pK_a of the acid closest to the targeted pH.

Solve
(a)   Acetic acid ($pK_a = 4.75$) would be the best buffer system to choose for pH = 5.2.
(b)   Hypobromous acid ($pK_a = 7.54$) would be the best buffer system to choose for pH = 8.0.

Think About It
When making a buffer we also have to consider toxicity. We would not want to prepare a buffer for biological work of pH = 9.00 with HCN ($pK_a = 9.25$). We might choose H_3BO_3 instead ($pK_a = 9.27$).

16.121. **Collect and Organize**

For three indicators, bromophenol blue, bromocresol green, and alizarin yellow R, we are to estimate the K_a values for the indicator given the pH at which each has a color transition.

Analyze

The color transition for indicators occurs around the value of the pK_a. To estimate the K_a of the indicator, we use $K_a = 1 \times 10^{-pH}$.

Solve

(a) $1 \times 10^{-3.8} = 1.6 \times 10^{-4}$
(b) $1 \times 10^{-4.3} = 5.0 \times 10^{-5}$
(c) $1 \times 10^{-10.9} = 1.3 \times 10^{-11}$

Think About It

Alizarin yellow R would be used to detect a pH change at pH > 10, whereas bromocresol green and bromophenol blue would be used to detect a pH change at pH < 6.

16.123. **Collect and Organize**

For solutions of formic acid and boric acid with NaOH, we are to calculate the pH at the equivalence point.

Analyze

The equivalence point is where the number of moles of NaOH added are equal to the moles of acid present. That reaction will produce 100% of the conjugate base, which will hydrolyze in the solution. Because both of those titrations will produce a weak conjugate base, we expect the pH at the equivalence point to be basic.

Solve

For the titration of 10.0 mL of 0.100 M formic acid (HCOOH, $K_a = 1.77 \times 10^{-4}$) with 0.100 M NaOH to the equivalence point, we will require 10.0 mL of titrant (for a total volume of 20.0 mL) and we will produce 0.00100 mol of HCOO$^-$. The concentration of HCOO$^-$ initially, before hydrolysis, at the equivalence point is 0.00100 mol/0.0200 L = 0.0500 M. We use a RICE table to now calculate the pH at the equivalence point.

Reaction	HCOO$^-$(aq) + H$_2$O(ℓ) $\rightleftharpoons$ HCOOH(aq)	+	OH$^-$(aq)
	[HCOO$^-$], M	[HCOOH], M	[OH$^-$], M
Initial	0.0500	0.0	0.0
Change	$-x$	$+x$	$+x$
Equilibrium	$0.0500 - x$	x	x

$$K_b = \frac{K_w}{K_a} = \frac{1 \times 10^{-14}}{1.77 \times 10^{-4}} = 5.65 \times 10^{-11} = \frac{[HCOOH][OH^-]}{[HCOO^-]} = \frac{(x)(x)}{0.0500 - x} \approx \frac{x^2}{0.0500}$$

$$x = 1.68 \times 10^{-6}$$

$$pOH = -\log(1.68 \times 10^{-6}) = 5.77$$

$$pH = 14.00 - 5.77 = 8.22$$

For the titration of 10.0 mL of 0.100 M boric acid (H$_3$BO$_3$, $K_a = 5.4 \times 10^{-10}$) with 0.100 M NaOH to the equivalence point, we will require 10.0 mL of titrant (for a total volume of 20.0 mL) and we will produce 0.00100 mol of H$_2$BO$_3^-$. The concentration of H$_2$BO$_3^-$ initially, before hydrolysis, at the equivalence point is 0.00100 mol/0.0200 L = 0.0500 M. We use a RICE table to now calculate the pH at the equivalence point.

Reaction	HBO$_3^-$(aq) + H$_2$O(ℓ) $\rightleftharpoons$ H$_2$BO$_3$(aq)	+	OH$^-$(aq)
	[HBO$_3^-$], M	[H$_2$BO$_3$], M	[OH$^-$], M
Initial	0.0500	0.0	0.0
Change	$-x$	$+x$	$+x$
Equilibrium	$0.0500 - x$	x	x

$$K_b = \frac{K_w}{K_a} = \frac{1\times10^{-14}}{5.4\times10^{-10}} = 1.85\times10^{-5} = \frac{[H_2BO_3][OH^-]}{[H_2BO_3^-]} = \frac{(x)(x)}{0.0500-x} \approx \frac{x^2}{0.0500}$$

$$x = 9.63\times10^{-4}$$
$$pOH = -\log(9.63\times10^{-4}) = 3.02$$
$$pH = 14.00 - 3.02 = 10.98$$

Because the pH for those titrations is so different (8.22 vs. 10.98), different indicators should be used.

Think About It
Our prediction that the pH at the equivalence point for both titrations would be greater than 7 was correct.

16.125. Collect and Organize
We are to answer several questions about the titration of a sample of H_2SO_3 with NaOH.

Analyze
Sulfurous acid is a diprotic acid and therefore two equivalence points will be present in the titration curve. At the equivalence point (reached here by adding 15.00 mL of NaOH), the moles of H_2SO_3 in the sample is equal to the moles of NaOH added.

Solve
(a) If 15.00 mL of titrant is needed to reach the first equivalence point, 15.00 mL more will be needed to reach the second equivalence point.
(b) When 7.50 mL of titrant has been added, the half-equivalence point has been reached. Here the pH = pK_{a1} = 1.77.
(c) Adding 22.50 mL of titration takes us to the second half-equivalence point. Here the pH = pK_{a2} = 7.21.
(d) At the first equivalence point H_2SO_3 has been deprotonated to form HSO_3^-, a small amount of which hydrolyzes to H_2SO_3. The added NaOH has reacted to form H_2O, leaving Na^+ in the solution. The major species present at the first equivalence point are Na^+ and HSO_3^-.
(e) At the second equivalence point HSO_3^- has been deprotonated to form SO_3^{2+-}, a small amount of which hydrolyzes to HSO_3^-. The added NaOH has reacted to form H_2O, leaving Na^+ in the solution. The major species present at the first equivalence point are Na^+ and SO_3^{2-}.

Think About It
At first reading, you might have thought that the statement of this problem was incomplete. We did not need the molar concentration of NaOH or of H_2SO_3 to answer any of these questions.

16.127. Collect and Organize
For the titration of 15.8 mL of 0.367 M pyridine with 0.500 M HCl, we are to answer several questions.

Analyze
Pyridine (C_5H_5N) is a base (pK_b = 8.77) that will react with 1 mol of acid and therefore the titration curve will have one equivalence point. At the equivalence point the moles of HCl required is equal to the moles of pyridine in the sample. The best indicator for a titration is one that changes color close to the pH of the equivalence point of the titration.

Solve
(a) The volume of HCl required to reach the equivalence point is

$$15.8 \text{ mL pyridine} \times \frac{0.367 \text{ mol pyridine}}{1000 \text{ mL}} \times \frac{1 \text{ mol HCl}}{1 \text{ mol pyridine}} \times \frac{1000 \text{ mL}}{0.500 \text{ mol HCl}} = 11.60 \text{ mL}$$

That gives a total volume of the solution at the equivalence point of 15.8 mL + 11.60 mL = 27.40 mL, or 0.02740 L. At the equivalence point pyridine has reacted completely with H^+ to give the conjugate acid, $C_5H_5NH^+$. The molarity of $C_5H_5NH^+$ produced is

$$15.8 \text{ mL pyridine} \times \frac{0.367 \text{ mol } C_6H_5N}{1000 \text{ mL}} \times \frac{1 \text{ mol } C_6H_5NH^+}{1 \text{ mol } C_6H_5N} \times \frac{1}{0.02740 \text{ L}} = 0.2116 \, M$$

Now we can use a RICE table to calculate the pH at the equivalence point:

Reaction	$C_5H_5NH^+(aq) + H_2O(\ell) \rightleftharpoons$	$C_5H_5N(aq)$ +	$H_3O^+(aq)$
	$[C_5H_5NH^+], M$	$[C_5H_5N], M$	$[H_3O^+], M$
Initial	0.2116	0.0	0.0
Change	$-x$	$+x$	$+x$
Equilibrium	$0.2116 - x$	x	x

$$K_a = \frac{K_b}{K_a} = \frac{1 \times 10^{-14}}{1 \times 10^{-8.77}} = 5.89 \times 10^{-6} = \frac{[C_5H_5N][H_3O^+]}{[C_5H_5NH^+]} = \frac{(x)(x)}{0.2116 - x} \approx \frac{x^2}{0.2116}$$

$$x = 1.12 \times 10^{-3}$$

$$pH = -\log(1.12 \times 10^{-3}) = 2.95$$

(b) Bromothymol blue would be an appropriate indicator; it changes from yellow to blue at about pH 3.50, which is close to the pH of the equivalence point of 2.95 for this titration.

(c) The initial pH of this solution is based on the concentration of pyridine and its K_b.

Reaction	$C_5H_5N(aq) + H_2O(\ell) \rightleftharpoons$	$C_5H_5NH^+(aq)$ +	$OH^-(aq)$
	$[C_5H_5N], M$	$[C_5H_5NH^+], M$	$[OH^-], M$
Initial	0.367	0.0	0.0
Change	$-x$	$+x$	$+x$
Equilibrium	$0.367 - x$	x	x

$$K_b = 1 \times 10^{-8.77} = 1.70 \times 10^{-9} = \frac{[C_5H_5NH^+][OH^-]}{[C_5H_5N]} = \frac{(x)(x)}{0.367 - x} \approx \frac{x^2}{0.367}$$

$$x = 2.50 \times 10^{-5}$$

$$pOH = -\log(2.50 \times 10^{-5}) = 4.60$$

$$pH = 14.00 - 4.60 = 9.40$$

After adding 1.0 mL of 0.500 *M* HCl:

$$\text{mol HCl added} = 1.0 \text{ mL} \times \frac{0.500 \text{ mol}}{1000 \text{ mL}} = 5.00 \times 10^{-4} \text{ mol}$$

$$\text{total volume of solution} = 15.8 \text{ mL} + 1.00 \text{ mL} = 16.8 \text{ mL, or } 0.0168 \text{ L}$$

$$\text{mol pyridine in sample} = 15.8 \text{ mL} \times \frac{0.367 \text{ mol}}{1000 \text{ mL}} = 5.80 \times 10^{-3} \text{ mol}$$

The added HCl will react with the pyridine to make the conjugate base:

Reaction	$C_5H_5N(aq)$ +	$H^+(aq)$ $\rightleftharpoons$	$C_5H_5NH^+(aq)$
	C_5H_5N (moles) +	H^+(mole)	$C_5H_5NH^+$(moles)
Initial	5.80×10^{-3}	5.00×10^{-4}	0.0
Change	-5.00×10^{-4}	-5.00×10^{-4}	$+5.00 \times 10^{-4}$
Final	5.3×10^{-3}	0.00	5.00×10^{-4}

The total volume of the sample is 0.0168 L and the concentrations of C_5H_5N and $C_5H_5NH^+$ are

$$\frac{0.00530 \text{ mol } C_5H_5N}{0.0168 \text{ L}} = 0.3155 \, M \, C_5H_5N$$

$$\frac{0.000500 \text{ mol } C_5H_5NH^+}{0.0168 \text{ L}} = 0.02976 \, M \, C_5H_5NH^+$$

Using the Henderson–Hasselbalch equation, we calculate the pH of this solution:

$$pK_a = 14.00 - 8.77 = 5.23$$

$$pH = 5.23 + \log\frac{0.3155\ M}{0.02976\ M} = 6.26$$

After adding 3.0 mL of 0.500 M HCl:

$$\text{mol HCl added} = 3.0\ \text{mL} \times \frac{0.500\ \text{mol}}{1000\ \text{mL}} = 1.50 \times 10^{-3}\ \text{mol}$$

$$\text{total volume of solution} = 15.8\ \text{mL} + 3.0\ \text{mL} = 18.8\ \text{mL, or } 0.0188\ \text{L}$$

$$\text{mol pyridine in sample} = 15.8\ \text{mL} \times \frac{0.367\ \text{mol}}{1000\ \text{mL}} = 5.80 \times 10^{-3}\ \text{mol}$$

The added HCl will react with the pyridine to make the conjugate base:

Reaction	$C_5H_5N(aq)$	+	$H^+(aq)$	$\rightleftharpoons$	$C_5H_5NH^+(aq)$
	C_5H_5N (moles)	+	H^+(mole)		$C_5H_5NH^+$(moles)
Initial	5.80×10^{-3}		1.50×10^{-3}		0.0
Change	-1.50×10^{-3}		-1.50×10^{-3}		$+1.50 \times 10^{-3}$
Final	4.3×10^{-3}		0.00		1.5×10^{-3}

The total volume of the sample is 0.0188 L and the concentrations of C_5H_5N and $C_5H_5NH^+$ are

$$\frac{0.00430\ \text{mol } C_5H_5N}{0.0188\ \text{L}} = 0.2287\ M\ C_5H_5N$$

$$\frac{0.00150\ \text{mol } C_5H_5NH^+}{0.0188\ \text{L}} = 0.07979\ M\ C_5H_5NH^+$$

Using the Henderson–Hasselbalch equation, we calculate the pH of this solution:

$$pH = 5.23 + \log\frac{0.2287\ M}{0.07979\ M} = 5.69$$

After adding 7.0 mL of 0.500 M HCl:

$$\text{mol HCl added} = 7.0\ \text{mL} \times \frac{0.500\ \text{mol}}{1000\ \text{mL}} = 3.50 \times 10^{-3}\ \text{mol}$$

$$\text{total volume of solution} = 15.8\ \text{mL} + 7.0\ \text{mL} = 22.8\ \text{mL, or } 0.0228\ \text{L}$$

$$\text{mol pyridine in sample} = 15.8\ \text{mL} \times \frac{0.367\ \text{mol}}{1000\ \text{mL}} = 5.80 \times 10^{-3}\ \text{mol}$$

The added HCl will react with the pyridine to make the conjugate base:

Reaction	$C_5H_5N(aq)$	+	$H^+(aq)$	$\rightleftharpoons$	$C_5H_5NH^+(aq)$
	C_5H_5N (moles)	+	H^+(mole)		$C_5H_5NH^+$(moles)
Initial	5.80×10^{-3}		3.50×10^{-3}		0.0
Change	-3.50×10^{-3}		-3.50×10^{-3}		$+3.50 \times 10^{-3}$
Final	2.3×10^{-3}		0.00		3.5×10^{-3}

The total volume of the sample is 0.0228 L and the concentrations of C_5H_5N and $C_5H_5NH^+$ are

$$\frac{0.00230\ \text{mol } C_5H_5N}{0.0228\ \text{L}} = 0.1009\ M\ C_5H_5N$$

$$\frac{0.00350\ \text{mol } C_5H_5NH^+}{0.0228\ \text{L}} = 0.1535\ M\ C_5H_5NH^+$$

Using the Henderson–Hasselbalch equation, we calculate the pH of this solution:

$$pH = 5.23 + \log\frac{0.1009\ M}{0.1535\ M} = 5.05$$

After adding 9.0 mL of 0.500 M HCl:

$$\text{mol HCl added} = 9.0\ \text{mL} \times \frac{0.500\ \text{mol}}{1000\ \text{mL}} = 4.50 \times 10^{-3}\ \text{mol}$$

$$\text{total volume of solution} = 15.8\ \text{mL} + 9.0\ \text{mL} = 24.8\ \text{mL, or } 0.0248\ \text{L}$$

$$\text{mol pyridine in sample} = 15.8\ \text{mL} \times \frac{0.367\ \text{mol}}{1000\ \text{mL}} = 5.80 \times 10^{-3}\ \text{mol}$$

The added HCl will react with the pyridine to make the conjugate base:

Reaction	$C_5H_5N(aq)$	+	$H^+(aq)$	$\rightleftharpoons$	$C_5H_5NH^+(aq)$
	C_5H_5N (moles)	+	H^+ (mole)		$C_5H_5NH^+$ (moles)
Initial	5.80×10^{-3}		4.50×10^{-3}		0.0
Change	-4.50×10^{-3}		-4.50×10^{-3}		$+4.50 \times 10^{-3}$
Final	1.3×10^{-3}		0.00		4.5×10^{-3}

The total volume of the sample is 0.0248 L and the concentrations of C_5H_5N and $C_5H_5NH^+$ are

$$\frac{0.00130\ \text{mol}\ C_5H_5N}{0.0248\ \text{L}} = 0.05242\ M\ C_5H_5N$$

$$\frac{0.00450\ \text{mol}\ C_5H_5NH^+}{0.0248\ \text{L}} = 0.1815\ M\ C_5H_5NH^+$$

Using the Henderson–Hasselbalch equation, we calculate the pH of this solution:

$$pH = 5.23 + \log\frac{0.07986\ M}{0.1215\ M} = 5.05$$

After adding 10.0 mL of 0.500 M HCl:

$$\text{mol HCl added} = 10.0\ \text{mL} \times \frac{0.500\ \text{mol}}{1000\ \text{mL}} = 5.00 \times 10^{-3}\ \text{mol}$$

$$\text{total volume of solution} = 15.8\ \text{mL} + 10.0\ \text{mL} = 25.8\ \text{mL, or } 0.0258\ \text{L}$$

$$\text{mol pyridine in sample} = 15.8\ \text{mL} \times \frac{0.367\ \text{mol}}{1000\ \text{mL}} = 5.80 \times 10^{-3}\ \text{mol}$$

The added HCl will react with the pyridine to make the conjugate base:

Reaction	$C_5H_5N(aq)$	+	$H^+(aq)$	$\rightleftharpoons$	$C_5H_5NH^+(aq)$
	C_5H_5N (moles)	+	H^+ (mole)		$C_5H_5NH^+$ (moles)
Initial	5.80×10^{-3}		5.000×10^{-3}		0.0
Change	-5.00×10^{-3}		-5.000×10^{-3}		$+5.00 \times 10^{-3}$
Final	8.0×10^{-4}		0.00		5.00×10^{-3}

The total volume of the sample is 0.0258 L and the concentrations of C_5H_5N and $C_5H_5NH^+$ are

$$\frac{0.00080\ \text{mol}\ C_5H_5N}{0.0258\ \text{L}} = 0.03101\ M\ C_5H_5N$$

$$\frac{0.00500\ \text{mol}\ C_5H_5NH^+}{0.0258\ \text{L}} = 0.1938\ M\ C_5H_5NH^+$$

Using the Henderson–Hasselbalch equation, we calculate the pH of this solution:

$$pH = 5.23 \ + \ \log\frac{0.03101 \ M}{0.1938 \ M} = 4.43$$

After adding 10.5 mL of 0.500 M HCl:

$$mol \ HCl \ added = 10.5 \ mL \times \frac{0.500 \ mol}{1000 \ mL} = 5.25 \times 10^{-3} \ mol$$

total volume of solution = 15.8 mL + 10.5 mL = 26.3 mL, or 0.0263 L

$$mol \ pyridine \ in \ sample = 15.8 \ mL \times \frac{0.367 \ mol}{1000 \ mL} = 5.80 \times 10^{-3} \ mol$$

The added HCl will react with the pyridine to make the conjugate base:

Reaction	$C_5H_5N(aq)$	+	$H^+(aq)$	$\rightleftharpoons$	$C_5H_5NH^+(aq)$
	C_5H_5N (moles)	+	H^+(mole)		$C_5H_5NH^+$(moles)
Initial	5.80×10^{-3}		5.25×10^{-3}		0.0
Change	-5.25×10^{-3}		-5.25×10^{-3}		$+5.25 \times 10^{-3}$
Final	5.5×10^{-4}		0.00		5.25×10^{-3}

The total volume of the sample is 0.0263 L and the concentrations of C_5H_5N and $C_5H_5NH^+$ are

$$\frac{0.00055 \ mol \ C_5H_5N}{0.0263 \ L} = 0.02091 \ M \ C_5H_5N$$

$$\frac{0.00525 \ mol \ C_5H_5NH^+}{0.0263 \ L} = 0.1996 \ M \ C_5H_5NH^+$$

Using the Henderson–Hasselbalch equation, we calculate the pH of this solution:

$$pH = 5.23 \ + \ \log\frac{0.02091 \ M}{0.1996 \ M} = 4.25$$

After adding 11.0 mL of 0.500 M HCl:

$$mol \ HCl \ added = 11.0 \ mL \times \frac{0.500 \ mol}{1000 \ mL} = 5.50 \times 10^{-3} \ mol$$

total volume of solution = 15.8 mL + 11.0 mL = 26.8 mL, or 0.0268 L

$$mol \ pyridine \ in \ sample = 15.8 \ mL \times \frac{0.367 \ mol}{1000 \ mL} = 5.80 \times 10^{-3} \ mol$$

The added HCl will react with the pyridine to make the conjugate base:

Reaction	$C_5H_5N(aq)$	+	$H^+(aq)$	$\rightleftharpoons$	$C_5H_5NH^+(aq)$
	C_5H_5N (moles)	+	H^+(mole)		$C_5H_5NH^+$(moles)
Initial	5.80×10^{-3}		5.50×10^{-3}		0.0
Change	-5.50×10^{-3}		-5.50×10^{-3}		$+5.50 \times 10^{-3}$
Final	3.0×10^{-4}		0.00		5.50×10^{-3}

The total volume of the sample is 0.0268 L and the concentrations of C_5H_5N and $C_5H_5NH^+$ are

$$\frac{0.00030 \ mol \ C_5H_5N}{0.0268 \ L} = 0.01119 \ M \ C_5H_5N$$

$$\frac{0.00550 \ mol \ C_5H_5NH^+}{0.0268 \ L} = 0.2052 \ M \ C_5H_5NH^+$$

Using the Henderson–Hasselbalch equation we calculate the pH of this solution

$$pH = 5.23 + \log\frac{0.01119\ M}{0.2052\ M} = 3.97$$

After adding 11.5 mL 0.500 M HCl:

$$\text{mol HCl added} = 11.5\ \text{mL} \times \frac{0.500\ \text{mol}}{1000\ \text{mL}} = 5.75 \times 10^{-3}\ \text{mol}$$

$$\text{total volume of solution} = 15.8\ \text{mL} + 11.5\ \text{mL} = 27.3\ \text{mL, or } 0.0273\ \text{L}$$

$$\text{mol pyridine in sample} = 15.8\ \text{mL} \times \frac{0.367\ \text{mol}}{1000\ \text{mL}} = 5.80 \times 10^{-3}\ \text{mol}$$

The added HCl will react with the pyridine to make the conjugate base:

Reaction	$C_5H_5N(aq)$	+	$H^+(aq)$	$\rightleftharpoons$	$C_5H_5NH^+(aq)$
	C_5H_5N (moles)	+	H^+ (mole)		$C_5H_5NH^+$ (moles)
Initial	5.80×10^{-3}		5.75×10^{-3}		0.0
Change	-5.75×10^{-3}		-5.75×10^{-3}		$+5.75 \times 10^{-3}$
Final	5.0×10^{-5}		0.00		5.75×10^{-3}

The total volume of the sample is 0.0273 L and the concentrations of C_5H_5N and $C_5H_5NH^+$ are

$$\frac{0.00005\ \text{mol } C_5H_5N}{0.0273\ \text{L}} = 0.001832\ M\ C_5H_5N$$

$$\frac{0.00575\ \text{mol } C_5H_5NH^+}{0.0273\ \text{L}} = 0.2106\ M\ C_5H_5NH^+$$

We are now starting to get out of the buffer zone, so the Henderson–Hasselbalch equation will no longer be valid. We set up a RICE table for the hydrolysis of $C_5H_5NH^+$ to solve for the pH of this solution.

Reaction	$C_5H_5NH^+(aq)$	+ $H_2O(\ell) \rightleftharpoons$	$C_5H_5N(aq)$	+	$H^+(aq)$
	$[C_5H_5NH^+]$, M		$[C_5H_5N]$, M		$[H^+]$, M
Initial	0.2106		0.001832		0.0
Change	$-x$		$+x$		$+x$
Equilibrium	$0.2106 - x$		$0.001832 + x$		x

$$K_a = \frac{K_w}{K_b} = \frac{1 \times 10^{-14}}{1.00 \times 10^{-8.77}} = 5.89 \times 10^{-6} = \frac{(0.001832 + x)x}{0.2106 - x}$$

$$5.89 \times 10^{-6}(0.2106 - x) = (0.001832 + x)x$$

$$1.240 \times 10^{-6} - 5.89 \times 10^{-6}x = 1.832 \times 10^{-3}x + x^2$$

$$x^2 + 1.838 \times 10^{-3}x - 1.240 \times 10^{-6} = 0$$

$$x = 0.0005248$$

$$pH = -\log(0.0005248) = 3.28$$

After adding 11.8 mL of 0.500 M HCl:

$$\text{mol HCl added} = 11.8\ \text{mL} \times \frac{0.500\ \text{mol}}{1000\ \text{mL}} = 5.90 \times 10^{-3}\ \text{mol}$$

$$\text{total volume of solution} = 15.8\ \text{mL} + 11.8\ \text{mL} = 27.6\ \text{mL, or } 0.0276\ \text{L}$$

$$\text{mol pyridine in sample} = 15.8\ \text{mL} \times \frac{0.367\ \text{mol}}{1000\ \text{mL}} = 5.80 \times 10^{-3}\ \text{mol}$$

The added HCl will react with the pyridine to make the conjugate base. We are now beyond the equivalence point, so the solution will have excess H^+ and all C_5H_5N will react.

Reaction	$C_5H_5N(aq)$	+	$H^+(aq)$	$\rightleftharpoons$	$C_5H_5NH^+(aq)$
	C_5H_5N (moles)	+	H^+ (mole)		$C_5H_5NH^+$ (moles)
Initial	5.80×10^{-3}		5.90×10^{-3}		0.0
Change	-5.80×10^{-3}		-5.80×10^{-3}		$+5.80 \times 10^{-3}$
Final	0.00		1.00×10^{-4}		5.80×10^{-3}

The total volume of the sample is 0.0276 L and the concentrations of $C_5H_5NH^+$ and H^+ are

$$\frac{0.00580 \text{ mol } C_5H_5NH^+}{0.0276 \text{ L}} = 0.2101 \ M \ C_5H_5NH^+$$

$$\frac{0.0001 \text{ mol } H^+}{0.0273 \text{ L}} = 0.003623 \ M \ H^+$$

We are out of the buffer zone, so the Henderson–Hasselbalch equation will no longer be valid. We set up a RICE table for the hydrolysis of $C_5H_5NH^+$ to solve for the pH of this solution.

Reaction	$C_5H_5NH^+(aq)$	$+ H_2O(\ell) \rightleftharpoons$	$C_5H_5N(aq)$	+	$H^+(aq)$
	$[C_5H_5NH^+], M$		$[C_5H_5N], M$		$[H^+], M$
Initial	0.2101		0.0		0.003623
Change	$-x$		$+x$		$+x$
Equilibrium	$0.2101 - x$		x		$0.003623 + x$

$$K_a = \frac{K_w}{K_b} = \frac{1 \times 10^{-14}}{1.00 \times 10^{-8.77}} = 5.89 \times 10^{-6} = \frac{(0.003623 + x)x}{0.2101 - x}$$

$$5.89 \times 10^{-6}(0.2101 - x) = (0.003623 + x)x$$

$$1.237 \times 10^{-6} - 5.89 \times 10^{-6}x = 3.623 \times 10^{-3}x + x^2$$

$$x^2 + 3.629 \times 10^{-3}x - 1.237 \times 10^{-6} = 0$$

$$x = 0.0003137$$

$$pH = -\log(0.0003137 + 0.003623) = 2.40$$

After adding 12.5 mL of 0.500 M HCl:

$$\text{mol HCl added} = 12.5 \text{ mL} \times \frac{0.500 \text{ mol}}{1000 \text{ mL}} = 6.25 \times 10^{-3} \text{ mol}$$

$$\text{total volume of solution} = 15.8 \text{ mL} + 12.5 \text{ mL} = 28.3 \text{ mL, or } 0.0283 \text{ L}$$

$$\text{mol pyridine in sample} = 15.8 \text{ mL} \times \frac{0.367 \text{ mol}}{1000 \text{ mL}} = 5.80 \times 10^{-3} \text{ mol}$$

The added HCl will react with the pyridine to make the conjugate base. We are now beyond the equivalence point, so the solution will have excess H^+ and all C_5H_5N will react.

Reaction	$C_5H_5N(aq)$	+	$H^+(aq)$	$\rightleftharpoons$	$C_5H_5NH^+(aq)$
	C_5H_5N (moles)	+	H^+ (mole)		$C_5H_5NH^+$ (moles)
Initial	5.80×10^{-3}		6.25×10^{-3}		0.0
Change	-5.80×10^{-3}		-5.80×10^{-3}		$+5.80 \times 10^{-3}$
Final	0.00		4.50×10^{-4}		5.80×10^{-3}

The total volume of the sample is 0.0283 L and the concentrations of $C_5H_5NH^+$ and H^+ are

$$\frac{0.00580 \text{ mol } C_5H_5NH^+}{0.0283 \text{ L}} = 0.2049 \ M \ C_5H_5NH^+$$

$$\frac{0.00045 \text{ mol } H^+}{0.0283 \text{ L}} = 0.01590 \ M \ H^+$$

We are out of the buffer zone, so the Henderson–Hasselbalch equation will no longer be valid. We set up a RICE table for the hydrolysis of $C_5H_5NH^+$ to solve for the pH of this solution.

Reaction	$C_5H_5NH^+(aq)$	$+ H_2O(\ell) \rightleftharpoons$	$C_5H_5N(aq) +$	$H^+(aq)$
	$[C_5H_5NH^+]$, M		$[C_5H_5N]$, M	$[H^+]$, M
Initial	0.2049		0.0	0.01590
Change	$-x$		$+x$	$+x$
Equilibrium	$0.2049 - x$		x	$0.01590 + x$

$$K_a = \frac{K_w}{K_b} = \frac{1 \times 10^{-14}}{1.00 \times 10^{-8.77}} = 5.89 \times 10^{-6} = \frac{(0.01590 + x)x}{0.2049 - x} \approx \frac{(0.01590)x}{0.2049}$$

$$x = 7.59 \times 10^{-5}$$
$$pH = -\log(0.01590 + 7.59 \times 10^{-5}) = 1.80$$

After adding 15.0 mL of 0.500 M HCl:

$$\text{mol HCl added} = 15.0 \text{ mL} \times \frac{0.500 \text{ mol}}{1000 \text{ mL}} = 7.50 \times 10^{-3} \text{ mol}$$

$$\text{total volume of solution} = 15.8 \text{ mL} + 15.0 \text{ mL} = 30.8 \text{ mL, or } 0.0308 \text{ L}$$

$$\text{mol pyridine in sample} = 15.8 \text{ mL} \times \frac{0.367 \text{ mol}}{1000 \text{ mL}} = 5.80 \times 10^{-3} \text{ mol}$$

The added HCl will react with the pyridine to make the conjugate base. We are now beyond the equivalence point, so the solution will have excess H^+ and all C_5H_5N will react.

Reaction	$C_5H_5N(aq) +$	$H^+(aq)$	$\rightleftharpoons$	$C_5H_5NH^+(aq)$
	C_5H_5N (moles) $+$	H^+(mole)		$C_5H_5NH^+$(moles)
Initial	5.80×10^{-3}	7.50×10^{-3}		0.0
Change	-5.80×10^{-3}	-5.80×10^{-3}		$+5.80 \times 10^{-3}$
Final	0.00	1.70×10^{-3}		5.80×10^{-3}

The total volume of the sample is 0.0308 L and the concentrations of $C_5H_5NH^+$ and H^+ are

$$\frac{0.00580 \text{ mol } C_5H_5NH^+}{0.0308 \text{ L}} = 0.1883 \ M \ C_5H_5NH^+$$

$$\frac{0.0017 \text{ mol } H^+}{0.0308 \text{ L}} = 0.05519 \ M \ H^+$$

We are out of the buffer zone, so the Henderson–Hasselbalch equation will no longer be valid. We set up a RICE table for the hydrolysis of $C_5H_5NH^+$ to solve for the pH of this solution.

Reaction	$C_5H_5NH^+(aq)$	$+ H_2O(\ell) \rightleftharpoons$	$C_5H_5N(aq) +$	$H^+(aq)$
	$[C_5H_5NH^+]$, M		$[C_5H_5N]$, M	$[H^+]$, M
Initial	0.1883		0.0	0.05519
Change	$-x$		$+x$	$+x$
Equilibrium	$0.1883 - x$		x	$0.05519 + x$

$$K_a = \frac{K_w}{K_b} = \frac{1 \times 10^{-14}}{1.00 \times 10^{-8.77}} = 5.89 \times 10^{-6} = \frac{(0.05519 + x)x}{0.1883 - x} \approx \frac{(0.05519)x}{0.1883}$$

$$x = 2.010 \times 10^{-5}$$
$$pH = -\log(0.05519 + 2.010 \times 10^{-5}) = 1.26$$

After adding 20.0 mL of 0.500 M HCl:

$$\text{mol HCl added} = 20.0 \text{ mL} \times \frac{0.500 \text{ mol}}{1000 \text{ mL}} = 1.00 \times 10^{-2} \text{ mol}$$

$$\text{total volume of solution} = 15.8 \text{ mL} + 20.0 \text{ mL} = 35.8 \text{ mL, or } 0.0308 \text{ L}$$

$$\text{mol pyridine in sample} = 15.8 \text{ mL} \times \frac{0.367 \text{ mol}}{1000 \text{ mL}} = 5.80 \times 10^{-3} \text{ mol}$$

The added HCl will react with the pyridine to make the conjugate base. We are now beyond the equivalence point, so the solution will have excess H^+ and all C_5H_5N will react.

Reaction	$C_5H_5N(aq)$	$+$	$H^+(aq)$	$\rightleftharpoons$	$C_5H_5NH^+(aq)$
	C_5H_5N (moles)	$+$	H^+ (mole)		$C_5H_5NH^+$ (moles)
Initial	5.80×10^{-3}		1.00×10^{-2}		0.0
Change	-5.80×10^{-3}		-5.80×10^{-3}		$+5.80 \times 10^{-3}$
Final	0.00		4.20×10^{-3}		5.80×10^{-3}

The total volume of the sample is 0.0308 L and the concentrations of $C_5H_5NH^+$ and H^+ are

$$\frac{0.00580 \text{ mol } C_5H_5NH^+}{0.0358 \text{ L}} = 0.1620 \text{ } M \text{ } C_5H_5NH^+$$

$$\frac{0.00420 \text{ mol } H^+}{0.0358 \text{ L}} = 0.1173 \text{ } M \text{ } H^+$$

We are out of the buffer zone, so the Henderson–Hasselbalch equation will no longer be valid. We set up a RICE table for the hydrolysis of $C_5H_5NH^+$ to solve for the pH of this solution.

Reaction	$C_5H_5NH^+(aq)$ $+ H_2O(\ell)$	$\rightleftharpoons$	$C_5H_5N(aq)$	$+$	$H^+(aq)$
	$[C_5H_5NH^+]$, M		$[C_5H_5N]$, M		$[H^+]$, M
Initial	0.1620		0.0		0.1173
Change	$-x$		$+x$		$+x$
Equilibrium	$0.1620 - x$		x		$0.1173 + x$

$$K_a = \frac{K_w}{K_b} = \frac{1 \times 10^{-14}}{1.00 \times 10^{-8.77}} = 5.89 \times 10^{-6} = \frac{(0.1173 + x)x}{0.1620 - x} \approx \frac{(0.1173)x}{0.1620}$$

$$x = 8.135 \times 10^{-6}$$

$$pH = -\log(0.1173 + 8.135 \times 10^{-6}) = 0.93$$

(d) The titration curve is:

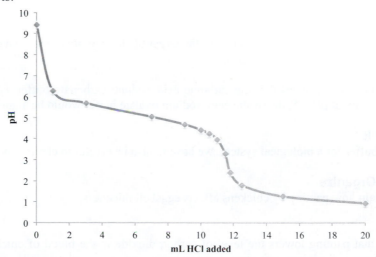

Where pH = pK_a (the half-equivalence point), the major species in solution are C_5H_5N and $C_5H_5NH^+$. At the equivalence point and when the titration is complete, the major species are Cl^- and $C_5H_5NH^+$.

Think About It
Notice in the titration that the presence of H^+ beyond the equivalence point reduces the acidity of $C_5H_5NH^+$ because of the common-ion effect.

16.129. **Collect and Organize**

We are asked whether $CoCO_3$ and $ZnCO_3$, which have nearly identical K_{sp} values, have the same molar solubility and solubility in grams per mole.

Analyze

Molar solubility is the moles of substance that dissolves per liter of solution.

Solve

Because $CoCO_3$ and $ZnCO_3$ have the same number of ions in their formula and they have nearly identical K_{sp} values, they will have the same molar solubilities. However, because of their different molar masses they will have different solubilities in grams per liter.

Think About It

$ZnCO_3$ has a higher molar mass, so it will be more soluble in grams per liter than $CoCO_3$.

16.131. **Collect and Organize**

Given their K_{sp} values we are asked which solution, $Mg(OH)_2$ or $Ca(OH)_2$, will have a higher pH.

Analyze

The greater the solubility of the hydroxide salt, the more OH^- is released, the more basic the solution and the higher the pH.

Solve

Calcium hydroxide has a greater solubility and therefore higher pH than magnesium hydroxide.

Think About It

You can calculate the actual pH of a saturated solution from the K_{sp} values given.

16.133. **Collect and Organize**

We are to choose an appropriate buffer for the action of trypsin and pepsin, whose activity is best at pH 6.5 and pH 1.5, respectively.

Analyze

The best buffer system will have a pK_a close to the target pH for activity of those enzymes.

Solve

For trypsin, with activity at pH 6.5, the carbonic acid–sodium carbonate buffer would be a good choice. For pepsin, with activity at pH 1.5, the oxalic acid–sodium oxalate buffer would be a good choice.

Think About It

In choosing a buffer for a biological system, we have to also be careful to choose one that is nontoxic.

16.135. **Collect and Organize**

We are to explain how panting by chickens affects eggshell thickness.

Analyze

We are given that panting lowers the level of carbon dioxide in the blood of chickens and that eggshells are composed primarily of calcium carbonate.

Solve

The lowering of CO_2 in the blood through panting reduces the amount of CO_2 available as CO_3^{2-} in the blood, which will lower the level of $CaCO_3$ in eggshells.

Think About It

Carbonate in the blood also precisely controls blood pH.

CHAPTER 17 | Electrochemistry: The Quest for Clean Energy

17.1. Collect and Organize

For the voltaic cell shown in Figure P17.1, we are to explain why a porous separator is not required.

Analyze

The porous separator serves to keep the reduction and oxidation half-reactions separate so that electrons are passed through the external circuit.

Solve

Through careful layering, two stacked half-cells are created, each consisting of a metal electrode in contact with a solution of its cations. Direct contact between the layered solutions eliminates the need for a porous bridge to allow ions to migrate between them.

Think About It

The half-reactions and overall reaction for this voltaic cell are

$$Zn(s) \rightarrow Zn^{2+}(aq) + 2\,e^- \qquad E^\circ_{anode} = -0.762 \text{ V}$$

$$Cu^{2+}(aq) + 2\,e^- \rightarrow Cu(s) \qquad E^\circ_{cathode} = 0.342 \text{ V}$$

$$Zn(s) + Cu^{2+}(aq) \rightarrow Zn^{2+}(aq) + Cu(s) \qquad E^\circ_{cell} = E^\circ_{cathode} - E^\circ_{anode}$$
$$= 0.342 - (-0.762) = 1.104 \text{ V}$$

17.3. Collect and Organize

For the voltaic cell shown in Figure P17.3, in which an Ag^+/Ag cell is connected to a standard hydrogen electrode (SHE), we are to determine which electrode is the anode and which is the cathode and indicate in which direction the electrons flow in the outside circuit.

Analyze

A voltaic cell runs spontaneously when E_{cell} is positive. By comparing the reduction potentials of each half-cell, we can write the reaction that is spontaneous for the cell. The half-cell where reduction (gain of electrons) occurs contains the cathode, and the half-cell where oxidation occurs contains the anode. Electrons flow from the anode, where they are produced by oxidation, towards the cathode, where they are required for reduction.

Solve

The spontaneous reaction for this cell is

$$2 \times (Ag^+ + e^- \rightarrow Ag) \qquad E^\circ_{cathode} = 0.7996 \text{ V}$$

$$H_2 \rightarrow 2\,H^+ + 2\,e^- \qquad E^\circ_{anode} = 0.000 \text{ V}$$

$$2\,Ag^+ + H_2 \rightarrow 2\,H^+ + Ag \qquad E^\circ_{cell} = E^\circ_{cathode} - E^\circ_{anode}$$
$$= 0.7996 - 0.000 = 0.7996 \text{ V}$$

Thus, Ag is the cathode, Pt in the SHE is the anode, and electrons flow from the SHE to Ag (to the left in the circuit shown in Figure P17.3).

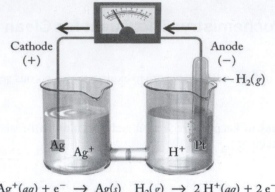

$$Ag^+(aq) + e^- \rightarrow Ag(s) \quad H_2(g) \rightarrow 2\,H^+(aq) + 2\,e^-$$

Think About It

The shorthand notation for this cell would be

$$Pt(s) \,|\, H_2(g) \,|\, H^+(aq) \,\|\, Ag^+(aq) \,|\, Ag(s)$$

where Pt is an inert electrode used in the SHE.

17.5. Collect and Organize

The graph of cell potential versus $[H_2SO_4]$ (Figure P17.5) shows four lines, some curved, some linear, some increasing, and some decreasing, as the concentration of H_2SO_4 decreases. From the shape of the curves and their trends, we are to choose the line that best represents the trend in potential versus $[H_2SO_4]$ in a lead–acid battery.

Analyze

The scale for $[H_2SO_4]$ is logarithmic, and voltage in the battery varies with the $[H_2SO_4]$ according to the Nernst equation:

$$E_{cell} = 2.04\text{ V} - \frac{0.0592}{2} \log \frac{1}{[H_2SO_4]^2}$$

Solve

From the Nernst equation, we see that the cell potential drops as the $\log 1/[H_2SO_4]^2$ decreases. So as $[H_2SO_4]$ decreases, the cell potential also decreases. The red line on the graph shows the opposite trend: the voltage increases as $[H_2SO_4]$ decreases. In considering which of the remaining lines might describe the lead–acid battery, we must consider that because the cell voltage drops as $\log 1/[H_2SO_4]^2$ we expect that the decrease in potential is linear. Therefore, the blue line best describes the potential as a function of $[H_2SO_4]$ concentration.

Think About It

Another characteristic of lead–acid batteries is that their cell voltage does not drop substantially until more than 90% of the battery has been discharged (see Figure 17.12).

17.7. Collect and Organize

For the electrolysis of water in which the two product gases, H_2 and O_2, are collected in burets (Figure P17.7), we are to write the half-reactions occurring at each electrode and discuss why a small amount of acid was added to the water to speed up the reaction.

Analyze

In the electrolysis of water, electricity is supplied to make the nonspontaneous oxidation and reduction reactions occur. From Figure P17.7, we notice that the left buret has collected twice the volume of gas as the right buret. From the overall balanced equation

$$2\,H_2O(\ell) \rightarrow 2\,H_2(g) + O_2(g)$$

we can identify the left buret as containing H_2 and the right buret as containing O_2. E°_{red} values are given in Appendix 6.

Solve

(a) The left electrode is the cathode, where reduction is occurring:
$$2\,H^+(aq) + 2\,e^- \rightarrow H_2(g) \qquad E^\circ_{cathode} = 0.000\ V$$
The right electrode is the anode, where oxidation is occurring:
$$2\,H_2O(\ell) \rightarrow O_2(g) + 4\,H^+(aq) + 4\,e^- \qquad E^\circ_{cathode} = 1.229\ V$$

Cathode reaction:
$$2\,H_2O(\ell) + 2\,e^-$$
$$\downarrow$$
$$H_2(g) + 2\,OH^-(aq)$$

Anode reaction:
$$2\,H_2O(\ell)$$
$$\downarrow$$
$$O_2(g) + 4\,H^+(aq) + 4\,e^-$$

Overall cell reaction
$$H_2O(\ell) \rightarrow H_2(g) + \tfrac{1}{2}\,O_2(g)$$

(b) The ions formed when sulfuric acid dissolves (H^+ and HSO_4^-) make the solution more conductive.

Think About It

The electrochemical potential for the overall process is
$$E^\circ_{cell} = E^\circ_{cathode} - E^\circ_{anode} = -0.828\ V - 1.229\ V = -2.057\ V$$

17.9. Collect and Organize

For the redox flow battery shown in Figure P17.9 and given the cell diagram for the battery, we are asked to write cell reactions for the battery, appropriately label the cells of the battery, and consider how pH changes the voltage of the battery.

Analyze

The cell diagram for the battery is
$$C(s)\ |\ V^{2+}(aq),\ V^{3+}(aq)\ ||\ VO_2^+(aq),\ VO^{2+}(aq)\ |\ C(s)$$
In this battery the inert electrodes are made of carbon. The convention in this cell diagram is that the reduction half-cell is on the right and the oxidation half-cell is on the left. Therefore, in the cell V^{2+} is oxidized to V^{3+} and VO_2^+ is reduced to VO^{2+}.

Solve

(a) At the anode, the following oxidation reaction is occurring:
$$V^{2+}(aq) \rightarrow V^{3+}(aq) + e^-$$
At the cathode, the following reduction reaction is occurring:
$$2\,H^+(aq) + VO_2^+(aq) + e^- \rightarrow VO^{2+}(aq) + H_2O(\ell)$$

(b) The overall redox reaction is
$$2\,H^+(aq) + VO_2^+(aq) + V^{2+}(aq) \rightarrow VO^{2+}(aq) + V^{3+}(aq) + H_2O(\ell)$$
with one electron being transferred in the reaction.

(c) Electrons flow from the oxidation half-cell to the reduction half-cell; from the anode to the cathode.

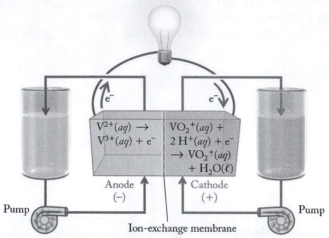

(d) From the Nernst equation we see that as the pH increases (becomes less acidic, $[H^+]$ decreases), the potential of the cell decreases because the value of the ln term becomes more positive.

$$E_{cell} = E^{\circ}_{cell} - \frac{RT}{nF} \ln \frac{[VO^{2+}][V^{3+}]}{[H^+]^2[VO_2^+][V^{2+}]}$$

Think About It
To keep the potential of this battery high, we would want to supply H^+ to the cathode for the oxidation reaction.

17.11. Collect and Organize
We are to explain what is meant by a half-reaction.

Analyze
A half-reaction is a part of how we describe oxidation–reduction reactions.

Solve
A half-reaction is half of an oxidation–reduction (redox) reaction. It can be either the reduction reaction or the oxidation reaction.

Think About It
A half-reaction is used in balancing electrochemical equations and in designing batteries.

17.13. Collect and Organize
We are to explain why a metal wire cannot function as a porous separator in an electrochemical cell.

Analyze
The porous separator allows charged ions to pass through to maintain electrical neutrality in each cell as the redox reaction progresses.

Solve
A wire can pass only electrons through it, not ions, so it cannot function as a porous separator.

Think About It
A wire is used as the conduit for electrons in the external circuit of the voltaic cell.

17.15. Collect and Organize

For each half-reaction, we are to balance it by adding the appropriate number of electrons to the reactant or product side and then identify each reaction as either a reduction or an oxidation reaction.

Analyze

All the reactions in this question are balanced for atoms, so we need to only add electrons to balance the total charge between the reactants side and the products side. A reduction reaction will have electrons on the reactants side; an oxidation reaction will have electrons on the products side.

Solve

(a) $Br_2(\ell) + 2\ e^- \rightarrow 2\ Br^-(aq)$; that is a reduction reaction.
(b) $Pb(s) + 2\ Cl^-(aq) \rightarrow PbCl_2(s) + 2\ e^-$; that is an oxidation reaction.
(c) $O_3(g) + 2\ H^+(aq) + 2\ e^- \rightarrow O_2(g) + H_2O(\ell)$; that is a reduction reaction.
(d) $H_2S(g) \rightarrow S(s) + 2\ H^+(aq) + 2\ e^-$; that is an oxidation reaction.

Think About It

In real chemical systems, oxidation and reduction reactions are always paired.

17.17. Collect and Organize

For the oxidation reaction of Fe_3O_4 to Fe_2O_3 in an acidic water environment, we are to balance the half-reaction describing that process.

Analyze

First, we will need to balance the iron atoms in the half-reaction and then the oxygen atoms by adding H_2O. Next, we balance hydrogen by adding an appropriate number of H^+ ions, followed by adding electrons (we expect that they will be added to the products side because this is an oxidation reaction) to balance the charge in the half-reaction.

Solve

Balancing the iron atoms:
$$2\ Fe_3O_4(s) \rightarrow 3\ Fe_2O_3(s)$$
Balancing the oxygen atoms:
$$2\ Fe_3O_4(s) + H_2O(\ell) \rightarrow 3\ Fe_2O_3(s)$$
Balancing the hydrogen atoms:
$$2\ Fe_3O_4(s) + H_2O(\ell) \rightarrow 3\ Fe_2O_3(s) + 2\ H^+(aq)$$
Balancing the charge:
$$2\ Fe_3O_4(s) + H_2O(\ell) \rightarrow 3\ Fe_2O_3(s) + 2\ H^+(aq) + 2\ e^-$$

Think About It

Indeed, this reaction does balance so that electrons are on the products side, consistent with our expectation that this is an oxidation reaction.

17.19. Collect and Organize

For the $Li(s)/Li_2S(s)$ oxidation and $FeS_2(s)/Fe(s)$ reduction used on a voltaic cell, we are to write half-reactions for the cell's anode and cathode, write the balanced overall cell reaction, and determine how many electrons are transferred in the cell reaction.

Analyze

We are given that the oxidation reaction is $Li(s)$ to $Li_2S(s)$, so that is the anode reaction, and we are given that the reduction reaction is $FeS_2(s)$ to $Fe(s)$, so that is the cathode reaction. In balancing the overall reaction, we must be sure to cancel all the electrons produced by oxidation with those used in reduction. To do so, we might have to multiply either half-reaction, or both, by some factor. The number of electrons transferred in the cell is the number of electrons used to obtain the overall balanced equation.

Solve

(a) $2 \text{Li}(s) + \text{S}^{2-}(non\text{-}aq) \rightarrow \text{Li}_2\text{S}(s) + 2 \text{e}^-$ anode, oxidation

$\text{FeS}_2(s) + 4 \text{e}^- \rightarrow \text{Fe}(s) + 2 \text{S}^{2-}(non\text{-}aq)$ cathode, reduction

(b) We have to multiply the first reaction above by 2 and then add the reactions to obtain the overall reaction:

$$4 \text{Li}(s) + \text{FeS}_2(s) \rightarrow 2 \text{Li}_2\text{S}(s) + \text{Fe}(s)$$

(c) The cell diagram is: $\text{Li}(s) \,|\, \text{LiS}_2(s) \,||\, \text{FeS}_2(s) \,|\, \text{Fe}(s)$

Think About It

As this reaction proceeds, the iron electrode gets heavier and the lithium electrode loses mass.

17.21. Collect and Organize

For the $\text{NiO(OH)}(s)/\text{Ni(OH)}_2(s)$ and $\text{Zn}(s)/\text{Zn(OH)}_2(s)$ voltaic cell, we are to write the appropriate half-reactions for the anode and the cathode, write the balanced overall cell reaction, and diagram the cell.

Analyze

Because we know that Ni(OH)_2 and Zn(OH)_2 are produced, NiO(OH)_2 is reduced and Zn is oxidized in this process. In balancing the overall reaction, we must be sure to cancel all the electrons produced by oxidation with those used in reduction. To do so, we might have to multiply either half-reaction, or both, by some factor. Finally, to diagram the cell we use the following convention:

anode | oxidation half-reaction species || reduction half-reaction species | cathode

making sure to indicate the phases of the species involved and to use vertical lines to separate phases.

Solve

(a) $\text{NiO(OH)}(s) + \text{H}_2\text{O}(\ell) + \text{e}^- \rightarrow \text{Ni(OH)}_2(s) + \text{OH}^-(aq)$ cathode, reduction

$\text{Zn}(s) + 2 \text{OH}^-(aq) \rightarrow \text{Zn(OH)}_2(s) + 2 \text{e}^-$ anode, oxidation

(b) We must multiply the first reaction by 2 and then add the above reactions to obtain the overall reaction:

$$2 \text{NiO(OH)}(s) + 2 \text{H}_2\text{O}(\ell) + \text{Zn}(s) \rightarrow 2 \text{Ni(OH)}_2(s) + \text{Zn(OH)}_2(s)$$

(c) $\text{Zn}(s) \,|\, \text{Zn(OH)}_2(s) \,||\, \text{NiO(OH)}(s) \,|\, \text{Ni(OH)}_2(s)$

Think About It

As this reaction proceeds, the zinc electrode loses mass.

17.23. Collect and Organize

We are asked to equate two different ways to calculate E°_{cell}.

Analyze

Equation 17.1 is

$$E^\circ_{cell} = E^\circ_{red}(\text{cathode}) - E^\circ_{red}(\text{anode})$$

Other textbooks express E°_{cell} in terms of E°_{red} and E°_{ox}:

$$E^\circ_{cell} = E^\circ_{red}(\text{cathode}) + E^\circ_{ox}(\text{anode})$$

Solve

Because $E^\circ_{red}(\text{anode}) = -E^\circ_{ox}(\text{anode})$, we can substitute $-E^\circ_{ox}(\text{anode})$ for $E^\circ_{red}(\text{anode})$ in the expression

$$E^\circ_{cell} = E^\circ_{red}(\text{cathode}) + E^\circ_{ox}(\text{anode})$$

to obtain

$$E^\circ_{cell} = E^\circ_{red}(\text{cathode}) + \left(-E^\circ_{red}(\text{anode})\right)$$
$$= E^\circ_{red}(\text{cathode}) - E^\circ_{red}(\text{anode})$$

That is equal to the expression for E°_{cell} in Equation 17.1.

Think About It

Either expression to calculate E°_{cell} is valid.

17.25. Collect and Organize
From the standard reduction potentials for Cu^{2+}, Co^{2+}, and Hg^{2+}, we are to find a combination of half-reactions that would give the highest positive and lowest positive value of E°_{cell}.

Analyze
The standard potentials for the half-reactions from Appendix 6 are

$$Cu^{2+}(aq) + 2\ e^- \rightarrow Cu(s) \quad E^{\circ}_{cell} = 0.342\ V$$
$$Co^{2+}(aq) + 2\ e^- \rightarrow Co(s) \quad E^{\circ}_{cell} = -0.277\ V$$
$$Hg^{2+}(aq) + 2\ e^- \rightarrow Hg(\ell) \quad E^{\circ}_{cell} = 0.851\ V$$

Solve
(a) The largest value of E°_{cell} is for the reduction of Hg^{2+} by Co.

$$Hg^{2+}(aq) + 2\ e^- \rightarrow Hg(\ell) \quad E^{\circ}_{cathode} = 0.851\ V$$
$$Co(s) \rightarrow Co^{2+}(aq) + 2\ e^- \quad E^{\circ}_{anode} = -0.277\ V$$
$$E^{\circ}_{cell} = E^{\circ}_{cathode} - E^{\circ}_{anode} = 0.851\ V - (-0.277\ V) = 1.128\ V$$

(b) The smallest positive value of E°_{cell} is for the reduction of Hg^{2+} by Cu.

$$Hg^{2+}(aq) + 2\ e^- \rightarrow Hg(\ell) \quad E^{\circ}_{cathode} = 0.851\ V$$
$$Cu(s) \rightarrow Cu^{2+}(aq) + 2\ e^- \quad E^{\circ}_{anode} = -0.342\ V$$
$$E^{\circ}_{cell} = E^{\circ}_{cathode} - E^{\circ}_{anode} = 0.851\ V - (0.342\ V) = 0.509\ V$$

Think About It
Another positive cell potential would be for the reduction of Cu^{2+} by Co with a standard cell potential of 0.619 V.

17.27. Collect and Organize
By calculating E°_{cell} for the possible redox reaction between Ag and Cu^{2+}, we can determine whether the reaction is spontaneous.

Analyze
Because we are told that $[Ag^+] = [Cu^{2+}] = 1.00\ M$, the reaction occurs under standard conditions, and we can calculate E°_{cell} from values of E°_{cell} in Appendix 6. If E°_{cell} is calculated as positive, the reaction is spontaneous.

Solve
The half-reactions, $E^{\circ}_{cathode}$ and E°_{anode}, and the overall reaction and cell potential are

$$2\ Ag(s) \rightarrow 2\ Ag^+(aq) + 2\ e^- \qquad E^{\circ}_{anode} = 0.7996\ V$$
$$\underline{Cu^{2+}(aq) + 2\ e^- \rightarrow Cu(s) \qquad\qquad E^{\circ}_{cathode} = 0.342\ V}$$
$$2\ Ag(s) + Cu^{2+}(aq) \rightarrow 2\ Ag^+(aq) + Cu(s) \quad E^{\circ}_{cell} = E^{\circ}_{cathode} - E^{\circ}_{anode} = -0.458\ V$$

No, the reaction is not spontaneous.

Think About It
The reverse reaction, Cu placed into Ag^+ to dissolve copper and deposit silver, is spontaneous.

17.29. Collect and Organize
For a Ni–Zn cell, we are asked to compare its E°_{cell} with that of a Cu–Zn cell that has $E^{\circ}_{cell} = 1.10\ V$.

Analyze
Using E°_{red} values in Appendix 6, we can identify the spontaneous reaction (having positive E°_{cell}) and calculate E°_{cell}.

Solve

The spontaneous reaction in the Ni–Zn cell is

$$\text{Ni}^{2+}(aq) + 2\,e^- \rightarrow \text{Ni}(s) \qquad\qquad E^\circ_{\text{cathode}} = -0.257\ \text{V}$$
$$\underline{\text{Zn}(s) \rightarrow \text{Zn}^{2+}(aq) + 2\,e^- \qquad\qquad E^\circ_{\text{anode}} = -0.762\ \text{V}}$$
$$E^\circ_{\text{cell}} = E^\circ_{\text{cathode}} - E^\circ_{\text{anode}} = 0.505\ \text{V}$$

The E°_{cell} for a Ni–Zn cell is less than E°_{cell} for a Cu–Zn cell.

Think About It

As we will see later in the chapter, because E°_{cell} for the Ni–Zn cell is less positive than that of a Cu–Zn cell, we also know that the cell reaction is less spontaneous.

17.31. Collect and Organize

For each pair of reduction reactions, we can use the reduction potentials in Appendix 6 to write the equation for the spontaneous cell reaction, identifying the reactions occurring at the anode and cathode.

Analyze

The overall cell potential, E°_{cell}, must be positive for the spontaneous voltaic cell reaction. To calculate E°_{cell} we use

$$E^\circ_{\text{cell}} = E^\circ_{\text{cathode}} - E^\circ_{\text{anode}}$$

The oxidation reaction occurs at the anode, and the reduction reaction occurs at the cathode.

Solve

(a)

Anode $\quad\quad\quad$ $\text{Zn}(s) \rightarrow \text{Zn}^{2+}(aq) + 2\,e^-$ $\quad$ $E^\circ_{\text{anode}} = -0.762\ \text{V}$

Cathode $\quad$ $\text{Hg}^{2+}(aq) + 2\,e^- \rightarrow \text{Hg}(\ell)$ $\quad$ $E^\circ_{\text{cathode}} = 0.851\ \text{V}$

$\overline{\text{Zn}(s) + \text{Hg}^{2+}(aq) \rightarrow \text{Zn}^{2+}(aq) + \text{Hg}(\ell) \quad E^\circ_{\text{cell}} = E^\circ_{\text{cathode}} - E^\circ_{\text{anode}} = 1.613\ \text{V}}$

(b)

Anode $\quad\quad\;$ $\text{Zn}(s) + 2\,\text{OH}^-(aq) \rightarrow \text{ZnO}(s) + \text{H}_2\text{O}(\ell) + 2\,e^-$ $\quad$ $E^\circ_{\text{anode}} = -1.25\ \text{V}$

Cathode $\quad$ $\text{Ag}_2\text{O}(s) + \text{H}_2\text{O}(\ell) + 2\,e^- \rightarrow 2\,\text{Ag}(s) + 2\,\text{OH}^-(aq)$ $\quad$ $E^\circ_{\text{cathode}} = 0.342\ \text{V}$

$\overline{\text{Zn}(s) + \text{Ag}_2\text{O}(s) \rightarrow \text{ZnO}(s) + 2\,\text{Ag}(s) \quad\quad\quad\quad E^\circ_{\text{cell}} = E^\circ_{\text{cathode}} - E^\circ_{\text{anode}} = 1.59\ \text{V}}$

(c)

Anode $\quad\quad\;$ $2 \times [\text{Ni}(s) + 2\,\text{OH}^-(aq) \rightarrow \text{Ni(OH)}_2(s) + 2\,e^-]$ $\quad$ $E^\circ_{\text{anode}} = -0.72\ \text{V}$

Cathode $\quad$ $\text{O}_2(g) + 2\,\text{H}_2\text{O}(\ell) + 4\,e^- \rightarrow 4\,\text{OH}^-(aq)$ $\quad$ $E^\circ_{\text{cathode}} = 0.401\ \text{V}$

$\overline{2\,\text{Ni}(s) + \text{O}_2(g) + 2\,\text{H}_2\text{O}(\ell) \rightarrow 2\,\text{Ni(OH)}_2(s) \quad E^\circ_{\text{cell}} = E^\circ_{\text{cathode}} - E^\circ_{\text{anode}} = 1.12\ \text{V}}$

Think About It

When we multiply a half-reaction to obtain the balanced cell reaction, the oxidation or reduction potential does not change because cell potential is an intensive, not extensive, property.

17.33. Collect, Organize, and Analyze

We are to explain how the negative sign in

$$w_{\text{elec}} = -QE_{\text{cell}}$$

is consistent with a positive cell voltage doing negative electrical work.

Solve

The sign of electrical work is negative because in passing voltage through a cell (or outside circuit), the redox reaction is doing work on the surroundings; therefore, the sign of work from the perspective of the system (the redox reaction) is negative.

Think About It

An example of positive electrical work would be charging a battery.

17.35. **Collect and Organize**

For each redox reaction given, we are to calculate the cell potential after calculating the value of free energy.

Analyze

We calculate $\Delta G°$ for each reaction by using free energy of formation values for the products and reactants from Appendix 4. The cell potential can then be calculated by using

$$\Delta G°_{cell} = -nFE°_{cell}$$

$$E°_{cell} = \frac{-\Delta G°}{nF}$$

Solve

(a) $\Delta G° = \left[(1\ mol\ Cu^{2+} \times 65.5\ kJ/mol) + (1\ mol\ Cu \times 0.0\ kJ/mol)\right] - (2\ mol\ Cu^{+} \times 50.0\ kJ/mol) = -34.5\ kJ$

$$E°_{cell} = \frac{+34,500\ J}{1\ mol \times 9.65 \times 10^4\ C/mol} = +0.358\ V$$

(b)

$$\Delta G° = \left[(1\ mol\ Ag^{+} \times 77.1\ kJ/mol) + (1\ mol\ Fe^{2+} \times -78.9\ kJ/mol)\right] -$$

$$\left[(1\ mol\ Ag \times 0.0\ kJ/mol) + (1\ mol\ Fe^{3+} \times -4.7\ kJ/mol)\right] = 2.9\ kJ$$

$$E°_{cell} = \frac{-2900\ J}{1\ mol \times (9.65 \times 10^4\ C/mol)} = -0.030\ V$$

Think About It

Reaction (a) is spontaneous and reaction (b) is nonspontaneous.

17.37. **Collect and Organize**

Using the relationship between free energy and cell potential,

$$\Delta G_{cell} = -nFE_{cell}$$

we can calculate ΔG_{cell} for a Zn–MnO$_2$ battery generating 1.50 V.

Analyze

To use the equation, we need the value of n, the number of electrons transferred in the overall balanced equation. From the cell reaction provided, we see that Zn is oxidized to Zn^{2+} and 2 mol of Mn^{4+} (in MnO_2) is reduced to Mn^{3+} (in Mn_2O_3), so 2 mol of electrons is transferred in the reaction. We should also be aware of the units for the quantities in this equation. The value of F is 9.65×10^4 C/mol and E is in volts, which, being equivalent to joules per coulomb (J/C), means that ΔG is calculated in joules, which we can convert to kilojoules, the usual units for free energy.

Solve

$$\Delta G_{cell} = -2\ mol \times \frac{9.65 \times 10^4\ C}{mol} \times \frac{1.50\ J}{C} \times \frac{1\ kJ}{1000\ J} = -290\ kJ$$

Think About It

This reaction is spontaneous because the cell potential is positive, making ΔG_{cell} negative.

17.39. **Collect and Organize**

For a battery based on $NiCl_2(s)$ and $Na(s)$ with the following overall cell reaction

$$2\ Na(s) + NiCl_2(s) \rightarrow Ni(s) + 2\ NaCl(s)$$

we are to assign oxidation numbers to the elements in both the reactants and the products, determine the number of electrons transferred in the cell reaction, and compute the free energy for the cell reaction.

Analyze
Using the relationship between free energy and cell potential,

$$\Delta G_{cell} = -nFE_{cell}$$

we can calculate ΔG_{cell} for this sodium nickel chloride cell generating 2.58 V. We should also be aware of the units for the quantities in this equation. The value of F is 9.65×10^4 C/mol and E is in volts, which, being equivalent to joules per coulomb (J/C), means that ΔG is calculated in joules, which we can convert to kilojoules, the usual units for free energy. To use the equation we need the value of n, the number of electrons transferred in the overall balanced equation. From the cell reaction provided we see that Ni^{2+} [in $NiCl_2$] is reduced to Ni and Na is oxidized to Na+ [in NaCl].

Solve
(a) Na has Na^0.
NaCl has Na^+ and Cl^-.
$NiCl_2$ has Ni^{2+} and Cl^-.
Ni has Ni^0.

(b) The reduction and oxidation half-reactions are:

$$Na(s) + Cl^-(nonaq) \rightarrow NaCl(s) + e^-$$
$$NiCl_2(s) + 2\,e^- \rightarrow Ni(s) + 2\,Cl^-(nonaq)$$

In the overall balanced reaction, 2 electrons are transferred.

$$2\,Na(s) + NiCl_2(s) \rightarrow 2\,NaCl(s) + Ni(s)$$

(c)

$$\Delta G_{cell} = -2\ \text{mol} \times \frac{9.65 \times 10^4\ \text{C}}{\text{mol}} \times \frac{2.58\ \text{J}}{\text{C}} \times \frac{1\ \text{kJ}}{1000\ \text{J}} = -498\ \text{kJ/mol}$$

Think About It
This reaction is spontaneous because the cell potential is positive, making ΔG_{cell} negative.

17.41. **Collect and Organize**
For each reaction, we can break it up into its appropriate half-reactions and, using E°_{red} values from Appendix 6, calculate E°_{cell}. From that cell potential, we can then calculate the free energy for each reaction.

Analyze
To calculate the cell potential, we use

$$E^\circ_{cell} = E^\circ_{cathode} - E^\circ_{anode}$$

Once we have calculated E°_{cell}, we use

$$\Delta G^\circ = -nFE^\circ_{cell}$$

to calculate the free energy of the reaction. Here, we have to remember that n is the number of moles of electrons transferred in the overall balanced equation.

Solve
(a)

$$Cu(s) \rightarrow Cu^{2+}(aq) + 2\,e^- \qquad E^\circ_{anode} = 0.342\ \text{V}$$
$$Sn^{2+}(aq) + 2\,e^- \rightarrow Sn(s) \qquad E^\circ_{cathode} = -0.136\ \text{V}$$
$$\overline{Cu(s) + Sn^{2+}(aq) \rightarrow Cu^{2+}(aq) + Sn(s) \qquad E^\circ_{cell} = E^\circ_{cathode} - E^\circ_{anode} = -0.478\ \text{V}}$$

$$\Delta G^\circ = -2\ \text{mol} \times \frac{9.65 \times 10^4\ \text{C}}{\text{mol}} \times \frac{-0.478\ \text{J}}{\text{C}} \times \frac{1\ \text{kJ}}{1000\ \text{J}} = 92.2\ \text{kJ}$$

(b)

$$
\begin{array}{ll}
Zn(s) \rightarrow Zn^{2+}(aq) + 2\,e^- & E^\circ_{anode} = -0.762 \text{ V} \\
Ni^{2+}(aq) + 2\,e^- \rightarrow Ni(s) & E^\circ_{cathode} = -0.257 \text{ V} \\
\hline
Zn(s) + Ni^{2+}(aq) \rightarrow Zn^{2+}(aq) + Ni(s) & E^\circ_{cell} = E^\circ_{cathode} - E^\circ_{anode} = 0.505 \text{ V}
\end{array}
$$

$$
\Delta G^\circ = -2 \text{ mol} \times \frac{9.65 \times 10^4 \text{ C}}{mol} \times \frac{0.505 \text{ J}}{C} \times \frac{1 \text{ kJ}}{1000 \text{ J}} = -97.5 \text{ kJ}
$$

Think About It

Reaction (b), with its positive cell potential, is spontaneous. Reaction (a) is nonspontaneous, so copper metal in contact with Sn^{2+} solution does not oxidize.

17.43. Collect, Organize, and Analyze

We are to describe the function of platinum in the standard hydrogen electrode.

Solve

The platinum electrode transfers electrons to the half-cell; because it is inert, it is not involved in the reaction.

Think About It

The inert electrode simply serves as a place for the oxidation or reduction reaction to occur, and as a conduit for electrons for the reaction.

17.45. Collect and Organize

We are to explain why the cell potential of most batteries changes little until the battery is almost discharged, when the voltage drops significantly.

Analyze

Voltage of a battery (a voltaic cell) is governed by the Nernst equation:

$$
E_{cell} = E^\circ_{cell} - \frac{RT}{nF} \ln Q
$$

As a battery discharges, the value of Q, the reaction quotient, changes:

$$
Q = \frac{[\text{products}]^x}{[\text{reactants}]^y}
$$

Solve

At the start of the reaction, Q is very small because [reactants] >> [products]. As the reaction proceeds [products] grows and Q increases but does not increase significantly until significant amounts of products form, that is, when the battery is nearly discharged.

Think About It

As an example, Figure 17.11 shows the cell potential of a lead–acid battery as a function of discharge. Notice that the voltage is relatively constant until the battery is approximately 90% discharged.

17.47. Collect and Organize

We can use the Nernst equation to calculate the cell potential when Fe^{3+} is combined with Cr^{2+} at nonstandard conditions.

Analyze

Because $T = 298$ K, we can use the following form of the Nernst equation:

$$
E_{cell} = E^\circ_{cell} - \frac{0.0592}{n} \log Q
$$

where E°_{cell} is the potential of the cell under standard conditions (calculated from tabulated E°_{red} values), n is the moles of electrons transferred in the overall balanced redox equation, and Q is the reaction quotient.

Solve

First, we need to calculate $E°_{cell}$ and determine the value of n. The half-reactions and overall cell reaction are

$$Fe^{3+}(aq) + e^- \rightarrow Fe^{2+}(aq) \qquad\qquad E°_{cathode} = 0.770 \text{ V}$$
$$\underline{Cr^{2+}(aq) \rightarrow Cr^{3+}(aq) + e^-} \qquad\qquad E°_{anode} = -0.41 \text{ V}$$
$$Fe^{3+}(aq) + Cr^{2+}(aq) \rightarrow Fe^{2+}(aq) + Cr^{3+}(aq) \qquad E°_{cell} = E°_{cathode} - E°_{anode} = 1.18 \text{ V}$$

We see that $n = 1$. Now we can use the Nernst equation to calculate E_{cell} when $[Fe^{3+}] = [Cr^{2+}] = 1.50 \times 10^{-3}$ M and $[Fe^{2+}] = [Cr^{3+}]$ 2.5×10^{-4} M.

$$E_{cell} = 1.18 \text{ V} - \frac{0.0592}{1} \log \frac{\left(2.5 \times 10^{-4}\right)^2}{\left(1.50 \times 10^{-3}\right)^2} = 1.27$$

Think About It

This reaction became more spontaneous (higher cell potential) under these conditions.

17.49. Collect and Organize

We can use the following equation to calculate the equilibrium constant for the given redox reaction:

$$\log K = \frac{nE°_{cell}}{0.0592}$$

Analyze

First, we have to determine the standard cell potential, $E°_{cell}$, using the $E°_{red}$ values in Appendix 6, and determine the value of n, the moles of electrons transferred in the overall balanced equation.

Solve

$$Fe^{3+}(aq) + e^- \rightarrow Fe^{2+}(aq) \qquad\qquad E°_{cathode} = 0.770 \text{ V}$$
$$\underline{Cr^{2+}(aq) \rightarrow Cr^{3+}(aq) + e^-} \qquad\qquad E°_{anode} = -0.41 \text{ V}$$
$$Fe^{3+}(aq) + Cr^{2+}(aq) \rightarrow Fe^{2+}(aq) + Cr^{3+}(aq) \quad E°_{cell} = E°_{cathode} - E°_{anode} = 1.18 \text{ V}$$
$$n = 1$$

$$\log K = \frac{1 \times 1.18 \text{ V}}{0.0592} = 19.9324$$
$$K = 1 \times 10^{19.9324} = 8.56 \times 10^{19}$$

Think About It

Because K for this reaction is very large, the reaction goes very far to the right.

17.51. Collect and Organize

We can use the Nernst equation to calculate the potential of the hydrogen electrode at pH = 7.00.

Analyze

We are reminded that under standard conditions [1 atm $H_2(g)$ and 1.00 M H^+ = pH = 0.00], the voltage of the hydrogen cell is zero. This is $E°_{cell}$. The overall reaction for the cell (against SHE) is

$$2 H^+(aq) + 2 e^- \rightarrow H_2(g) \qquad\qquad ([H^+] = 1 \times 10^{-7} M)$$
$$\underline{H_2(g) \rightarrow 2 H^+(aq) + 2 e^- \qquad (SHE)}$$
$$2 H^+(aq) \, (1 \times 10^{-7} M) + H_2(g) \, (1 \text{ atm}) \rightarrow 2 H^+(aq) \, (1.00 M) + H_2(g) \, (1 \text{ atm})$$

The form of Q for the Nernst equation is

$$Q = \frac{\left(1.00 M\right)^2 \times \left(1 \text{ atm}\right)}{\left(1 \times 10^{-7} M\right)^2 \times \left(1 \text{ atm}\right)} = \frac{1}{\left(1 \times 10^{-7}\right)^2}$$

Solve

$$E_{cell} = 0.000 \text{ V} - \frac{0.0592}{2} \log \frac{1}{\left(1 \times 10^{-7}\right)^2} = -0.414 \text{ V}$$

Think About It

The spontaneous reaction actually is the reverse reaction:

$$2 \text{ H}^+(aq) \ (1.00 \ M) + \text{H}_2(g) \ (1 \text{ atm}) \rightarrow 2 \text{ H}^+(aq) \ (1 \times 10^{-7} \ M) + \text{H}_2(g) \ (1 \text{ atm}) \qquad E^\circ_{cell} = 0.414 \text{ V}$$

In this redox cell, acid in the SHE will be reduced and H_2 in the cell where pH = 7.00 will be oxidized.

17.53. Collect and Organize

We can use the Nernst equation to calculate the potential for the reduction of MnO_4^- to MnO_2 in the presence of SO_3^{2-} when $[MnO_4^-] = 0.150 \ M$, $[SO_3^{2-}] = 0.256 \ M$, $[SO_4^{2-}] = 0.178 \ M$, and $[OH^-] = 0.0100 \ M$. We are also to assess whether the potential increases or decreases as reactants become products in the reaction.

Analyze

To use the Nernst equation, we need to know the value of E°_{cell} and n. We can determine both of those by writing the half-reactions and balancing the redox reaction.

Solve

$$2 \text{ MnO}_4^-(aq) + 4 \text{ H}_2\text{O}(\ell) + 6 \text{ e}^- \rightarrow 2 \text{ MnO}_2(s) + 8 \text{ OH}^-(aq) \qquad E^\circ_{cathode} = 0.59 \text{ V}$$

$$3 \text{ SO}_3^{2-}(aq) + 6 \text{ OH}^-(aq) \rightarrow 3 \text{ SO}_4^{2-}(aq) + 3 \text{ H}_2\text{O}(\ell) + 6 \text{ e}^- \qquad E^\circ_{anode} = -0.92 \text{ V}$$

$$\overline{2 \text{ MnO}_4^-(aq) + 3 \text{ SO}_3^{2-}(aq) + \text{H}_2\text{O}(\ell) \rightarrow 2 \text{ MnO}_2(s) + 3 \text{ SO}_4^{2-}(aq) \ 2 \text{ OH}^-(aq)} \qquad \begin{array}{l} E^\circ_{cell} = E^\circ_{cathode} - E^\circ_{anode} = 1.51 \text{ V} \\ n = 6 \end{array}$$

$$E_{cell} = 1.51 \text{ V} - \frac{0.0592}{6} \log \frac{(0.178 \ M)^3 (0.0100 \ M)^2}{(0.150 \ M)^2 (0.256 \ M)^3}$$

$$E_{cell} = 1.54 \text{ V}$$

As the reaction proceeds, the concentrations of the reactants decrease and the concentrations of the products increase, so Q increases and $\log Q$ becomes more positive. When a more positive $\log Q$ is multiplied by $0.0592/6$ and then subtracted from the E°_{cell}, E_{cell} decreases.

Think About It

Be sure to use half-reactions to determine the correct value of n.

17.55. Collect and Organize

For the reaction of copper pennies with nitric acid, we are to calculate E°_{cell} and then E_{cell} when $[\text{H}^+] = 0.100 \ M$, $[\text{NO}_3^-] = 0.0250 \ M$, $[\text{Cu}^{2+}] = 0.0375 \ M$, and $P_{NO} = 0.00150$ atm.

Analyze

E°_{cell} is calculated by adding E°_{anode} and $E^\circ_{cathode}$. The reaction is spontaneous, so our calculated E°_{cell} must be positive. In balancing the reaction, we can also determine the value of n for the Nernst equation.

Solve

(a)

$$3 \text{ Cu}(s) \rightarrow 3 \text{ Cu}^{2+}(aq) + 6 \text{ e}^- \qquad E^\circ_{anode} = 0.3419 \text{ V}$$

$$\underline{2 \text{ NO}_3^- + 8 \text{ H}^+(aq) + 6 \text{ e}^- \rightarrow 2 \text{ NO}(g) + 4 \text{ H}_2\text{O}(\ell)} \qquad E^\circ_{cathode} = 0.96 \text{ V}$$

$$3 \text{ Cu}(s) + 2 \text{ NO}_3^- + 8 \text{ H}^+(aq) \rightarrow 3 \text{ Cu}^{2+}(aq) + 2 \text{ NO}(g) + 4 \text{ H}_2\text{O}(\ell) \qquad E^\circ_{cell} = E^\circ_{cathode} - E^\circ_{anode} = 0.62 \text{ V}$$

(b) $\quad E_{cell} = 0.6181 \text{ V} - \dfrac{0.0592}{6} \log \dfrac{(0.0375 \ M)^3 (0.00150 \text{ atm})^2}{(0.0250 \ M)^2 (0.100 \ M)^8} = 0.61 \text{ V}$

Think About It

We may mix concentration units of atmospheres and molarity, as we do in this calculation of Q.

17.57. Collect and Organize

For the reaction of NH_4^+ with O_2 in water, we are to calculate E_{cell}°. We can then use the Nernst equation to determine $[NO_3^-]/[NH_4^+]$ for $P_{O_2} = 0.21$ atm and pH = 5.60 at 298 K.

Analyze

E_{cell}° is calculated by adding E_{anode}° and $E_{cathode}^\circ$. The reaction is spontaneous, so our calculated E_{cell}° must be positive. In balancing the reaction, we can determine the value of n. Because the system is at equilibrium in part b, we know that $E_{cell} = 0$. Therefore, $[NO_3^-]/[NH_4^+]$ can be determined through the equation

$$0 = E_{cell}^\circ - \frac{0.0592}{n} \log \frac{[NO_3^-][H^+]^2}{[NH_4^+]\left(P_{O_2}\right)^2}$$

Solve

(a)

$$
\begin{array}{ll}
NH_4^+(aq) + 3\,H_2O(\ell) \rightarrow NO_3^-(aq) + 10\,H^+ + 8\,e^- & E_{anode}^\circ = 0.88\ V \\
2\,O_2(g) + 8\,H^+(aq) + 8\,e^- \rightarrow 4\,H_2O(\ell) & E_{cathode}^\circ = 1.229\ V \\
\hline
NH_4^+(aq) + 2\,O_2(g) \rightarrow NO_3^-(aq) + H_2O(\ell) + 2\,H^+(aq) & E_{cell}^\circ = E_{cathode}^\circ - E_{anode}^\circ = 0.349\ V
\end{array}
$$

(b)

$$0 = 0.349\ V - \frac{0.0592}{8} \log \frac{[NO_3^-]\left(2.512 \times 10^{-6}\ M\right)^2}{[NH_4^+](0.21\ \text{atm})^2}$$

$$-0.349\ V = -\frac{0.0592}{8} \log \frac{[NO_3^-]\left(2.512 \times 10^{-6}\ M\right)^2}{[NH_4^+](0.21\ \text{atm})^2}$$

$$47.16 = \log\left(\left(1.431 \times 10^{-10}\right) \times \frac{[NO_3^-]}{[NH_4^+]} \right)$$

$$1 \times 10^{47.16} = 1.453 \times 10^{47} = 1.43 \times 10^{-10} \times \frac{[NO_3^-]}{[NH_4^+]}$$

$$\frac{[NO_3^-]}{[NH_4^+]} = 1.02 \times 10^{57}$$

Think About It

This ratio is consistent with a spontaneous reaction as indicated by the positive E_{cell}°.

17.59. Collect and Organize

We are to compare a 12 V lead–acid battery with one that has a lower ampere-hour rating.

Analyze

An ampere-hour is a unit of electrical charge and is defined as the electric charge transferred by 1 A of current for 1 h. It is used to describe the life of a battery.

Solve

The total masses of the electrode materials (c) and the combined surface areas of the electrodes (f) are likely to be different.

Think About It

Both batteries use the same components (b and e) and have the same voltage (a and d).

17.61. **Collect and Organize**

We are to compare two voltaic cells to determine which produces more charge per gram of anode material.

Analyze

For each cell we must first identify which species is the anode and the number of electrons transferred when 1 mol of anode is consumed in the reaction. The charge generated by the reaction is

$$C = nF$$

where C = charge in coulombs, n = moles of electrons transferred in the balanced equation, and $F = 9.65 \times 10^4$ C/mol. That gives the charge per mole of anode. To convert into charge per gram, we use the molar mass:

$$\frac{\text{charge}}{\text{gram}} = \frac{\text{coulombs/mol}}{\text{molar mass of anode material}}$$

Solve

For Cell A, the NiO(OH)–Cd voltaic cell, Cd is the anode material:

$$\frac{C}{g} = \frac{2 \text{ mol e}^- \times 9.65 \times 10^4 \text{C / mol}}{112.41 \text{ g/mol}} = 1.72 \times 10^3 \text{ C/g Cd}$$

For Cell B, the Al–O_2 voltaic cell, Al is the anode material:

$$\frac{C}{g} = \frac{12 \text{ mol e}^- \times 9.65 \times 10^4 \text{C / mol}}{26.98 \text{ g/mol}} = 4.29 \times 10^4 \text{ C/g Al}$$

Therefore, Cell B, the Al–O_2 cell, produces a greater charge per gram.

Think About It

The number of electrons transferred in the oxidation of Al to Al^{3+} is 4 mol × 3 e$^-$/mol = 12 e$^-$.

17.63. **Collect and Organize**

We are to compare two voltaic cells to determine which produces more energy per gram of anode material.

Analyze

The energy of a voltaic cell is the force to move electrons from the anode to the cathode. The unit of volts is energy per unit charge, so

$$\text{Energy} = \text{volts} \times \text{charge (in units of V} \cdot \text{C)}$$

where 1 V = 1 J/C. The charge in the cell generated by 1 g of anode material is

$$\text{Charge} = 1 \text{ g} \times \text{molar mass of anode material} \times \frac{\text{mol e}^-}{\text{mol anode}} \times \frac{9.65 \times 10^4 \text{C}}{\text{mol}}$$

Solve

For Reaction E, the Zn–Ni(OH)$_2$ cell, Zn is the anode material:

$$\text{Charge} = 1 \text{ g} \times \frac{1 \text{ mol}}{65.38 \text{ g}} \times \frac{2 \text{ mol e}^-}{1 \text{ mol Zn}} \times \frac{9.65 \times 10^4 \text{ C}}{\text{mol e}^-} = 2.952 \times 10^3 \text{ C}$$

$$\text{Energy} = 1.20 \frac{\text{J}}{\text{C}} \times 2.952 \times 10^3 \text{ C} = 3.54 \times 10^3 \text{ J}$$

For Reaction F, the Li–MnO$_2$ cell, Li is the anode material:

$$\text{Charge} = 1 \text{ g} \times \frac{1 \text{ mol}}{6.941 \text{ g}} \times \frac{1 \text{ mol e}^-}{1 \text{ mol Li}} \times \frac{9.65 \times 10^4 \text{ C}}{\text{mol e}^-} = 1.390 \times 10^4 \text{ C}$$

$$\text{Energy} = 3.15 \frac{\text{J}}{\text{C}} \times 1.390 \times 10^4 \text{C} = 4.38 \times 10^4 \text{ J}$$

Therefore, Reaction F, the Li–MnO$_2$ cell, generates more energy per gram of anode.

Think About It

Although the charge generated per mole of Li versus that of a mole of Zn in those cells is lower, the high voltage of the Li–MnO$_2$ cell means that this cell generates more energy.

17.65. Collect and Organize

We are to explain the differences in the signs of the cathode in a voltaic versus an electrolytic cell.

Analyze

The signs of the electrodes in a cell indicate the direction of electron flow.

Solve

In a voltaic cell, the electrons are produced at the anode, so a negative (–) charge builds up there; in an electrolytic cell, electrons are being forced onto the cathode so that it builds up negative (–) charge. The flow of electrons in the outside circuit is reversed in an electrolytic cell in comparison with the flow in a voltaic cell.

Think About It

An electrolytic cell uses an outside source of electrical energy to cause a nonspontaneous reaction to occur.

17.67. Collect and Organize

In a mixture of molten Br^- and Cl^- salts, we are to predict which product, Br_2 or Cl_2, forms first in an electrolytic cell as the voltage is increased.

Analyze

The oxidations of Br^- and Cl^- are expressed as

$$2\,Br^-(\ell) \rightarrow Br_2(\ell) + 2\,e^-$$
$$2\,Cl^-(\ell) \rightarrow Cl_2(\ell) + 2\,e^-$$

Solve

The halide that is first to be oxidized is the one with the lowest ionization energy. Br^-, being larger and less electronegative than Cl^-, loses its electron more readily; therefore, Br_2 forms first in the cell as the voltage is increased.

Think About It

If the molten salt also contains F^-, F_2 would form after Br_2 and Cl_2.

17.69. Collect and Organize

For the electrolysis of a 1.0 $M\,Cu^{2+}$ solution, we are to determine whether the potential at the cathode where the reduction of Cu^{2+} occurs needs to be more negative or less negative than 0.34 V to quantitatively reduce the Cu^{2+} in solution to Cu.

Analyze

We are given that E°_{red} for Cu^{2+} is 0.34 V. That is under standard conditions when $[Cu^{2+}]$ = 1.0 M, the concentration of the solution at the start of the electrolysis. As $[Cu^{2+}]$ decreases as Cu is deposited on the electrode, E_{cell} can be calculated by using the Nernst equation:

$$E_{cell} = 0.34\text{ V} - \frac{0.0592}{2}\log\frac{1}{[Cu^{2+}]}$$

Solve

As the reaction proceeds and $[Cu^{2+}]$ decreases, the value of log $(1/[Cu^{2+}])$ becomes more positive. As a result, E_{cell} decreases, so the cathode potential must be more negative than 0.34 V to complete the reduction.

Think About It

A slight overpotential might be required to accomplish the electrolysis, however, because of a kinetic barrier to the reduction reaction.

17.71. Collect and Organize

In an electroplating process, we are to calculate the mass of silver deposited on an object when a charge of 1.7 A · h is delivered from a battery. To determine that, we need to relate the ampere-hours to the total number of electrons generated. Then we can relate the number of electrons to the mass of silver deposited from a solution of Ag^+.

Analyze

We can convert ampere-hours to coulombs:

$$A \cdot h \times \frac{1\ C}{A \cdot s} \times \frac{3600\ s}{h}$$

We can calculate the moles of electrons used in the process from that result, knowing that 1 mol of $e^- = 9.65 \times 10^4$ C. To calculate the mass of Ag deposited, we also need to know that the reduction of Ag^+ to Ag is a one-electron process.

Solve

$$1.7\ A \cdot h \times \frac{1\ C}{A \cdot s} \times \frac{3600\ s}{h} = 6120\ C$$

$$6120\ C \times \frac{1\ mol\ e^-}{9.65 \times 10^4\ C} = 6.342 \times 10^{-2}\ mol\ e^-$$

$$6.342 \times 10^{-2}\ mol\ e^- \times \frac{1\ mol\ Ag}{1\ mol\ e^-} \times \frac{107.87\ g\ Ag}{mol} = 6.8\ g$$

Think About It

The higher the amps for the battery, the faster an object can be electroplated.

17.73. **Collect and Organize**

From knowing that 0.446 g of gold is produced when the same amount of charge is passed through the solution as when 0.732 g of silver is deposited from $AgNO_3(aq)$, we are to determine the oxidation state of the gold.

Analyze

From the mass of silver deposited, we can calculate the moles of electrons needed to deposit that amount of silver, as 1 mol of electrons would be needed to reduce 1 mol of Ag^+ ions. We can then calculate the moles of Au deposited and divide the number of electrons used to deposit that gold by the moles of Au deposited to give the charge on the gold in the solution.

Solve

Moles of electrons used to deposit 0.732 g of silver:

$$0.732\ g \times \frac{1\ mol\ Ag}{107.87} \times \frac{1\ mol\ e^-}{1\ mol\ Ag} = 6.79 \times 10^{-3}\ mol$$

Moles of gold deposited:

$$0.466\ g \times \frac{1\ mol}{196.97\ g} = 2.26 \times 10^{-3}\ mol$$

Charge of the gold ions in the solution:

$$\frac{6.79 \times 10^{-3}\ mol\ e^-}{2.26 \times 10^{-3}\ mol\ Au} = 3$$

Think About It

For the same amount of charge delivered to the silver nitrate solution, one-third the molar amount of gold is deposited because of the difference in the charge of their ions (Au^{3+} versus Ag^+).

17.75. **Collect and Organize**

We are to calculate how long it will take to recharge a battery that contains 4.10 g of NiO(OH) and is 50% discharged. Therefore, 2.05 g of NiO(OH) has been depleted from the battery. The charger for the battery operates at 2.00 A and 1.3 V.

Analyze

We first need to calculate the moles of electrons needed to recover 2.05 g of NiO(OH), which in the NiMH battery forms $Ni(OH)_2$ in a one-electron process. Next, we will convert the moles of electrons to coulombs. Because $1\ C = 1\ A \cdot s$ and we know the amperes at which the charger operates, we can then calculate the time it takes the charger to deliver the electrons to recharge the battery.

Solve

$$2.05 \text{ g NiO(OH)} \times \frac{1 \text{ mol}}{91.70 \text{ g}} \times \frac{1 \text{ mol e}^-}{1 \text{ mol NiO(OH)}} = 0.0224 \text{ mol e}^-$$

$$0.0224 \text{ mol e}^- \times \frac{9.65 \times 10^4 \text{C}}{\text{mol e}^-} \times \frac{\text{A} \cdot \text{s}}{\text{C}} \times \frac{1}{2.00 \text{ A}} = 1080 \text{ s}$$

In minutes that is $1080 \text{ s} \times 1 \text{ min}/60 \text{ s} = 18.0 \text{ min}$.

Think About It

The larger the battery, the more of the reactant is needed to be regenerated and the longer it takes to recharge it.

17.77. Collect and Organize

We are to calculate the amount of O_2 that could be generated in 1 h on a submarine by using electrolysis and then consider the practicality of using seawater as the source of oxygen for the submarine.

Analyze

We can calculate the moles of O_2 produced by the electrolytic cell through

$$\text{Moles } O_2 = \text{time in seconds} \times \text{amperes} \times \frac{1 \text{ C}}{\text{A} \cdot \text{s}} \times \frac{\text{mol e}^-}{9.65 \times 10^4 \text{C}} \times \frac{1 \text{ mol } O_2}{4 \text{ mol e}^-}$$

Notice that the oxidation of water to O_2 is a 4 e$^-$ process. We can then use the ideal gas law to calculate the volume of O_2 produced.

Solve

(a)
$$\text{Moles } O_2 = 1 \text{ hr} \times \frac{3600 \text{ s}}{1 \text{ hr}} \times 0.025 \text{ A} \times \frac{1 \text{ C}}{\text{A} \cdot \text{s}} \times \frac{1 \text{ mol e}^-}{9.65 \times 10^4 \text{C}} \times \frac{1 \text{ mol } O_2}{4 \text{ mol e}^-} = 2.332 \times 10^{-4} \text{ mol } O_2$$

$$V = \frac{2.332 \times 10^{-4} \text{ mol} \times 0.08206 \text{ L} \cdot \text{atm} / \text{mol} \cdot \text{K} \times 298 \text{ K}}{0.98692 \text{ atm}/1 \text{ bar}} = 5.78 \times 10^{-3} \text{ L, or } 5.8 \text{ mL}$$

(b) Seawater contains a fairly high concentration of Cl$^-$ and Br$^-$ that can be oxidized, so the direct electrolysis of seawater would not be useful as an oxygen source.

Think About It

Submarines probably purify their water, perhaps through a reverse osmosis process, to remove the chloride and bromide and other ions before the electrolysis process.

17.79. Collect and Organize

For the process that electroplates nickel, we are to calculate the lowest potential required to deposit Ni onto a piece of iron by using a 0.35 M Ni^{2+} solution.

Analyze

To solve this problem we need to use the Nernst equation:

$$E_{\text{cell}} = E_{\text{cell}}^{\circ} - \frac{0.0592}{n} \log Q$$

where E_{cell}° is the reduction potential of Ni^{2+} versus the SHE (-0.257 V), n is the number of electrons needed to reduce Ni^{2+} to Ni^{+0}, and Q is 1/[Ni^{2+}] on the basis of the reduction reaction

$$\text{Ni}^{2+}(aq) + 2 \text{ e}^- \rightarrow \text{Ni}(s)$$

Solve

$$E_{\text{cell}} = -0.257 \text{ V} - \frac{0.0592}{2} \log \frac{1}{0.35 \, M} = -0.270 \text{ V}$$

Think About It

For this electrolysis reaction, using a more dilute solution of the metal cation necessitates an increase in the potential needed to cause Ni^{2+} to deposit on the iron.

17.81. Collect and Organize

We are to consider the advantages and disadvantages of hybrid power systems versus all-electric fuel cell systems.

Analyze

A parallel hybrid power system uses traditional petroleum-based fuel for high power demands and an electric motor for lower power demands. An all-electric fuel cell system uses only electrochemical power based on combustion half-reactions to supply power.

Solve

A hybrid vehicle uses a relatively inexpensive fuel (gasoline) in the internal combustion engine and has good fuel economy but still gives off emissions. A fuel-cell vehicle does not give off emissions (the reaction produces H_2O) but requires a more expensive and explosive fuel (hydrogen); moreover, current battery technologies incorporate materials that are still very expensive and bulky.

Think About It

The factor that determines whether alternate fuels and power systems get used will be the cost of petroleum used to produce gasoline, which traditionally has been much less expensive than alternative energy sources.

17.83. Collect and Organize

Given the information that methane may be used in fuel cells, we are to consider why these fuel cells are likely to produce less CO_2 emissions per mile than an internal combustion engine fueled by methane.

Analyze

Both the methane fuel cell and the combustion of methane in a combustion reaction have the balanced equation

$$CH_4(g) + 2\,O_2(g) \rightarrow 2\,H_2O(g) + CO_2(g)$$

Solve

Fuel cells burn methane fuel more efficiently. Electric engines are more efficient by converting more of the energy into motion instead of losing it as heat. Therefore, less CO_2 is produced per mile with fuel cells.

Think About It

The bulkiness and short range of fuel cells currently limit their use in transportation.

17.85. Collect and Organize

For the reactions of CH_4 and CO with water, we are to assign oxidation numbers to the C and H atoms in all the species in the reactions and calculate ΔG°_{rxn} for each and for the overall reaction.

$$CH_4(g) + 2\,H_2O(g) \rightarrow 4\,H_2(g) + CO_2(g)$$

Analyze

We can use the usual rules of assigning oxidation states from Chapter 4. To calculate ΔG°_{rxn} we use ΔG°_f values from Appendix 4 in the following equation:

$$\Delta G^{*}_{rxn} = \sum n\Delta G^{*}_{f,products} - \sum n\Delta G^{*}_{f,reactants}$$

Solve

(a) $\overset{-4\,+1}{C\,H_4}(g) + \overset{+1}{H_2O}(g) \rightarrow \overset{+2}{C\,O}(g) + 3\,\overset{0}{H_2}(g)$

$\overset{+2}{C\,O}(g) + \overset{+1}{H_2O}(g) \rightarrow \overset{0}{H_2}(g) + \overset{+4}{C\,O_2}(g)$

(b) For the reaction of CH_4 with H_2O:

$\Delta G^{\circ} = \left[(1\text{ mol }CO \times -137.2\text{ kJ/mol}) + (3\text{ mol }H_2 \times 0.0\text{ kJ/mol}) \right] -$

$\left[(1\text{ mol }CH_4 \times -50.8\text{ kJ/mol}) + (1\text{ mol }H_2O \times -228.6\text{ kJ/mol}) \right] = 142.2\text{ kJ}$

For the reaction of CO with H_2O:

$$\Delta G° = \left[(1 \text{ mol } CO_2 \times -394.4 \text{ kJ/mol}) + (1 \text{ mol } H_2 \times 0.0 \text{ kJ/mol})\right] -$$
$$\left[(1 \text{ mol } CO \times -137.2 \text{ kJ/mol}) + (1 \text{ mol } H_2O \times -228.6 \text{ kJ/mol})\right] = -28.6 \text{ kJ}$$

For the overall reaction:

$$\Delta G°_{overall} = \Delta G°_{rxn_1} + \Delta G°_{rxn_2} = 113.6 \text{ kJ}$$

Think About It

The spontaneity of the second reaction is not enough to overcome the positive free energy of the first reaction, so the overall reaction is nonspontaneous.

17.87. Collect, Organize, and Analyze

The values of $E°_{red}$ in Appendix 6 use the standard hydrogen electrode (SHE):

$$2 H^+(aq) + 2 e^- \rightarrow H_2(g)$$

as $E°_{red} = 0.000$ V. We are asked how the values in Appendix 6 would change if

$$2 H_2O(\ell) + 2 e^- \rightarrow H_2(g) + 2 OH^-(aq)$$

were defined as $E°_{red} = 0.000$ V.

Solve

The $E°_{red}$ of the reduction of water to give H_2 versus the SHE is -0.8278 V. If that reduction reaction were redefined to have $E°_{red} = 0.000$ V, all the values in Appendix 6 would increase by $+0.828$ V.

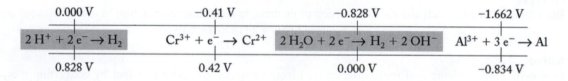

Think About It

Any reduction half-reaction could have been chosen as the standard half-reaction and the voltage of its reduction defined as 0.000 V.

17.89. Collect and Organize

We are asked to write the overall cell reaction for the reaction of Li with SO_2 in the lithium sulfur dioxide battery to form $Li_2S_2O_4(s)$, to determine the number of electrons transferred in the cell reaction, and to draw the Lewis structure for the $S_2O_4^{2-}$ ion.

Analyze

For each half-reaction, we need to have OH^- or H_2O as a reactant. We balance each for both atoms and charge before adding the two half-reactions.

Solve

(a)

Anode	$[Li(s) \rightarrow Li^+(aq) + e^-] \times 2$
Cathode	$2 SO_2(\ell) + 2 e^- \rightarrow S_2O_4^{2-}(nonaq)$

$$2 Li(s) + 2 SO_2(\ell) \rightarrow Li_2S_2O_4(s)$$

(b) Two electrons are transferred in the cell reaction.

(c)

:O: :O:⁻
 ‖   /
 :S——S:
 /   ‖
:O: :O:
⁻

Think About It
The lithium sulfur dioxide battery has a high energy density and has found some military applications.

17.91. Collect and Organize
To find E_{cell} for a concentration cell composed of 0.25 M copper(II) nitrate and 0.00075 M copper(II) nitrate, we will use the Nernst equation.

Analyze
Because the cell is composed of the same Cu/Cu^{2+} half cells, the $E°_{cell} = 0.00$ V. In order for the cell potential to be positive (for the spontaneous cell reaction), the log ratio in the Nernst equation will have to be negative, and that will occur when the ratio is 0.00075 M/0.25 M. The number of electrons transferred in the cell reaction is 2.

Solve

$$E_{cell} = 0.00 \text{ V} - \frac{0.0592}{2} \log \frac{0.00075 \ M}{0.25 \ M} = 0.075 \text{ V}$$

Think About It
This reaction is spontaneous with a ΔG value of

$$\Delta G° = -2 \text{ mol} \times \frac{9.65 \times 10^4 \text{ C}}{\text{mol}} \times \frac{0.075 \text{ J}}{\text{C}} \times \frac{1 \text{ kJ}}{1000 \text{ J}} = -14 \text{ kJ}$$

17.93. Collect and Organize
Using the two equations below, we are to derive an expression that relates $E°_{cell}$ to K.

$$\Delta G = \Delta G° + RT \ln Q$$

$$\Delta G° = -nFE°_{cell}$$

Analyze
Because at equilibrium $\Delta G = 0$ and $Q = K$, we can substitute K for Q and set ΔG as 0 in the first equation. Then we can substitute $-nFE°_{cell}$ for $\Delta G°$ from the second equation in that first equation and rearrange to solve for $E°_{cell}$.

Solve
Setting $\Delta G = 0$ and $Q = K$, we obtain

$$0 = \Delta G° + RT \ln K$$

Substituting $-nFE°_{cell}$ for $\Delta G°$ in that equation gives

$$0 = -nFE°_{cell} + RT \ln K$$

Rearranging to solve for $E°_{cell}$ gives

$$E°_{cell} = \frac{RT \ln K}{nF}$$

Think About It
Using this relationship, we can experimentally obtain equilibrium constants from measuring the potential of an electrochemical cell.

17.95. Collect and Organize
For a $Mg–Mo_3S_4$ battery for which we are given the half-reaction potential of the anode reaction (2.37 V) and the overall cell potential (1.50 V), we are to calculate $E°_{red}$ of Mo_3S_4. We are also to consider why Mg^{2+} is added to the battery's electrolyte and determine the oxidation states and electron configurations of Mo in Mo_3S_4 and $MgMo_3S_4$.

Analyze
Because $E°_{cell} = E°_{cathode} - E°_{anode}$ the reduction potential for Mo_3S_4 will be $E°_{cell} + E°_{anode}$.

Solve

(a) $E^\circ_{cathode} = 1.50 \text{ V} + (-2.37 \text{ V}) = -0.87 \text{ V}$

(b) Sulfur usually carries a 2– charge, so each Mo atom in Mo_3S_4 has a calculated charge of 2.67+. That, therefore, is a mixed oxidation state compound in which two of the Mo atoms probably have a 3+ charge and one Mo atom has a 2+ charge. In $MgMo_3S_4$, the Mg atom has a 2+ charge, so the Mo atoms have a 2+ charge. The electron configurations for the two oxidation states of Mo are

$$\text{Mo in +2 oxidation state } [Kr]4d^4$$
$$\text{Mo in +3 oxidation state } [Kr]4d^3$$

(c) Mg^{2+} is added to the electrolyte to better carry the charge in the cell. That cation is produced at the anode and consumed at the cathode.

Think About It

This battery resembles the lithium–ion battery in that a migrating cation in the cell generates the electrical current.

17.97. Collect and Organize

We consider the thermodynamic and electrochemical properties of the synthesis of F_2 both chemically and in an electrolysis reaction.

Analyze

(a) We can use the usual rules described in Chapter 4 to assign oxidation numbers to all the elements in the reactants and products for the chemical synthesis of F_2. From the change in oxidation numbers we can deduce the number of electrons involved in the process.

(b) To calculate ΔH°_{rxn} we use

$$\Delta H^\circ_{rxn} = \sum n\Delta H^\circ_{f,products} - \sum n\Delta H^\circ_{f,reactants}$$

(c) When we assume $\Delta S \approx 0$

$$\Delta G = \Delta H - T\Delta S \approx \Delta H$$

so we can use ΔH for ΔG in the equation to calculate E°_{cell}.

Solve

(a) In K_2MnF_6: K = +1, Mn = +4, F = –1
In SbF_5: Sb = +5, F = –1
In $KSbF_6$: K = +1, Sb = +5, F = –1
In MnF_3: Mn = +3, F = –1
In F_2: F = 0
This is a one-electron process.

(b) $\Delta H^\circ_{rxn} = \left[\left(2 \text{ mol } KSbF_6 \times -2080 \text{ kJ/mol}\right) + \left(1 \text{ mol } MnF_3 \times -1579 \text{ kJ/mol}\right) + \left(\frac{1}{2} \text{ mol } F_2 \times 0.0 \text{ kJ/mol}\right) \right]$
$- \left[\left(1 \text{ mol } K_2MnF_6 \times -2435 \text{ kJ/mol}\right) + \left(2 \text{ mol } SbF_5 \times -1324 \text{ kJ/mol}\right) \right] = -656 \text{ kJ}$

(c) $E^\circ_{cell} \approx \dfrac{6.56 \times 10^5 \text{ J}}{1 \text{ mol } e^- \times 9.65 \times 10^4 \text{ C/mol}} = 6.80 \text{ V}$

(d) If ΔS is positive, then the ΔG estimate is too positive; ΔG would be more negative, giving a more positive E°_{cell}. Therefore, our calculated E°_{cell} using only the ΔH°_{rxn} value would be too low.

(e) In H_2: H = 0
In F_2: F = 0
In KF: K = +1, F = –1
In KHF_2: K = +1, H = +1, F = –1
This is a two-electron process.

Think About It

The chemical synthesis of F_2 relies on the oxidation of F^- by the strong oxidant MnF_6^{2-}. That process, however, is not very practical, and fluorine is prepared industrially by the electrolysis reaction described in part e.

17.99. **Collect and Organize**

Given the balanced reactions of UO_2 with HF and UF_4 with Mg to ultimately produce $U(s)$, we are to identify the reducing agent and the element that is reduced. We will also determine the highest possible value for the reduction of UF_4 and determine whether 1.00 g of Mg could produce 1.00 g of solid uranium.

Analyze

(a) and (b) A species that is reduced is one that decreases (makes less positive) its oxidation state. A reducing agent is that species that causes some other species to be reduced; it is itself oxidized.

(b) The reduction of UF_4 must be positive when combined with the oxidation of magnesium.

(c) We can determine whether 1.00 g of Mg is enough to give 1.00 g of U through the stoichiometry of the given balanced equation for the process.

Solve

(a) Mg is the reducing agent; it is oxidized to Mg^{2+}.

(b) Uranium in UF_4 is reduced from a +4 to a 0 oxidation state.

(c) The half-cell reduction potential for Mg^{2+} is –2.37 V. Therefore, the reduction of UF_4 must have a reduction potential greater than –2.37 V for the reaction to have a positive cell potential and be spontaneous.

(d) Yes, 1.00 g of Mg will be enough to produce 1.00 g of U:

$$1.00 \text{ g Mg} \times \frac{1 \text{ mol Mg}}{24.305 \text{ g}} \times \frac{1 \text{ mol U}}{2 \text{ mol Mg}} \times \frac{238.03 \text{ g U}}{1 \text{ mol}} = 4.90 \text{ g U could be produced.}$$

Think About It

The first reaction in this process where UF_4 is produced is not a redox reaction, but an acid–base reaction.

17.101. **Collect and Organize**

We consider electrolysis of a molten Mg^{2+} salt from evaporated seawater (so it may contain NaCl).

Analyze

The possible reactions are (with $E°$ values when listed in Appendix 6):

$$Mg^{2+}(\ell) + 2 \text{ e}^- \rightarrow Mg(s)$$
$$Na^+(\ell) + \text{e}^- \rightarrow Na(s)$$
$$2 H_2O(\ell) + 2 \text{ e}^- \rightarrow H_2(g) + 2 OH^-(aq) \qquad E°_{red} = -0.8277 \text{ V}$$
$$2 H_2O(\ell) \rightarrow O_2(g) + 4 H^+(aq) + 4 \text{ e}^- \qquad E°_{red} = 1.229 \text{ V}$$
$$Mg^{2+}(aq) + 2 \text{ e}^- \rightarrow Mg(s) \qquad E°_{red} = -2.37 \text{ V}$$
$$Na^+(aq) + \text{e}^- \rightarrow Na(s) \qquad E°_{red} = -2.71 \text{ V}$$

Solve

(a) Mg^{2+} undergoes a reduction reaction that occurs at the cathode. Mg forms at the cathode.

(b) No. Mg^{2+}, with a higher positive charge, has a lower (less negative) reduction potential than Na^+, so the Mg^{2+} would not need to be separated from the NaCl in seawater first.

(c) No. The electrolysis of $MgCl_2(aq)$ would not produce $Mg(s)$ because water, with a less negative reduction potential, would be electrolyzed.

(d) H_2 and O_2 gases would be produced.

Think About It

Because different components are reduced at different potentials, electrolysis of molten salts is one way to separate components (such as metals).

CHAPTER 18 | The Solid State: A Particulate View

18.1. Collect and Organize

From the drawings shown in Figure P18.1, we are to choose which represents a crystalline solid and which represents an amorphous solid.

Analyze

In crystalline solids, atoms or molecules arrange themselves in regular, repeating three-dimensional patterns. In an amorphous solid, the atoms or molecules are arranged randomly, with no defined repeating pattern.

Solve

Drawings (b) and (d) are analogous to crystalline solids because they show a definite pattern, whereas drawings (a) and (c) are amorphous.

Think About It

Drawings (b) and (d) have two kinds of atoms; if the drawings represent metals, those two substances would be alloys.

18.3. Collect and Organize

Using Figure P18.3, we are to determine the unit cell and write the chemical formula for the compound in which element A is represented as red spheres and element B is represented as blue spheres.

Analyze

Once the unit cell is determined, the chemical formula can be deduced. Spheres on the corner are shared by eight unit cells, so they are counted as $\frac{1}{8}$ in the unit cell. Likewise, spheres on the edges are shared by four unit cells and are counted as $\frac{1}{4}$ in the unit cell, and spheres on the faces of the unit cell are shared by two unit cells and are counted as $\frac{1}{2}$ in the unit cell. Any atom completely inside a unit cell belongs entirely to it and counts as one.

Solve

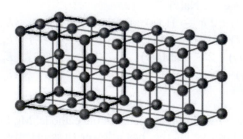

The unit cell has $8(\frac{1}{8}) + 6(\frac{1}{2}) = 4$ red spheres, or four A atoms. The unit cell has $12(\frac{1}{4}) + 1(1) = 4$ blue spheres, or four B atoms. The chemical formula is A_4B_4 or AB.

Think About It

The empirical formula for that compound would be AB because that is the lowest whole-number ratio of elements in the substance.

18.5. Collect and Organize

From Figure P18.5, in which a portion of the unit cell has six face A atoms and eight corner B atoms, we are to determine the number of equivalent atoms of A and B.

Analyze

Atoms on the corners contribute $\frac{1}{8}$ of their volume to the unit cell and atoms on the faces contribute $\frac{1}{2}$ of their volume to the unit cell.

Solve

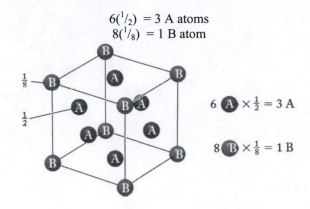

$$6(^1/_2) = 3 \text{ A atoms}$$
$$8(^1/_8) = 1 \text{ B atom}$$

$$6 \, \text{\normalsize A} \times \tfrac{1}{2} = 3 \text{ A}$$

$$8 \, \text{\normalsize B} \times \tfrac{1}{8} = 1 \text{ B}$$

Think About It

The empirical formula for that substance, on the basis of the unit cell, is BA_3.

18.7. **Collect and Organize**

From the portion of a unit cell shown in Figure P18.7, we are to deduce the chemical formula of the ionic compound.

Analyze

Atoms on the corners contribute $^1/_8$ of their volume to the unit cell and atoms on the faces contribute $^1/_2$ of their volume to the unit cell. Atoms completely inside the unit cell contribute all their volume to the unit cell.

Solve

$$8\left(\tfrac{1}{8}\right) = 1 \text{ A cation}$$
$$6\left(\tfrac{1}{2}\right) = 3 \text{ X cations}$$
$$1(1) = 1 \text{ B anion}$$

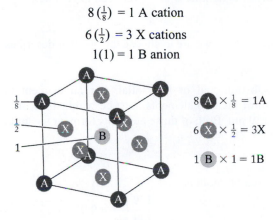

$$8 \, \text{\normalsize A} \times \tfrac{1}{8} = 1\text{A}$$

$$6 \, \text{\normalsize X} \times \tfrac{1}{2} = 3\text{X}$$

$$1 \, \text{\normalsize B} \times 1 = 1\text{B}$$

The chemical formula is ABX_3.

Think About It

In that unit cell, the X anion is located in the center, occupying the body-centered position, and the B cations occupy face-centered positions.

18.9. **Collect and Organize**

Knowing the distance between the phosphorus atoms in its cubic form, we are to calculate the density.

Analyze

Density is mass per unit volume. The mass of phosphorus in one unit cell will be the number of atoms of P in one unit cell multiplied by the molar mass of P and divided by Avogadro's number. The volume of the unit cell is the cube of the length of the side of the unit cell. Density is usually expressed as grams per cubic centimeter (g/cm^3), so we have to convert picometers to centimeters. The unit cell is one cube of eight P atoms:

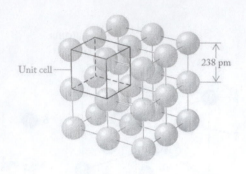

Unit cell

238 pm

Solve

Mass of phosphorus in one unit cell:

$$8\left(\tfrac{1}{8}\right)\times\frac{30.97\text{ g}}{1\text{ mol}}\times\frac{1\text{ mol}}{6.022\times10^{23}\text{ atoms}}=5.143\times10^{-23}\text{ g}$$

Volume of the unit cell:

$$\left(238\text{ pm}\times\frac{1\times10^{-10}\text{ cm}}{1\text{ pm}}\right)^{3}=1.35\times10^{-23}\text{ cm}^{3}$$

Density of cubic phosphorus:

$$\frac{5.143\times10^{-23}\text{ g}}{1.35\times10^{-23}\text{ cm}^{3}}=3.81\text{ g/cm}^{3}$$

Think About It

The density of phosphorus in another form would be different. If we measure the density of a crystal of known composition, we may be able to determine what its unit cell looks like.

18.11. Collect and Organize

From the drawings in Figure P18.11, we are to choose the one that describes an alloy of gold and platinum.

Analyze

An alloy may be either substitutional or interstitial. For an element to substitute for another atom in a substitutional alloy, the radii of the two elements must be within 15% of each other. The radius of gold is 144 pm and that of platinum is 139 pm. Both of those elements form an fcc unit cell, as shown in Figure 18.10 in the textbook.

Solve

The ratio of the radii of platinum to gold is

$$\frac{139\text{ pm}}{144\text{ pm}}=0.97$$

That is about a 3% difference in their radii, so this forms a substitutional alloy, which is depicted in drawing (b), which has an fcc unit cell of 11 platinum atoms and 2 gold atoms.

Think About It

Drawing (a) shows an interstitial alloy for an fcc unit cell, drawing (c) shows a substitutional alloy for a bcc unit cell, and drawing (d) shows a substitutional alloy for a diamond structure.

18.13. Collect and Organize

We are to choose the drawing from Figure P18.13 that best explains the electrical conductivity of copper metal, magnesium metal, silicon, and sodium chloride.

Analyze

Drawing (a) describes an element/substance with a partially filled valence band (a metal), drawing (b) describes an element/substance with overlapping valence and conduction bands (a metal), drawing (c) describes an element/substance with a relatively small band gap between the valence and conduction bands (a semiconductor), drawing (d) describes a *p*-doped element/substance, drawing (e) describes an *n*-doped element/substance, and drawing (f) describes an element/substance with a very large band gap (an insulator).

Solve

(a) Copper metal has partially filled valence bands because of its $[Ar]3d^{10}4s^1$ electron configuration; drawing (a) best describes copper metal.

(b) Magnesium metal has filled valence bands because of its $[Ne]3s^2$ electron configuration; drawing (b) best describes magnesium metal.

(c) Silicon is a semiconductor (undoped); drawing (c) best describes silicon.

(d) Sodium chloride is an ionic solid, which, if pure, is not conductive as a solid and is an insulator; drawing (f) best describes sodium chloride.

Think About It

Sodium chloride dissolved in a solution, however, is electrically conducting because the ions are free to move in the solution.

18.15. Collect and Organize

From Figure P18.15 showing the unit cell of magnesium boride, we are to determine its formula.

Analyze

We are given that one B atom is inside the trigonal prismatic unit cell. Each corner of the unit cell has a Mg atom. Because the unit cell is trigonal prismatic, each corner atom is shared with 11 other unit cells and so has $\frac{1}{12}$ of its volume inside the unit cell.

Solve

The number of B atoms in the unit cell is 1. The number of Mg atoms in the unit cell is $6 \times \frac{1}{12} = \frac{1}{2}$. The formula from the unit cell is MgB_2 (in the lowest whole-number ratio).

Think About It

It is tricky to see that the Mg atoms here are shared between 12 unit cells. Consider the center atom in the figure on the right-hand side. That atom is shared with six unit cells as shown, but six more unit cells are stacked on top of it that are not shown in the diagram.

18.17. Collect and Organize

From the structure shown in Figure P18.17 we are to determine the number of large (blue) and small (red) spheres that are assignable to the unit cell and to calculate the maximum radius of the small spheres if the radius of the large spheres is 140 pm.

Analyze

Atoms on the corners of the unit cell contribute $\frac{1}{8}$ of their volume to the unit cell. Atoms on the unit cell edges contribute $\frac{1}{4}$ of their volume to the unit cell. Atoms on the face of a unit cell contribute $\frac{1}{2}$ of their volume to the unit cell. Atoms inside the unit cell contribute all their volume to the unit cell. In the structure shown, the red spheres occupy the tetrahedral holes in the lattice of blue spheres. A tetrahedral hole will accommodate an atom with the ratio $r_{nonhost}/r_{host}$ between 0.22 and 0.41. The maximum size of the small red spheres will be

$$\frac{r_{nonhost}}{r_{host}} = 0.41$$

$$r_{nonhost} = 0.41 \times r_{host}$$

Solve

(a) 8(1/8 large blue) + 6(1/2 large blue) = 4 large blue spheres.

All the small red spheres are completely inside the unit cell = 8 small red spheres.

(b) The maximum radius of the small red spheres is

$$r_{nonhost} = 0.41 \times r_{host} = 0.41 \times 140 \text{ pm}$$
$$r_{nonhost} = 57 \text{ pm}$$

Think About It
If the small spheres were larger than 57 pm, they would be better accommodated in octahedral holes in the lattice.

18.19. Collect and Organize
We are asked to differentiate between cubic closest-packed (ccp) and hexagonal closest-packed (hcp) structures.

Analyze
Both structures contain layers of close-packed atoms and differ only in how the layers are stacked.

Solve
Cubic closest-packed structures have an *abcabc* . . . pattern, and hexagonal closest-packed structures have an *abab* . . . pattern.

Think About It
The unit cell for ccp is face-centered cubic (Figure 18.11), and the unit cell for hcp is hexagonal (Figure 18.8).

18.21. Collect and Organize
We are asked which has the greater packing efficiency, the simple cubic or the body-centered cubic structure.

Analyze
Packing efficiency is the fraction of space within a unit cell that is occupied by the atoms.

Solve
We read in Section 18.3 of the textbook that the simple cubic cell has the lowest packing efficiency of all the unit cells, so the body-centered cubic structure has a greater packing efficiency than the simple cubic structure.

Think About It
The packing efficiency of the unit cell structures can be calculated. For simple cubic, the packing efficiency is only 52%, whereas for body-centered cubic it is 68% and for face-centered and hexagonal it is 74%.

18.23. Collect and Organize
Iron can adopt either the bcc unit cell structure (at room temperature) or the fcc unit cell structure (at 1070°C). We are asked whether those two forms are allotropes.

Analyze
Allotropes are defined as different *molecular* forms of an element.

Solve
Iron is not molecular, and the bcc and fcc unit cell structures describe only a difference in atom packing in the metal. Therefore, those structural forms are not allotropes.

Think About It
Elements that do have allotropes include phosphorus, sulfur, and carbon. Metals that take on alternative lattice structures at different temperatures and pressures exhibit *polytypism*.

18.25. Collect and Organize
Using the information in Figure 18.11, we are to derive the expression for the edge length ℓ of the bcc and fcc unit cells in terms of r, the radius of the atoms.

Analyze

For the bcc unit cell, the atoms touch along the body diagonal, for a distance of $\ell\sqrt{3}$. For the fcc unit cell, the atoms touch along the face diagonal, for a distance of $\ell\sqrt{2}$.

Solve

In the bcc unit cell, the body diagonal is $4r$. From the Pythagorean theorem, the body diagonal is

$$4r = \sqrt{(\text{edge length})^2 + (\text{face diagonal})^2}$$

where the edge length $= \ell$ and the face diagonal is

$$\sqrt{\ell^2 + \ell^2} = \sqrt{2\ell^2} = \ell\sqrt{2}$$

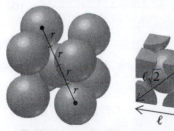

Body-centered cubic:
Atoms touch along body diagonal

Therefore,

$$4r = \sqrt{\ell^2 + \left(\ell\sqrt{2}\right)^2} = \sqrt{\ell^2 + 2\ell^2} = \sqrt{3\ell^2}$$

$$4r = \ell\sqrt{3}$$

$$\ell = \frac{4r}{\sqrt{3}} = 2.309r$$

In the fcc unit cell, the face diagonal is $4r$:

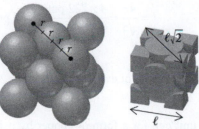

Face-centered cubic:
Atoms touch along face diagonal

$$\sqrt{\ell^2 + \ell^2} = \sqrt{2\ell^2} = \ell\sqrt{2} = 4r$$

$$\ell = \frac{4r}{\sqrt{2}} = 2.828r$$

Think About It

For a given atom, the edge length of the bcc unit cell is less than that of atoms packed in an fcc unit cell.

18.27. **Collect and Organize**

Knowing that $\ell = 240.6$ pm for the bcc structure of europium, we are to calculate the radius of one atom of europium.

Analyze

For the bcc structure, the body diagonal contains two atoms of europium, or $4r$, and the body diagonal is equal to $\ell\sqrt{3}$ (Figure 18.16), where ℓ is the unit cell edge length, 240.6 pm.

Solve

$$4r = \ell\sqrt{3}$$
$$r = \frac{\ell\sqrt{3}}{4} = \frac{240.6 \text{ pm} \times \sqrt{3}}{4} = 104.2 \text{ pm}$$

Think About It

If the structure of the unit cell were simple cubic, the radius of the europium atom would simply be
$$2r = \ell$$
$$r = \frac{\ell}{2} = 120.0 \text{ pm}$$

18.29. **Collect and Organize**

We are to calculate the edge length of the unit cell of Ba, knowing that it crystallizes in a bcc unit cell and that $r_{\text{Ba}} = 222$ pm.

Analyze

The body diagonal of a bcc unit cell has the relationship $\ell\sqrt{3} = 4r$. An edge has the length ℓ, so all we need to do is rearrange the body diagonal expression to solve for ℓ:

$$\ell = \frac{4r}{\sqrt{3}}$$

Solve

$$\ell = \frac{4r}{\sqrt{3}} = \frac{4 \times 222 \text{ pm}}{\sqrt{3}} = 513 \text{ pm}$$

Think About It

Be careful to not assume that the edge length of every unit cell is $\ell = 2r$, as in a simple cubic.

18.31. **Collect and Organize**

We are to determine the type of unit cell for a form of copper by using the density of the crystal (8.95 g/cm^3) and the radius of the Cu atom (127.8 pm).

Analyze

For each type of unit cell (simple cubic, body-centered cubic, and face-centered cubic), we can compare the calculated density with that of the actual density given for the crystalline form of copper. Density is mass per unit volume. The mass of each unit cell is the mass of the copper atoms contained in each unit cell. To find that, we first have to determine the number of atoms of Cu in each unit cell on the basis of its structure. Then, we multiply by the mass of one atom of Cu:

$$1 \text{ Cu atom} \times \frac{1 \text{ mol}}{6.022 \times 10^{23} \text{ atoms}} \times \frac{63.55 \text{ g}}{1 \text{ mol}} = 1.055 \times 10^{-22} \text{ g}$$

The volume of each unit cell is ℓ^3, which is related to r in a way that depends on the type of unit cell.

Solve

For a simple cubic unit cell in which all the Cu atoms are at the corners of the cube:

$$\text{Number of Cu atoms} = 8 \times \tfrac{1}{8} = 1 \text{ Cu atom}$$
$$\ell = 2r = 2 \times 127.8 \text{ pm} = 255.6 \text{ pm}$$

Converting to centimeters for the calculation of density (grams per cubic centimeter) gives

$$255.6 \text{ pm} \times \frac{1 \times 10^{-10} \text{ cm}}{1 \text{ pm}} = 2.556 \times 10^{-8} \text{ cm}$$

$$\text{Volume} = \ell^3 = (2.556 \times 10^{-8} \text{ cm})^3 = 1.670 \times 10^{-23} \text{ cm}^3$$

$$\text{Density} = \frac{1 \text{ Cu atom} \times 1.055 \times 10^{-22} \text{ g/atom}}{1.670 \times 10^{-23} \text{ cm}^3} = 6.32 \text{ g/cm}^3$$

For a body-centered cubic unit cell with one Cu atom in the center of the unit cell and eight Cu atoms at the corners of the cube:

$$\text{Number of Cu atoms} = (8 \times \tfrac{1}{8}) + 1 = 2 \text{ Cu atoms}$$

From the body diagonal $4r = \ell\sqrt{3}$ or $\ell = 4r/\sqrt{3}$,

$$\ell = \frac{4 \times 127.8 \text{ pm}}{\sqrt{3}} = 295.1 \text{ pm}$$

Converting to centimeters for calculation of density (grams per cubic centimeter) gives

$$295.1 \text{ pm} \times \frac{1 \times 10^{-10} \text{ cm}}{1 \text{ pm}} = 2.951 \times 10^{-8} \text{ cm}$$

$$\text{Volume} = \ell^3 = (2.951 \times 10^{-8} \text{ cm})^3 = 2.570 \times 10^{-23} \text{ cm}^3$$

$$\text{Density} = \frac{2 \text{ Cu atoms} \times 1.055 \times 10^{-22} \text{ g/atom}}{2.570 \times 10^{-23} \text{ cm}^3} = 8.21 \text{ g/cm}^3$$

For a face-centered cubic unit cell with eight Cu atoms at the corners of the unit cell and six on the faces:

$$\text{Number of Cu atoms} = (8 \times \tfrac{1}{8}) + (6 \times \tfrac{1}{2}) = 4 \text{ Cu atoms}$$

From the face diagonal $4r = \ell\sqrt{2}$ or $\ell = 4r/\sqrt{2}$,

$$\ell = \frac{4 \times 127.8 \text{ pm}}{\sqrt{2}} = 361.5 \text{ pm}$$

Converting to centimeters for calculation of density (grams per cubic centimeter) gives

$$361.5 \text{ pm} \times \frac{1 \times 10^{-10} \text{ cm}}{1 \text{ pm}} = 3.615 \times 10^{-8} \text{ cm}$$

$$\text{Volume} = \ell^3 = (3.615 \times 10^{-8} \text{ cm})^3 = 4.724 \times 10^{-23} \text{ cm}^3$$

$$\text{Density} = \frac{4 \text{ Cu atoms} \times 1.055 \times 10^{-22} \text{ g/atom}}{4.724 \times 10^{-23} \text{ cm}^3} = 8.93 \text{ g/cm}^3$$

The fcc unit cell gives a density closest to the actual density, so we predict that in this crystalline form of Cu the atoms pack in an fcc unit cell (c).

Think About It

Even though the fcc unit cell has the longest edge, the unit cell contains more atoms, so it yields the densest structure.

18.33. ## Collect and Organize

We can use the definitions of solid solution and homogeneous alloy to determine whether a difference exists between those terms.

Analyze

A solid solution is a homogeneous mixture of solids. An alloy is a mixture of two or more metallic elements in solution with each other.

Solve

The terms have no real difference. Both are homogeneous mixtures of solids. A subtle difference may exist, however. An alloy is a mixture of two or more metallic elements, and a solid solution is a mixture of two or more solids, which may or may not be metallic elements.

Think About It

All alloys are solid solutions, but not all solid solutions are alloys.

18.35. **Collect and Organize**

We are to explain why an alloy of Cu and Ag melts at a lower temperature than either pure Ag or pure Cu.

Analyze

When other metals are alloyed into a pure metal, the atomic structure of the metal is changed so that the atoms are not in so regular of an arrangement as in the pure metal. From Sample Exercise 18.3, we know that Cu and Ag form substitutional alloys because their atomic radii are so similar.

Solve

When those two metals form alloys with each other, the metallic bonding is weakened because of their slightly different sizes and different ionization energies. That weaker metallic bonding means that the melting points of the alloys are lower than those of the pure metals.

Think About It

Lowering the melting point of an alloy allows for easier forming, processing, and workability than for a pure metal.

18.37. **Collect and Organize**

For a substitutional alloy composed of Mo and W (both having bcc structure and same atomic radius) we are to determine whether the alloy has a greater, lesser, or equal density than that of pure molybdenum.

Analyze

For a substitutional alloy, the atoms of the alloying metal (here W) substitute for the host metal (here Mo) in the structure. For this alloy the same size atoms are used to make the alloy and the resultant alloy will have a bcc structure.

Solve

The size of the unit cell will not change when replacing Mo with W in the structure because those atoms are of the same size. However, W has a higher atomic mass (183.84 g/mol) than Mo (95.96 g/mol), and so replacing Mo with W will add more mass to the unit cell. That alloy, therefore, will have a greater density than pure Mo because of the increased mass of the unit cell.

Think About It

On the other hand, making a substitutional alloy by replacing W with Mo will result in an alloy of lower density than Mo.

18.39. **Collect and Organize**

Given that a unit cell of an alloy of X and Y consists of an fcc arrangement of X atoms at each corner and Y atoms on each face, we are to write the formula of the alloy and then consider the formula if the two atom positions were reversed.

Analyze

Atoms at the corner of a unit cell count for $1/8$ and atoms on the faces count for $1/2$ of the total volume of the unit cell.

Solve

(a) For eight X atoms at the corners and six Y atoms on the faces of the unit cell:

$$1/8(8 \text{ X atoms}) + 1/2(6 \text{ Y atoms}) = \text{XY}_3$$

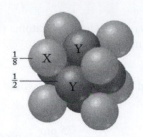

(b) If the positions were reversed, we would have

$$\tfrac{1}{8}(8 \text{ Y atoms}) + \tfrac{1}{2}(6 \text{ X atoms}) = \text{YX}_3$$

Think About It
The fcc lattice is closest-packed. If X and Y may be in either the corners or faces in the unit cell, that alloy is a substitutional alloy.

18.41. Collect and Organize
Given the atomic radii of C and V (77 and 135 pm, respectively), we can calculate the radius ratio to determine which holes, octahedral or tetrahedral, C occupies in the V closest-packed structure.

Analyze
The radius ratios for tetrahedral, octahedral, and cubic holes are given in Table 18.4 as 0.22–0.41, 0.41–0.73, and 0.73–1.00, respectively.

Solve

$$\frac{r_C}{r_V} = \frac{77 \text{ pm}}{135 \text{ pm}} = 0.57$$

That ratio fits the range for carbon's occupying octahedral holes in the vanadium lattice.

Think About It
Carbon can also fit into the larger cubic holes (radius ratio, 0.73–1.00) but would occupy the smallest hole it can in the lattice to maximize its interactions with the host (V) atoms in the closest-packed structure.

18.43. Collect and Organize
By calculating the radius ratio of Sn to Ag, we can determine whether dental alloys of Sn and Ag are substitutional or interstitial alloys.

Analyze
It would be possible for an alloy to be both interstitial and substitutional as long as the voids within the lattice are about the same size as the metal atom. In particular, if the host metal has a simple cubic structure, the alloying metal may fit in a cubic void (radius ratio > 0.73) as well as substitute for the host metal in the lattice structure.

Solve

$$\frac{r_{Sn}}{r_{Ag}} = \frac{140 \text{ pm}}{144 \text{ pm}} = 0.972$$

Because those radii are within 15% of each other, an alloy of silver and tin is a substitutional alloy.

Think About It
Remember that the metals in substitutional alloys must crystallize in the same closest-packed (hcp, ccp, simple cubic, bcc) structures.

18.45. Collect and Organize

Using the radius ratio rule, we can determine whether N atoms (radius, 75 pm) fit into the octahedral holes of closest-packed Fe atoms (radius, 126 pm).

Analyze

If the radius ratio falls between 0.41 and 0.73, the nitrogen would fit into the octahedral holes.

Solve

$$\frac{r_N}{r_{Fe}} = \frac{75 \text{ pm}}{126 \text{ pm}} = 0.60$$

That ratio does lie in the range for octahedral holes. Yes, N occupies the octahedral holes in Fe.

Think About It

The radius ratio would need to drop below 0.41 before N could occupy the tetrahedral holes. That would occur for metals with atomic radii greater than 183 pm, such as in the lanthanides.

18.47. Collect and Organize

We are to predict the formula of alloys formed when atom B variously occupies octahedral and tetrahedral holes in an fcc lattice of A atoms.

Analyze

The fcc arrangement of A atoms has four atoms of A in the unit cell. That fcc unit cell also contains four octahedral holes and eight tetrahedral holes.

Solve

(a) If all the octahedral holes are occupied, the unit cell has four B atoms and four A atoms. The formula of the alloy is AB.
(b) If half the octahedral holes are occupied, the unit cell has two B atoms and four A atoms. The formula of the alloy is A_2B.
(c) If half the tetrahedral holes are occupied, the unit cell has four B atoms and four A atoms. The formula of the alloy is AB.

Think About It

If all the tetrahedral holes were filled, the formula of the alloy would be AB_2.

18.49. Collect and Organize

For an alloy in which the A atoms are arranged in an fcc structure and in which one B atom is present for every five A atoms, we are to determine what fraction of the octahedral holes in the structure are occupied by B atoms.

Analyze

An fcc unit cell of A atoms contains four A atoms and four octahedral holes. If five A atoms exist for every one B atom, we can consider five unit cells to give a whole number of A atoms, B atoms, and unit cells.

Solve

Five unit cells contain 20 A atoms and 20 octahedral holes. From the 5:1 ratio of A to B atoms, the five unit cells contain four B atoms. Those four B atoms, therefore, occupy $^4/_{20}$, or one-fifth, of the octahedral holes in the fcc structure.

Think About It

If the ratio of A to B atoms were 2:1, that would mean that half the octahedral holes would be filled with B atoms in the fcc lattice structure of A atoms.

18.51. Collect and Organize

Given that the unit cell edge length for an alloy of Cu and Sn is the same as that for pure Cu, we are to determine whether the alloy would be denser than pure Cu.

Analyze

Density is mass per unit volume. If the unit cell edge length is the same for both the alloy and pure copper, the volumes of the unit cells of both substances are the same. The density would be greater only if the mass of the unit cell alloy is greater than that of the pure copper.

Solve

Because tin has a greater molar mass than copper, the mass of the unit cell in the alloy would be greater and therefore, yes, the density of the alloy would be greater than the density of pure copper.

Think About It

The unit cell volume does not change much for substitutional alloys because the alloying atoms are close in size (within 15%) to the atoms being replaced.

18.53. **Collect and Organize**

We are to explain how the electron-sea model accounts for the electrical conductivity of gold.

Analyze

In the electron-sea model the metal atoms exist in the structure as cations with the ionized electrons not associated with particular metal ions.

Solve

Because the electrons are free to distribute themselves throughout the metal, the application of an electrical potential across the metal causes the electrons to travel freely towards the positive terminal. That movement of electrons in the structure is electrical conductivity.

Think About It

Metals have low ionization energies. That makes the electron sea easy to achieve in the lattice of metals.

18.55. **Collect and Organize**

Given that the melting and boiling points of Na are lower than those of NaCl, we are to predict the relative strengths of metallic and ionic bonds.

Analyze

Solid sodium is held together by metallic bonds, whereas sodium chloride is held together by ionic bonds. The higher the melting point or boiling point of a substance, the stronger the forces between particles of that substance.

Solve

Because the melting point and boiling point of NaCl are higher than those of Na, ionic bonds are stronger than metallic bonds.

Think About It

Some metallic bonds are strong, as evidenced by the melting points of some metals. Whereas Na has a melting point of 97.72°C, tungsten melts at 3422°C.

18.57. **Collect and Organize**

We are to consider whether band theory can explain why hydrogen at very low temperatures and high pressures might act like a metal and conduct electricity.

Analyze

Band theory is a model of bonding in which orbitals on many atoms are combined, as in molecular orbital theory, to form a filled or partially filled valence band and an empty conduction band.

Solve

Yes, by combining the 1*s* orbitals on many hydrogen atoms at very low temperatures and at high pressures, we could construct a molecular orbital diagram that would show a partially filled valence band (as for sodium in Figure 18.25) in which the electrons can move into the unfilled portion of the band, where they can migrate freely.

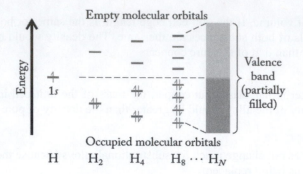

Empty molecular orbitals
Energy
1s
Valence
band
(partially
filled)
Occupied molecular orbitals
H
H2
H4
H8 ··· HN

Think About It

Metallic liquid hydrogen was discovered with hydrogen at very high pressures, but at high temperatures, in 1996 at the Lawrence Livermore Laboratory by accident.

18.59. **Collect and Organize**

We are asked to identify which groups in the periodic table contain metals with filled valence bands.

Analyze

The metals in the periodic table are in groups 1–12 with some in group 13 (Al, Ga, In, Tl), group 14 (Si, Ge, Sn, Pb), group 15 (As, Sb, Bi), group 16 (Te, Po), and group 17 (At). Filled valence bands for metals occur for electron configurations of ns^2 and nd^{10}.

Solve

Metals of groups 2 and 12 have filled valence bands.

Think About It

The orbital energy diagram for those metals will be similar to that of zinc in Figure 18.26.

18.61. **Collect and Organize**

We are asked why excluding phosphorus contaminants in the manufacture of silicon chips might be important.

Analyze

The difference between silicon and phosphorus is that phosphorus has five valence electrons and silicon has four valence electrons.

Solve

Phosphorus, with one more valence electron than silicon, would give pure silicon a higher conductivity and so would change the electrical nature of the silicon chips.

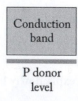

Conduction
band
P donor
level
Valence
band
n-type
·Si· + ·P·

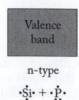

Think About It
Phosphorus, with one more electron than silicon, would form an n-type semiconductor if it were present as a dopant in silicon.

18.63. **Collect and Organize**
We are asked to identify in which group of the periodic table we might find elements that would be useful to form a p-type semiconductor with Sb_2S_3.

Analyze
Antimony has five valence electrons, and sulfur has six valence electrons. To form a p-doped semiconductor, we would choose elements with fewer than five valence electrons.

Solve
Group 14 elements, with four valence electrons, would form p-type semiconductors when doped into Sb_2S_3.

Think About It
Similarly, group 13 elements, with three valence electrons, might also be used to form p-doped Sb_2S_3 semiconductors.

18.65. **Collect and Organize**
We consider the properties of nitrogen-doped diamond.

Analyze
(a) Nitrogen, with five valence electrons, has one more valence electron than carbon, with four valence electrons.
(b) In diamond, the valence band and the conduction band are separated by a large band gap, making diamond an insulator. Adding nitrogen to the diamond structure adds a partially filled band above diamond's valence band and below its conduction band.
(c) Because $E = hc/\lambda$, we can calculate the energy of the band gap when $\lambda = 4.25 \times 10^{-7}$ m.

Solve
(a) Because nitrogen has one more valence electron than carbon, nitrogen serves as an n-type dopant in diamond.
(b)

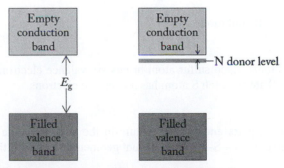

Diamond insulator   Nitrogen-doped diamond

(c) $E_g = \dfrac{6.626 \times 10^{-34}\,\text{J}\cdot\text{s} \times 3.00 \times 10^8\,\text{m/s}}{4.25 \times 10^{-7}\,\text{m}} = 4.68 \times 10^{-19}\,\text{J}$

Think About It
Because nitrogen-doped diamonds absorb violet light, we observe a yellow color (the complement of violet) in the diamond.

18.67. **Collect and Organize**

By calculating the wavelength associated with the band gap energies for AlN, GaN, and InN, we can determine which energies correspond to radiation in the visible part of the electromagnetic spectrum.

Analyze

The wavelength of light absorbed or emitted by a semiconductor is inversely related to the band gap energy:

$$E_g = \frac{hc}{\lambda}$$

To calculate the wavelength, we can rearrange that equation to

$$\lambda = \frac{hc}{E_g}$$

The wavelength is expressed in terms of a single photon, so we have to convert energy units (from kilojoules per mole to joules per photon).

Solve

For AlN: $\lambda = \dfrac{6.626 \times 10^{-34} \text{ J} \cdot \text{s} \times 3.00 \times 10^8 \text{m/s}}{\left(\dfrac{580.6 \text{ kJ}}{\text{mol}} \times \dfrac{1000 \text{ J}}{\text{kJ}} \times \dfrac{1 \text{ mol}}{6.022 \times 10^{23} \text{ photons}} \right)} = 2.06 \times 10^{-7}$ m, or 206 nm

For GaN: $\lambda = \dfrac{6.626 \times 10^{-34} \text{ J} \cdot \text{s} \times 3.00 \times 10^8 \text{m/s}}{\left(\dfrac{322.1 \text{ kJ}}{\text{mol}} \times \dfrac{1000 \text{ J}}{\text{kJ}} \times \dfrac{1 \text{ mol}}{6.022 \times 10^{23} \text{ photons}} \right)} = 3.72 \times 10^{-7}$ m, or 372 nm

For InN: $\lambda = \dfrac{6.626 \times 10^{-34} \text{ J} \cdot \text{s} \times 3.00 \times 10^8 \text{m/s}}{\left(\dfrac{192.9 \text{ kJ}}{\text{mol}} \times \dfrac{1000 \text{ J}}{\text{kJ}} \times \dfrac{1 \text{ mol}}{6.022 \times 10^{23} \text{ photons}} \right)} = 6.21 \times 10^{-7}$ m, or 621 nm

Only InN (E_g = 192.9 kJ/mol) emits in the visible region of the spectrum, which ranges from 390 to 700 nm.

Think About It

The color of InN's emission is red.

18.69. **Collect and Organize**

We are to explain why S_8 is not a flat molecule.

Analyze

S_8 is a ring of eight sulfur atoms. Each sulfur atom brings six valence electrons to the structure. Because each sulfur is bound to two other S atoms, each S atom has an octet of electrons.

Solve

Each S atom has two bonding pairs and two lone pairs in the structure. The sp^3 hybridization on S and the presence of the lone pairs give the S—S—S bonds a bent geometry; therefore, the ring is not flat.

Think About It

The hybridization at each S atom is sp^3, giving bond angles of ~109.5°.

18.71. **Collect and Organize**

We are to consider the structure of graphite in which the C atoms have been replaced by B and N atoms and decide whether that replacement would pucker the rings in the structure (refer to Figure P18.8).

Analyze

Boron has three valence electrons and nitrogen has five. Those atoms are replacing two carbon atoms in graphite (Figure P18.8b), in which each carbon atom brings four valence electrons to the structure.

Solve

To pucker the ring, lone pairs would have to be introduced onto the atoms in the structure. Because B and N, with eight total valence electrons, replace two C atoms, also with eight valence electrons, the ring of BN atoms remains flat.

Think About It

BN is boron nitride, a very tough, hard, high-temperature ceramic material.

18.73. **Collect and Organize**

For the ionic form of ice in which the O^{2-} ions are bcc and the H^+ ions occupy the holes, we are to determine the number of H^+ and O^{2-} ions in each unit cell and then draw the Lewis structure for ionic ice.

Analyze

A body-centered cubic arrangement of O^{2-} ions would have four O^{2-} ions at the corners of a cube and one O^{2-} ion in the center of the unit cell.

Solve

(a) The bcc unit cell has two O^{2-} ions. To balance the charge, four H^+ ions must be present in the unit cell.

(b)

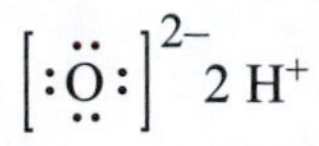

$$\left[:\ddot{O}:\right]^{2-} 2\,H^+$$

Think About It

The bcc unit cell has twice as many holes as there are ions in the unit cell. Therefore, the bcc unit cell has four holes. In ionic ice all four of those holes are filled with H^+ ions.

18.75. **Collect and Organize**

We are to determine the bond angles in cyclic S_6.

Analyze

Each sulfur atom in S_6 has two bonding pairs and two lone pairs and is sp^3 hybridized.

Solve

Each sulfur atom has a bent geometry with an ideal bond angle of 109.5°.

Think About It

The S_6 ring is not flat but, rather, is puckered and is flexible.

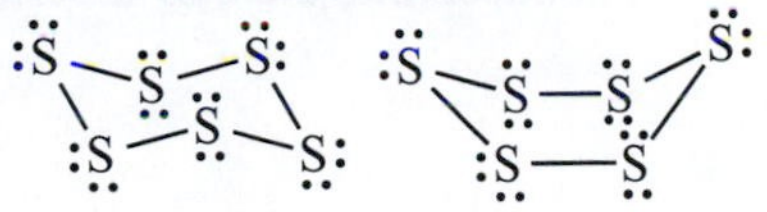

18.77. **Collect and Organize**

We are to explain why the Cl^- ions in the rock salt structure of LiCl touch in the unit cell, but those in KCl do not.

Analyze

In the fcc rock salt lattice, the alkali metal, Li^+ or K^+, is placed into the octahedral holes of the unit cell. From the radius ratio rule applied to LiCl and KCl (ionic radii are given in Figure 3.35),

$$\frac{r_{Li^+}}{r_{Cl^-}} = \frac{76\ pm}{181\ pm} = 0.42 \qquad \frac{r_{K^+}}{r_{Cl^-}} = \frac{138\ pm}{181\ pm} = 0.76$$

Solve

The radius ratios show that K^+ is large and so does not fit well into the octahedral holes (radius ratio between 0.41 and 0.73). Therefore, the rock salt structure for KCl has Cl^- ions that may not touch in order to accommodate the large K^+ ions in the fcc lattice.

Think About It

The radius ratio rule is simply a guide to predict in which type of hole cations will fit within the anion lattice. You will find that sometimes we would predict an octahedral hole for a cation when the actual crystal places the cation into a cubic hole, for example.

18.79. Collect and Organize

We are to describe how CsCl could be viewed as both a simple cubic and a body-centered cubic structure.

Analyze

A simple cubic structure consists of atoms located only at the corners of a cube. A body-centered cubic structure consists of an atom inside a cube of atoms.

Solve

The radius of Cl^- is 181 pm and the radius of Cs^+ is 170 pm, and so their radii are similar. The Cs^+ ion at the center of Figure P18.79 occupies the center of the cubic cell, so CsCl could be viewed as a body-centered cubic structure when taking into account the ions' slight difference in size. However, if we look at the ions as roughly equal in size, the unit cell becomes two interpenetrating simple cubic unit cells.

Think About It

In the body-centered cubic unit cell, the cubic hole in the simple cubic unit cell is filled.

18.81. Collect and Organize

We consider the rock salt structure to determine whether we could describe it as an fcc array of Na^+ with Cl^- in octahedral holes.

Analyze

For Cl^- to fit into the octahedral holes, the radius ratio

$$\frac{r_{Cl^-}}{r_{Na^+}} = \frac{181 \text{ pm}}{102 \text{ pm}}$$

would have to be between 0.41 and 0.73.

Solve

Although the Na^+ ions, when viewed without the Cl^- ions, are in an fcc array, the Na^+ ions are not closest-packed. Also, the radius ratio r_{Cl^-}/r_{Na^+} is greater than 1, indicating that the Cl^- does not fit well into the octahedral holes in the Na^+ lattice. It is not accurate, then, to describe NaCl as an fcc array of Na^+ with Cl^- in the octahedral holes.

Think About It

It is because of the large difference in radii that the "fcc array" of Na^+ ions is not closest-packed. The Cl^- radius is so large in comparison that it has greatly expanded the fcc lattice of Na^+ ions.

18.83. Collect and Organize

We are to predict the effect on density of an ionic compound of rock salt structure as the cation–anion radius ratio increases.

Analyze

When the cation–anion ratio increases, the cation is increasing relatively in size. If the anions remain in the rock salt structure, they might have to expand the fcc structure to accommodate larger cations.

Solve

As the cation–anion radius ratio increases, the closest-packed anions would expand to maintain the cations in the octahedral holes of the fcc lattice. The cell volume, therefore, increases and thus the calculated density would be less than the measured density.

Think About It
Another factor in density is the molar mass of the ions of the structure.

18.85. **Collect and Organize**
For a ccp array of O^{2-} ions with $^1/_4$ of the octahedral holes containing Fe^{3+}, $^1/_8$ of the tetrahedral holes containing Fe^{3+}, and $^1/_4$ of the octahedral holes containing Mg^{2+}, we are asked to write the formula of the ionic compound.

Analyze
A ccp array of ions has four octahedral holes and eight tetrahedral holes. The ccp array has an fcc unit cell, which has four closest-packed anions.

Solve
The unit cell will have the following:
    4 O^{2-} anions in the fcc unit cell
    1 Fe^{3+} in octahedral holes
    1 Fe^{3+} in tetrahedral holes
    1 Mg^{2+} in octahedral holes
The formula is $MgFe_2O_4$.

Think About It
The salt has charge balance as well: 4 O^{2-} gives a charge of 8–, which is balanced by 2 Fe^{3+} plus 1 Mg^{2+}, or a charge of 8+.

18.87. **Collect and Organize**
Given that a uranium oxide packs into a ccp structure and the oxide ions are located in all the tetrahedral holes, we are to write the formula for that U_xO_y salt.

Analyze
The unit cell for the ccp structure is the face-centered cube. That unit cell would contain four uranium atoms with eight tetrahedral holes.

Solve
If all the tetrahedral holes are filled, the unit cells have eight O^{2-} ions. That gives four U and eight O atoms in the unit cell, for a formula of UO_2.

Think About It
The oxidation number of uranium in that oxide is +4.

18.89. **Collect and Organize**
We consider the structure of anatase, a form of TiO_2, found on a map of Vinland believed to date from the 1400s.

Analyze
We can predict the type of hole Ti^{4+} is likely to occupy by calculating the radius ratio $r_{Ti^{4+}}/r_{O^{2-}}$. The radius ratios for tetrahedral, octahedral, and cubic holes are 0.22–0.41, 0.41–0.73, and 0.73–1.00, respectively (Table 18.4). In the ccp structure the unit cell is fcc. That unit cell contains four octahedral and eight tetrahedral holes.

Solve
(a) From the radius ratio,

$$\frac{r_{Ti^{4+}}}{r_{O^{2-}}} = \frac{60.5 \text{ pm}}{140 \text{ pm}} = 0.432$$

Ti^{4+} is expected to occupy octahedral holes.
(b) To give charge balance, one Ti^{4+} ion is present for every two O^{2-} ions in the lattice. Because the unit cell contains four O^{2-} ions, two Ti^{4+} ions must be in the unit cell. Because the unit cell has four octahedral holes, half the octahedral holes in the unit cell must be occupied.

Think About It
If the Ti^{4+} were small enough to fit into the tetrahedral holes, $^1/_4$ of the tetrahedral holes would be occupied.

18.91. Collect and Organize
For two forms of CdS, we consider why the rock salt structure would be denser than the sphalerite form.

Analyze
In the rock salt structure, the S^{2-} ions would be in an fcc array with Cd^{2+} occupying all the octahedral holes. In the sphalerite structure, the S^{2-} ions are also in an fcc array, but with Cd^{2+} ions occupying half the tetrahedral holes. From the radii of S^{2-} and Cd^{2+} given, we can calculate the radius ratio to determine which type of hole Cd^{2+} fits into.

$$\frac{r_{Cd^{2+}}}{r_{S^{2-}}} = \frac{95 \text{ pm}}{184 \text{ pm}} = 0.52$$

Solve
The radius ratio predicts that Cd^{2+} fits into the octahedral holes in the ccp lattices of S^{2-} ions. That rock salt arrangement is denser than the sphalerite arrangement because in sphalerite the lattice of S^{2-} ions must expand to accommodate the Cd^{2+} ions in the smaller tetrahedral holes.

Think About It
This example clearly shows that the radius ratio rule is only a guide to the structure of ionic solids. Here we predict CdS to be like rock salt, but in nature it is found as sphalerite.

18.93. Collect and Organize
Given parameters for the structure of the unit cell of ReO_3 and the radii of Re and O atoms (given as 137 and 73 pm, respectively), we are asked to sketch the unit cell for ReO_3, calculate the density of ReO_3, and calculate the percent empty space in a unit cell of ReO_3.

Analyze
We are given that Re atoms are present at each corner of the cubic unit cell, $(8 \times ^1/_8) = 1$ Re atom is in the unit cell. Likewise, given that 12 edge O atoms are present, $(12 \times ^1/_4) = 3$ O atoms are in the unit cell. We are told that the atoms touch (Re–O–Re) along an edge of the unit cell; therefore, the cell edge length is
$$137 + 73 + 73 + 137 = 420 \text{ pm}$$
To calculate the density of ReO_3, we use

$$d = \frac{\text{mass of atoms in unit cell}}{\text{volume of unit cell}}$$

To calculate the percent empty space, we subtract the volume that the atoms in the unit cell occupy from the total volume of the unit cell and then divide by the total volume of the unit cell.

$$\% \text{ empty space} = \frac{\ell^3 - \left[\left(1 \text{ Re}^{6+} \text{ ion} \times \frac{4}{3}\pi r^3\right) + \left(3 \text{ O}^{2-} \text{ ions} \times \frac{4}{3}\pi r^3\right)\right]}{\ell^3} \times 100$$

Solve
(a) ReO_3 has the following structure

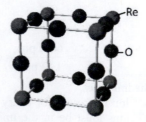

(b) The mass, m, of atoms in a unit cell of ReO_3 is

$$m = \frac{1 \text{ Re atom} \times \dfrac{186.21 \text{ g}}{\text{mol}} + 3 \text{ O atoms} \times \dfrac{15.999 \text{ g}}{\text{mol}}}{6.022 \times 10^{23} \text{ atoms/mol}} = 3.889 \times 10^{-22} \text{ g}$$

The volume, V, of the unit cell (in cubic centimeters) is

$$V = \left(420 \text{ pm} \times \frac{1 \times 10^{-10} \text{ cm}}{1 \text{ pm}} \right)^3 = 7.409 \times 10^{-23} \text{ cm}^3$$

The density, d, of the solid is then

$$d = \frac{3.889 \times 10^{-22} \text{ g}}{7.409 \times 10^{-23} \text{ cm}^3} = 5.25 \text{ g/cm}^3$$

(c) The percent empty space in the unit cell of ReO_3 is

$$\% \text{ empty space} = \frac{(420 \text{ pm})^3 - \left[\left(1 \text{ Re}^{6+} \text{ ion} \times \dfrac{4}{3}\pi(137 \text{ pm})^3 \right) + \left(3 \text{ O}^{2-} \text{ ions} \times \dfrac{4}{3}\pi(73 \text{ pm})^3 \right) \right]}{(420 \text{ pm})^3} \times 100 = 78.9\%$$

Think About It
Remember that this is a calculated, or theoretical, density. The actual density, even if the structure is correctly described, may differ slightly from the theoretical value.

18.95. Collect and Organize
Given the structure (rock salt) and density (3.60 g/cm^3) of MgO, we are asked to calculate the length of the edge of the unit cell (ℓ).

Analyze
The volume of the unit cell is simply ℓ^3. From the density we can calculate the volume, V, as

$$V = \frac{\text{mass of unit cell}}{\text{density}}$$

The fcc unit cell has four Mg atoms and four O atoms to give a mass, m, of

$$m = \frac{4 \text{ Mg atoms} \times \dfrac{24.305 \text{ g}}{\text{mol}} + 4 \text{ O atoms} \times \dfrac{15.999 \text{ g}}{\text{mol}}}{6.022 \times 10^{23} \text{ atoms/mol}} = 2.677 \times 10^{-22} \text{ g}$$

Solve

$$V = \frac{2.677 \times 10^{-22} \text{ g}}{3.60 \text{ g/cm}^3} = 7.436 \times 10^{-23} \text{ cm}^3$$

$$\ell = \sqrt[3]{V} = \sqrt[3]{7.436 \times 10^{-23} \text{ cm}^3} = 4.21 \times 10^{-8} \text{ cm, or } 421 \text{ pm}$$

Think About It
That value seems reasonable. The radius of an oxygen atom is 73 pm and that of a manganese atom is 127 pm (Appendix 3). If we assume that the O–Mn–O atoms touch along the fcc unit edge length (a good first approximation), the edge length would be

$$73 + 127 + 127 + 73 = 400 \text{ pm}$$

18.97. Collect and Organize
From a list of properties, we are to identify which describe ceramics and which are associated with metals.

Analyze
Metallic bonding in metals allows atoms to slip past each other and be easily deformed. Ceramics have either covalent or ionic bonding (or sometimes a mixture of both) that does not allow the atoms to easily slip past each other under stress. Metals have either partially filled valence bands or overlapping valence and conduction

bands that allow electrons to flow through the bands. Ceramics, however, have large band gaps between their valence and conduction bands that make them insulators and nonconductive to the flow of electrons.

Solve

Ductility, electrical and thermal conductivity, and malleability describe metals. Ceramics can be characterized as being both electrical and thermal insulators and brittle. Therefore:
(a) Metals are ductile.
(b) Ceramics are thermal insulators.
(c) Metals are electrically conductive.
(d) Metals are malleable.

Think About It

The bonding differences between metals and ceramics account for their different properties and, therefore, unique applications.

18.99. Collect and Organize

We are to write the formula for the mineral formed when Mg^{2+} replaces Al^{3+} in kaolinite $[Al_2(Si_2O_5)(OH)_4]$.

Analyze

When replacing Al^{3+} with Mg^{2+}, we must be careful to put in as many Mg^{2+} ions as needed to balance the charge of the Al^{3+} removed. Here, because we replace two Al^{3+} ions (6+ charge), we need 3 Mg^{2+} ions to balance the charge.

Solve

$Mg_3(Si_2O_5)(OH)_4$

Think About It

Several other ions (such as Fe^{3+} or Be^{2+}) could replace the Al^{3+} in kaolinite.

18.101. Collect and Organize

For the reaction of $KAlSi_3O_8$ with water and carbon dioxide, we are asked to determine whether it is a redox reaction and to balance the reaction to give $Al_2(Si_2O_5)(OH)_4$, SiO_2, and K_2CO_3 as products.

Analyze

In a redox reaction, the oxidation state of atoms must change. All species for that reaction contain O^{2-}, K^+, Si^{4+}, H^+, Al^{3+}, and C^{4+}.

Solve

$$2\ KAlSi_3O_8(s) + 2\ H_2O(\ell) + CO_2(aq) \rightarrow Al_2(Si_2O_5)(OH)_4(s) + 4\ SiO_2(s) + K_2CO_3(aq)$$

Because none of the atoms change oxidation state, that is not a redox reaction.

Think About It

In that weathering process, water breaks the larger silicate anion $Si_3O_8^{4-}$ down to $Si_2O_5^{2-}$ and SiO_2.

18.103. Collect and Organize

For the transformation of anorthite $(CaAl_2Si_2O_8)$ to a mixture of grossular $[Ca_3Al_2(SiO_4)_3]$, kyanite (Al_2SiO_5), and quartz (SiO_2) under high pressure, we are asked to write the balanced equation and to determine the charges on the silicate anions.

Analyze

The charge on the silicate ions is found, knowing that the cations are Ca^{2+} and Al^{3+}.

Solve

(a) $3\ CaAl_2Si_2O_8(s) \rightarrow Ca_3Al_2(SiO_4)_3(s) + 2\ Al_2SiO_5(s) + SiO_2(s)$
(b) In anorthite, the silicate anion is $Si_2O_8^{8-}$.
In grossular, the silicate anion is SiO_4^{4-}.
In kyanite, the silicate anion is SiO_5^{6-}.

Think About It
Many silicate minerals exist in nature.

18.105. **Collect and Organize**
Given the radii of Ba^{2+}, Ti^{4+}, and O^{2-} (135, 60.5, and 140 pm, respectively), we can use the radius ratio rule to determine which holes Ba^{2+} and Ti^{4+} occupy in a closest-packed arrangement of O^{2-} anions.

Analyze
The radius ratio can help us predict whether the smaller ion would fit into a cubic (radius ratio, 0.73–1.00), octahedral (0.41–0.73), or tetrahedral (0.22–0.41) hole in the crystal lattice.

Solve
Cubic holes can accommodate Ba^{2+}:

$$\frac{r_{Ba^{2+}}}{r_{O^{2-}}} = \frac{135 \text{ pm}}{140 \text{ pm}} = 0.964$$

Octahedral holes (maybe tetrahedral) can accommodate Ti^{4+}:

$$\frac{r_{Ti^{4+}}}{r_{O^{2-}}} = \frac{60.5 \text{ pm}}{140 \text{ pm}} = 0.432$$

Think About It
Ti^{4+} has a radius ratio at the edge of the tetrahedral–octahedral ranges, so predicting which of those holes it will occupy is difficult.

18.107. **Collect and Organize**
We are asked to explain why an amorphous solid does not give sharp peaks when scanned by X-ray diffraction.

Analyze
X-ray diffraction gives sharp peaks for crystalline materials that have a regular repeating array of atoms. Amorphous materials have no long-range order in their arrangement of atoms.

Solve
An amorphous solid has no regular repeating lattice to diffract the X-rays and therefore cannot give rise to the distinct constructive and destructive interference needed to produce sharp peaks in the X-ray diffraction scan.

Think About It
Amorphous materials in X-ray diffraction either give no signal or may show very broad peaks.

18.109. **Collect and Organize**
X-rays and microwaves are both electromagnetic radiation. We are to consider why X-rays, not microwaves, are used to determine structure.

Analyze
X-rays have wavelengths of the order of 10,000 to 10 pm. Microwaves have wavelengths of the order of 1 m to 1 mm.

Solve
The separation of atoms in crystal lattices is on the order of 10^{-10} m, or 100 pm. X-rays have wavelengths on the order of the separation of atoms in crystals, so constructive and destructive interference occurs. Microwaves have wavelengths too long for crystal lattices to diffract.

Think About It
X-ray diffraction is a useful technique, but it has difficulty "seeing" light elements, such as H atoms. Neutron-diffraction experiments can yield complementary information on structure.

18.111. Collect and Organize

We are to consider why different wavelengths of X-rays might be used to determine crystal structures.

Analyze

From the Bragg equation

$$n\lambda = 2d \sin \theta$$

we see that if λ is smaller, θ is smaller.

Solve

If a crystallographer uses a shorter wavelength, the data set can be collected over a smaller scanning range.

Think About It

With that approach, the crystallographer can see more data over a smaller scanning range so that if the camera or source can obtain only $2\theta = 30°$, for example, a shorter λ would reveal more peaks than a longer λ.

18.113. Collect and Organize

Given that the lattice spacing in sylvite (KCl) is larger than in halite (NaCl), we can use the Bragg equation to predict which crystal will diffract X-rays of a particular wavelength through higher 2θ values.

Analyze

Rearranging the Bragg equation to solve for d gives

$$d = \frac{n\lambda}{2\sin \theta}$$

That equation shows that $\sin \theta$ is inversely proportional to d.

Solve

The smaller the distance d (the lattice spacing), the larger the $\sin \theta$ and the larger the 2θ. Therefore, halite, with smaller lattice spacing, diffracts X-rays through larger 2θ values.

Think About It

Therefore, the pattern for NaCl is more spread out than that of KCl.

18.115. Collect and Organize

For galena, which shows reflections in X-ray diffraction ($\lambda = 71.2$ pm) at $2\theta = 13.98°$ and $21.25°$, we are asked to determine the value of n for those reflections and to calculate the spacing between the layers of the lattice (d).

Analyze

To determine n, we must find a pattern in the angles of reflection. We notice here that for $\theta = 6.99°$ and $\theta = 10.62°$ we have a factor of 1.52. The spacing of the layers in the lattice is calculated from the Bragg equation:

$$d = \frac{n\lambda}{2\sin \theta}$$

Solve

Because the ratio of the reflection angles (θ) is $10.62°/6.99° = 1.52$, the values of n are 2 ($\theta = 6.99°$) and 3 ($\theta = 10.62°$). The lattice spacings are

$$d = \frac{2 \times 71.2 \text{ pm}}{2\sin(6.99°)} = 585 \text{ pm}$$

$$d = \frac{3 \times 71.2 \text{ pm}}{2\sin(10.62°)} = 580 \text{ pm}$$

Average lattice spacing $(585 + 580)/2 = 582$ pm.

Think About It
The calculated lattice spacings from the reflections may not be exactly equal, but they will be close, as in the example above.

18.117. Collect and Organize
For a lattice distance of 1855 pm, we are to calculate the smallest angle of diffraction (2θ) for 154 pm wavelength X-rays diffracted in pyrophyllite.

Analyze
We can rearrange the Bragg equation to solve for $\sin \theta$:

$$\sin \theta = \frac{n\lambda}{2d}$$

where $n = 1$ for the smallest angle of diffraction. The 2θ value is twice the calculated value of θ.

Solve

$$\sin \theta = \frac{1 \times 154 \text{ pm}}{2 \times 1855 \text{ pm}} = 0.04151$$

$$\theta = 2.38°$$

$$2\theta = 4.76°$$

Think About It
For $n = 3$, the reflection would appear at

$$\sin \theta = \frac{3 \times 154 \text{ pm}}{2 \times 1855 \text{ pm}} = 0.1245$$

$$\theta = 7.15°$$

$$2\theta = 14.3°$$

18.119. Collect and Organize
For a unit cell with X at the eight corners of the cubic unit cell, Y at the center of the cube, and Z at the center of each face, we are to write the formula of the compound.

Analyze
Each atom at the corner of a cube counts as $1/8$ in the unit cell. Each atom on a face counts as $1/2$ in the unit cell. Each atom in the center of the unit cell counts as one.

Solve
Element X = $1/8 \times 8 = 1$ X
Element Y = $1 \times 1 = 1$ Y
Element Z = $6 \times 1/2 = 3$ Z
The formula for the compound is XYZ_3.

Think About It
An example of that kind of structure is perovskite, $CaTiO_3$.

18.121. Collect and Organize
Given the density and radius of silicon (2.33 g/mL and 117 pm, respectively), we can determine the volume of a silicon atom and the mass of 1.00 cm³ of silicon to calculate the packing efficiency of the atoms in the solid by using the equation

$$\text{Packing efficiency (\%)} = \frac{\text{volume occupied Si atoms}}{\text{volume of unit cell}} \times 100$$

Analyze

Assuming 1 cm³ of silicon that weighs 2.33 g, we can calculate the number of atoms in that 1 cm³. From the radius of one silicon atom, we can calculate the volume of that silicon atom and multiply by the number of atoms in the 1 cm³ sample. That gives the volume of space occupied by all the atoms in 1.00 cm³ of silicon. The percentage of space occupied, then, is the volume of Si divided by 1 cm³.

Solve

Number of Si atoms in 1 cm³:

$$2.33 \text{ g} \times \frac{1 \text{ mol}}{28.09 \text{ g}} \times \frac{6.022 \times 10^{23} \text{ atoms}}{1 \text{ mol}} = 4.995 \times 10^{22} \text{ atoms}$$

Volume of 1 Si atom:

$$V = \frac{4}{3}\pi r^3 = \frac{4}{3}\pi \left(117 \text{ pm} \times \frac{1 \times 10^{-10} \text{ cm}}{1 \text{ pm}} \right)^3 = 6.709 \times 10^{-24} \text{ cm}^3$$

Volume of Si atoms in 1.00 cm³ of Si:

$$6.709 \times 10^{-24} \frac{\text{cm}^3}{\text{atom}} \times 4.995 \times 10^{22} \text{ atoms} = 0.3351 \text{ cm}^3$$

Packing efficiency:

$$\frac{0.3351 \text{ cm}^3}{1.00 \text{ cm}^3} \times 100 = 33.5\%$$

Think About It

Therefore, 66.5% of the silicon block is empty space.

18.123. Collect and Organize

We consider nanocubes of Mo (4.8 nm on each side) and are asked to calculate the radius of Mo in the nanocubes, the density, and the number of Mo atoms in each nanocube.

Analyze

(a) We are given that the unit cells are body-centered cubic. In that arrangement the unit-cell edge length (ℓ) is $4r/\sqrt{3}$, where r is the radius of the atoms. The length of a unit cell is 4.8 nm/15 = 0.32 nm, or 320 pm, because we are given that the nanocube is 15 unit cells on an edge.

(b) The density of the nanocube can be calculated, knowing that each bcc unit cell has two Mo atoms. Once we calculate the mass of those two atoms of Mo, we divide by the volume (ℓ^3) of the unit cell.

(c) Because we know that 15 unit cells are on an edge for the nanocubes and that two Mo atoms are in each unit cell, we can calculate the number of atoms in each nanocube.

Solve

(a) Rearranging the equation for the unit edge length for a bcc unit cell gives the effective radius for Mo:

$$r = \frac{\ell\sqrt{3}}{4} = \frac{320 \text{ pm}\sqrt{3}}{4} = 139 \text{ pm}$$

(b) The density of the bcc array of Mo in the nanocubes is

$$d = \frac{\left(2 \text{ atoms} \times \dfrac{95.96 \text{ g}}{\text{mol}} \times \dfrac{1 \text{ mol}}{6.022 \times 10^{23} \text{ atoms}} \right)}{\left(320 \text{ pm} \times \dfrac{1 \times 10^{-10} \text{ cm}}{1 \text{ pm}} \right)^3} = 9.73 \frac{\text{g}}{\text{cm}^3}$$

(c) The number of unit cells and Mo atoms in a nanocube are

$$(15 \text{ unit cells on an edge})^3 = 3375 \text{ unit cells}$$

$$3375 \text{ unit cells} \times \frac{2 \text{ Mo atoms}}{\text{unit cell}} = 6750 \text{ Mo atoms}$$

Think About It

Even though nanocubes are very small, they still contain thousands of Mo atoms.

18.125. Collect and Organize

We consider different possible crystal structures of Fe (radius, 126 pm) and compare the densities of bcc and hcp iron and calculate the density of a crystal of 96% Fe and 4% Si.

Analyze

For each structure, the density is the mass of atoms in the unit cell divided by the volume of the cubic unit cell. For the bcc unit cell, two atoms exist per unit cell and the edge length is $\ell = 4r/\sqrt{3}$. Each hcp unit cell has two atoms.

Solve

(a) The edge length of a bcc unit cell is

$$\ell = \frac{4r}{\sqrt{3}} = \frac{4 \times 126 \text{ pm}}{\sqrt{3}} = 291.0 \text{ pm, or } 2.91 \times 10^{-8} \text{ cm}$$

The volume of the unit cell is

$$V = \left(291.0 \times 10^{-8} \text{ cm}\right)^3 = 2.464 \times 10^{-23} \text{ cm}^3$$

The mass of the unit cell is

$$2 \text{ atoms Fe} \times \frac{55.85 \text{ g}}{\text{mol}} \times \frac{1 \text{ mol}}{6.022 \times 10^{23} \text{ atoms}} = 1.855 \times 10^{-22} \text{ g}$$

The density is

$$d = \frac{1.855 \times 10^{-22} \text{ g}}{2.464 \times 10^{-23} \text{ cm}^3} = 7.53 \text{ g/cm}^3$$

(b) The hcp unit cell also has two atoms. We are given that the volume of the unit cell is $5.414 \times 10^{-23} \text{ cm}^3$, so the density is

$$d = \frac{1.855 \times 10^{-22} \text{ g}}{5.414 \times 10^{-23} \text{ cm}^3} = 3.42 \text{ g/cm}^3$$

(c) If 4% of the mass is due to Si replacing Fe, the molar mass of the alloy would be

$$0.04 \times \frac{28.09 \text{ g}}{\text{mol}} + 0.96 \times \frac{55.85 \text{ g}}{\text{mol}} = 54.74 \text{ g/mol}$$

The mass of two atoms in the unit cell would be

$$2 \text{ atoms} \times \frac{54.74 \text{ g}}{\text{mol}} \times \frac{1 \text{ mol}}{6.022 \times 10^{23} \text{ atoms}} = 1.818 \times 10^{-22} \text{ g}$$

The density is

$$d = \frac{1.818 \times 10^{-22} \text{ g}}{5.414 \times 10^{-23} \text{ cm}^3} = 3.36 \text{ g/cm}^3$$

Think About It

Replacing Fe with an element of lower molar mass (silicon) gives a lower-density alloy.

18.127. **Collect and Organize**

Given that substitutional alloys form when the difference in the radii of the alloying metals is no greater than 15%, we are to predict which alloy of AuZn, AgZn, and CuZn has the greatest mismatch of atomic radii. Then, using the atomic radii in Appendix 3, we are to calculate the percent difference in the atomic radii for the three alloys to determine which will form substitutional alloys. Finally, we are to determine whether the atoms in the AuAg alloy touch along the face diagonal of an fcc unit cell.

Analyze

(a) We can use periodic trends to determine which metal atoms have the greatest mismatch in size. As we go across a row in the periodic table, atoms generally get smaller; as we go down a group in the periodic table, atoms generally get larger.

(b) The atomic radii of those metals are Cu, 128 pm; Ag, 144 pm; Au, 144 pm; and Zn, 134 pm.

(c) In the fcc structure (Figure 18.11b) for atoms of the same size, the atoms touch along the face diagonal.

Solve

(a) From periodic trends for Cu, Ag, Au, and Zn, we would expect zinc to have the smallest atomic radius and gold to have the largest atomic radius. Therefore, we would expect the largest mismatch in size for AuZn.

(b) The percent differences in radii for AuZn, AgZn, and CuZn are

$$\text{For AuZn:} \quad \frac{144 \text{ pm} - 134 \text{ pm}}{\left(\dfrac{144 \text{ pm} + 134 \text{ pm}}{2} \right)} \times 100 = 7.19\%$$

$$\text{For AgZn:} \quad \frac{144 \text{ pm} - 134 \text{ pm}}{\left(\dfrac{144 \text{ pm} + 134 \text{ pm}}{2} \right)} \times 100 = 7.19\%$$

$$\text{For CuZn:} \quad \frac{134 \text{ pm} - 128 \text{ pm}}{\left(\dfrac{128 \text{ pm} + 134 \text{ pm}}{2} \right)} \times 100 = 4.58\%$$

All those alloys are expected to form substitutional alloys.

(c) Because gold and silver have the same atomic radii, yes, the atoms touch along the face diagonal of an fcc unit cell.

Think About It

The unit cell for AuZn and AgZn will have the same edge length and therefore the same volume. Because of the greater molar mass of Au versus Ag, however, AuZn will be denser than AgZn.

18.129. **Collect and Organize**

We are to determine the formula of a transition metal sulfide for which the unit cell is shown in Figure P18.129. Then we are to count the M^{2+} and S^{2-} ions in the unit cell, determine which spheres represent the S^{2-} ions, and determine the number of yellow spheres that the brown spheres touch.

Analyze

Atoms on the corners contribute $\frac{1}{8}$ of their volume to the unit cell. Atoms on the edges of the unit cell contribute $\frac{1}{4}$ of their volume to the unit cell. Atoms completely inside the unit cell contribute all their volume to the unit cell.

Solve

(a) Following from the results in part (b) below, the formula of the transition metal sulfide is MS.

(b) The number of yellow spheres (anions) in the unit cell is

$$(8 \times \tfrac{1}{8}) + 1 = 2$$

The number of metal cations (brown spheres) in the unit cell is 8. The formula for the compound is MS.

$$(4 \times \tfrac{1}{4}) + 1 = 2$$

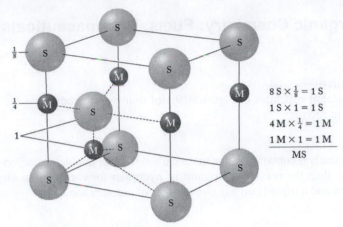

$8\,S \times \frac{1}{8} = 1\,S$

$1\,S \times 1 = 1\,S$

$4\,M \times \frac{1}{4} = 1\,M$

$\underline{1\,M \times 1 = 1\,M}$

MS

(c) The yellow spheres are S^{2-} ions and the brown spheres are M^{2+} ions.
(d) Each brown sphere touches four yellow spheres.

Think about It
The metal ions are located in the tetrahedral holes of the S^{2-} lattice.

18.131. **Collect and Organize**
From the formula Cu_3Al for the alloy with a bcc structure, we are to determine how the copper and aluminum atoms are distributed between the unit cells.

Analyze
The bcc unit cell contains two atoms. The radii of Al and Cu are 143 and 128 pm, respectively. The radius ratio is 0.895.

Solve
The radius ratio indicates that Cu atoms fit into the cubic holes of the Al lattice. Another way to look at that is as a simple cubic arrangement of Cu atoms with an Al atom in the center of the unit cell. To be consistent with the formula (Cu_3Al), we would have to consider three unit cells, one of which would have one Al atom in the center.

Think About It
It is probably also possible to have an alloy of formula Al_3Cu because their atomic radii are so similar.

CHAPTER 19 | Organic Chemistry: Fuels, Pharmaceuticals, and Modern Materials

19.1. Collect and Organize

Given the carbon-skeleton structures in Figure P19.1 of four hydrocarbons, we are to determine the degrees of unsaturation present in each.

Analyze

An unsaturated hydrocarbon contains one or more carbon–carbon double or carbon–carbon triple bonds. Those compounds have less than the maximum amount of hydrogen for each carbon atom. A double bond has one degree of unsaturation and a triple bond has two degrees of unsaturation.

Solve

(a) This structure has one double bond. It has one degree of unsaturation.
(b) This structure has two double bonds. It has two degrees of unsaturation.
(c) This structure has neither double nor triple bonds. It is a saturated hydrocarbon with no degrees of unsaturation.
(d) This structure has one double and one triple bond. It has three degrees of unsaturation.

Think About It

Because the structure in (d) has three degrees of unsaturation, it will combine with three molecules of hydrogen (H_2) to form a saturated hydrocarbon.

19.3. Collect and Organize

From the structures of fragrant oils in Figure P19.3, we are to identify those that contain the alkene functional group.

Analyze

An alkene functional group is a carbon–carbon double bond.

Solve

Both pine oil and oil of celery contain a $C = C$ bond and so are classified as alkenes.

Think About It

Camphor is not an alkene, but it does have a $C = O$ double bond. That functional group is a ketone.

19.5. Collect and Organize

Of the four hydrocarbons shown in Figure P19.5, we are to identify those that are aromatic.

Analyze

Aromatic compounds have planar, hexagonal rings of six sp^2-hybridized carbon atoms with alternating single and double bonds.

Solve

Compounds b and d are aromatic.

Think About It

Aromatic compounds have a special stability due to resonance.

19.7. Collect and Organize

Given the structures of polypropylene in Figure P19.7, we are to determine which is soft and rubbery and which is rigid and resists deformation.

Analyze

Molecules that have a repeating structure stack more regularly and have greater crystallinity than those with irregularly repeating units.

Solve

Polypropylene (b) is soft and rubbery because its methyl groups are randomly positioned, which does not allow for the polymer chains to crystallize well. Polypropylene (a) is stiff and resistant to deformation because its regular arrangement of methyl groups allows greater crystallinity.

Think About It

Polymers like (a) with regular arrangement of their side chains are called *isotactic*; polymers like (b) are called *atactic*.

19.9. Collect and Organize

From the structure of dihydroxydimethylsilane (Figure P19.9), we are to draw the condensed structure for the repeating monomeric unit in Silly Putty.

Analyze

The condensation reaction of dihydroxydimethylsilane combines the monomers of $Si(CH_3)_2(OH)_2$ to create a larger molecule, losing water as a by-product. The –OH groups on the monomer will combine to produce the –O–Si–O– linkage for the polymer backbone.

Solve

$$HO-\underset{\underset{CH_3}{|}}{\overset{\overset{CH_3}{|}}{Si}}-OH \;+\; HO-\underset{\underset{CH_3}{|}}{\overset{\overset{CH_3}{|}}{Si}}-OH \;\xrightarrow[-H_2O]{}\; HO-\underset{\underset{CH_3}{|}}{\overset{\overset{CH_3}{|}}{Si}}-O-\underset{\underset{CH_3}{|}}{\overset{\overset{CH_3}{|}}{Si}}-OH$$

Continuing that reaction produces a long-chain polymer with the condensed structure

$$\left[-O-\underset{\underset{CH_3}{|}}{\overset{\overset{CH_3}{|}}{Si}}-O- \right]_n$$

Think About It

Those types of siloxane polymers have found uses in soft contact lenses, oils, and greases.

19.11. Collect and Organize

We are asked to draw monomeric units of *cis*- and *trans*-polyisoprene.

Analyze

The structures of the two polyisoprenes are shown in Figure P19.11. Those polymers form through addition reactions of an alkene. The monomeric unit is the smallest repeating unit in the polymer.

Solve

For *cis*-polyisoprene, the monomeric unit is

For *trans*-polyisoprene, the monomeric unit is

Think About It
The difference in those polymers is the orientation of how the monomeric units are joined, which gives a different orientation of the C $=$ C double bond in the polymer backbone.

19.13. Collect and Organize
After listing the different set of hybrid orbitals in the bonding in organic compounds, we are to determine what combinations of single and multiple bonds are possible for each set.

Analyze
A carbon atom has available the 2*s* and 2*p* orbitals for hybridization.

Solve
The possible hybridized orbital sets and the combinations of multiple bonds possible for each are as follows:
sp^3, four single bonds
sp^2, one double bond and two single bonds
sp, one triple bond and one single bond or two double bonds

Think About It
For hydrocarbons (molecules composed of only carbon and hydrogen bonds), sp^3-hybridized carbons are in alkanes, sp^2-hybridized carbons are in alkenes, and sp-hybridized carbons are in alkynes.

19.15. Collect and Organize
Tungsten carbide contains carbon. We are asked whether that compound is considered an organic compound.

Analyze
Organic compounds are compounds that have a carbon–element bond.

Solve
The carbon atoms in WC occupy the interstices of the closest-packed tungsten lattice and therefore are not formally bonded to the tungsten at all. No, tungsten carbide is not considered an organic compound.

Think About It
We generally consider organic compounds to contain carbon covalently bonded to carbon, hydrogen, oxygen, nitrogen, phosphorus, sulfur, and other atoms.

19.17. Collect and Organize
From the functional groups shown in Table 19.1 of the textbook, we are to determine which are polar.

Analyze
A functional group will be polar if, owing to differences in the electronegativity of the bonded elements and the symmetry of the molecule, the molecule has a permanent dipole moment.

Solve
The functional groups in Table 19.1 that are polar are amine, alcohol, ether, aldehyde, ketone, carboxylic acid, ester, and amide.

Think About It
By identifying the functional groups in an organic compound as polar, chemists can predict some of its properties and reactivities.

19.19. Collect and Organize
Given that sample A of polyethylene has a molar mass twice that of a polyethylene sample B, we are to predict which polymer sample would soften at a higher temperature.

Analyze
We know from intermolecular forces between molecules that samples with higher molar masses have more intermolecular forces and so have higher melting points. Softening in polymers indicates the same type of molecular movement (although restricted) seen in melting, so the two are analogous.

Solve
Sample A, with a higher molar mass, will soften at a higher temperature than sample B, with a lower molar mass.

Think About It
Thermoplastics are polymers that can be softened at high temperatures, molded, and then returned to solid form by cooling.

19.21. Collect and Organize
We are to compare the empirical formulas for linear and branched alkanes that have the same number of carbon atoms.

Analyze
For any alkane, whether branched or linear, the empirical formula is C_nH_{2n+2}.

Solve
Yes, linear and branched alkanes with the same number of carbon atoms have the same empirical formula.

Think About It
For example, consider the following alkanes with four carbon atoms and the empirical formula C_4H_{10}:

n-Butane   2-Methylpropane

19.23. Collect and Organize
By considering the number of covalent bonds that carbon forms in alkanes, we can determine the hybridization of the carbon atoms.

Analyze
In alkanes, carbon is singly bonded to four other atoms, either to other carbon atoms or to hydrogen atoms.

Solve
Carbon forms four single bonds when it undergoes sp^3 hybridization.

Think About It
When the carbon is sp^2 hybridized, it may form double bonds to other carbon atoms (alkenes) or to oxygen (ketones, aldehydes, carboxylic acids).

19.25. Collect and Organize
Given that cyclohexane (C_6H_{12}) is not planar, we can draw the Lewis structure to help explain why.

Analyze
The Lewis structure for cyclohexane is

Solve
Each carbon atom in the six-membered ring is sp^3 hybridized with ideal bond angles of 109.5°. The ring, therefore, is not planar.

Think About It
The cyclohexane ring can adopt two structures; the chair structure is more stable.

Chair   Boat

19.27. Collect and Organize
By considering the definition of a saturated hydrocarbon, we can determine whether cycloalkanes are saturated.

Analyze
By the textbook definition, a saturated hydrocarbon has the maximum ratio of hydrogen to carbon atoms in its structure and has an empirical formula of C_nH_{2n+2}.

Solve
No. A cycloalkane has a formula of C_nH_{2n} and therefore is not a saturated hydrocarbon.

Think About It
That is a tricky definition, however. Another definition of a saturated hydrocarbon is that every carbon in the molecule is bonded to four other atoms. That definition would classify cycloalkanes as saturated hydrocarbons.

19.29. Collect and Organize
By considering the definition of constitutional isomers, we can determine whether they always have the same chemical properties.

Analyze
Two compounds are constitutional isomers if they have the same formula but different arrangements of the atoms.

Solve
No. Constitutional isomers have different chemical properties because of their different arrangement of atoms.

Think About It
Constitutional isomers also have different physical properties, such as different melting points and vapor pressures.

19.31. Collect and Organize
For the molecular formula C_5H_{12}, we are asked to draw and name all the constitutional isomers.

Analyze

Constitutional isomers all have the same molecular formula, so all our structures must contain 5 C atoms and 12 H atoms, just arranged differently. Naming of alkanes is described in Section 19.2, with examples in Table 19.3 for the linear alkanes.

Solve

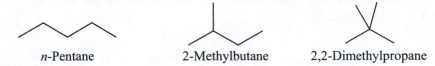

n-Pentane   2-Methylbutane   2,2-Dimethylpropane

Think About It

The naming convention used for those compounds uniquely describes the structure of each isomer.

19.33. Collect and Organize

Of the five molecules shown in Figure P19.33, we are to identify and name those that are constitutional isomers of *n*-octane.

Analyze

Any constitutional isomer of *n*-octane has to have the molecular formula C_8H_{18}.

Solve

(a) This molecule is C_8H_{18} and is therefore a constitutional isomer.
(b) This molecule is C_9H_{20} and is not a constitutional isomer of *n*-octane.
(c) This molecule is C_8H_{18} and is therefore a constitutional isomer.
(d) This molecule is C_8H_{18} and is therefore a constitutional isomer.
(e) This molecule is C_9H_{20} and is not a constitutional isomer of *n*-octane.

The constitutional isomers of *n*-octane are (a) 2,3-dimethylhexane, (c) 2-methylheptane, and (d) 2-methylheptane.

Think About It

Molecules c and d are not constitutional isomers of each other because they have the same arrangement of atoms (just drawn differently), but molecules b and e are constitutional isomers of each other.

19.35. Collect and Organize

Using the examples for hydrocarbons described in Section 19.2, we can convert the line drawings of each molecule in Problem 19.33 to chemical formulas.

Analyze

In line drawings, only the carbon skeleton is shown. The end of a line is understood to be the –CH_3 group and the intersection of two lines is understood to be a –CH_2– group. If more than two lines intersect, the number of hydrogen atoms at that carbon is 4 minus the number of intersecting lines because carbon makes bonds to four atoms to satisfy its octet.

Solve

(a) The condensed formula is $(CH_3)_2CHCH(CH_3)CH_2CH_2CH_3$, for a molecular formula of C_8H_{18}.
(b) The condensed formula is $(CH_3)_2CHCH_2CH_2CH_2CH_2CH_2CH_3$, for a molecular formula of C_9H_{20}.
(c) The condensed formula is $(CH_3)_2CHCH_2CH_2CH_2CH_2CH_3$, for a molecular formula of C_8H_{18}.
(d) The condensed formula is $(CH_3)_2CHCH_2CH_2CH_2CH_2CH_3$, for a molecular formula of C_8H_{18}.
(e) The condensed formula is $CH_3CH_2CH(CH_3)CH(CH_3)CH(CH_3)_2$, for a molecular formula of C_9H_{20}.

Think About It

The names of the isomers of C_9H_{20} shown in Problem 19.33 are (b) 2-methyloctane and (e) 2,3,4-trimethylhexane.

19.37. **Collect and Organize**

We are to place C_6H_{14}, $C_{18}H_{38}$, $C_{12}H_{26}$, and C_9H_{20} in the order in which they would appear in the distillation of crude oil.

Analyze

All those compounds are nonpolar. The larger the dispersion forces between molecules, the higher the boiling point. Dispersion forces are greater for nonpolar molecules with more atoms and for those that are less branched or those that have longer chains.

Solve

Appearing earliest to latest in the distillation: $C_6H_{14} < C_9H_{20} < C_{12}H_{26} < C_{18}H_{38}$.

Think About It

If any of those alkanes are branched, it is the branched isomers that have lower boiling points than the linear isomer.

19.39. **Collect and Organize**

By looking at the molecular formulas of hexane and cyclohexane, we are to determine which has the higher H:C ratio.

Analyze

Hexane has the molecular formula C_6H_{14}. Cyclohexane has the molecular formula C_6H_{12}.

Solve

With 14 hydrogen atoms per 6 carbon atoms (versus 12 hydrogen atoms to 6 carbon atoms), hexane has a higher H:C ratio.

Think About It

Those hexanes are not constitutional isomers since they do not have the same molecular formula.

19.41. **Collect and Organize**

By considering the empirical formula of an alkene and a cycloalkane with the same number of carbon atoms, we can determine whether combustion analysis could distinguish between the two.

Analyze

Both cycloalkanes and alkenes have the formula C_nH_{2n}.

Solve

Because no difference exists in the number of either C atoms or H atoms between the alkene and the cycloalkane, no, we cannot distinguish them by combustion analysis.

Think About It

We could, however, distinguish between an alkane of n C atoms (C_nH_{2n+2}) and an alkene (C_nH_{2n}).

19.43. **Collect and Organize**

We are to explain why alkenes in which the double bond involves the first (or last) carbon do not exhibit cis and trans isomerism.

Analyze

A cis isomer has two like groups on the same side of a line drawn through the double bond. A trans isomer has two like groups on opposite sides of a line drawn through the double bond.

Solve

When the double bond is "terminal" (occurs at the end or beginning of the carbon chain), two like groups (H) are present, so no cis and trans isomers are possible.

Think About It
No cis and trans isomers would likewise be possible for a double bond for which two of the substituents on a single carbon atom are the same, as shown in the following example:

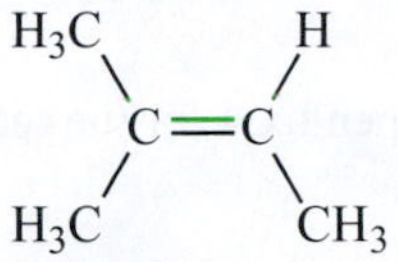

$$H_3C \quad\quad H$$
$$C = C$$
$$H_3C \quad\quad CH_3$$
2-Methyl-2-butene

19.45. Collect and Organize
From the structure of carvone, we are asked to explain why that molecule does not have cis and trans isomers.

Analyze
A cis isomer has two like groups on the same side of a line drawn through the double bond. A trans isomer has two like groups on opposite sides of a line drawn through the double bond.

Solve
The C=C double bond outside the ring does not show cis–trans isomerism because the terminal carbon atom does not have two dissimilar groups. The C=C double bond in the ring of carbon atoms is cis in the structure of carvone. That bond cannot be trans, or the ring of six carbon atoms would not be possible.

Think About It
A related molecule that would show trans and cis isomers would be

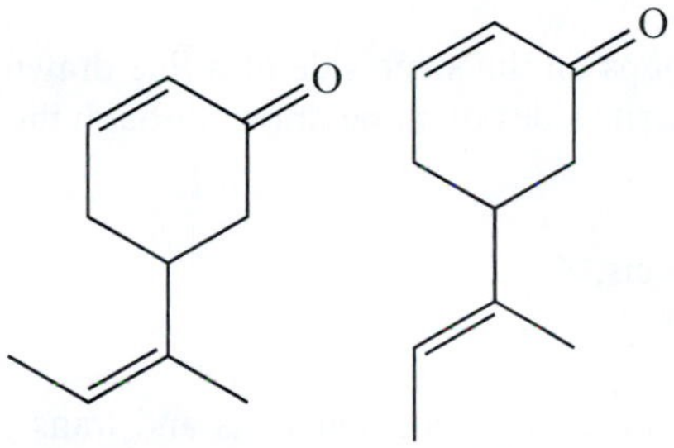

Trans   Cis

19.47. Collect and Organize
By comparing the structures of ethylene and polyethylene, we can explain why ethylene reacts with HBr but polyethylene does not.

Analyze
The structures of ethylene and polyethylene are as follows:

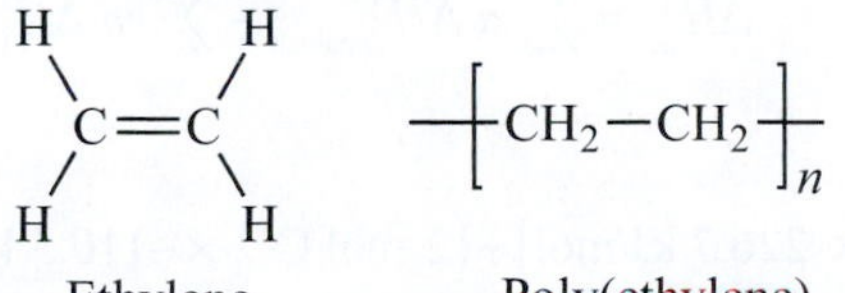

$$H \quad\quad H$$
$$C = C$$
$$H \quad\quad H$$
Ethylene

$$+CH_2-CH_2+_n$$
Poly(ethylene)

Solve
Ethylene has a C=C bond with which HBr is reactive, but polyethylene has only saturated C—C bonds.

Think About It
Polyethylene is produced from ethylene under high temperature and pressure.

19.49. **Collect and Organize**

By adding the energy of the bonds that form ($-\Delta H$) and the energy of the bonds broken ($+\Delta H$) in the reaction of C_2H_4 with H_2 to give C_2H_6, we can estimate $\Delta H_{hydrogenation}$.

Analyze

The relevant bond strengths from Appendix 4 for the reactants and products are as follows:

    C—H   413 kJ/mol
    C=C   614 kJ/mol
    H—H   436 kJ/mol
    C—C   348 kJ/mol

In the reactants, a C=C bond and a H—H bond are broken. In forming the products, a C—C bond and two C—H bonds are formed.

Solve

$$\Delta H_{rxn} = \sum \Delta H_{bond\ breaking} + \Delta H_{bond\ forming}$$
$$= [614 + 436] + [-348 + (2 \times -413)] = -124\ kJ$$

Think About It

Because that reaction is exothermic, it is favored by enthalpy.

19.51. **Collect and Organize**

For the two structures shown in Figure P19.51, we are asked to label the isomers as cis or trans and *E* or *Z*.

Analyze

A cis or *Z* isomer has two like groups on the same side of a line drawn through the double bond. A trans or *E* isomer has two like groups on opposite sides of a line drawn through the double bond.

Solve

Isomer a is trans, *E* and isomer b is cis, *Z*.

Think About It

In designating those isomers, we more often encounter cis and trans rather than *E* and *Z*, but knowing both designations is helpful in studying organic chemistry.

19.53. **Collect and Organize**

Using ΔH_f° values for the reactants and products, we can calculate the ΔH_{rxn}° for the controlled combustion of methane.

Analyze

To calculate ΔH_{rxn}° we use

$$\Delta H_{rxn}^\circ = \sum n\ \Delta H_{f,products}^\circ - \sum m\ \Delta H_{f,reactants}^\circ$$

Solve

$$\Delta H_{rxn}^\circ = \left[(2\ mol\ C_2H_2 \times 226.7\ kJ/mol) + (2\ mol\ CO \times -110.5\ kJ/mol) + (10\ mol\ H_2 \times 0\ kJ/mol) \right]$$
$$- \left[(6\ mol\ CH_4 \times -74.8\ kJ/mol) + (1\ mol\ O_2 \times 0\ kJ/mol) \right]$$
$$= 681.2\ kJ$$

That reaction, a controlled combustion of methane, is endothermic.

Think About It

The uncontrolled combustion of methane to give CO_2 and H_2O, however, is exothermic.

19.55. Collect and Organize

From the balanced hydrogenation reactions of acetylene and ethylene to form ethane, we are to estimate the $\Delta H^{\circ}_{\text{rxn}}$ for the hydrogenation of acetylene to ethylene.

Analyze

To obtain the hydrogenation of acetylene to ethylene, we can reverse the second chemical reaction and then add to the first.

$$HC\equiv CH(g) + 2\ H_2(g) \rightarrow H_3C-CH_3(g)$$
$$H_3C-CH_3(g) \rightarrow H_2C=CH_2(g) + H_2(g)$$
$$\overline{HC\equiv CH(g) + H_2(g) \rightarrow H_2C=CH_2(g)}$$

The heat of a reaction may be calculated by using $\Delta H^{\circ}_{\text{f}}$ values for the reactants and products according to the equation

$$\Delta H^{\circ}_{\text{rxn}} = \sum n\ \Delta H^{\circ}_{\text{f,products}} - \sum n\ \Delta H^{\circ}_{\text{f,reactants}}$$

Solve

The enthalpy of the hydrogenation of acetylene is:

$$\Delta H^{\circ}_{\text{rxn}} = \left[\left(1\ \text{mol}\ C_2H_6\right)\left(-84.67\ \text{kJ/mol}\right)\right] - \left[\left(1\ \text{mol}\ C_2H_2\right)\left(226.7\ \text{kJ/mol}\right) + \left(2\ \text{mol}\ H_2\right)\left(0.0\ \text{kJ/mol}\right)\right]$$
$$= -311.4\ \text{kJ/mol}$$

The enthalpy of the hydrogenation of ethylene is:

$$\Delta H^{\circ}_{\text{rxn}} = \left[\left(1\ \text{mol}\ C_2H_6\right)\left(-84.67\ \text{kJ/mol}\right)\right] - \left[\left(1\ \text{mol}\ C_2H_4\right)\left(52.4\ \text{kJ/mol}\right) + \left(1\ \text{mol}\ H_2\right)\left(0.0\ \text{kJ/mol}\right)\right]$$
$$= -137.1\ \text{kJ/mol}$$

For the overall hydrogenation reaction of acetylene to ethylene:

$$HC\equiv CH(g) + 2\ H_2(g) \rightarrow H_3C-CH_3(g) \qquad \Delta H^{\circ}_{\text{rxn}} = -311.4\ \text{kJ}$$
$$H_3C-CH_3(g) \rightarrow H_2C=CH_2(g) + H_2(g) \qquad \Delta H^{\circ}_{\text{rxn}} = 137.1\ \text{kJ}$$
$$\overline{HC\equiv CH(g) + H_2(g) \rightarrow H_2C=CH_2(g) \qquad \Delta H^{\circ}_{\text{rxn}} = -174.3\ \text{kJ}}$$

That is an exothermic reaction.

Think About It

The partial hydrogenation of acetylene to ethylene in practice, however, is difficult because ethylene will easily hydrogenate to ethane under the conditions of the hydrogenation reaction.

19.57. Collect and Organize

From the carbon-skeleton structure of vinyl acetate (Figure P19.57), we are to draw the structure of poly(vinyl acetate).

Analyze

Poly(vinyl acetate) is an addition polymer in which the $C=C$ double bonds link to form the polymer backbone. In that polymer, the acetate group is a side chain off the polymer chain.

Solve

Think About It

Poly(vinyl acetate), or PVA, is used in bookbinding because of its flexibility and, as an emulsion in water, as an adhesive for wood, paper, and cloth.

19.59. Collect and Organize

We are asked to explain why the changes to the structure of polyacetylene upon reaction with O_2 shown in Figure P19.59 decrease the conductivity.

Analyze

Upon reaction, oxygen is incorporated into the structure of polyacetylene along the carbon chain.

Solve

The conductivity of polyacetylene decreases because the conjugated C—C double bonds that give polyacetylene its conductivity are disrupted.

Think About It

The discoverers of polyacetylene's conductivity were awarded the Nobel Prize in Chemistry in 2000.

19.61. Collect and Organize

By looking at the Lewis structure of benzene, we can explain why it is a planar molecule.

Analyze

Benzene has a six-membered carbon ring with alternating single and double bonds.

Solve

The line structure of benzene is

In benzene, each C atom is sp^2 hybridized with bond angles of 120°. That geometry at each carbon atom in the ring makes benzene a planar molecule.

Think About It

Benzene's π electrons are delocalized by resonance, lending benzene a special stability.

19.63. Collect and Organize

By drawing the line structures of tetramethylbenzene and pentamethylbenzene, we can determine whether those molecules have any constitutional isomers.

Analyze

The methyl groups on those compounds are ring substituents. That is, they take the place of H atoms in the benzene structure.

Solve

Tetramethylbenzene has three constitutional isomers:

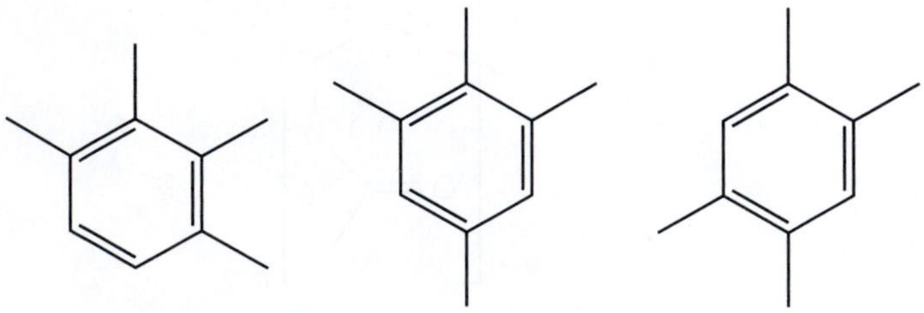

Pentamethylbenzene has no constitutional isomers:

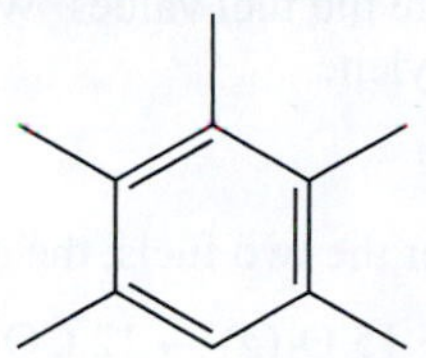

Think About It
Remember that constitutional isomers have distinct chemical and physical properties.

19.65. Collect and Organize
By examining pyridine's structure (Figure P19.65), we can determine whether that compound is aromatic.

Analyze
Aromatic structures have alternating single and double bonds through which resonance delocalizes the electrons.

Solve
Yes. Pyridine is an aromatic compound because the π electrons are delocalized over all the atoms in the ring through resonance.

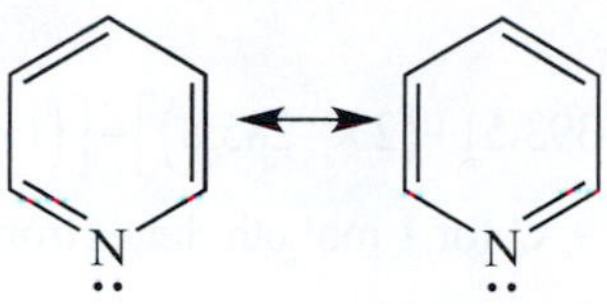

Think About It
In that aromatic compound, the alternating single and double bonds include C—N bonds.

19.67. Collect and Organize
We are to draw all the constitutional isomers of trimethylbenzene.

Analyze
The methyl groups on the benzene ring are substituents that have replaced the H atoms on benzene.

Solve

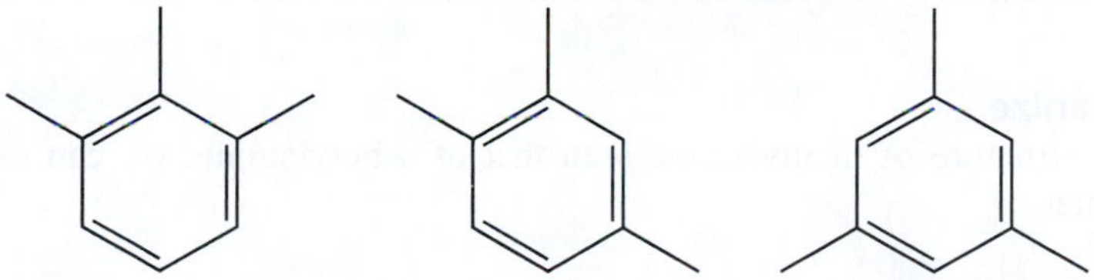

Think About It
The structures below are not additional structural isomers. They are the same as the first isomer above. All we need to do is rotate each one below to give the isomer above.

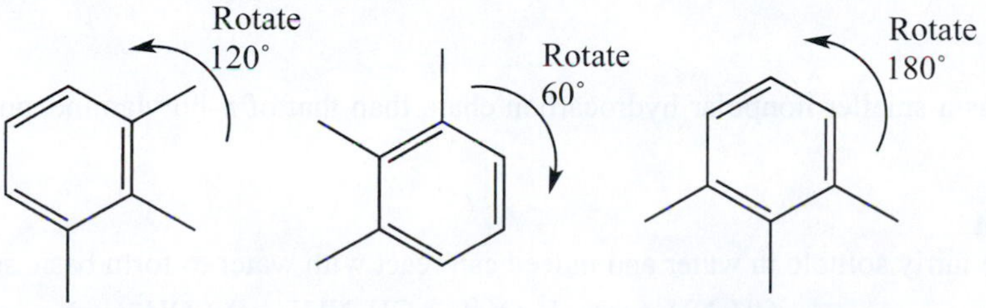

19.69. **Collect and Organize**

For benzene and ethylene, we are to calculate the fuel values. We are also asked whether 1 mol of benzene has a higher or lower fuel value than 3 mol of ethylene.

Analyze

The fuel value is ($-\Delta H_{comb}$/mass of fuel). For the two fuels, the combustion reactions are

$$2\,C_6H_6(\ell) + 15\,O_2(g) \rightarrow 12\,CO_2(g) + 6\,H_2O(\ell)$$
$$C_2H_4(g) + 3\,O_2(g) \rightarrow 2\,CO_2(g) + 2\,H_2O(\ell)$$

Solve

The ΔH°_{comb} for 1 mol of benzene:

$$\Delta H^\circ_{comb} = \left[(12 \times -393.5) + (6 \times -285.8)\right] - \left[(2 \times 49.0) + (15 \times 0)\right]$$
$$= -6534.8 \text{ kJ for 2 mol benzene (from balanced equation)}$$
$$= -3267.4 \text{ kJ for 1 mol benzene}$$

Fuel value:

$$\frac{3267.4 \text{ kJ/mol}}{78.11 \text{ g/mol}} = 41.83 \text{ kJ/g}$$

The ΔH°_{comb} for 3 mol of ethylene:

$$\Delta H^\circ_{comb} = \left[(2 \times -393.5) + (2 \times -285.8)\right] - \left[(1 \times 52.3) + (3 \times 0)\right]$$
$$= -1410.9 \text{ kJ for 1 mol ethylene (from balanced equation)}$$

Fuel value:

$$= \frac{1410.9 \text{ kJ/mol}}{28.05 \text{ g/mol}} = 50.30 \text{ kJ/g}$$

When we compare those findings:

1 mol of benzene has an energy content of 3267.4 kJ.

3 mol of ethylene has an energy content of 3 × 1410.9 = 4232.7 kJ.

Therefore, 1 mol of benzene has a lower energy content than 3 mol of ethylene.

Think About It

Ethylene has six additional C—H bonds in 3 mol compared with 1 mol of benzene. Breaking of those bonds and the formation of additional H—O bonds in water must account for the difference in energy content.

19.71. **Collect and Organize**

By comparing the structure of methylamine with that of *n*-butylamine, we can explain why methylamine is more soluble in water.

Analyze

The line structures of those two amines are

Solve

Methylamine has a smaller nonpolar hydrocarbon chain than that of *n*-butylamine, and so it is more soluble in water.

Think About It

Both amines are fairly soluble in water and indeed can react with water to form basic solutions:

$$CH_3NH_2(aq) + H_2O(\ell) \rightarrow CH_3NH_3^+(aq) + OH^-(aq)$$

19.73. Collect and Organize

In the structures of serotonin and amphetamine (Figure P19.73), we are to identify the primary and secondary amine functional groups.

Analyze

Primary amines (1°) have one R group bonded to the nitrogen atom, whereas secondary amines (2°) have two R groups bonded to the nitrogen atom.

Solve

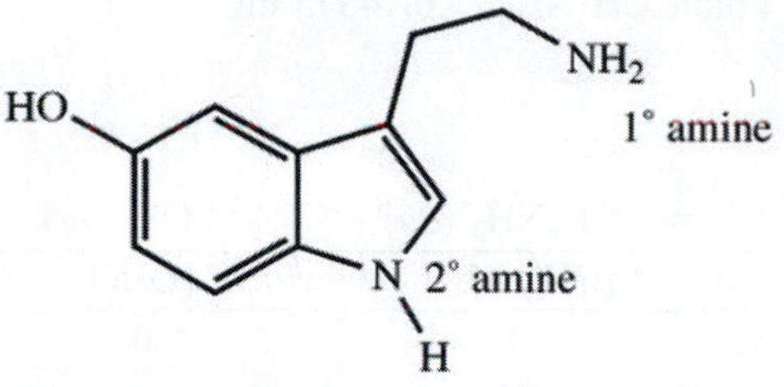

Serotonin

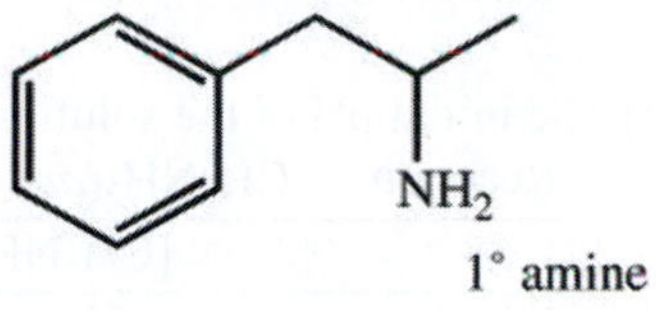

Amphetamine

Think About It

By analogy, tertiary amines (3°) have three R groups bonded to the nitrogen atom and quaternary amines (4°) have four R groups bonded to the nitrogen atom.

19.75. Collect and Organize

We can use the ΔH_f° values in Appendix 4 and the given ΔH_f° for methylamine (–23.0 kJ/mol) to calculate the enthalpy of the reaction

$$4\ CH_3NH_2(g) + 2\ H_2O(\ell) \rightarrow 3\ CH_4(g) + CO_2(g) + 4\ NH_3(g)$$

Analyze

The heat of a reaction may be calculated by using ΔH_f° for the reactants and products according to the equation

$$\Delta H_{rxn}^\circ = \sum n\Delta H_{f,products}^\circ - \sum m\Delta H_{f,reactants}^\circ$$

Solve

$$\Delta H_{rxn}^\circ = \left[\left(3\ mol\ CH_4\right)\left(-74.8\ kJ/mol\right) + \left(1\ mol\ CO_2\right)\left(-393.5\ kJ/mol\right) + \left(4\ mol\ NH_3\right)\left(-46.1\ kJ/mol\right) \right]$$
$$- \left[\left(4\ mol\ CH_3NH_2\right)\left(-23.0\ kJ/mol\right) + \left(2\ mol\ H_2O\right)\left(-285.8\ kJ/mol\right) \right]$$
$$\Delta H_{rxn}^\circ = -138.7\ kJ$$

Think About It

Be careful to use the correct enthalpy of formation for the phases shown in the equation. Here, ammonia and carbon dioxide are in the gas phase and water is in the liquid phase.

19.77. Collect and Organize

Knowing that methylamine is a weak base, we are to sketch the curve for the titration of 125 mL of 0.015 M CH_3NH_2 with 0.100 M HCl, labeling the titration curve with the pH for the initial solution and at the half-equivalence and equivalence points, and then draw the structure of the species present at the equivalence point and, finally, draw the structures of the species present at the equivalence point.

Analyze

We will set up RICE tables to determine the pH at the various points in the titration curve. The K_b for methylamine is 4.4×10^{-4}. The equivalence point is reached when the moles of added HCl equals the moles of CH_3NH_2 in the solution:

$$125 \text{ mL} \times \frac{0.015 \text{ mol CH}_3\text{NH}_2}{1000 \text{ mL}} \times \frac{1 \text{ mol HCl}}{1 \text{ mole CH}_3\text{NH}_2} \times \frac{1000 \text{ mL}}{0.100 \text{ mol HCl}} = 18.75 \text{ mL}$$

Total volume at equivalence point = 125 + 18.75 = 143.75 mL

At the equivalence point, the weak base, CH_3NH_2 has been converted to $CH_3NH_3^+$. The concentration of that species at the equivalence point, then, is

$$125 \text{ mL} \times \frac{0.015 \text{ mol CH}_3\text{NH}_2}{1000 \text{ mL}} \times \frac{1 \text{ mol CH}_3\text{NH}_3^+}{1 \text{ mole CH}_3\text{NH}_2} \times \frac{1}{0.14375 \text{ mL}} = 0.0130 \; M$$

Solve

(a) and (b) The initial pH of the solution:

Reaction	$CH_3NH_2(aq) + H_2O(\ell)$	$\rightleftharpoons$	$CH_3NH_3^+(aq)$	+	$OH^-(aq)$
	$[CH_3NH_2]$		$[CH_3NH_3^+]$		$[OH^-]$
Initial	0.015		0		0
Change	$-x$		$+x$		$+x$
Equilibrium	$0.015 - x$		x		x

$$K_b = 4.4 \times 10^{-4} = \frac{(x)(x)}{0.015 - x}$$

$$x^2 + 4.4 \times 10^{-4} x - 6.6 \times 10^{-6} = 0$$

$$x = 0.00236 = [OH^-]$$

$$pOH = -\log(0.00236) = 2.63$$

$$pH = 14 - pOH = 14 - 2.63 = 11.37$$

The pH at the half-titration point is where pH = pK_a.

$$pK_a = 14 - pK_b = 14 - (-\log 4.4 \times 10^{-4}) = 10.64$$

The pH at the equivalence point:

Reaction	$CH_3NH_2^+(aq)$	$\rightleftharpoons$	$CH_3NH_2(aq)$	+	$H^+(aq)$
	$[CH_3NH_2^+]$		$[CH_3NH_2]$		$[H^+]$
Initial	0.013		0		0
Change	$-x$		$+x$		$+x$
Equilibrium	$0.013 - x$		x		x

$$K_a = \frac{1.00 \times 10^{-14}}{4.4 \times 10^{-4}} = 2.273 \times 10^{-11} = \frac{(x)(x)}{0.013 - x} \approx \frac{x^2}{0.013}$$

$$x = 5.436 \times 10^{-7} = [H^+]$$

$$pH = -\log(5.436 \times 10^{-7}) = 6.26$$

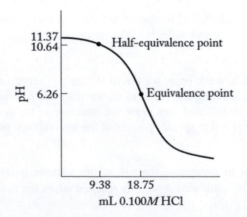

(a) The species at the equivalence point is the methylammonium cation and chloride anion.

$$H-\overset{\overset{\displaystyle H}{|}}{\underset{\underset{\displaystyle H}{|}}{C}}-\overset{\overset{\displaystyle H}{|}}{\underset{\underset{\displaystyle H}{|}}{\overset{+}{N}}}-H \quad \left[:\overset{..}{\underset{..}{Cl}}:\right]^{-}$$

Think About It
Recall from Chapter 15 that when titrating a weak base, we expect the pH at the equivalence point to be below 7 because of the reaction of the conjugate acid with water.

19.79. **Collect and Organize**
We are asked why the fuel values of ethanol and dimethyl ether are lower than that of ethane. In comparing the fuel values, we must take into account the oxygenation of the compounds.

Analyze
The more oxygenated a fuel, the lower its fuel value. All three compounds have two carbons and six hydrogens.

Solve
Because dimethyl ether and ethanol both contain oxygen in their structures but ethane does not, the fuel values of both dimethyl ether and ethanol are lower than that of ethane.

Think About It
We might expect, though, that the fuel values of dimethyl ether and ethanol will not differ much from each other since they are isomers.

19.81. **Collect and Organize**
Ethers have the general structure R—O—R′ and alcohols have the general structure R—OH. We are asked to explain their general difference in boiling point.

Analyze
The lower the boiling point, the weaker the intermolecular forces between the molecules. Because ethers boil at lower temperatures than isomeric alcohols, as stated in the problem, ethers must have weaker intermolecular forces than alcohols.

Solve
Ethers have lower boiling points than alcohols because they have weaker dipole–dipole forces than the alcohols, which have hydrogen bonding between the molecules.

Think About It
Both ethers and alcohols also have dispersion forces that attract their molecules to each other. Those forces get stronger as the R groups on the ether or alcohol get longer.

19.83. **Collect and Organize**
We need to consider the evaporation of ethanol to explain why our skin feels cold after wiping with ethanol.

Analyze
When ethanol comes in contact with your skin, it begins to evaporate.

Solve
Evaporation of ethanol from the skin is an endothermic process (phase change from liquid to vapor). The heat transfers from the skin to the ethanol, so the skin feels cold.

Think About It
The reverse process, condensation, is an exothermic process, as you learned from Chapter 5.

19.85. **Collect and Organize**

From the structures shown in Figure P19.85, we are to identify which are ethers and which are alcohols and place them in order of increasing boiling point.

Analyze

Ethers have the general formula R—O—R′ and alcohols have the general formula R—OH. The greater the intermolecular forces between molecules, the higher the boiling point. Because of hydrogen bonding, alcohols generally have higher boiling points than ethers. The larger the molecule (more atoms) and the less branching it has, the greater the boiling point.

Solve

Compounds a and d are alcohols; compounds b and c are ethers. In order of increasing boiling point, b < c < a < d.

Think About It

Our prediction for boiling point order is nearly correct: (b) diethyl ether, 35°C < (c) isobutyl methyl ether, 59°C < (d) 2,5-dimethylcyclohexanol, 170°C < (a) 3-methyl-4-heptanol, 174°C. Compounds a and d have nearly equal boiling points.

19.87. **Collect and Organize**

We are to calculate the fuel values of butanol and diethyl ether and indicate which has the higher fuel value.

Analyze

The fuel value is equal to $-\Delta H^{\circ}_{comb}$ divided by the mass of fuel. For those fuels, the combustion reactions are

$$(C_2H_5)_2O(\ell) + 6\ O_2(g) \rightarrow 4\ CO_2(g) + 5\ H_2O(\ell)$$
$$C_4H_9OH(\ell) + 6\ O_2(g) \rightarrow 4\ CO_2(g) + 5\ H_2O(\ell)$$

Solve

The fuel value for diethyl ether:

$$\Delta H^{\circ}_{comb} = \left[(4\times-393.5)+(5\times-285.8)\right]-\left[(1\times-279.6)+(6\times0.0)\right]$$

$$= -2723.4 \text{ kJ/mol}$$

$$\text{Fuel value} = \frac{2723 \text{ kJ/mol}}{74.12 \text{ g/mol}} = 36.74 \text{ kJ/g}$$

The fuel value for *n*-butanol:

$$\Delta H^{\circ}_{comb} = \left[(4\times-393.5)+(5\times-285.8)\right]-\left[(1\times-327.3)+(6\times0.0)\right]$$

$$= -2675.7 \text{ kJ}$$

$$\text{Fuel value} = \frac{2675.7 \text{ kJ/mol}}{74.12 \text{ g/mol}} = 36.10 \text{ kJ/g}$$

Diethyl ether has a slightly higher fuel value than *n*-butanol.

Think About It

We would expect those compounds to have similar fuel values because they are isomers.

19.89. **Collect and Organize**

After calculating the fuel values of ethanol (C_2H_5OH) and methanol (CH_3OH), we can decide the validity of our prediction (Problem 19.80) that the fuel value increases as carbon atoms are added to the alcohol alkyl chain.

Analyze

Fuel values are computed from $-\Delta H^{\circ}_{comb}$. For methanol and ethanol, the balanced combustion reactions are

$$2\ CH_3OH(\ell) + 3\ O_2(g) \rightarrow 2\ CO_2(g) + 4\ H_2O(\ell)$$
$$C_2H_5OH(\ell) + 3\ O_2(g) \rightarrow 2\ CO_2(g) + 3\ H_2O(\ell)$$

We need ΔH°_f values from the textbook (Appendix 4, Table A4.3) to compute $-\Delta H^{\circ}_{comb}$.

Solve
For methanol:

$$\Delta H^{\circ}_{comb} = \left[\left(2 \text{ mol CO}_2\right)\left(-393.5 \text{ kJ/mol}\right) + \left(4 \text{ mol H}_2O\right)\left(-285.8 \text{ kJ/mol}\right)\right]$$
$$- \left[\left(2 \text{ mol CH}_3OH\right)\left(-238.7 \text{ kJ/mol}\right) + \left(3 \text{ mol O}_2\right)\left(0.0 \text{ kJ/mol}\right)\right]$$

$$\Delta H^{\circ}_{comb} = -1452.8 \text{ kJ for 2 mol CH}_3OH \text{ burned}$$

$$\text{Fuel value} = \frac{1452.8 \text{ kJ}}{2 \text{ mol} \times 32.04 \text{ g/mol}} = 22.67 \text{ kJ/g}$$

For ethanol:

$$\Delta H^{\circ}_{comb} = \left[\left(2 \text{ mol CO}_2\right)\left(-393.5 \text{ kJ/mol}\right) + \left(3 \text{ mol H}_2O\right)\left(-285.8 \text{ kJ/mol}\right)\right]$$
$$- \left[\left(1 \text{ mol C}_2H_5OH\right)\left(-277.7 \text{ kJ/mol}\right) + \left(3 \text{ mol O}_2\right)\left(0.0 \text{ kJ/mol}\right)\right]$$

$$\Delta H^{\circ}_{comb} = -1366.7 \text{ kJ for 1 mol CH}_3OH \text{ burned}$$

$$\text{Fuel value} = \frac{1366.7 \text{ kJ}}{1 \text{ mol} \times 46.07 \text{ g/mol}} = 29.67 \text{ kJ/g}$$

Yes, the answer supports the prediction made in Problem 19.80 that fuel values of alcohols increase as the number of C atoms increases.

Think About It
For comparing alcohols with alkanes, however, alcohols have lower fuel values.

19.91. Collect and Organize
By looking at the structures of carboxylic acids and aldehydes, we can explain why carboxylic acids are generally more soluble in water.

Analyze
Carboxylic acids have the general molecular structure

Aldehydes have the general molecular structure

Solve
Both carboxylic acids and aldehydes have polar functional groups. Carboxylic acids, however, are more soluble in water because they form strong hydrogen bonds with water.

Think About It
Remember that hydrogen bonds between a species and water are stronger than dipole–dipole interactions between a species and water.

19.93. Collect and Organize
Given the structures for butanal (an aldehyde) and 2-butanone (a ketone) (Figure P19.93), we are asked whether those compounds are constitutional isomers.

Analyze
Constitutional isomers have the same molecular formula but different connectivity of their atoms. The molecular formula of butanal is C_4H_8O, and so is the formula of 2-butanone, C_4H_8O.

Solve

Yes, those compounds are constitutional isomers.

Think About It

n-Butanol ($C_4H_{10}O$), however, is not a constitutional isomer of those compounds because it has a different molecular formula.

19.95. **Collect and Organize**

We can compare the molecular formulas of aldehydes and ketones to determine whether we could distinguish them by combustion analysis.

Analyze

The molecular formula for aldehydes is $C_nH_{2n}O$ and for ketones it is also $C_nH_{2n}O$.

Solve

Because the empirical formulas for a ketone and for an aldehyde with the same number of carbon atoms are the same, no, we cannot distinguish between those compounds by combustion analysis.

Think About It

We would be able, however, to distinguish between a ketone or aldehyde and an alcohol, which has the molecular formula $C_nH_{2n+2}O$.

19.97. **Collect and Organize**

By assigning formal charges to the atoms in the two resonance structures for acetic acid (Figure P19.97), we can determine which form contributes more to the bonding.

Analyze

The resonance form that has the lowest formal charges and/or the negative formal charges on the most electronegative atoms (oxygen for acetic acid) contributes the most to the structure.

Solve

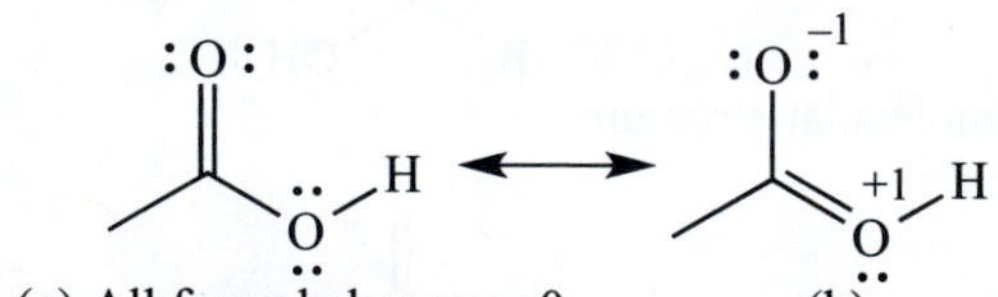

(a) All formal charges = 0   (b)

Structure (a) contributes more to the bonding in acetic acid because all the formal charges are zero.

Think About It

Once deprotonated, acetic acid forms the acetate anion, which has two equivalent resonance forms:

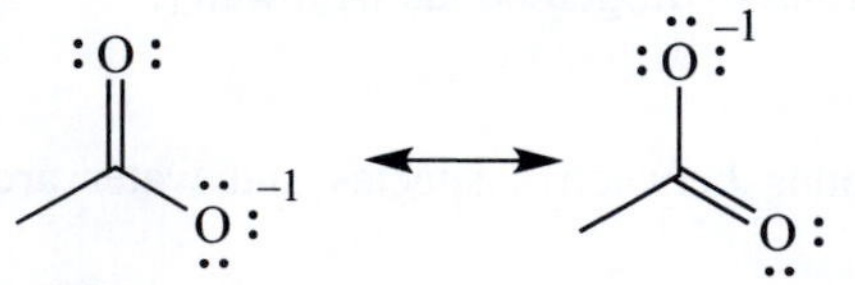

19.99. **Collect and Organize**

By comparing the general structures of an amine and an amide, we can distinguish between the two functional groups.

Analyze

An amine has the general structure

$$R-NH_2$$

An amide has the general structure

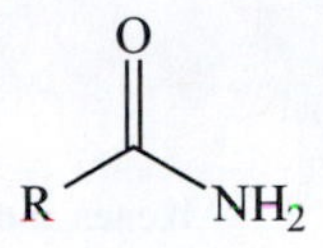

Solve

An amide includes a carbonyl (C=O) group as part of its functional group in addition to the –NH$_2$ group.

Think About It

Amides are structurally similar to carboxylic acids:

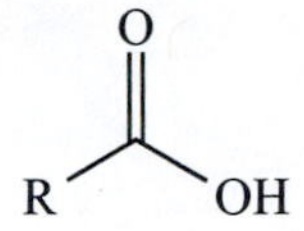

19.101. Collect and Organize

Given the molecular formula of an aldehyde (C$_5$H$_{10}$O), we are to choose which of four structures (Figure P19.101) are constitutional isomers of it.

Analyze

Constitutional isomers have the same molecular formula but different connectivity of their atoms.

Solve

Structure (a) has a formula of C$_5$H$_{10}$O, so it is a constitutional isomer of the given aldehyde. Structure (b) has a formula of C$_5$H$_{10}$O, so it is a constitutional isomer. Structure (c) has a formula of C$_4$H$_8$O, so it is not a constitutional isomer of aldehyde. Structure (d) has a formula of C$_5$H$_{10}$O, so it is also a constitutional isomer of the given aldehyde.

Think About It

The names of the three constitutional isomers shown are (a) 3-methylbutanal, (b) 2-methylbutanal, and (d) pentanal.

19.103. Collect and Organize

From the structures shown in Figure P19.103, we are to determine which compound is a ketone.

Analyze

Ketones have the general structure

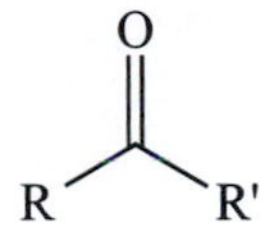

Solve

Compound (b) is a ketone.

Think About It

Structure (a) represents an aldehyde, structure (c) represents a carboxylic acid, and structure (d) represents an alcohol.

19.105. Collect and Organize

After plotting the carbon–hydrogen ratio in aldehydes as a function of the number of carbon atoms and comparing it with the ratios of alkanes and alkenes, we can find the better correlation for aldehydes.

Analyze

Aldehydes have the general formula C$_n$H$_{2n}$O, so the C:H ratio is always 0.5. Alkanes have the general formula C$_n$H$_{2n+2}$, so the C:H ratio changes as the number of C atoms increases. Alkenes have the general formula C$_n$H$_{2n}$, so the C:H ratio is always 0.5.

Solve

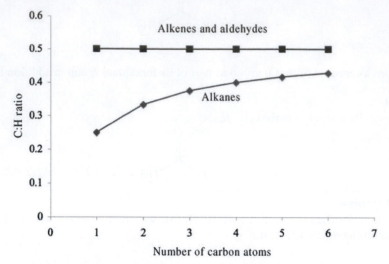

The plot of C:H ratio versus number of C atoms for aldehydes correlates exactly to that of alkenes and poorly to that of alkanes.

Think About It
Although that correlation exists, aldehydes are not structural isomers of alkenes. Aldehydes have an oxygen in their structure; alkenes do not.

19.107. **Collect and Organize**
Esters are formed in condensation reactions between carboxylic acids and alcohols. For each ester structure shown in Figure P19.107, we can write an equation to identify the carboxylic acid and alcohol used to synthesize them.

Analyze
The general reaction for the formation of esters is

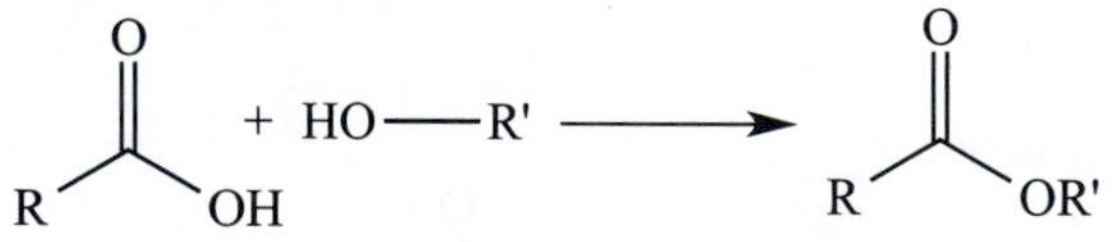

Solve
(a) Pineapples

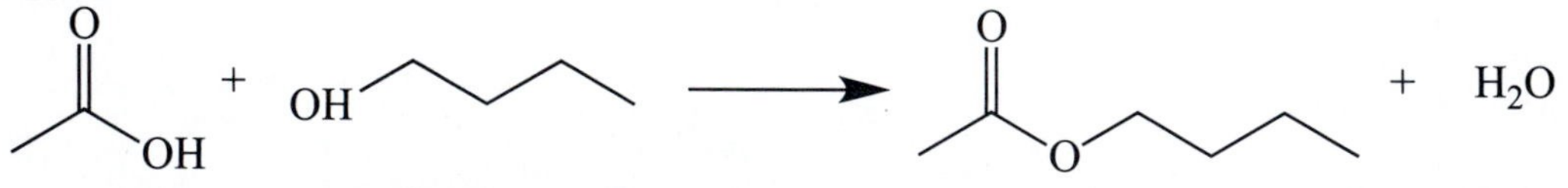

      Acetic acid   *n*-Butanol

(b) Bananas

      2-Methylbutanoic acid   Ethanol

(c) Apples

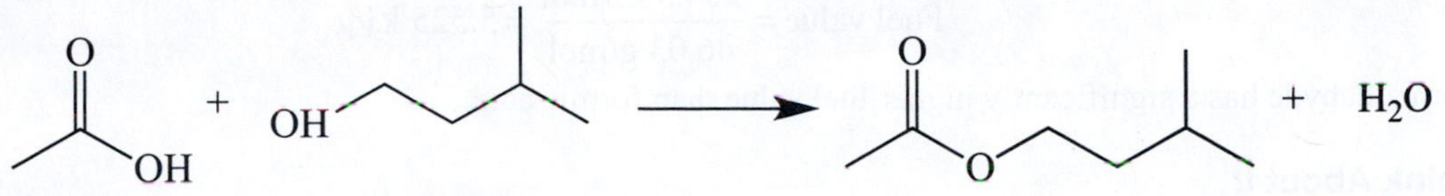

Acetic acid   3-Methylbutanol

Think About It
Those are all condensation reactions because they give a small molecule, H_2O, as the other product.

19.109. **Collect and Organize**
For the highlighted N atoms in nicotine and Valium (Figure P19.109), we are to identify the associated functional group as either an amine or an amide.

Analyze
An amine has the following possible structures:

$$R{-}NH_2 \qquad R_2{-}NH \qquad R_3N$$
Primary   Secondary   Tertiary

An amide has the following possible structures:

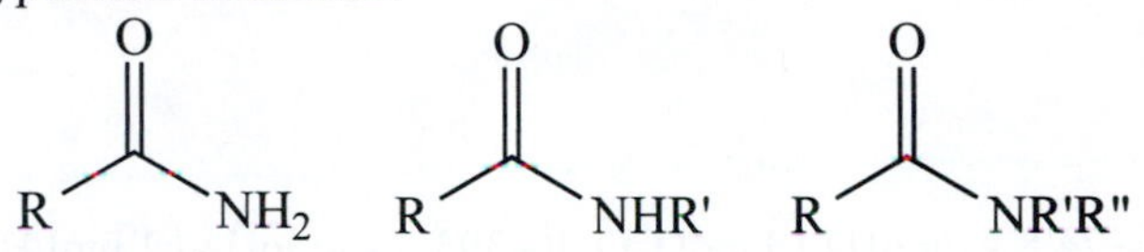

Solve
(a) Nicotine's highlighted N atom is a tertiary amine group.
(b) Valium's highlighted N atom, having an adjacent $C{=}O$ double bond, is an amide group.

Think About It
Both nicotine and Valium also contain aromatic rings in their structures.

19.111. **Collect and Organize**
After calculating the fuel values of formaldehyde and formic acid (Figure P19.111), we can determine which has the higher fuel value.

Analyze
In general, the more oxygenated the fuel, the lower the fuel value. We predict that formaldehyde has the higher fuel value because it has less oxygen in its structure. The fuel value is equal to $-\Delta H^{\circ}_{comb}$ divided by the mass of fuel. For those two fuels, the combustion reactions are

$$CH_2O(g) + O_2(g) \rightarrow CO_2(g) + H_2O(\ell)$$
$$CH_2O_2(\ell) + \tfrac{1}{2}O_2(g) \rightarrow CO_2(g) + H_2O(\ell)$$

Solve
For formaldehyde:

$$\Delta H^{\circ}_{comb} = \left[(-393.5)+(-285.8)\right]-\left[(-108.6)+(0.0)\right]$$

$$= -570.7 \text{ kJ/mol}$$

$$\text{Fuel value} = \frac{570.7 \ \text{kJ/mol}}{30.03 \ \text{g/mol}} = 19.00 \text{ kJ/g}$$

For formic acid:

$$\Delta H^{\circ}_{comb} = \left[(-393.5)+(-285.8)\right]-\left[(-425.0)+(0.0)\right]$$

$$= -254.3 \text{ kJ/mol}$$

$$\text{Fuel value} = \frac{254.3 \text{ kJ/mol}}{46.03 \text{ g/mol}} = 5.525 \text{ kJ/g}$$

Formaldehyde has a significantly higher fuel value than formic acid.

Think About It
Our prediction, based on the level of oxygenation of the two fuels, was correct.

19.113. Collect and Organize
For two reactions involving methanogenic bacteria, we are to calculate ΔH°_{rxn}.

Analyze
We determine the enthalpy of each reaction by using the given value of the enthalpy of formation of formic acid and the values in Appendix 4.

$$\Delta H^{\circ}_{rxn} = \sum n \Delta H^{\circ}_{f,products} - \sum n \Delta H^{\circ}_{f,reactants}$$

Solve
For reaction 1:

$$\Delta H^{\circ}_{rxn} = \left[\left(1 \text{ mol } CH_4\right)\left(-74.8 \text{ kJ/mol}\right) + \left(1 \text{ mol } CO_2\right)\left(-393.5 \text{ kJ/mol}\right) \right] - \left[\left(1 \text{ mol } CH_3COOH\right)\left(-485.8 \text{ kJ/mol}\right) \right]$$

$$\Delta H^{\circ}_{rxn} = 17.5 \text{ kJ}$$

For reaction 2:

$$\Delta H^{\circ}_{rxn} = \left[\left(1 \text{ mol } CH_4\right)\left(-74.8 \text{ kJ/mol}\right) + \left(3 \text{ mol } CO_2\right)\left(-393.5 \text{ kJ/mol}\right) + \left(2 \text{ mol } H_2O\right)\left(-285.8 \text{ kJ/mol}\right) \right]$$
$$- \left[\left(4 \text{ mol } HCOOH\right)\left(-425.0 \text{ kJ/mol}\right) \right]$$

$$\Delta H^{\circ}_{rxn} = -126.9 \text{ kJ}$$

Think About It
The breakdown of acetic acid to methane and carbon dioxide is endothermic and therefore not favored by enthalpy, but the process that gives methanol, carbon dioxide, and water is exothermic and is favored by enthalpy.

19.115. Collect and Organize
For each polymer shown in Figure P19.115, which were synthesized through the condensation reaction of $H_2N(CH_2)_6NH_2$ with $HO_2C(CH_2)_nCO_2H$, we are asked to determine the number of carbon atoms in the chain (n) of the dicarboxylic acids used.

Analyze
The number of carbon atoms in the dicarboxylic acid chains includes the two carbon atoms double-bonded to oxygen.

Solve
(a) 6
(b) 8
(c) 10

Think About It
Because n is defined in the problem as the number of carbon atoms in the dicarboxylic acid formula $HO_2C(CH_2)nCO_2H$, we do not count the carboxylic acid carbon atoms.

19.117. Collect and Organize
Given the reaction between dimethyl terephthalate and 1,4-di(hydroxymethyl)cyclohexane to form Kodel (Figure P19.117), we are asked to classify the reaction as either a condensation or an addition reaction and to compare the properties of Kodel with those of Dacron, which is prepared using ethylene glycol.

Analyze

(a) In addition polymerization reactions, the atoms are joined in the monomers to form the polymeric backbone without loss of atoms. In condensation polymerization reactions, the two monomers react to form the polymeric backbone while a small molecule like water is formed.

(b) The structure of ethylene glycol is

$$HO\diagup\diagdown\diagup^{OH}$$

Solve

(a) Kodel is a condensation polymer. Methanol (CH_3OH) is the by-product of the polymerization reaction.

(b) Kodel has in its backbone a carbon six-membered ring, not a straight chain as in Dacron. Therefore, Kodel might be better able to accept organic dyes that are nonpolar.

Think About It

Kodel, being fairly polar because of the saturated six-membered carbon ring, is fairly resistant to water and has been used to make clothing.

19.119. Collect and Organize

Using the definitions of *enantiomer*, *achiral*, and *optically active*, we are to determine whether all three terms may be applied to a single compound.

Analyze

An *enantiomer* is a molecule that is a nonsuperimposable mirror image of another molecule and is *optically active*. An *achiral* molecule is superimposable on its mirror image.

Solve

No, all three terms cannot describe a single compound. Whereas *enantiomer* and *optically active* describe the same chiral molecule, *achiral* does not.

Think About It

Many biomolecules are chiral.

19.121. Collect and Organize

We are asked whether cis–trans isomers can also be chiral.

Analyze

Simple cis or trans isomers are superimposable on their mirror images and so are not chiral. The R group, however, may contain a chiral carbon center.

Solve

Yes. If R contains a chiral center, the cis or trans isomer of RCH==CHR would have optical isomers.

Think About It

Remember the definition of a chiral carbon center: A carbon atom is chiral when it is bonded to four different groups.

19.123. Collect and Organize

For the structure of myristicin shown in Figure P19.123, we are to determine whether cis and trans isomers are possible and to determine the number of structural isomers possible on the basis of the arrangement of the groups around the aromatic ring.

Analyze

Cis and trans isomers are possible where two different groups are present on each of the carbon atoms that are doubly-bonded to each other. For the aromatic ring in that structure, four positions are available in which to place the two substituents.

Solve

(a) No, cis and trans isomers are not possible for that compound. The double bonds in the aromatic ring are locked into their geometry and the terminal alkene has two H atoms bonded to the carbon, so no cis or trans isomers are possible.

(b) T Myristicin has six possible structural isomers:

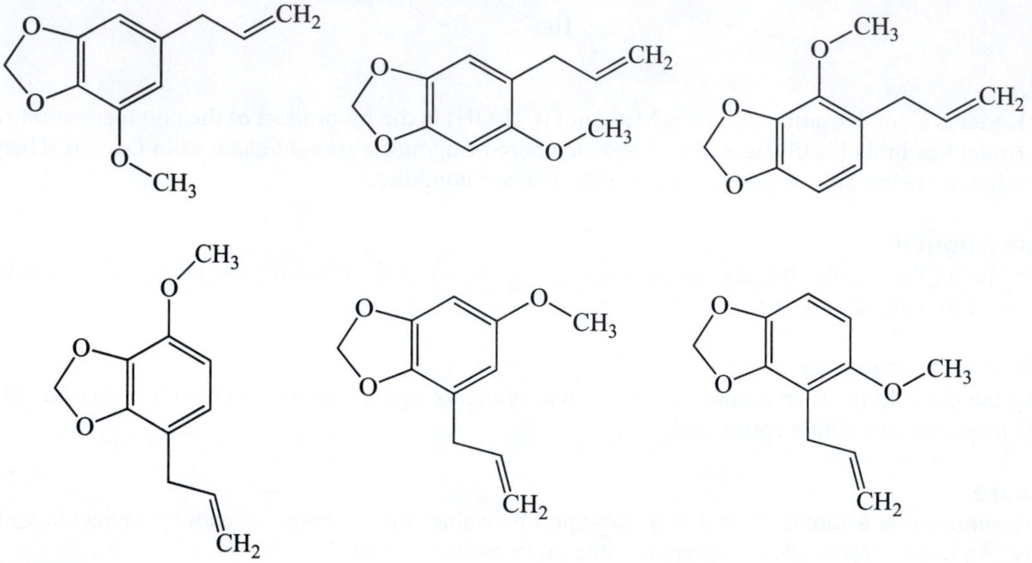

Think About It

The effects of myristicin poisoning include headache, dizziness, anxiety, hallucinations, vomiting, and collapse.

19.125. **Collect and Organize**

For the three carboxylic acids in Figure P19.125, we are to determine which, if any, have no possible constitutional isomers.

Analyze

Constitutional isomers have the same molecular formula but have a different arrangement of their bonds.

Solve

All those acids have possible constitutional isomers because at least one isomer may be drawn for each one:

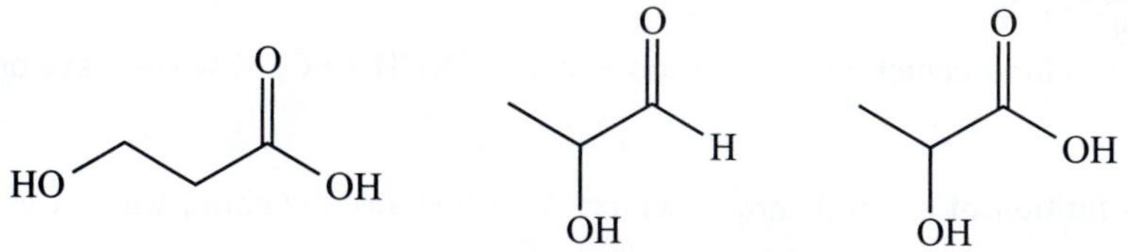

(a)   (b)   (c)

Think About It

Carboxylic acid (a) is chiral. The chiral center is shown here:

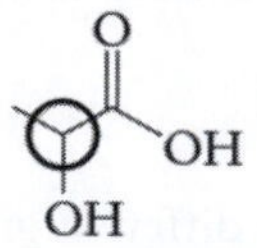

19.127. Collect and Organize

For the combustion reaction of methanol

$$CH_3OH(\ell) + \tfrac{3}{2} O_2(g) \rightarrow CO_2(g) + 2\,H_2O(\ell)$$

we are to calculate how many grams of methanol would be needed to raise the temperature of 454 g of water from 20.0°C to 50.0°C. We are also to calculate the mass of CO_2 produced in that reaction.

Analyze

If all the heat from burning the methanol is used to heat the water, then

$$q_{water} = -q_{comb}$$

The heat needed to raise the temperature of the water is given by

$$q_{water} = mc_s\Delta T$$

where m is the mass of the water, c_s is the specific heat capacity of water (4.184 J/g · °C), and ΔT is the change in temperature of the water (30.0°C). We can use ΔH_f° values from Appendix 4 to calculate the enthalpy of combustion of 1 mol of methanol. The molar amount of methanol needed to heat the water, then, can be calculated through

$$\frac{q_{comb}}{\Delta H_{comb}} = \text{mol } CH_3OH \text{ to heat the water}$$

Once we know the moles of methanol required for the reaction, we can calculate the mass of CH_3OH needed and the mass of CO_2 produced from the balanced equation.

Solve

The heat generated by the combustion reaction is

$$q_{water} = -q_{comb} = 454\ g \times 4.184\ J/g \cdot °C \times 30.0°C = 57{,}000\ J, \text{ or } 57.0\ kJ$$

$$\Delta H_{rxn}^\circ = \big[(1\ mol\ CO_2)(-393.5\ kJ/mol) + (2\ mol\ H_2O)(-285.8\ kJ/mol)\big]$$
$$- \big[(1\ mol\ CH_3OH)(-238.7\ kJ/mol) + (\tfrac{3}{2}\ mol\ O_2)(0.0\ kJ/mol)\big]$$

$$\Delta H_{rxn}^\circ = -726.4\ kJ \text{ for 1 mol methanol}$$

Moles of methanol to heat the water:

$$\frac{-57.0\ kJ}{-726.4\ kJ/mol} = 7.85 \times 10^{-2}\ mol\ methanol$$

Mass of methanol needed to heat the water:

$$7.85 \times 10^{-2}\ mol \times \frac{32.04\ g}{mol} = 2.52\ g\ methanol$$

Mass of CO_2 produced:

$$7.85 \times 10^{-2}\ mol\ CH_3OH \times \frac{1\ mol\ CO_2}{1\ mol\ CH_3OH} \times \frac{44.01\ g}{mol} = 3.45\ g\ CO_2$$

Think About It

That mass of CO_2 would occupy 1.92 L at 25°C and 1 atm pressure:

$$V = \frac{nRT}{P} = \frac{7.85 \times 10^{-2}\ mol \times 0.0821\ L \cdot atm/mol \cdot K \times 298\ K}{1\ atm} = 1.92\ L$$

19.129. Collect and Organize

We are asked to explain the very different melting points of *o*-xylene (13°C) and *p*-xylene (−48°C) despite their having similar boiling points.

Analyze

The structures of those xylenes are:

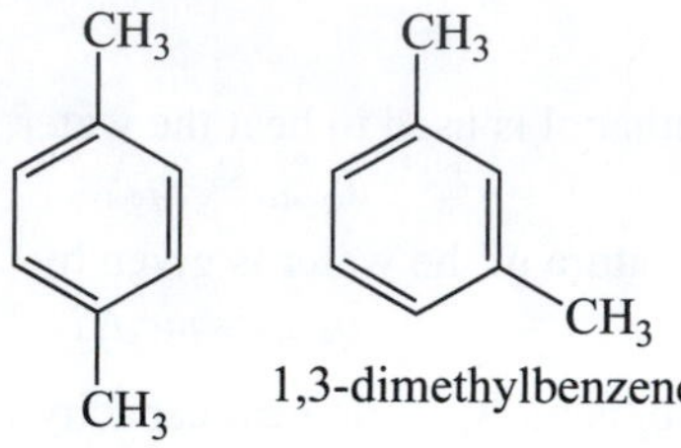

1,4-dimethylbenzene

1,3-dimethylbenzene

Solve

1,4-Dimethylbenzene (*p*-xylene) has a symmetrical structure, so it packs efficiently in the solid state and, therefore, has a much higher melting point than 1,3-dimethylbenzene.

Think About It

Dimethylbenzene has one more structural isomer: 1,2-dimethylbenzene.

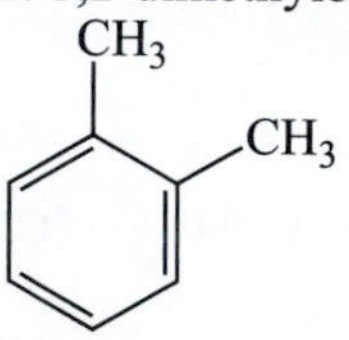

19.131. Collect and Organize

From the structure of curcumin (Figure P19.131), we can determine whether the substituents on the double bonds are cis or trans. We are to then draw two other isomers of that compound and list the hybridization of all the carbon atoms in the molecule.

Analyze

Substituents that are cis have two like groups on the same side of a line drawn through the double bond. Trans substituents have two like groups on opposite sides of a line drawn through the double bond.

Solve

(a) Two C=C bonds are present in the structure of curcumin, and they are both in the trans configuration.

(b) If both C=C bonds were in the cis configuration, the structure we would have is isomer A. If one C=C bond were in the trans configuration and the other in the cis configuration, we would have isomer B.

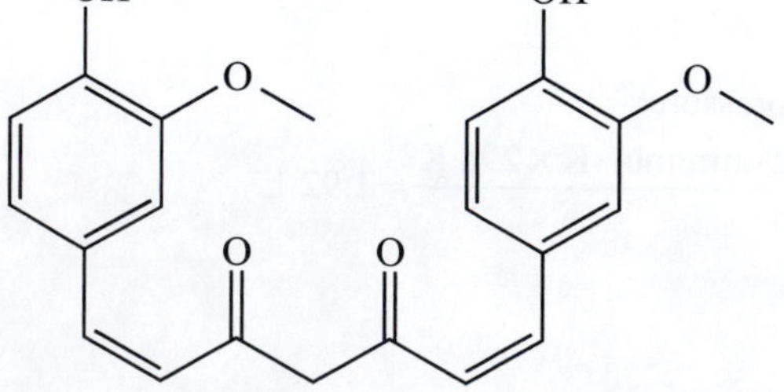

Isomer A

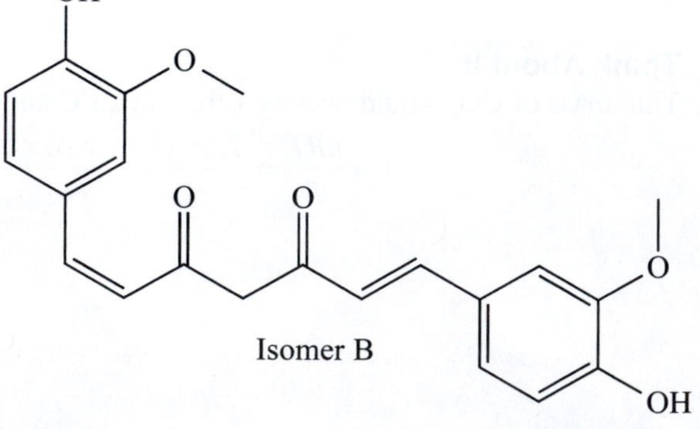

Isomer B

(c) The hybridizations on the carbon atoms in curcumin are sp^3 for the carbon of the −OCH$_3$ group and sp^2 for all other carbon atoms.

Think About It
Only three geometric isomers of that compound exist because the two ends of the molecule are the same. The cis–trans and the trans–cis isomers are identical.

19.133. Collect and Organize
For each polymer shown in Figure P19.133, we are asked to write the reactants that form it.

Analyze
Both polymers are formed from condensation reactions of difunctional monomers. Polymer (a) is a polyaramide that results from the reaction of a carboxylic acid group with an amine group. Polymer (b) is a polyester that results from the reaction of a carboxylic acid group with an alcohol.

Solve
For polymer (a):

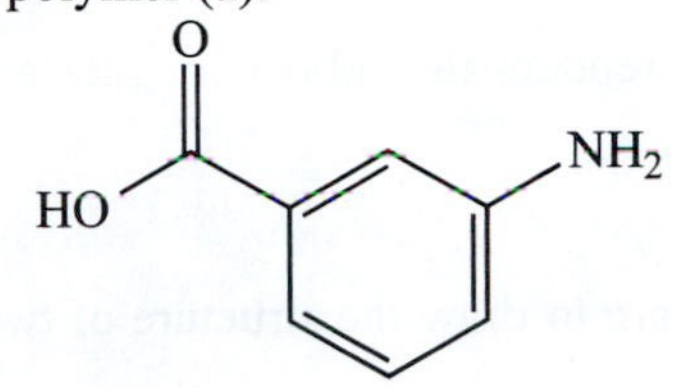

For polymer (b):

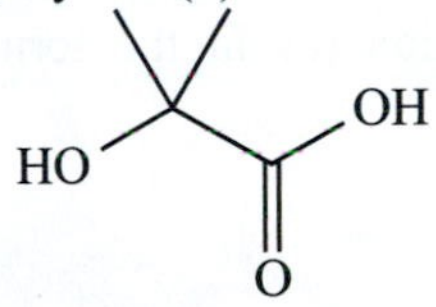

Think About It
The polyaramide here contains an aromatic ring (the benzene ring), but we may also call polymer (a) a polyamide because it contains amides.

19.135. Collect and Organize
We are to draw the carbon-skeleton structure of the condensation polymer of $H_2N(CH_2)_6COOH$ and compare that structure with that of nylon-6 (Figure 19.45).

Analyze
In a condensation polymerization reaction, a small molecule is formed. When the reaction is between an amine and a carboxylic acid, that small molecule is water. In the condensation of $H_2N(CH_2)_6COOH$ with itself, the polymer has a repeating unit of just that monomer. Nylon-6 is discussed in Sample Exercise 19.12.

Solve
The condensation reaction and the line structure of the polymer are

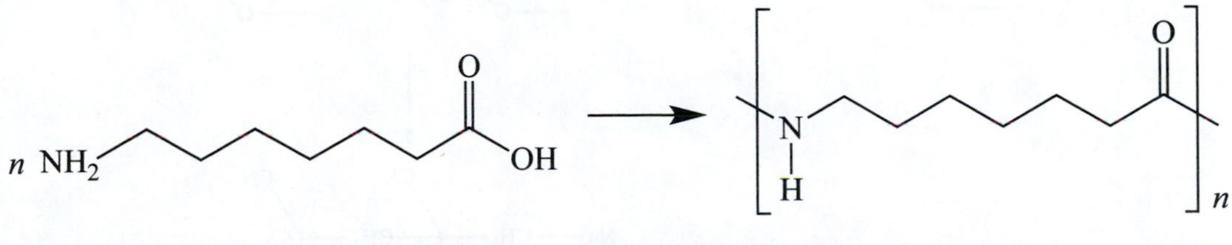

That polymer has one monomeric repeating unit with seven carbon atoms because it is prepared from the difunctional $H_2N(CH_2)_6COOH$ monomer. In nylon-6 the polymer also has a single monomeric unit but with six carbon atoms.

Think About It

We can make different nylons with different properties by changing the monomers used in the condensation reaction.

19.137. Collect and Organize

We consider how cross-linking of a polymer affects its physical properties. We are asked to predict how the properties of a styrene–divinylbenzene copolymer might differ from those of 100% polystyrene.

Analyze

Cross-linking "ties" individual polymer strands together with covalent bonds.

Solve

Cross-linking increases a polymer's strength, hardness, melting (softening) point, and chemical resistance.

Think About It

Synthetic rubber for tires is a cross-linked polymer. The cross-linking renders the rubber tougher and more elastic.

19.139. Collect and Organize

For the polymerization of methyl 2-cyanoacrylate (Figure P19.139), we are to draw the structure of two of the repeating units for the polymer.

Analyze

The polymerization of methyl 2-cyanoacrylate is through addition polymerization (as in the formation of polystyrene).

Solve

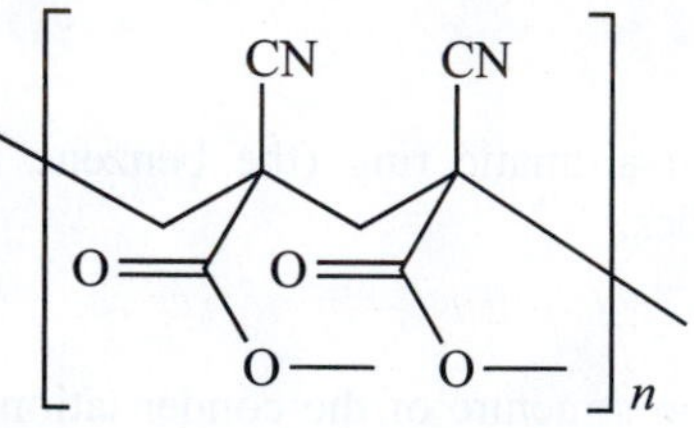

Think About It

The presence of water, –OH, or –NHR functioning as OH⁻ and RNH⁻ (Nu:⁻) groups serves to start the polymerization reaction.

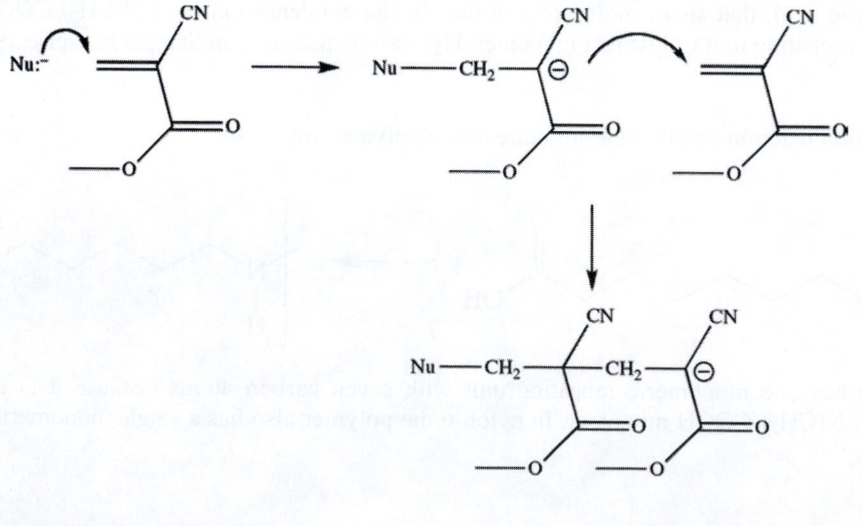

19.141. **Collect and Organize**

From the structures given for piperine and capsaicin (Figure P19.141), we are asked to draw the amine and the carboxylic acid that would react to form the amides. We are also to determine whether the C=C bonds are cis or trans and name the functional groups containing oxygen atoms in each compound.

Analyze

(a) The amide functional group is formed when water is eliminated in the reaction of R—NH$_2$ with R—COOH.

(b) Substituents that are cis have two like groups on the same side of a line drawn through the double bond. Trans substituents have two like groups on opposite sides of a line drawn through the double bond.

(c) The functional groups that contain oxygen are ethers (R—O—R), alcohols (R—OH), carboxylic acids (R—COOH), aldehydes (R—COH), ketones (RCOR), and amides [RN(H)COR].

Solve

(a)

For piperine:

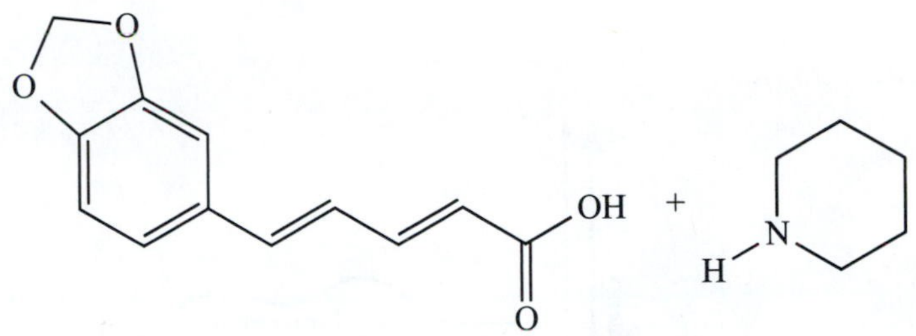

For capsaicin:

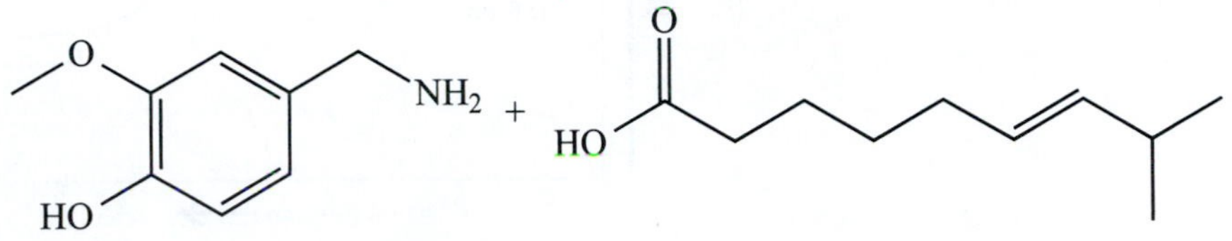

(b) All C=C bonds are trans.

(c) Piperine has two ether groups (in the five-membered ring) and the amide group. Capsaicin has an ether group, an alcohol group, and the amide group.

Think About It

Capsaicin is also the ingredient used in pepper spray and can be used to deter squirrels from eating bird seed because the compound does not affect birds.

CHAPTER 20 | Biochemistry: The Compounds of Life

20.1. Collect and Organize

By examining the reaction in Figure P20.1, in which α-glucose is converted to β-glucose, we are asked to sketch a reaction energy diagram showing the energy changes for the transformation.

Analyze

The α form of glucose is different from the β form in that the position of the –OH group at carbon 1 is in the axial position in the α form but in the equatorial position in the β form. To make that transformation, the C1—O bond must break, to form the open-chain sugar as an intermediate that is less stable than either cyclic form, and then re-forms after a bond rotation to give the other form. The β form is slightly more stable than the α form because the bulky —OH group is more stable in the roomier equatorial position. The transformation is facile and fast in solution, so the activation energy for the transformation (both forward and reverse) is small.

Solve

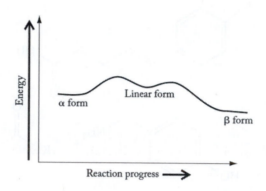

Think About It

As the temperature decreases, the transformation will be slower because fewer molecules will have enough energy to go over the activation energy barrier.

20.3. Collect and Organize

For the triglycerides shown in Figure P20.3, we are to identify which of the fatty acids that make up the triglycerides are saturated and then decide whether the unsaturated fatty acids are likely to be cis or trans isomers.

Analyze

A saturated fatty acid has no C=C double bonds in its structure. Common fatty acids are shown in Table 20.2.

Solve

(a) Triglyceride a contains palmitic acid (saturated), oleic acid (unsaturated), and linoleic acid (unsaturated). Triglyceride b contains oleic acid (unsaturated) and stearic acid (saturated). The saturated fatty acids in those triglycerides are palmitic acid and stearic acid.

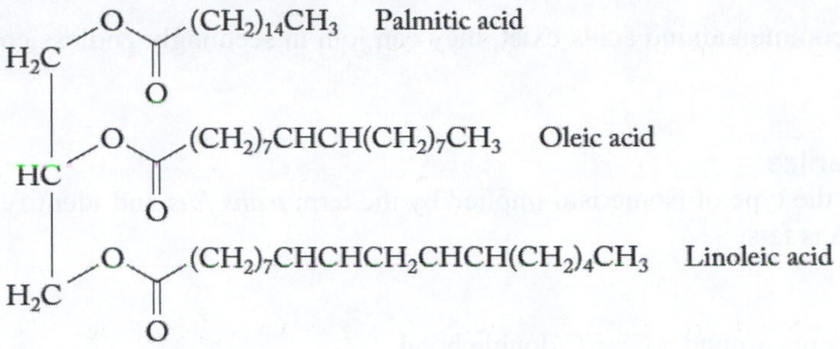

(a)

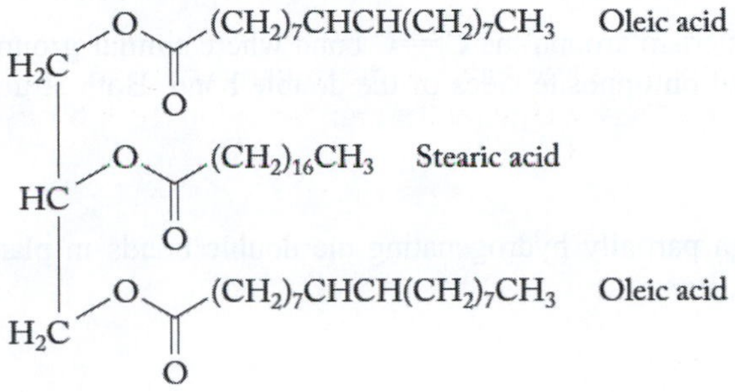

(b)

(b) In nature, fatty acids are more likely to be cis.

Think About It

The triglycerides are formed through a condensation reaction between glycerol and three fatty acids.

20.5. Collect and Organize

For the structure of enkephalin shown in Figure P20.5, we are to name the five amino acids that make up the polypeptide.

Analyze

Table 20.1 shows the 20 common amino acids. A peptide bond is formed between two amino acids in an acid–base reaction:

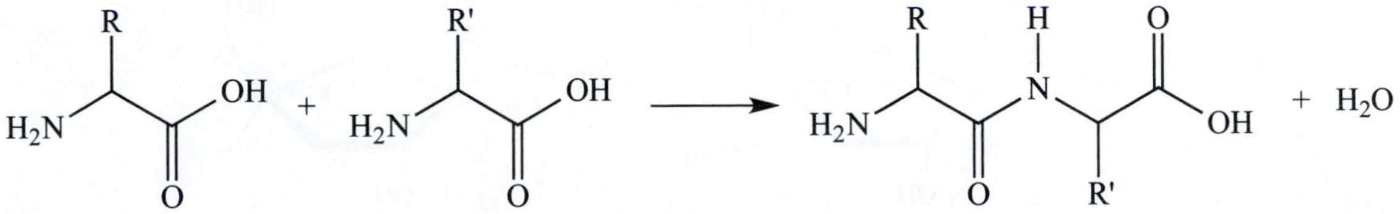

Therefore, the C—NH linkage is where two amino acids are joined.

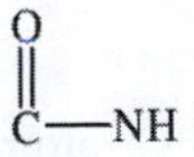

Solve

The amino acids that make up the structure are (from left to right in the molecule) methionine, phenylalanine, glycine, glycine, and tyrosine.

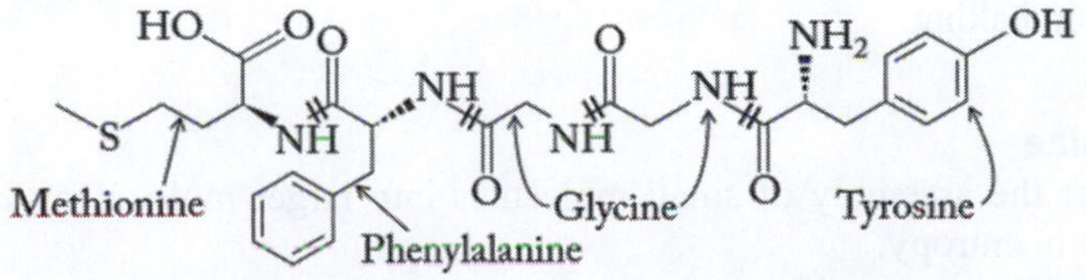

Think About It

Although only 20 common amino acids exist, they can join in seemingly endless combinations to make a huge variety of peptides.

20.7. Collect and Organize

We are to describe the type of isomerism implied by the term *trans fats* and identify which molecules shown in Figure P20.7 are trans fats.

Analyze

Trans isomerism occurs around a $C=C$ double bond.

Solve

Trans fats exhibit geometric isomerism around the $C=C$ bond where similar groups (such as the H atoms) on the two carbon atoms are situated on opposite sides of the double bond. Both structures a and c contain trans fats.

Think About It

Most trans fats are made through partially hydrogenating the double bonds in plant oils to give them higher melting points and a longer shelf life.

20.9. Collect and Organize

From the structure given for sucralose in Figure P20.9, we are to identify the sugar from which the sweetener is prepared and then comment on the implications that, that sweetener is "natural."

Analyze

Figures 20.19, 20.21, and 20.22 show some of the common natural sugars (glucose, fructose, and sucrose, respectively).

Solve

The structure of sucralose appears to be derived from sucrose. The difference in the structures is that in sucralose, Cl atoms have replaced three OH groups on sucrose.

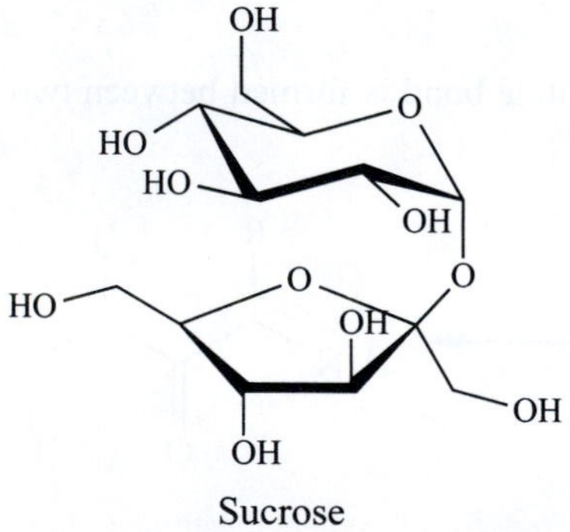

Sucrose 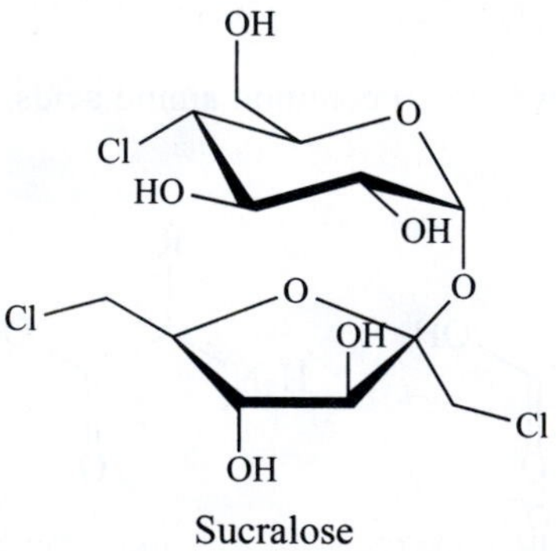   Sucralose

Being derived from sucrose implies that the sugar is natural, but the presence of Cl atoms on sugars is not natural.

Think About It

Sucralose is different from other artificial sweeteners such as aspartame or saccharin in that it is heat stable and therefore can be used in baking.

20.11. Collect and Organize

We are asked whether the assembly of small molecules into larger molecules in a cell is accompanied by an increase or a decrease in entropy.

Analyze

An increase in entropy is an increase in the number of particles in a process, whereas a decrease in entropy is a decrease in the number of particles in a process.

Solve

Because the number of particles (molecules) is decreasing as small molecules are bonded to form a larger molecule, the entropy decreases.

Think About It

Remember that a decrease in entropy is not favorable for a process. For the process with negative ΔS to be spontaneous, the enthalpy of the reaction must be negative so that the free energy is negative for the process.

20.13. **Collect and Organize**

We are to explain the meaning of alpha in "α-amino acid."

Analyze

The α refers to the placement of the $-NH_2$ and $-COOH$ groups on the structure of the amino acid.

Solve

The α refers to the single carbon atom in amino acids to which both $-NH_2$ and $-COOH$ groups are bonded.

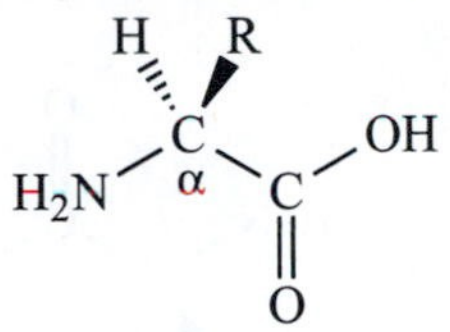

Think About It

A β-amino acid has its amino group bonded to the β-carbon instead of the α-carbon. An example is β-alanine.

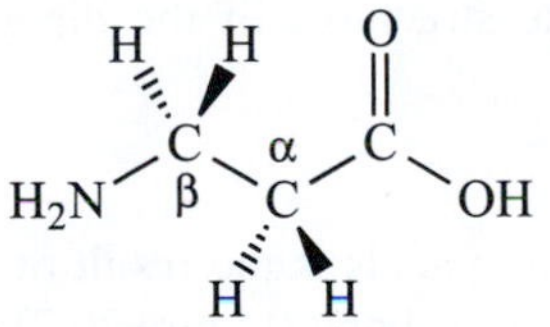

20.15. **Collect, Organize, and Analyze**

We are asked what is meant by the prefixes D and L in D-amino acids and L-amino acids.

Solve

D- and L- refer to how the four groups on a chiral carbon atom are oriented in three-dimensional space. They do not indicate how plane-polarized light is rotated; that is indicated by (+) and (−).

Think About It

Very few biological systems on Earth contain D-amino acids; they have been found in bacterial cell walls, in fungal toxins, and in some exotic sea animals.

20.17. **Collect and Organize**

Of the structures shown in Figure P20.17, we are asked which are not α-amino acids.

Analyze

An α-amino acid has both the amine and carboxylic acid groups bonded to the same carbon.

Solve

Compounds a and c have their amine and carboxylic acid groups bonded to different carbons, and so they are not α-amino acids. Only compound b has the amine group bonded to the same carbon as the carboxylic acid, and so it is the only α-amino acid in Figure P20.17.

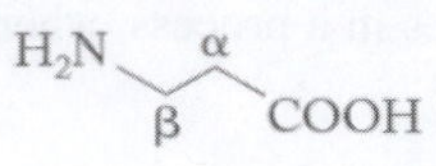

(a) 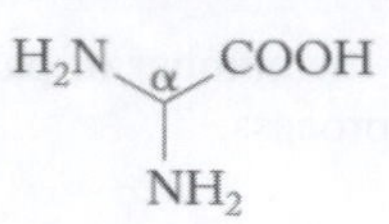   (b) 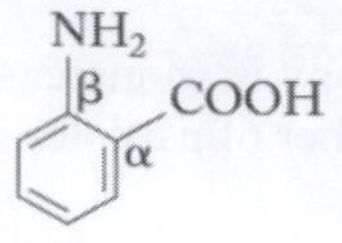   (c)

Think About It
Compound b actually has two amine groups in its structure.

20.19. Collect, Organize, and Analyze
We are asked to explain why most amino acids exist in their zwitterionic form (contain both a positive and a negative group on the same molecule) at a slightly basic pH.

Solve
The amine group of an amino acid is basic (becomes protonated) and the carboxylic acid group is acidic (deprotonates). Those combine to form the zwitterion at pH 7.4. That near-neutral pH is basic enough to deprotonate the carboxylic acid and yet acidic enough to protonate the amine group.

Think About It
For example, the zwitterion of alanine is

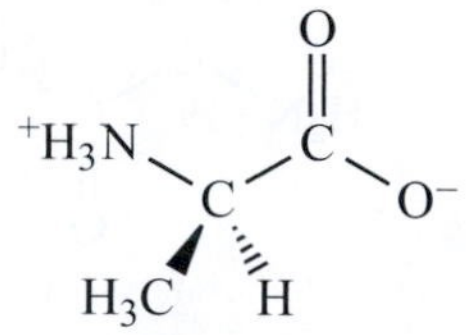

20.21. Collect and Organize
We are asked to draw all the possible structures of the dipeptides formed when two amino acids undergo condensation.

Analyze
The condensation reaction between amino acids is the result of the carboxylic acid of one amino acid reacting with the amine of another to give the peptide bond (Figure 20.7) and releasing water.

Solve
(a) Alanine + serine:

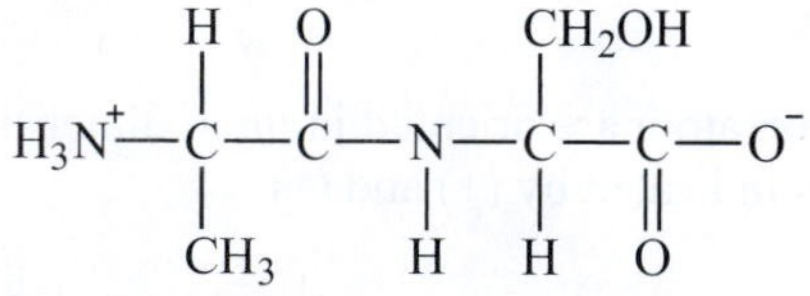

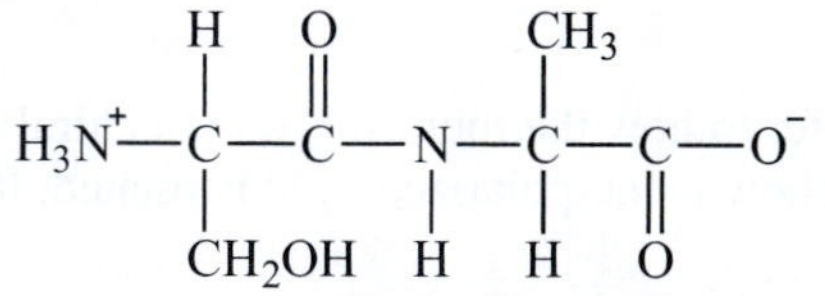

(b) Alanine + phenylalanine:

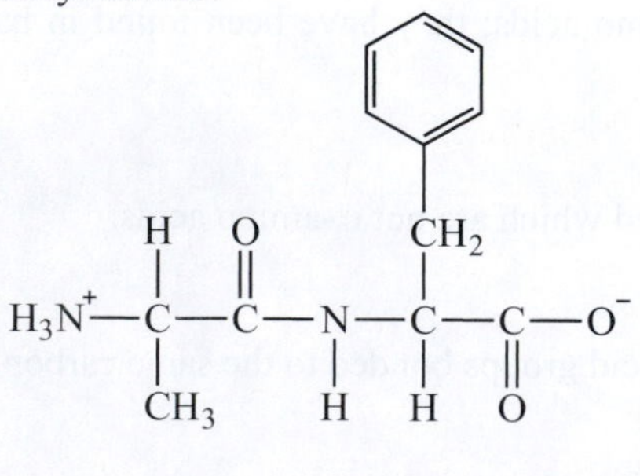

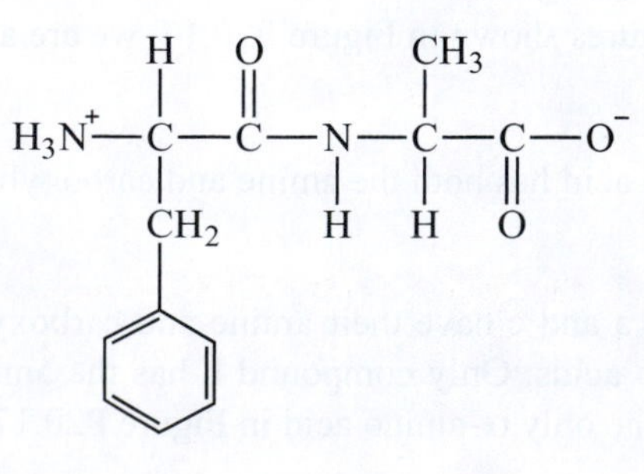

(c) Alanine + valine:

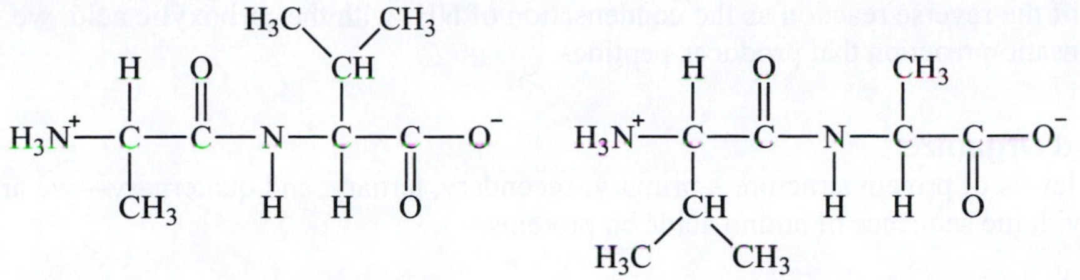

Think About It

By the naming rules described in the textbook, those dipeptides are (a) alanylserine and serylalanine, (b) alanylphenylalanine and phenylalylalanine, and (c) alanylvaline and valylalanine.

20.23. Collect and Organize

For each dipeptide structure shown in Figure P20.23, we are to identify the parent amino acids.

Analyze

The peptide bond formed between two amino acids is

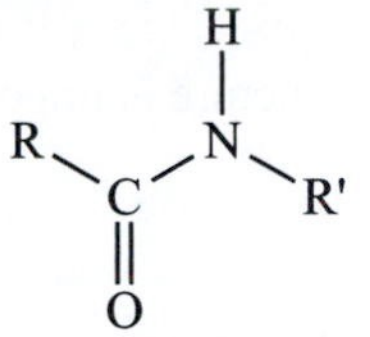

If we imagine the C—N bond rehydrated, we would regenerate the carboxylic acid containing R and the amine containing R′. From the structures of R and R′, we can identify the amino acids.

Solve

(a) Glycine + alanine
(b) Leucine + leucine
(c) Phenylalanine + tyrosine

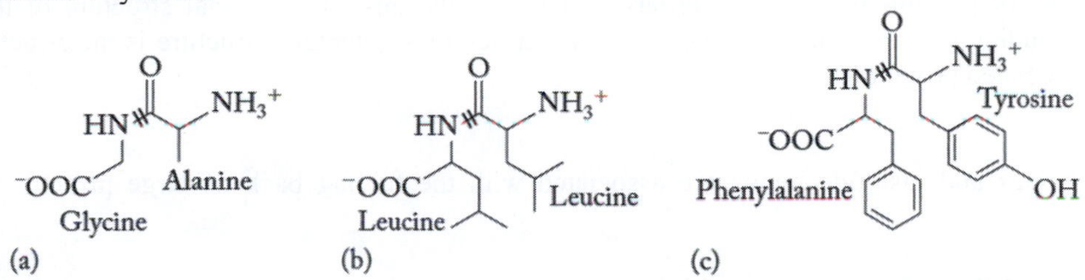

Think About It

For longer-chain peptides that make up proteins, the order in which the amino acids are linked is important to their structure, which is key to their function in the cell.

20.25. Collect and Organize

We are asked to identify the second product in the reaction shown in Figure P20.25.

Analyze

The only difference in the structures of the products and the reactants is the replacement of the –NH_2 group in the reactant with –OH in the product.

Solve

Upon reaction with water, the amide group forms a carboxylic acid group and ammonia. The second product in that reaction is simply NH_3.

Think About It

If we think of the reverse reaction as the condensation of NH_3 with the carboxylic acid, we see that we have the same condensation reaction that produces peptides.

20.27. **Collect and Organize**

Of the four levels of protein structure—primary, secondary, tertiary, and quaternary—we are to decide which is associated with the sequence of amino acids on proteins.

Analyze

Primary structure is at the level of the ordering of the amino acids as they are joined in the protein chain. Secondary structure is associated with the first stage of folding of a protein and is the pattern of the arrangement of segments of proteins in a chain. Tertiary structure is the three-dimensional structure of the protein, fully folded, including the segments from the secondary structure. Quaternary structure is the structure of the entire unit of associated proteins.

Solve

The sequence of amino acids in a protein is associated with primary structure.

Think About It

The sequence of amino acids in the protein structure is important to the associations between protein strands for secondary and tertiary structure.

20.29. **Collect and Organize**

Of the four levels of protein structure—primary, secondary, tertiary, and quaternary—we are to decide which is associated with ion–ion interactions and disulfide bond formation.

Analyze

Primary structure is at the level of the ordering of the amino acids as they are joined in the protein chain. Secondary structure is associated with the first stage of folding of a protein and is the pattern of the arrangement of segments of proteins in a chain. Tertiary structure is the three-dimensional structure of the protein, fully folded, including the segments from the secondary structure. Quaternary structure is the structure of the entire unit of associated proteins.

Solve

Ion–ion bonds and disulfide bonds are associated with the folding back of large proteins to form tertiary structures.

Think About It

Ion–ion bonds are fairly strong, as are the covalent S—S bonds, and help to maintain the tertiary structure of folded proteins.

20.31. **Collect, Organize, and Analyze**

We are to describe the *induced-fit* theory of enzyme activity.

Solve

In the induced-fit model, the binding site in the enzyme changes slightly upon substrate binding to the active site so that it better matches the transition state for the reaction being catalyzed.

Think About It

That model is different from a lock-and-key model, in which the substrate key fits perfectly without changes to the active site lock.

20.33. Collect, Organize, and Analyze
We are asked to describe how molecular structure affects the specificity of enzyme reactivity.

Solve
The structure of the enzyme—in particular, of the active site—is specific to the shape and intermolecular interactions of a particular substrate. If the molecule does not "fit," the reaction cannot be catalyzed.

Think About It
The matching of structure and interactions is precise, giving each enzyme a specialized function.

20.35. Collect and Organize
In the folding of proteins we are asked why lysine often pairs with glutamic acid.

Analyze
Table 20.1 gives the structures of the two amino acids.

Solve
Lysine contains two amino groups, one of which is on a long carbon tail. That can react with the carboxylic acid on the carbon tail of glutamic acid to form a salt bridge.

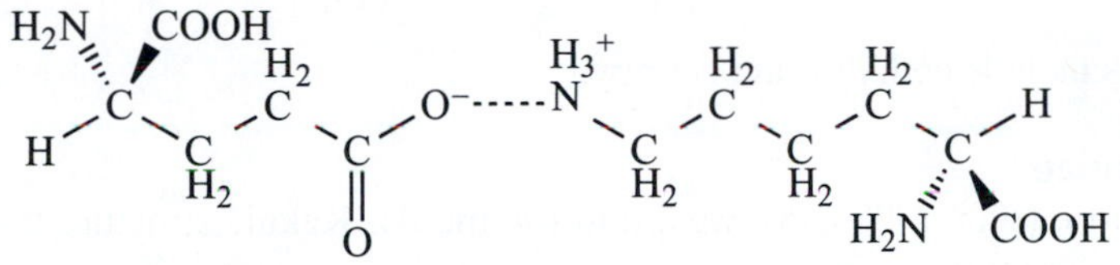

Think About It
That interaction is formed without much crowding of the amino acids and so is facile.

20.37. Collect, Organize, and Analyze
By looking at the structures in Figure 20.23 in the textbook, we can describe the structural differences between starch and cellulose.

Solve
Starch has α-glycosidic bonds and cellulose has β-glycosidic bonds. Starch coils into granules, and cellulose forms linear molecules.

Think About It
That subtle difference in how the glucose chain is oriented gives starch and cellulose different properties. We can easily digest starch, but we cannot digest cellulose.

20.39. Collect and Organize
We are asked whether the fuel value of glucose in its cyclic form is the same as that in its linear form.

Analyze
In the cyclic form of glucose we have an additional C—O bond not present in the linear form. We also have the formation of a C—O single bond from a C=O.

Solve
No. Because the bonding is different in the cyclic form of glucose, we do not expect the fuel values of the cyclic and linear forms to be the same.

Think About It
We might expect, though, that the fuel values are close to each other because the structures differ by only a few bonds.

20.41. Collect, Organize, and Analyze

We are to describe the function of carbohydrates in the diet.

Solve

Carbohydrates supply energy to all the cells of the body and are required for its functioning, including physical activity, brain function, and operation of organs and tissues.

Think About It

Carbohydrates also store energy, spare fats and proteins for other needs, and build macromolecules.

20.43. Collect and Organize

We can recall from Chapter 12 how to calculate ΔG for an overall process made up of two steps.

Analyze

ΔG is a state function; therefore, we can apply Hess's law to calculating the free-energy change for a multistep process.

Solve

To calculate the free-energy change for a two-step process, we need to only sum the ΔG values for each reaction.

Think About It

Other state functions include enthalpy and entropy.

20.45. Collect and Organize

For the sugar galactose (Figure P20.45), we are to use the flat Kekulé structure to draw the isomers that form when that linear structure forms a ring.

Analyze

Galactose is a six-carbon sugar, and the OH group on the C5 carbon reacts with the aldehyde group at C1 to form a six-membered ring. Depending on the orientation of the OH group attached in relation to the aldehyde, we obtain either the α or β isomer.

Solve

Using Figure 20.21 as a reference, we see that the α and β forms of galactose are

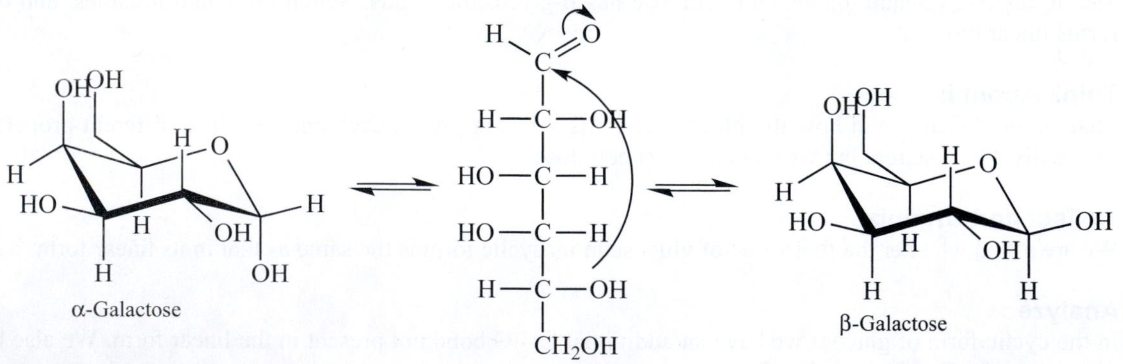

Think About It

Galactose is not as water soluble as glucose and is less sweet.

20.47. Collect and Organize

Given the cyclic structures of sugars shown in Figure P20.47, we are to identify those that are β isomers.

Analyze

Structures that are β forms have the OH group at C1 pointing up.

Solve

Structure c shows a β isomer.

Think About It

Because the orientation of the OH group on carbon atom 3 in structure b is different from the others, we can say that structure b is a different sugar.

20.49. **Collect and Organize**

Given the cyclic structures shown in Figure P20.49, we are to determine which are α monosaccharides.

Analyze

Structures that are α forms have the OH group at C1 pointing down.

Solve

Structure a shows an α isomer.

Think About It

Both structures b and c are β isomers.

20.51. **Collect and Organize**

Given the structures shown in Figure P20.51, we are to choose the saccharides digestible by humans.

Analyze

Humans can digest α-glycosidic linkages between sugars, but not β linkages. Structures that are α forms have the OH group at C1 pointing down.

Solve

Only structure b is fully digestible by humans.

Think About It

Structure c has both α and β linkages, so it is only partially digestible.

20.53. **Collect and Organize**

Given values for the free energy of formation of maltose, glucose, and water, we are to calculate the free-energy change for the reaction of maltose with water to produce glucose.

Analyze

To calculate the free-energy change, we use the equation

$$\Delta G^{\circ}_{\text{rxn}} = \sum n \Delta G^{\circ}_{\text{f,products}} - \sum n \Delta G^{\circ}_{\text{f,reactants}}$$

Solve

$$\Delta G^{\circ}_{\text{rxn}} = \left(2 \text{ mol} \times -1274.4 \text{ kJ/mol}\right) - \left[\left(1 \text{ mol} \times -2246.6 \text{ kJ/mol}\right) + \left(1 \text{ mol} \times -285.8 \text{ kJ/mol}\right)\right] = -16.4 \text{ kJ}$$

Think About It

That reaction is spontaneous under standard conditions, as indicated by the negative free-energy change for the reaction.

20.55. **Collect, Organize, and Analyze**

We can use definitions to discriminate between a saturated and an unsaturated fatty acid.

Solve

Saturated fatty acids have all C—C single bonds in their structure; unsaturated fatty acids have C═C double bonds in their structure.

Think About It

Table 20.2 shows the common fatty acids. Of those listed, oleic, linoleic, and linolenic acids have C═C double bonds in their structures.

20.57. Collect and Organize

We consider why Arctic explorers would eat sticks of butter.

Analyze

Butter is composed mostly of fatty acids.

Solve

Fatty acids have a high fuel value (see Problem 20.56), and eating sticks of butter affords Arctic explorers more energy per gram of food than would carbohydrates or proteins.

Think About It

Arctic explorers need many more calories not only to move around the environment but also to keep warm.

20.59. Collect and Organize

By examining the structure of a triglyceride, we can determine whether those molecules have a chiral center.

Analyze

A triglyceride is formed when glycerol reacts with long-chain fatty acids to form esters.

Solve

If the two fatty acids linked to the glycerol at C1 and C3 are different, then, yes, the triglyceride has a chiral center. That configuration will give a structure in which the central carbon on the triglyceride has four different groups attached, and the fatty acid is therefore chiral.

Think About It

If all three fatty acid chains are different, the triglyceride must be chiral.

20.61. Collect and Organize

Given the three triglycerides shown in Figure P20.61, we are to choose those that are unsaturated fats.

Analyze

Unsaturated fats contain C═C bonds in their structures.

Solve

Fats b and c are unsaturated fats.

Think About It

However, structure b is not as unsaturated as structure c, so we might be more likely to observe that fat b has a higher melting temperature than fat c.

20.63. Collect and Organize

For the reaction of glycerol with octanoic acid, decanoic acid, and dodecanoic acid, we are to draw the structures of the resulting triglycerides.

Analyze

The reaction of glycerol with the carboxylic acid functional group of the fatty acid makes an ester bond and releases water as a by-product.

Solve

(a)

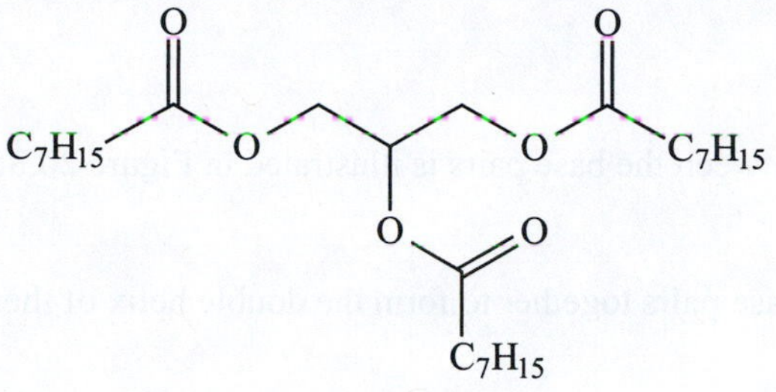

Glycerol with octanoic acid

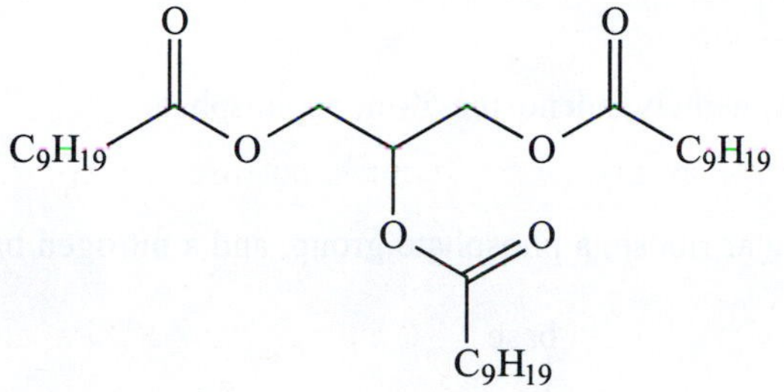

Glycerol with decanoic acid

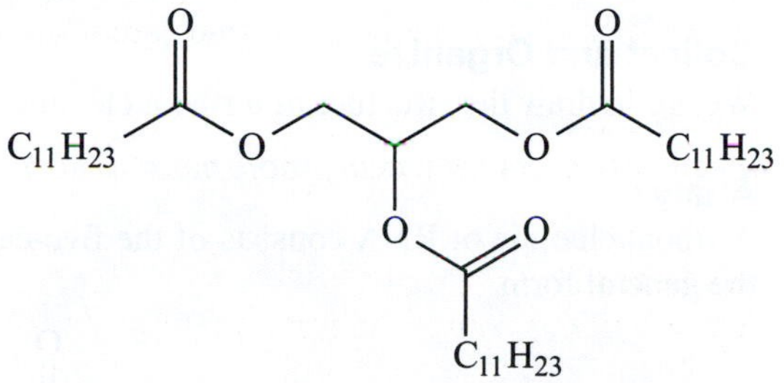

Glycerol with dodecanoic acid

Think About It

You may have read this problem as asking for the reaction of glycerol with one equivalent each of the fatty acids. In that case you would have the following triglycerides:

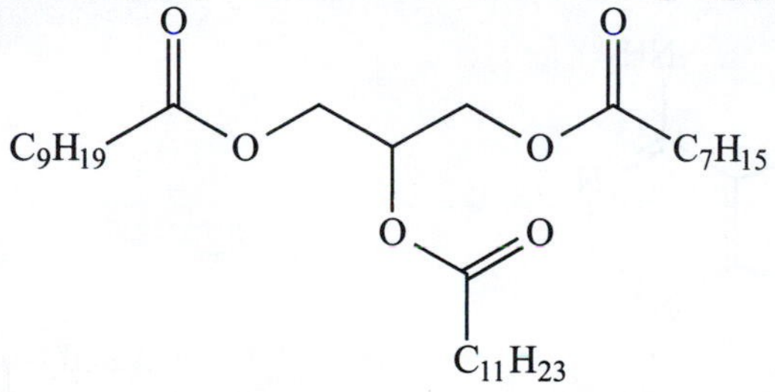

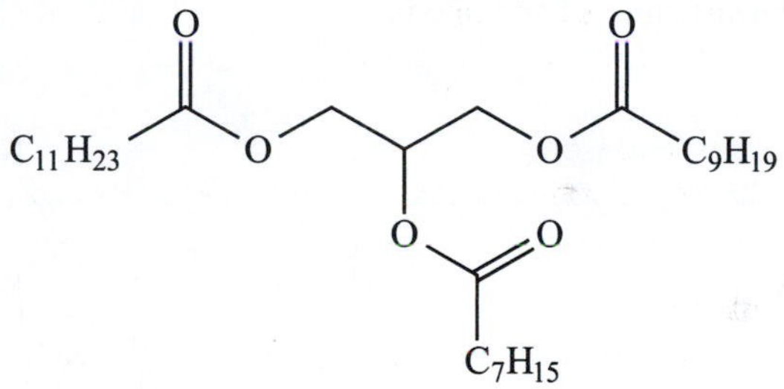

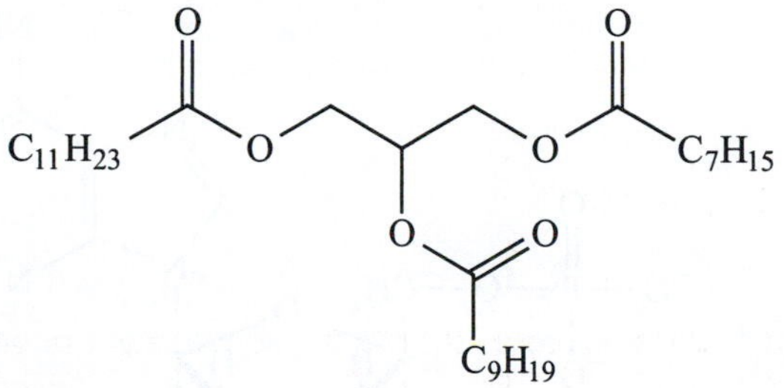

20.65. Collect, Organize, and Analyze

We are to name the three subunits present in DNA and identify the two subunits that form the backbone of DNA strands.

Solve

The three subunits of the DNA structure are a phosphate group, a five-carbon sugar, and a nitrogen base. As shown in Figure 20.34, the backbone of DNA is composed of alternating sugar residues and phosphate groups.

Think About It

The entire unit of the three subunits is called a *nucleotide*.

20.67. **Collect and Organize**

We are to identify the kind of intermolecular force that holds together the strands of DNA and forms the double-helix configuration.

Analyze

The intermolecular force between the base pairs is illustrated in Figure 20.35.

Solve

Hydrogen bonds hold the base pairs together to form the double helix of the DNA molecule.

Think About It

Recall that hydrogen bonds are strong, and so the integrity of the DNA molecule is fairly secure.

20.69. **Collect and Organize**

We are to draw the structure of a ribonucleotide of RNA, namely, adenosine 5′-monophosphate.

Analyze

A ribonucleotide of RNA consists of the five-carbon sugar ribose, a phosphate group, and a nitrogen base with the general form

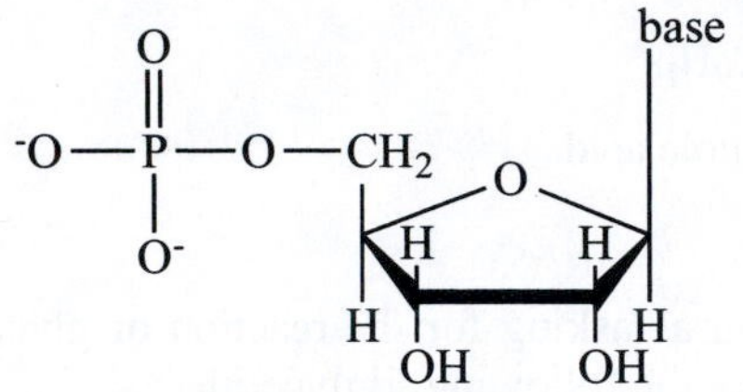

The structure of adenine is

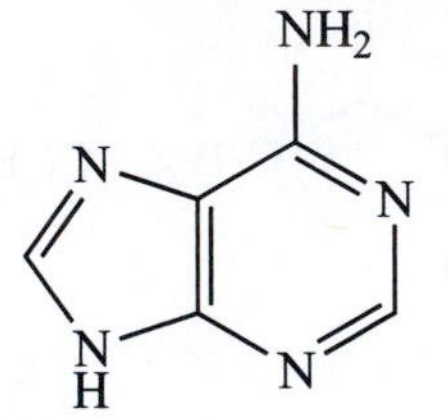

Solve

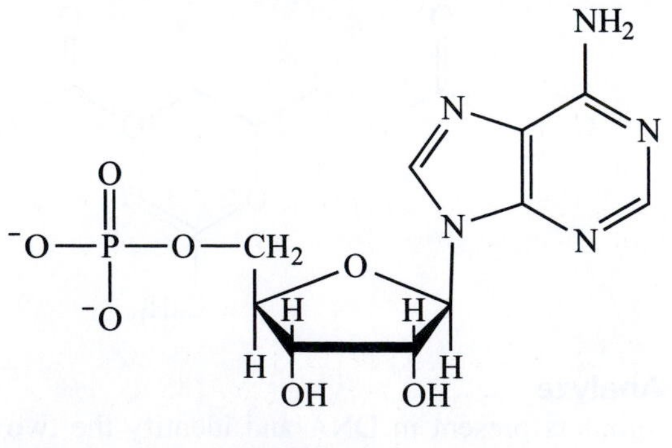

Think About It

That nucleoside plays an important role in energy transfer as ATP and ADP.

20.71. **Collect and Organize**

In the replication of DNA, we are to determine the sequence of the double-stranded helix formed when the original strand has the sequence T-C-G-G-T-A.

Analyze

In replication, DNA copies information from a DNA strand by making a sequence of bases complementary to that on the DNA segment. For DNA, thymine (T) complements adenine (A), and guanine (G) complements cytosine (C).

Solve

The sequence of the double-stranded helix containing the complementary base sequence of T-C-G-G-T-A is
A-G-C-C-A-T

Think About It

DNA is a crucial molecule for life on our planet. It contains the instructions used in the development and functioning of all known living organisms and even of some viruses.

20.73. **Collect and Organize**

We consider the structure of olestra shown in Figure P20.73.

Analyze

Olestra is a disaccharide in which the –OH groups on the sugars have been converted to esters through reaction with long-chain carboxylic acids.

Solve

(a) The disaccharide core of olestra is composed of sucrose.
(b) Esters have replaced the hydroxyl groups on the disaccharide as functional groups.
(c) The carboxylic acid used to make olestra is $C_{15}H_{31}COOH$.

Think About It

The fatty acid that reacted with the disaccharide in this problem is palmitic acid.

20.75. **Collect and Organize**

We are to compare the structure of cysteine with that of homocysteine shown in Figure P20.75. We are also asked whether homocysteine is chiral.

Analyze

The structure of cysteine from Table 20.1 is

Solve

(a) Homocysteine's sulfur-containing side chain has an extra –CH_2– group that is not present in cysteine.
(b) Yes, homocysteine is chiral because it contains a carbon bonded to four different groups.

Think About It

Because of the extra –CH_2– group in homocysteine, it can form a five-membered ring. That configuration, however, does not allow the amino acid to form stable peptide bonds.

20.77. **Collect and Organize**

Given the structure of hypoglycin shown in Figure P20.77, we are to determine whether that compound is an α-amino acid.

Analyze

An α-amino acid has both the carboxylic acid group and the amine group bonded to the same carbon atom.

Solve

Yes, hypoglycin is an α-amino acid. The carbon to which the carboxylic acid and amine group are bonded is indicated below.

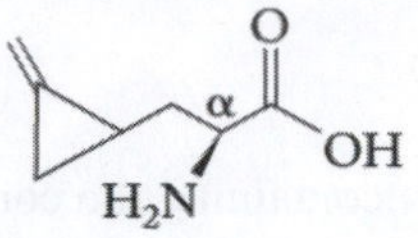

Think About It

Hypoglycin is chemically related to lysine, and hypoglycin's toxic effects may include coma and death.

20.79. **Collect and Organize**

Given the structure of creatine shown in Figure P20.79, we are asked whether creatine is an α-amino acid. We are also to draw the two dipeptides that may form when glycine reacts with creatine.

Analyze

When the –NH$_2$ group and the –COOH group are bonded to the same carbon atom, the amino acid is an α-amino acid. When glycine reacts with creatine, either (1) the –NH$_2$ group of glycine reacts with the –COOH group of creatine or (2) the –NH$_2$ group of creatine reacts with the –COOH group of glycine.

Solve

(a) No, creatine is not an α-amino acid.
(b) When the –NH$_2$ group of glycine reacts with the –COOH group of creatine, we get the following dipeptide:

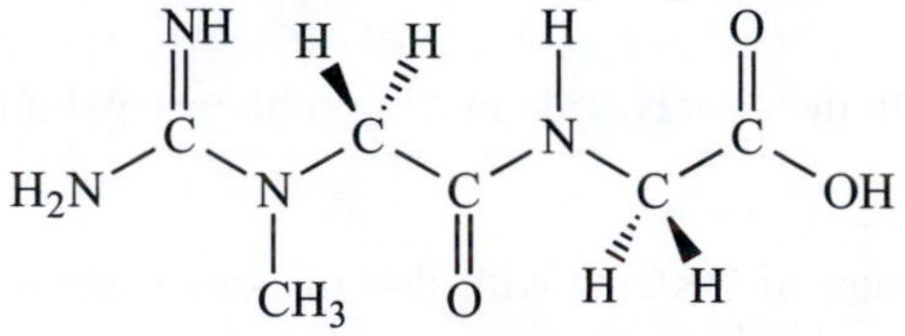

When the –NH$_2$ group of creatine reacts with the –COOH group of glycine, we get this dipeptide:

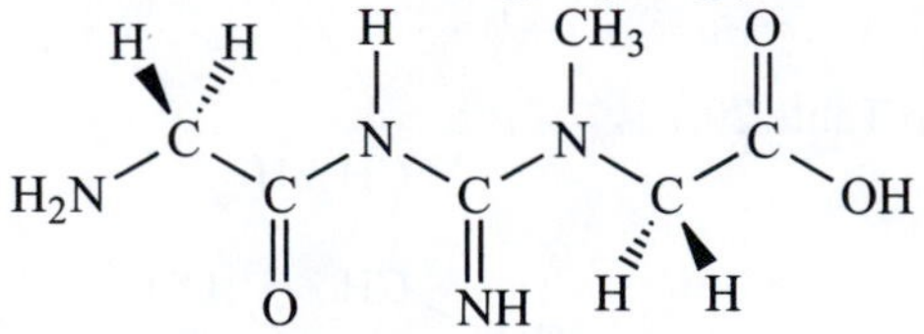

Think About It

The two dipeptides would be named glycylcreatine and creatylglycine, respectively.

20.81. **Collect and Organize**

From the structure of glutathione (Figure P20.81), we are to identify the three amino acids that combine to make it up.

Analyze

Table 20.1 shows the 20 common naturally occurring amino acids. Glutathione is prepared from the condensation of three amino acids to make the two peptide bonds in the structure.

Solve

Glutamic acid, cysteine, and glycine are the amino acids that make up glutathione.

Think About It

Glutathione is an antioxidant and, as such, protects cells from damage due to free radicals.

20.83. **Collect and Organize**

By considering only the average bond energies, we are asked whether the relative fuel values of leucine and isoleucine should be the same. We are also to explain why calorimetric measurements show isoleucine to have a lower fuel value than leucine.

Analyze

The structures of leucine and isoleucine show that they are isomers with similar structures and bonding between the atoms. They differ only in location of the methyl ($-CH_3$) group on the three-carbon side chain.

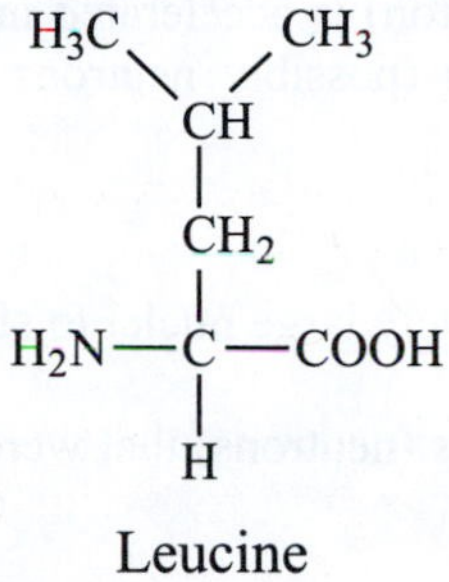

$$
\begin{array}{cc}
\text{Leucine} & \text{Isoleucine} \\
\text{(Leu)} & \text{(Ile)}
\end{array}
$$

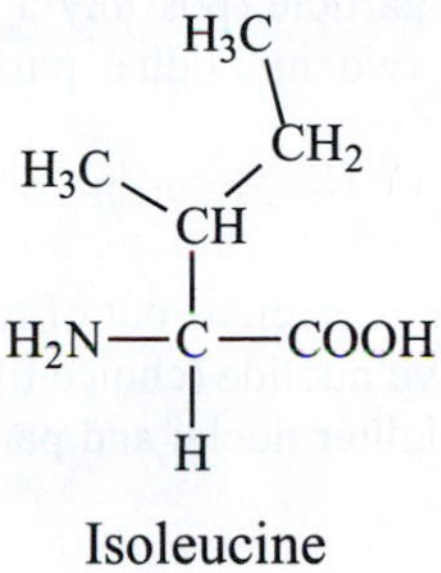

Solve

Yes. Because no difference exists in the number of C—C, C—H, C=O, C—O, or N—H bonds between the two compounds, we expect, on the basis of average bond energies, that the fuel values of leucine and isoleucine should be identical. Isoleucine might have a lower fuel value because the CH_3 group is closer to the COOH and NH_2 groups, and that difference in shape must contribute to the slightly different fuel values.

Think About It

Isoleucine and leucine are not enantiomers of each other, but they are structural isomers.

CHAPTER 21 | Nuclear Chemistry: The Risks and Benefits

21.1. **Collect and Organize**
Of the five nuclear processes named, we are to determine which two are represented in Figure P21.1(a) and (b).

Analyze
(a) In this process a smaller nucleus is fusing with a larger nucleus to give one larger nucleus accompanied by the emission of three small particles (probably neutrons).
(b) In this process a single particle (possibly a neutron) is accelerated into a large nucleus, and two nuclei of about equal mass plus two individual particles (possibly neutrons again) are the result of the nuclear reaction.

Solve
(a) This process makes a large nucleus out of relatively large nuclei to start, and so it is a process of fusion to synthesize a supermassive nuclide (choice b).
(b) This process produces lighter nuclei and particles (neutrons) that were once in the nuclei, so it is a process of fission (choice d).

Think About It
A β decay representation would show a neutron becoming a proton with the emission of the negatively charged β particle. Primordial nucleosynthesis would be represented by the fusing of protons with neutrons to produce deuterium. Solar fusion would be represented by the fusing of protons with the emission of positrons to produce helium.

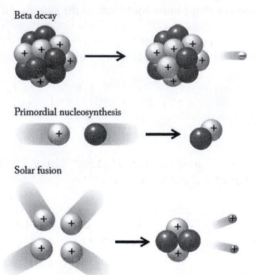

21.3. **Collect and Organize**
Of the highlighted elements in the periodic table shown in Figure P21.2, we are to choose those that are products of the natural radioactive decay of uranium.

Analyze
The decay series of uranium is shown in Figure 21.2 of the textbook.

Solve
Radium (red-shaded element in Figure P21.2) is one of the products of the radioactive decay.

Think About It
Uranium is a naturally occurring radioactive element. It was until recently believed that bismuth was the last stable isotope, but theory had predicted that, indeed, bismuth is unstable. In 2003 researchers in France measured the half-life of bismuth-209 to be 1.9×10^{19} yr. That is a billion times longer than the age of the universe, so scientists still treat ^{209}Bi as the last stable element in the periodic table.

21.5. Collect and Organize
From the two graphs shown in Figure P21.5, we are to determine which one describes β decay.

Analyze
In β decay, a neutron is changed into a proton and an electron is produced. In β decay, then, the number of neutrons decreases and the number of protons increases.

Solve
Because β decay results in the emission of an electron from the nucleus, which increases the atomic number of the nucleus, graph (a) illustrates β decay.

Think About It
Figure P21.5(b) represents proton emission from the nucleus, in which the number of protons is reduced by 1 but the number of neutrons remains the same.

21.7. Collect and Organize
Given the five curves shown in Figure P21.7, we are to choose the one that represents a nuclear process with $t_{1/2} = 2.0$ days.

Analyze
The half-life of a nuclear decay process is the time it takes for half the original concentration to decay. Starting from the original isotope quantity of 100%, it is the time it takes for that concentration to decrease to 50%.

Solve
The blue line (b) represents a decay process with $t_{1/2} = 2.0$ days. The quantity of the isotope is 50% after 2 days after starting at 100%.

Think About It
The value of $t_{1/2}$ does not change. On line (b) the time it takes for the quantity of the isotope to drop from 50% to 25% also is 2.0 days.

21.9. Collect and Organize
From the processes depicted in Figure P21.9, we are to assign each as either fission or fusion.

Analyze
In fission, the nucleus splits into smaller nuclei. In fusion, smaller nuclei combine to form a heavier nucleus.

Solve
Process 1 represents fission, where smaller nuclei are generated (along with some nuclear particles, shown as gray spheres). Process 2 represents fusion, where lighter nuclei are fused into a heavier nucleus.

Think About It
In fusion, some nuclear particles (neutrons, α particles, or β particles) may be released to stabilize the larger nucleus formed.

21.11. **Collect and Organize**

Using Figure P21.11 and our knowledge of the charges and the masses of α and β particles and neutrons/gamma radiation, we are to determine whether the radiation represented by the red arrow penetrates solids better than the radiation represented by the blue and green arrows.

Analyze

Figure P21.11 represents each particle's behavior as it moves through an electric field. The red arrow represents α particles that are positively charged and will be attracted to the negative pole of the electric field. The green arrow represents the trajectory of lighter and negatively charged β particles, which are attracted to the positive pole of the electric field. Notice, too, that the β particles travel less far and with a smaller arc in their deflection than the heavier α particles. The blue arrow represents the trajectory of either neutrons or gamma radiation whose trajectory is not affected by the electric field.

Solve

No, both neutrons and gamma radiation, being neutral in charge, and β particles, being negatively charged but lighter, penetrate solid objects better than the heavier, positively charged α particles.

Think About It

The penetration of the different forms of radiation are shown in Figure 21.19; α particles are stopped by clothing but are dangerous if ingested or inhaled.

21.13. **Collect and Organize**

We consider here what happens to the neutron-to-proton ratio when a nucleus undergoes β decay.

Analyze

Beta decay converts a neutron into a proton.

Solve

In a nucleus that undergoes β decay, a neutron is "lost" and a proton is "gained." The neutron-to-proton ratio will decrease.

Think About It

Remember that, although the β particle is an electron, it is emitted from the nucleus of an atom. It is not simply lost by the atom in the ionization process to become a cation.

21.15. **Collect and Organize**

We are to describe how the belt of stability (shown in Figure 21.1) can be used to predict the possible decay modes of radioactive nuclides.

Analyze

The belt of stability in Figure 21.1 plots the number of neutrons versus the number of protons.

Solve

If the nuclide lies in the belt of stability (green dots on the plot in Figure 21.1), it is not radioactive and is stable. If it lies above the belt of stability, it is neutron rich and tends to undergo β decay to increase the number of protons and reduce the number of neutrons in its nucleus. If it lies below the belt of stability, it is neutron poor and tends to undergo positron emission or electron capture to increase the number of neutrons and reduce the number of protons in its nucleus.

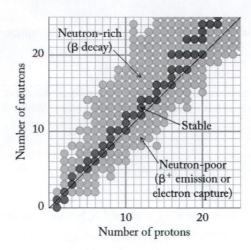

Think About It
From Figure 21.1 we can see that an element often has several radioactive isotopes (light gray dots).

21.17. Collect and Organize
Using the decay series for ^{238}U (shown in Figure 21.2), we are to explain why several α decays are often followed by β decay.

Analyze
We know, too, from the statement of the problem that the neutron-to-proton ratio of decay products must decrease to form stable fission products.

Solve
Alpha decay increases the neutron-to-proton ratio to produce less stable isotopes, which can then be made more stable through β emission to decrease the neutron-to-proton ratio.

Think About It
Ultimately, ^{238}U decays to form stable ^{206}Pb.

21.19. Collect and Organize
Knowing that ^{64}Cu decays to ^{64}Zn and ^{64}Ni, we are to identify the modes of decay for both processes.

Analyze
In the decay of ^{64}Cu to ^{64}Zn, the number of protons increases by 1 and the number of neutrons decreases by 1. In the decay of ^{64}Cu to ^{64}Ni, the number of protons decreases by 1 and the number of neutrons increases by 1.

Solve
The decay of ^{64}Cu to ^{64}Zn occurs by beta decay.

$$^{64}_{29}\text{Cu} \rightarrow {}^{64}_{30}\text{Zn} + {}^{0}_{-1}\beta$$

The decay of ^{64}Cu to ^{64}Ni occurs by either electron capture or positron emission.

$$^{64}_{29}\text{Cu} + {}^{0}_{-1}\text{e} \rightarrow {}^{64}_{28}\text{Ni}$$

$$^{64}_{29}\text{Cu} \rightarrow {}^{64}_{28}\text{Ni} + {}^{0}_{+1}\beta$$

Think About It
Copper-64 does decay by all three of those processes. The decay of ^{64}Cu produces ^{64}Ni 18% of the time by positron emission and 43% of the time by electron capture while zinc-64 is produced by ^{64}Cu beta decay of ^{64}Cu 39% of the time.

21.21. **Collect and Organize**

We consider the neutron-to-proton ratio for an isotope with a mass number, A, more than two times the atomic number, Z.

Analyze

When the neutron-to-proton ratio is 1.00, the nucleus has equal numbers of protons and neutrons. When the ratio is greater than 1, the nucleus has more neutrons than protons. When the ratio is less than 1, the nucleus has more protons than neutrons.

Solve

When $A > 2Z$, the neutron-to-proton ratio is greater than 1.

Think About It

For light elements, the neutron-to-proton ratio is about 1.0, but for heavier stable nuclei the ratio is greater than 1.0 and up to 1.5 for the heaviest elements.

21.23. **Collect and Organize**

For each given radioactive isotope, we are to calculate the neutron-to-proton ratio and predict the mode of decay.

Analyze

The neutron-to-proton ratio can be calculated by first determining the number of protons in each nuclide (the atomic number in the symbol) and the number of neutrons in each nuclide (the mass number – atomic number).

$$\text{}^{\text{mass number}}_{\text{atomic number}} X$$

When the neutron-to-proton ratio is too high (neutron-rich), the radionuclide may undergo beta decay. When the neutron-to-proton ratio is too low (neutron-poor), the radionuclide may undergo positron emission or electron capture. Alpha decay occurs for massive nuclides and results in a higher neutron-to-proton ratio. We can use Figure 21.1 to help us determine whether the radionuclide lies above or below the belt of stability.

Solve

(a) ^{47}Sc has 21 protons and 26 neutrons, giving a neutron-to-proton ratio of 1.24. That radionuclide is neutron-rich and will undergo beta decay.

(b) ^{89}Zr has 40 protons and 49 neutrons, giving a neutron-to-proton ratio of 1.23. That radionuclide is neutron-poor and will undergo positron emission or electron capture.

(c) ^{230}Th has 90 protons and 140 neutrons, giving a neutron-to-proton ratio of 1.56. That radionuclide is a massive nuclide and will undergo alpha decay.

Think About It

The decay product of ^{230}Th alpha decay is ^{226}Ra.

21.25. **Collect and Organize**

Given that ^{192}Ir can undergo both beta decay and electron capture, we are to write the balanced nuclear equations for those processes.

Analyze

Beta decay results in a neutron converting into a proton, and electron capture results in a proton converting into a neutron.

Solve

$$^{192}_{77}\text{Ir} \rightarrow \,^{192}_{78}\text{Pt} + \,^{0}_{-1}\beta$$
$$^{192}_{77}\text{Ir} + \,^{0}_{-1}\text{e} \rightarrow \,^{192}_{76}\text{Os}$$

Think About It

As shown in the answer to Problem 21.18, both ^{192}Pt and ^{192}Os are stable nuclides.

21.27. Collect and Organize

For ^{56}Co and ^{44}Ti we are to predict the mode of decay.

Analyze

If the nuclide lies in the belt of stability (green dot in the plot in Figure 21.2), it is not radioactive and is stable. If it lies above the belt of stability, it is neutron rich and tends to undergo β decay to increase the number of protons and reduce the number of neutrons in its nucleus. If it lies below the belt of stability, it is neutron poor and it tends to undergo positron emission or electron capture to increase the number of neutrons and reduce the number of protons in its nucleus.

Solve

^{56}Co has 27 protons and 29 neutrons and is neutron poor; it may undergo electron capture or positron emission. ^{44}Ti has 22 protons and 22 neutrons and is neutron poor; it may undergo electron capture or positron emission.

Think About It

Both radioisotopes, then, are expected to be positron emitters.

21.29. Collect and Organize

We are to explain why radiocarbon dating is reliable only for artifacts younger than 50,000 years.

Analyze

The half-life of ^{14}C is 5730 yr. The number of half-lives that 50,000 years represents is 50,000/5730 = 8.726 half-lives.

Solve

After 8.726 half-lives, the ratio of ^{14}C present to that originally in an artifact is

$$\frac{N_t}{N_0} = 0.50^{8726} = 0.00236, \text{ or } 0.236\%$$

That is too little to detect.

Think About It

Longer-lived isotopes would be better suited for dating old artifacts. For example, ^{40}K (Problem 21.75) is useful for objects older than 300,000 years.

21.31. Collect and Organize

We consider why ^{40}K is a useful isotope for dating only objects older than 300,000 years.

Analyze

The half-life of ^{40}K is 1.28×10^9 yr; 300,000 years represents $300,000/1.28 \times 10^9 = 0.00023$ half-life.

Solve

After 0.00023 half-life, the ratio of ^{40}K present to that originally in a sample is

$$\frac{N_t}{N_0} = 0.50^{0.00023} = 0.9998, \text{ or } 99.98\%$$

That level is just when we can detect the difference in amounts of ^{40}K.

Think About It

For objects younger than 300,000 years, an isotope with a shorter half-life must be used.

21.33. **Collect and Organize**

We are to determine what percentage of radioactivity of a sample remains after two half-lives. A half-life is defined as the time at which half the radioactivity remains.

Analyze

After each half-life, radioactivity decreases by 50%. We can start with 100% and decrease by 50% for each half-life.

Solve

$$100\% \xrightarrow[\text{half-life}]{\text{first}} 50\% \xrightarrow[\text{half-life}]{\text{second}} 25\% \xrightarrow[\text{half-life}]{\text{third}} 12.5\%$$

Think About It

Alternatively, we can use Equation 21.9

$$\frac{N_t}{N_0} = 0.5^n$$

where N is the number of half-lives:

$$\frac{N_t}{N_0} = 0.5^3 = 0.125, \text{ or } 12.5\%$$

21.35. **Collect and Organize**

For ^{199}Au that in 168 hours decays 16.5%, we are to calculate its half-life.

Analyze

The half-life of a radionuclide can be found by rearranging the equation

$$\frac{N_t}{N_0} = 0.5^{t/t_{1/2}}$$

$$\ln\frac{N_t}{N_0} = -0.693\frac{t}{t_{1/2}}$$

$$t_{1/2} = \frac{-0.693t}{\left(\ln\dfrac{N_t}{N_0}\right)}$$

Solve

For $N_0 = 100$ and $N_t = (100 - 16.5) = 83.5$, and $t = 168$ h, the half-life of this radionuclide is

$$t_{1/2} = \frac{-0.693\times 168 \text{ h}}{\left(\ln\dfrac{83.5}{100}\right)} = 646 \text{ h}$$

Think About It

The number of half-lives that this radionuclide has undergone in 646 hours is

$$\frac{N_t}{N_0} = 0.5^n$$

$$\ln\frac{N_t}{N_0} = n\times\ln\ 0.5$$

$$n = \frac{\left(\ln\dfrac{N_t}{N_0}\right)}{-0.693} = \frac{\left(\ln\dfrac{83.5}{100}\right)}{-0.693} = 0.26$$

21.37. **Collect and Organize**

For cesium-137 that was released at the Fukushima power plant in 2011 and that has a half-life of 30.2 yr, we are to calculate the time it will take for the radioactivity to decay to 5.0% of the level released.

Analyze

The time of decay of a radionuclide can be found by rearranging the equation

$$\frac{N_t}{N_0} = 0.5^{t/t_{1/2}}$$

$$\ln \frac{N_t}{N_0} = -0.693 \frac{t}{t_{1/2}}$$

$$t = \frac{t_{1/2} \times \left(\ln \frac{N_t}{N_0} \right)}{-0.693}$$

Solve

For $N_0 = 100$ and $N_t = 5.00$, and $t_{1/2} = 30.2$ yr, the time needed for cesium-137 to decay to 5.0% is

$$t = \frac{30.2 \text{ yr} \times \left(\ln \frac{5.0}{100} \right)}{-0.693} = 131 \text{ yr}$$

Think About It

Therefore, the year that the level of radioactivity will be 5% will be $2011 + 131 = 2142$.

21.39. Collect and Organize

For a piece of charcoal that is 8700 years old, we are to calculate the fraction of ^{14}C remaining.

Analyze

We can use Equation 21.10 to solve for N_t/N_0:

$$t = -\frac{t_{1/2}}{0.693} \ln \frac{N_t}{N_0}$$

Solve

$$8700 \text{ yr} = -\frac{5730 \text{ yr}}{0.693} \ln \frac{N_t}{N_0}$$

$$\frac{N_t}{N_0} = 0.35, \text{ or } 35\%$$

Think About It

Alternatively, this problem can be solved by first calculating n, the number of half-lives:

$$n = \frac{8700}{5730} = 1.518 \text{ half-lives}$$

$$\frac{N_t}{N_0} = 0.5^{1.518} = 0.35, \text{ or } 35\%$$

21.41. Collect and Organize

For a wood sample from a giant sequoia tree that in 1891 was 1342 years old, we are to compare the fraction of ^{14}C in the innermost ring with that in the outermost ring.

Analyze

We can use Equation 21.10 to solve for N_t/N_0:

$$t = -\frac{t_{1/2}}{0.693} \ln \frac{N_t}{N_0}$$

Solve

$$1342 \text{ yr} = -\frac{5730 \text{ yr}}{0.693} \ln \frac{N_t}{N_0}$$

$$\frac{N_t}{N_0} = 0.850, \text{ or } 85.0\%$$

Think About It

Alternatively, this problem can be solved by first calculating n, the number of half-lives:

$$n = \frac{1342}{5730} = 0.2342 \text{ half-life}$$

$$\frac{N_t}{N_0} = 0.5^{0.2342} = 0.850, \text{ or } 85.0\%$$

21.43. **Collect and Organize**

Given that the ^{14}C to ^{12}C ratio is only 1.19% for the ancient mammoth tusk compared with the ratio in elephants today, we are to calculate the age of the mammoth tusk.

Analyze

We can use Equation 21.10 to solve for t:

$$t = -\frac{t_{1/2}}{0.693} \ln \frac{N_t}{N_0}$$

Solve

$$t = -\frac{5730 \text{ yr}}{0.693} \ln \frac{1.19}{100} = 36,640 \text{ yr}$$

Think About It

The number of half-lives that represents for ^{14}C is

$$n = \frac{36,640}{5730} = 6.39 \text{ half-lives}$$

21.45. **Collect, Organize, and Analyze**

We are asked to explain why all nuclear reactions produce heat.

Solve

Nuclear reactions produce heat because they result in a loss of mass from reactants to products, which is released as energy, some of it in the form of heat.

Think About It

Einstein's equation allows us to calculate the change in energy of a nuclear reaction from the change in mass.

21.47. **Collect and Organize**

For the radioactive ^{11}C nucleus, we are to write the balanced nuclear reaction and calculate the binding energy.

Analyze

From Figure 21.4 we see that ^{11}C is neutron poor; it is likely to be a positron emitter. We can calculate the binding energy of ^{11}C by adding the masses of the individual protons, neutrons, and electrons for ^{11}C and subtracting that result from the actual mass of 1.82850×10^{-26} kg to find the mass defect. We then apply Einstein's equation to calculate the energy difference.

Solve

(a) The decay of ^{11}C by positron emission is given by the nuclear reaction

$$^{11}_{6}C \rightarrow {}^{11}_{5}B + {}^{0}_{+1}\beta$$

(b) The mass defect for ^{11}C is

$$\Delta m = \left[\left(6 \times 1.67262 \times 10^{-27}\,\text{kg} \right) + \left(5 \times 1.67493 \times 10^{-27}\,\text{kg} \right) + \left(6 \times 9.10938 \times 10^{-31}\,\text{kg} \right) \right] - \left[1.82850 \times 10^{-26}\,\text{kg} \right]$$

$$= 1.3084 \times 10^{-28}\,\text{kg}$$

The binding energy of ^{11}C is

$$\Delta E = \left(\Delta m \right) c^2 = 1.3084 \times 10^{-28}\,\text{kg} \times \left(2.998 \times 10^{8}\,\text{m/s} \right)^2$$

$$= 1.1760 \times 10^{-11}\,\text{J}$$

Think About It

The binding energy calculated above is per atom of ^{11}C. To calculate the binding energy per mole, we multiply by Avogadro's number.

21.49. Collect and Organize

We are asked to calculate the binding energy in kilojoules per mole of ^{35}Cl and ^{37}Cl with exact masses of 34.9689 amu and 36.9659 amu, respectively.

Analyze

The binding energy of each nuclide can be found using Einstein's equation and the mass defect. The mass defect is found by adding the masses of the individual protons, neutrons, and electrons and subtracting that result from the actual mass of each nuclide.

Solve

The binding energy of ^{35}Cl is

$$\Delta m = \left[\left(17 \times 1.67262 \times 10^{-27}\,\text{kg} \right) + \left(18 \times 1.67493 \times 10^{-27}\,\text{kg} \right) + \left(17 \times 9.10938 \times 10^{-31}\,\text{kg} \right) \right] -$$

$$\left[34.9689\,\text{amu} \times \frac{1.66054 \times 10^{-27}\,\text{kg}}{1\,\text{amu}} \right] = 5.3151 \times 10^{-28}\,\text{kg}$$

$$\Delta E = \left(\Delta m \right) c^2 = 5.3151 \times 10^{-28}\,\text{kg} \times \left(2.998 \times 10^{8}\,\text{m/s} \right)^2$$

$$= 4.7772 \times 10^{-11}\,\text{J}$$

In kilojoules per mole, that energy is

$$\Delta E = \frac{4.7772 \times 10^{-11}\,\text{J}}{\text{nuclide}} \times \frac{6.022 \times 10^{23}\,\text{nuclides}}{\text{mol}} \times \frac{1\,\text{kJ}}{1000\,\text{J}} = 2.8768 \times 10^{10}\,\text{kJ/mol}$$

The binding energy of ^{37}Cl is

$$\Delta m = \left[\left(17 \times 1.67262 \times 10^{-27}\,\text{kg} \right) + \left(20 \times 1.67493 \times 10^{-27}\,\text{kg} \right) + \left(17 \times 9.10938 \times 10^{-31}\,\text{kg} \right) \right] -$$

$$\left[36.9659\,\text{amu} \times \frac{1.66054 \times 10^{-27}\,\text{kg}}{1\,\text{amu}} \right] = 5.6527 \times 10^{-28}\,\text{kg}$$

$$\Delta E = \left(\Delta m \right) c^2 = 5.6527 \times 10^{-28}\,\text{kg} \times \left(2.998 \times 10^{8}\,\text{m/s} \right)^2$$

$$= 5.0807 \times 10^{-11}\,\text{J}$$

In kilojoules per mole, that energy is

$$\Delta E = \frac{5.0807 \times 10^{-11}\,\text{J}}{\text{nuclide}} \times \frac{6.022 \times 10^{23}\,\text{nuclides}}{\text{mol}} \times \frac{1\,\text{kJ}}{1000\,\text{J}} = 3.0596 \times 10^{10}\,\text{kJ/mol}$$

Think About It

The binding energy *per nucleon* could also be determined by dividing ΔE for each nuclide by the total number of nucleons per nuclide (35 for ^{35}Cl and 37 for ^{37}Cl).

21.51. **Collect and Organize**
We are asked to write balanced nuclear equations to describe the bombardment of ^{209}Bi to form ^{211}At.

Analyze
Bismuth and astatine differ in atomic number by 2. The mass numbers for the two isotopes for this problem differ only by 2. Therefore, an appropriate particle with which to bombard ^{209}Bi is the α particle with the emission of two neutrons.

Solve

$$^{209}_{83}\text{Bi} + ^{4}_{2}\alpha \rightarrow ^{211}_{85}\text{At} + 2\,^{1}_{0}\text{n}$$

Think About It
The emission of neutrons does not change the atomic number, only the mass number.

21.53. **Collect and Organize**
For the nuclear reactions used to prepare some isotopes used in medicine, we are to complete them by supplying the missing nuclide or nuclear particle.

Analyze
For those reactions to balance, the sum of the mass numbers of the reactants must equal that of the products. Likewise, the sum of the atomic numbers of the reactants must equal that of the products.

Solve

(a) $^{197}_{79}\text{Au} + 2\,^{1}_{0}\text{n} \rightarrow ^{199}_{80}\text{Hg} + \underline{^{0}_{-1}\beta}$

(b) $^{64}_{28}\text{Ni} + ^{1}_{1}\text{H} \rightarrow ^{64}_{29}\text{Cu} + \underline{^{1}_{0}\text{n}}$

(c) $^{63}_{29}\text{Cu} + ^{4}_{2}\alpha \rightarrow ^{66}_{31}\text{Ga} + \underline{^{1}_{0}\text{n}}$

(d) $^{67}_{30}\text{Zn} + ^{1}_{0}\text{n} \rightarrow ^{67}_{29}\text{Cu} + \underline{^{1}_{1}\text{H}}$

Think About It
All those nuclides are produced through the collision of light particles (neutron or hydrogen) with nuclides of larger mass.

21.55. **Collect and Organize**
For the nuclear reactions given, we are to complete them by supplying the missing nuclide or nuclear particle.

Analyze
For those reactions to balance, the sum of the mass numbers of the reactants must equal that of the products. Likewise, the sum of the atomic numbers of the reactants must equal that of the products.

Solve

(a) $^{131}_{52}\text{Te} \rightarrow ^{131}_{53}\text{I} + \underline{^{0}_{-1}\beta}$

(b) $^{122}_{53}\text{I} \rightarrow ^{122}_{54}\text{Xe} + \underline{^{0}_{-1}\beta}$

(c) $^{10}_{5}\text{B} + ^{4}_{2}\text{He} \rightarrow ^{13}_{7}\text{N} + \underline{^{1}_{0}\text{n}}$

(d) $\underline{^{68}_{30}\text{Zn}} + ^{1}_{1}\text{H} \rightarrow ^{67}_{31}\text{Ga} + 2\,^{1}_{0}\text{n}$

Think About It
Adding a neutron to a nucleus, as in reaction c, does not change the atomic number, only the mass number.

21.57. Collect and Organize
Given a list of nuclear particles, we are to arrange them in order of increasing mass.

Analyze
For this problem we need to recognize that a β particle has the same mass and charge as an electron, that a positron has the same mass as but opposite charge of the electron, that a neutron is slightly more massive than a proton, that an α particle is equivalent in mass to a helium nucleus with two protons and two neutrons, and that a deuteron consists of one proton and one neutron.

Solve
In order of increasing mass: electron = β particle = positron < proton < neutron < deuteron < α particle.

Think About It
A positron is an antielectron. When it encounters an electron, both are immediately annihilated.

21.59. Collect and Organize
We are to describe how antihydrogen differs from hydrogen.

Analyze
An antiparticle has the same mass but the opposite charge of its partner particle.

Solve
Antihydrogen has the same mass as hydrogen, but its nucleus has a negative charge with a positively charged electron. It contains the antiproton in the nucleus and a positron in place of the electron in the $1s$ orbital.

Think About It
Antihydrogen is immediately destroyed when it encounters the walls of the reactor in which it is generated.

21.61. Collect and Organize
We are asked to describe how the formation of α particles in fusion in primordial synthesis differs from that in the sun.

Analyze
In the sun, the processes to make α particles rely on hydrogen as fuel. In primordial synthesis, protons and neutrons were the fuel.

Solve
In primordial synthesis, a proton and neutron fused to form deuterium that then fused with another deuterium to give the α particle. The sun's fusion process involves more steps. With a lower concentration of neutrons, it relies ultimately on the fusion of four protons (hydrogen) to give the α particle and two positrons.

Think About It
Accomplished either way, the production of α particles through fusion releases huge amounts of energy.

21.63. Collect, Organize, and Analyze
We are asked to identify the ions of the most abundant element after hydrogen that flow out of the sun on the solar wind.

Solve
Our sun is composed mostly of hydrogen burning to make helium. Therefore, helium is the next-most-abundant element in the solar wind after hydrogen.

Think About It

The sun burns hydrogen to make helium with the release of energy through the following overall fusion reaction:

$$4\,{}_1^1\text{H} \rightarrow {}_2^4\text{He} + 2\,{}_{+1}^0\beta$$

21.65. Collect, Organize, and Analyze

We are asked to explain the presence of carbon in our star's core despite the sun's not being hot enough to support the triple-alpha process (Figure 21.11).

Solve

The carbon must have come from the explosion of other stars in the distant past that were hot enough to nucleosynthesize carbon.

Think About It

Elements through iron can be synthesized through fusion on stars if they are hot enough, but elements beyond iron must be made during supernovae.

21.67. Collect and Organize

For the annihilation of a proton and an antiproton when they collide, we can use Einstein's relation between mass and energy to calculate the energy released and the relationship between energy and wavelength to calculate the wavelength of the gamma rays emitted from this process.

Analyze

We first have to find Δm for the annihilation. Because no particle mass is left over after the collision, Δm is the sum of the mass of the proton and the antiproton. That is simply twice the mass of the proton since the mass of the proton equals the mass of the antiproton. We can then use Einstein's equation to calculate ΔE. The wavelength of the gamma ray emitted is found by the relationship

$$\lambda = \frac{hc}{E}$$

Solve

$$\Delta m = 2 \times 1.67262 \times 10^{-24}\ \text{g} \times \frac{1\ \text{kg}}{1000\ \text{g}} = 3.34524 \times 10^{-27}\ \text{kg}$$

$$\Delta E = 3.34524 \times 10^{-27}\ \text{kg} \times \left(2.998 \times 10^8\ \text{m/s}\right)^2 = 3.007 \times 10^{-10}\ \text{J}$$

$$\lambda = \frac{hc}{E} = \frac{6.626 \times 10^{-34}\ \text{J} \cdot \text{s} \times 2.998 \times 10^8\ \text{m/s}}{3.007 \times 10^{-10}\ \text{J/2 photons}} = 1.321 \times 10^{-15}\ \text{m, or } 1.321 \times 10^{-6}\ \text{nm}$$

Think About It

If a mole of protons collided with a mole of antiprotons, the energy released would be

$$3.007 \times 10^{-10}\ \text{J/photon} \times \frac{6.022 \times 10^{23}}{1\ \text{mol}} = 1.849 \times 10^{14}\ \text{J/mol}$$

21.69. Collect and Organize

For the four fusion reactions given, we can use Einstein's equation to calculate the energy released in each.

Analyze

In the equation

$$\Delta E = \Delta mc^2$$

c is the speed of light and Δm is the mass defect. The mass defect is the mass lost in the reaction and must be expressed in kilograms. Because the masses in this problem are expressed in atomic mass units (amu), we need to use the conversion factor

$$1\ \text{amu} = 1.6605402 \times 10^{-27}\ \text{kg}$$

Solve

(a) For the production of ^{28}Si from ^{14}N + ^{14}N:

$$\Delta m = 27.97693 \text{ amu} - (2 \times 14.00307 \text{ amu}) = -0.02921 \text{ amu}$$

$$0.02921 \text{ amu} \times \frac{1.6605402 \times 10^{-27} \text{ kg}}{1 \text{ amu}} = 4.8504379 \times 10^{-29} \text{ kg}$$

$$\Delta E = 4.8504379 \times 10^{-29} \text{ kg} \times (3.00 \times 10^8 \text{ m/s})^2 = 4.37 \times 10^{-12} \text{ J released}$$

(b) For the production of ^{28}Si from ^{10}B + ^{16}O + ^{2}H:

$$\Delta m = 27.97693 \text{ amu} - (10.0129 \text{ amu} + 15.99491 \text{ amu} + 2.0146 \text{ amu}) = -0.04548 \text{ amu}$$

$$0.04548 \text{ amu} \times \frac{1.6605402 \times 10^{-27} \text{ kg}}{1 \text{ amu}} = 7.552 \times 10^{-29} \text{ kg}$$

$$\Delta E = 7.552 \times 10^{-29} \text{ kg} \times (3.00 \times 10^8 \text{ m/s})^2 = 6.80 \times 10^{-12} \text{ J released}$$

(c) For the production of ^{28}Si from ^{16}O + ^{12}C:

$$\Delta m = 27.97693 \text{ amu} - (15.994915 \text{ amu} + 12.000 \text{ amu}) = -0.01798 \text{ amu}$$

$$0.01798 \text{ amu} \times \frac{1.6605402 \times 10^{-27} \text{ kg}}{1 \text{ amu}} = 2.98565 \times 10^{-29} \text{ kg}$$

$$\Delta E = 2.98565 \times 10^{-29} \text{ kg} \times (3.00 \times 10^8 \text{ m/s})^2 = 2.69 \times 10^{-12} \text{ J released}$$

(d) For the production of ^{28}Si from ^{24}Mg + ^{4}He:

$$\Delta m = 27.97693 \text{ amu} - (23.98504 \text{ amu} + 4.00260 \text{ amu}) = -0.01071 \text{ amu}$$

$$0.01071 \text{ amu} \times \frac{1.6605402 \times 10^{-27} \text{ kg}}{1 \text{ amu}} = 1.77844 \times 10^{-29} \text{ kg}$$

$$\Delta E = 1.77843 \times 10^{-29} \text{ kg} \times (3.00 \times 10^8 \text{ m/s})^2 = 1.60 \times 10^{-12} \text{ J released}$$

Think About It

Those reactions all release energy (a negative mass defect occurs in proceeding from reactants to products). In solving for the *energy released*, we have used the positive value of the mass defect.

21.71. Collect and Organize

We are to calculate the change in energy for the fusion reaction between a neutron and a lithium-6 nucleus to form helium and tritium.

Analyze

In the equation

$$\Delta E = \Delta mc^2$$

c is the speed of light and Δm is the mass defect. The mass defect is the mass lost in the reaction and must be expressed in kilograms. The exact masses of tritium and lithium-6 are given as 5.00827×10^{-27} kg and 9.98841×10^{-27} kg, respectively. The exact mass of the neutron and of helium are given in the textbook (Appendices 2 and 3) as 1.67493×10^{-27} kg and 4.002603 amu, respectively.

Solve

The mass of the helium nucleus must be converted to kilograms, and we must remove the mass of the two electrons.

$$\text{mass of } ^4_2\text{He} = 4.002603 \text{ amu} \times \frac{1.66054 \times 10^{-27} \text{ kg}}{\text{amu}} - (2 \times 9.10938 \times 10^{-31} \text{ kg}) = 6.64466 \times 10^{-27} \text{ kg}$$

The mass and energy change released in the nuclear reaction to produce tritium and helium from the collision of a neutron with lithium-6 is

$$\Delta m = \left[6.64466 \times 10^{-27}\,\text{kg} + 5.00827 \times 10^{-27}\,\text{kg}\right] - \left[9.98841 \times 10^{-27}\,\text{kg} + 1.67493 \times 10^{-27}\,\text{kg}\right]$$

$$= -1.04100 \times 10^{-29}\,\text{kg}$$

$$\Delta E = \left(-1.04100 \times 10^{-29}\,\text{kg}\right) \times \left(2.998 \times 10^{8}\,\text{m/s}\right)^{2} = -9.357 \times 10^{-13}\,\text{J released}$$

Think About It

In kilojoules per mole, that is 5.634×10^{8} kJ/mol.

21.73. Collect and Organize

For the fusion reactions in the core of a giant star, we are to predict the nuclide formed.

Analyze

We are to assume that each reaction has only one product. For those reactions to balance, the sum of the mass numbers of the reactants must equal that of the product. Likewise, the sum of the atomic numbers of the reactants must equal that of the product.

Solve

(a) $^{12}_{6}\text{C} + ^{4}_{2}\text{He} \rightarrow ^{16}_{8}\text{O}$

(b) $^{20}_{10}\text{Ne} + ^{4}_{2}\text{He} \rightarrow ^{24}_{12}\text{Mg}$

(c) $^{32}_{16}\text{S} + ^{4}_{2}\text{He} \rightarrow ^{36}_{18}\text{Ar}$

Think About It

Each of those fusion reactions with helium resulted in an increase of atomic number of 2.

21.75. Collect and Organize

For the fusion reactions in the core of a collapsing giant star, we are to predict the nuclide formed.

Analyze

For those reactions to balance, the sum of the mass numbers of the reactants must equal that of the product. Likewise, the sum of the atomic numbers of the reactants must equal that of the product.

Solve

(a) $^{96}_{42}\text{Mo} + 3\,^{1}_{0}\text{n} \rightarrow \underline{^{99}_{43}\text{Tc}} + \underline{^{0}_{-1}\beta}$

(b) $^{118}_{50}\text{Sn} + 3\,^{1}_{0}\text{n} \rightarrow \underline{^{121}_{51}\text{Sb}} + \underline{^{0}_{-1}\beta}$

(c) $^{108}_{47}\text{Ag} + ^{1}_{0}\text{n} \rightarrow \underline{^{109}_{48}\text{Cd}} + \underline{^{0}_{-1}\beta}$

Think About It

Each of those fusion reactions with neutrons resulted in an increase of atomic number of 1 with a release of a beta particle.

21.77. Collect and Organize

We are asked to describe how the rate of energy release is controlled in nuclear reactors.

Analyze

Nuclear reactions that power the reactors release neutrons, which promote more nuclear fission processes, as shown in Figure 21.14. The reaction can be controlled by absorbing some of the neutrons.

Solve

Control rods made of boron or cadmium are used to absorb the excess neutrons to control the rate of energy release in a nuclear reactor.

Think About It
When the control rods are removed, the reactor core may go critical. That is what occurred in the reactor at Chernobyl in the Soviet Union on April 26, 1986.

21.79. Collect and Organize
Figure 21.1 shows the belt of stability of the radionuclides. Using that, we are to explain why neutrons are by-products in fission reactions.

Analyze
In a fission reaction a heavier, unstable nucleus splits into two lighter nuclei.

Solve
The neutron-to-proton ratio for heavy nuclei is high, and when the nuclide undergoes fission to form smaller nuclides, it must emit neutrons because the fission products require a lower neutron-to-proton ratio for stability.

Think About It
In fusing nuclei, more neutrons are needed for the heavier nuclei, and so we might expect that β decay often accompanies those processes.

21.81. Collect and Organize
For the incomplete fission reactions given, we are to determine the missing nuclides.

Analyze
Before we can balance those reactions, the sum of the mass numbers of the reactants must equal that of the products. Likewise, the sum of the atomic numbers of the reactants must equal that of the products.

Solve
(a) $^{235}_{92}U + ^{1}_{0}n \rightarrow ^{96}_{40}Zr + ^{138}_{52}Te + 2\,^{1}_{0}n$

(b) $^{235}_{92}U + ^{1}_{0}n \rightarrow ^{99}_{41}Nb + ^{133}_{51}Sb + 4\,^{1}_{0}n$

(c) $^{235}_{92}U + ^{1}_{0}n \rightarrow ^{90}_{37}Rb + ^{143}_{55}Cs + 3\,^{1}_{0}n$

Think About It
All those fission products (^{96}Zr, ^{138}Te, ^{99}Nb, ^{133}Sb, ^{90}Rb, and ^{143}Cs) come from the fission of ^{235}U.

21.83. Collect and Organize
For the incomplete fission reactions given, we are to determine the missing nuclides.

Analyze
Before we can balance these reactions, the sum of the mass numbers of the reactants must equal that of the products. Likewise, the sum of the atomic numbers of the reactants must equal that of the products.

Solve
(a) $^{235}_{92}U + ^{1}_{0}n \rightarrow ^{131}_{53}I + ^{103}_{39}Y + 2\,^{1}_{0}n$

(b) $^{235}_{92}U + ^{1}_{0}n \rightarrow ^{103}_{44}Ru + ^{130}_{48}Cd + 3\,^{1}_{0}n$

(c) $^{235}_{92}U + ^{1}_{0}n \rightarrow ^{95}_{40}Zr + ^{138}_{52}Te + 3\,^{1}_{0}n$

Think About It
All those fission products (^{131}I, ^{103}Y, ^{103}Ru, ^{130}Cd, ^{95}Zr, and ^{138}Te) come from the fission of ^{235}U.

21.85. **Collect, Organize, and Analyze**

We can use the definitions of the terms to describe the difference between the *level* of radioactivity and the *dose* of radioactivity.

Solve

The level of radioactivity is the amount of radioactive particles present in a given instant of time. The dose is the accumulation of exposure over a length of time.

Think About It

A person can get a high dose of radioactivity either from a brief exposure to a highly intense radioactive source or through prolonged exposure to low levels of radiation.

21.87. **Collect and Organize**

We are to describe the dangers of ^{222}Rn.

Analyze

Radon is a colorless, odorless gas that results from the natural decay of uranium in the earth. It is an α emitter that decays to ^{218}Po, also an α emitter.

Solve

When radon-222 decays to polonium-218 while in the lungs, the ^{218}Po, a reactive solid chemically similar to oxygen, lodges in the lung tissue, where it continues to emit α radiation. Alpha radiation is one of the most damaging kinds of radiation when in contact with biological tissues. The result of exposure to high levels of radon is an increased risk of lung cancer.

Think About It

Most hardware stores sell radon detection kits for homeowners to learn whether the level of radon-222 in their homes is unusually high.

21.89. **Collect and Organize**

We are asked to calculate the grays that are equal to 5 μSv and how much energy 5 μSv corresponds to for a 50 kg person.

Analyze

From Table 21.5, we see that

$$1 \text{ Sv} = 1 \text{ Gy} \times \text{RBE}$$

The relative biological effectiveness (RBE) of X-rays is given in the problem as 1, so for dental X-rays

$$1 \text{ Sv} = 1 \text{ Gy} \times 1, \text{ or } 1 \text{ μSv} = 1 \text{ μGy} \times 1$$

Also from Table 21.5, we see that

$$1 \text{ Gy} = 1 \text{ J/kg of tissue mass, or } 1 \text{ μGy} = 1 \text{ μJ/kg}$$

Solve

$$5 \text{ μSv} = 5 \text{ μGy}$$

$$5 \text{ μGy} \times \frac{1 \text{ μJ/kg}}{1 \text{ μGy}} \times 50 \text{ kg} = 250 \text{ μJ}$$

Think About It

This dosage of 5 μSv is well below the dose that may cause toxic effects (Table 21.5).

21.91. **Collect and Organize**

For the radioactive isotope ^{90}Sr, we are to write a balanced nuclear equation corresponding to its β decay, calculate the atoms of ^{90}Sr in 200 mL of milk that has 1.25 Bq/L of ^{90}Sr radioactivity, and give a reason why ^{90}Sr would be more concentrated in milk than in other foods.

Analyze

(a) Beta decay increases the atomic number, leaving the mass number unchanged.

(b) To calculate the atoms of ^{90}Sr in the milk, we can use the equation

$$\text{milliliters of milk} \times \frac{1.25\ \text{Bq}}{1000\ \text{mL}} \times \frac{1\ \text{disintegration/s}}{1\ \text{Bq}} = \text{disintegrations per second}$$

That is the rate of the first-order decay of ^{90}Sr that follows the rate law

$$\text{Rate} = k[^{90}\text{Sr}]$$

where $k = 0.693/t_{1/2}$. Given that the $t_{1/2}$ for ^{90}Sr is 28.8 yr, we need to convert from years to seconds.

Solve

(a) $^{90}_{38}\text{Sr} \rightarrow ^{0}_{-1}\beta + ^{90}_{39}\text{Y}$

(b) The number of disintegrations in 200 mL of milk is

$$200\ \text{mL} \times \frac{1.25\ \text{Bq}}{1000\ \text{mL}} \times \frac{1\ \text{disintegration/s}}{1\ \text{Bq}} = 0.250\ \text{disintegration/s}$$

The rate constant k in reciprocal seconds is

$$k = \frac{0.693}{28.8\ \text{yr}} \times \frac{1\ \text{yr}}{365\ \text{d}} \times \frac{1\ \text{d}}{24\ \text{hr}} \times \frac{1\ \text{hr}}{60\ \text{min}} \times \frac{1\ \text{min}}{60\ \text{s}} = 7.63 \times 10^{-10}\ \text{s}^{-1}$$

The concentration of ^{90}Sr in the milk from the first-order rate law is

$$\left[^{90}\text{Sr}\right] = \frac{0.250\ \text{disintegration/s}}{7.63 \times 10^{-10}\ \text{s}^{-1}} = 3.28 \times 10^{8}\ ^{90}\text{Sr atoms}$$

(c) Strontium-90 is found in milk and not other foods because it is chemically similar to calcium, and milk is rich in calcium.

Think About It

In more familiar chemical concentration terms, the concentration of ^{90}Sr in these samples is

$$3.28 \times 10^{8}\ ^{90}\text{Sr atoms} \times \frac{1\ \text{mol}}{6.022 \times 10^{23}\ \text{atoms}} \times \frac{1}{0.200\ \text{L}} = 2.72 \times 10^{-15}\ M$$

21.93. Collect and Organize

For drinking water we are to calculate the number of decay events per second in 1.0 mL with a radon level of 4.0 pCi/mL. We are then to calculate the number of Rn atoms in 1.0 mL given that $t_{1/2}$ of ^{222}Rn = 3.8 d.

Analyze

(a) To calculate the number of decay events per second from picocuries, we need the conversions

$$1\ \text{pCi} = 1 \times 10^{-12}\ \text{Ci}$$
$$1\ \text{Ci} = 3.70 \times 10^{10}\ \text{Bq}$$
$$1\ \text{Bq} = 1\ \text{disintegration/s}$$

(b) The decay of ^{222}Rn follows the first-order rate law

$$\text{Rate} = k[^{222}\text{Rn}]$$

where the rate is the number of decay events per second and k, the rate constant, is

$$k = \frac{0.693}{t_{1/2}}$$

Here $t_{1/2} = 3.8$ d, which must be converted to seconds.

Solve

(a) $4.0\ \text{pCi} \times \dfrac{1 \times 10^{-12}\ \text{Ci}}{1\ \text{pCi}} \times \dfrac{3.70 \times 10^{10}\ \text{Bq}}{1\ \text{Ci}} \times \dfrac{1\ \text{decay/s}}{1\ \text{Bq}} = 0.15\ \dfrac{\text{decay}}{\text{s}}$

(b) $[^{222}\text{Rn}] = \dfrac{0.148\ \text{decay/s}}{2.11 \times 10^{-6}\ \text{s}^{-1}} = 7.0 \times 10^{4}\ \text{atoms}$

Think About It

Even though that seems like a large number of ^{222}Rn atoms, the percentage of ^{222}Rn atoms in 1.0 mL of water is very low:

$$1.0 \text{ mL} \times \frac{1 \text{ g}}{\text{mL}} \times \frac{1 \text{ mol}}{18 \text{ g}} \times \frac{6.022 \times 10^{23} \text{ H}_2\text{O molecules}}{\text{mole}} = 3.35 \times 10^{22} \text{ molecules of H}_2\text{O}$$

$$\% \ ^{222}\text{Rn atoms} = \frac{7.0 \times 10^4 \text{ atoms of }^{222}\text{Rn}}{3.35 \times 10^{22} \text{ molecules of H}_2\text{O}} \times 100$$

$$= 2.1 \times 10^{-16} \ \%$$

21.95. Collect, Organize, and Analyze

We are asked to consider how a radioactive isotope for radiotherapy is selected on the basis of its half-life, decay mode, and properties of its products.

Solve

(a) The half-life should be long enough to effect treatment of the cancerous cells but not so long as to damage healthy tissues.

(b) Because α radiation does not penetrate far beyond a tumor, the α decay mode is best.

(c) Products should be nonradioactive, if possible, or have short half-lives and be able to be flushed from the body by normal cellular and biological processes.

Think About It

Researchers continue to investigate and develop new radioisotopes.

21.97. Collect and Organize

For each isotope given, we can use the belt of stability (Figure 21.2) to predict the mode of decay.

Analyze

If the nuclide has a neutron-to-proton ratio that places it below the belt of stability, either electron capture or positron emission is likely. If a nuclide lies above the belt of stability, β emission is likely.

Solve

(a) $^{197}_{80}$Hg has 80 protons and 117 neutrons. That nuclide lies below the belt of stability, so it is likely to decay by positron emission or electron capture.

(b) $^{75}_{34}$Se has 34 protons and 41 neutrons. That nuclide lies below the belt of stability, so it is likely to decay by positron emission or electron capture.

(c) $^{18}_{9}$F has nine protons and nine neutrons. That nuclide lies below the belt of stability, so it is likely to decay by positron emission or electron capture.

Think About It

Those imaging agents have relatively short half-lives: ^{197}Hg, 64 h; ^{75}Se, 120 d; ^{18}F, 110 min.

21.99. Collect and Organize

For a 1.00 mg sample of ^{192}Ir, 0.756 mg remains after 30 days. From that information, we are to calculate the half-life of ^{192}Ir.

Analyze

We need to rearrange Equation 21.10 to solve this problem:

$$t = -\frac{t_{1/2}}{0.693} \ln \frac{N_t}{N_0}$$

$$t_{1/2} = -\frac{0.693t}{\ln\left(N_t/N_0\right)}$$

Solve

$$t_{1/2} = -\frac{0.693 \times 30\text{ d}}{\ln(0.756\text{ mg}/1.00\text{ mg})} = 74.3\text{ d}$$

Think About It

After 30 days, then, that isotope has decayed through not even one half-life (Equation 21.10):

$$\frac{0.756}{1.00} = 0.5^n$$

$$0.756 = 0.5^n$$

$$n = 0.404\text{ half-lives}$$

21.101. Collect and Organize

Using the information given about the half-life of ^{131}I ($t_{1/2}$ = 8.1 d) and the initial and residual activity of the isotope after 30 days (108 and 4.1 counts/min, respectively), we are to determine whether the brain cells took up any of the ^{131}I.

Analyze

We need to calculate the expected activity of the ^{131}I if it were not taken up by the cells. We find that by using Equation 21.10 to solve for N_t.

$$t = -\frac{t_{1/2}}{0.693} \ln \frac{N_t}{N_0}$$

If the activity of the sample after 30 days is less than the calculated activity, the cell did take up ^{131}I.

Solve

$$30\text{ days} = -\frac{8.1\text{ d}}{0.693} \ln \frac{N_t}{108}$$

$$-2.567 = \ln \frac{N_t}{108}$$

$$0.07679 = \frac{N_t}{108}$$

$$N_t = 8.29\text{ counts/min}$$

Because that is more than the counts per minute in the sample, yes, the brain cells must have incorporated some of the ^{131}I.

Think About It

The sample was taken after the ^{131}I had decayed through 3.7 half-lives:

$$\frac{8.29}{108} = 0.5^n$$

$$0.0768 = 0.5^n$$

$$n = 3.7\text{ half-lives}$$

21.103. Collect and Organize

Using the half-life of ^{105}Rh ($t_{1/2}$ = 35.4 h), we are to determine what percentage of a shipment of 250 mg remains after an overnight shipment lasting 12 h and calculate how long it will take for 95% of the ^{109}Rh to decay.

Analyze

(a) To determine the percentage of the shipment remaining, we need to rearrange Equation 21.10 to solve for N_t/N_0:

$$t = -\frac{t_{1/2}}{0.693} \ln \frac{N_t}{N_0}$$

$$\frac{-0.693t}{t_{1/2}} = \ln \frac{N_t}{N_0}$$

$$\frac{N_t}{N_0} = e^{-0.693t/t_{1/2}}$$

(b) To calculate the time for 95% of the ^{105}Rh to decay, we can use

$$t = -\frac{t_{1/2}}{0.693} \ln \frac{N_t}{N_0}$$

where $N_t/N_0 = 5/100$.

Solve

(a) The amount in milligrams and the percentage of the shipment remaining is

$$\frac{N_t}{250 \text{ mg}} = e^{-0.693 \times 12 \text{ h}/35.4 \text{ h}} = 0.791, \text{ or } 79.1\%$$

(b) The time for 95% of the ^{105}Rh to decay is

$$t = -\frac{35.4 \text{ h}}{0.693} \ln \frac{5}{100} = 153 \text{ h}$$

Think About It

We did not need to know the starting amount of ^{109}Rh to solve either question for this problem.

21.105. **Collect and Organize**

For the decay of ^{11}B formed during boron neutron-capture therapy (BNCT) for cancers and which decays to ^{7}Li, we are to write balanced nuclear equations for the neutron absorption and α decay; then calculate the energy released in the process by using Einstein's equation; and finally consider why this process, in which an α emitter is produced by neutron capture, is an effective cancer treatment.

Analyze

(a) In writing the balanced equation we need to balance the sum of the mass numbers of the products with that of the reactants. Likewise, the sum of the atomic numbers of the products must equal that of the reactants.

(b) In the Einstein equation

$$\Delta E = \Delta mc^2$$

c is the speed of light and Δm is the mass defect. The mass defect is the mass lost in the reaction and must be expressed in kilograms. Because the masses in this problem are expressed in atomic mass units, we need to use the conversion factor

$$1 \text{ amu} = 1.6605402 \times 10^{-27} \text{ kg}$$

Solve

(a) $^{10}_{5}\text{B} + ^{1}_{0}\text{n} \rightarrow ^{11}_{5}\text{B}$

$^{11}_{5}\text{B} \rightarrow ^{7}_{3}\text{Li} + ^{4}_{2}\alpha$

Overall process: $^{10}_{5}\text{B} + ^{1}_{0}\text{n} \rightarrow ^{7}_{3}\text{Li} + ^{4}_{2}\alpha$

(b) $\Delta m = (7.01600 \text{ amu} + 4.00260) - (10.0129 \text{ amu} + 1.008665 \text{ amu}) = -0.002965 \text{ amu}$

$$0.002965 \text{ amu} \times \frac{1.6605402 \times 10^{-27} \text{ kg}}{1 \text{ amu}} = 4.9235 \times 10^{-30} \text{ kg}$$

$$\Delta E = 4.9235 \times 10^{-30} \text{ kg} \times (3.00 \times 10^8 \text{ m/s})^2 = 4.43 \times 10^{-13} \text{ J}$$

(c) Alpha particles have a high RBE, and they do not penetrate into healthy tissue if the radionuclide is placed inside a tumor.

Think About It
BNCT is especially useful in the treatment of brain tumors. However, the only source of neutrons is from nuclear reactors, and only a few sites worldwide have that treatment available.

21.107. Collect and Organize
We are asked to explain why antihydrogen would have been a suitable fuel for the starship *Enterprise* and to describe the challenges to storing such a fuel.

Analyze
In Problem 21.67 we calculated that a large amount of energy (1.813×10^{14} J) is released when 1 mol of hydrogen collides with 1 mol of antihydrogen (the proton and antiproton).

Solve
(a) Besides releasing a large amount of energy to power the starship *Enterprise*, hydrogen is an abundant fuel in the universe and therefore could easily react with any antihydrogen produced.
(b) Antimatter of any sort will react with matter, so storing it in an ordinary container made of ordinary matter is impossible. It might be contained within a magnetic or energy field, however.

Think About It
Antihelium, although much more difficult to produce, might also have been a good choice.

21.109. Collect and Organize
We are asked to verify that ^{8}Be nuclei require no energy to decompose into ^{4}He nuclei by calculating the binding energy of ^{8}Be and comparing it with the binding energy of ^{4}He.

Analyze
The binding energy can be calculated using Einstein's equation and the mass defect. The mass defect is found by adding up the masses of the individual protons, neutrons, and electrons and subtracting that result from the actual mass of each nuclide. The actual masses of ^{8}Be and ^{4}He (looking them up in Appendix 2 and on the Internet) are 8.00530510 and 4.002603 amu, respectively. The pertinent nuclear equation for the decomposition of ^{8}Be is

$$_4^8\text{Be} \rightarrow 2\,_2^4\text{He}$$

Solve
The binding energy of ^{8}Be is

$$\Delta m = \left[\left(4 \times 1.67262 \times 10^{-27}\,\text{kg}\right) + \left(4 \times 1.67493 \times 10^{-27}\,\text{kg}\right) + \left(4 \times 9.10938 \times 10^{-31}\,\text{kg}\right)\right] -$$

$$\left[8.00530510\,\text{amu} \times \frac{1.66054 \times 10^{-27}\,\text{kg}}{1\,\text{amu}}\right] = 1.00714 \times 10^{-28}\,\text{kg}$$

$$\Delta E = (\Delta m)c^2 = 1.00714 \times 10^{-28}\,\text{kg} \times \left(2.998 \times 10^8\,\text{m/s}\right)^2$$

$$= 9.0522 \times 10^{-12}\,\text{J}$$

The binding energy of ^{4}He

$$\Delta m = \left[\left(2 \times 1.67262 \times 10^{-27}\,\text{kg}\right) + \left(2 \times 1.67493 \times 10^{-27}\,\text{kg}\right) + \left(2 \times 9.10938 \times 10^{-31}\,\text{kg}\right)\right] -$$

$$\left[4.002603\,\text{amu} \times \frac{1.66054 \times 10^{-27}\,\text{kg}}{1\,\text{amu}}\right] = 5.04395 \times 10^{-28}\,\text{kg}$$

$$\Delta E = (\Delta m)c^2 = 5.04395 \times 10^{-28}\,\text{kg} \times \left(2.998 \times 10^8\,\text{m/s}\right)^2$$

$$= 4.5335 \times 10^{-12}\,\text{J}$$

That is for one ^{4}He nucleus; for two ^{4}He produced $\Delta E = 2 \times (4.5335 \times 10^{-12}\,\text{J}) = 9.0670110^{-12}\,\text{J}$.
The binding energy for two ^{4}He nuclei is close to the binding energy (and is slightly greater, indicating that ^{4}He is more stable) of ^{8}Be.

Think About It
Beryllium-8 is produced in the universe through spallation, in which larger atoms collide with cosmic rays.

21.111. Collect and Organize
We consider the ^{241}Am isotope in smoke detectors, which decays by α emission.

Analyze
(a) A Geiger counter functions by conducting electricity when ions are in the gas owing to the presence of ionizing radiation. A scintillation counter detects radiation through the detection of light caused by the fluorescence of the material when ionizing radiation strikes it.
(b) To calculate the time it takes for the activity of ^{241}Am to decay to 1% of its original activity, we can use a rearrangement of Equation 21.8:

$$t = -\frac{t_{1/2}}{0.693} \ln \frac{N_t}{N_0}$$

Solve
(a) Because the ^{241}Am ionizes the air and the smoke detector registers a change in current, a smoke detector resembles a Geiger counter in its operation.
(b)

$$t = -\frac{433 \text{ yr}}{0.693} \ln \frac{1}{100} = 2877 \text{ yr}$$

(c) Smoke detectors are safe to handle because the ^{241}Am is an α emitter, and α particles do not travel more than a few inches in air and cannot penetrate the first layer of skin.

Think About It
Even though the half-life of ^{241}Am is long, the useful life of a smoke detector is 10 years.

21.113. Collect and Organize
For the synthesis of ^{294}Og, we are to describe its decay reactions with balanced equations and, by looking at its position in the periodic table, select another element that has similar properties to Og.

Analyze
To balance those reactions, the sum of the mass numbers of the reactants must equal that of the products. Likewise, the sum of the atomic numbers of the reactants must equal that of the products. By balancing the equations, we can identify the nuclides in parts b, c, and d of this problem.

Solve
(a) $^{249}_{98}\text{Cf} + ^{48}_{20}\text{Ca} \rightarrow ^{294}_{118}\text{Og} + 3^{1}_{0}\text{n}$

(b) $^{294}_{118}\text{Og} \rightarrow ^{4}_{2}\alpha + ^{290}_{116}\text{Lv}$

(c) $^{290}_{116}\text{Lv} \rightarrow ^{4}_{2}\alpha + ^{286}_{114}\text{Fl}$

(d) $^{286}_{114}\text{Fl} \rightarrow ^{4}_{2}\alpha + ^{282}_{112}\text{Cn}$

(e) Because ^{294}Og is a member of the noble gas family, it has chemical and physical properties similar to those of naturally occurring radon.

Think About It
Even though those superheavy elements are short-lived, their half-lives are long enough (milliseconds) to allow some of their chemical and physical properties to be experimentally determined.

21.115. Collect, Organize, and Analyze
We are asked to write Lewis structures for H_2O^+, H_3O^+, and OH.

Solve

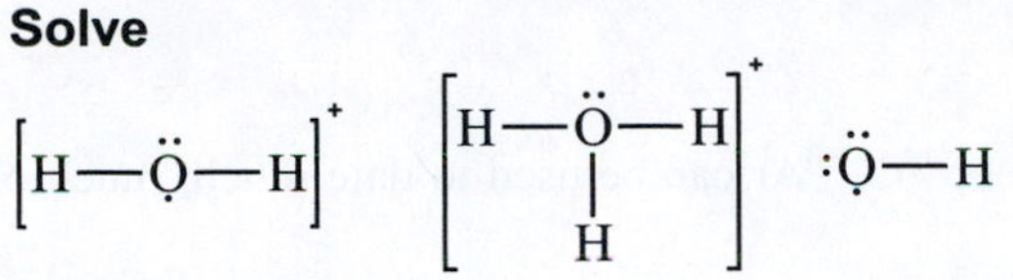

$$\left[H-\ddot{\underset{}{O}}-H \right]^{+} \qquad \left[H-\overset{\displaystyle\ddot{O}}{\underset{\displaystyle |}{\underset{\displaystyle H}{|}}}-H \right]^{+} \qquad :\ddot{\underset{}{O}}-H$$

Think About It

Both H_2O^+ and OH are highly reactive radical species.

21.117. Collect and Organize

For the synthesis of Pt by two fusion reactions, we can write the balanced nuclear equations for their formation to determine which isotopes of Pt are formed.

Analyze

Before we can balance those reactions, the sum of the mass numbers of the reactants must equal that of the products. Likewise, the sum of the atomic numbers of the reactants must equal that of the products.

Solve

(a) $\,^{64}_{28}\text{Ni} + \,^{124}_{50}\text{Sn} \rightarrow \,^{188}_{78}\text{Pt}$

(b) $\,^{64}_{28}\text{Ni} + \,^{132}_{50}\text{Sn} \rightarrow \,^{196}_{78}\text{Pt}$

Think About It

Both those isotopes of Pt lie well above the belt of stability (Figure 21.2), so we would expect that they might decay by β or α emission.

21.119. Collect and Organize

We are to determine the ratio of ^{14}C present in a bone sample that is 15,000 years old to another sample that is 25,000 years old.

Analyze

We can determine the ratio by using

$$\frac{N_t/N_0 (15{,}000 \text{ years old})}{N_t/N_0 (25{,}000 \text{ years old})} = \frac{0.5^n}{0.5^n}$$

where n for 15,000 years ago is 15,000 yr/5730 yr and n for 25,000 years ago is 25,000 yr/5730 yr.

Solve

$$\frac{N_t/N_0 (15{,}000 \text{ years ago})}{N_t/N_0 (25{,}000 \text{ years ago})} = \frac{0.5^{15{,}000/5730}}{0.5^{25{,}000/5730}} = 3.35$$

Think About It

The number of half-lives of ^{14}C that have passed for the 25,000-year-old sample is 25,000 yr/5730 yr = 4.4 half-lives.

21.121. Collect and Organize

For the dating of ancient human skulls in Ethiopia on the basis of the amount of ^{40}Ar, we are to propose a decay mechanism for that nuclide to form from ^{40}K and explain why researchers used ^{40}Ar to date the skulls and not ^{14}C.

Analyze

(a) The mass number in the decay of ^{40}K to ^{40}Ar does not change, which indicates either a β decay or positron emission.

(b) To explain why researchers used ^{40}Ar instead of ^{14}C dating, we must compare the half-lives of the parent isotopes: ^{14}C, 5730 yr; ^{40}K, 1.28×10^9 yr.

Solve

(a) $^{40}_{19}\text{K} \rightarrow \,^{40}_{18}\text{Ar} + \,^{0}_{+1}\beta$

(b) Because the half-life of ^{40}K is so much longer than that of ^{14}C, ^{40}Ar can be used to date much, much older objects.

Think About It

The amount of ^{14}C remaining after 154,000 yr would be too small to measure:

$$\frac{N_t}{N_0} = 0.5^{154,000/5730} = 8.1 \times 10^{-9}$$

21.123. Collect and Organize

Given the counts per minute over time for the decay of ^{208}Tl, we can use a first-order plot to find the value of the rate constant k and then compute the half-life of the decay.

Analyze

Radioactive decay follows first-order kinetics. A plot of ln(counts per minute) versus time gives a straight line with slope $= -k$. The half-life of the first-order decay is equal to $0.693/k$.

Solve

The first-order plot gives a straight line with slope $= -5.87 \times 10^{-3}$, so $k = 5.87 \times 10^{-3}$ s^{-1}. The half-life of the decay is

$$t_{1/2} = \frac{0.693}{5.87 \times 10^{-3}} = 118 \text{ s}$$

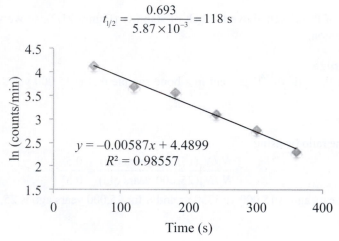

$y = -0.00587x + 4.4899$
$R^2 = 0.98557$

Think About It

If a radioactive decay is fast enough, experiments such as those described in Chapter 15 can be performed to measure with confidence the half-life of the decay.

21.125. Collect and Organize

For nuclear reactions in which ^{11}B absorbs a neutron with later decay by either α or β decay, we are to write the balanced nuclear equations.

Analyze

To write and balance those nuclear reactions, the sum of the mass numbers of the reactants must equal that of the products. Likewise, the sum of the atomic numbers of the reactants must equal that of the products. Upon absorption of a neutron, ^{11}B is converted to ^{12}B.

Solve

$$^{12}_{5}B \rightarrow \, ^{8}_{3}Li + \, ^{4}_{2}He$$

$$^{12}_{5}B \rightarrow \, ^{12}_{6}C + \, ^{0}_{-1}\beta$$

Think About It

Of the decay products in the absorption of a neutron by ^{11}B, ^{12}C is stable, whereas ^{8}Li is not stable.

21.127. Collect and Organize

We are to write balanced nuclear equations for the synthesis of ^{10}B from collisions between protons and ^{12}C and ^{14}N.

Analyze

To write and balance a nuclear reaction, the sum of the mass numbers of the reactants must equal that of the products. Likewise, the sum of the atomic numbers of the reactants must equal that of the products.

Solve

$$^{12}_{6}C + \, ^{1}_{1}p \rightarrow \, ^{10}_{5}B + \, ^{3}_{2}He$$

$$^{14}_{7}N + \, ^{1}_{1}p \rightarrow \, ^{10}_{5}B + \, ^{5}_{3}Li$$

Think About It

Boron-10 is used in radiation shielding, in thermal control rods in nuclear reactors, and in neutron capture therapy for cancer.

21.129. Collect and Organize

We are to write the balanced nuclear reaction for the formation of ^{105}Rh from the bombardment of ^{104}Rh with neutrons and, given the half-life of ^{105}Rh as 4.4 h, we are to calculate the time it will take for 99% of a sample of ^{105}Rh to decay.

Analyze

(a) To write and balance a nuclear reaction, the sum of the mass numbers of the reactants must equal that of the products. Likewise, the sum of the atomic numbers of the reactants must equal that of the products.

(b) We can use the equation from the rearrangement of Equation 21.8 to solve this problem:

$$\ln \frac{N_t}{N_0} = -\frac{0.693t}{t_{1/2}}$$

$$t = -\frac{t_{1/2}}{0.693} \ln \frac{N_t}{N_0}$$

where $N_t/N_0 = 1/100$.

Solve

(a) $\quad ^{104}_{44}Ru + \, ^{1}_{0}n \rightarrow \, ^{105}_{45}Rh + \, ^{0}_{-1}\beta$

(b) $\quad t = -\dfrac{4.4 \text{ h}}{0.693} \ln \dfrac{1}{100} = 29 \text{ h}$

Think About It

Rhodium-105 undergoes β decay to give palladium-105.

CHAPTER 22 | The Main Group Elements: Life and the Periodic Table

22.1. Collect and Organize

From the periodic trend representations showing increasing radii in Figure P22.1, we are to choose the one that best describes the trend for monatomic cation radii.

Analyze

Cations are produced when an atom loses an electron. A monatomic cation is one with +1 charge. As we move across a period, more protons in the nucleus hold electrons more tightly, decreasing the radii. As we move down a group in the periodic table, electrons have been added to shells farther from the nucleus.

Solve

Part (d) represents the periodic trend in monatomic cation radii. Larger cations will be at the beginning of a period and at the bottom of a group.

Think About It

Cations are always smaller than the neutral atom because there is reduced electron–electron repulsion upon removal of an electron.

22.3. Collect and Organize

From the two groups highlighted in the periodic table in Figure P22.3, we are to choose which forms ions with radii larger than those of the neutral atoms.

Analyze

The two groups highlighted are the alkali metals (green), which typically form +1 ions, and the chalcogens (pink), which typically form –2 anions.

Solve

Because adding an electron increases the size of an atom, whereas removing an electron decreases the size of an atom, the chalcogen (pink) elements typically forms ions with larger radii than those of the neutral atoms.

Think About It

Elements that are metals tend to form cations; elements that are nonmetals tend to form anions.

22.5. Collect and Organize

For two solutions of 150 mM and 10 mM Na^+ separated by a semipermeable membrane, we are to calculate the value of ΔG for the transport of Na^+ ions from the side of higher concentration to the side of lower concentration.

Analyze

For this problem we use the Nernst equation to calculate the potential difference between the separated Na^+ solutions

$$E_{cell} = E^{\circ}_{cell} - \frac{0.0592}{n} \log Q$$

where Q is the ratio of the lower concentration of Na^+ to the higher concentration of Na^+, and $n = 1$ since both cations are Na^+. $E^{\circ}_{cell} = 0.00$ V because both solutions are composed of the same cation, Na^+. We can then use the calculated value of E_{cell} to calculate ΔG from

$$\Delta G = -nFE_{cell}$$

where $n = 1$ and $F = 96,486$ C/mol.

Solve

$$E_{cell} = 0.00 \text{ V} - \frac{0.0592}{1} \log \frac{10 \text{ m}M}{150 \text{ m}M} = 0.0696 \text{ V}$$

$$\Delta G = -1 \times \frac{96{,}485 \text{ C}}{\text{mol}} \times 0.0696 \text{ V} = -6.72 \times 10^3 \text{ J/mol, or } -6.72 \text{ kJ/mol}$$

Think About It
The transport of Na^+ from the higher concentration region to the lower concentration region is spontaneous.

22.7. Collect and Organize
For the molecule shown in Figure P22.7, we are to describe the geometry around each germanium atom.

Analyze
Each germanium atom in this compound is bonded to three atoms and does not have any lone pairs.

Solve
By VSEPR theory, the geometry around both germanium atoms is trigonal planar.

Think About It
Ignoring any possible participation of d orbitals in the bonding, we could describe the hybridization of the Ge atoms in this compound as sp^2.

22.9. Collect and Organize
Using Figure P22.9, we are to predict which polyatomic anion would be the largest: sulfate, phosphate, or perchlorate.

Analyze
Sulfate (SO_4^{2-}), phosphate (PO_4^{3-}), and perchlorate (ClO_4^-) all have the same number of oxygen atoms. What will determine the size of the anion, then, is the size of the central atom.

Solve
Because P > S > Cl in atomic radii, phosphate, PO_4^{3-}, will be the largest of the three polyatomic anions.

Think About It
However, because their atomic radii are close together, we might expect little difference in the sizes of sulfate, phosphate, and perchlorate anions.

22.11. Collect, Organize, and Analyze
We are to differentiate between an essential element and a nonessential element.

Solve
An essential element has beneficial physiological effects, whereas a nonessential element has no known function.

Think About It
Essential elements include hydrogen, sodium, potassium, magnesium, calcium, carbon, oxygen, chlorine, nitrogen, phosphorus, and sulfur.

22.13. Collect, Organize, and Analyze
We are to describe the main criterion that distinguishes major, trace, and ultratrace essential elements from one another.

Solve
The criterion that distinguishes major, trace, and ultra trace essential elements from one another is the amount present in the body.

Think About It
Major essential elements include oxygen and calcium; trace essential elements include silicon and iodine; selenium is an ultratrace essential element.

22.15. **Collect and Organize**

Given the concentration of fluorine, silicon, and iodine in milligrams or grams per kilogram, we are to express the concentration of each in parts per million.

Analyze

A part per million is one part in 10^6.

Solve

(a) $\dfrac{110 \text{ mg}}{70 \text{ kg}} \times \dfrac{1 \text{ g}}{1000 \text{ mg}} \times \dfrac{1 \text{ kg}}{1000 \text{ g}} = 1.6 \times 10^{-6}$, or 1.6 ppm

(b) $\dfrac{525 \text{ mg}}{\text{kg}} \times \dfrac{1 \text{ g}}{1000 \text{ mg}} \times \dfrac{1 \text{ kg}}{1000 \text{ g}} = 5.25 \times 10^{-4}$, or 525 ppm

(c) $\dfrac{0.043 \text{ g}}{100 \text{ kg}} \times \dfrac{1 \text{ kg}}{1000 \text{ g}} = 4.3 \times 10^{-7}$, or 0.43 ppm

Think About It

Ultratrace essential elements are present in part-per-billion levels.

22.17. **Collect and Organize**

For each pair of elements, we are to choose the one more abundant in the human body.

Analyze

We can use information from the text (Figure 22.1) to determine which elements are major, trace, or ultratrace essential elements.

Solve

(a) Oxygen is more abundant as a major essential element than silicon, a trace essential element.
(b) Oxygen is more abundant as a major essential element than iron, a trace essential element.
(c) Carbon is more abundant as a major essential element than aluminum, a nonessential element.

Think About It

Aluminum is present in the human body at about 65 mg and is a nonessential element. It is not considered toxic at normal levels.

22.19. **Collect, Organize, and Analyze**

We are to explain why chemists refer to groups 1 and 2 on the periodic table as *s*-block elements and groups 13–18 as *p*-block elements.

Solve

Groups 1 and 2 have valence electrons in the *s* orbitals, so they are referred to as *s*-block elements. Groups 13–18 have valence electrons in the *p* orbitals, so they are referred to as *p*-block elements.

Think About It

The lanthanoid and actinoid elements are likewise referred to as *f*-block elements.

22.21. **Collect, Organize, and Analyze**

We are to explain why, despite their nearly same molar masses, Li_2O is a solid, whereas CO is a gas.

Solve

Lithium oxide is a solid because it is an ionic compound with very strong inter-ion forces, whereas carbon monoxide is a gas because it is a covalent compound with much weaker intermolecular dipole–dipole forces.

Think About It

The intermolecular forces decrease in strength in order: ionic >> hydrogen bonds > dipole-dipole > dispersion.

22.23. **Collect and Organize**

Given the properties of atomic radius, electrical conductivity, and molar mass, we are to determine which would allow us to distinguish a metallic element from a semimetallic element.

Analyze

Semimetals lie along the border of metals and nonmetals on the periodic table.

Solve

Electrical conductivity will distinguish between metals and semimetals. Metals have high electrical conductivity and semimetals have lower conductivities. Molar mass and atomic radii change smoothly from metals to nonmetals back to metals again, so these would not be useful in distinguishing between metals and semimetals.

Think About It

Semimetals have only a small overlap in their valence and conduction bands, which limits their conductivity. Metals have high conductivities because they have partially filled conduction bands, which gives them high conductivities.

22.25. **Collect and Organize**

We are to explain why Be^{2+} is more likely than Ca^{2+} to displace Mg^{2+} in biomolecules.

Analyze

Beryllium and calcium are in the same group in the periodic table as magnesium, so their chemistries should both be similar to magnesium's. Be^{2+}, however, has a much smaller ionic radius than Ca^{2+}.

Solve

Be^{2+} is more likely to displace Mg^2 in biomolecules than Ca^{2+} is because of beryllium's smaller size. Ca^{2+} ions must be too large to fit into magnesium-containing biomolecules.

Think About It

Long-term exposure to low levels of beryllium results in berylliosis, a lung disease.

22.27. **Collect and Organize**

Of the two ion channels, sodium or potassium, we are to determine which must accommodate a larger cation.

Analyze

Potassium ions are larger than sodium ions.

Solve

Potassium ion channels must accommodate larger cations than sodium ion channels because K^+ is larger than Na^+.

Think About It

Potassium cations are larger than sodium ions by 36 pm.

22.29. **Collect and Organize**

We are to place the ions Mg^{2+}, Li^+, Al^{3+}, and Cl^- in order of increasing ionic radius.

Analyze

For isoelectronic ions (Mg^{2+} and Al^{3+}), ionic size decreases as we go across a period. As we add electrons, atoms increase in radius (Cl^- vs. Cl). As we add electrons to the outer shell of atoms as we go down a group in the periodic table, ions get larger.

Solve

$Li^+ < Al^{3+} < Mg^{2+} < Cl^-$

Think About It

Chloride ion has an electron configuration of [Ar] and thus is the largest ion of this group.

22.31. **Collect and Organize**
We are to place the elements K, S, F, and Mg in order of increasing electronegativity.

Analyze
Electronegativity increases as we go across a period and up a group on the periodic table.

Solve
K < Mg < S < F

Think About It
Fluorine, with a Pauling electronegativity of 4.0, is the most electronegative element.

22.33. **Collect and Organize**
We are asked how the electron affinity of Cl atoms is related to the ionization energy of Cl^- ions.

Analyze
The electron affinity of Cl is described by the following reaction:
$$Cl(g) + e^- \rightarrow Cl^-(g)$$
The ionization energy of Cl^- is described by the following reaction:
$$Cl^-(g) \rightarrow Cl(g) + e^-$$

Solve
The electron affinity of Cl is the reverse reaction of the ionization of Cl^-.

Think About It
The magnitude of the electron affinity of Cl and ionization energy of Cl^- will be the same, but they will have opposite signs; the ionization energy will be positive and the electron affinity will be negative.

22.35. **Collect, Organize, and Analyze**
We are asked to describe three ways that ions may enter and exit cells.

Solve
Ions may enter and exit cells through diffusion, by ion transport through ion channels, and through ion pumps.

Think About It
An ion pump exchanges ions from inside the cell with those in intracellular fluid, and that process requires energy.

22.37. **Collect, Organize, and Analyze**
We are to explain why crossing the cell membrane is hard for ions.

Solve
Ions, being polar, have difficulty crossing the cell membrane because the inner portion of the phospholipid bilayer of the cell membrane is nonpolar.

Think About It
Ion channels, which can have favorable ion–dipole interactions, help the ions cross the cell membrane to enter and exit the cell.

22.39. **Collect and Organize**
We are asked to identify the alkali metal ion that Rb^+ is most likely to substitute for.

Analyze
For an ion to substitute for another, it ideally should have the same charge and similar chemistry. The ion probably substitutes for an ion from the same group on the periodic table with similar size.

Solve

Rubidium ions are most likely to substitute for K^+.

Think About It

The human body does treat Rb^+ the same as it does K^+; the ions are not toxic, if present in less than 50% of the amount of normal K^+ concentrations.

22.41. **Collect, Organize, and Analyze**

We are asked to consider why nature chose $CaCO_3$, not $CaSO_4$, as a major exoskeleton shell material.

Solve

Because of respiration, carbonates are more abundant in nature than sulfates are, so nature made use of carbonate ions preferentially over sulfate ions.

Think About It

Carbonate minerals are ubiquitous in nature in the form of limestone and marble, for example.

22.43. **Collect and Organize**

For red blood cells in which the concentration of NaCl is 11 mM, we are to calculate the osmotic pressure at 37°C.

Analyze

Osmotic pressure is calculated through

$$\Pi = iMRT$$

where i is the number of ions in the salt (the van t' Hoff factor), M is the molarity of the solution, R is the ideal gas constant (in L·atm/mol·K), and T is the temperature in kelvin.

Solve

$$\Pi = 2 \times \left(\frac{11 \text{ mmol}}{L} \times \frac{1 \text{ mol}}{1000 \text{ mmol}} \right) \times \frac{0.08206 \text{ L·atm}}{\text{mol·K}} \times 310 \text{ K} = 0.56 \text{ atm}$$

Think About It

This pressure is about 430 mm Hg.

22.45. **Collect and Organize**

For two solutions of 11 mM Na^+ in red blood cells and 160 mM Na^+ in blood plasma, separated by a membrane to form a concentration cell, we are to calculate the cell potential due to the unequal concentrations of Na^+ at 37°C.

Analyze

For this problem we use the Nernst equation to calculate the potential difference between the separated Na^+ solutions

$$E_{cell} = E_{cell}^\circ - \frac{RT}{nF} \ln Q$$

where Q is the ratio of the lower concentration of Na^+ to the higher concentration of Na^+, $n = 1$ since both cations are Na^+, R is the gas constant, and F is Faraday's constant. $E_{cell}^\circ = 0.00$ V because both solutions are composed of the same cation, Na^+.

Solve

$$E_{cell} = 0.00 \text{ V} - \frac{8.314 \text{ J/mol·K} \times 310 \text{ K}}{1 \times 96,485 \text{ C/mol}} \ln \frac{11 \text{ m}M}{160 \text{ m}M} = 0.072 \text{ V}$$

Think About It

The transport of Na^+ from the higher concentration region (the blood plasma) to the lower concentration region (red blood cells) is spontaneous.

22.47. **Collect and Organize**

The reaction of ATP^{4-} with water to produce ADP^{3-} and HPO_4^{2-} releases 30.5 kJ of free energy. If the transport of K^+ across a cell membrane requires 5 kJ/mol, we are to calculate the mol of ATP^{4-} that must be hydrolyzed.

Analyze

We need only to divide the energy required for the transport of K^+ by the energy released in the hydrolysis reaction to find the energy necessary for the transport of K^+ across the cell membrane.

Solve

$$\text{Moles required} = \frac{5 \text{ kJ}}{30.5 \text{ kJ/mol}} = 0.16 \text{ mol}$$

Think About It

One hydrolysis reaction could transport 5 K^+ ions across the cell membrane.

22.49. **Collect and Organize**

Given a concentration of sulfate ion in seawater of 0.028 M, we are to calculate the solubility of strontium sulfate in moles per liter at 25°C.

Analyze

The solubility constant of $SrSO_4$ at 25°C from Appendix A is 3.44×10^{-7}.

Solve

$$K_{sp} = [Sr^{2+}][SO_4^{2-}]$$

$$3.44 \times 10^{-7} = [Sr^{2+}] \times 0.028 \, M$$

$$[Sr^{2+}] = 1.2 \times 10^{-5} \, M$$

Thus, the molar solubility of $SrSO_4$ in the seawater is $1.2 \times 10^{-5} \, M$.

Think About It

As the concentration of sulfate ion in seawater increases, $SrSO_4$ will be less soluble.

22.51. **Collect, Organize, and Analyze**

We are to describe the danger ^{137}Cs poses to human health.

Solve

Cesium-137 is a radioactive beta emitter (which could cause cancer) and can interfere with K^+-dependent functions in the body.

Think About It

Cesium-137 is a common fission product from uranium and plutonium. Because of its solubility, ^{137}Cs spreads quickly in nature.

22.53. **Collect and Organize**

We are to predict the signs of entropy and free energy for the dissolution of tooth enamel.

Analyze

The dissolution of tooth enamel (composed of hydroxyapatite) is described by the equation
$$Ca_5(PO_4)_3OH(s) \rightleftharpoons OH^-(aq) + 3 \, PO_4^{3-}(aq) + 5 \, Ca^{2+}(aq)$$

Solve

From the dissolution equation, it is clear that ΔS is positive (favored by entropy). The reaction is spontaneous (even though it is slow), so ΔG is negative.

Think About It

Acidic substances react with the OH^- released upon dissolution of hydroxyapatite. The equilibrium is shifted to the right, dissolving more hydroxyapatite.

22.55. Collect and Organize

We are asked to explain why peroxide ions are strong oxidizing agents.

Analyze

A strong oxidizing agent is one that will be easily reduced, gaining electrons to give a more negative oxidation state.

Solve

Peroxide ions (O_2^{2-}), in which each oxygen atom has -1 charge, are strong oxidizing agents because the formation of O^{2-} (the most stable oxidation state for oxygen) is formed upon peroxide's reduction.

Think About It

Hydrogen peroxide, however, can function as both an oxidizing agent and a reducing agent.

22.57. Collect and Organize

We are asked to predict the products of ^{137}Cs decay.

Analyze

^{137}C decays by β decay to change a neutron into a proton.

Solve

The products of ^{137}C decay are a beta particle and ^{137}Ba.

$$^{137}_{55}Cs \rightarrow {}^{0}_{-1}\beta + {}^{137}_{56}Ba$$

Think About It

Most of a sample of ^{137}Cs decays to metastable ^{137}Ba, which then decays to stable ^{137}Ba upon emission of a gamma ray.

22.59. Collect and Organize

For a 1.00×10^{-3} M solution of selenocysteine, which has a pK_{a1} of 2.21 and pK_{a2} of 5.43, we are to calculate the pH.

Analyze

Selenocysteine is a diprotic acid, but the acidity of the solution will be due almost entirely to the first acid dissociation. We can solve this problem by setting up a RICE table as we did in Chapters 14, 15, and 16.

Solve

Reaction	selenocysteine(aq) $\rightleftharpoons$	H_3O^+(aq) +	selenocysteine$^-$(aq)
	[selenocysteine], M	[H_3O^+], M	[selenocysteine$^-$], M
Initial	0.00100	0	0
Change	$-x$	$+x$	$+x$
Equilibrium	$0.00100 - x$	x	x

$$K_{a1} = 10^{-2.21} = 6.17 \times 10^{-3}$$

$$6.17 \times 10^{-3} = \frac{x^2}{0.00100 - x}$$

$$x^2 + 6.17 \times 10^{-3} x - 6.17 \times 10^{-6} = 0$$

$$x = 8.88 \times 10^{-4}$$

$$pH = -\log(8.88 \times 10^{-4}) = 3.05$$

Think About It
Selenocysteine is a fairly acidic amino acid.

22.61. **Collect and Organize**
Given the equation for the reaction of fluoride with hydroxyapatite and the equilibrium constant, we are to write the equilibrium constant expression and determine whether the reaction lies toward reactants or products.

Analyze
The equilibrium constant expression is the concentration of the products (raised to their stoichiometric powers) divided by the concentration of the reactants (raised to the stoichiometric powers). That applies for all species in solution but not solids. An equilibrium constant greater than 1 favors product formation.

Solve
$$K = \frac{[OH^-]}{[F^-]} = 8.48$$
Because $K > 1$, that reaction favors products.

Think About It
This is the reaction that occurs from use of fluoride toothpastes and fluoridated water to strengthen teeth and reduce the solubility of hydroxyapatite.

22.63. **Collect and Organize**
Given the K_{sp} for the calcium mineral in tooth enamel, we are to determine whether tooth enamel is more or less soluble than hydroxyapatite, calculate the molar solubility of hydroxyapatite, and explain why the presence of weak acids produced by bacteria increases the solubility of hydroxyapatite.

Analyze
The larger the value of K_{sp}, the more soluble a substance is. The equilibrium expression for the molar solubility of hydroxyapatite is
$$K_{sp} = [Ca^{2+}]^5[PO_4^{3-}]^3[OH^-]$$
where, if $[OH^-]$ is x, then $[Ca^{2+}] = 5x$, and $[PO_4^{3-}] = 3x$.

Solve
(a) The K_{sp} for the dissolution of $Ca_8(HPO_4)_2(PO_4)_4 \cdot 6H_2O$ is larger than that of hydroxyapatite, $Ca_5(PO_4)_3(OH)$, and therefore, that mineral is more soluble than hydroxyapatite.
(b) The molar solubility of hydroxyapatite (x) is
$$K_{sp} = [Ca^{2+}]^5[PO_4^{3-}]^3[OH^-]$$
$$2.3 \times 10^{-59} = (5x)^5(3x)^3(x) = 8.44 \times 10^4 x^9$$
$$x = 8.7 \times 10^{-8} \, M$$
(c) The weak acids produced by the bacteria react with the OH^- released upon dissolution of hydroxyapatite. The equilibrium is shifted to the right, dissolving more hydroxyapatite.

Think About It
The equilibrium would be shifted in the opposite direction (to the left) in an alkaline environment.

22.65. **Collect and Organize**
Given that the K_{sp} of $Ca_5(PO_4)_3(OH)$ is 2.3×10^{-59}, we are to calculate the K_{sp} of $Ca_{10}(PO_4)_6(OH)_2$.

Analyze
For $Ca_{10}(PO_4)_6(OH)_2$, the formula of hydroxyapatite has been doubled. Therefore, all ions formed will be doubled, which will square the value of K_{sp}.

Solve
$K_{sp} = (2.3 \times 10^{-59})^2 = 5.3 \times 10^{-118}$

Think About It
The value of K_{sp} for $Ca_{10}(PO_4)_6(OH)_2$ is much less than the K_{sp} for $Ca_5(PO_4)_3(OH)$.

22.67. Collect and Organize
For the group of compounds H_2O, H_2S, H_2Se, and H_2Te, we are to determine the most polar and the least polar compound.

Analyze
Polarity depends on the structure (those are all linear) and the difference in electronegativity between the atoms. The greater the difference in electronegativity, the more polar. Periodic trends tell us that electronegativity will decrease down the group: O > S > Se > Te.

Solve
The most polar of that group is H_2O. The least polar is H_2Te.

Think About It
H_2O is so polar that it forms a special kind of dipole–dipole bond—the hydrogen bond—with itself and many other molecules.

22.69. Collect, Organize, and Analyze
We are asked to explain why it is important to consider the decay mode as well as the half-life for an isotope to be used for imaging.

Solve
While the half-life has to be long enough to obtain images and short enough to be eliminated by the body, the mode of decay also must be considered. Different decay modes (such as α, β, γ) have different penetrating powers, so we must pick the one that enables the imaging of the organ or body part of interest, and they have different RBE (relative biological effectiveness) values.

Think About It
Using radioisotopes for imaging is like imaging from the inside out because it is the radiation from within the body that is imaged.

22.71. Collect, Organize, and Analyze
We are asked to consider why a β emitter is better for imaging than an α emitter.

Solve
For tissues to be imaged, the radiation must exit the body to be detected. Beta particles have much greater tissue penetration (to reach the imaging detector) than α particles, which are stopped within millimeters of traveling through tissue. Beta particles are also less damaging to tissues overall.

Think About It
In PET scanning, the β emitter in the form of a positron travels only a little way in the tissue to interact with an electron, which then generates γ radiation, which the scanner then detects.

22.73. Collect and Organize
We are to explain why ^{231}Bi undergoes β decay, whereas ^{111}In undergoes electron capture.

Analyze
Electron capture has the overall effect of converting a proton into a neutron; β decay has the overall effect of converting a neutron into a proton.

Solve
Bismuth-213 undergoes β decay because it is neutron-rich; indium-111 undergoes electron capture because it is neutron-poor.

Think About It

The balanced nuclear reactions for those conversions are:

$$^{213}_{83}\text{Bi} \rightarrow {}^{0}_{-1}\beta + {}^{213}_{84}\text{Po}$$

$$^{111}_{49}\text{In} + {}^{0}_{-1}\text{e} \rightarrow {}^{111}_{48}\text{Cd}$$

22.75. Collect and Organize

We are to identify the intermolecular forces accounting for the solubility of Xe and Ar in blood.

Analyze

Xenon is a noble monatomic gas and is thus nonpolar. Blood, we can assume, is composed mostly of water, which is polar.

Solve

Dipole–induced dipole forces account for the solubility of nonpolar monatomic gases such as Xe and Ar in blood.

Think About It

Xenon doping is believed to enhance performance by increasing the amount of the protein responsible for producing more red blood cells.

22.77. Collect and Organize

For a patient injected with a 5 μM solution ^{68}Ga, which has a half-life of 9.4 h, we are to calculate the time it would take for the activity of the ^{68}Ga to drop to 5%.

Analyze

We can use Equation 21.10 to solve for t:

$$t = -\frac{t_{1/2}}{0.693} \ln \frac{N_t}{N_0}$$

and where $N_t/N_0 = 0.05$ and $t_{1/2} = 9.4$ h.

Solve

$$t = -\frac{9.4\text{ h}}{0.693} \ln 0.05 = 41\text{ h}$$

Think About It

That time amounts to 4.4 half-lives.

22.79. Collect and Organize

We are to draw the Lewis structure for BiO^+.

Analyze

The bismuth atom brings 5 valence electrons and the oxygen brings 6 valence electrons to that ion. We have to account for the positive charge by removing one electron. Therefore, that ion has 10 total electrons in its Lewis structure.

Solve

$$\left[\text{:Bi}\equiv\text{O:} \right]^+$$

Think About It

Bismuth is about twice as abundant as gold in the earth's crust and is produced in the process of refining lead.

22.81. Collect and Organize

We are asked to write a balanced net ionic equation for the reaction of $Al(OH)_3$ with HCl.

Analyze

That is an acid–base reaction in which the hydroxide ions on aluminum will be neutralized by the strong acid, HCl.

Solve

Ionic equation: $Al(OH)_3(s) + 3\ H^+(aq) + 3\ Cl^-(aq) \rightarrow Al^{3+}(aq) + 3\ H_2O(\ell) + 3\ Cl^-(aq)$

Net ionic equation: $Al(OH)_3(s) + 3\ H^+(aq) \rightarrow Al^{3+}(aq) + 3\ H_2O(\ell)$

Think About It

That reaction indicates that solid aluminum hydroxide dissolves in strong acid solutions.

22.83. Collect and Organize

In comparing $Mg(OH)_2$ and $Al(OH)_3$, we are to determine which will neutralize more acid on a per-mole basis and then determine whether that substance also neutralizes more acid than the other on a per-gram basis.

Analyze

We must consider that on a per-mole basis, $Mg(OH)_2$ has 2 mol of OH^- in its formula, whereas $Al(OH)_3$ has 3 mol of OH^- in its structure. We will need the molar masses of each to consider the per-gram basis: $Mg(OH)_2$ = 58.32 g/mol and $Al(OH)_3$ = 78.00 g/mol.

Solve

Because $Al(OH)_3$ has more OH^- in its formula, $Al(OH)_3$ will neutralize more acid on a per-mole basis. It also neutralizes more acid on a per-gram basis:

$$1.00\ \text{g Al(OH)}_3 \times \frac{1\ \text{mol}}{78.00\ \text{g}} \times \frac{3\ \text{mol OH}^-}{1\ \text{mol Al(OH)}_3} = 0.038\ \text{mol OH}^-\ \text{per gram Al(OH)}_3$$

$$1.00\ \text{g Mg(OH)}_2 \times \frac{1\ \text{mol}}{58.32\ \text{g}} \times \frac{2\ \text{mol OH}^-}{1\ \text{mol Al(OH)}_3} = 0.035\ \text{mol OH}^-\ \text{per gram Mg(OH)}_2$$

Think About It

The acid neutralized per gram, though, is close, so on a weight basis $Mg(OH)_2$ is not that much worse in acid neutralizing capability than $Al(OH)_3$.

22.85. Collect and Organize

We are to calculate the grams of $Mg(OH)_2$ required to neutralize 115 mL of 0.75 *M* stomach acid.

Analyze

We can assume for this problem that stomach acid is the strong acid HCl. The molar mass of $Mg(OH)_2$ is 58.32 g/mol.

Solve

$$115\ \text{mL acid} \times \frac{0.75\ \text{mol}}{1000\ \text{mL}} \times \frac{1\ \text{mol OH}^-}{1\ \text{mole acid}} \times \frac{1\ \text{mol Mg(OH)}_2}{2\ \text{mol OH}^-} \times \frac{58.32\ \text{g}}{1\ \text{mol Mg(OH)}_2} = 2.5\ \text{g}$$

Think About It

Milk of magnesia has a dosage strength of about 400 mg/5 mL. To neutralize the 115 mL of stomach acid in this problem, a patient would have to take 31 mL. For acid indigestion, it is recommended that patients take 5–15 mL every 4 h (with no more than four doses in 24 h).

CHAPTER 23 | Transition Metals: Biological and Medical Applications

23.1. Collect and Organize

From the highlighted elements in Figure P23.1, we are to choose those whose chlorides are colored.

Analyze

Chloride compounds are colored for the transition elements that have incomplete d shells. The chlorides of the highlighted elements and the electron configurations of the transition metal ions are

$CaCl_2$	Ca^{2+}	$[Ar]$
$CrCl_2$	Cr^{2+}	$[Ar]3d^4$
$CrCl_3$	Cr^{3+}	$[Ar]3d^3$
$CoCl_2$	Co^{2+}	$[Ar]3d^7$
$CoCl_3$	Co^{3+}	$[Ar]3d^6$
$ZnCl_2$	Zn^{2+}	$[Ar]3d^{10}$

Solve

Chromium (green) and cobalt (yellow) have colored chloride salts.

Think About It

Remember that we remove the s electrons first in forming transition metal cations.

23.3. Collect and Organize

Of the elements highlighted in Figure P23.2, we are to identify which have M^{2+} ions that form colorless tetrahedral complex ions.

Analyze

Transition metal ions that are colorless have either filled or empty d orbitals. The electron configurations for the M^{2+} cations are

Red	V^{2+}	$[Ar]d^3$
Purple	Mn^{2+}	$[Ar]d^5$
Yellow	Co^{2+}	$[Ar]d^7$
Blue	Zn^{2+}	$[Ar]d^{10}$

Solve

Zinc (blue) forms colorless tetrahedral complex ions.

Think About It

Because nearly all tetrahedral complex ions are high spin with the d orbital splitting diagram shown below,

the numbers of unpaired electrons for the other ions in this problem are as follows: V^{2+}, three unpaired e^-; Mn^{2+}, five unpaired e^-; and Co^{2+}, three unpaired e^-.

23.5. Collect and Organize

For the structure of a chelating ligand shown in Figure P23.5, we are to count the electron-pair donor groups for the ligand when the –COOH groups are ionized.

Analyze

The electron-pair donor groups on the neutral ligand include the –SH groups. When the carboxylic acid groups are ionized, the –OH group becomes –O⁻ and so these are added to the –SH electron-donor groups.

Solve

When the two carboxylic acid groups are ionized, two more electron-donor groups are added to give a total of four electron-pair donors.

Think About It

When ionized, this ligand goes from being a bidentate to being a tetradentate ligand.

23.7. **Collect and Organize**

In Figure P23.7, three solutions of different colors for three Co^{3+} ions are shown. Using the spectrochemical series (Table 23.6), we can identify which solution is yellow, blue, and orange. The three ligands are F^-, NH_3, and CN^-.

Analyze

First, we have to recognize that the observed color is the complementary color of the light absorbed. The yellow solution, therefore, absorbs violet light, the blue solution absorbs orange light, and the orange solution absorbs blue light. The order of those wavelengths by increasing energy is

$$orange < blue < violet$$

The higher the ligand (F^-, NH_3, CN^-) on the spectrochemical series, the higher is the energy absorbed because of increased $d–d$ splitting on the transition metal ion.

Solve

From the spectrochemical series, the order of splitting of the d orbitals on Co^{3+} for those ligands is

$$F^- < NH_3 < CN^-$$

The solutions therefore are
(a) Yellow solution absorbing violet = $[Co(CN)_6]^{3-}$
(b) Blue solution absorbing orange = $[CoF_6]^{3-}$
(c) Orange solution absorbing blue = $[Co(NH_3)_6]^{3+}$

Think About It

Ligands high on the spectrochemical series are called strong field ligands and are usually low spin.

23.9. **Collect and Organize**

In each of three pairs of complexes represented as ball-and-stick models, we are to identify whether the pairs are identical to each other, isomers of each other, or neither (different compounds).

Analyze

For two molecules to be identical, they must have their atoms connected exactly the same way in the same orientation. For compounds to be isomers, they must have the same chemical composition, but a different bonding between the atoms. If the compounds are different, they will have different chemical compositions.

Solve

(a) These compounds are identical. Their chemical composition and the way in which the atoms are bonded to each other are the same. We need to only rotate the second molecule by 90° to obtain the first.

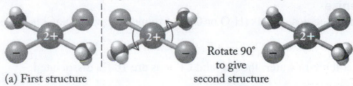

(a) First structure   Rotate 90° to give second structure

(b) These compounds are isomers. Their chemical compositions are the same, but the connectivity of their atoms is different. If we rotate the second structure to line up the chelating ligand, we see that the green anionic ligands are not trans to each other, but rather cis.

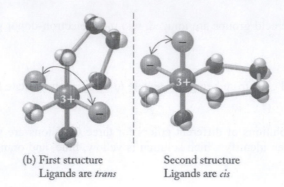

(b) First structure   Second structure
Ligands are *trans*   Ligands are *cis*

(c) These compounds are identical. Their chemical composition and the way in which the atoms are bonded to each other are the same. We need to only rotate the second molecule by 90° to obtain the first.

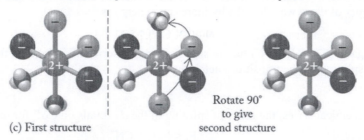

(c) First structure   Rotate 90° to give second structure

Think About It
To make isomers identical to each other, we would have to break bonds and rearrange the atoms.

23.11. Collect and Organize
Given that the periodic trend for the halide ions in the spectrochemical series if $Cl^- > Br^- > I^-$, we are to determine which curve in Figure P23.11 represents $CoCl_4^{2-}$ and which represents CoI_4^{2-}.

Analyze
The curves in Figure P23.11 shift from short to longer wavelength, or from higher to lower energy. In the spectrochemical series Cl^- splits the *d* orbitals more than I^-, so $CoCl_4^{2-}$ will absorb at a higher energy and shorter wavelength than CoI_4^{2-}, with the absorption of $CoBr_4^{2-}$ in between those absorbances.

Solve
Curve c represents $CoCl_4^{2-}$ and curve a represents CoI_4^{2-}.

Think About It
If CoF_4^{2-} were included in the series, we would expect its absorption spectrum to be shifted to higher energy/shorter wavelength than $CoCl_4^{2-}$.

23.13. Collect and Organize
We are asked which molecules or ions (H_2O or Cl^-) surround Na^+ when NaCl is dissolved in water.

Analyze
NaCl is completely soluble in water; the Na^+ and Cl^- ions are 100% dissociated.

Solve
Water molecules occupy the inner coordination sphere of Na^+ ions.

Think About It
The oxygen atoms, which carry partial negative charge, are pointed towards the Na^+ ion in the coordination sphere.

23.15. Collect and Organize

We are asked which species surround Ni^{2+} when $Ni(NO_3)_2$ dissolves in water.

Analyze

The compound dissolved in water consists of Ni^{2+} and NO_3^- ions. We also have a lot of H_2O that may bind to the metal cation. The cation attracts species that are negative or that are polar. Therefore, potentially Ni^{2+} might be surrounded by NO_3^- or H_2O. Which one of those surrounds the Ni^{2+} depends, then, on their Lewis basicity toward Ni^{2+}. From our previous study of acid–base behavior, we know that in order of basicity, $NO_3^- < H_2O$.

Solve

H_2O molecules in solution surround the Ni^{2+} ion because H_2O is the strongest base.

Think About It

Therefore, the NO_3^- ions are spectator ions in the solution of $Ni(NO_3)_2$.

23.17. Collect and Organize

We are to identify the counter ion present in $Na_2[Zn(CN)_4]$.

Analyze

A counter ion is the ion of opposite charge to the complex ion and is not directly bonded to the metal cation.

Solve

Na^+, sodium ion

Think About It

The counter ion balances the charge to give a neutral salt. Here we know that the charge on the complex ion is 2– because two Na^+ counter ions are present.

23.19. Collect and Organize

From the information in the table giving the number of moles of silver chloride precipitated per mole of Ag^+ added for each of five platinum coordination compounds, we are asked to write the formulas for each compound.

Analyze

In writing the formulas for each compound, the number of uncoordinated Cl^- ions equals the number of moles of AgCl precipitated. Those counter ions are written outside the brackets in the formula, and the coordination compound of platinum is written inside the brackets.

Solve

$Pt(NH_3)_6Cl_4 = [Pt(NH_3)_6]Cl_4$
$Pt(NH_3)_5Cl_4 = [Pt(NH_3)_5Cl]Cl_3$
$Pt(NH_3)_4Cl_4 = [Pt(NH_3)_4Cl_2]Cl_2$
$Pt(NH_3)_3Cl_4 = [Pt(NH_3)_3Cl_3]Cl$
$Pt(NH_3)_2Cl_4 = [Pt(NH_3)_2Cl_4]$

Think About It

The coordinated Cl^- ions (those bonded to the Pt^{4+} ion in all those compounds) do not precipitate with added Ag^+, but rather stay bound to the metal cation.

23.21. Collect and Organize

We are to name the compounds we identified in Problem 23.19.

Analyze

To name those coordination compounds, we separately name the cation and the anion, with the name of the cation being written first. All these have the transition metal complex as the cation. For each complex, we first name the ligands in alphabetical order, indicating with a prefix how many of each ligand are bonded to the metal ion. Then we add the name of the metal and indicate the charge on the metal ion with a Roman numeral.

Solve

$[Pt(NH_3)_6]Cl_4$ = hexaammineplatinum(IV) chloride

$[Pt(NH_3)_5Cl]Cl_3$ = pentaamminechloroplatinum(IV) chloride

$[Pt(NH_3)_4Cl_2]Cl_2$ = tetraamminedichloroplatinum(IV) chloride

$[Pt(NH_3)_3Cl_3]Cl$ = triamminetrichloroplatinum(IV) chloride

$[Pt(NH_3)_2Cl_4]$ = diamminetetrachloroplatinum(IV) chloride

Think About It

We do not indicate the number of chloride counter anions; those are inferred from the number of charged ligands and the oxidation state on the metal cation in the complexes.

23.23. Collect and Organize

We can use the conventions for naming given in Section 23.2 to name three transition metal complex ions: $Cr(NH_3)_6^{3+}$, $Co(H_2O)_6^{3+}$, and $[Fe(NH_3)_5Cl]^{2+}$.

Analyze

For each cation, we first name the ligands in alphabetical order, indicating with a prefix how many of each ligand are bonded to the metal ion. Then we add the name of the metal, indicating the charge on the metal ion with a Roman numeral.

Solve

(a) Hexaamminechromium(III)

(b) Hexaaquacobalt(III)

(c) Pentaamminechloroiron(III)

Think About It

Be sure to correctly account for the charge on the metal ion by considering the charge on the ligands and the charge on the overall complex. In part c the chloro ligand has a 1– charge. With an overall charge on the complex of 2+, the iron ion must have a 3+ charge.

23.25. Collect and Organize

We can use the conventions for naming given in Section 23.2 to name three transition metal complex ions: $CoBr_4^{2-}$, $Zn(H_2O)(OH)_3^-$, and $Ni(CN)_5^{3-}$.

Analyze

For each anion, we first name the ligands in alphabetical order, indicating with a prefix how many of each ligand are bonded to the metal ion. Then we add the name of the metal, using *-ate* as the ending and indicate the charge on the metal ion with a Roman numeral.

Solve

(a) Tetrabromocolbaltate(II)

(b) Aquatrihydroxozincate(II)

(c) Pentacyanonickelate(II)

Think About It

Be sure to correctly account for the charge on the metal ion by considering the charge on the ligands and the charge on the overall complex. In part c the cyanide ligands have a 1– charge each. With an overall charge on the complex of 3–, the Ni metal ion must have a 2+ charge.

23.27. Collect and Organize

We can use the conventions for naming given in Section 23.2 to name three transition metal coordination compounds: $[Zn(en)]SO_4$, $[Ni(NH_3)_5(H_2O)]Cl_2$, and $K_4Fe(CN)_6$.

Analyze

To name those coordination compounds, we separately name the cation and the anion, with the name of the cation being written first.

Solve

(a) Bis(ethylenediamine)zinc(II) sulfate

(b) Pentaammineaquanickel(II) chloride

(c) Potassium hexacyanoferrate(II)

Think About It

We need not indicate the number of sulfate, chloride, or potassium ions in the name. They are understood to be counter ions. When writing the formulas we would indicate how many are needed to balance the charge to make a neutral compound.

23.29. **Collect, Organize, and Analyze**

After first defining *sequestering agent*, we are to describe the properties that make a compound an effective sequestering agent.

Solve

A sequestering agent is a multidentate ligand that separates metal ions from other substances so that they can no longer react. Properties that make a sequestering agent effective include strong bonds formed between the metal and the ligand and large formation constants.

Think About It

EDTA is an example of a good sequestering agent.

23.31. **Collect and Organize**

By examining the structure of an aminocarboxylate ligand, we can predict how its chelating ability changes as pH changes.

Analyze

The general structure of an aminocarboxylate is

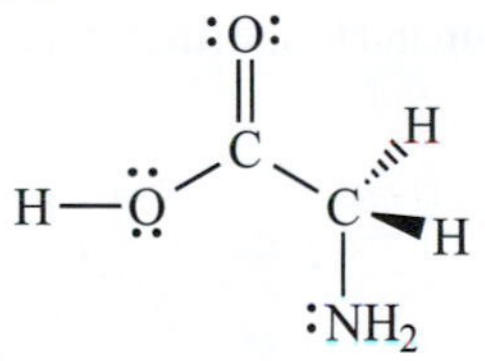

Solve

At low pH the aminocarboxylate ligand contains $-COOH$ and $-NH_2^+$ groups. As the pH increases, the chelating ability increases because OH^- removes the H on the amine and carboxylic acid groups, providing additional sites for binding to the metal cation.

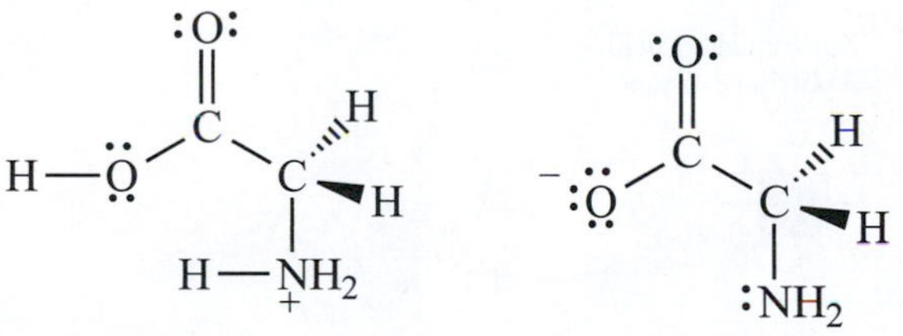

At low pH   At high pH

Think About It

EDTA is an example of an aminocarboxylate ligand.

23.33. **Collect and Organize**

We are to explain why most compounds of Sc through Zn are colored.

Analyze

Most of the first-row transition metals form ions with electrons in $3d$ orbitals.

Solve

When the transition metals bond to ligands, the *d* orbitals split in energy. If a *d*-to-*d* transition is possible for the ion, the compound is likely to be colored. The transitions for an octahedral d^2 complex are shown as an example below.

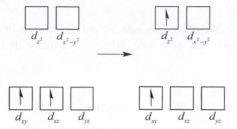

Think About It

Not all the first-row transition elements give colored complexes. For example, if no electrons are in the *d* orbitals, as in Sc^{3+}, or if the *d* orbitals are filled, as in Zn^{2+}, the compounds are not colored.

23.35. Collect and Organize

For a square planar crystal field geometry, we are to explain why d_{xy} is higher in energy than either the d_{xz} or d_{yz} orbitals.

Analyze

In a crystal field, *d* orbitals pointed directly at the ligands are raised in energy, and those not pointed directly at ligands are lowered in energy.

Solve

The repulsions due to the ligands in a square planar crystal field are highest for the d_{xy} orbital, and so it is raised in energy because that orbital lies in the plane of the ligands. The d_{xz} and d_{yz} orbitals, however, are perpendicular to the plane of the four ligands, and therefore those orbitals are lower in energy than the d_{xy} orbital.

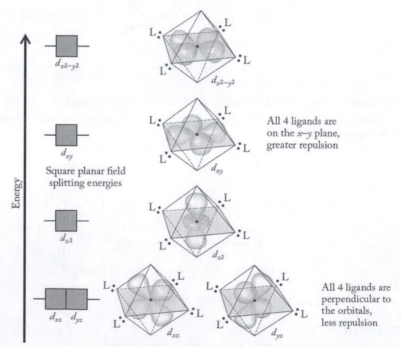

Think About It

In a square planar geometry the $d_{x^2-y^2}$ orbital points directly at the four ligands. It has the most repulsions and therefore is the highest orbital in energy.

23.37. **Collect and Organize**
Given yellow and violet aqueous solutions of Cr^{3+} complex ions, we are to determine which solution contains $Cr(H_2O)_6^{3+}$ and which contains $Cr(NH_3)_6^{3+}$.

Analyze
From the spectrochemical series we know that NH_3 splits the d orbitals more than H_2O (Table 16.5). $Cr(NH_3)_6^{3+}$ therefore absorbs a higher energy (shorter wavelength) of light than $Cr(H_2O)_6^{3+}$.

Solve
The wavelength (color) absorbed is complementary to the wavelength (color) observed. The yellow aqueous solution absorbs violet light, and the violet aqueous solution absorbs yellow light. Because violet light has a higher energy and shorter wavelength than yellow light, the yellow solution contains (b) $Cr(NH_3)_6^{3+}$. The violet solution contains (a) $Cr(H_2O)_6^{3+}$.

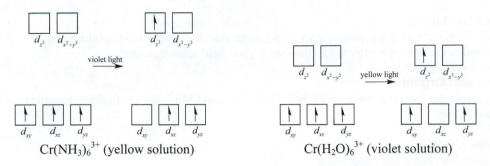

Think About It
A Cr^{3+} compound giving a red solution would have a d orbital splitting energy between that of $Cr(NH_3)_6^{3+}$ and $Cr(H_2O)_6^{3+}$.

23.39. **Collect and Organize**
Given the octahedral crystal field splitting for $Co(phen)_3^{3+}$ (5.21×10^{-19} J/ion), we can determine the color of the solution of $Co(phen)_3^{3+}$ by calculating the wavelength of light absorbed and correlating that to the color of light reflected or transmitted.

Analyze
To calculate the wavelength of light absorbed, we rearrange $E = hc/\lambda$ to solve for the wavelength of light:

$$\lambda = \frac{hc}{E}$$

We can use the visible spectrum (400–700 nm) to determine the color of the absorbed wavelength and then the color wheel (Figure 23.13) to choose the complementary color (the color we observe).

Solve

$$\lambda = \frac{6.626 \times 10^{-34} \text{ J} \cdot \text{s} \times 3.00 \times 10^8 \text{ m/s}}{5.21 \times 10^{-19} \text{ J/ion}} = 3.82 \times 10^{-7} \text{ m, or 382 nm}$$

That wavelength is in the UV region, so the solution is colorless.

Think About It
If a complex absorbs in the UV, sometimes it appears slightly yellow because its absorption "tails" into the violet region of the visible spectrum.

23.41. **Collect and Organize**
For the complexes $NiCl_4^{2-}$ and $NiBr_4^{2-}$, whose solutions absorb light at 702 and 756 nm, respectively, we are asked which ion has the greater d-orbital energy splitting.

Analyze
The shorter the wavelength absorbed, the larger the split of the d-orbital energies, Δ_o.

Solve

Because $NiCl_4^{2-}$ absorbs at a shorter wavelength than $NiBr_4^{2-}$, $NiCl_4^{2-}$ has a greater split of the *d*-orbital energies.

Think About It

A solution of $NiCl_4^{2-}$ appears blue-green and a solution of $NiBr_4^{2-}$ appears yellow-green.

23.43. **Collect, Organize, and Analyze**

We can use the crystal field model to explain how a transition metal may be either high spin or low spin.

Solve

The magnitude of the crystal field splitting energy in comparison with the pairing energy of the electrons in a lower energy *d* orbital determines whether a transition metal cation is high spin or low spin.

Think About It

For a high-spin complex the pairing energy is greater than the crystal field splitting energy, whereas for a low-spin complex the pairing energy is less than the crystal field splitting energy.

23.45. **Collect and Organize**

For high-spin octahedral complexes of Fe^{2+}, Cu^{2+}, Co^{2+}, and Mn^{3+}, we are to determine the number of unpaired electrons.

Analyze

The octahedral crystal field diagram is

Since all those complexes are high spin, the pairing energy is greater than the crystal field splitting, Δ_t.

Solve

Fe^{2+} has a d^6 electron configuration, giving four unpaired electrons.

Cu^{2+} has a d^9 electron configuration, giving one unpaired electron.

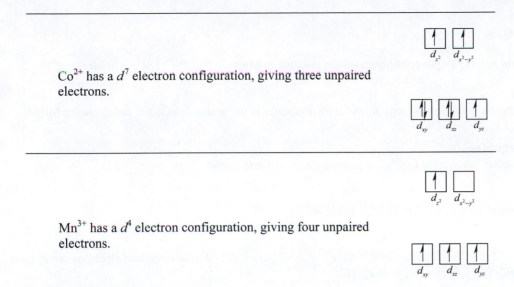

Co^{2+} has a d^7 electron configuration, giving three unpaired electrons.

Mn^{3+} has a d^4 electron configuration, giving four unpaired electrons.

Think About It
Remember that for high-spin complexes, electrons are "promoted" to the higher-lying d orbitals before being paired into the lower-lying d orbitals.

23.47. Collect and Organize
Of the transition metal cations Co^{2+}, Cr^{3+}, Ni^{2+}, and Zn^{2+}, we are to determine which of their electron configurations could result in either a high-spin or low-spin complex in a tetrahedral field.

Analyze
The tetrahedral field diagram is

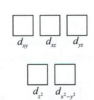

and it may hold 10 e⁻. The possibility for high-spin or low-spin configurations occurs when, depending on the magnitude of Δ_t, electrons may be placed in either the low-lying orbital set or the high-lying orbital set after the d^2 configuration. Those situations may occur for metal ions that have three, four, five, or six d electrons.

Solve
Co^{2+} has the electron configuration $[Ar]4s^0 3d^7$, so it has seven electrons in its $3d$ orbitals.
Cr^{3+} has the electron configuration $[Ar]4s^0 3d^3$, so it has three electrons in its $3d$ orbitals.
Ni^{2+} has the electron configuration $[Ar]4s^0 3d^8$, so it has eight electrons in its $3d$ orbitals.
Zn^{2+} has the electron configuration $[Ar]4s^0 3d^{10}$, so it has 10 electrons in its $3d$ orbitals.
Only Cr^{3+} may have either high-spin or low-spin configurations.

Think About It
Because $\Delta_t \approx 4/9\Delta_o$, nearly all tetrahedral complexes are high spin.

23.49. Collect and Organize
For the minerals MnO_2 and Mn_3O_4, where the Mn ions are surrounded by six O^{2-} ions (and are therefore in an octahedral crystal field), we are to determine the charges of the Mn ions in each mineral and which mineral might have a possibility of high-spin and low-spin Mn ions.

Analyze
High-spin and low-spin complexes are possible in an octahedral field when four, five, six, or seven electrons are occupying the d orbitals.

Solve

(a) Mn^{4+} in MnO_2

    2 Mn^{3+} and 1 Mn^{2+} in Mn_3O_4

(b) Both low-spin and high-spin configurations are possible in Mn_3O_4 (d^4 and d^5) but not in MnO_2 (d^3).

Think About It

For minerals to have a metal ion present in two oxidation states is not unusual, as is the case here for Mn_3O_4.

23.51. Collect and Organize

We are asked whether tetrahedral $CoCl_4^{2-}$ is paramagnetic or diamagnetic.

Analyze

The electron configuration of Co^{2+} in $CoCl_4^{2-}$ is $[Ar]4s^0 3d^7$.

Solve

In a tetrahedral field the Co^{2+} ions in $CoCl_4^{2-}$ have a d^7 configuration, and three unpaired electrons are in its d_{xy}, d_{xz}, and d_{yz} orbitals. The complex is paramagnetic.

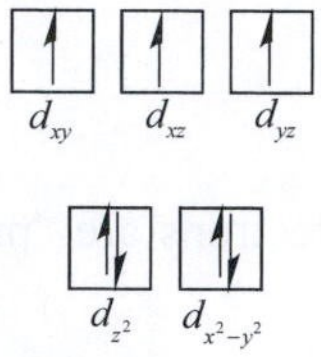

Think About It

Because of its partially filled d orbitals, we also expect the compound to be colored.

23.53. Collect, Organize, and Analyze

We are asked to differentiate between cis- and trans- for an octahedral complex ion.

Solve

For an octahedral geometry, cis means that two ligands are side by side and have a 90° bond angle between them. Ligands that are trans to each other have a 180° bond angle between them.

Cis   Trans

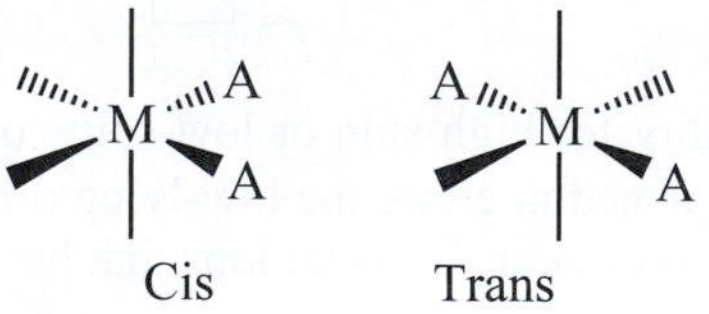

Think About It

Complexes that have cis- and trans-placed ligands are isomers of each other.

23.55. Collect and Organize

In considering stereoisomers of a square planar complex, we are to determine the minimum number of different types of donor groups required.

Analyze

A square planar complex has four coordination sites.

Solve

To have stereoisomers, we must have at least two ligands that are different from the others.

Think About It
Some geometric isomer possibilities for square planar complexes are as follows:

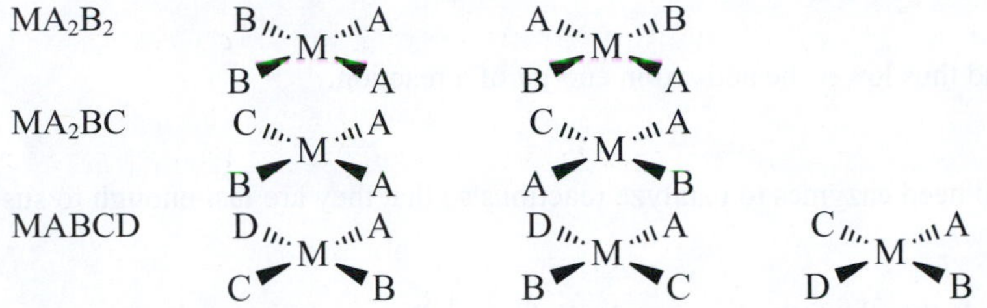

MA_2B_2

MA_2BC

$MABCD$

23.57. Collect and Organize
By looking at the structure of $Co(en)(H_2O)_2Cl_2$, we can determine whether it can have geometric isomers.

Analyze
Ethylenediamine (en) must bind in a cis fashion to the Co^{4+} metal ion.

Solve
Yes, $Co(en)(H_2O)_2Cl_2$ has geometric isomers because the H_2O (or Cl) ligands may be either cis or trans to each other and in relation to the ethylenediamine ligand:

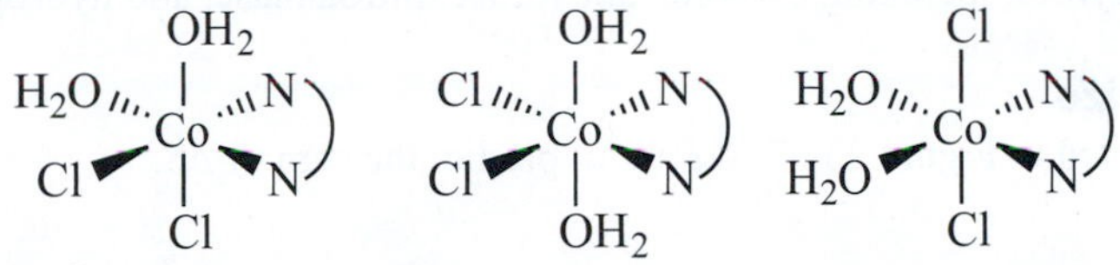

Think About It
Isomer (a) also has an optical isomer (nonsuperimposable mirror image) as a possibility.

23.59. Collect and Organize
For square planar $CuCl_2Br_2{}^{2-}$ we are asked to sketch the possible geometric isomers and asked whether any of the isomers drawn are chiral.

Analyze
The possible arrangements of the two different ligands on a square planar metal cation are cis and trans.

Solve

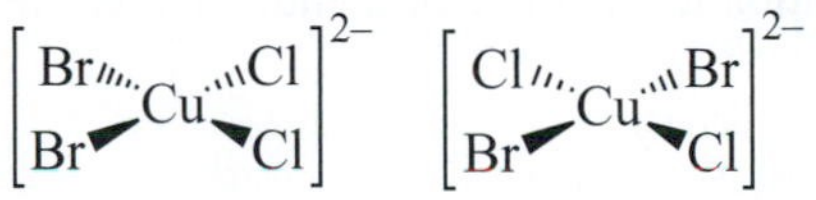

Cis   Trans

No, neither isomer is chiral because both are superimposable on their mirror images.

Think About It
If that complex were tetrahedral, it also would not have any geometric isomers and would not be chiral.

23.61. Collect, Organize, and Analyze
We consider the definitions of enzymes and proteins.

Solve
(a) Enzymes catalyze biological reactions by lowering the activation energy for the reaction.
(b) No, not all proteins are enzymes.

Think About It
Proteins can also play a role in replication of DNA, transporting molecules, and responding to stimuli.

23.63. Collect, Organize, and Analyze

We consider how an enzyme affects the activation energy of a biochemical reaction.

Solve

Enzymes are catalysts and thus lower the activation energy of a reaction.

Think About It

Most processes in the cell need enzymes to catalyze reactions so that they are fast enough to sustain life.

23.65. Collect and Organize

We are asked to predict whether the formation constant of a metalloenzyme is much greater or much less than 1.

Analyze

An enzyme consists of proteins that have multiple sites that may bond to the metal cation; those are in essence chelate complexes, which form very stable compounds.

Solve

The formation constant for the formation of a metalloenzyme is much greater than 1.

Think About It

Several important enzymes, including carbonic anhydrase, nitrogenase, and hydrogenase, contain metal cations.

23.67. Collect and Organize

For the reaction depicted in Figure 23.67, we are to predict the sign of ΔS.

Analyze

Entropy generally increases when more particles (in the form of molecules, for example) are formed as products than those present as reactants.

Solve

In that reaction one molecule is split into two molecules, so we expect the sign of ΔS to be positive.

Think About It

A positive change in entropy for a reaction means that the reaction is favored by entropy.

23.69. Collect and Organize

Given the activation energies for the catalyzed and uncatalyzed decomposition of hydrogen peroxide, we are to use a form of the Arrhenius equation to calculate how much larger the rate constant for the uncatalyzed reaction is than the catalyzed reaction.

Analyze

In the equation given, k_1 = reaction rate constant for the uncatalyzed reaction, k_2 = reaction rate constant for the catalyzed reaction, E_{a1} = activation energy for the uncatalyzed reaction, and E_{a2} = activation energy for the catalyzed reaction.

$$RT \ln\left(\frac{k_1}{k_2}\right) = E_{a_2} - E_{a_1}$$

Solve

$$\frac{8.314 \text{ J}}{\text{mol} \cdot \text{K}} \times 293.15 \text{ K} \times \ln\left(\frac{k_1}{k_2}\right) = \left(\frac{75.3 \text{ kJ}}{\text{mol}} - \frac{29.3 \text{ kJ}}{\text{mol}}\right) \times \frac{1000 \text{ J}}{\text{kJ}}$$

$$\ln\left(\frac{k_1}{k_2}\right) = 18.87 \text{ J/mol}$$

$$\frac{k_1}{k_2} = 1.57 \times 10^8$$

Think About It
By reducing the activation energy by about 60%, the rate was enhanced by 10^8, which is a very large rate enhancement. That allows the catalyzed reaction to reach equilibrium much faster than an uncatalyzed reaction.

23.71. Collect and Organize
Using the graph in Figure P23.71, we are to determine the apparent reaction order for high concentrations of the substrate (far right of the graph).

Analyze
The graph shows that initially as concentration increases, the rate of the reaction also increases; however, at high concentrations, the rate of the reaction levels off.

Solve
At high concentrations of substrate (far right of the graph), the rate does not change with a change in concentration. Therefore, the apparent reaction order is zero.

Think About It
At high concentrations of the substrate, the enzyme becomes saturated and the reaction reaches its maximum rate.

23.73. Collect, Organize, and Analyze
We are asked to describe under what circumstances a coordination complex of a toxic metal might be useful in medicine.

Solve
A coordination compound of a toxic metal might be useful to kill targeted disease cells such as cancer.

Think About It
Platinum coordination complexes, such as cisplatin, are used to treat a variety of cancers.

23.75. Collect and Organize
We are to predict what radiation $^{153}_{64}\text{Gd}$ produces when undergoing electron capture that makes it useful for imaging.

Analyze
Electron capture in $^{153}_{64}\text{Gd}$ would convert a proton into a neutron to produce $^{153}_{63}\text{Eu}$. That isotope then decays to produce radiation for imaging.

Solve
Possible radiation emitted might be beta particles, positrons, alpha particles, or gamma radiation. Of those, gamma radiation is most penetrating and would be useful for imaging.

Think About It
Europium-153 is a stable nuclide, and so it would not undergo further decay through beta, positron, or alpha emission.

23.77. Collect, Organize, and Analyze
We are to explain how platinum and ruthenium drugs fight cancer.

Solve
Platinum- and ruthenium-containing drugs fight cancer by binding to DNA to prevent the replication of tumor cells.

Think About It
Those drugs, however, are toxic because they cannot discriminate well between normal cells and cancer cells.

23.79. Collect and Organize

We are asked whether the glucose tolerance factor that contains Cr^{3+} is paramagnetic or diamagnetic.

Analyze

Chromium(III) has an electron configuration of $[Ar]4s^0 3d^3$.

Solve

The three electrons in the d orbitals of Cr^{3+} will be unpaired; the glucose tolerance factor, therefore, will be paramagnetic.

Think About It

Chromium(III) solutions absorb in the visible range, so their solutions are colored.

23.81. Collect and Organize

Given the value of the formation equation and constant for the formation of $Hg(methionine)^{2+}$ and the formation equation and constant for $Hg(penicillamine)^{2+}$, we are to calculate the equilibrium constant for the formation of $Hg(penicillamine)^{2+}$ from $Hg(methionine)^{2+}$ reacting with penicillamine.

Analyze

The two formation reaction can be combined to give the formation equation of $Hg(penicillamine)^{2+}$ from $Hg(methionine)^{2+}$ reacting with penicillamine:

$$Hg(methionine)^{2+} \rightleftharpoons Hg^{2+} + methionine \qquad K_{f1} = 1/10^{14.2}$$
$$Hg^{2+} + penicilliamine \rightleftharpoons Hg(penicillamine)^{2+} \qquad K_{f2} = 10^{16.3}$$

In reversing the equation for the formation of $Hg(methionine)^{2+}$, we take the inverse of the formation constant. Those equations add up to

$$Hg(methionine)^{2+} + penicilliamine \rightleftharpoons Hg(penicillamine)^{2+} + methionine$$

and where the overall equilibrium constant is

$$K_{eq} = K_{f1} \times K_{f2}$$

Solve

$$K_{eq} = K_{f1} \times K_{f2} = \frac{1}{10^{14.2}} \times 10^{16.3} = 130$$

Think About It

The formation of $Hg(methionine)^{2+}$ from $Hg(penicillamine)^{2+}$, the reverse reaction, would favor the reactants, not the products.

23.83. Collect and Organize

Given the equilibrium constant of 0.633 for the formation of $CH_3Hg(cysteine)^+$ from the reaction of $CH_3Hg(penicillamine)^+$ with cysteine, we are to calculate the equilibrium concentrations of cysteine and penicillamine if the initial concentrations of cysteine and $CH_3Hg(penicillamine)^+$ are 1.00 M and 1.00 mM, respectively.

Analyze

To solve this we set up a RICE table in which the initial concentration of cysteine is 1.00 M and the initial concentration of $CH_3Hg(penicillamine)^+$ is 1.00 mM.

Solve

Reaction	$CH_3Hg(penicillamine)^+$	+ cysteine	$\rightleftharpoons$	$CH_3Hg(cysteine)^+$	+ penicillamine
Initial	0.001	1.00		0	0
Change	$-x$	$-x$		$+x$	$+x$
Equilibrium	$0.001 - x$	$1.00 - x$		x	x

Solving the equilibrium constant expression by the quadratic equation gives

$$0.633 = \frac{x^2}{(0.001-x)(1.00-x)}$$

$$0.367x^2 + 0.633633x - 0.000633 = 0$$

$$x = 9.98 \times 10^{-4}$$

Therefore, [penicillamine] $= 9.99 \times 10^{-4}$ M and [cysteine] $= 1.00$ M.

Think About It

That reaction favors the reactants, not the products.

23.85. Collect and Organize

Given the structures of two platinum compounds used in cancer therapy, we are to describe the difference between the compounds, name the complexes, determine whether the two compounds have the same orbital diagram, and sketch the orbital diagram for the compounds.

Analyze

Both complexes have an octahedral geometry and have the same ligands (N_3^-, OH^-, and NH_3). With two of each ligand, the charge on the platinum metal cation at the center of the complex is +4.

Solve

(a) The complexes differ in the placement of the ligands around the central metal cation. In both the OH ligands are trans, but in the structure on the left the N_3 ligands are cis and so are the NH_3 ligands, whereas in the structure on the right, the N_3 ligands are trans and so are the NH_3 ligands. Those complexes are stereoisomers.

(b) The complex on the left is *cis*-diammine-*cis*-diazido-*trans*-hydroxoplatinum(IV). The complex on the right is *trans*-diammine-*trans*-diazido-*trans*-hydroxoplatinum(IV).

(c) Yes, both compounds have the same orbital diagram because they both contain Pt^{4+} and have octahedral geometry.

(d) Platinum in both compounds is +4 and therefore is a d^6 transition metal ion. We will assume here that the complexes are both low spin because NH_3 is a fairly high-field ligand.

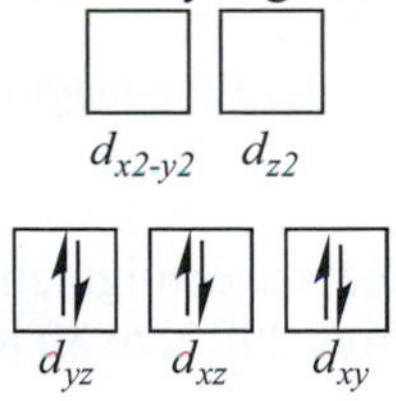

Think About It

Neither of those coordination compounds is chiral.

23.87. Collect and Organize

For two complexes, one of platinum and the other of gold, both of which are square planar, we are to determine whether the two complexes have the same number of valence electrons in their orbital diagrams and whether the complexes are paramagnetic or diamagnetic.

Analyze

Platinum is +2 and Au is +1 in those compounds.

Solve

(a) No, those complexes do not have the same number of d electrons. When platinum is +2 it has 8 electrons in its d orbitals; when gold is +1 it has 10 electrons in its d orbitals.

(b) Both complexes are diamagnetic, having an even number of electrons to fill the square planar d orbital diagram (Figure 23.15).

Think About It

Both those complexes are used to treat cancer.

23.89. Collect and Organize

Given the observed colors and magnetic properties of two cobalt complexes, we are asked which has the largest Δ_o.

Analyze

We can presume that both complexes have octahedral geometry. Oxidation converts Co^{2+} to Co^{3+}. Co^{2+} has a d^7 configuration. Because that complex is observed to be purple, it is absorbing relatively low-energy yellow light. For that complex to have three unpaired e^- (high spin), Δ_o must be small. Co^{3+} has a d^6 configuration. Because that complex is observed to be yellow, it is absorbing relatively high-energy purple light. For that complex to have no unpaired e^- (low spin), Δ_o must be large.

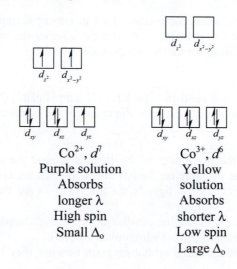

Co^{2+}, d^7
Purple solution
Absorbs longer λ
High spin
Small Δ_o

Co^{3+}, d^6
Yellow solution
Absorbs shorter λ
Low spin
Large Δ_o

Solve

The yellow complex containing Co^{3+} in aqueous ammonia has the larger Δ_o.

Think About It

Remember that the spin of a complex is a result of the magnitude of Δ_o.

23.91. Collect and Organize

We are to explain how Fe^{2+}, with a d^6 electron configuration, changes from paramagnetic (4 unpaired e^-) to diamagnetic (0 unpaired e^-) upon cooling from 298 K to 80 K.

Analyze

The d^6 electron configuration is paramagnetic if Δ_o is low field, but it is diamagnetic if Δ_o is high field.

Solve

As the sample is cooled, the Δ_o must increase and a transition from high spin to low spin occurs.

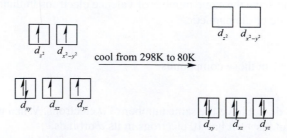

cool from 298K to 80K

Think About It

A larger Δ_o at low temperature must mean that at low temperature the ligands are interacting more strongly with the d orbitals. That effect might be due to lower thermal vibrations in the complex, which bring the ligands closer to the metal ions.

23.93. Collect and Organize

We are to explain why $Fe(bipy)_2(SCN)_2$ is paramagnetic but $Fe(bipy)_2(CN)_2$ is diamagnetic.

Analyze

Both complexes have d^6 Fe^{2+} ions. The only difference between them is the ligands CN^- and SCN^-.

Solve

CN^- is a strong-field ligand with a very large Δ_o, which leaves d^6 Fe^{2+} with no unpaired electrons and diamagnetic. SCN^-, however, must be a weak-field ligand, which leaves d^6 Fe^{2+} with four unpaired electrons and paramagnetic.

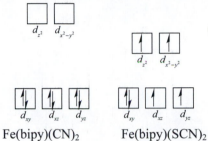

Fe(bipy)(CN)$_2$ Fe(bipy)(SCN)$_2$

Think About It

Both compounds have possible d–d transitions, being d^6, and so we expect both to be colored.

23.95. Collect and Organize

For the complex $MnCl_2(H_2O)_4$, we are to determine the number of unpaired electrons.

Analyze

Manganese has a +2 charge in that coordination compound and therefore has five electrons in the d orbitals. The geometry of the complex is octahedral. Both water and chloride ligands are relatively weak field ligands, so the complex is expected to be high spin.

Solve

For a high-spin d^5 complex, we expect five unpaired electrons.

Think About It

If the ligands on that complex were replaced with stronger field ligands, we would expect the complex to be low spin with one unpaired electron.